# Space Optics

Proceedings of the
Ninth International Congress of the
International Commission for Optics

(ICO IX), SANTA MONICA, CALIFORNIA
October 9–13, 1972

*Editors*

B. J. THOMPSON   The Institute of Optics
University of Rochester

R. R. SHANNON   Optical Sciences Center
University of Arizona

NATIONAL ACADEMY OF SCIENCES
Washington, D.C. 1974

**Library of Congress Cataloging in Publication Data**

International Commission for Optics.
    Space optics; proceedings of the ninth international congress of the International Commission for Optics.

    Held in Santa Monica, Calif., Oct. 9–13, 1972.
    Includes bibliographies.
    1. Astronomical instruments—Congresses.
2. Optical instruments—Congresses. I. Thompson, Brian J., ed. II. Shannon, Robert Rennie, 1932– ed. III. Title.
QB86.I59      1973      522      73–19588
ISBN 0–309–02144–8

*Available from*
Printing and Publishing Office, National Academy of Sciences
2101 Constitution Avenue, Washington, D.C. 20418

Printed in the United States of America

# Preface

It has been a privilege to act as Editors of this volume on behalf of the U.S. National Committee for the International Commission for Optics (ICO) of the National Academy of Sciences. This volume is a record of the proceedings of ICO IX, held in Santa Monica, California, October 9–13, 1972. The material has been reorganized slightly to provide for better groupings of the papers and to aid the reader.

This volume is a record of the subject of space optics as of late 1972 and provides, under a single cover, a state-of-the-art review of this important field.

On behalf of the participants we would like to express our thanks to the following for their efforts to ensure a successful meeting: the ICO Bureau (Appendix A), the USNC for ICO and their organizing subcommittees (Appendix B), the National Aeronautics and Space Administration for their support under contract NSR 09-012-105 and other organizations (Appendix C) for their contributions, especially the Jet Propulsion Laboratory for arranging the live reception and presentation of Mars data from Mariner 9 and both the National Academy of Sciences and the University of California at Los Angeles for their help in organizing the meeting.

B. J. THOMPSON<br>
R. R. SHANNON

# Contents

**SPACE SYSTEMS**

| | |
|---|---|
| Skylab Astronomy Instrumentation | 3 |
|    *Karl G. Henize* | |
| Photometric Calibration of an Extreme-Ultraviolet Spectroheliometer for the Skylab Mission | 33 |
|    *M. C. E. Huber, E. M. Reeves,* and *J. G. Timothy* | |
| Image Quality Criteria for the Large Space Telescope | 55 |
|    *William B. Wetherell* | |
| Calibration and Performance Evaluation of Systems and Devices for Space Astronomy: Discussion of a Test Facility | 104 |
|    *R. H. Meier* | |
| Design of Athermal Lens Systems | 116 |
|    *H. Köhler* and *F. Strähle* | |
| Space Optics—A Review of Telescope Designs and Technology for the Space-Science Disciplines | 154 |
|    *John D. Mangus* | |
| Terrestrial Engineering of Space-Optical Elements | 171 |
|    *W. P. Barnes, Jr.* | |
| Elastic Analysis of Large Spacebound Mirrors | 184 |
|    *K. R. Maser* and *K. Soosaar* | |

Vibrational Modes of the Primary Mirror Structure in the
Large Space Telescope System 208
    *Changhwi Chi, Pravin Mehta,* and *Aaron Ostroff*
Recent Advances in Optical Control for
Large Space Telescopes 239
    *W. E. Howell*
Large Space Telescope Support Systems Module and Orbital
Operations 259
    *Fred Steputis* and *Hal Nylander*

## ULTRAVIOLET INSTRUMENTS

Ultraviolet Solar Eclipse Spectroscopy from Space 293
    *R. J. Speer*
Astronomical Observations in the Middle Ultraviolet with
High Spectral Resolution 319
    *B. Bates, C. D. McKeith, G. R. Courts,* and *J. K. Conway*
Observations on Degradation of Ultraviolet Systems on Nimbus
Spacecraft 340
    *Donald F. Heath* and *James B. Heany*
An All-Reflection Interferometer with Possible Applications
in the Vacuum Ultraviolet 355
    *F. L. Roesler, R. A. Kruger,* and *L. W. Anderson*
Development and Fabrication of Large-Area Extreme-Ultraviolet
Filters for the Apollo Telescope Mount 367
    *Gordon N. Steele*

## INFRARED METHODS

Sensitive Pyroelectric Detectors 393
    *C. B. Roundy* and *R. L. Byer*
Hadamard-Transform Instrumentation for Infrared Space Optics 405
    *John A. Decker, Jr.*
An Infrared Pneumatic Transducer with Capacitive Detection 419
    *Georges Gauffre* and *Michel Chatanier*
A Tunable Filter for Calibrating Radiometers in the 11–19-Micrometer Spectral Region 429
    *C. R. Munnerlyn* and *J. W. Balliett*
Near-Infrared Heterodyne Interferometer for the Measurement
of Stellar Diameters 442
    *H. van de Stadt, Th. de Graauw, J. C. Shelton,* and *C. Veth*

## COMMUNICATIONS AND RADIOMETRY

Matched-Filter Detection of Mode-Locked Laser Signals 461
    *R. J. D'Orazio* and *Nicholas George*

*Contents* vii

Scintillation and Polarization Compensation Techniques
for Optical Communications 476
    *W. N. Peters*
Image Deterioration Due to Atmospheric Turbulence 490
    *A. Consortini, P. Pandolfini, L. Ronchi,* and *R. Vanni*
Absolute Radiometry and the Solar Constant 502
    *Richard C. Willson*
Absolute Intensity Calibration of a High-Resolution Rocket
Spectrometer 511
    *J. L. Kohl* and *W. H. Parkinson*

THIN FILMS

New Developments in Vacuum-Ultraviolet Reflecting Coatings
for Space Astronomy 525
    *G. Hass* and *W. R. Hunter*
Multilayer Antireflection Coatings for Ultraviolet (200–400 nm)
and Application 554
    *Hideo Ikeda, Hideki Akasaka,* and *Zenji Wakimoto*
An Automatic Thin-Film Interference Filter Design Program
Based on the Use of Minus Filters 570
    *J. A. Dobrowolski*
Multilayer Interference Coatings for the Vacuum Ultraviolet 581
    *Eberhard Spiller*

IMAGE PROCESSING AND HOLOGRAPHY

A New Method of Optical Processing Applied to Detection of
Differences between Two Images 601
    *S. Debrus, M. Françon,* and *C. P. Grover*
Subtraction (or Addition) of Illuminance 613
    *Y. Belvaux* and *S. Lowenthal*
Holographic Microscopy in Exobiology: Transmission of a
Holomicrogram over a Limited Telemetry Channel 622
    *J. T. Winthrop* and *R. F. van Ligten*
Contouring from Holographic Stereomodels 635
    *N. Balasubramanian*
Holographic Testing of Aspheric Optical Elements 643
    *J. C. Wyant*

OPTICAL TECHNOLOGY

Practical Applications of the Electromagnetic Theory of Gratings 667
    *R. Petit, D. Maystre,* and *M. Nevière*
Thin-Film Geodesic Lenses 682
    *G. C. Righini, V. Russo, S. Sottini,* and *G. Toraldo di Francia*

Determination of the In-Flight Optical Transfer Function of
Orbital Earth Resources Sensors ........................... 693
   *R. A. Schowengerdt* and *P. N. Slater*
Fabrication of an Aspheric Surface for Observing Airflow
in a Rocket Nozzle ....................................... 704
   *Masaharu Kawai*

## OPTICAL METHODS

Scattering from Mirror Surfaces Used in Space Applications ... 717
   *H. E. Bennett, J. L. Stanford,* and *J. M. Bennett*
Recent Developments in Selective Modulation Spectrometry .... 739
   *A. Maréchal* and *G. Fortunato*
Use of Lateral Waves in the Study of Surface Films ............ 750
   *O. S. Heavens* and *S. K. Sharma*
Focusing Systems Adaptable to Integrated Optics ............. 757
   *R. Boulay, J. W. Y. Lit,* and *R. Tremblay*
A Wall-Stabilized Doubled Arc as a Standard Source in the
Vacuum Ultraviolet ...................................... 772
   *G. L. Weissler* and *Santosh K. Srivastava*

## INSTRUMENTATION

Method and Equipment for Localizing Satellites by Laser
Range-and-Direction Finding .............................. 789
   *Pierre Weber* and *Pierre Durrenberger*
Spectrometric Device for the Analysis of Light Emission
from a Venus Orbiter .................................... 806
   *A. Monfils, J. P. Macau,* and *S. Gardier*
The Utrecht Orbiting Stellar Spectrophotometer S 59 ......... 825
   *R. Hoekstra, K. A. van der Hucht, Th. Kamperman,* and
   *H. J. Lamers*

## APPENDIXES

A   ICO Bureau 1969–1972 ................................ 837
B   U.S. National Committee for ICO ....................... 838
C   Sponsors ............................................ 840

# SPACE SYSTEMS

KARL G. HENIZE

# Skylab astronomy instrumentation

## Introduction

On April 30, 1973, the United States will launch its first manned orbiting space laboratory—Skylab. The spacecraft is designed for a total mission duration of 8 months, including 5 months of manned operation and 3 months of unmanned operation. Manned operation will be conducted by three separate crews of three men each, who will visit the laboratory on successive missions. The first mission is planned for a duration of 28 days, the second and third missions for 56 days each. The general configuration of Skylab is illustrated in Figure 1.

The Skylab program will carry out more than 50 individual scientific experiments, which may be grouped into four major categories: (1) medical experiments, (2) solar-physics observations with the Apollo Telescope Mount (ATM) instruments, (3) earth observations with the Earth Resources Experiment Package (EREP), and (4) corollary experiments of which a major subgroup is astronomically oriented and operates through the scientific air locks in the Orbiting Workshop area. The aim of this paper is to describe the objectives and the instrumentation of the ATM experiments and of the astronomy-oriented corollary experiments.

---

The author is at the NASA Manned Spacecraft Center, Houston, Texas 77058.

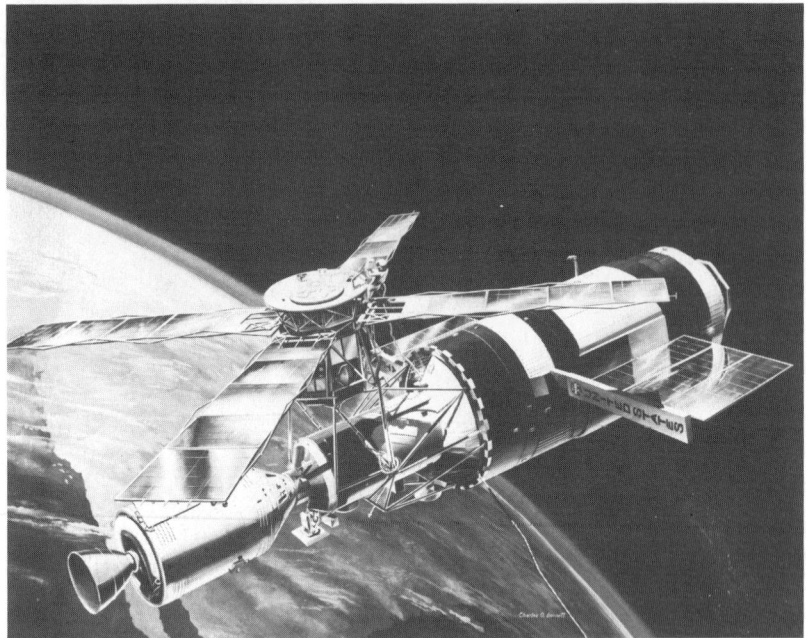

FIGURE 1  The Skylab orbital configuration. To the left is a command module docked to the Multiple Docking Adapter (MDA), which leads to the Orbiting Workshop on the right. The structure above the MDA, from which four panels of solar cells radiate, is the ATM.

## Apollo Telescope Mount

The general configuration of ATM is illustrated in Figure 2. The instrument package consists of eight major instruments mounted on a spar within a pointable canister. The canister, which is 7 ft in diameter and 11 ft long, is attached to the Multiple Docking Adapter (MDA) by a larger support structure. The overall weight including the support structure is 22,000 lb, while the weight of the instruments is 2200 lb. Four large panels of solar cells provide power for the ATM. The canister has a thermal control system maintaining it at $50 \pm 3 \,°F$, while the individual instruments have heaters giving thermal stability to $\pm 1°F$. The ATM contains three Control Moment Gyros, which are a part of the primary Skylab attitude-control system. The canister has a secondary stabilization system and may be moved independently of the spacecraft through $\pm 2°$ in pitch and yaw and through $\pm 120°$ in roll. Fine pointing on the sun accurate to $\pm 2.5$ sec of arc is achieved through use of sun sensors, and a star tracker is used to provide reference for canister roll.

FIGURE 2 A view of the ATM during a film-retrieval EVA. The astronaut seen at the solar end of the canister is retrieving film from the xuv spectroheliograph and uv spectrograph. Film from the other experiments is retrieved through the door visible in the upper side of the ATM.

All but one of the instruments use photographic film as the primary detector. As is illustrated in Figure 2, it is necessary for astronauts to undertake extravehicular activity (EVA) in order to retrieve and replace film magazines. Six EVAs are planned for this purpose—one during the first mission, three during the second, and two during the final mission.

The ATM instruments are operated at a control and display panel located in the MDA. This is illustrated in Figure 3. This panel has both video and numerical displays that depict solar activity at several wavelengths. These aid the astronaut in monitoring solar activity, in detecting flares, and in selecting features worthy of detailed study. The video displays include images of the solar disk in H-$\alpha$, in the extreme ultraviolet (170 to 550 Å), and in x rays (2 to 10 Å), as well as an image of the corona in white light. A white-light image of a 3 × 3 min of arc portion of the solar disk reflected from the slit jaws of the ultraviolet spectrograph may also be displayed. Numerical displays include x-ray fluxes in three wavelength ranges (0 to 2 Å, 2 to 8 Å, and 8 to 20 Å), as well as

FIGURE 3  The ATM control and display panel mounted in the MDA. (NASA photo.)

the 6-cm radio flux. The 2 to 8 Å and 8 to 20 Å x-ray bands and the 6-cm radio data may be displayed on a plotter.

## The ATM Experiments

Figure 4 gives a summary of the ATM experiments. There are eight major instruments mounted on the spar. Six are primary research instruments designated as "experiments," while two—the H-$\alpha$ telescopes—are support instruments to aid the astronaut in pointing the other instruments.

As the wavelength ranges indicate, these instruments, as a group, cover almost the entire wavelength range from 0.2 Å to visible light. One small gap from 60 to 150 Å is covered by experiment S-020, which, strictly speaking, is not an ATM experiment, since it is operated through the scientific air lock in the Orbiting Workshop area. It is logical that space-

The diameter of the solar image is 3.95 mm. At this plate scale the angular resolution of the system is limited to 8 sec of arc by the resolution of the film (Kodak 026-02). The instrument is designed to detect light levels on the order of $10^{-10}$ × the mean solar radiance. Because of vignetting, caused by the external disk assembly, the effective coronal radiance is "flattened" over the field of view so that the radiance in the film plane varies by a factor of only 5 from 1.5 to 6 solar radii even though the actual radiance varies by a factor of $10^3$.

Photometric calibration is provided on each frame by a 15-step calibration wedge illuminated by the sun. An image of this wedge is projected by an auxiliary lens onto the film, where it is superposed on the image of the occulting disks. The appearance of a typical photo is shown in Figure 8.

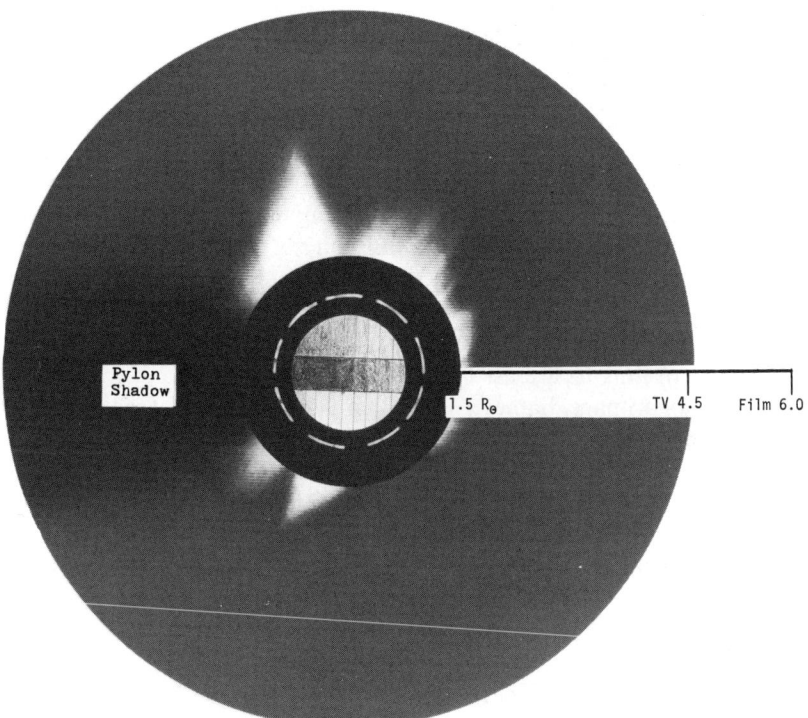

FIGURE 8 Schematic representation of a white-light coronagraph photo. The dark area to the left is the shadow of the pylon, which supports the inner occulting disk. The inner light area represents the image of the calibration wedge. The dashed white circle represents the occulted solar disk. (NASA photo.)

A four-position filter wheel contains a clear aperture plus three linear polaroids that allow determination of line-of-sight electron densities in the corona. The polaroids are normally used during synoptic observations of the corona. However, when transient events are being observed, the requirement for high time resolution may dictate the use of a sequence utilizing the clear aperture.

An auxiliary mirror may be used to divert the light beam into a TV camera, which transmits a TV image of the corona to the ATM control and display panel. The TV image has a resolution of 30 sec of arc and a field diameter of 4.5 solar diameters.

This instrument was constructed by Ball Brothers Research Corporation. The PI is Robert M. MacQueen of the High Altitude Observatory of the National Center for Atmospheric Research.

### X-Ray Spectroheliograph, Skylab Experiment S-054

This instrument utilizes grazing-incidence mirrors to form high-resolution images of the sun in x-ray wavelengths from 3.5 to 60 Å. Filtergrams obtained using five selectable filters will allow studies of the spatial, spectral, and temporal distribution of x-ray emission from solar flares and the solar corona. A typical x-ray filtergram obtained with a similar instrument in a sounding rocket is shown in Figure 9. The basic optical configuration of the grazing-incidence mirror system is shown in Figure 10. Two reflections from confocal paraboloidal and hyperboloidal surfaces give an image at $F_2$ corrected for coma. In the S-054 telescope, two nested sets of mirrors are used to give a total light gathering area of 42 cm$^2$. The solar image diameter is 1.92 cm, and the field of view is 48 min of arc. The on-axis resolution is limited to a value of about 2.5 sec of arc. The emulsion used is Kodak SO-212.

A functional diagram of the x-ray spectroheliograph is given in Figure 11. Spectral resolution within the 3.5 to 60 Å band is achieved primarily by five filters mounted in a wheel just forward of the focal plane. These consist of Teflon, parylene, and beryllium in three thicknesses. The narrowest band [3 Å full width at half-maximum (FWHM)] is produced by a 51-μm thickness of beryllium. A sixth slot in the wheel is blank and allows the imaging of the entire 3.5 to 60 Å band. The 3.5 Å limit is set by the incidence angles of the light on the Kanigen-coated beryllium mirrors, and the 60 Å limit is set by an aluminum–polypropylene filter located elsewhere in the optical system. A 1440 lines/mm transmission grating may also be placed in the light beam and produces x-ray spectroheliograms with a wavelength resolution of 0.15 Å at 7 Å.

*Karl G. Henize* 13

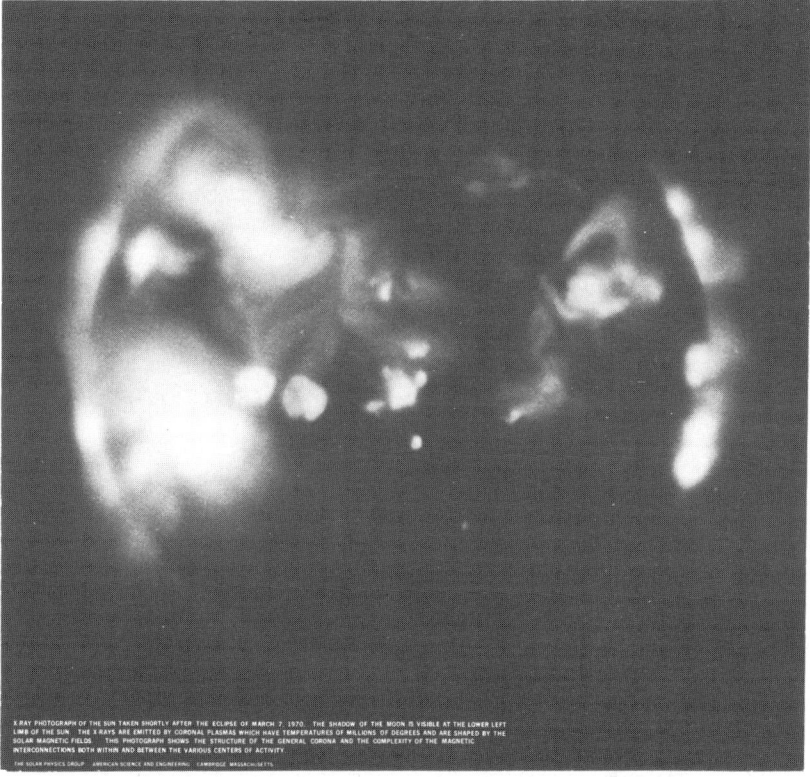

FIGURE 9 X-ray filtergram of the sun similar to those to be obtained by the x-ray spectroheliograph. (Courtesy of American Science and Engineering, Inc.)

Three auxiliary optical systems are also illustrated in Figure 11. These include: a visible light lens, which provides a white-light image of the sun to record the centering and the orientation of each x-ray frame; a small x-ray mirror, which provides the image for the astronaut's x-ray "finder"; and an uncollimated scintillation detector, which serves several purposes.

The x-ray "finder" forms an image on a scintillation crystal optically coupled to an image dissector, the output of which is displayed on a cathode-ray tube on the control and display panel. This display has a resolution of 1 min of arc and allows the astronaut to observe and point at bright x-ray features, such as a flare in its early stages, which might not be correlated with H-$\alpha$ phenomena.

The x-ray scintillation detector utilizes a photomultiplier tube coupled

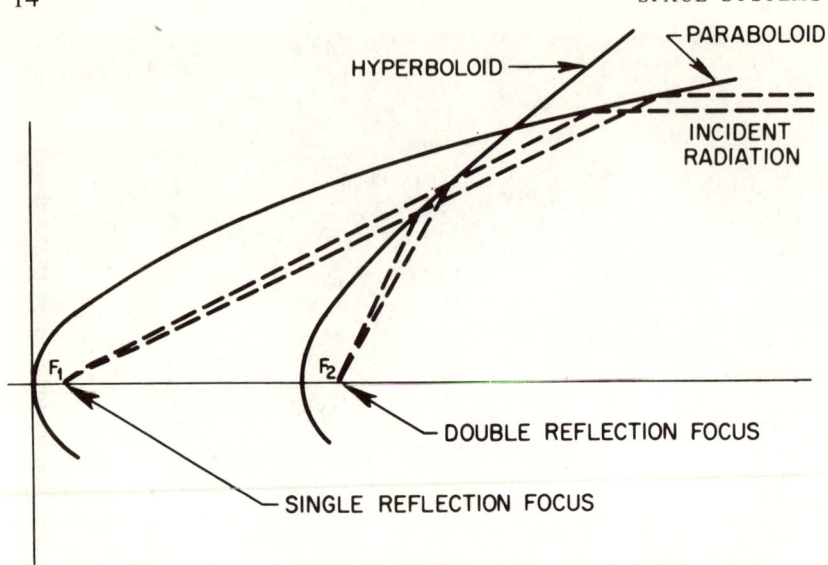

FIGURE 10 Optical configuration of the x-ray grazing-incidence mirror. (NASA photo.)

to a Na I scintillation crystal to monitor the 0 to 8 Å solar x-ray flux. The output from the photomultiplier is used: (a) to trigger an alarm designed to alert the astronauts to the occurrence of a flare, (b) to provide exposure control to the main instrument, and (c) to provide eight-channel pulse-height spectra of hard x rays in the range from approximately 0.2 to 2 Å. These spectra are recorded and telemetered to the ground.

The x-ray spectroheliograph was manufactured by American Science and Engineering, Incorporated. The PI is R. Giacconi of the same company.

## Ultraviolet Spectrometer–Spectroheliometer, Skylab Experiment S-055

This instrument is designed to obtain both spectra and spectroheliograms in the ultraviolet wavelength range from 280 to 1350 Å with an angular resolution of 5 sec of arc. It utilizes seven independent, open-channel, electron-multiplier detection systems and is the only ATM instrument with completely electronic detectors.

The optical configuration of the uv spectrometer–spectroheliometer is shown in Figure 12. The iridium-coated primary mirror is an off-axis paraboloid that may be rotated in two directions to scan the solar image across the square spectrometer entrance slit. This mirror has a focal ratio of $f/12$ and a focal length of 2.3 m, producing a solar image 21.4 mm in

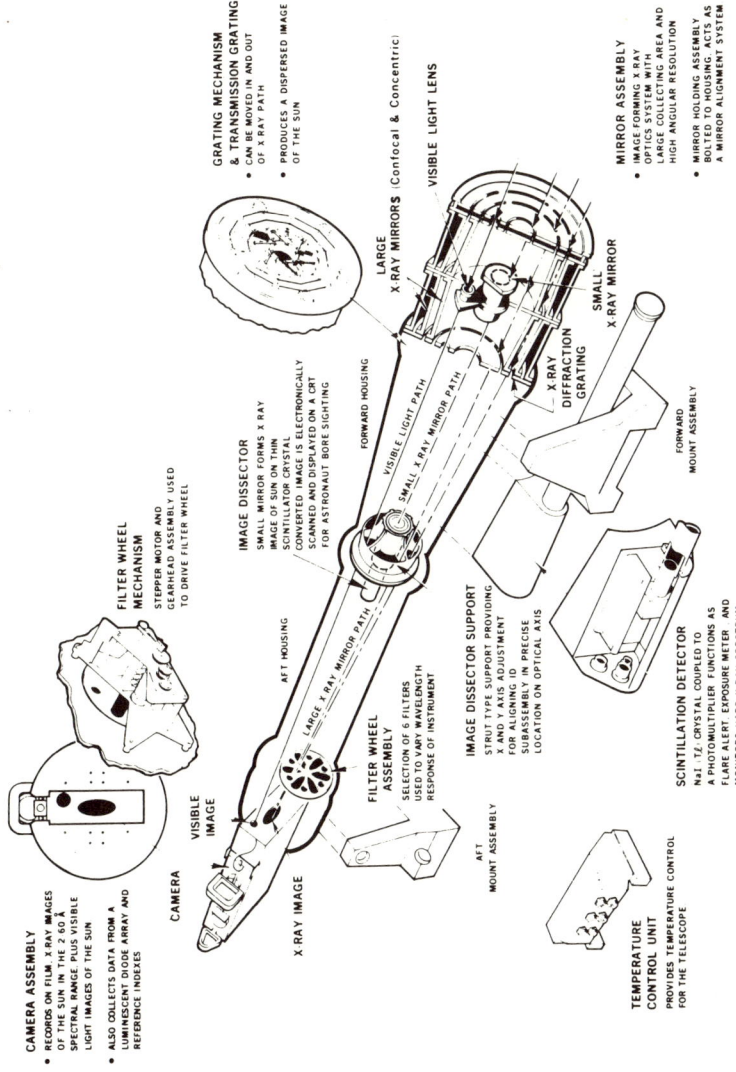

FIGURE 11 Functional diagram of the American Science and Engineering x-ray spectroheliograph. (NASA photo.)

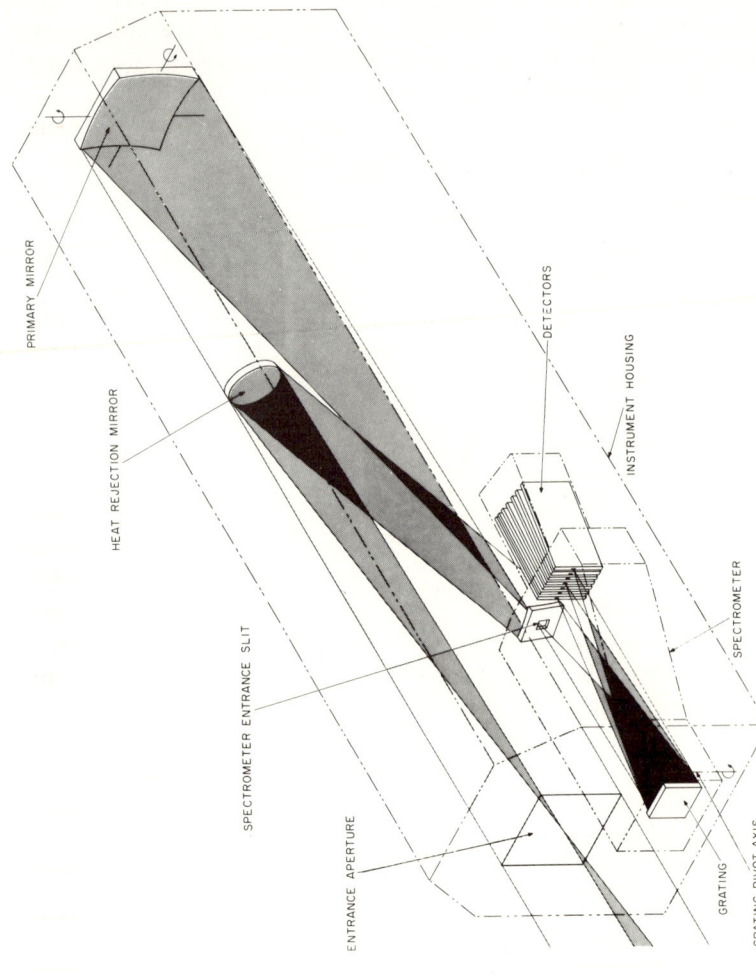

FIGURE 12  Optical configuration of the Harvard College Observatory uv spectrometer-spectroheliometer. (NASA photo.)

diameter in the plane of the entrance slit. An 1800 lines/mm gold-coated concave grating disperses the light and focuses images of the entrance slit on the seven detectors. Heat produced by the unused portion of the solar image is reflected out of the instrument by means of heat-rejection mirrors.

In the spectrometer mode, a 5 × 5 sec of arc portion of the solar image enters the spectrometer, where it is dispersed and scanned across detector number one by rotation of the grating. A complete spectrum with 1.6 Å resolution (FWHM) is obtained in 3.8 min.

In the spectroheliometer mode, the grating is positioned at a predetermined angle at which important wavelengths fall on each of the seven detectors. Then the primary mirror is rotated so that a 5 min of arc square section of the solar image is scanned over the entrance slit. The output from the detectors can then be reconstructed into spectroheliograms at each of seven wavelengths. Each spectroheliogram covers a 5 × 5 min of arc section of the sun with 5 sec of arc resolution. A complete scan requires 5 min of time. It is also possible to scan a single line 5 min of arc in length once every 5 sec in order to obtain better time resolution for rapidly evolving phenomena. A 40-msec time resolution can be achieved for one 5 sec of arc square region of the solar disk by stopping all scanning motions.

The detectors are positioned so that they coincide with seven important chromospheric and coronal emission lines in the solar spectrum when the grating is set at a predetermined angle. These lines are listed in Table 1 and are also indicated on the spectrum shown in Figure 13. These lines span a temperature range from $10^4$ to $2 \times 10^6$ K and thus yield information at various levels in the solar atmosphere from the chromosphere, through the transition region, into the corona.

TABLE 1 Wavelengths Measured in the Primary and Secondary Grating Positions of the UV Spectrometer–Spectroheliometer

| Detector | Position 1 | | | Position 2 | |
|---|---|---|---|---|---|
| | Line | | $T_{max}$ (K)[a] | Line | |
| 1 | C II | λ1335 | $5 \times 10^4$ | — | |
| 2 | Ly-α | λ1216 | $10^4$ | Ly-α | λ1216 |
| 3 | O VI | λ1032 | $3 \times 10^5$ | Ly-β | λ1025 |
| 4 | C III | λ977 | $10^5$ | Ly-γ | λ972 |
| 5 | Ly-Cont | λ896 | $10^4$ | Ly-Cont | λ890 |
| 6 | Mg X | λ625 | $1.5 \times 10^6$ | — | |
| 7 | O IV | λ554 | $2 \times 10^5$ | — | |

[a] $T_{max}$ is the temperature at which the line is of maximum intensity.

Several other positions of the grating give chance coincidences of interesting groups of lines. One of these is indicated as "Position 2" in Table 1. The grating can also be positioned to select any desired single wavelength in the 280 to 1350 Å region.

Figure 13 shows a spectral scan from a similar Harvard College Observatory spectrometer aboard OSO-4, which illustrates the character of the solar spectrum in the wavelength region of interest to S-055.

The instrument was constructed by Ball Brothers Research Corporation. The PI is Leo Goldberg, currently on leave from Harvard College Observatory. The acting PI is E. M. Reeves of Harvard College Observatory.

### X-Ray Telescope, Skylab Experiment S-056

This instrument will produce x-ray filtergrams in the 5 to 40 Å wavelength region for the purpose of studying solar x-ray radiation from flares and from the corona. The basic optical system is a paraboloidal-hyperboloidal pair of grazing incidence mirrors such as is used in the S-054 x-ray spectroheliograph. In this case, only one set of mirrors is used and the effective light collecting area is 14.8 cm$^2$. The reflecting surfaces are highly polished quartz and are designed to minimize scattered radiation. The solar image diameter is 1.92 cm. The on-axis angular resolution of 2.5 sec of arc is limited by the emulsion, Kodak SO-212. The field diameter is 38 min of arc.

A system diagram for the x-ray telescope is shown in Figure 14. Spectral resolution within the 5 to 40 Å band is achieved by means of five filters mounted in a filter wheel just forward of the focal plane. These consist of titanium, two thicknesses of beryllium, and two thicknesses of aluminum. The narrowest bandpass produced by these filters is about 3 Å (FWHM).

Two additional detectors shown in Figure 14 are proportional counters. One, with a beryllium window, is sensitive from 2 to 8 Å; the other, with an aluminum window, is sensitive from 8 to 20 Å. Pulses from these counters are analyzed into a ten-channel pulse-height spectrum, which is recorded and telemetered to the ground. The integrated output of each counter may also be displayed to the astronaut either as a number or as a graph on the x-ray/radio-frequency history plotter.

This instrument was constructed at Marshall Space Flight Center. The PI is James Milligan of that organization. Operational and data-analysis support are provided by Aerospace Corporation.

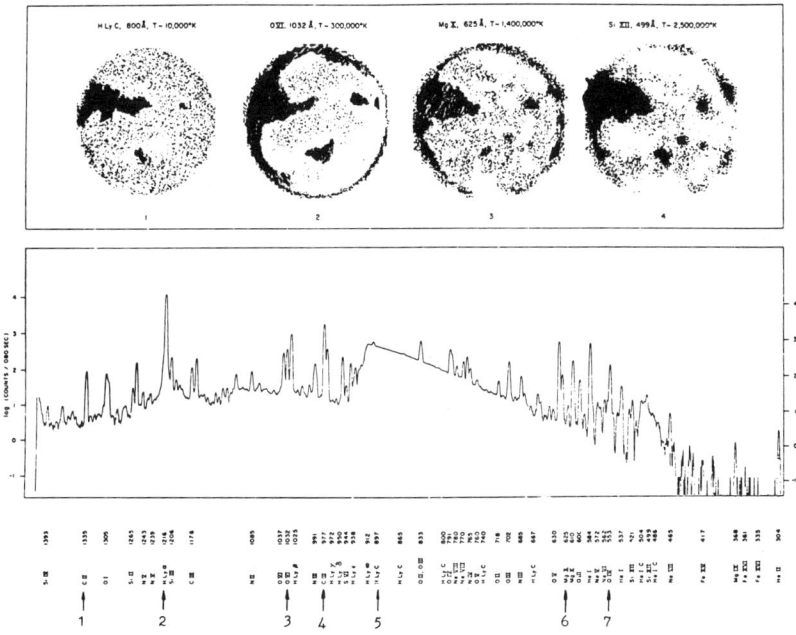

FIGURE 13  Ultraviolet solar spectrum and solar spectroheliograms observed on OSO-4. This instrument was similar, in principle, to the Skylab uv spectrometer–spectroheliometer, but the Skylab instrument will have greater resolution. For example, the OSO-4 spectroheliograms give 1 min of arc resolution over the entire solar disk, whereas the Skylab instrument will give 5 sec of arc resolution over a 5 min of arc square section of the solar disk. The seven wavelengths noted below the spectrum are those to be observed in the Skylab instrument when the grating is in its primary reference position. (Courtesy of Harvard College Observatory.)

## Extreme Ultraviolet (xuv) Spectroheliograph, Skylab Experiment S-082A

The xuv spectroheliograph is a slitless objective-grating spectrograph operating over the 150 to 630 Å wavelength range. Spectroheliograms (sometimes overlapping) are thus formed at the wavelengths of the stronger emission lines in this range. The angular resolution is 2 to 10 sec of arc depending on the wavelength and is best in the central portion of the wavelength range. For a sharp feature 10 sec of arc in diameter the spectral resolution is 0.13 Å. These data will be used to study temperature, density, and composition at various heights in the solar chromosphere and lower corona.

The optical schematic shown in Figure 15 illustrates the optical sys-

20        SPACE SYSTEMS

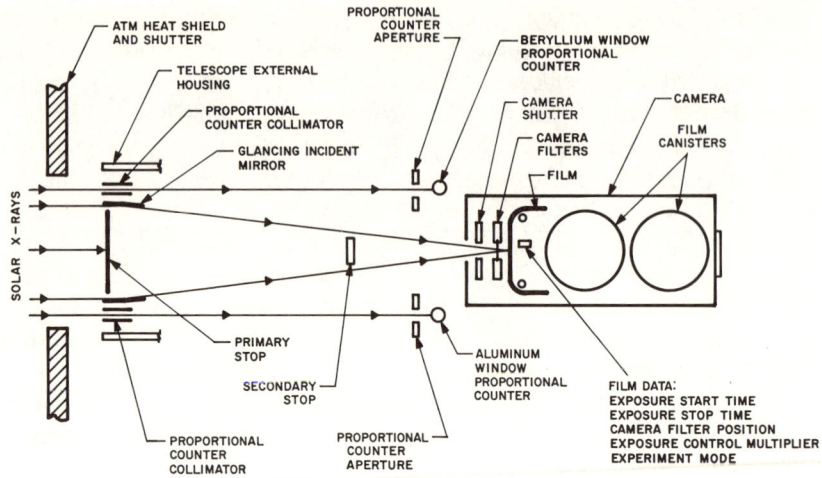

FIGURE 14  System diagram for the Marshall Space Flight Center x-ray telescope. (NASA photo.)

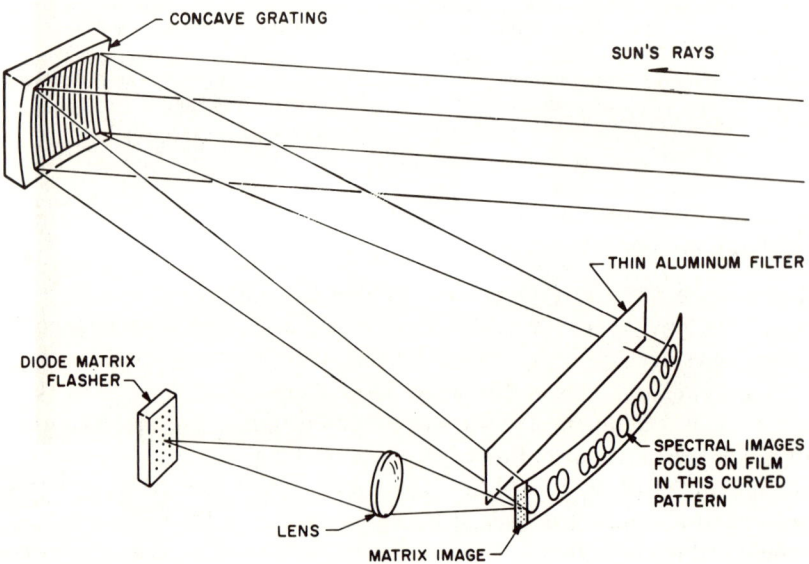

FIGURE 15  Optical schematic for the Naval Research Laboratory xuv spectroheliograph. (Courtesy of Ball Brothers Research Corporation.)

tem. A 3600 lines/mm concave grating collects, disperses, and focuses the solar radiation on a curved focal plane. The images are recorded on Kodak 104 (formerly SWR) emulsion, which is shielded from stray light of wavelengths longer than 835 Å by a 0.1-μm thickness of aluminum directly in front of the focal plane. The field of view is 56 min of arc in diameter. Only half of the wavelength range may be recorded on a given frame. Rotation of the grating allows the astronaut to select either the short (150 to 335 Å) or the long (321 to 630 Å) wavelength range.

Exposure times may be varied from 2.5 sec to prolonged manual exposures, the longest of which is not planned to exceed 48 min. The large frame format allows only 200 exposures per film canister, and thus this experiment is the most film-critical of the ATM experiments.

The diode matrix flasher is used to imprint the vital data for each exposure on the film. Most of the other experiments use similar systems for recording such data.

Figure 16 shows a portion of a frame taken with a prototype instru-

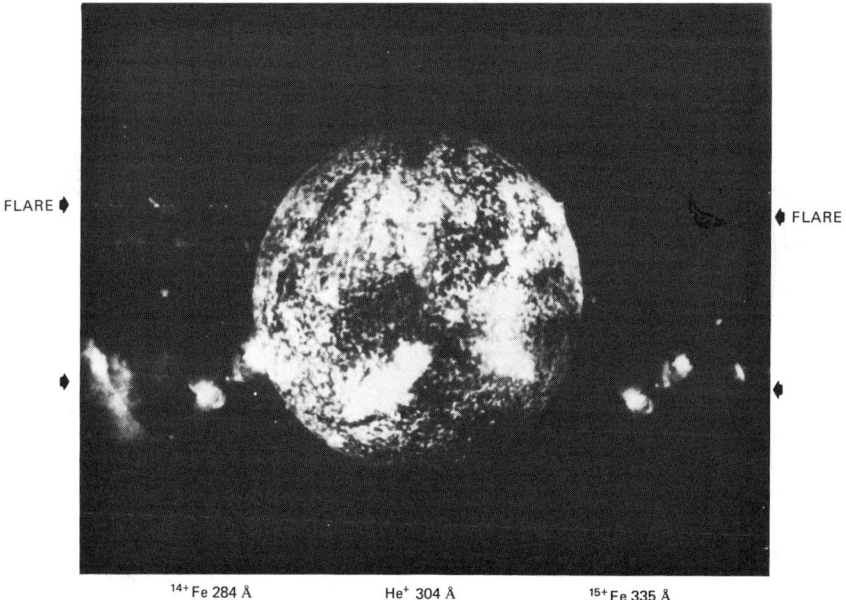

FIGURE 16 Objective-grating spectroheliogram of the sun in the light of He II λ 304 obtained with a prototype of the S-082A instrument flown on a rocket on November 4, 1969. The helium image at the center comes from the sun's chromosphere; the highly ionized iron images at either side come from the corona. Note the tiny dots that are the spectrum of the flare and the butterfly-like region of intense solar activity in the corona. (Courtesy of the Naval Research Laboratory.)

ment flown on a rocket. The dominant image is that of He II λ304. The spectrum of a Class 1B flare appears as a string of dots in the upper part of the frame.

This instrument was constructed by Ball Brothers Research Corporation. The PI is Richard Tousey of the Naval Research Laboratory.

### Ultraviolet Spectrograph, Skylab Experiment S-082B

The ultraviolet spectrograph is designed to give high-resolution spectra of the solar atmosphere, primarily the chromosphere, in the 970 to 3970 Å wavelength range. The optical system (see Figure 17) uses a predisperser grating to attenuate stray light, to correct for astigmatism, and to select either the short (970 to 1970 Å) or the long (1940 to 3940 Å) wavelength range by interchange of the predisperser. Partial correction for astigmatism is achieved by ruling the predisperser gratings in ten strips of differing dispersion to approximate a continuously changing dispersion. This reduces the residual astigmatism to 1 min of arc thus producing a significant increase in instrument speed. However, the remaining astigmatism is sufficient to eliminate any spatial resolution along the slit.

The resolution is 0.08 Å in the short-wavelength range and 0.16 Å

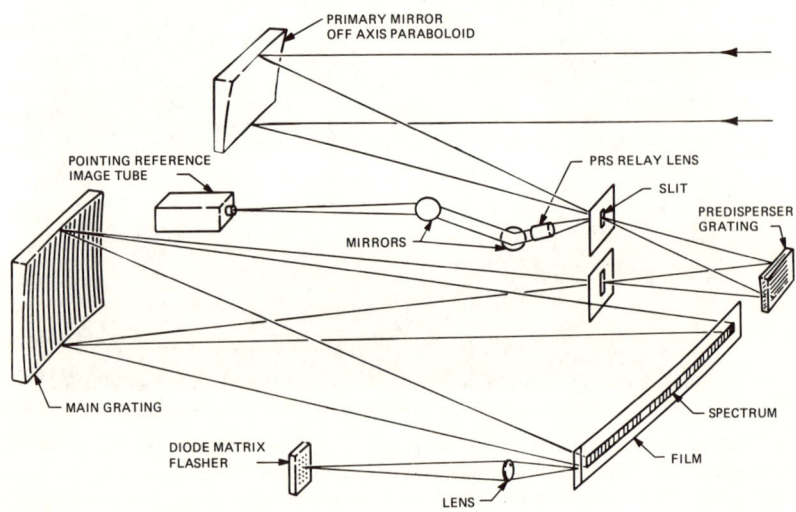

FIGURE 17 Optical schematic of the Naval Research Laboratory uv spectrograph. (Courtesy of Ball Brothers Research Corporation.)

in the long-wavelength range. This is due to a change in dispersion caused by the predisperser gratings. Eight spectra are recorded on each film strip. The emulsion is Kodak 104.

The entrance slit defines the angular resolution of 2 X 60 sec of arc. The astronaut can monitor the position of the slit on the solar disk by means of a video system using an image dissector tube that views the white-light image reflected from the slit jaws. This display makes possible the accurate coalignment of this spectrograph with the uv spectrometer–spectroheliometer and with the H-$\alpha$ video display. The video system is also used to operate a servo system controlling the primary mirror, so that spectra taken with the slit parallel to the solar limb can be made at automatically selected positions relative to the limb accurate to ±1 sec of arc.

This instrument also incorporates an xuv monitor, which presents an image of the sun in the wavelength range from 170 to 550 Å on the monitor screens of the control and display panel. The optical system consists of a platinum-coated off-axis paraboloid that images the solar disk on an SEC vidicon detector. Thin aluminum filters are used to filter out the longer wavelengths. Images from this instrument may also be telemetered to the ground.

The ultraviolet spectrograph was constructed by Ball Brothers Research Corporation. The PI is Richard Tousey of the Naval Research Laboratory.


## The ATM Observing Program

In most cases exposure sequences and observing modes were not discussed in the above description of the ATM instruments. However, in each case a number of operating modes are possible, each with its particular exposure sequence. Let us consider for example one of the simpler instruments in this respect—the xuv spectroheliograph. This instrument has three automatic modes and one manual mode of operation. In automatic mode "AUTO 1" three frames are taken with different exposure times the lengths of which depend on whether the short- or the long-wavelength range is being observed. For the short-wavelength range the exposures will be 10, 40, and 160 sec each; and for the long, the exposures will be 20, 80, and 320 sec each. An additional control switch makes it possible to multiply or divide these exposures by a factor of 4.

In automatic mode "AUTO 2" six frames are taken automatically: three in the short-wavelength region and three in the long-wavelength region with exposures as noted above. In the "FLARE" mode, 15 exposures will be taken in the short-wavelength range and 9 in the long. In the manual or "TIME" mode, one exposure at a time is taken and its length is controlled by the astronaut.

It becomes obvious that interweaving the operation of six such instruments in a way that will give the maximum scientific data return is a complex task requiring considerable preplanning. However, such interweaving is highly desirable, since simultaneous observations at several wavelengths will give a more complete description of any particular phenomenon. In other cases, the observing requirements of the experiments may conflict with one another, and it is necessary to foresee and avoid such conflicts.

As a consequence, the ATM experimenters have developed 23 basic observing sequences called "building blocks." Each provides for a specific type of observation such as high-angular-resolution observations of faint or bright areas or high-time-resolution observations. Each block consists of prescribed switch settings for the various instruments plus a few additional settings to be specified by the ATM experimenters or by the astronaut just before execution (for example, whether the xuv spectroheliograph should operate at long or short wavelengths). Since the building blocks represent standard observing patterns that may be thoroughly rehearsed before the mission, astronaut operations in orbit are simplified and the astronauts may concentrate on telescope pointing or other scientific decisions.

These building blocks are, in turn, combined into Joint Observing Programs (JOP's) of which 13 have been defined. These are listed in Table 2. One of the more obvious advantages of the building blocks and

TABLE 2  ATM Joint Observing Programs

| Number | Title |
|---|---|
| 1 | Study of the Chromospheric Network and Its Coronal Extension |
| 2 | Active Regions |
| 3 | Flares |
| 4 | Prominances and Filaments |
| 5 | Constant Latitude Studies |
| 6 | Synoptic Observations of the Sun |
| 7 | Atmospheric Extinction |
| 8 | Coronal Transients |
| 9 | Solar Wind |
| 10 | Lunar Libration Points, Lunar Calibration |
| 11 | Chromospheric Oscillations and Heating |
| 12 | Program Calibration |
| 13 | Nightside Observations |

JOP's is the way in which they simplify communications between the astronauts and the group of supporting scientists on the ground. In general, the supporting scientists work out the observing "strategy" for one or two days at a time, while the astronauts decide the "tactics" by which the strategy should be executed. Strategy can be communicated to the astronauts in a relatively simple way by specifying the JOP's that should be carried out.

It is also evident that the value of the Skylab observations can be greatly enhanced if they can be coordinated with ground observations made at the many solar observatories around the world. An organization has been formed to plan such coordinated observations. Scientists interested in participating are requested to contact Robert O. Doyle at Harvard College Observatory, Cambridge, Massachusetts.

## The Astronomy Corollary Experiments

A number of the "corollary" experiments are of interest to astronomers. These include Nuclear Emulsion, Experiment S-009, which measures cosmic-ray fluxes; UV Airglow Horizon Photography, Experiment S-063, which obtains visual and ultraviolet photographs of night airglow and of the daytime ozone layer; Gegenschein/Zodiacal Light, Experiment S-073, which uses a photometer to measure the brightness and polarization of the visible background of the sky; Particle Collection, Experiment S-149, which collects interplanetary dust particles; and Galactic X-Ray Mapping, Experiment S-150, which will conduct a survey for faint x-ray sources in the 0.2- to 12-keV range. However, in the interest of keeping this paper within manageable bounds, I have elected to describe only those that pertain to solar observations (X-Ray/UV Solar Photography, Experiment S-020) or to stellar observations (UV Stellar Astronomy, Experiment S-019, and UV Panorama, Experiment S-183).

## X-Ray/UV Solar Photography, Skylab Experiment S-020

This instrument uses a grazing-incidence concave grating to obtain high-resolution spectra of the integrated solar disk in the wavelength range from 10 to 200 Å. By making exposures as long as 60 min it will be possible to record solar x-ray and extreme uv emission lines that are too faint to be detected from sounding rockets.

The optics are illustrated in Figure 18. The primary element is a concave grazing-incidence grating that images the solar spectrum on a strip

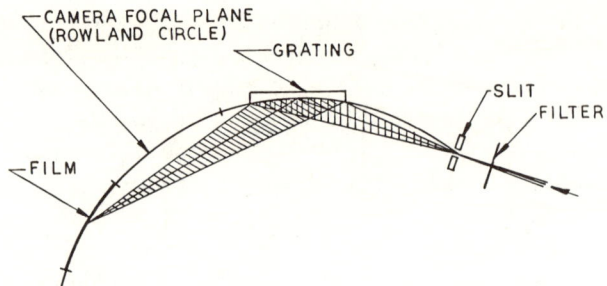

FIGURE 18  Optical schematic for the Naval Research Laboratory x-ray/uv solar spectrograph.

of Kodak 101-01 film. The grating has a double ruling, one ruling lying above the other, so that it produces two spectra simultaneously. One spectrum covers the 10 to 100 Å spectral range with a resolution of 0.05 Å, the other covers the 20 to 200 Å range with a resolution of 0.08 Å. A two-part metallic filter is positioned before the slit to reject stray light in the longer wavelengths and to suppress unwanted spectral orders.

The spectrograph views the sun through the solar-oriented scientific airlock in the Orbiting Workshop. All functions are manually controlled by the astronaut.

This instrument was constructed at the Naval Research Laboratory. The PI is Richard Tousey of the same organization.

**Ultraviolet Stellar Astronomy, Skylab Experiment S-019**

This instrument is an objective-prism spectrograph designed to obtain stellar spectra in the 1300 to 5000 Å wavelength range. Each field covers a 4° × 5° region of the sky, and the basic purpose is to conduct a survey of ultraviolet stellar spectra over as large an area of the sky as possible. The basic telescope (see Figure 19) is an $f/3$ Ritchey-Chrétien system of 15-cm aperture. The field flattener–astigmatism corrector consists of both $CaF_2$ and LiF elements in order to reduce chromatic aberration. The on-axis angular image diameter is 15 sec of arc.

All functions of the spectrograph are manually controlled. The astronaut points the line of sight; verifies the star field in a small finder telescope; advances film; and starts, stops, and times exposures all in a manual mode. The film magazine contains 164 slides of Kodak 101-06 film of which about 14 will be used for preflight and postflight calibration.

The 4° $CaF_2$ objective prism produces spectra with a resolution of

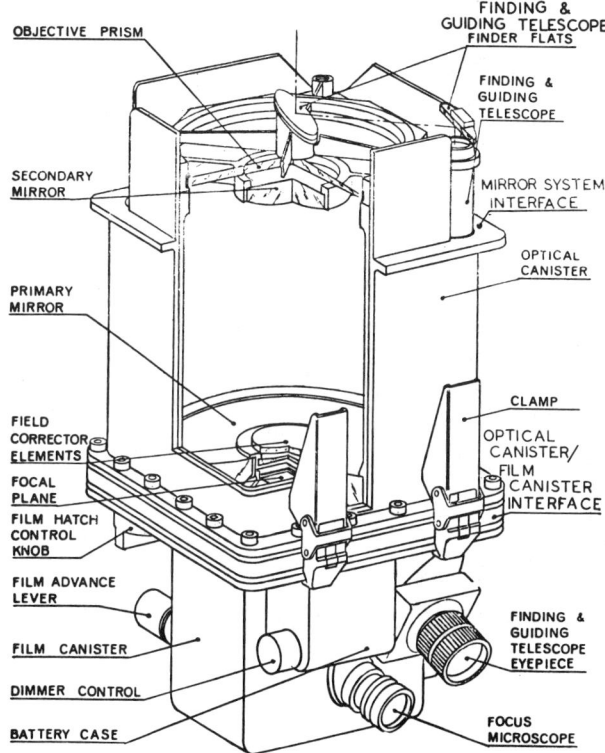

FIGURE 19  Cutaway view of the Northwestern University–University of Texas uv stellar spectrograph.

1.5 Å at 1400 Å and 31 Å at 2800 Å. The limiting visual magnitudes for B0 stars at these wavelengths are expected to be 6.0 and 8.0, respectively. Figure 20 illustrates a ground-based photograph taken with the prototype instrument.

This instrument operates through the antisolar scientific airlock in the Orbiting Workshop. Since the spacecraft is not readily maneuverable, an articulated mirror system is first extended through the airlock (see Figure 21) to allow quick pointing to any area within a 30° band around the sky. This system also incorporates a spectral widening mechanism that produces a spectrum width of 0.6 mm.

Exposures of 30, 90, and 270 sec (each with its appropriate widening rate) may be taken on each field. The limiting magnitudes are based on 270-sec exposures widened at the rate of 1 sec of arc per sec. Since spacecraft gyro drift rates on the nightside of the earth are expected to

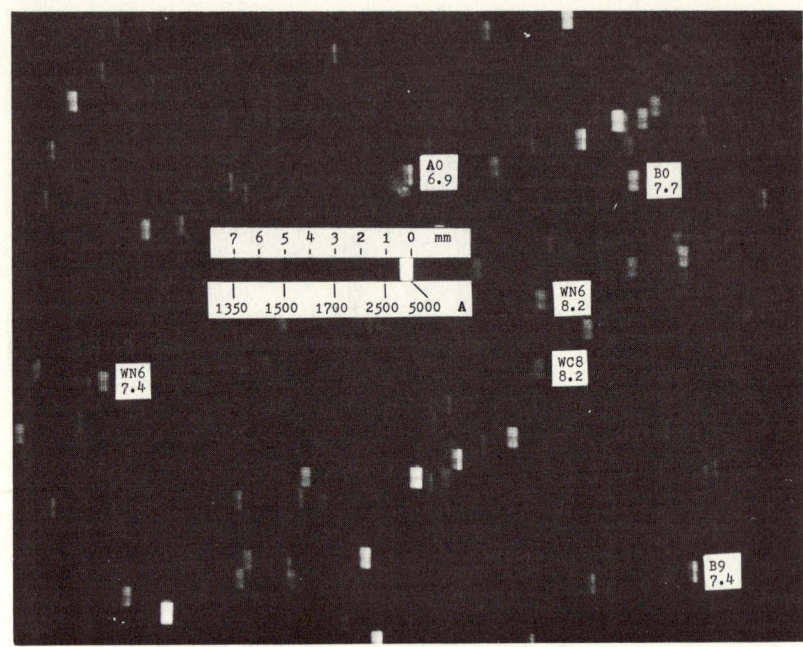

FIGURE 20 Ground-based spectra of a field of stars in Cygnus obtained with the uv stellar spectrograph prototype instrument. These spectra extend from 5000 Å at the right to 3000 Å at the left, where they are cut off by atmospheric absorption. The extent of the potential uv spectrum is indicated for one star. Note the sharp emission lines in the Wolf-Rayet stars and the Balmer absorption continuum in the stars of type B8 and later. (Courtesy of Northwestern University.)

be about one tenth of this rate, it is planned to take occasional exposures widened only by the spacecraft drift. In this case, we may extend the limiting magnitude by up to 2.5 magnitudes.

The spectrograph was designed and constructed by Cook Electric Company, with later modifications by the Boller & Chivens Division of The Perkin-Elmer Corporation. The articulated mirror system was designed at Northwestern University and constructed by Boller & Chivens. The PI is Karl G. Henize, formerly of Northwestern University but now affiliated with the University of Texas at Austin.

**Ultraviolet Panorama, Skylab Experiment S-183**

This experiment, which might be more appropriately titled "Ultraviolet Stellar Photometer," will conduct a survey of stellar colors at uv wave-

FIGURE 21  The uv stellar spectrograph mounted on the articulated mirror system. The cylinder between the rotation and tilt counters is the spectrum-widening mechanism. (Courtesy of Northwestern University.)

lengths. The primary instrument (see Figure 22) is a wide-field grating spectrograph system producing Fabry-type images in two bands with 600 Å full width at half-intensity centered on 1800 and 3000 Å. An auxiliary camera records direct photographs in a third band centered on 2500 Å.

In basic concept the primary instrument is a Fabry system in which an array of small lenses near the photographic plate forms an array of images of the rectangular entrance aperture. However, the entrance aperture is masked to form two apertures, and a grating is used to feed different wavelength ranges into each aperture. Thus each Fabry image is broken into two images, each covering a given wavelength range.

The spherical mirror at the bottom of the photometer (see Figure 22) is, in effect, the entrance aperture at which the mask is located. The true entrance aperture at the top of the photometer is imaged onto the spherical mirror by an auxiliary elliptical mirror midway up the tube. However, the elliptical mirror also images the sky onto a 610 lines/mm grating blazed at 1750 Å. Thus the spectrally dispersed star images on the grating become the "object" for the spherical mirror. Therefore, each of the two masked apertures at the spherical mirror sees the sky in only a restricted wavelength range, and the Fabry images thus represent the same restricted wavelength ranges.

The angular resolution of the system is set by the size of the Fabry

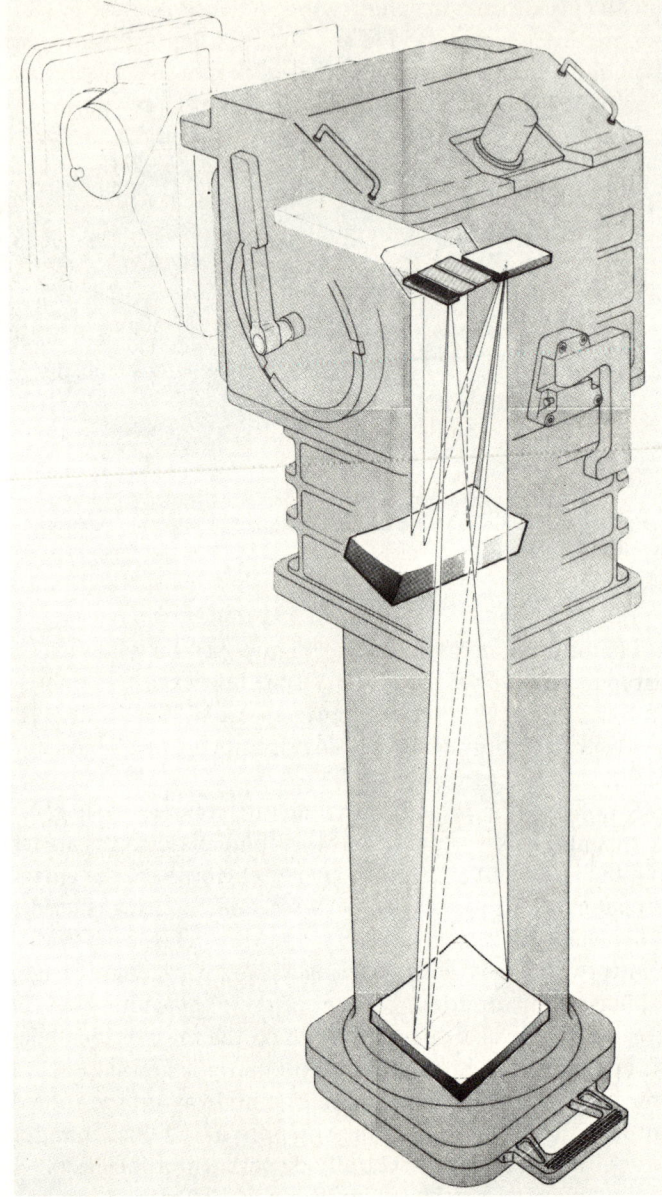

FIGURE 22 Optical configuration of the French Space Astronomy Laboratory uv stellar photometer. (Courtesy of the French Space Astronomy Laboratory.)

lenses. Both cylindrical and spherical Fabry lenses are used. The cylindrical lenses are 0.9 mm wide and give a 7 min of arc resolution (in one dimension only). The spherical lenses are 2.0 mm in diameter and give a resolution of 15 min of arc. As long as the beam from a particular star is passing through the same Fabry lens, the shape and photometric quality of its image are unaffected by instrument or spacecraft pointing. Thus pointing excursions of a few minutes of arc are of little concern. In general, only those lenses on which a beam of star light falls will produce a Fabry image. However, on long exposures the sky background will also produce Fabry images at each lens.

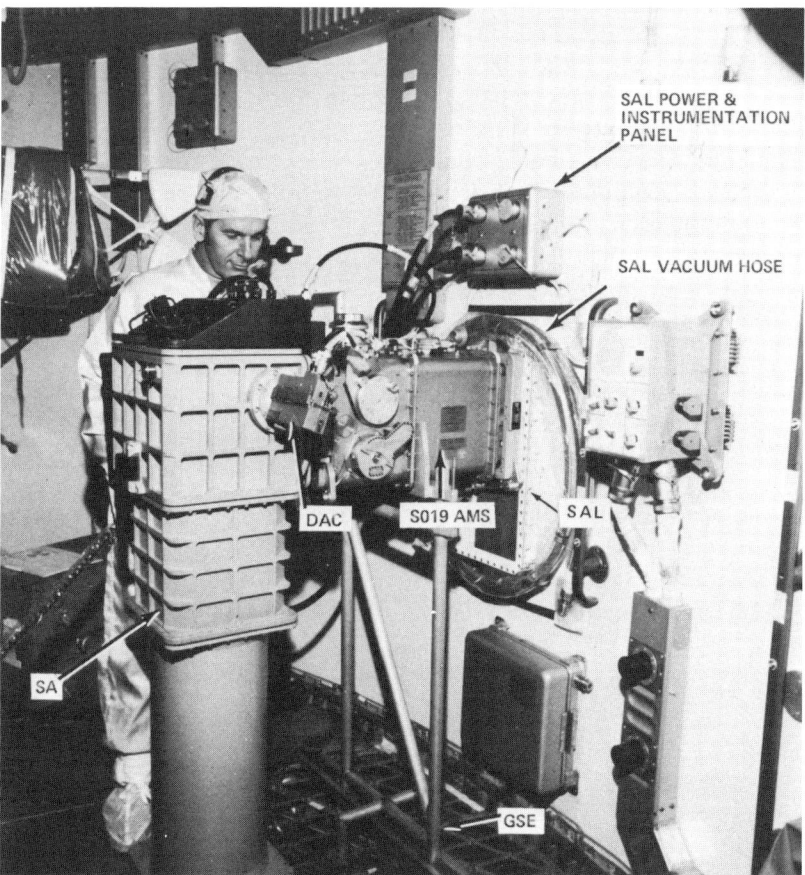

FIGURE 23  The uv stellar photometer mounted on the articulated mirror system at the antisolar scientific airlock. (NASA photo.)

The field of view of the system is 7° X 9°, and the limiting magnitude for B0 stars is about 7.0. An automatic exposure system will take exposures ranging from 0.3 to 21 min on Kodak Pathe SC5 film.

The auxiliary camera is a 16-mm movie camera with a special $f/1.5$ catadioptric objective of 60-mm focal length. The mirrors are reflection filters that isolate a 600 Å wide passband centered at 2500 Å. These data are useful not only for their astrophysical significance but also for documenting instrument pointing and image motion.

The ultraviolet stellar photometer operates through the antisolar airlock in the Orbiting Workshop and makes use of the S-019 articulated mirror system in order to point the line of sight about the sky. Figure 23 shows the uv stellar photometer and the articulated mirror system mounted on the airlock.

This instrument has the distinction of being the only Skylab instrument constructed outside the United States. The prototype instrument was designed and constructed at the Space Astronomy Laboratory, and the flight units were constructed by the SAGEM and CROUZET French industries. The PI is George Courtes of the Space Astronomy Laboratory of the National Center for Scientific Research (CNRS) in Marseille, France. The project is supported by the National Center of Space Research (CNES), the French national space agency.

In reviewing the work of other scientists, I have relied heavily on both verbal and documentary information from the PI's, their staffs, and their support contractors. I extend my gratitude to these persons for their continuing courteous assistance. Information has been taken from NASA documents to which it is difficult to give proper reference. Particular credit should be given to the *Skylab ATM Classroom Training Manuals* and to the *Experiment Operations Handbooks.* I have also borrowed heavily from "Observing Programs in Solar Physics during the 1973 ATM Skylab Program" by E. M. Reeves, R. W. Noyes, and G. L. Withbroe, which is to be published in *Solar Physics,* and from "Instruments, Systems, and Manned Operations of the Apollo Telescope Mount" by O. K. Garriott, D. L. Forsythe, and E. H. Cagle, which appeared in the June 1971 issue of *Astronautics and Aeronautics.*

M. C. E. HUBER, E. M. REEVES,
and J. G. TIMOTHY

# Photometric Calibration of an Extreme-Ultraviolet Spectroheliometer for the Skylab Mission

## Introduction

The absolute photometric calibration of the Harvard extreme ultraviolet (euv) spectroheliometer (S-055A) to be flown on Skylab in 1973 presents some unique problems, owing not only to the lack of standard euv sources available for use in space but also to the complexity of the overall mission, which required completion of final laboratory measurements 2½ years before launch. This paper briefly describes the construction and modes of operation of the flight instrument and details the procedures adopted for laboratory calibration, functional testing during the prolonged integration period prior to launch, and final calibration while the experiment is operating in orbit.

## Flight Instrument

The flight instrument is a spectrometer–spectroheliometer designed to obtain up to seven simultaneous spectroheliograms in the wavelength

---

All authors were at the Center for Astrophysics—Harvard College Observatory and Smithsonian Astrophysical Observatory, Cambridge, Massachusetts 02138 when this work was performed. M. C. E. Huber's permanent address is Atomic Physics and Astrophysics Group, Federal Institute of Technology (ETH-Z), CH-8006, Zurich, Switzerland.

range 280–1340 Å with a spatial resolution of 5 × 5 (sec of arc)$^2$. The instrument also obtains spectral scans with a resolution of 1.6 Å from any selected 5 × 5 (sec of arc)$^2$ area on the sun over the same wavelength range. It will be flown on the Apollo Telescope Mount (ATM) of the orbiting Skylab space station scheduled for launch in May 1973; owing to its large size it is expected to provide data of a quality far superior to that of any photoelectric euv spectroheliometer flown to date.

A schematic of the instrument is shown in Figure 1. An iridium-coated Cer-Vit telescope mirror forms an image of the sun on the entrance slit plate, a small portion of this image being admitted through the entrance slit into the spectrometer. A gold concave reflection grating then diffracts this radiation and focuses the euv spectrum onto seven exit slits, each equipped with an open-structure channel electron multiplier as the detector. In addition, a photocell that is sensitive to visible light and monitors the zeroth-order image of the grating is used to define a reference position in the spectral scan (optical reference). The spectrometer is mounted in a Johnson-Onaka arrangement in which the grating is rotated about an axis lying outside its surface, in a way such that the first of the seven exit slits is in exact focus for two different grating angles and hence for two different wavelengths. The exit slits are chosen and set so that the radiation of seven strong emission features, originating

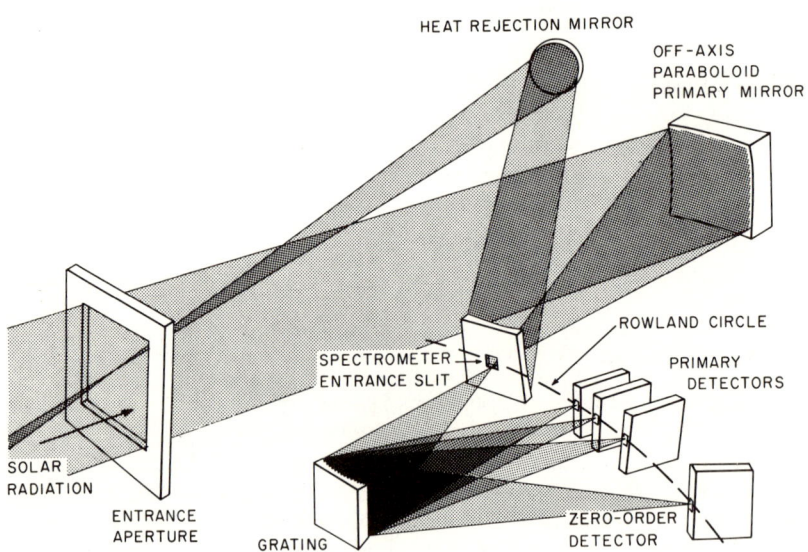

FIGURE 1  Optical schematic of the S-055A euv spectroheliometer.

at widely varying heights in the outer solar atmosphere, falls on the detectors when the zeroth-order image falls on the photocell at the optical reference position. The wavelengths monitored in this first polychromatic position are listed in Table 1. The exit slit widths are narrow enough to separate the majority of the solar euv emission lines but wide enough to admit the wavelengths of the most prominent features in the spectrum of neutral hydrogen (namely the L-$\alpha$, L-$\beta$, and L-$\gamma$ lines, as well as part of the Lyman continuum) through four of the seven exit slits in a second polychromatic position at a slightly different grating angle. Additional polychromatic positions are available in which two or more solar emission features happen to be coincident with exit slits. At this time over 50 such positions have been identified. All the exit slits are wider than the entrance slit and also significantly wider than the spectral lines generated by laboratory light sources or the solar plasma. The exit slits are also wider than the blurred spectral image of the entrance slit produced at wavelengths where the Johnson-Onaka mounting results in some defocus. The instrument line profiles, therefore, have a trapezoidal shape with a flat top, so that a detector receives the full intensity of a given spectral line for several steps of the grating scan. Wavelength scans are always taken from long to short wavelengths; at the short wavelength end of the scan the grating retraces rapidly to the long wavelength end and also actuates a microswitch that provides a mechanical reference of the grating position. The grating is driven by a stepping motor, moving through an angle equivalent to 0.2 Å per step

TABLE 1  Bandpass and Nominal Wavelength of euv Detectors and Solar Lines To Be Observed in the Two Polychromatic Modes

| Detector Number | Nominal Bandpass (Å) | Nominal Location of Slit Center in First Polychromatic Position, $\lambda$(Å) | Solar Line in First Polychromatic Position, $\lambda$(Å) | | Solar Line in Second Polychromatic Position, $\lambda$(Å) |
|---|---|---|---|---|---|
| 1 | 1.6 | 1335.7 | C II | 1335.7 | |
| 2[a] | 8.3 | 1219.0 | H I (L-$\alpha$) | 1215.7 | H I (L-$\alpha$) 1215.7 |
| 3 | 4.0 | 1031.9 | O VI | 1031.9 | H I (L-$\beta$) 1025.7 |
| 4 | 4.0 | 977.8 | C III | 977.0 | H I (L-$\gamma$) 972.5 |
| 5 | 8.3 | 896.0 | H L-cont. | 896 | H L-cont. 890 |
| 6 | 4.0 | 625.3 | Mg X | 625.3 | |
| 7 | 4.0 | 554.6 | O IV | $\begin{cases}553.3\\554.1\\554.5\\555.3\end{cases}$ | |

[a] An attenuator mounted in front of exit slit 2 reduces the flux seen by this detector by a factor of 25.

at a rate of 24 steps per sec. A complete wavelength scan thus requires a time of 3.8 min. The grating drive can also be commanded to move in single steps or to return to either mechanical or optical reference position.

The photoelectric detectors employed in the spectrometer are two-stage channel electron multipliers developed by Bendix Research Laboratories specifically for this experiment. The channel electron multiplier (see Figure 2) together with its associated charge-sensitive amplifier is potted in a module, the front face of which carries an exit slit of the spectrometer. These multipliers differ in an important respect from conventional units in that a low-resistance ($\sim 10^8$-$\Omega$) channel section has been added in series with the standard channel, which has a nominal resistance of the order of $10^9$ $\Omega$. This has the effect of increasing the wall current at the output end of the multiplier, thereby increasing the overall gain and providing a linear response to signal levels in excess of $5 \times 10^5$ counts/sec. To accommodate the astigmatic exit image of the spectrometer, a slot 8 mm long and 1 mm wide has been cut in the channel wall at the input end of the multiplier. Radiation entering this slot strikes the opposite wall of the channel at normal incidence, the photocathode being simply the semiconducting inner surface of the channel. The multiplier is operated in a pulse counting mode and has a life expectancy of greater than $10^{11}$ accumulated counts.

The telescope mirror, which is mounted in gimbals (see Figure 3), is an off-axis paraboloid with a focal length of 2.3 m and a collection area of 340 cm². Coarse pointing is achieved by orienting the complete Apollo Telescope Mount (ATM) [Garriott et al., 1971], and fine pointing is achieved by offsetting the optical axis of the telescope within a 5 X 5 (min of arc)² field of view. A telescope for H-$\alpha$, which is coaligned with the optical axis of the euv instrument provides a photographic record of the pointing. When the mirror is driven around the two-gimbal axes, the slit of the spectrometer can be made to scan part of the solar image in a raster pattern. This pattern covers the 5 X 5 (min of arc)² field of view and consists of an array of 60 lines that are scanned at a continuous speed of 1 min of arc per sec (see Figure 4). The mirror-drive system can be commanded to perform complete rasters, scan one line continuously, or fix the mirror at any position in the field of view.

The three alternative modes of operation of the instrument are as follows:

1. Spatial scans, i.e., line scan or raster, at fixed wavelengths with any grating position;
2. Wavelength scans with fixed pointing;

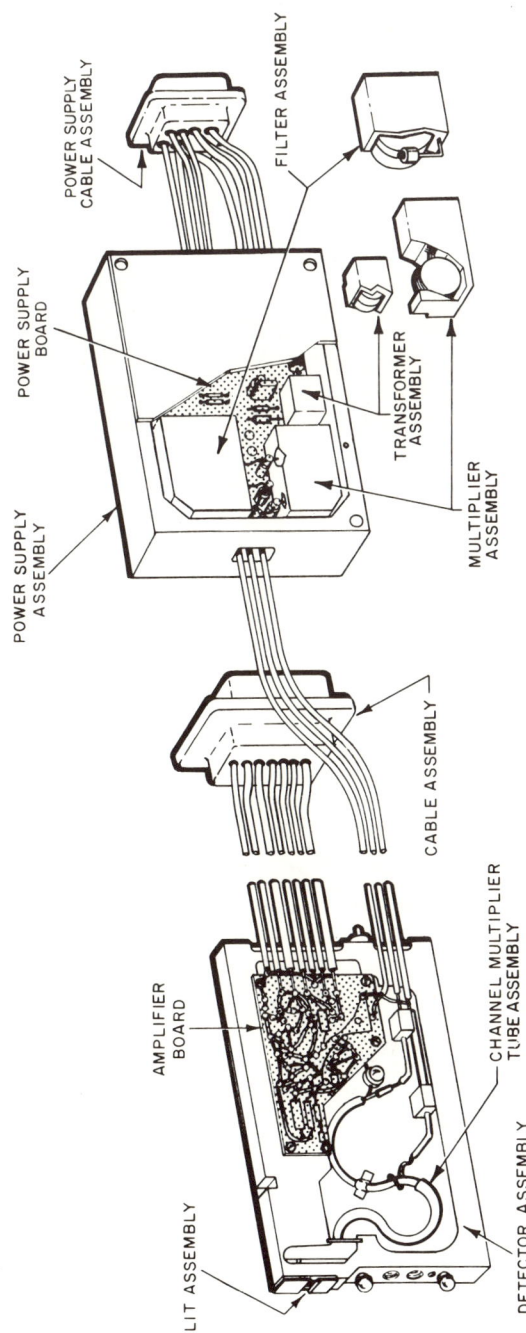

FIGURE 2  Construction of the channel electron multiplier detector unit.

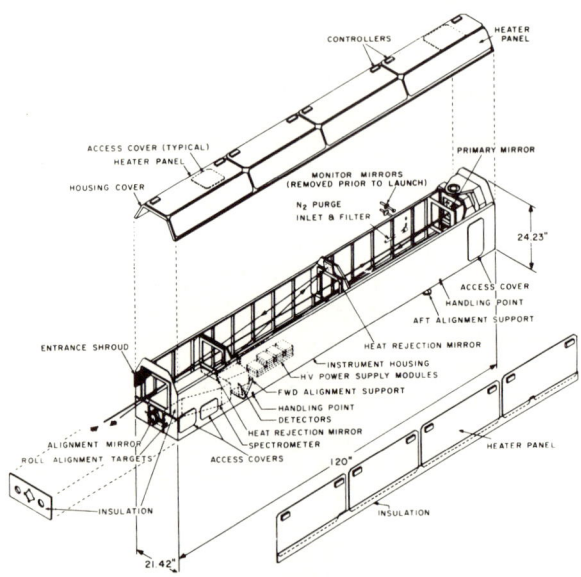

FIGURE 3 Assembly of the euv spectroheliometer.

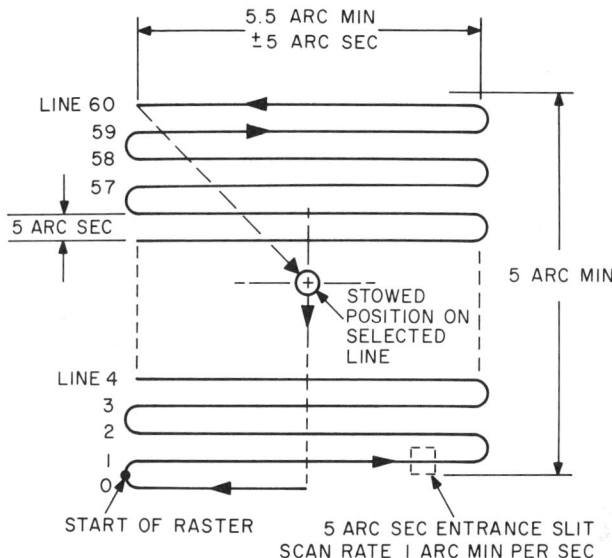

FIGURE 4 Raster pattern executed by telescope mirror.

3. Monitoring of line intensities as a function of time with both grating and telescope mirror in fixed positions.

**Photometric Calibration**

The aim of the experiment is to obtain measurements of the absolute intensity emitted by a given area of the solar disk, at a given wavelength in the case of a strong emission line or over a narrow wavelength interval in the case of continuum emission. The goal of the calibration program is to attain a photometric accuracy of ±10% over the range 280 to 1340 Å and to maintain this accuracy throughout the orbital lifetime.

The output count rate of the photoelectric spectroheliometer set to monitor radiation from a given area of the solar disk with a given wavelength bandpass can be expressed as

$$N(\lambda) = A \frac{a}{f^2} \int_{\lambda_1}^{\lambda_2} R_\lambda E_\lambda \Phi_\lambda \, d\lambda,$$

where

$N(\lambda)$ = output count rate from the spectroheliometer monitoring radiation in the wavelength range $\lambda_1$ to $\lambda_2$ (counts sec$^{-1}$ Å$^{-1}$),

$A$ = collection area of telescope mirror (cm²),
$a$ = area of spectrometer entrance slit (cm²),
$f$ = focal length of telescope mirror (cm),
$R_\lambda$ = reflectance of telescope mirror at wavelength $\lambda$ (photons/photon),
$E_\lambda$ = overall spectrometer efficiency at wavelength $\lambda$ (counts/photon),
$\Phi_\lambda$ = solar intensity in wavelength range $\lambda_1$ to $\lambda_2$ (photons cm⁻² sec⁻¹ sr⁻¹ Å⁻¹).

The overall spectrometer efficiency can be further expressed as

$$E_\lambda = \Delta_\lambda \, \epsilon_\lambda,$$

where

$\Delta_\lambda$ = diffraction efficiency of the grating at wavelength $\lambda$ (photons/photon),
$\epsilon_\lambda$ = detection efficiency of the photomultiplier at wavelength $\lambda$ (counts/photon).

Finally, the photomultiplier detection efficiency can be expressed as

$$\epsilon_\lambda = \gamma_\lambda P,$$

where

$\gamma_\lambda$ = photoelectric yield of the cathode material at wavelength $\lambda$ (photoelectrons/photon),
$P$ = probability of producing an output pulse greater than the set amplifier threshold: for the wavelength range 280 Å to 1340 Å this function may reasonably be expected to be independent of wavelength (counts/photoelectron).

The two most critical parts of the laboratory calibration are the determination of the telescope mirror reflectance and the absolute efficiency of the spectrometer. Because of the lack of high-focal-ratio standard light sources for the euv wavelength region, these measurements must be carried out independently, the spectrometer efficiency and the mirror reflectance being determined in separate experiments. The measurement of the mirror reflectance is, in principle, a relatively simple task in that any linear detector, or correlated pair of linear detectors, may be used without any knowledge of their absolute response at a given wavelength. Great care must be taken, however, to minimize errors arising both from the use of a geometry of illumination in the laboratory different from that to be expected during flight and from the effects of non-uniform illumination. The measurement of the spectrometer efficiency is the most difficult task since its response must be determined in absolute units of energy. Two approaches to the problem are available, namely, the use of an absolute radiation source or comparison with the response of an absolute detector. In the laboratory calibration of the ATM spectrometer, the output count rates of the seven detectors were compared with the signal from a standard photodiode or photomultiplier mounted behind the entrance slit when the spectrometer was illuminated with monochromatic radiation. The absolute efficiencies of these standards were determined before and after spectrometer calibration in a separate series of experiments.

The aim at Harvard is to develop a series of photometric standards having pedigrees directly traceable to the National Bureau of Standards (NBS). The calibration paths from the NBS to the ATM spectrometer are outlined in Figure 5. The standard light sources used in the laboratory are Eppley total-irradiance tungsten-filament lamps emitting in the visible and Penray mercury lamps used in conjunction with interference filters at 2537 Å. A rare-gas double-ionization chamber [Samson, 1964] is used as the absolute standard detector at euv wavelengths, and a Reeder gold black thermopile [Johnston and Madden, 1965] is used as a primary standard detector to transfer the calibration from the visible standard sources to the euv detectors. The principal transfer standard used in the spectrometer calibration was an open tungsten diode. This was independently calibrated against a rare-gas ionization chamber by Samson at GCA Corporation and against both the thermopile and a rare-gas ionization chamber at Harvard prior to the spectrometer calibration. After the spectrometer calibration, the design of the Harvard ionization chamber was improved and a calibrated, sealed, Spicer CsTe diode having a $MgF_2$ window was acquired from NBS with a calibration over the wavelength range 1164–2537 Å. Open $Al$–$Al_2O_3$ diodes that have a high stability were also obtained from NBS with calibrations over the wavelength range 584–1216 Å. The values of the efficiency of the ATM tungsten diode obtained by comparison with all these standards are summarized in Figure 6. The estimated error is ±10% below 1020 Å and ±15% at longer wavelengths. A Bendix cone-channel electron multiplier was also calibrated against the tungsten diode for use as a transfer stan-

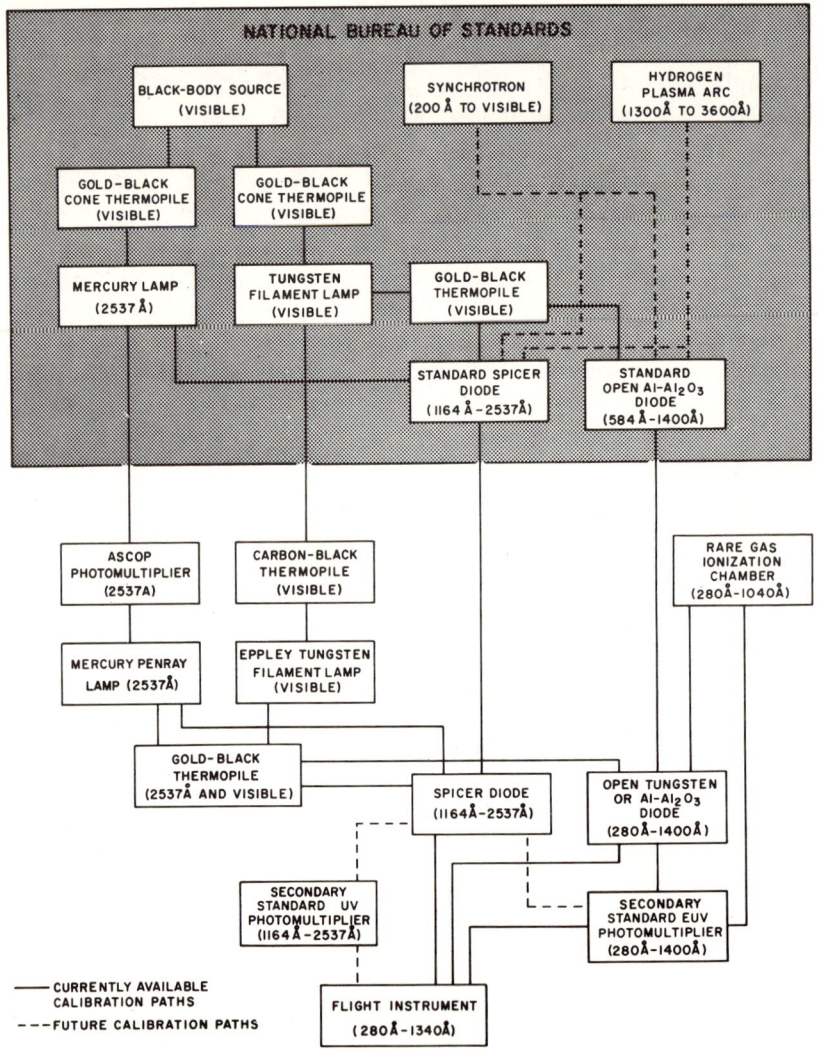

FIGURE 5 Calibration paths to the flight spectrometer.

dard for weak lines in the range 304–1216 Å. However, the cathode nonuniformity of this detector was high, and it cannot be considered to be as reliable a standard as the tungsten diode.

The layout of the laboratory system used for the calibration of the spectrometer is shown in Figure 7. A concave grating monochromator dispersed the radiation from a windowless dc hollow-cathode discharge

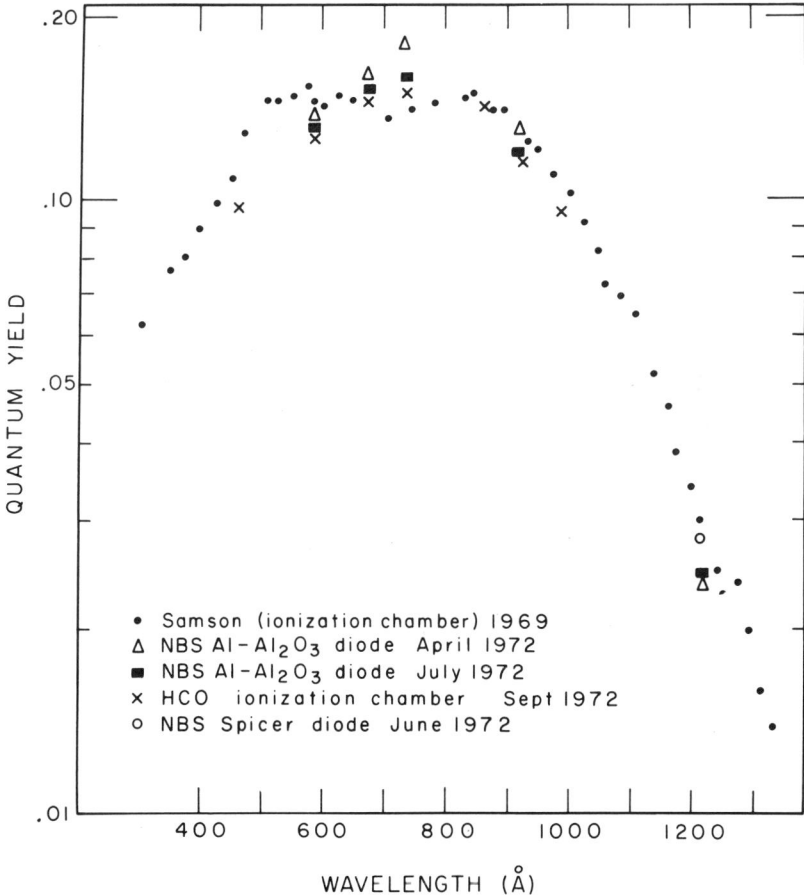

FIGURE 6  Efficiencies of standard tungsten diode.

tube [Newburgh *et al.*, 1962; Newburgh, 1963] to provide a beam of monochromatic radiation that just filled the spectrometer of the flight instrument. The planes of dispersion of the laboratory monochromator and the flight spectrometer were set at right angles to reduce effects due to the nonuniform beams diffracted by the concave gratings. The light source was operated with helium, neon, argon, krypton, xenon, or carbon dioxide to produce intense line spectra in the wavelength range 460–1335 Å. The reference tungsten diode was connected to a calibrated electrometer amplifier and was mounted on a movable arm behind the entrance slit so that it could be inserted into the beam for determining

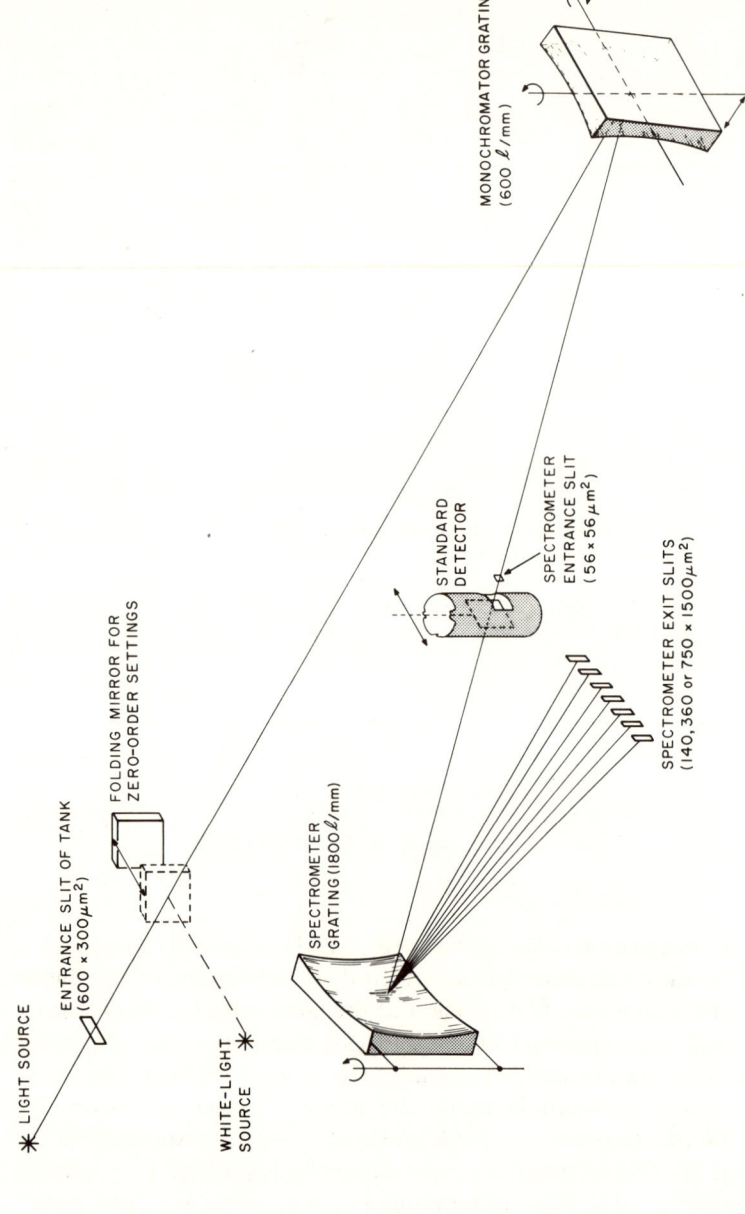

FIGURE 7 Laboratory facility for spectrometer calibration.

the photon flux entering the spectrometer. The saturation current was measured before and after the output count rates from the detectors were recorded on magnetic tape by the minicomputer in the calibration system; this allowed the elimination of errors induced by drifts in the light-source intensity. Calibration errors resulting from a nonuniform efficiency of the predispersing grating were estimated by using the laboratory monochromator in second order at selected strong emission lines. The resulting efficiency curve for the flight spectrometer, together with the associated calibration errors, is shown in Figure 8 (top). No measurements were made of polarization in the calibration beam. However, earlier measurements made on the similar OSO-6 spectrometer [Huber, 1972] indicated no measurable effect, as expected for normal-incidence optical systems. Errors arising from these effects can thus be assumed to be negligible.

The reflectance of the telescope mirror was measured in a similar laboratory monochromator. The detector used was an RCA 1P21 visible photomultiplier coupled to a sodium salicylate conversion phosphor. This was mounted on a movable arm so that the intensity of the monochromatic exit beam could be measured before and after reflection from the mirror. Because of the large size of the telescope mirror, the reflectance was measured for about 40% of the surface at one time. The average reflectance was calculated from the data for several measurements covering different areas of the surface. The measured reflectivity, together with the associated measurement errors, for the flight mirror is shown in Figure 8 (bottom). Monitor mirrors, coated and calibrated at the same time as the flight mirror, accompany the flight instrument through the integration period and provide a means of easily detecting any contamination arising from these tests.

**Extreme Ultraviolet Functional Tests**

The laboratory calibrations were completed in December 1970, i.e., about 2½ years before launch. In order to monitor changes in the sensitivity of the assembled instrument during the prolonged integration period of the ATM, it was necessary periodically to conduct functional tests with extreme ultraviolet (euv) radiation up to a time as late as possible before launch. The last opportunity to conduct such a test occurred after the thermal vacuum test of the ATM in September 1972. The light source for these tests consisted of a dc hollow-cathode discharge tube coupled to a Cassegrain collimator and mounted on a cart inside the thermal vacuum chamber (see Figure 9). The circular light-

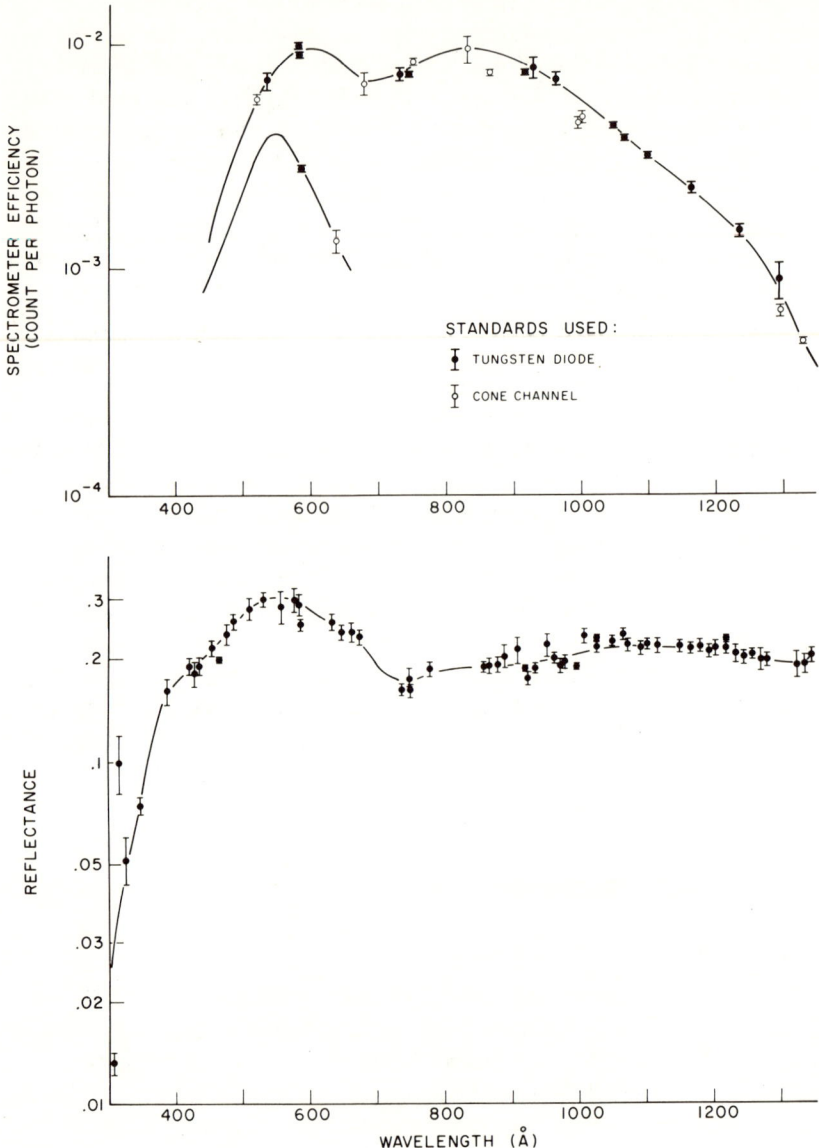

FIGURE 8 *Top:* Efficiency of flight spectrometer (detector #1) in first order (upper curve) and second order (lower curve). *Bottom:* Reflectance of flight mirror.

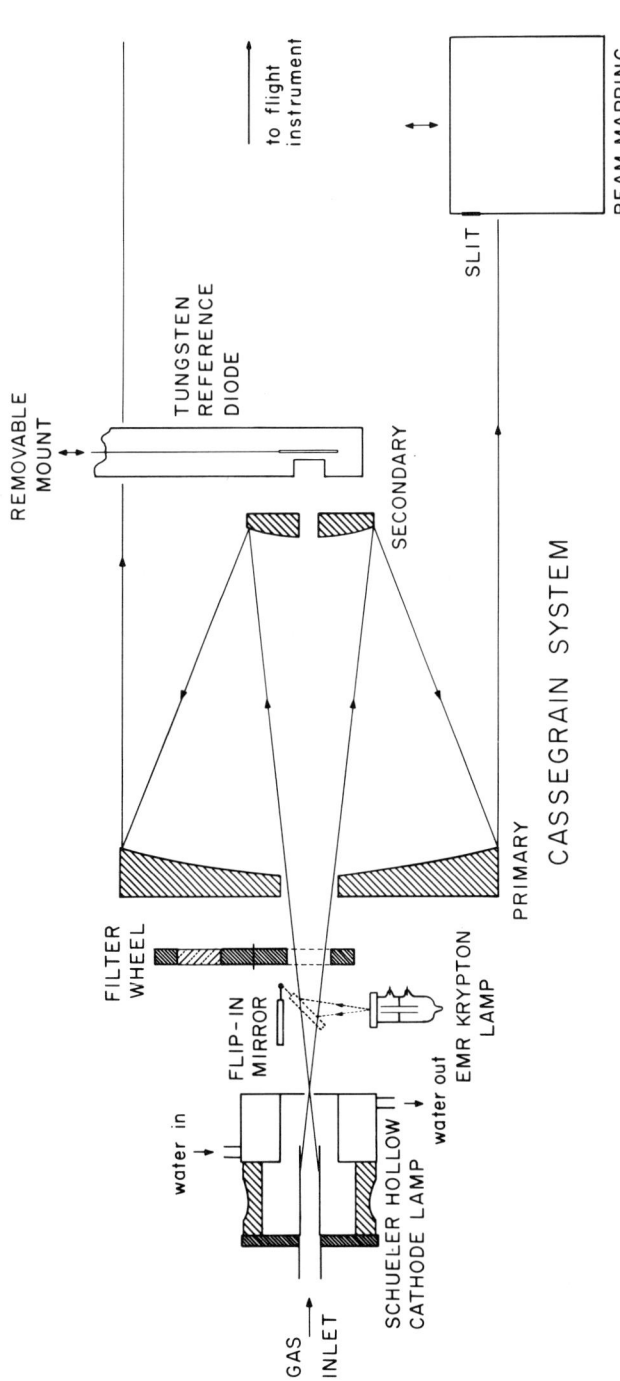

FIGURE 9 Light source for euv functional tests for assembled spectroheliometer.

source aperture was placed at the focus of the Cassegrain system. With an effective focal length of 3.3 m, the 1-mm-diam aperture produced a light beam from the primary having an angular divergence of 1 min of arc. For the thermal vacuum test of the complete ATM, the light source was mounted beneath the ATM inside the 24-m vacuum chamber at the Manned Spacecraft Center in Houston, as shown in Figure 10. Co-alignment between the optical axes of the light source and flight instrument was effected by means of jacking motors on the light-source cart. The data from the euv functional tests supplemented that obtained during the laboratory photometric calibration, and the specific aims of the functional tests were

1. To verify that the first polychromatic position coincided with optical reference,

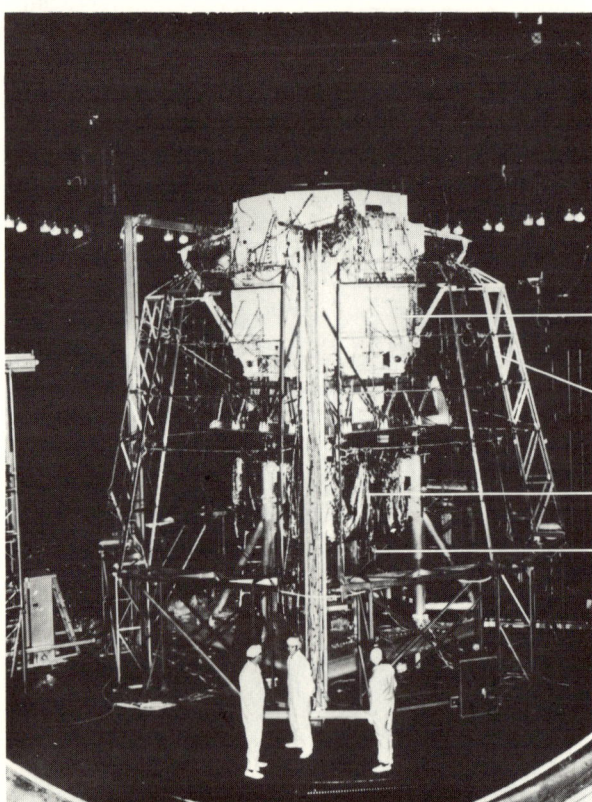

FIGURE 10  ATM and calibration light source installed in the thermal vacuum chamber at Houston.

2. To determine the high-voltage characteristics of each detector both before and after the tests,

3. To determine the noise level in the detection system by counting the noise (or dark) pulses that are present in the absence of euv illumination,

4. To determine instrument line profiles and verify detector separations from wavelength scans of different rare-gas spectra,

5. To check the absolute photometric response of the optical system for at least one wavelength.

The coincidence of optical reference and first polychromatic position was set and checked in the following way. First, the grating was set at optical reference, while the instrument was at atmosphere pressure, using either the solar image from a coelostat or a sun simulator. Then, with the instrument under vacuum and illuminated with radiation at the C II 1336 Å wavelength, the grating was stepped off the peak of the line profile. At the end of the vacuum test, the grating was set to polychromatic position and the operation of optical reference verified after repressurization.

The channel electron multipliers exhibit high-voltage characteristics as shown in Figure 11 when illuminated with a constant euv flux. The

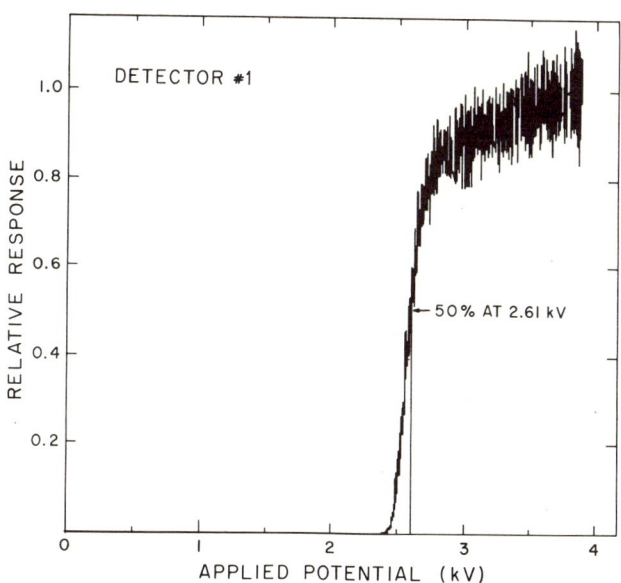

FIGURE 11   High-voltage characteristic for channel electron multiplier.

voltage at which the count rate is 50% of that obtained with the nominal operating voltage ($\sim 4$ kV) is a sensitive monitor of the overall gain of the multiplier. Any increase in the voltage at which the 50% count level is attained will, therefore, be indicative of gain fatigue in the multiplier; furthermore, any change in the overall shape of the characteristic will indicate contamination or physical damage to the channel. Thus, monitoring this characteristic in conjunction with measurements of the dark count rate (typically 1 count per 200 sec) provides the data necessary to verify the correct performance of the multiplier.

Spectral scans taken with polychromatic radiation allowed the exit-slit separations to be determined accurately when two emission lines are monitored in near coincidence on the relevant detectors. The separations determined for the flight instrument are shown in Figure 12. The same

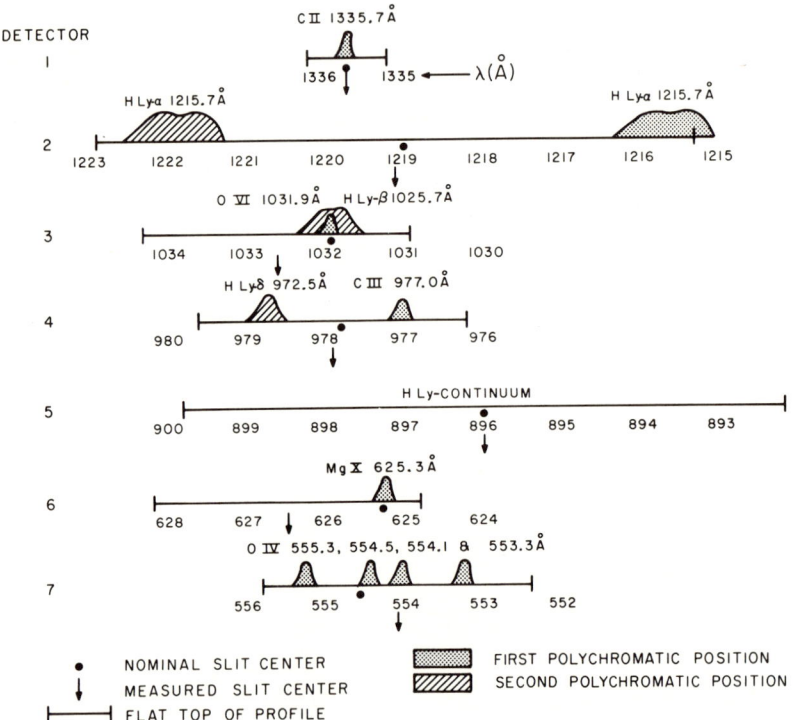

FIGURE 12 Exit slit positions in flight instrument. Solar lines are indicated with their estimated widths.

scans were also used to define the instrumental line profiles; a sample profile for detector #1 is shown in Figure 13. The profiles are narrower than predicted by ray traces of the Johnson-Onaka mounting, implying that the effective focal ratio is smaller than expected. This probably results from effects of nonuniform sensitivity in the diffraction grating at euv wavelengths.

Finally, the absolute efficiency of the instrument was measured by use of a sealed krypton discharge tube with a $MgF_2$ window mounted on the calibration light source (see Figure 9). The efficiency cannot be determined with the hollow-cathode source since this lamp emits radiation over a wide wavelength range. The sealed lamp, however, emits solely at the wavelengths of the krypton resonance lines, namely, 1164.9 and 1235.8 Å. LiF, $MgF_2$, and $CaF_2$ filters mounted on the optical system were used to define the relative intensities of the two lines, and the intensity was measured both on a calibrated tungsten diode and a calibrated channel electron multiplier, which was also employed to map the beam. The calibration carried out at the wavelengths in September 1972 indicated no detectable loss of efficiency outside the ±20% measurement error.

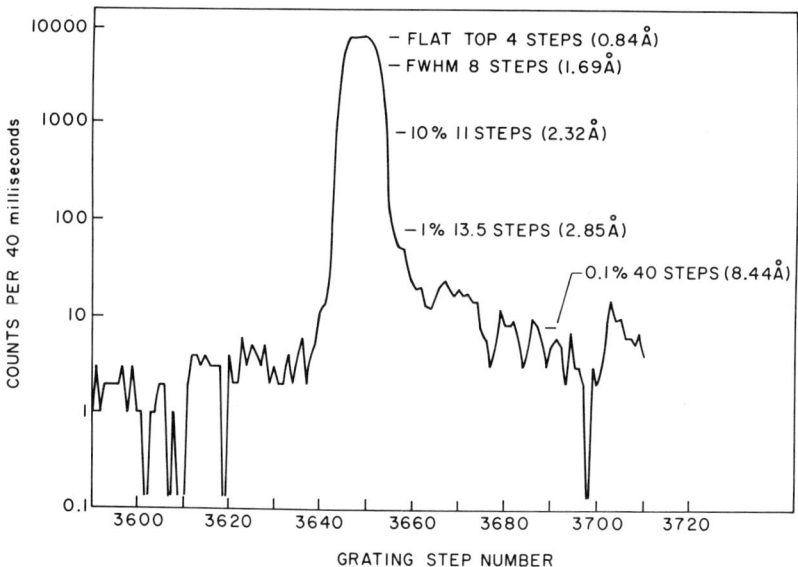

FIGURE 13 Instrument line profile at 584 Å for detector #1.

## Calibration Rocket Program

Since the final check with euv radiation took place at least 9 months before launch, and past experience with satellite instruments of this type indicates that dramatic changes in sensitivity occur in orbit [Reeves and Parkinson, 1970], recalibration after launch is vital if the desired photometric accuracy is to be maintained. At this time, no standard euv sources are available for use in satellites, and the development of such sources cannot be expected in the near future. Accordingly, the only currently feasible means of calibrating in orbit is to compare the ATM instrument with a calibrated spectrometer flown on a sounding rocket using a quiet region on the sun itself as the transfer standard.

The layout of the calibration rocket instrument is virtually identical to that of the ATM flight instrument, except that the telescope section employs a mirror having a collection area of 52 cm$^2$ and a focal length of 90 cm. The entrance slit of the spectrometer collects radiation from an area of the solar disk 20 sec of arc by 4 min of arc, a total field of 4 X 4 (min of arc)$^2$ being mapped in 12 steps by means of a single axis scan of the telescope mirror. Two-channel electron multipliers are used to cover the wavelength ranges 290–850 Å and 790–1340 Å. The resolution of the spectrometer is 1.6 Å, identical to that of the ATM instrument, and the complete instrument is calibrated both before and after flight. This calibration is transferred to the satellite instrument when both spectrometers simultaneously map the emission from a 16 (min of arc)$^2$ area on the solar disk (see Figure 14). The area observed must be free of active regions, filamentary structure, and coronal holes and also be near the disk center. Data obtained with previous flights of euv spectroheliometers suggest that the average emission from a quiet region a few minutes of arc in area on the disk was sufficiently constant with time over a 6- to 9-month period at solar maximum to serve as an intermediate standard; thus, it should be possible, even at a different part of the solar cycle, to monitor short-term changes in the relative efficiency of the satellite instrument by observing such regions on a day-to-day basis, the absolute efficiency being remeasured at regular intervals by means of the calibration rocket instrument.

## Conclusions

The data of Figure 8 show that after extensive effort the calibration accuracy attained in the laboratory for the ATM spectroheliometer is

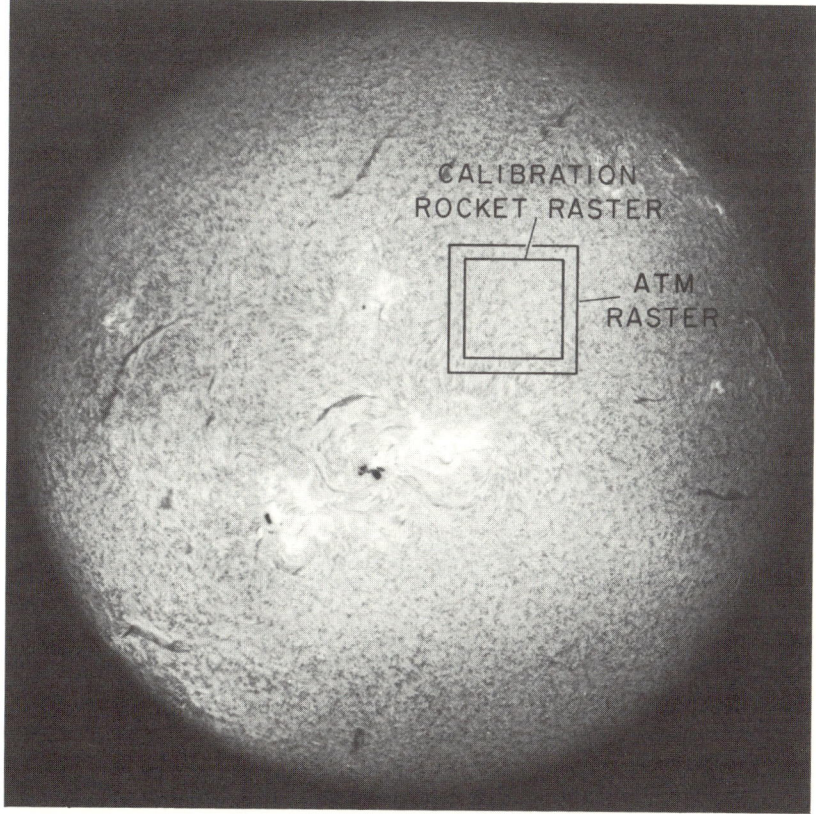

FIGURE 14 Areas of solar disk monitored by the ATM and the calibration rocket spectroheliometers.

of the order of ±20% over the wavelength range 460–1340 Å. Furthermore, this calibration is directly traceable to the NBS. These data were recorded some 2½ years before launch, and while information on the performance of the instrument has been obtained to within 9 months of launch and reflectance data will continue to be obtained from the monitor mirrors to within a few days of launch, recalibration in orbit will be essential to establish the photometric accuracy of the solar data. The quality of these data, thus, depends critically on the success of the calibration rocket program. The goal of a ±10% photometric calibration at euv wavelengths has thus still to be realized in satellite or sounding-rocket instruments.

The tests described in this paper were performed using facilities at the Harvard College Observatory (HCO) in Cambridge, Massachusetts, at Ball Brothers Research Corporation (BBRC) in Boulder, Colorado, and at the Manned Spacecraft Center (MSC) in Houston, Texas. Since over 50 people were directly involved in these tests, a complete list of acknowledgments is impractical. However, we wish to thank the staff of the HCO program at BBRC (prime contractor for this experiment) for their resourceful, courteous, and efficient support. Furthermore, the outstanding contribution of the Marshall Space Flight Center (MSFC) program office and of the MSC and MSFC personnel to the faultless operation of the functional tests in the Houston vacuum chamber is gratefully acknowledged. Finally, of the many people at HCO who contributed to the calibration effort with unfailing enthusiasm, we wish to thank especially M. L. Alford, E. S. Allen, R. M. Chambers, A. M. d'Entremont, J. C. Flagg, H. B. Freeman, W. Harby, G. U. Nystrom, and E. E. Thomson.

The work was supported by the National Aeronautics and Space Administration under contract NAS 5-3969 with the Marshall Space Flight Center.

## References

Garriott, O. K., D. L. Forsythe, and E. H. Cagle, *Astronaut. Aeronaut. 9,* 50 (1971).
Huber, M. C. E., "The Harvard Experiment on OSO-6," Harvard College Observatory Special Report (1972).
Johnston, R. G., and R. P. Madden, *Appl. Opt. 4,* 1574 (1965).
Newburgh, R. G., L. Heroux, and H. E. Hinteregger, *Appl. Opt. 1,* 733 (1962).
Newburgh, R. G., *Appl. Opt. 2,* 864 (1963).
Reeves, E. M., and W. H. Parkinson, *Appl. Opt. 9,* 1201 (1970).
Samson, J. A. R., *J. Opt. Soc. Am. 54,* 6 (1964).

WILLIAM B. WETHERELL

# IMAGE QUALITY CRITERIA FOR THE LARGE SPACE TELESCOPE

## I. Introduction

The 3-m-aperture Large Space Telescope (LST) will offer the astronomical community a tool for examining the universe that has unprecedented levels of performance. The absence of an atmosphere makes it possible to utilize fully the resolution capabilities of the 3-m aperture, within limits set only by the data sensors and by the optics of the telescope itself. In large ground-based telescopes, atmospheric seeing reduces image resolution to a level corresponding to that of a diffraction-limited, 20- to 25-cm-aperture telescope. A 3-m LST must therefore have substantially higher image quality than its ground-based counterpart. As a result, more sophisticated image-quality criteria are needed during the design definition stages, when design parameters are being chosen to achieve a balance between desired performance goals, available production technology, and costs.

In evaluating a telescope, the astronomer wishes to know system performance characteristics, such as the minimum brightness star or galaxy that can be detected, or the finest detail that can be recorded in a planetary image. The design engineer wishes to know what optical de-

---

The author is at Itek Corporation, Lexington, Massachusetts 02173.

sign parameters he must choose, how accurately he must maintain pointing during an exposure interval, and what error budgets he must establish for fabrication, alignment, and environmental effects. Traditional image-quality criteria such as two-point resolution or the Rayleigh quarter-wave criterion are not adequate for either the astronomer or the design engineer. What is needed is a criterion that will predict performance in terms meaningful to the astronomer, starting from telescope design characteristics that can be controlled by the design engineer. Given such a criterion, it is possible to show what any change in the choice of design parameters will cost in terms of loss in astronomical performance.

Three image quality criteria applicable to the LST are discussed in this paper. Two are based on signal-to-noise ratio and are concerned with point-source detection and continuous-tone imagery. The third is Maréchal's criterion for diffraction-limited image quality and is based on Strehl definition. The two criteria based on signal-to-noise ratio allow analysis of performance for the complete system, including both optics and image sensor, in terms relating astronomical performance directly to optical design parameters and the optical error budgets. Maréchal's criterion is included primarily for historical reasons, since it gives a basis for comparing the other criteria to a rigorous definition of the term "diffraction-limited performance." Strehl definition, the basis of Maréchal's criterion, also gives a convenient example to show how a general image-quality criterion can be calculated from basic design characteristics such as wavefront error.

The purposes of this paper are threefold: first, to define the image-quality criteria in a usable form; second, to show how these criteria relate performance in astronomical terms to optical image quality in terms useful to the design engineer; third, to develop elementary calculation techniques and to provide sufficient background data for the reader to use in his own preliminary design analysis. To this end, the following sections will develop the optical image-quality criteria, starting with a description of the factors affecting image quality in large, nearly diffraction-limited optics, and carrying it through to a description of their application to the point-source detection and continuous-tone imagery problems in astronomy.

## II. Image Quality and the Factors Affecting It

Image quality measures the degree to which the optical system and image sensor degrade the geometrical distribution of irradiance in collecting light from an object and forming its recorded image. The most complete measures of image quality for an optical system are its point-spread function and its optical transfer function. Image quality for the image sensor can also be measured in terms of a point-spread function and transfer function, but an additional factor relating to signal-to-noise ratio is also required.

The point-spread function measures the distribution of light in the image of a point source. The distribution of light in the image of objects other than a point source can be determined by convolving the object distribution of light with the point-spread function. The optical transfer function is the complex Fourier transform of the point-spread function. It measures the alteration in amplitude (modulation) and phase (lateral position) of the sine-wave spatial frequency components of the object in transferring light from object to image. The image distribution is computed from the object distribution and optical-transfer function by simple multiplication. This simple multiplicative relationship makes the optical transfer function particularly convenient. The image-quality criteria discussed here are all computed from the optical transfer function.

The optical transfer function $O(\nu,\phi)$ can be divided into two separate functions, the modulation transfer function $T(\nu,\phi)$ and the phase transfer function $P(\nu,\phi)$. The image-quality criteria we deal with here are concerned only with the modulation transfer function (MTF). In this paper, the MTF symbol $T(\nu,\phi)$ should always be assumed to be

$$T(\nu,\phi) = |O(\nu,\phi)|$$
$$= [O(\nu,\phi)O^\star(\nu,\phi)]^{1/2}, \qquad (1)$$

where the star ($\star$) indicates the complex conjugate. The MTF will be given in cylindrical coordinates, with the modulation transfer $T$ being expressed as a function of the sine-wave spatial frequency $\nu$ and the orientation angle $\phi$.

Six factors affect the optical system MTF:

1. Entrance pupil diameter, $D_p$
2. System focal ratio or $f$/number, $F$
3. Wavelength, $\lambda$
4. Root-mean-square wavefront error, $\omega$
5. Central obstruction diameter ratio, $\epsilon$
6. Root-mean-square image motion, $\sigma$

The first three determine the performance limit for a perfect lens and how performance scales in object and image coordinates. The last three determine how much the image quality is degraded below that of a perfect lens. In addition to these six factors, the image sensor degrades the system MTF, and scattered light can raise the background image irradiance. Scattering does affect the MTF, but its effect is negligible when compared to those of $\omega$, $\epsilon$, and $\sigma$. These latter three factors will be termed the image-quality parameters, and models will be developed for estimating their effect on image quality.

In large, well-corrected reflecting optics, such as the Ritchey-Chrétien design expected to be used for the LST, surface irregularities in the primary and secondary mirrors are expected to be the principal sources of wavefront error. These irregularities are a result of manufacturing errors, for the most part. Environmentally introduced thermal gradients will be a significant cause of surface deformations as well. Other sources of wavefront error will include defocus and misalignment of the secondary mirror. These can be caused by assembly errors, environmental effects, and, if servo control systems are used for focus and alignment control, servo loop dither. Any limitation on the maximum permissible value of $\omega$ will go primarily into defining error budgets for fabrication, thermal gradients, and secondary-mirror alignments. The telescope design parameter most likely to be restricted by these error budgets is the primary-mirror focal ratio $F_p$, principally through its influence on sensitivity to secondary mirror misalignment [see Wetherell, 1972 or Wetherell and Rimmer, 1972]. The restriction will be to place a minimum value boundary on $F_p$.

The central obstruction consists of the secondary mirror and any baffle cones attached to it. Techniques for computing the central obstruction diameter ratio are described in the literature [Young, 1967; Prescott, 1968; Wetherell, 1972]. $\epsilon$ is a function solely of the choice of telescope design parameters, of which the most significant are $F_p$ and the angular field-of-view diameter. The secondary mirror magnification $m$ also is a factor. The principal effect of placing an upper limit on $\epsilon$ would be to place an upper limit on $F_p$, since the field-of-view diameter will be defined by external considerations.

Image motion will come primarily from servo dither in the pointing stabilization servo system and from vibration of the secondary-mirror support structure. If offset tracking is accomplished on a star imaged through the telescope, magnification changes in the telescope image can cause image movement. Setting a maximum tolerable limit on $\sigma$ will affect the design of the image stabilization servo system, will establish stiffness requirements for the main structure holding the primary and secondary mirrors in alignment, and will be an input to the thermal error budget.

The relationship of the image-quality parameters to the telescope design parameters is discussed in more detail in Wetherell [1972] and Wetherell and Rimmer [1972]. For now, it is sufficient to know that they can be related. It is more important to establish their relationship to the three image-quality criteria mentioned above.

Strehl definition is computed from the volume integral under the MTF. The criterion related to point-source detection, the reciprocal equivalent sampling area, is computed from the volume integral under the square of the MTF. The criterion for continuous-tone imagery, change in the signal-to-noise ratio at different spatial frequencies, is directly proportional to the MTF. The first step in relating $\omega$, $\epsilon$, and $\sigma$ to these criteria, then, is to show how they can be used in computing the MTF.

## III. Computing MTF from the Three Image-Quality Parameters

In any lens, the MTF falls to zero at a cutoff spatial frequency $\nu_0$, and is zero for $\nu > \nu_0$. The cutoff spatial frequency is given by

$$\nu_0 = D_p/\lambda \quad \text{(cycles per radian)} \tag{2}$$

for object space and by

$$\nu_0 = 1/\lambda F \quad \text{(cycles per unit length)} \tag{3}$$

for image space. For most cases discussed in this paper, it is convenient to use the normalized spatial frequency

$$\nu_n = \nu/\nu_0, \tag{4}$$

so that results can be discussed in terms independent of either aperture diameter or focal ratio.

For an ideal lens with a circular aperture, the MTF, $T_I(\nu_n)$, is given by

$$T_I(\nu_n) = (2/\pi) \ [\arccos \nu_n - \nu_n \sin(\arccos \nu_n)] \tag{5}$$

and is invariant with orientation $\phi$. For analytical purposes, it is convenient to define the effects of $\omega$, $\epsilon$, and $\sigma$ on the MTF in terms of three MTF degradation functions, $T_\omega(\nu,\phi)$, $T_\epsilon(\nu)$, and $T_\sigma(\nu,\phi)$. The real lens MTF $T(\nu,\phi)$ is then given by

$$T(\nu,\phi) = T_\omega(\nu,\phi) \times T_\epsilon(\nu) \times T_\sigma(\nu,\phi) \times T_I(\nu). \tag{6}$$

$T_\omega$ and $T_\epsilon$ are not completely independent when the wavefront error involved is a systematic one, such as spherical aberration. However, if aberration is small and is expressed by its rms deviation measured over the clear aperture of the telescope, Eq. (6) is approximately correct. O'Neill [1963] has shown this to be valid for random wavefront errors taken as a statistical mean.

In particular cases, the MTF varies with both $\nu$ and $\phi$ and must be shown as a three-dimensional plot. Figure 1 shows the MTF of a lens having a central obstruction $\epsilon = 0.35$ and a "fixed" random wavefront for which $\omega \cong 0.1$ wavelength rms. Such fixed random wavefronts are characteristic of mirrors having surface figure errors. The lumps in the MTF of Figure 1 are due to individual irregularities in the fixed random wavefront and have no general significance. It is convenient to eliminate such irregularities from consideration by using statistical models for the wavefront error and image motion. These models are rotationally symmetric (invariant in $\phi$) and can be represented by sections $T(\nu)$. This also sets $P(\nu,\phi) \equiv 0$ for all spatial frequencies.

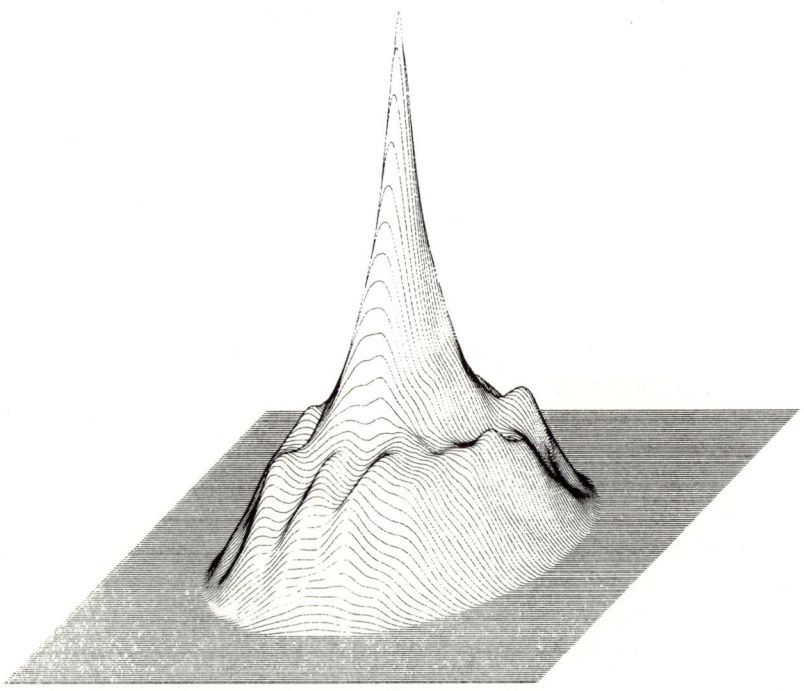

FIGURE 1   MTF for lens with random wavefront error and a central obstruction.

All of our calculations will involve the MTF. The point-spread function is useful in comparing the differences in how wavefront error, the central obstruction diameter ratio, and image motion affect image quality. Figure 2 shows the point-spread and encircled energy functions for a perfect lens and will be used as a standard for comparison to similar curves for lenses degraded by the three-image quality parameters.

### A. ROOT-MEAN-SQUARE WAVEFRONT ERROR ($\omega$)

Wavefront error is measured in terms of the optical path difference $\omega$ between the real wavefront and a reference sphere. The center of the reference sphere is placed at best focus, so that $\omega$, averaged over the entire wavefront passing through the exit pupil, is zero. The mean-square deviation between the real wavefront and the reference sphere is $E_0$. The rms wavefront error $\omega$, as used here, is

$$\omega = (E_0/\lambda^2)^{1/2}. \tag{7}$$

O'Neill [1963] has shown that the MTF degradation function $T_\omega(\nu)$ for random wavefront error is

$$T_\omega(\nu) = \exp\left(-\left\{(2\pi\omega)^2[1-\Phi_{11}(\nu)]\right\}\right), \tag{8}$$

where $\Phi_{11}(\nu)$ is the autocorrelation function for the wavefront. It is generally assumed that the autocorrelation function for random wavefronts is Gaussian, based on studies of atmospheric turbulence and ionospheric disturbance [Hufnagel and Stanley, 1964; Hufnagel, 1965; Ratcliffe, 1956; Barakat, 1971]. The form used here is attributed to Hufnagel, although no specific reference has been found:

$$\Phi_{11}(\nu_n) = \exp(-2N_H^2 \nu_n^2). \tag{9}$$

The Hufnagel constant $N_H$ is the reciprocal of the spatial frequency at which $\Phi_{11} = 0.135$. Above this frequency, $\Phi_{11}$ rapidly decreases to zero, and $T_\omega(\nu) = \exp[-(2\pi\omega)^2]$.

Figure 3 shows $T_\omega(\nu)$ for three values of $N_H$, 3, 7, and 12, and the same value for wavefront error $\omega = 0.10$. The larger the value of $N_H$, the more rapidly the MTF will drop at low spatial frequencies. $N_H$ also affects the point-spread function, as can be seen by comparing Figures 4 and 5. These represent the two extremes of Figure 3, $N_H = 3$ and 12. As can be seen, substantially more energy has been projected out into the ring structure when $N_H = 12$.

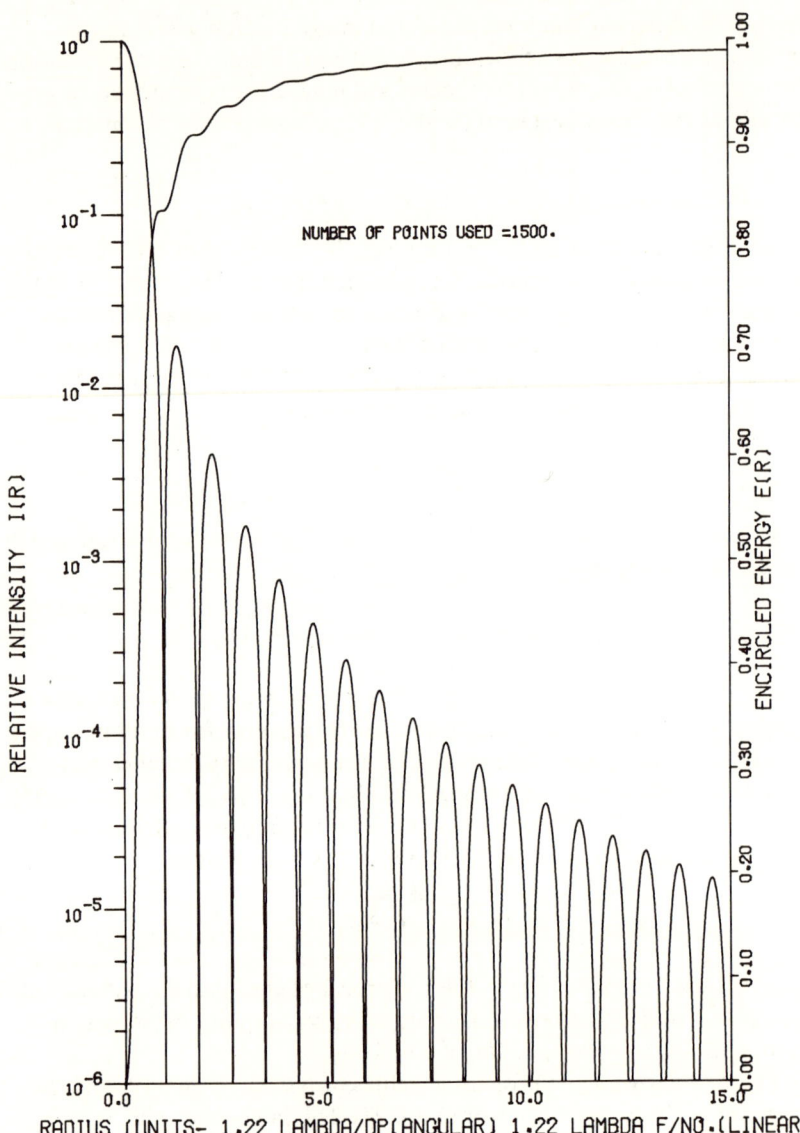

FIGURE 2 Point-spread function for a perfect lens.

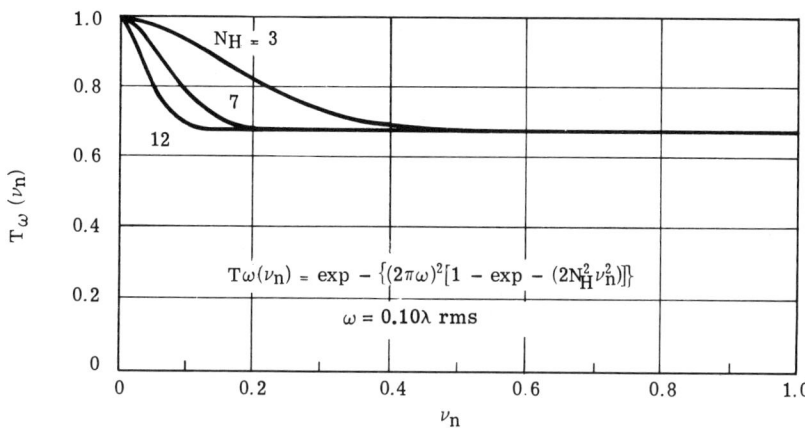

FIGURE 3  MTF degradation function for Hufnagel random wavefront error, $\omega = 0.10$ wavelength rms.

Figures 4 and 5 are examples of a general rule: reductions in the MTF at low spatial frequencies are associated with energy transfer well out into the ring structure of the point-spread function. Both are caused by small-scale (i.e., small fraction of the pupil diameter) irregularities in the wavefront. In finishing mirrors for the LST, it is necessary to be concerned with reducing $N_H$ as much as with reducing $\omega$. Rough measurements of the autocorrelation function of a high-quality 48-in.-aperture parabola suggests that $N_H = 7$ to $9$ may be typical for the LST mirrors.

A comparison of Figures 2, 4, and 5 shows how random wavefront error affects the point-spread function. The positions of the rings remain about the same. The minima (dark rings) are filled in, and maxima (bright rings) near the center of the pattern are increased. Beyond a certain radius, about 15 Airy radii in these cases, the maxima are about the same as for a perfect lens. The general impression is that a Gaussian point spread has been added to the normal diffraction pattern at the cost of removing energy from the center of the pattern. [Note that the relative intensity has been normalized to that for the perfect lens, so that the relative intensity at $r = 0$ represents the Strehl definition (see Section IV).]

B. CENTRAL OBSTRUCTION DIAMETER RATIO ($\epsilon$)

Analytical equations for the MTF of a perfect lens with a central obstruction have been given by O'Neill [1956] and Levi [1968]. The

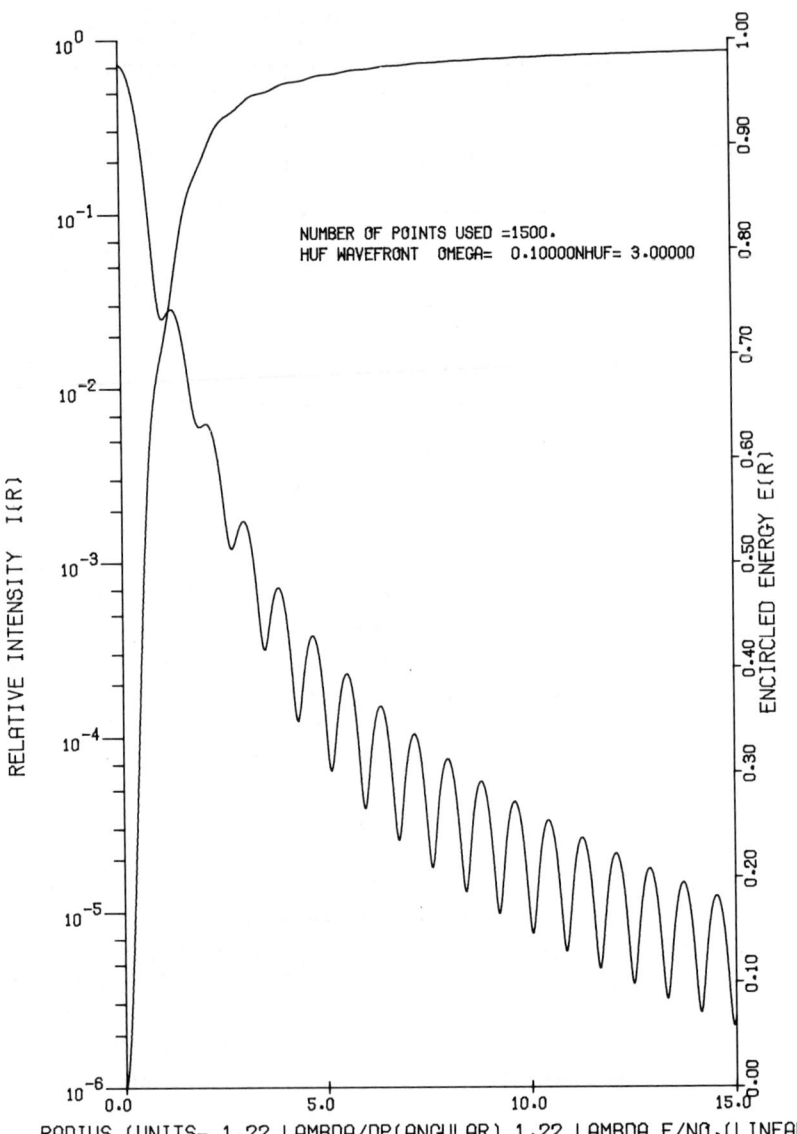

FIGURE 4  Point-spread function, $\omega = 0.10, N_H = 3$.

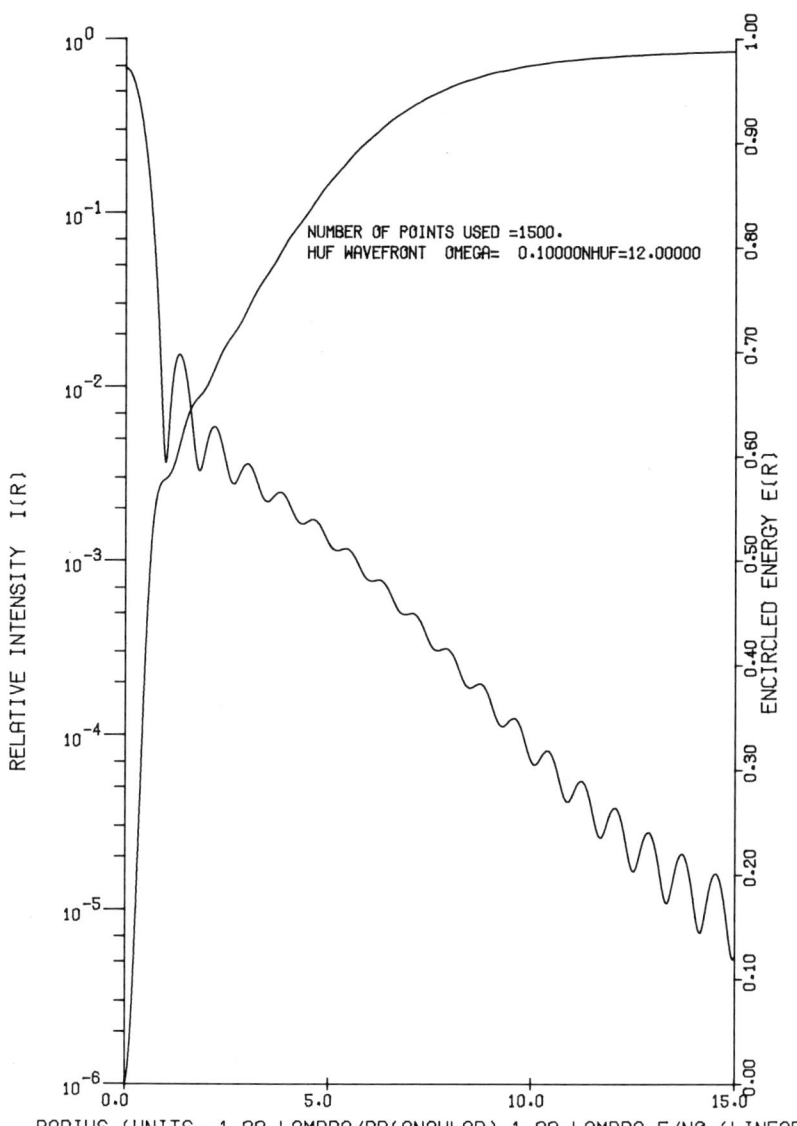

FIGURE 5 Point-spread function, $\omega = 0.10, N_H = 12$.

MTF degradation function $T_\epsilon(\nu)$ can be computed from these by dividing them by the perfect lens MTF. This has been done, and $T_\epsilon(\nu_n)$ has been plotted for $\epsilon = 0.10$ to $\epsilon = 0.70$ in Figure 6. A more complete set of values for $T_\epsilon(\nu_n)$ has been listed in Table 1, at $\epsilon$ and $\nu_n$ increments of 0.05 from 0.05 to 0.95. $T_\epsilon$ is given rather than $T_\epsilon \times T_I$, which is the more common practice, because $T_\epsilon$ will be used as a separate function in discussing continuous-tone imagery.

In general, the analytical equation for $T_\epsilon(\nu)$ is too complex to be worth repeating here. If $\nu_n \geqslant (1 + \epsilon)/2$, however,

$$T_\epsilon = (1 - \epsilon^2)^{-1}. \tag{10}$$

Since Eq. (10) produces values of $T_\epsilon$ greater than 1.0, it signifies that a central obstruction increases the MTF at high spatial frequencies. This is sometimes considered to be an artifact of the convention of normalizing the MTF to 1.0 at $\nu = 0$. It will be shown, however, that it has a real significance in signal-to-noise calculations.

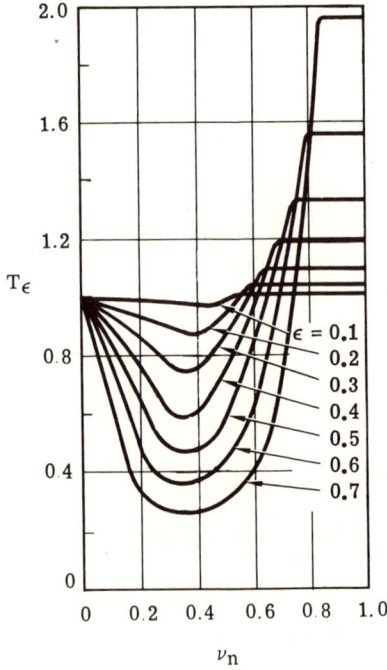

FIGURE 6  MTF degradation function $T_\epsilon(\nu_n)$.

TABLE 1  $T_\epsilon(\nu_n)$ as a Function of $\nu_n$ and $\epsilon$

| $\nu_n$ \ $\epsilon$ | 0.05 | 0.10 | 0.15 | 0.20 | 0.25 | 0.30 | 0.35 | 0.40 | 0.45 | 0.50 | 0.55 | 0.60 | 0.65 | 0.70 | 0.75 | 0.80 | 0.85 | 0.90 | 0.95 |
|---|---|---|---|---|---|---|---|---|---|---|---|---|---|---|---|---|---|---|---|
| 0.05 | 0.9972 | 0.9927 | 0.9882 | 0.9832 | 0.9775 | 0.9710 | 0.9635 | 0.9548 | 0.9445 | 0.9321 | 0.9170 | 0.8981 | 0.8738 | 0.8415 | 0.7962 | 0.7282 | 0.6149 | 0.3884 | 0.1740 |
| 0.10 | 0.9968 | 0.9870 | 0.9761 | 0.9649 | 0.9525 | 0.9384 | 0.9223 | 0.9035 | 0.8814 | 0.8549 | 0.8225 | 0.7820 | 0.7299 | 0.6606 | 0.5635 | 0.4178 | 0.2906 | 0.1874 | 0.0921 |
| 0.15 | 0.9963 | 0.9852 | 0.9662 | 0.9462 | 0.9255 | 0.9024 | 0.8761 | 0.8456 | 0.8098 | 0.7668 | 0.7144 | 0.6490 | 0.5650 | 0.4530 | 0.3565 | 0.2767 | 0.2035 | 0.1339 | 0.0665 |
| 0.20 | 0.9958 | 0.9831 | 0.9614 | 0.9301 | 0.8975 | 0.8631 | 0.8246 | 0.7802 | 0.7282 | 0.6661 | 0.5904 | 0.4960 | 0.4145 | 0.3459 | 0.2827 | 0.2230 | 0.1655 | 0.1095 | 0.0545 |
| 0.25 | 0.9952 | 0.9806 | 0.9558 | 0.9200 | 0.8720 | 0.8217 | 0.7677 | 0.7066 | 0.6353 | 0.5504 | 0.4767 | 0.4135 | 0.3552 | 0.3000 | 0.2471 | 0.1958 | 0.1458 | 0.0966 | 0.0481 |
| 0.30 | 0.9945 | 0.9777 | 0.9492 | 0.9081 | 0.8529 | 0.7818 | 0.7062 | 0.6239 | 0.5523 | 0.4903 | 0.4331 | 0.3791 | 0.3274 | 0.2775 | 0.2291 | 0.1818 | 0.1354 | 0.0898 | 0.0447 |
| 0.35 | 0.9936 | 0.9743 | 0.9413 | 0.8938 | 0.8301 | 0.7480 | 0.6632 | 0.5918 | 0.5296 | 0.4723 | 0.4182 | 0.3665 | 0.3168 | 0.2687 | 0.2218 | 0.1761 | 0.1311 | 0.0869 | 0.0432 |
| 0.40 | 0.9926 | 0.9701 | 0.9318 | 0.8765 | 0.8180 | 0.7560 | 0.6867 | 0.6078 | 0.5387 | 0.4785 | 0.4226 | 0.3698 | 0.3192 | 0.2703 | 0.2229 | 0.1767 | 0.1314 | 0.0870 | 0.0432 |
| 0.45 | 0.9913 | 0.9649 | 0.9321 | 0.8945 | 0.8503 | 0.7983 | 0.7374 | 0.6658 | 0.5812 | 0.5080 | 0.4454 | 0.3878 | 0.3335 | 0.2816 | 0.2317 | 0.1832 | 0.1361 | 0.0899 | 0.0446 |
| 0.50 | 0.9962 | 0.9848 | 0.9660 | 0.9396 | 0.9052 | 0.8621 | 0.8092 | 0.7449 | 0.6669 | 0.5719 | 0.4910 | 0.4233 | 0.3616 | 0.3039 | 0.2490 | 0.1963 | 0.1453 | 0.0958 | 0.0474 |
| 0.55 | 1.0025 | 1.0089 | 1.0089 | 0.9965 | 0.9743 | 0.9422 | 0.8991 | 0.8438 | 0.7740 | 0.6865 | 0.5764 | 0.4841 | 0.4086 | 0.3406 | 0.2774 | 0.2176 | 0.1605 | 0.1054 | 0.0520 |
| 0.60 | 1.0025 | 1.0101 | 1.0230 | 1.0417 | 1.0446 | 1.0308 | 1.0031 | 0.9611 | 0.9032 | 0.8263 | 0.7258 | 0.5946 | 0.4859 | 0.3991 | 0.3219 | 0.2508 | 0.1839 | 0.1202 | 0.0591 |
| 0.65 | 1.0025 | 1.0101 | 1.0230 | 1.0417 | 1.0667 | 1.0989 | 1.1069 | 1.0899 | 1.0523 | 0.9929 | 0.9077 | 0.7899 | 0.6283 | 0.4961 | 0.3931 | 0.3027 | 0.2201 | 0.1429 | 0.0699 |
| 0.70 | 1.0025 | 1.0101 | 1.0230 | 1.0417 | 1.0667 | 1.0989 | 1.1396 | 1.1905 | 1.2045 | 1.1804 | 1.1236 | 1.0298 | 0.8896 | 0.6824 | 0.5149 | 0.3881 | 0.2784 | 0.1790 | 0.0868 |
| 0.75 | 1.0025 | 1.0101 | 1.0230 | 1.0417 | 1.0667 | 1.0989 | 1.1396 | 1.1905 | 1.2539 | 1.3333 | 1.3548 | 1.3138 | 1.2150 | 1.0436 | 0.7666 | 0.5432 | 0.3796 | 0.2401 | 0.1151 |
| 0.80 | 1.0025 | 1.0101 | 1.0230 | 1.0417 | 1.0667 | 1.0989 | 1.1396 | 1.1905 | 1.2539 | 1.3333 | 1.4337 | 1.5625 | 1.5922 | 1.5065 | 1.2973 | 0.9015 | 0.5813 | 0.3561 | 0.1673 |
| 0.85 | 1.0025 | 1.0101 | 1.0230 | 1.0417 | 1.0667 | 1.0989 | 1.1396 | 1.1905 | 1.2539 | 1.3333 | 1.4337 | 1.5625 | 1.7316 | 1.9608 | 1.9924 | 1.7553 | 1.1354 | 0.6232 | 0.2810 |
| 0.90 | 1.0025 | 1.0101 | 1.0230 | 1.0417 | 1.0667 | 1.0989 | 1.1396 | 1.1905 | 1.2539 | 1.3333 | 1.4337 | 1.5625 | 1.7316 | 1.9608 | 2.2857 | 2.7778 | 2.7304 | 1.6146 | 0.6186 |
| 0.95 | 1.0025 | 1.0101 | 1.0230 | 1.0417 | 1.0667 | 1.0989 | 1.1396 | 1.1905 | 1.2539 | 1.3333 | 1.4337 | 1.5625 | 1.7316 | 1.9608 | 2.2857 | 2.7778 | 3.6036 | 5.2632 | 3.0712 |

A central obstruction affects the distribution of energy in the point-spread function differently than wavefront error. Figure 7 shows point-spread function for $\epsilon = 0.50$. The point-spread function for a lens with a central obstruction is formed by the "beating" of two Bessel functions, one representing the diffraction pattern for the aperture $D_p$ and the other the diffraction pattern for the obstruction. Energy is shifted from bright ring to bright ring, and the positions of the minima and maxima are shifted. The minima always go to zero, however. A central obstruction will transfer energy further out into the ring structure than wavefront error, if $N_H$ is small.

### C. ROOT-MEAN-SQUARE IMAGE MOTION ($\sigma$)

A Gaussian model for image motion will be used. That is to say, image motion is presumed to make the nominal delta-function object look like a Gaussian intensity distribution $I(r)$.

$$I(r) = I(0) \exp\left[-(r^2/2\sigma_{mg}^2)\right], \qquad (11)$$

where $\sigma_{mg}$ is the rms deviation in angular or linear units and $r$ is the radius. Taking the Fourier transform gives the MTF

$$T_{mg} = \exp\left[-(2\pi^2 \sigma_{mg}^2 \nu^2)\right]. \qquad (12)$$

To combine this function with the lens MTF, it is necessary to normalize $\sigma_{mg}$. A convenient form for normalization is

$$\begin{aligned}\sigma &= \sigma_{mg}\nu_0 \\ &= \sigma_{mg}/\lambda F \quad (\sigma_{mg} \text{ in linear units}) \\ &= \sigma_{mg} D_p/\lambda \quad (\sigma_{mg} \text{ in angular units}).\end{aligned} \qquad (13)$$

In these normalized units, the MTF degradation function $T_\sigma(\nu_n)$ is given by

$$T_\sigma(\nu_n) = \exp\left[-(2\pi^2 \sigma^2 \nu_n^2)\right]. \qquad (14)$$

Figures 8 and 9 are point-spread functions for $\sigma = 0.10$ and $0.50$. Image motion degrades the point-spread function by spreading energy from the center core and the bright rings into the adjacent dark rings. For $\sigma \geqslant 0.50$, the ring structure completely vanishes. A comparison of Figures 9 and 2 shows the average value of the smeared point-spread

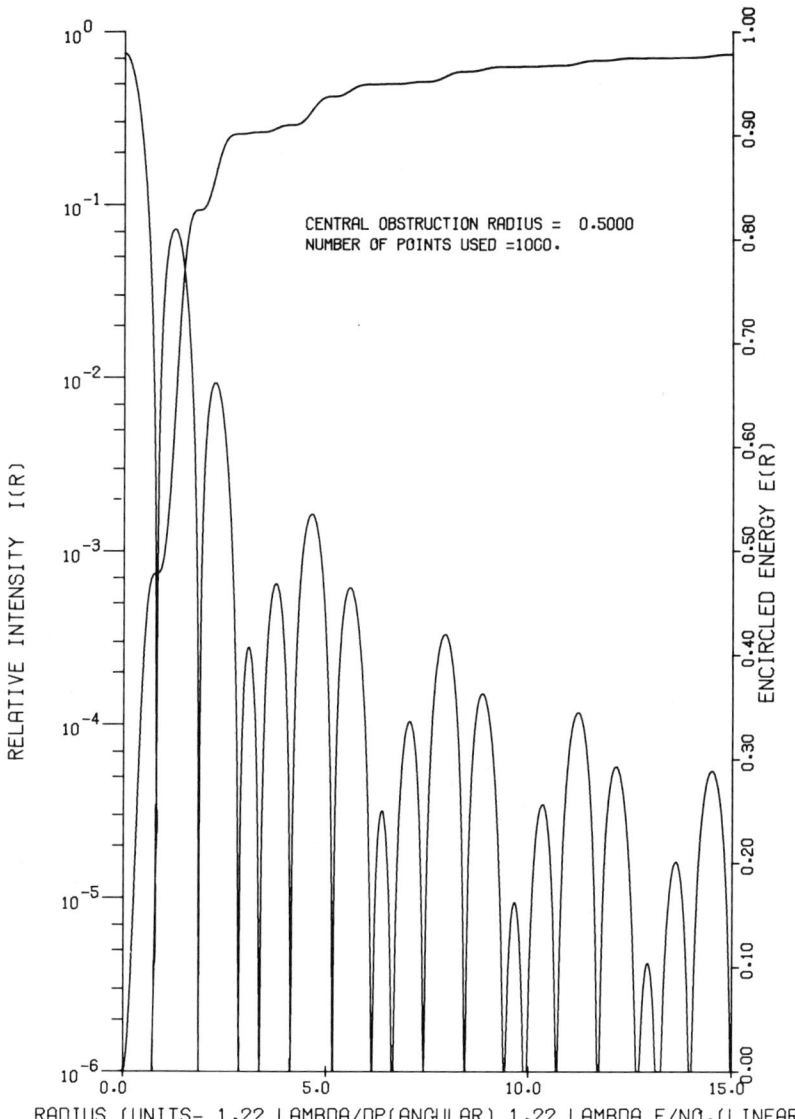

FIGURE 7  Point-spread function, $\epsilon = 0.50$.

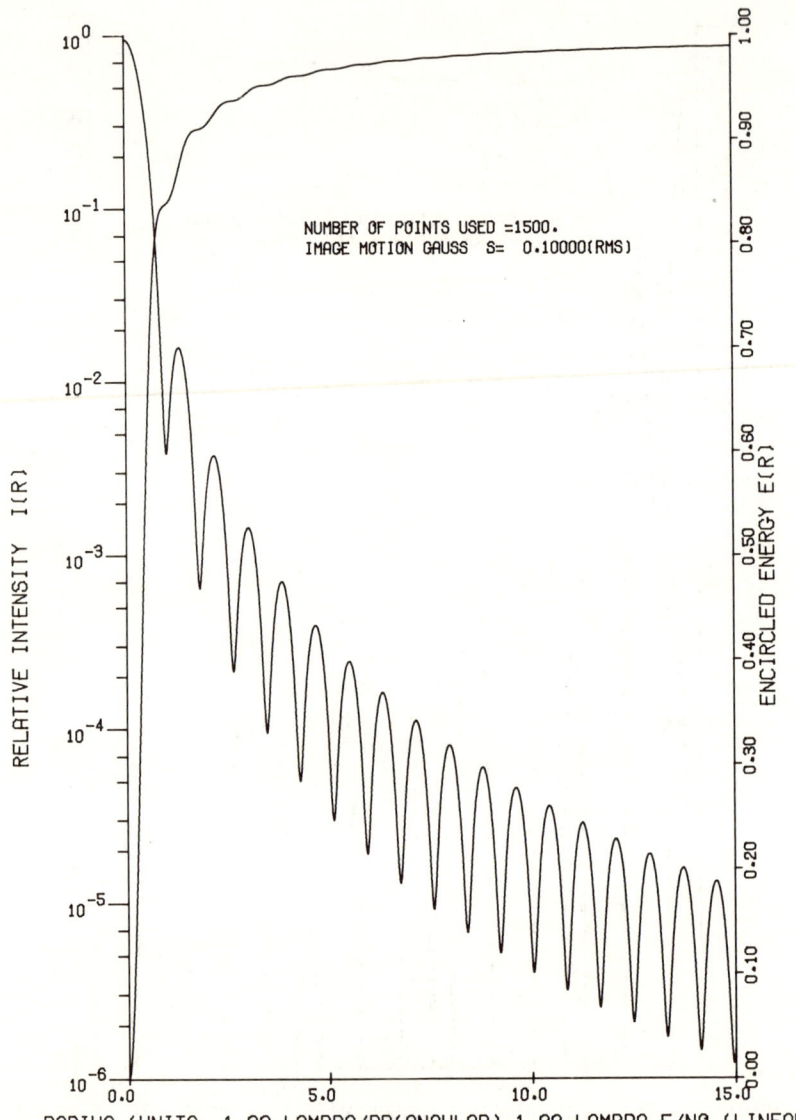

FIGURE 8 Point-spread function, $\sigma = 0.10$.

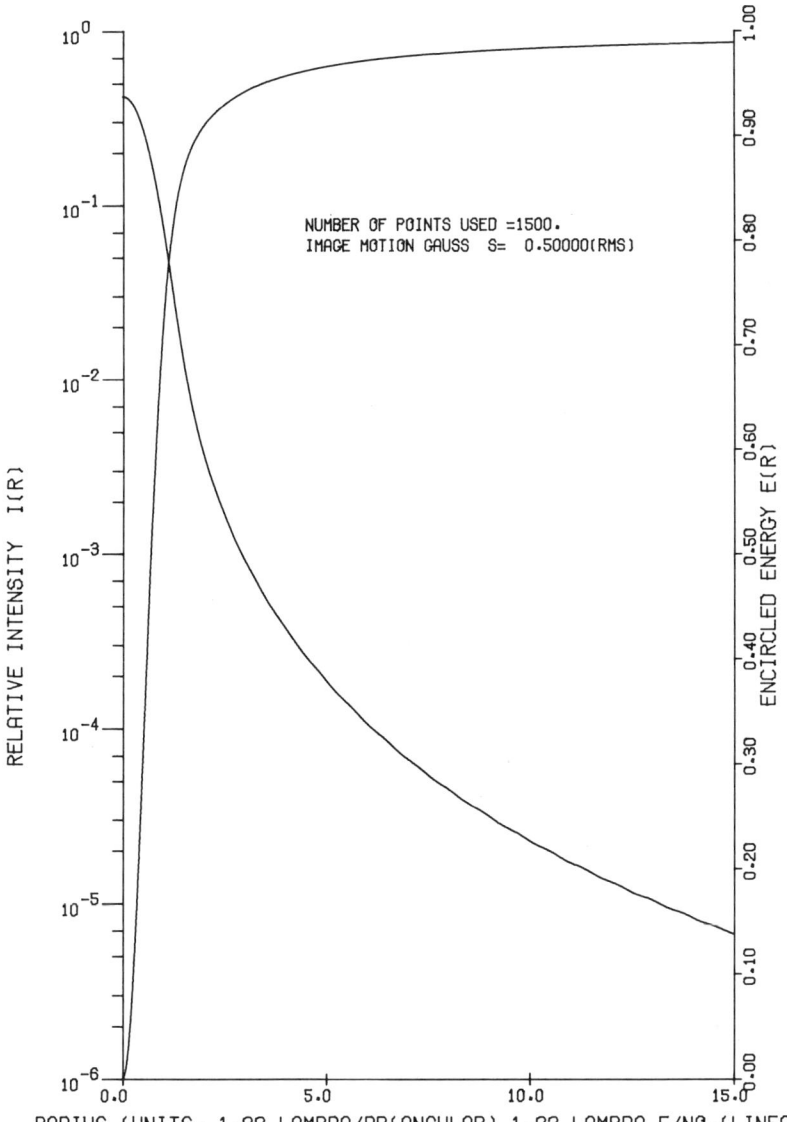

FIGURE 9  Point-spread function, $\sigma = 0.50$.

function to be one half of the maxima in the ring structure. There is no significant transfer of energy beyond the first bright ring unless $\sigma \gg 0.50$.

## IV. Strehl Definition ($\mathcal{D}$)

The Strehl definition $\mathcal{D}$ is the ratio of the peak intensity of the point-spread function of a lens to the peak intensity of the point-spread function for a perfect lens of the same aperture diameter and focal ratio. In its classic form, the intensity ratio is defined for a point-source object on a black background. Maréchal's equation for Strehl definition as a function of rms wavefront error [Born and Wolf, 1959] is

$$\mathcal{D} \geq (1 - 2\pi^2 \omega^2)^2. \tag{15}$$

A common variation of Eq. (15) is

$$\mathcal{D} \geq 1 - 4\pi^2 \omega^2. \tag{16}$$

In more general terms, Strehl definition is the normalized volume integral under the MTF,

$$\mathcal{D} = \int_0^{2\pi} \int_0^{\nu_0} T(\nu,\phi)\, \nu d\nu d\phi \Big/ \int_0^{2\pi} \int_0^{\nu_0} T_I(\nu)\, \nu d\nu d\phi. \tag{17}$$

Equation (17) can be used as the basis for extending Strehl definition to include the effects of image motion and the central obstruction diameter ratio. The latter raises some questions concerning the forms of normalization used for Strehl definition and the MTF, however. A central obstruction lowers the peak intensity by redistributing light in the point-spread function and by reducing the total light flux reaching the image plane. If both effects are considered, then the peak intensity is reduced by $(1 - \epsilon^2)^2$. Substituting a conventionally normalized MTF into Eq. (17), however, will lead to the conclusion that $\mathcal{D}_\epsilon = (1 - \epsilon^2)$.

The practical astronomical point source is a point object against a background continuum, as in Figure 10. In imaging this point source with a lens having a central obstruction diameter ratio $\epsilon$, the background irradiance is suppressed by a factor of $(1 - \epsilon^2)$, in comparison with the background irradiance with an ideal lens, because of the reduction in transmittance. In an astronomical experiment, the critical factor is exposure per unit area accumulated during an exposure interval, not irradiance. A loss in transmittance is frequently interchangeable with exposure

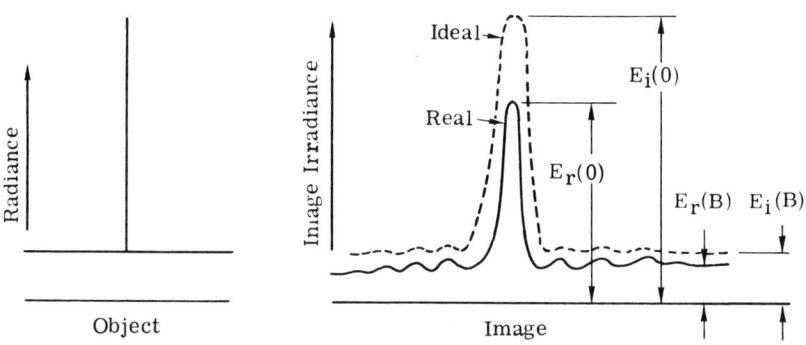

FIGURE 10 Astronomical point-source object and its image.

time, and in the situation shown in Figure 10, the common practice would be to bring the background exposure up to the same level in both cases. Under these conditions, it is more realistic to say

$$\mathcal{D} = [E_r(0)/E_i(0)] \; [E_i(B)/E_r(B)]. \tag{18}$$

Normalizing $T(\nu,\phi)$ to 1.0 at $\nu = 0$ means that the point-spread function contains unit energy. This is consistent with the practice of compensating for loss in transmittance by increasing exposure time. Equations (17) and (18) will therefore give the same answer. In effect, this form of normalizing $T(\nu,\phi)$ distinguishes between factors affecting the distribution of energy reaching the image plane and factors affecting the rate at which energy reaches the image plane. This distinction will be found to be important in calculating signal-to-noise ratio, even when exposure times are not adjusted to compensate for the loss in transmittance.

Equation (17) and the MTF degradation factors of Section III have been used to derive simplified equations for Strehl definition. In the case of Gaussian random wavefront error,

$$\mathcal{D}_\omega = \exp\left[-(C_{DH}\,\omega^2)\right], \tag{19}$$

where $C_{DH}$ is a function of $N_H$ whose value can be taken from Figure 11. For extremely large $N_H$,

$$\mathcal{D}_\omega \to \exp\left[-(2\pi\omega)^2\right], \quad N_H \to \infty. \tag{20}$$

Expanding (20) in a power series will lead to Eqs. (15) and (16). For

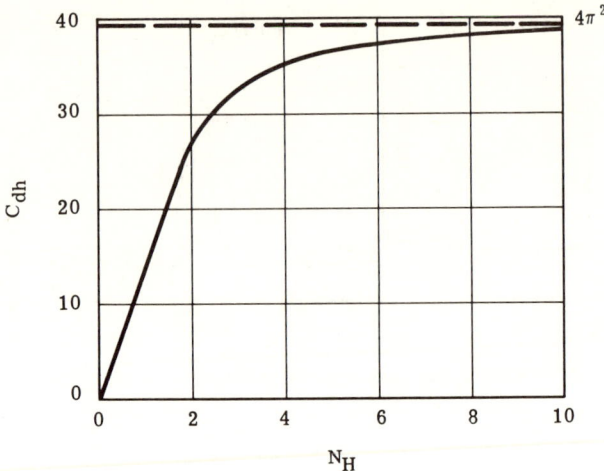

FIGURE 11  $C_{DH}$ as a function of $N_H$ [see Eq. (19)].

our purposes, expression (20) produces adequate accuracy, as long as $N_H > 6$, $\omega < 0.12$.

For the central obstruction diameter ratio,

$$\mathcal{D}_\epsilon = (1 - \epsilon^2). \tag{21}$$

For image motion $\sigma$,

$$\mathcal{D}_\sigma \simeq \exp[-(4.3\sigma^2)] \tag{22}$$

if $\sigma \leq 0.30$.

Equations (20), (21), and (22) can be considered to be independent functions, as long as $\omega$, $\epsilon$, and $\sigma$ are reasonably small. A general equation for Strehl definition can therefore be written

$$\mathcal{D} \simeq (1 - \epsilon^2) \exp\left\{-[(2\pi\omega)^2 + 4.3\sigma^2]\right\}, \tag{23}$$

$$\omega \leq 0.12, \quad \sigma \leq 0.30.$$

Maréchal's criterion for diffraction-limited performance is that $\mathcal{D} \geq 0.80$. If this value is substituted into expression (20), a graph can be produced that will show the combinations of $\omega$, $\epsilon$, and $\sigma$ just meeting Maréchal's criterion. Such an $\omega$–$\epsilon$–$\sigma$ chart for $\mathcal{D} = 0.80$ is shown in Figure 12. Curves of $\omega$ versus $\epsilon$ are plotted for constant values of $\sigma$. When $\epsilon = \sigma = 0$, $\omega = 0.075$ wavelength rms. This corresponds to Ray-

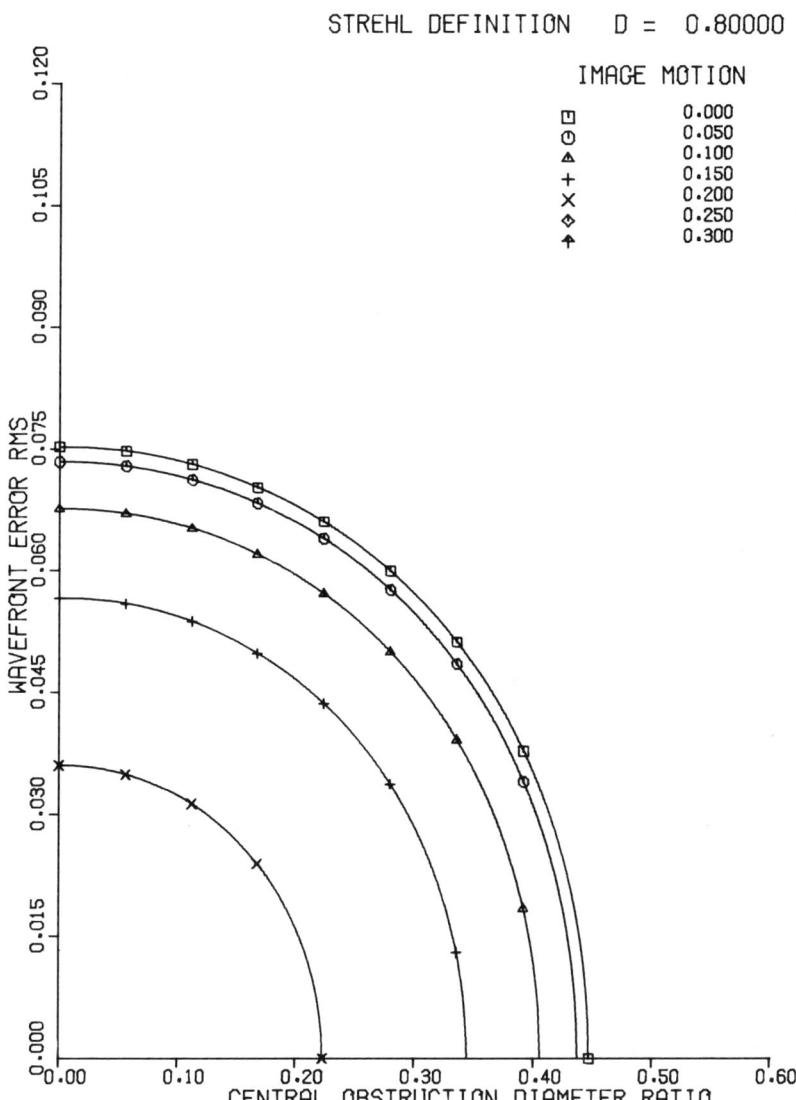

FIGURE 12  $\omega$-$\epsilon$-$\sigma$ chart for Maréchal's criterion.

leigh's quarter-wavelength limit for spherical aberration. If $\mathscr{D} \geqslant 0.80$ is taken as a rigorous definition for diffraction-limited performance, it will be seen that the optical path difference must be considerably less than one-quarter wavelength for a telescope having a typical central obstruction ($\epsilon \simeq 0.30$ to $0.40$) and image motion ($\sigma \simeq 0.10$).

## V. Normalized Reciprocal Equivalent Sampling Area ($Q$)

The concept of reciprocal equivalent sampling area derives from work by Schade [1951, 1952, 1953, 1954, 1955] in defining image-quality criteria for television cameras and other image sensors. Schade was looking for a criterion that would agree with the visual concept of image sharpness but that could be calculated from objectively measured data. Schade defined a quantity he termed "equivalent passband," $N_e^*$, which measures the amount of white noise passed by the optical system and image sensor. The concept of normalized reciprocal equivalent sampling area ($Q$), or normalized equivalent passband, is a direct descendant of $N_e^*$.

Schade pictured image sharpness as a measure of the total harmonic content of a scene. The eye is viewed as scanning the scene repeatedly and measuring its harmonic content in the same way as an ac wattmeter measures the mean-square deviation in a modulated current. Given two recorded images of the same scene, which are identical in all respects except that each was made with a system having a different spatial frequency passband, the one having the greatest harmonic content (largest mean-square deviation) will appear the sharpest.

Quantitatively speaking, scanning an image with a detector (sampling area) will produce a complex waveform $\psi$. If passed through a spectrum analyzer, it will be found to consist of a series of sine-wave components $\psi_N$, where $N$ is the associated spatial frequency in line numbers (half-cycles). In normalized form,

$$r_{\psi N} = \psi_N/\psi_0, \qquad (24)$$

where $\psi_0$ is the sine-wave component for $N \to 0$. (Schade's original notation is used here.)

Consider a uniformly illuminated plane. The modulation of this image is due to photon noise, the randomness in the position of arriving photons. If scanned with an infinitely small sampling area, it will produce a spectrum $r_{\psi N} = 1.0$, to very high spatial frequencies. If the scanning aperture has a finite size, the spectrum will fall off as in Figure 13. If the output of this detector is measure with a dc meter, an average value $\bar{\psi} =$

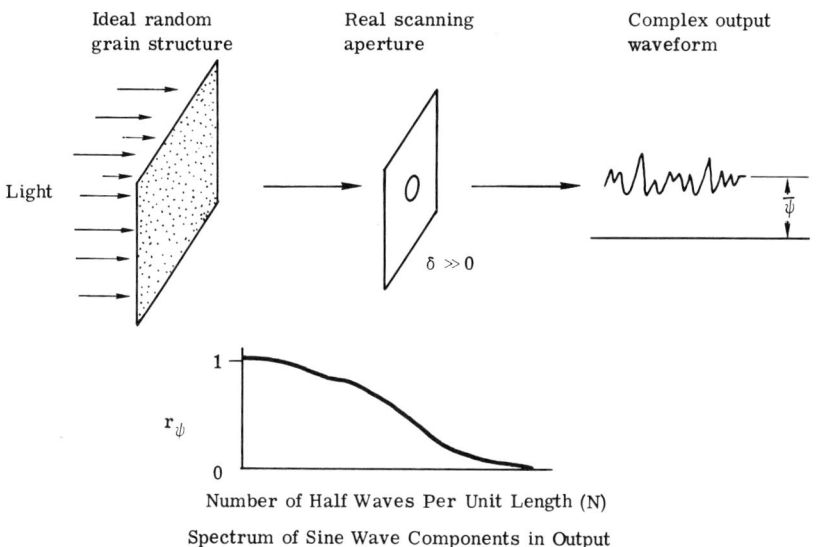

FIGURE 13  Finite-aperture detector examining random grain structure.

$\psi_0$ will be produced. Measured with a broadband ac meter, output waveform $\psi$ will be found to have an rms deviation $[\psi]$, where

$$[\psi]^2 = \psi_0^2 \int_0^\infty (r_\psi)_N{}^2 \, dN. \tag{25}$$

Suppose the original detector is replaced by one for which $r_{\psi N} = 1.0$ for $N \leqslant N_e{}^*$, and $r_{\psi N} = 0$ for $N > N_e{}^*$. The mean-square deviation would then be

$$[\psi]^2 = \psi_0^2 \, N_e{}^* \tag{26}$$

for a white-noise source. We can then define

$$N_e{}^* = \int_0^\infty (r_\psi)_N{}^2 \, dN \tag{27}$$

as the equivalent passband, or, more accurately, the white-noise-equivalent passband for a system with the response function $r_{\psi N}$. [The asterisk (*) is used here and in subsequent equations to indicate that this equivalency is for white noise only.] Since $[\psi]^2$ is directly proportional to $N_e{}^*$, the latter is a good measure of an imaging system's effect on image sharpness, by Schade's analogy.

In defining the signal-to-noise ratio for detection of a point source against a background continuum, we must measure the noise resulting from the background. This noise is photon noise, which is white and is, in fact, [$\psi$] from the above derivation of $N_e{}^*$. Bradley [1972] has considered the point-source detection problem in detail and has extended Schade's equivalent passband to two dimensions:

$$N_{e2}{}^* = \int_0^{2\pi}\int_0^\infty T^2(\nu,\phi)\,\nu d\nu d\phi \tag{28}$$

using the current notation for MTF.

$N_{e2}{}^*$ has the dimensions of reciprocal area. The reciprocal of $N_{e2}{}^*$ is an equivalent sampling area

$$A_s{}^* = 1/N_{e2}{}^*. \tag{29}$$

For a uniform background, then, the number of photons used to calculate background noise is simply

$$n_b = N_b A_s{}^* t, \tag{30}$$

where $N_b$ is the flux in detected photons (or photoelectrons) per unit area per second, and $t$ is the exposure time. The procedures for calculating signal-to-noise ratio will be discussed in more detail in Section VIII.

The design engineer is interested in comparing the performance of a real system to that of one with a perfect lens. The important quantity, then, is the normalized value of $N_{e2}{}^*$, or the normalized reciprocal equivalent sampling area $Q$:

$$Q = \int_0^{2\pi}\int_0^\infty T^2(\nu,\phi)\nu d\nu d\phi \bigg/ \int_0^{2\pi}\int_0^\infty T_l{}^2(\nu)\nu d\nu d\phi. \tag{31}$$

If we substitute the perfect lens MTF, Eq. (5), into the denominator,

$$Q = \frac{12\pi}{(3\pi^2 - 16)r_0{}^2}\int_0^{2\pi}\int_0^{\nu_0} T^2(\nu,\phi)\nu d\nu d\phi. \tag{32}$$

For rotationally symmetric MTF's,

$$Q = \frac{24\pi^2}{(3\pi^2 - 16)r_0{}^2}\int_0^{\nu_0} T^2(\nu,\phi)\nu d\nu. \tag{33}$$

$Q$ is our fundamental measure of image quality for optical systems used for point-source detection, and $A_s{}^*$ is used in the actual signal-to-noise ratio calculations. For a perfect lens

$$A_s^* \text{ (perf)} = 2.77\lambda^2 F^2. \tag{34}$$

For a real lens, then, the equivalent sampling area is

$$A_s^* = 2.77\lambda^2 F^2 Q^{-1}, \tag{35}$$

and the diameter $d_s^*$ of an equivalent circular aperture is

$$d_s^* = 1.88\lambda F Q^{-1/2}. \tag{36}$$

For a perfect lens, this is 77% of the diameter of the first dark ring.

Equation (32) or (33) can be used to calculate $Q$ for a real system, where the actual MTF is known. For general analysis of near-diffraction-limited systems, approximations for calculating $Q$ directly from $\omega$, $\epsilon$, and $\sigma$ are desired. Approximate figures have been derived numerically, using the MTF degradation functions of Section III with Eq. (33).

For random wavefront error, if $N_H = \infty$,

$$Q_\omega = \exp[-(8\pi^2\omega^2)]. \tag{37}$$

For lesser values of $N_H$,

$$Q_\omega = \exp[-(C_{QH}\omega^2)], \tag{38}$$

where $C_{QH}$ is taken from Figure 14. For $N_H = 8$, $C_{QH} = 74$, the value most commonly used in this paper. Figure 14 was derived for a series of calculations in which $\omega = 0.10$ and $N_H$ was varied from 1 to 10. The form of Eq. (38) was derived from a series in which $N_H = 8$, and $\omega$ was varied from 0.01 to 0.12 wavelength rms. In these series, Eq. (38) and the numerical integration agreed within 3 in the third place. This level of precision should hold well for large $N_H$ but may deteriorate for low $N_H$. Calculations were not made for $\omega > 0.12$.

The comparable equation for the central obstruction diameter ratio $\epsilon$ is

$$Q_\epsilon = \exp[-(2.95\epsilon^2)]. \tag{39}$$

For rms image motion $\sigma$, it is

$$Q_\sigma = 1/(1 + 6\sigma^2). \tag{40}$$

In both cases, the fit to numerical analysis is comparable to that for

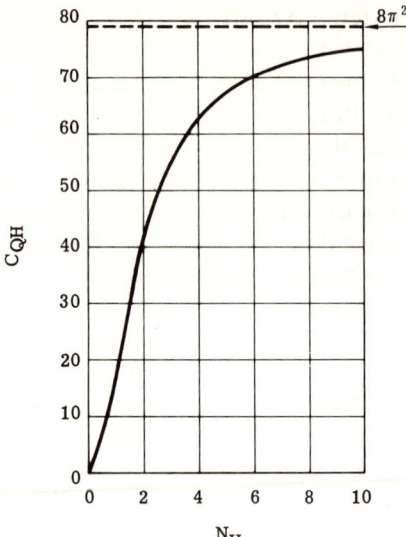

FIGURE 14 $C_{QH}$ as a function of $N_H$ [see Eq. (38)].

wavefront error, over the ranges

$$\epsilon \leqslant 0.60 \text{ and } \sigma \leqslant 0.60. \tag{41}$$

If $\sigma \leqslant 0.30$, a Gaussian fit may be used.

$$Q_\sigma \simeq \exp[-(5.50\sigma^2)], \tag{42}$$

but it is of somewhat lesser accuracy.

For the combination of $\omega$, $\epsilon$, and $\sigma$, we can say

$$Q \simeq (1 + 6\sigma^2)^{-1} \exp[-(74\omega^2 + 2.95\epsilon^2)], \tag{43}$$

when $\omega \leqslant 0.12$, $\epsilon$, $\sigma \leqslant 0.60$. If $\sigma \leqslant 0.30$, this can be reduced to

$$Q \simeq \exp[-(74\omega^2 + 2.95\epsilon^2 + 5.5\sigma^2)]. \tag{44}$$

These equations have been spot checked against numerical analysis and found to produce differences that are generally less than 5%.

For Maréchal's criterion, $\mathscr{D} = 0.80$, the wavefront error in a lens free of a central obstruction and image motion should not exceed $\omega = 0.075$ wavelength rms. The corresponding reciprocal equivalent sampling area is $Q_\omega = 0.66$. Figure 15 is an $\omega$-$\epsilon$-$\sigma$ chart for this case. A comparison to Figure 12 indicates that $Q$ is less sensitive to image motion (curves for different values of $\sigma$ are more closely spaced) and more sensitive to the size of the central obstruction diameter ratio. Any practical system for which $Q \geq 0.66$ would have to have an obstruction smaller than $\epsilon \simeq 0.25$. By this criterion as well, diffraction-limited performance would be difficult to achieve in an LST.

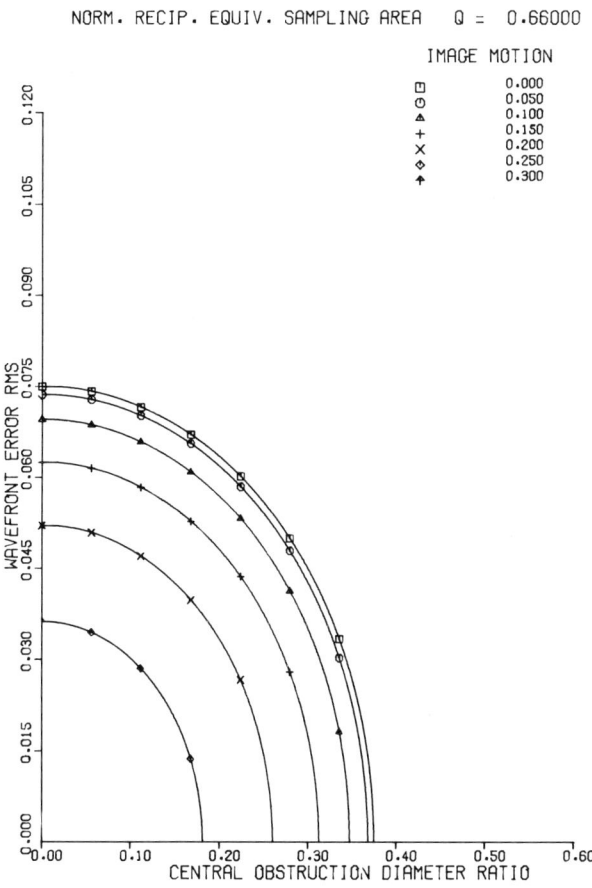

FIGURE 15  $\omega$-$\epsilon$-$\sigma$ chart for $Q = 0.66$.

## VI. Image-Sensor Models

At the present time, the LST is expected to use only TV-type image sensors. The leading candidates are advanced versions of the SEC vidicon and silicon target vidicon (SIT vidicon) now being developed under the guidance of J. Lowrance of Princeton University. Both of these image sensors are expected to operate in a photon-noise-limited mode, although this may be only marginally true of the SEC vidicon, when operating near its minimum illumination level. Both have rather similar MTF's. For the purposes of this paper, only one needs to be considered, and the SIT has been selected. The information used in modeling its performance characteristics was supplied by W. Bradley of Itek, after consultation with J. Lowrance.

The SIT vidicon consists of a photocathode, an electron optics image relay, a silicon storage target, and an electron-beam readout mechanism. The most significant characteristics of the SIT vidicon for this analysis are the photocathode sensitivity and thermionic emission rate and the MTF and saturation level of the silicon target. Other sources of image degradation do exist within the tube, but it is assumed that they are negligible or can be compensated for. There will be fixed pattern noise from variations in sensitivity across the photocathode and variations in gain across the storage target. These can be removed in data processing. The MTF of the electron optics should be high enough to be ignorable, in comparison with that of the silicon target. Similarly, noise introduced by the scanning electron beam and by the output preamplifiers should be negligible in comparison with photon noise.

Photocathode sensitivity $S$ is selectable at will, within the limits of presently available photocathode materials. The highest sensitivity available with trialkali photocathodes is $S \simeq 2 \times 10^{-4}$ A/lm. This will be adequate for the present analysis, where detailed radiometric calculations are not critical. An appropriate spectral sensitivity curve in amperes per watt per wavelength increment can be substituted for more detailed analyses.

Thermionic dark current for the above photocathode typically ranges from 100 electrons $cm^{-2}$ $sec^{-1}$ at room temperature down to 3 electrons $cm^{-2}$ $sec^{-1}$ at temperatures below $-10$ °C. The latter figure will be used here since it allows thermionic emission to be discounted. Methods for inclusion of thermionic emission will be discussed briefly in Section VII.A for the interested reader.

The MTF of the silicon target is given by the approximation

$$T_{SIT}(\nu) = \exp[-(2\pi t_t \nu)], \tag{45}$$

where $t_t$ is the characteristic target thickness. For this case, it is assumed that $t_t = 0.007$ mm (see Figure 16). To match this equation to the normalized spatial frequency scale for the lens, the constant $C_s = t_t/\lambda F$ must be introduced and Eq. (45) rewritten

$$T_{SIT}(\nu_n) = \exp[-(2\pi C_s \nu_n)]. \tag{46}$$

Equations (46) and (33) can be used to define an approximation for $Q_{SIT}$. For completeness, two equations are needed:

$$Q_{SIT} = (1 + 1.63 C_s)^{-2.72}, \quad C_s < 2; \tag{47}$$

and

$$Q_{SIT} = 0.0924 C_s^{-2}, \quad C_s > 2. \tag{48}$$

For $\lambda = 633$ nm, $C_s < 2$ if $F > 5.5$. Therefore, Eq. (47) is adequate for all cases here. Equation (48) should define the system $Q$ (including optics) adequately enough for readers interested in fast systems, where $F < 5.5$.

The silicon target has a hard saturation limit corresponding to an exposure of $1.25 \times 10^{12}$ photoelectrons/m². Details in continuous-tone images producing a greater exposure will be clipped at about this level. The maximum permissible average exposure level should be below this cutoff. Continuous-tone astronomical objects are generally low-contrast

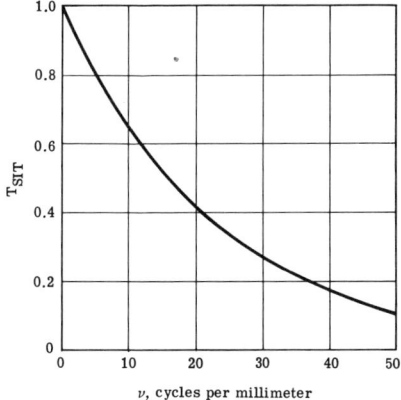

FIGURE 16  MTF of model SIT vidicon.

objects. Setting the maximum permissible exposure $K_s$ to

$$K_s = 2.5 \times 10^{11} \text{ photoelectrons/m}^2, \tag{49}$$

one fifth of the saturation limit, should give an adequate safety margin.

## VII. Point-Source Detection

In astronomy, stars always appear against a background continuum. Unless the star is so much brighter than the background that the latter is below the detection threshold of the image sensor, detection is a matter of sorting out the signal photoelectrons from variations in the background due to photon statistics. Detection will occur when the peak signal exceeds the rms noise by some threshold ratio. For simple detection, the signal-to-noise ratio $SNR_p$ must exceed a threshold of about 3. For a photometric accuracy of 10%, $SNR_p = 10$. The latter figure is commonly used in astronomy and will be used here.

Three aspects of point-source detection are examined here. First, a model will be developed to compute $SNR_p$ for photon-noise-limited image sensors and to determine the exposure time needed to reach the detection threshold. Second, the loss in detection threshold limit caused by image degradation in the optical system and image sensor will be computed. Third, the point-source case will be extended to include small finite sources, such as distant galaxies.

### A. SIGNAL-TO-NOISE RATIO MODEL

The principal noise sources ahead of the storage target are photon statistics and thermionic emission statistics. Both are white noise, whose rms noise can be determined by taking the square root of the average number of (photo) electrons involved. In the general case, where there are enough thermionic electrons to count,

$$SNR_p = n_s/(n_s + n_b + n_d)^{1/2}, \tag{50}$$

where $n_s$ is the number of signal photoelectrons, $n_b$ the number of background photoelectrons, and $n_d$ the number of dark-current (thermionic) electrons. When the photocathode is cooled, or a low dark-current photocathode is used (e.g., blue sensitive only),

$$SNR_p = n_s/(n_s + n_b)^{1/2}. \tag{51}$$

The latter equation is assumed to apply here, so that dark current can be ignored. The limitations on this assumption will be discussed below.

The signal consists of all light from the star that reaches the image sensor and is converted to photoelectrons. In the most rigorous radiometric terms,

$$n_s = \frac{\pi D_p^2 t}{4e} \int_{\lambda_1}^{\lambda_2} \tau(\lambda) S_e(\lambda) E_{ae}(\lambda) d\lambda, \qquad (52)$$

where

$e \equiv$ electron charge $= 1.6 \times 10^{-19}$ coulomb/electron,
$E_{ae}(\lambda) \equiv$ spectral irradiance of the aperture due to the star (W/m$^2$ in wavelength interval $d\lambda$),
$S_e(\lambda) \equiv$ spectral sensitivity of photocathode (A/W at wavelength $\lambda$),
$t \equiv$ exposure time (sec),
$\tau(\lambda) \equiv$ effective transmittance.

For an all-reflecting optical system,

$$\tau(\lambda) = (1 - \epsilon^2) \rho_\lambda^n, \qquad (53)$$

where $\rho_\lambda$ is the spectral reflectivity (assumed to be the same for all mirrors) and $n$ is the number of mirrors.

The background photoelectrons can be computed by Eq. (30). If the image irradiation $N_b$ (photoelectrons m$^{-2}$ sec$^{-1}$) is expanded in rigorous radiometric terms, using standard lens equations,

$$n_b = \frac{2.77\pi t}{4e} \int_{\lambda_1}^{\lambda_2} \frac{\tau(\lambda) S_e(\lambda) L_0^2(\lambda) \lambda^2}{Q(\lambda)} d\lambda, \qquad (54)$$

where $L_0(\lambda)$ is the object spectral radiance in W m$^{-2}$ sr$^{-1}$ in wavelength increment $d\lambda$.

The above equations are necessary for exact radiometric analysis, where photocathode type and spectral classification of the star may both vary. A simplified set of photometric equations may be used if we are interested only in the relative effects of image degradation, as is the case in this paper. The luminous sensitivity, $S$, in amperes per lumen can be substituted for $S_e(\lambda)$, and average values can be used for $\tau$, $Q$, and $\lambda$.

If $m_v$ is the apparent visual magnitude of the star, the aperture illumination $E_{av}$, in lumens per square meter, is

$$E_{av} = 2.65 \times 10^{-(6 + 0.4 m_v)}. \qquad (55)$$

Similarly, if $\mathcal{M}_v$ is the brightness of the background (or any continuous source) in apparent visual magnitude per square second of arc, the object luminance, $L_{ov}$ in lm m$^{-2}$ sr$^{-1}$, is

$$L_{ov} = 1.13 \times 10^{(5 - 0.4\mathcal{M}_v)}. \tag{56}$$

Using these values, Eqs. (52) and (54) become, respectively,

$$n_s = 1.30\tau St D_p^2 \times 10^{(13 - 0.4 m_v)} \tag{57}$$

and

$$n_b = (1.54\tau St \lambda^2/Q) \times 10^{(24 - 0.4\mathcal{M}_v)}. \tag{58}$$

The thermionic contribution $n_d$ is given by

$$n_d = N_d A_s^* t, \tag{59}$$

where $N_d$ is the thermionic emission rate in electrons per square meter. Since $A_s^*$ is proportional to $F^2$, the number of thermionic electrons per resolution element will be proportional to $F^2$. The error introduced by using Eq. (51) instead of Eq. (50) will be 10% or less when

$$n_d \leqslant 0.21 n_b, \tag{60}$$

neglecting $n_s$.

If we consider the minimum brightness background likely to be encountered, $\mathcal{M}_v \simeq 23$, and a typical trialkali photocathode sensitivity $S = 2 \times 10^{-4}$ A/lm, we can determine at what focal ratio the error can reach 10%. For room-temperature operation, where $N_d \simeq 10^6$ electrons m$^{-2}$ sec$^{-1}$, the error is greater than 10%, if $F \geqslant 77$. For $-10°$C or cooler operation, where $N_d \simeq 30{,}000$ electrons m$^{-2}$ sec$^{-1}$, the error is greater than 10% when $F \geqslant 443$. The LST is expected to operate at $F \simeq 100$ ($F = 200$ at most), with a cooled photocathode. Therefore, Eq. (51) can be used for $SNR_p$ here.

The signal-to-noise ratio can be determined by substituting Eqs. (57) and (58) into Eq. (51). In general, we are more interested in knowing the exposure time $t$ required to achieve a fixed $SNR_p$:

$$t = [0.77\,(SNR_p)^2/\tau SD_p^2] \times 10^{(0.4 m_v - 13)}$$

$$+ [1.18\lambda^2 (SNR_p)^2/\tau SD_p^4 Q] \times 10^{(0.8 m_v - 0.4\mathcal{M}_v - 2)}. \tag{61}$$

In principle, the threshold $SNR_p$ can be achieved for any combination of $m_v$ and $\mathcal{M}_v$, given enough time. In practice, there is a maximum permissible exposure time set by the saturation limit of the image sensor storage target. If the maximum allowable average exposure is $K_s$ photoelectrons/m², 

$$t(\max) = (1.81 K_s F^2/\tau S) \times 10^{(0.4 \mathcal{M}_v - 24)}. \tag{62}$$

Consider a specific example: Set $K_s = 2.5 \times 10^{11}$ photoelectrons/m², $S = 2 \times 10^{-4}$ A/lm, and $\mathcal{M}_v = 23$. From Eq. (62),

$$t(\max) = 3.59 F^2/\tau \quad (\text{sec}). \tag{63}$$

If we further stipulate that $D_p = 3.0$ m, $\lambda = 6.33 \times 10^{-7}$ m, and $SNR_p = 10$, Eq. (61) leads to

$$m_v(\max) = 25.36 + 2.5 \log F + 1.25 \log Q, \tag{64}$$

for exposure to one fifth of target saturation.

The LST is expected to operate at about $F = 100$. If we take a typical case in which $\omega = 0.05$ wavelength rms, $\epsilon = 0.35$, and $\sigma = 0.10$ and compute $Q$ for the lens and SIT vidicon from Eqs. (44) and (48), $m_v$ (max) = 29.8. The mirror reflectivity will be about 0.88, and there will be six mirrors in the $F = 100$ optics. The transmittance, $\tau = 0.408$, from Eq. (53). The exposure time will therefore be 88,000 sec, or 24.4 h.

### B. THRESHOLD LOSS FOR FIXED EXPOSURE TIME

If the exposure time is fixed, both $\tau$ and $Q$ affect $SNR_p$. To investigate the loss in threshold magnitude due to $\tau$ and $Q$, it is worthwhile to identify two special cases. If we are looking for bright stars with very short exposure times, $n_s \gg n_b$, and $n_b$ can be neglected. Thus

$$SNR_p = \sqrt{n_s} = 3.61(\tau St)^{1/2} D_p \times 10^{(6 - 0.2 m_v)}. \tag{65}$$

If, on the other hand, we are looking for the dimmest possible stars with a very long exposure time, $n_b \gg n_s$, and only $n_b$ need be considered a noise source. Thus

$$SNR_p = n_s/\sqrt{n_b} = 1.05 \, (D_p^2/\lambda)(\tau St Q)^{1/2} \times 10^{(0.2 \mathcal{M}_v - 0.4 m_v)}. \tag{66}$$

If a perfect optical system and image sensor are used ($\tau = Q = 1.0$), let the threshold detectable magnitude be $m_v$. If an imperfect lens and sensor are used in its place ($\tau < 1.0; Q < 1.0$), the threshold magnitude will drop to $m_v - \Delta m$, assuming $D_p$, $S$, $\lambda$, $\tau$, $\mathcal{M}_v$, and $SNR_p$ are the same. From Eqs. (65) and (66), it can be shown that

$$\tau = 10^{-0.4\Delta m} \quad \text{(bright star)} \tag{67}$$

and

$$\tau Q = 10^{-0.8\Delta m} \quad \text{(dim star)}. \tag{68}$$

The dim-star case is the more significant since it corresponds to the limit astronomers wish to improve upon by using spaceborne telescopes.

Equations (67) and (68) form a very simple and useful image criterion relating performance loss in astronomical terms to the choice of design parameters and error budgets. They allow inclusion of both image sensor and optics. Exact calculations can be made using MTF integration to calculate $Q$. Preliminary analysis can be carried out using the approximations for $Q$ from Sections V and VI. A specific performance criterion can be set by defining a maximum permissible value for $\Delta m$.

If we expand Eq. (67), we get a bright-star detection threshold loss

$$\Delta m = -2.50 \left[ n \log \rho + \log (1 - \epsilon^2) \right]. \tag{69}$$

If we set $\Delta m \leqslant 1.0$, $n = 6$, and $\rho = 0.88$, then $\epsilon \leqslant 0.38$. Thus we cannot meet our performance criterion for bright stars unless the central obstruction diameter ratio is within an upper bound set by $n$ and $\rho$.

The dim-star case expands into a series of equations

$$\Delta m_\omega = 0.543 C_{QH} \, \omega^2 \simeq 40\omega^2, \tag{70}$$

$$\Delta m_\sigma = 1.25 \log (1 + 6\sigma^2), \tag{71}$$

$$\Delta m_\epsilon = 1.6\epsilon^2 - 1.25 \log (1 - \epsilon^2), \tag{72}$$

$$\Delta m_\rho = -1.25 \, n \log \rho, \tag{73}$$

$$\Delta m_{SIT} = 3.4 \log \left[ 1 + (1.63 t_t / \lambda F) \right]. \tag{74}$$

Equations (70), (71), and (72) have been plotted in Figure 17, and Eq. (74) has been plotted in Figure 18. For the reflectivity $\rho$, $\Delta m_\rho = 0.07$ per surface for $\rho = 0.88$. For six mirrors, $\Delta m_\rho = 0.42$. This is the largest single loss in fixed exposure time threshold. For $\epsilon = 0.35$ and $F = 100$, the central obstruction and image tube tie for second largest performance degradation source.

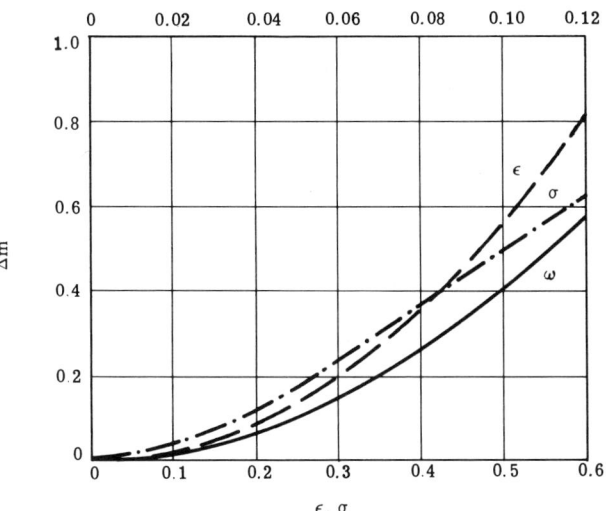

FIGURE 17 Threshold loss for point-source detection, $\Delta m$, as a function of $\omega$, $\epsilon$, and $\sigma$, for fixed exposure time.

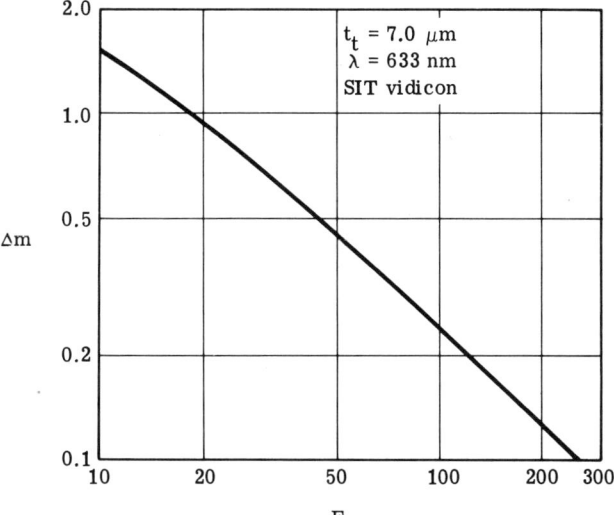

FIGURE 18 Threshold loss for point-source detection with an SIT vidicon, as a function of focal ratio $F$, for fixed exposure time.

Equations (67) and (68) have one more interesting feature. If $\tau = Q$, Eq. (68) reduces to Eq. (67). The threshold loss is therefore the same for both extremes, bright and dim star, and one surmises that the threshold loss remains constant for intermediate cases. To check this, Eq. (61) was used to plot exposure time versus $m_v$ for an ideal 3-m-aperture telescope ($\tau = Q = 1.0$) and for a telescope for which $\tau = Q = 0.398$. The latter corresponds to $\Delta m = 1.0$ magnitude at each extreme. The results are plotted in Figure 19. These curves confirm the surmised constancy of $\Delta m$ through the intermediate range, where $n_s \simeq n_b$. They also give some idea of the performance to be expected from the LST, if it can be designed to meet the criterion $\Delta m = 1.0$.

Equation (68) can be used as the basis for generating an $\omega$–$\epsilon$–$\sigma$ chart for use in trading off image-quality parameters, once a value for $\Delta m$ has been chosen. Such a chart has been generated for $\Delta m = 1.0$ due to the

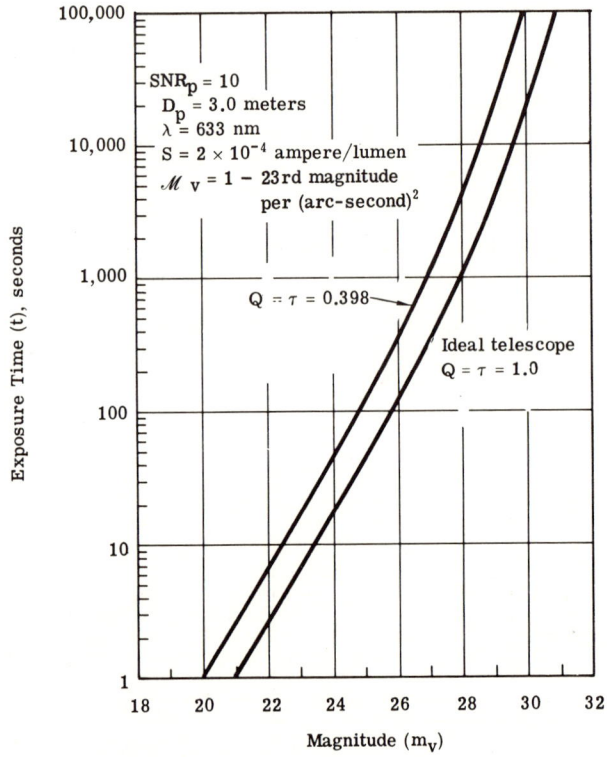

FIGURE 19 Variation in the detection threshold for point sources as a function of exposure time.

optical system alone (the SIT vidicon is not included) and is shown in Figure 20. The criterion $\Delta m = 1.0$ for the optics is substantially looser than Maréchal's criterion. For $\epsilon = 0.35$ and $\sigma = 0.10$, the criterion $\Delta m = 1.0$ leads to $\omega \simeq 0.084$, which allows more wavefront error than is presently expected for the LST. Inclusion of the SIT vidicon within the $\Delta m = 1.0$ would tighten this limit considerably, at least at $F = 100$.

C. DETECTION OF SMALL FINITE OBJECTS

The concept of $Q$ can be extended to include small finite objects, such as distant galaxies. Astronomers customarily define the brightness of such sources in apparent magnitude, in the same manner as for stars. That is,

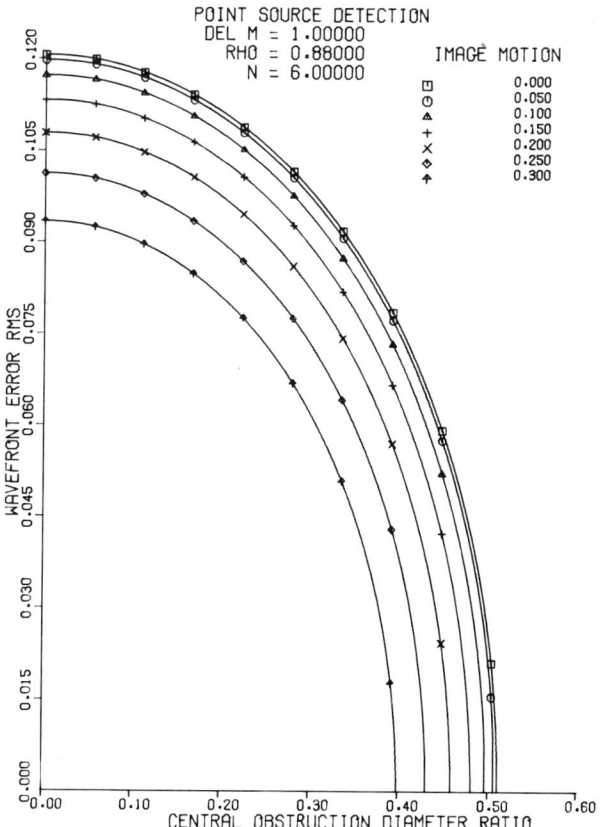

FIGURE 20 $\omega$–$\epsilon$–$\sigma$ chart for a point-source detection threshold loss of $\Delta m = 1.0$ due to optical system.

a galaxy of apparent visual magnitude $m_v$ will provide the same aperture illumination $E_{av}$ as a star of the same magnitude. Its signal can therefore be computed by Eq. (57). The intensity distribution in the galaxy can be treated as if it were a point-spread function and convolved with that for the telescope. If the relative intensity distribution of the object is Fourier transformable, it can be converted into a spatial frequency distribution, and multiplied by the system MTF, for use with Eqs. (67) and (68) in computing $Q$. The equations of Section VII.A can then be used to calculate $SNR_p$ or the exposure time required to achieve the detection threshold. Equation (68) can be used to calculate $\Delta m$, which in this case compares the detection threshold for the galaxy with the perfect lens threshold for a star of the same magnitude.

If the characteristic diameter of the galaxy is substantially larger than the Airy disk diameter, it can be assumed that the effective spread-function approximates the relative intensity distribution of the galaxy itself. In this case, Parseval's theorem can be used to compute $A_s^*$ directly from the relative intensity distribution.

Parseval's theorem states that, if $f(r,\phi)$ and $T(\nu,\phi)$ are a Fourier transform pair, then

$$\int_0^{2\pi}\int_0^{\infty} |f(\nu,\phi)|^2\, \nu d\nu d\phi = |T(\nu,\phi)|^2\, \nu d\nu d\phi. \tag{75}$$

Thus $N_{e2}^*$ can be computed directly from the point-spread function $f(r,\phi)$, if it has been normalized so that it is a transform of the MTF. Since $T(0,0) = 1.0$, the volume integral under $f(r,\phi)$ must equal unity. If $f(r,\phi)$ is the relative intensity, the normalization can be accomplished in calculating $N_{e2}^*$:

$$N_{e2}^* = \frac{1}{A_s^*} = \int_0^{2\pi}\int_0^{\infty} |f(r,\phi)|^2 r dr d\phi \bigg/ \left[\int_0^{2\pi}\int_0^{\infty} f(r,\phi)\, rdrd\phi\right]^2. \tag{76}$$

J. Greenstein (personal conversation with the author) has provided a model for the cross section of a galaxy:

$$I(r) = I_0\, [1 + (r^2/d_g^2)]^{-5/2}, \tag{77}$$

when $d_g$ is the characteristic diameter of the galaxy. [When $r = d_g/2$, $I(r) = 0.5724 I_0$.] For rotationally symmetric galaxies, Eq. (76) gives

$$A_s^* = 16\pi d_g^2/9. \tag{78}$$

The equivalent sampling aperture diameter $d_s^*$,

$$d_s^* = 2.667 d_g. \tag{79}$$

Dividing $A_s^*$ for a perfect lens [Eq. (34)] by $A_s^*$ from Eq. (78) gives $Q$,

$$Q = 0.496\lambda^2 F^2 / d_g^2. \qquad (80)$$

In terms of the Airy disk diameter $d_A = 2.44\lambda F$,

$$Q = 0.0833 d_A^2 / d_g^2. \qquad (81)$$

If $d_g$ is converted to the equivalent sample area $d_s^*$,

$$Q = 0.592 d_A^2 / (d_s^*)^2. \qquad (82)$$

If the small finite source were a uniform radiance disk of diameter $d_{gc}$, then $d_{gc} \simeq d_s^*$ for $d_{gc} \gg d_A$. To examine the range of values for $d_{gc}$ over which the spread function of the lens will cause significant reduction in $Q$, the circular disk was Fourier transformed and multiplied by the lens MTF. $Q$ was then computed by numerical analysis using Eq. (33) for varying-size disks with a perfect lens and with a lens for which $\omega = 0.050$ wavelength rms, $\epsilon = 0.35$, and $\sigma = 0.10$. The results are plotted in Figure 21. The solid curve is from Eq. (82). The dashed line is for the disk with a perfect lens, and the singly broken line is for the disk with the degraded lens. The doubly broken line is for the Greenstein model galaxy, based on Eq. (81).

Figure 21 indicates that for $d_s^* > 10 d_A$, the effects of the lens can be ignored, and $Q$ can be computed directly from Eq. (82), for circular disks. The same should be true for the Greenstein model, so that Eq. (81) is valid for $d_g > 4 d_A$. Equations (68), (81), and (82) can then be combined to indicate the loss in detection threshold magnitude for galaxies. Assuming an all-reflecting lens system of $n$ mirrors, we see that for the Greenstein model,

$$\Delta m = 1.35 - 1.25 n \log \rho - 1.25 \log(1 - \epsilon^2) - 2.5 \log d_A + 2.5 d_g, \qquad (83)$$

where $d_A$ and $d_g$ must be in the same units. For a uniform circular object, or for any object expressed by its equivalent circular sampling aperture diameter,

$$\Delta m = 0.28 - 1.25 n \log \rho - 1.25 \log(1 - \epsilon^2) - 2.5 \log d_A + 2.5 \log d_s^*. \qquad (84)$$

Equations (83) and (84) are for fixed exposure times. If the exposure can be adjusted to compensate for transmission losses, the terms in $\rho$ and $\epsilon$ can be omitted.

For the 3-m-aperture LST, $n = 6$, $\rho = 0.88$, $\epsilon = 0.35$, and $d_A = 0.1062$

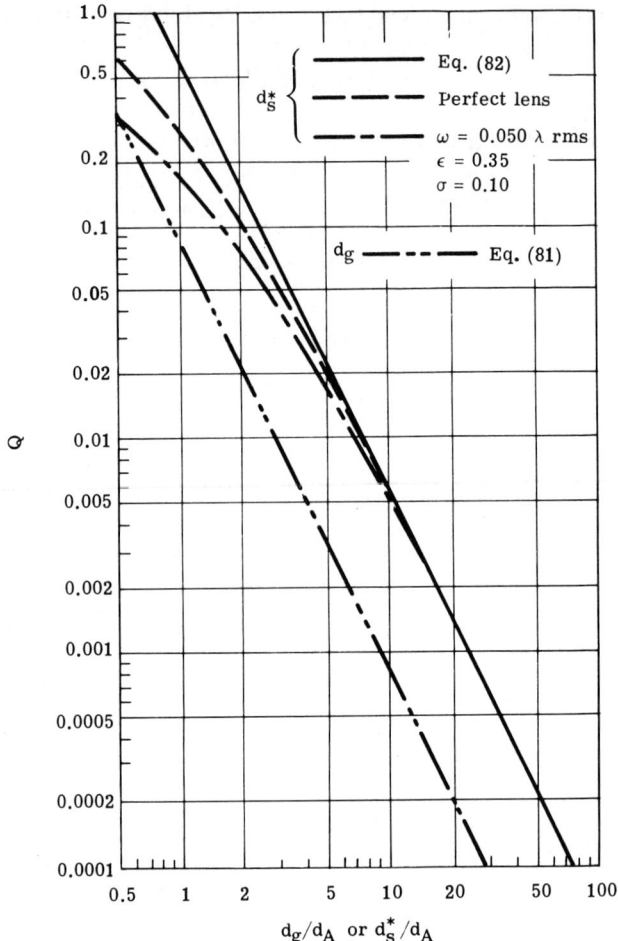

FIGURE 21  $Q$ as a function of diameter for disk-shape and Greenstein-model objects.

sec of arc at $\lambda$ = 633 nm. If $d_g$ = 1.0 sec of arc, $\Delta m$ = 4.27. If $d_g$ = 10 sec of arc, $\Delta m$ = 6.77. Thus, for any exposure time $t$, the detection threshold for galaxies of these sizes will be the point-source detection threshold for $\tau = Q = 1.0$, minus the above figures.

For the LST, the limiting threshold for detecting point sources with the SIT target exposed to one fifth of its saturation limit is given by Eq. (64), for the example cited. For Greenstein-model galaxies,

$$m_v \text{ (max)} \simeq 24.01 + 2.5 \log F + 2.5 \log d_A - 2.5 \log d_g. \tag{85}$$

For any source with an equivalent circular sampling aperture $d_s^* \gg 10 d_A$,

$$m_v \text{ (max)} \simeq 25.08 + 2.5 \log F + 2.5 \log d_A - 2.5 \log d_s^*. \tag{86}$$

[See discussion related to Eq. (64) for assumptions concerning sensor characteristics.] If $F = 100$ and $d_A = 0.1062$ min of arc, then $m_v$ (max) = 26.6 for $d_g = 1$ sec of arc, and 24.1 for $d_g = 10$ sec of arc.

## VIII. Continuous-Tone Imagery

The conventional means for specifying image quality for continuous-tone imagery is by stating the threshold resolution, which is the maximum spatial frequency in a test target whose lines can be resolved. Resolution occurs when the signal-to-noise ratio $SNR_c(\nu)$ at spatial frequency $\nu$ exceeds some threshold value. The exact threshold value depends on the type of target. For three-bar targets, $SNR_c = 1.2$ is considered adequate. For astronomical purposes, the signal-to-noise ratios quoted earlier may be appropriate. Thus we may say $SNR_c = 3$ for the detection of small detail, and $SNR_c = 10$ for radiometric measurements to 10% accuracy.

No specific threshold will be used here. Rather, signal-to-noise ratio will be discussed as a function of spatial frequency and how it is affected by image degradation factors. A model for $SNR_C$ will be described, based on Rosell [1971]. Based on this model, curves will be developed showing the maximum signal-to-noise ratio that can be achieved with the SIT vidicon exposed to one-fifth saturation and how these curves are affected by $\omega$, $\epsilon$, $\sigma$, and the choice of focal ratio $F$ in typical cases relevant to the LST. Finally, the effects of using loss in signal-to-noise ratio at a given spatial frequency on the choice of $\omega$, $\epsilon$, and $\sigma$ will be discussed.

### A. SIGNAL-TO-NOISE RATIO MODEL

The model used here for signal-to-noise ratio differs from that of Rosell only in the definition of contrast and in the use of cycles (line pairs or optical lines) rather than TV lines in defining spatial frequency. Here, the object contrast $C_0$ is a modulation,

$$C_0 = (\text{max} - \text{min})/(\text{max} + \text{min}) \tag{87}$$

rather than Rosell's $(\text{max} - \text{min})/\text{max}$.

Consider two adjacent resolution elements (½ cycle each), one receiving $n_{\text{max}}$ detected photons on average in an exposure interval and the other receiving $n_{\min}$ detected photons. The peak-to-peak signal seen in examining the image is $n_{\max} - n_{\min}$. The rms noise will be the square root of the mean-square noises in each element, $(n_{\max} + n_{\min})^{1/2}$. Assuming the sensor to be photon noise limited,

$$SNR_c = (n_{\max} - n_{\min})/(n_{\max} + n_{\min})^{1/2}. \tag{88}$$

From this basic model the full equation for $SNR_c$ can be derived:

$$SNR_c(\alpha) = (\pi\tau S t L_o^{1/2}/8e) D_p \alpha C_o(\alpha) T(\alpha), \tag{89}$$

where $\tau$, $S$, $t$, $e$, and $D_p$ are as defined before, and

$\alpha \equiv$ angle subtended by one cycle at spatial frequency $\nu$,
$C_o(\alpha) \equiv$ object modulation at given spatial frequency,
$T(\alpha) \equiv$ system MTF at given spatial frequency,
$L_o \equiv$ object radiance in units appropriate to $S$.

The quantities $\tau$, $S$, $L_o$, $C_o(\alpha)$, and $T(\alpha)$ vary with wavelength. This variance must be considered when analyzing spectroscopic or narrow-band images. For our purposes, it is convenient to use photometric units, with the object radiance specified by $\mathcal{M}_\nu$ in apparent visual magnitudes per square second of arc. Rewriting Eq. (89) in terms of $\mathcal{M}_\nu$ and the normalized spatial frequency $(\nu_n)$, and substituting for the numerical constants,

$$SNR_c = 5.27\lambda(\tau S t)^{1/2} C_o(\nu_n)[T(\nu_n)/\nu_n] \times 10^{(11 - 0.2\mathcal{M}_\nu)}. \tag{90}$$

### B. $SNR_c$ VERSUS SPATIAL FREQUENCY FOR A SATURATED TARGET

The exposure time required to reach one fifth of the saturation limit was defined in Eq. (62). Since continuous-tone astronomical objects tend to be low in contrast, this should be a good safe limit for the maximum exposure time. If Eq. (62) is substituted into Eq. (90),

$$SNR_c \text{ (max)} = 0.71 \lambda F \sqrt{K_s} C_o(\nu_n) T(\nu_n)/\nu_n, \tag{91}$$

where $K_s$ is, as before, the average exposure at one fifth of saturation for the target, in electrons per square meter. ($\lambda$ must also be in meters.)

The optical system characteristics enter this equation in two terms, $F$ and $T(\nu_n)$. Changing $F$ will change the signal-to-noise ratio in two respects. First, the MTF degradation function $T_{SIT}(\nu_n)$, due to the image sensor, is a function of $F$. Second, the size of the basic resolution element scales with $F$, increasing the number of photoelectrons that can be stored in a resolution element at the normalized spatial frequency $\nu_n$. (The exposure time must increase as $F^2$, however.)

The effects of focal ratio scaling and image sensor MTF are shown in Figure 22, in which $SNR_c$ curves for a perfect lens with and without an SIT vidicon are plotted for the cases $F = 12$, 100, and 200. In all cases $K_s = 2.5 \times 10^{11}$ electrons/m², corresponding to one-fifth saturation for the SIT target. The differences in position of the three solid lines, for the perfect lens alone, represent the effects of focal ratio and exposure-time scaling. The differences between solid and dashed lines at each focal ratio represent the loss due to image sensor MTF. The losses due to image sensor MTF reduce with the larger focal ratio but may be significant (20% at high spatial frequencies) even at $F = 200$.

If it were necessary to hold the exposure time fixed at, say, the time

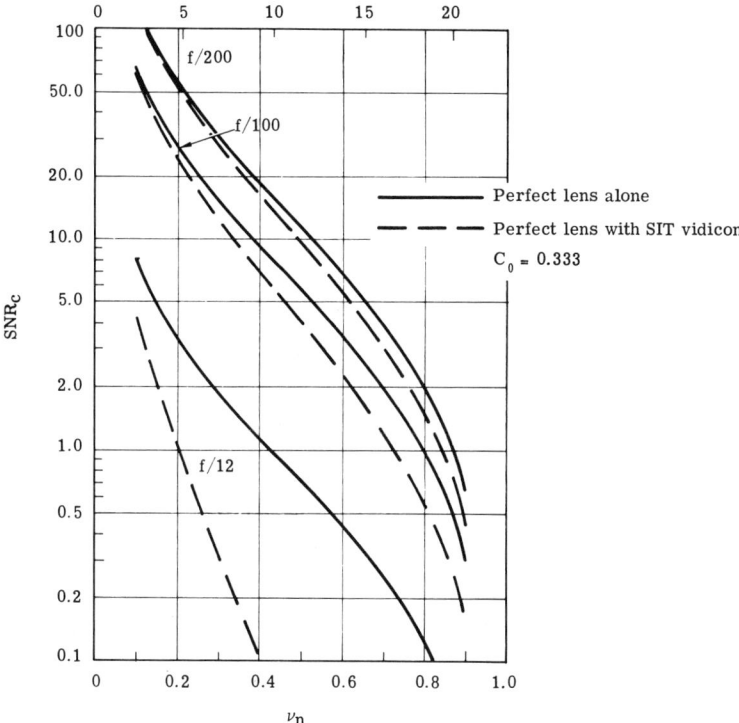

FIGURE 22 $SNR_c$ for a perfect lens alone and for a perfect lens with an SIT vidicon, when target is exposed to one fifth of saturation limit ($\tau t/F^2$ fixed).

corresponding to saturation for the $F = 12$ case, each pair of curves would shift downward (on this logarithmic scale) until the three solid lines coincided. There would still be a gain in $SNR_c$ due to the MTF change in going to the largest focal ratio possible, provided the exposure level on the target is not so low that readout noise will add a significant contribution to the total noise.

The change in $SNR_c$ due to wavefront error $\omega$, central obstruction diameter ratio $\epsilon$, and image motion $\sigma$ is illustrated in Figure 23, for the $F = 100$ and $F = 200$ cases. Typical values have been chosen for the image-quality parameters. Curves have been plotted with and without the central obstruction diameter ratio, to emphasize its unique effects. The circles represent the SIT vidicon with a perfect lens.

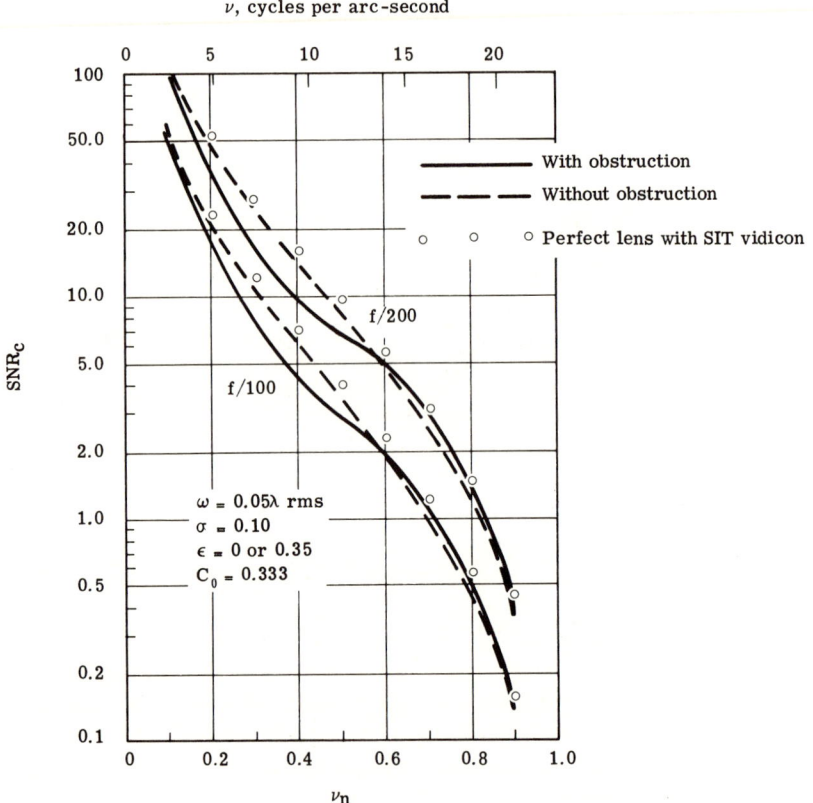

FIGURE 23 Effect of $\omega$, $\epsilon$, and $\sigma$ on $SNR_C$ when SIT target is exposed to one fifth of saturation limit ($\tau t/F^2$ fixed).

The loss in performance due to $\omega$ and $\sigma$ is on the order of 20% at higher spatial frequencies and corresponds to that due to the SIT vidicon alone at $F = 200$. Below $F = 200$, the SIT vidicon is the chief source of MTF-induced loss in signal-to-noise ratio.

The central obstruction boosts $SNR_c$ at high spatial frequencies, somewhat. This is offset by a larger suppression of $SNR_c$ at low spatial frequencies. Whether the central obstruction is useful or objectionable will depend on the threshold value for $SNR_c$ required for a particular experiment. The major limitation on $SNR_c$ is the exposure-time limitation set by target saturation, however. This can be overcome by increasing $F$ or by taking multiple exposures and integrating them in data processing.

C. LIMITATIONS ON CHOICE OF IMAGE-QUALITY PARAMETERS

Since the signal-to-noise ratio at a given spatial frequency determines whether it can be resolved, the degradation of $SNR_c$ at that spatial frequency is one measure of optical image quality. We can define a degradation constant D.C., which is the ratio of the actual and perfect lens signal-to-noise ratios at a given spatial frequency,

$$\text{D.C.}(v_n) = SNR_c \text{ (real)}/SNR_c \text{ (perf)}. \tag{92}$$

For a fixed exposure time, Eq. (90) leads to

$$\text{D.C.}(v_n) = \tau \times T_\omega(v_n) \times T_\epsilon(v_n) \times T_\sigma(v_n) \times T_{\text{SIT}}(v_n). \tag{93}$$

For the saturated target case, there is some ambiguity in how to handle the exposure-time scaling effect due to scaling $F$. If we resolve this by saying that the exposure time can only be scaled by a factor of $1/\tau$, to offset transmission loss, then

$$\text{D.C.}(v_n) = T_\omega(v_n) \times T_\epsilon(v_n) \times T_\sigma(v_n) \times T_{\text{SIT}}(v_n). \tag{94}$$

This will hold regardless of the value of $F$.

The only difference between Eqs. (93) and (94) is in the presence or absence of $\tau = \rho^n (1 - \epsilon^2)$. In the fixed-exposure-time case, the high spatial frequency boost due to $T_\epsilon(v_n)$ will be offset by a factor $(1 - \epsilon^2)^{1/2}$ due to transmission loss. It is interesting to note that if $v_n \geq (1 + \epsilon)/2$, where $T_\epsilon(v_n) = (1 - \epsilon^2)^{-1}$, there is a net boost

$$\text{D.C.}_\epsilon(v_n) = (1 - \epsilon^2)^{-1/2} \tag{95}$$

even when the loss in transmittance due to $\epsilon$ is not compensated for. This boost is not great, being about 6.7% for $\epsilon = 0.35$, but its presence is one reason why it is important to distinguish between factors affecting transmission and factors affecting energy distribution in the image when normalizing the MTF.

If we confine our attention to the lens system itself, either Eq. (93) or (94) can be made the basis of an image-quality criterion for use in trading $\omega$, $\epsilon$, and $\sigma$ against each other. D.C.$(\nu_n)$ is set to a constant, and an $\omega$–$\epsilon$–$\sigma$ chart generated. The form taken by the resultant chart will be strongly dependent on the chosen spatial frequency $\nu_n$. Figure 24 illustrates this. $\omega$–$\epsilon$–$\sigma$ charts are presented for the fixed-exposure-time case, for $\nu_n = 0.30, 0.55$, and $0.70$. D.C. was set to 0.631, which corresponds to the difference in $SNR_c$ between two objects differing by $\Delta \mathcal{M} = 1.0$, and it was assumed that the optical system has two mirrors of reflectivity $\rho = 0.88$.

Neither the particular value of D.C. nor the number of mirrors used in generating the figures is important, however. What is important to note here is the relative weighting of $\epsilon$ and $\sigma$ at different spatial frequencies. The relative weighting of these two image-quality parameters is reversed at spatial frequencies well below and well above $\nu_n = 0.55$. In general, the object in continuous-tone imagery is to obtain threshold performance at the highest spatial frequency possible. Any criterion based on continuous-tone imagery must therefore stress reducing image motion.

## IX. Conclusions

Two of the three image-quality criteria discussed here relate performance in astronomical terms directly to quantities useful to the design engineer. One of these, the normalized reciprocal equivalent sampling area $Q$, is particularly convenient from an analytical point of view. It can be calculated exactly by numerical integration from the system modulation transfer function, or it can be approximated by elementary equations based on the three image-quality parameters. It represents the complete system performance with a single number, which is useful for tradeoff analysis. It relates directly to the loss in performance for point-source detection due to imperfections in the optical system. Image sensor characteristics can be included in the calculations in a very simple manner.

Examination of the image-quality criteria described here has pointed out a number of factors affecting the choice of telescope design parameters and error budget allocations, which are worth summarizing.

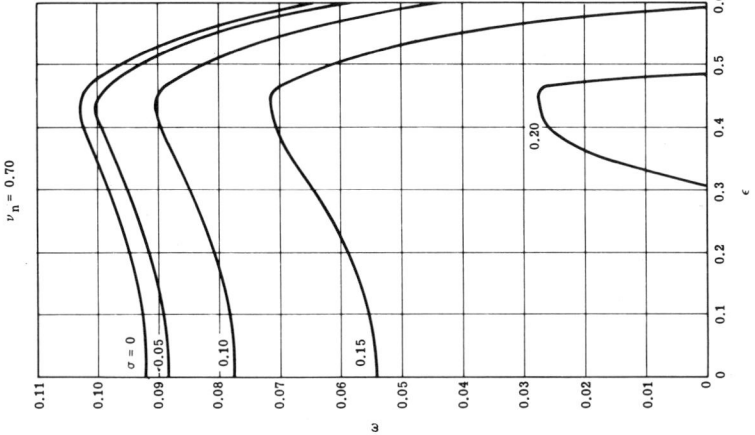
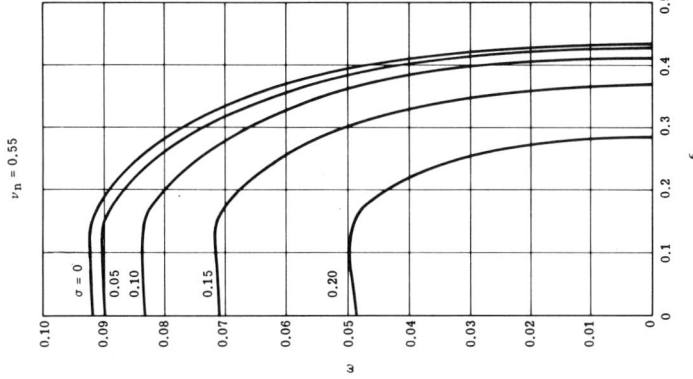
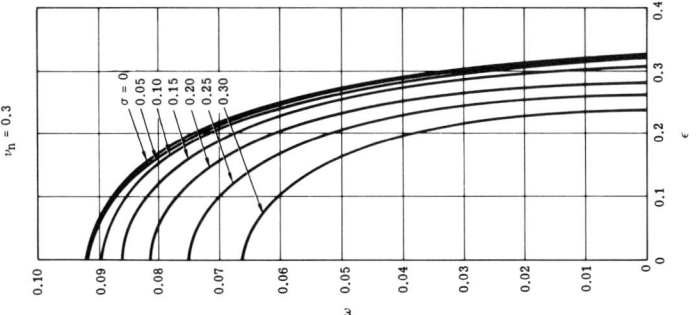

FIGURE 24 $\omega$–$\epsilon$–$\sigma$ charts for continuous-tone imagery, based on a fixed reduction in $SNR_C$ at $\nu_n = 0.30$, 0.55, and 0.70.

1. A rigorous interpretation of Maréchal's criterion for diffraction-limited performance, $\mathscr{D} \geq 0.80$, leads to very tight tolerances on wavefront error if the effects of image motion and the central obstruction diameter ratio are included. For $\sigma = 0.10$ and $\epsilon = 0.30$ to $0.38$, the wavefront error must be in the range $\omega = 0.047$ to $0.038$ wavelength rms.

2. Use of an image-quality criterion based on a loss in the point-source detection threshold for fixed exposure times may allow a substantial relaxation of the wavefront error tolerance. In a six-mirror system, a criterion of $\Delta m = 1.0$ for the optics alone leads to wavefront error limits in the range $\omega = 0.095$ to $0.078$ wavelength rms, when $\sigma = 0.10$ and $\epsilon = 0.30$ to $0.38$.

3. The largest single factor affecting the point-source detection threshold for fixed exposure times is transmittance, including the transmission loss due to the central obstruction. In the bright star case, if there are six mirrors in the optics train, an image quality criterion of $\Delta m = 1.0$ can be met only if $\epsilon \leq 0.38$.

4. During early design stages for the LST, a system focal ratio $F \approx 100$ was chosen in hopes of minimizing performance loss due to the image sensor MTF. The present analysis indicates that losses due to the tube are still very high at $F = 100$. For point-source detection, $\Delta m_{\text{SIT}} = 0.24$ at $F = 100$. For continuous-tone imagery, the loss in $SNR_c$ at $\nu_n = 0.75$ is about 40%. Both findings indicate that it would be desirable to operate at substantially larger focal ratios than $F = 100$, for optics-limited performance.

5. To maintain $SNR_c$ at high values for large spatial frequencies $(0.6 < \nu_n < 1.0)$, it is essential to reduce the image motion as much as possible. At $\nu_n = 0.75$, $\sigma = 0.10$ produces an 11% reduction in $SNR_c$, and $\sigma = 0.20$ produces a 36% reduction. Restrictions on wavefront error are not so strict, e.g., $\omega = 0.075$ produces a 20% reduction in $SNR_c$. A central obstruction actually increases $SNR_c$ at high spatial frequencies. At $\nu_n = 0.75$, a central obstruction of diameter ratio $\epsilon = 0.35$ will boost $SNR_c$ by 14% if the loss in transmittance is compensated for by an increase in exposure time, or by 7% if the exposure time is fixed.

As a general summary, it appears that point-source detection considerations restrict the central obstruction to a diameter ratio $\epsilon \leq 0.38$, continuous-tone imagery restricts the image motion to $\sigma \leq 0.10$, while the wavefront error should be in the range $\omega = 0.07$ to $0.08$. In both cases it would be desirable to use the SIT vidicon at a system focal ratio nearer to $F = 200$ than $F = 100$.

The author would like to express special thanks to W. Bradley for bringing the work of Otto Schade to his attention and for aid and advice in the development of some of the material presented here. Thanks are also due to Mark Egdall, for aid in computer programming.

The work reported in this paper was sponsored by the National Aeronautics and Space Administration, under contract No. NASw-2313.

## References

Barakat, R., *Opt. Acta 18*, 683 (1971).
Born, M., and E. Wolf, *Principles of Optics*, 1st ed., p. 468, Pergamon, New York (1959).
Bradley, W. C., Itek Corp., unpublished paper.
Hufnagel, R. E., *Photogr. Sci. Eng. 9*, 244 (1965).
Levi, L., *Applied Optics*, Vol. 1, p. 505, Wiley, New York (1968).
O'Neill, E. L., *J. Opt. Soc. Am. 46*, 258 (1956).
O'Neill, E. L., *Introduction to Statistical Optics*, p. 99, Addison-Wesley, Reading, Mass. (1963).
Prescott, R., *Appl. Opt. 7*, 479 (1968).
Ratcliffe, J. A., *Rep. Prog. Phys. XIX*, 188 (1956).
Rosell, F. A., in *Photoelectronic Imaging Devices*, Vol. 1, p. 307, Plenum, New York (1971).
Schade, O. H., *Soc. Motion Pict. Telev. Eng. 56*, 137 (1951).
Schade, O. H., *Soc. Motion Pict. Telev. Eng. 58*, 181 (1952).
Schade, O. H., *Soc. Motion Pict. Telev. Eng. 61*, 97 (1953).
Schade, O. H., in *Optical Image Engineering*, Natl. Bur. Stand. Circ. 526, p. 231, U.S. Govt. Printing Office, Washington, D.C. (1954).
Schade, O. H., *Soc. Motion Pict. Telev. Eng. 64*, 593 (1955).
Wetherell, W. B., in *Instrumentation in Astronomy*, SPIE Seminar Proc., Vol. 28 (1972).
Wetherell, W. B., and M. P. Rimmer, *Appl. Opt. 11*, 2817 (1972).
Young, A. T., *Appl. Opt. 6*, 1063 (1967).

R. H. MEIER

# Calibration and Performance Evaluation of Systems and Devices for Space Astronomy: Discussion of a Test Facility

## Introduction

For any optical system intended to be operated from a space platform, a comprehensive prelaunch calibration and performance evaluation is a prerequisite for its ultimate usefulness. This is particularly true for astronomical instruments designed to sense and measure radiant energy from sources emitting in the infrared and submillimeter region of the electromagnetic spectrum. A meaningful calibration and evaluation of such sensors is quite difficult, and aside from the present lack of generally accepted primary calibration standards, the necessary realistic simulation of their operational environment requires great care and expense.

An existing large-scale facility, which appears to be well suited for the testing and calibration of a variety of infrared astronomical sensors and telescopes, is described in the following. It was developed by the McDonnell Douglas Astronautics Company in Huntington Beach, California, as part of its Space Simulation Laboratory.

The author is at McDonnell Douglas Astronautics Company, Huntington Beach, California 92647.

## Description of Test Facility

GENERAL

The facility, as shown photographically in Figure 1 and schematically in Figure 2, consists basically of a vacuum chamber with an inner cryoshroud housing an on-axis parabolic collimator, a radiant-energy-source assembly, a calibration monitor that occupies one portion of the collimated beam, a pair of scanning mirrors directing radiant energy from another portion of the collimated beam into the sensor under test, and an optional star background generator. Sensors to be tested can be mounted either inside the chamber or downward-looking on top of their saddle portion, as shown in Figure 2. All optical elements consist of Kanigen-coated aluminum and have their active surfaces gold coated. The vacuum chamber is a stainless steel horizontal cylinder, 2.4 m in diameter and 4.3 m long, with a vertical penetration 2 m in diameter. The latter can be extended on demand by stacking spools to the required length. Vacuum levels of $10^{-8}$ Torr or better can be achieved with a multistage contamination-free

FIGURE 1   View of test facility.

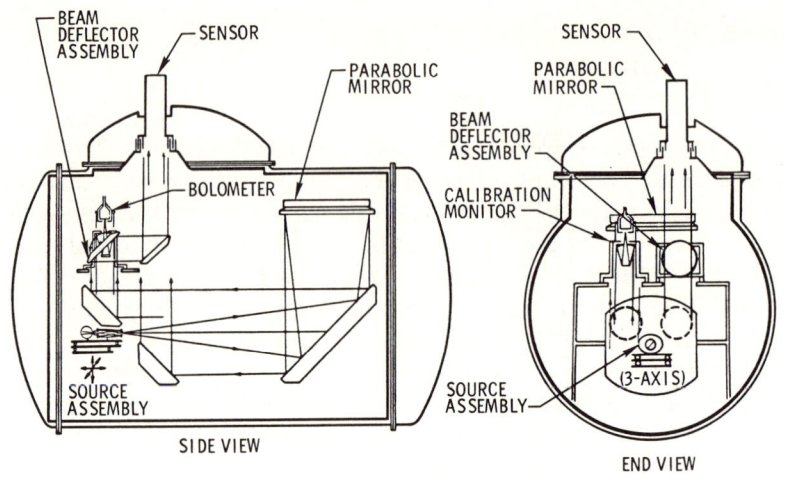

FIGURE 2  Layout of test facility.

pumping system. A gaseous helium, closed-cycle cooling system cools the cryoshroud and all other interior structures, components, and optical elements to a temperature of about 20 K, thereby providing an extremely low radiation background.

COLLIMATOR

The collimator assembly, shown in Figure 2, consists of a 0.81-m-diameter parabola having a focal length of 3.02 m and two 0.81 m × 1.06 m plano folding mirrors. Its performance was found to be diffraction-limited for wavelengths longer than 8 $\mu$m. The spatial distribution of the collimated beam's flux density was assessed by measuring the distribution of the radiant flux in a plane across the diverging beam about 15 cm from the source, where it could be measured with a bare detector. Tests have shown no measurable performance degradation of the collimator at cryogenic temperatures.

SOURCE ASSEMBLIES

Located behind the second folding mirror of the collimator is a three-axis source table with the source assemblies mounted on it. The sources radiate through a 0.2-m-diameter central circular aperture in this mirror.

The standard source assembly is shown in Figure 3. It consists of a blackbody with a chopper and a set of stepped limiting apertures in

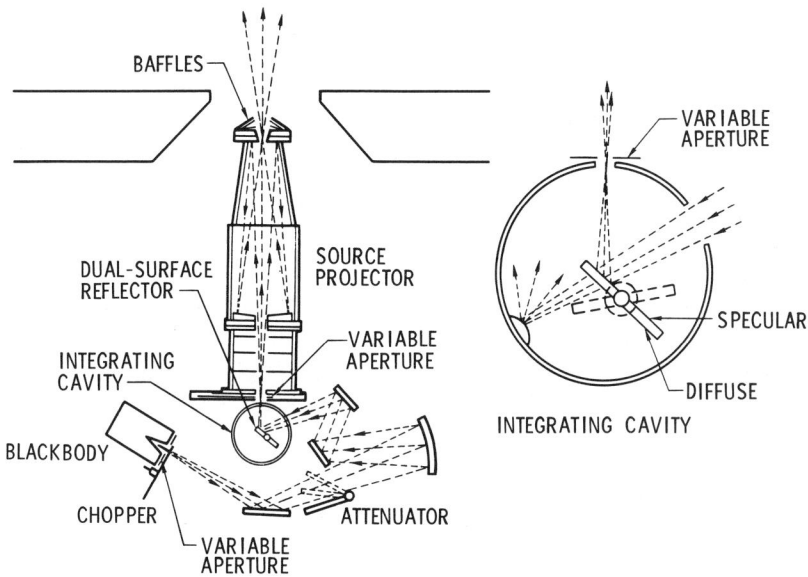

FIGURE 3 Standard source assembly.

front of its cavity, a set of transfer optics, a variable attenuator, an integrating cavity with a second set of stepped variable apertures located at its exit hole, a special dual-surface reflector mounted inside the integrating cavity, and a source projector. One side of the dual-surface reflector is highly polished for specular reflection. Its other side is selectively glass-ball peened for diffuse reflection, as is the interior wall of the integrating cavity. The selected aperture in front of the blackbody cavity is imaged onto the selected exit aperture of the integrating cavity with unity magnification by a toric mirror via four folding reflectors, which include the dual-surface reflector inside the integrating cavity. This aperture is located at the back focus of the source projector, which projects its 2.4 times reduced image into the focal plane of the parabolic collimator.

The source assembly can be operated in three different modes, depending on the angular position of the dual-surface reflector inside the cavity. The specular reflection mode is used when high irradiance levels are desired. In the diffuse reflection mode, the mirror is rotated 180° so that its diffuse surface deflects the radiant energy onto the integrating cavity's exit aperture. The lowest irradiance levels are achieved by rotating it into a position where it does not intercept the beam entering the cavity and where the diffuse side is facing the cavity's exit aperture. In this test mode, called the integrating cavity mode of operation, the beam strikes the strongly convex and diffuse reflecting surface of a button-type insert at the opposite wall of the integrating cavity. The insert reflects and diverges the radiant flux toward the front half of the integrating cavity, from where it is again diffusely reflected, mainly toward the diffuse surface of the dual-surface reflector. This surface now provides a uniform Lambertian radiant background within the field of view of the source projector.

In the specular mode of operation, the 8-mm blackbody aperture is used exclusively; only the exit apertures of the integrating cavity are varied. In the two other modes, all combinations of the two sets of variable apertures—five in the blackbody set and seven in the integrating cavity set—can be used to control the collimator irradiance. Furthermore, in the integrating cavity mode, the attenuator, which is a simple metal plate that can be rotated into the exit beam of the blackbody to block any desired portion of it with a measurement accuracy of about 1 percent, can be utilized. In total, at any given blackbody temperature, this source assembly can provide a range of irradiance values within the collimated beam of the facility that spans more than nine orders of magnitude. This is achieved without filters or impractical aperture sizes.

Another available source assembly that can be mounted on the three-axis source table is an all-reflective grating monochromator, shown in Figure 4. Its circular exit aperture is located at the focal point of the collimator. At present, three gratings are available with rulings of 20, 40, and 80 lines/mm, respectively. The spectrum is scanned across the exit aperture by rotating the grating from its specular (zero-order) position ($\phi = 0$) through angles $\phi$, yielding the wavelength according to

$$m\lambda = 2d \sin \phi \cos \delta,$$

with $\delta$ being the angle of incidence at $\phi = 0$, $d$ the grating constant, and $m$ the spectral order. The gratings are blazed for $\lambda = 4.3$, 8.6, and 17.2 $\mu$m, respectively. Using the first-order spectrum of each grating only, the linear dispersion of the monochromator at its exit aperture amounts to 0.05 $\mu$m/mm for the 80 lines/mm grating and to 0.22 $\mu$m/mm for the 20 lines/mm grating. At any wavelength setting, the calibration monitor is subjected to the same spectral irradiance as the sensor under test. Since the spectral response of the monitor's bolometer-type detector is essentially flat, the absolute spectral response of the sensor under test is obtained by simply dividing its normalized output signal by the calibration monitor's normalized output signal. A set of typical system response curves of the monochromator is plotted in Figure 5. The curves were

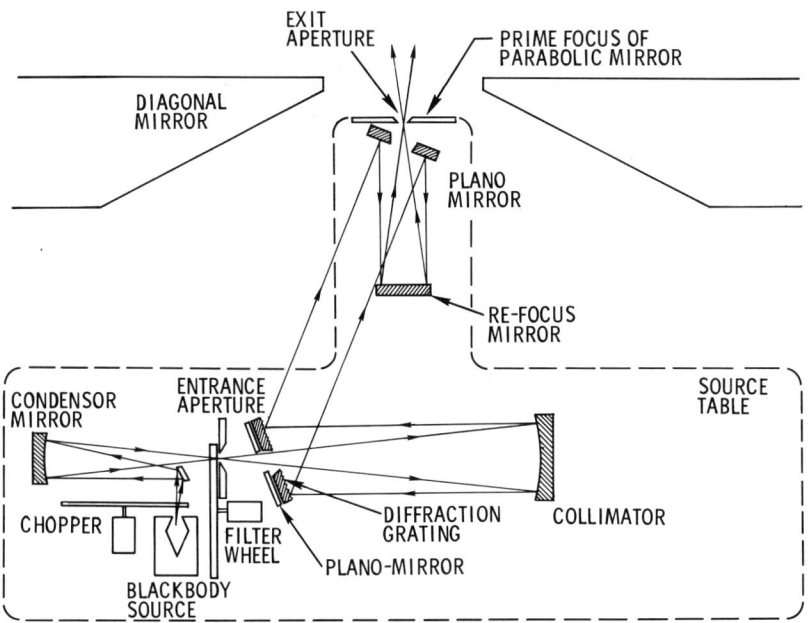

FIGURE 4  Monochromator source.

obtained with a 900 K source during two test cycles; one with the 80 lines/mm grating, the other with the 20 lines/mm grating. The filters used were commercial products from Optical Coating Laboratory, Inc., Santa Rosa, California.

Since a changeover from one source assembly to the other or a change of gratings during a monochromator test requires very costly and time-consuming warm-up periods of the facility, a new universal source assembly has been designed that combines all the characteristics and capabilities of both sources.

Figure 6 is a schematic drawing of this universal source. With the mirror side of the rotatable cube in the position shown in the drawing, the assembly functions exactly like the described standard source with its wide dynamic range of irradiance levels. With any one of the three grating sides rotated into the collimated portion of the beam, the assembly is changed into a monochromator. The exit aperture of the integrating cavity now acts as the monochromator exit aperture. Again, the spectrum is scanned across the exit aperture by rotating the selected grating about its normal (zero-order) position.

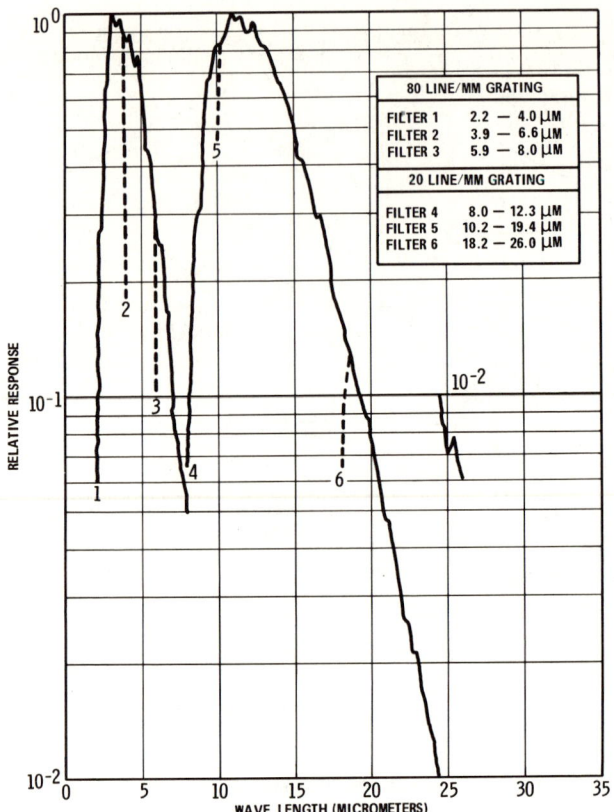

FIGURE 5  System response of monochromator.

SCANNING MIRRORS

As illustrated in Figure 2, a pair of scanning mirrors deflect a 0.3-m-diameter portion of the collimated beam into the aperture of the sensor under test. A magnetically driven positional servo system can rotate each mirror out of its rest position by at least ±5°, one in elevation, the other in azimuth, with a positional precision of about 20 sec of arc. In addition, the mirrors can perform periodic scanning motions at varying angular velocities ranging from 0.1 mrad/sec to 0.3 rad/sec.

STAR BACKGROUND SIMULATOR

Another capability that can easily be added to the facility is the simulation of a superimposed star background within the instantaneous field of

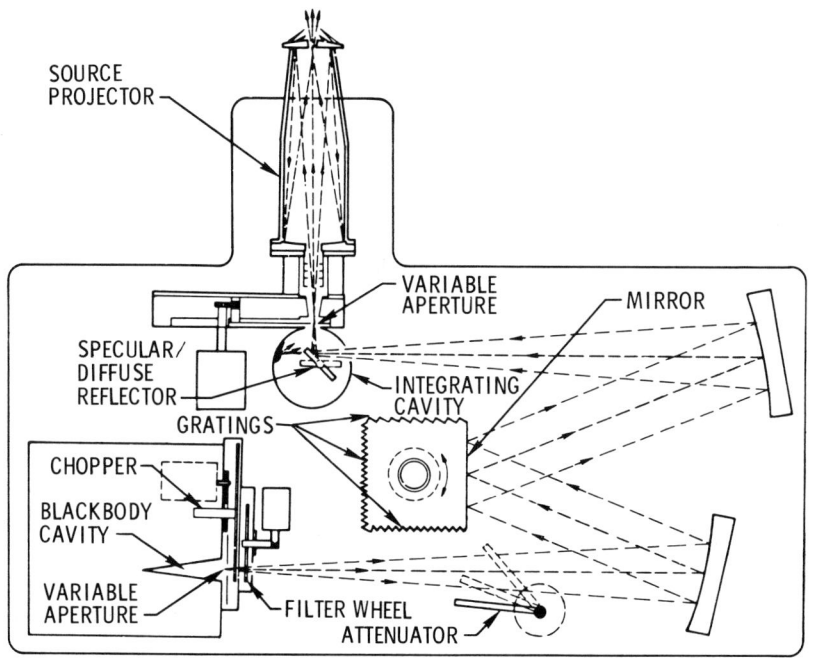

FIGURE 6  Universal source assembly.

view of the sensor. A schematic drawing that illustrates how this or any other background can be generated is presented in Figure 7. An extended source, typically a large blackbody cavity, irradiates uniformly a templet with numerous small pinholes. The pinholes may have different sizes and may, if desired, be covered by selected filters. A simple Newtonian optical system projects the punctured templet toward infinity, parallel to the axis of the collimated beam and within the clear aperture of the sensor under test. The sensor then forms an image of the punctured templet in its focal plane.

## Calibration of Test Facility

GENERAL

An attempt to predict the prevailing irradiance in the collimated beam from such data as blackbody temperature and geometry, source assembly geometry, and reflectance values of all utilized surfaces including the

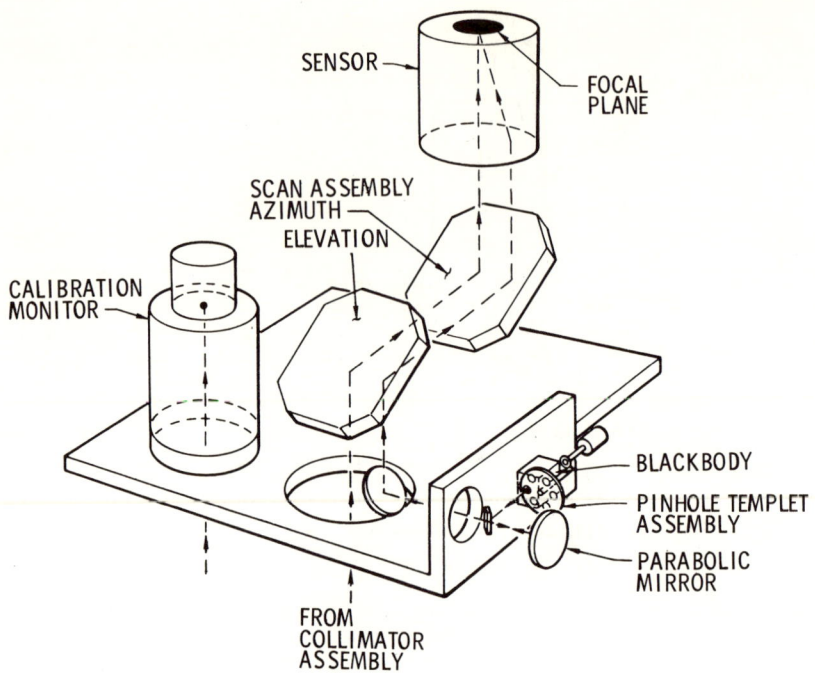

FIGURE 7 Concept of star background generator.

glass-ball-peened diffuse interior of the integrating cavity, is, to say the least, hazardous. For this reason, the irradiance is continuously measured with the permanently installed calibration monitor, which occupies a generally unused portion of the collimated beam. The detector of the calibration monitor, a gallium-doped germanium bolometer, supplied by Infrared Laboratories, Inc., of Tucson, Arizona, has exhibited remarkably stable, repeatable characteristics ever since its selection as a working transfer standard. Its calibration is repeated whenever convenient and consists of the determination of its responsivity to radiant energy from a blackbody source at selected temperatures and chopping frequencies. The responsivity is defined by

$$R_{BB} = V_{S,B,N}/\phi_{BB} \quad \text{[Volts (rms)/Watt]},$$

where $V_{S,B,N}$ is the signal resulting from the blackbody and background irradiance and from bolometer noise; $\phi_{BB}$ is the total radiant power impinging on the bolometer, according to

$$\phi_{BB} = \epsilon \sigma T_{BB}^4 A_{BB} A_D / \pi r^2 \quad [\text{Watt}],$$

with $\epsilon$ being the emissivity of the blackbody cavity, $\sigma$ the Stefan-Boltzmann constant, $T_{BB}$ the measured temperature of the blackbody cavity, $A_{BB}$ the area of the blackbody cavity, $A_D$ the area of the bolometer, and $r$ the distance between blackbody and detector.

At an operating bolometer temperature of 2.03 K, a blackbody temperature of 300 K, and a chopping frequency of 25.5 Hz, the bolometer responsivity was measured to be

$$R_{BB}(1\sigma) = (1.54 \pm 0.03) \times 10^5 \quad [\text{Volts (rms)/Watt}].$$

The utilized blackbody cavity was fabricated from ultra-pure aluminum by Electro-Optical Industries, Inc., Santa Barbara, California. Its temperature is monitored by three copper–constantan thermocouples embedded in its wall. Although the precision to which the temperature can be set and maintained is better than 0.2 K, an uncertainty of ±5 K has been assigned to its accuracy. This estimate is primarily based on observed departures of the measured values from those expected according to the $T^4$ law of blackbody radiation.

CALIBRATION MONITOR

The calibration monitor is essentially a calibrated radiometer, permanently installed inside the facility, where it intercepts a 0.2-m-diameter portion of the collimated beam. Its optical layout is shown in Figure 8. It has an effective focal length of 0.8 m. The area of its clear aperture is 212 cm$^2$. The bolometer assembly is mounted on a three-axis table. The selected bolometer operating temperature of 2.03 K is carefully controlled by maintaining the pressure in the helium Dewar constant to within ±10$^{-3}$ Torr.

CALIBRATION RESULTS

Since the calibration monitor and the scanning mirror assembly, which directs radiant flux into the sensor under test, occupy areas of the collimated beam that are about equally spaced from its centerline, it can be assumed that the irradiance in both locations will not differ significantly. Therefore, the irradiance values measured with the calibration monitor in its location can be applied to the portion of the beam that is occupied by the sensor under test. For sensors whose entrance aperture exceeds

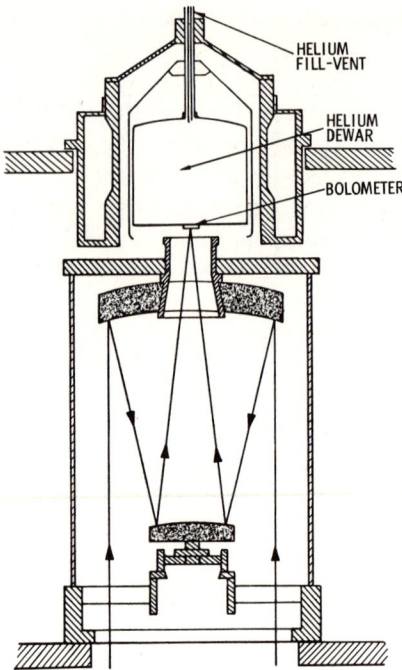

FIGURE 8  Calibration monitor.

0.3 m, a careful mapping of the amount of blockage caused by the collimated beam's central obscuration must be performed. Sensors approaching the diameter of the collimated beam, 0.81 m, are mounted centrally. In this case, the calibration monitor must be either removed after calibration or mounted outside the beam with a retractable mirror system.

Figure 9 is a typical plot of beam irradiance versus aperture settings, measured with the calibration monitor at a blackbody temperature of 300 K. It shows sizable overlaps in the dynamic ranges of the three modes of source operation. In the specular mode, only the exit aperture of the integrating cavity was varied. The blackbody aperture was kept constant at 8 mm. In both the diffuse mode and the integrating cavity mode, the normalized effective aperture area values are obtained by multiplying every normalized blackbody aperture area with every normalized integrating cavity aperture area. Values below the bolometer's noise-equivalent irradiance are extrapolated. This is considered permissible since the data obtained at higher irradiance levels strongly indicate the validity of the linear relationship between aperture settings and irradiance levels.

The error bars at the measured data points indicate the existing uncer-

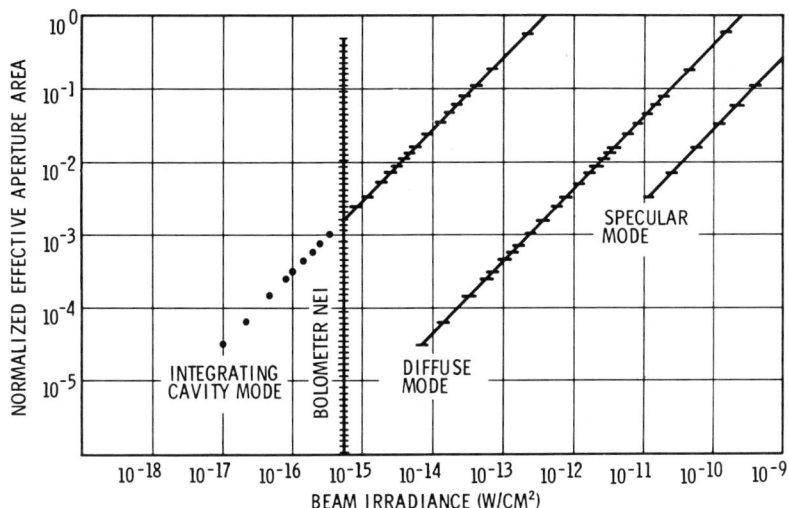

FIGURE 9 Collimated-beam irradiance with standard source assembly.

tainty in the total systems calibration. From a simple error analysis it can be shown that this uncertainty is roughly $\pm [2.2 \times 10^{-3} + (4\Delta T)^2 \, T^{-2}]^{1/2}$ percent. The first term of this expression represents the sum of all uncertainties and measurement errors that are independent of the blackbody temperature. Some are by necessity derived from only a limited number of measurements. They also include the somewhat arbitrary assumption of a 2 percent uncertainty in the blackbody's effective emissivity. The second term represents the uncertainty caused by the assumed ±5 K uncertainty in the blackbody temperature determination. At 300 K—the temperature at which the responsivity of the bolometer was determined— this term amounts to about $4.5 \times 10^{-3}$. Thus, the existing uncertainty in the total systems calibration is about 8.2 percent. With an essentially wavelength-independent bolometer response, this uncertainty should hold for all blackbody temperatures. Under the present circumstances, the quality of this calibration can be considered good. Prospects for refining it depend to some extent on the future availability of more accurate methods for determining blackbody temperatures.

The author wishes to acknowledge the efforts and contributions made by R. P. Day and C. J. C. Thompson during the design and development phase and by A. B. Dauger, W. A. Fraser, H. Popick, and J. D. Shoore during the calibration phase. Also, the support given by J. R. Averett, F. A. Eckhardt, J. T. Morrow, and E. H. Schwiebert to the solution of several unusual engineering problems in mechanics, cryogenics, and vacuum technology has greatly contributed to the successful completion of this unique facility. Finally, the author would like to express his appreciation for many fruitful discussions with R. H. McFee and S. Zwerdling.

H. KÖHLER and F. STRÄHLE

# DESIGN OF ATHERMAL LENS SYSTEMS

## I. Introduction

The temperature behavior of optical lens systems is governed primarily by the following two thermal glass constants:

1. The linear coefficient of expansion

$$\alpha^* = \frac{1}{l}\frac{dl}{T}$$

2. The temperature coefficient of the refractive index

$$\beta^* = dn/dT$$

[It should be mentioned that the numerical values of $\beta^*$ are quite different if the optical element is in air ($\beta^*_{rel}$) or in a vacuum ($\beta^*_{abs}$).]

In the practical application of optical systems that are affected by temperature changes, the following two cases must be distinguished:
(a) External influence of temperature causes—after a certain transient time—a locally stationary *homogeneous* temperature change of all parts

---

The authors are at Carl Zeiss, 7082 Oberkochen/West Germany.

of the optical system. (b) Instead of a homogeneous temperature distribution in the stationary temperature equilibrium there are temperature gradients. We would like to concentrate here on the specific case that all optical elements of a system have the same *radial temperature gradient*. This is the most important case for practical work.

In case of homogeneous temperature changes, the temperature effects, such as changes in focal length and Gaussian image scale, are reduced by a factor of about 10 compared with the corresponding changes in the case of radial temperature gradients. There is a difference in the relationship between the changes in the optical image-forming parameters as a function of the temperature of the thermal glass constants $\alpha^*$ and $\beta^*$, depending on whether *homogeneous* or *inhomogeneous* temperature changes are concerned.

Changes in the image-forming properties are governed by a thermal glass constant

$$\gamma^* = [\beta^*/(n-1)] - \alpha^*,$$

in case of *homogeneous* temperature changes, and by a glass constant

$$\Gamma^* = [\beta^*/(n-1)] + \alpha^*,$$

different from the above in case of *inhomogeneous* temperature changes.

## II. Homogeneous Temperature Changes

Although the influence of radial temperature gradients on the image quality of optical lens systems and the design of lens systems that are insensitive to such radial temperature gradients are more important for practical work than the influence of homogeneous changes, we would like to deal first with the latter.

Perry [1943], Volosov [1958], and Strähle [1969] have contributed to the theory of the influence of homogeneous temperature changes on the image quality of optical lens systems. Figure 1 shows part of the results that have been obtained—the change of the refractive power or of the focal length is proportional to the thermal glass constant $\gamma^*$.

In a system with thin-lens elements with finite air spaces, the change of the back focal length for the individual lens elements of the system is also determined with good approximation by the values of the thermal glass constant $\gamma^*$. As shown by the last equation in Figure 1, the change of the refractive power of a thick lens may be assumed to be

1.) Thin lens: $\quad \dfrac{d\phi}{dT} = \phi \gamma^*$ ; $\quad \dfrac{df'}{dT} = -\gamma^* f'$

2.) System consisting of k thin lenses:

$$\dfrac{d\phi'_k}{dT} = -\dfrac{{\sigma'_k}^2}{n'_k}\left(\dfrac{h_1}{h_k}\right)^2 \left\{\sum_{m=1}^{k}\left(\dfrac{h_m}{h_1}\right)^2 \phi_m \gamma^*_m + \sum_{m=2}^{k}\left(\dfrac{h_m}{h_1}\right)^2 \dfrac{1}{\sigma^2_{m-1}} \dfrac{d(d_{m-1})}{dT}\right\}$$

for $\sigma_1 = \infty$, $\quad \dfrac{d(d_{m-1})}{dT} = 0$, $n'_k = 1$

$$\boxed{\dfrac{d\sigma'_k}{dT} = -{\sigma'_k}^2 \left(\dfrac{h_1}{h_k}\right)^2 \left\{\sum_{m=1}^{k}\left(\dfrac{h_m}{h_1}\right)^2 \phi_m \gamma^*_m\right\}}$$

3.) Thick lens:

$$\phi_{\Delta T} = \phi_0 (1 + \gamma^* \Delta T) + C$$

$$C = \dfrac{(n-1)d}{(n + \beta^* \Delta T) r_{10} \cdot r_{20}} (1 + 2\gamma^* \Delta T)$$

FIGURE 1

proportional to $\gamma^*$. The additional parameter $C$ in this equation is in most cases negligible, especially for telescope lenses.

As shown by Strähle (see Figure 2), the constant $\gamma^*$ determines changes not only in the optical parameters in the paraxial region but also in the direction of a wavefront element at a finite distance from the axis, which are proportional to the constant $\gamma^*$ in the case of temperature changes.

For a number of optical glasses produced by Jenaer Glaswerk Schott & Gen., Mainz, the values of our glass constant $\gamma^*$, valid for lenses in air, are given as a function of the Abbe number $\nu$, as shown in Figure 3. Similar values were obtained for lenses in a vacuum. The values used for the temperature variation of the refractive index correspond to the measurements carried out by Schott in Mainz. In this connection, we would like to emphasize that the $\gamma^*$ values vary considerably with the wavelengths. This is demonstrated in Figure 4. Strictly speaking, athermalization can be achieved for only one wavelength. However, according to our computations the gain in temperature constancy for an achromat made from athermal glasses is still considerable in the entire visual wavelength range.

As illustrated on Figure 3, there are only a few glass pairs that, at considerable differences in the Abbe number $\nu$, have values of $\gamma^*$ near zero.

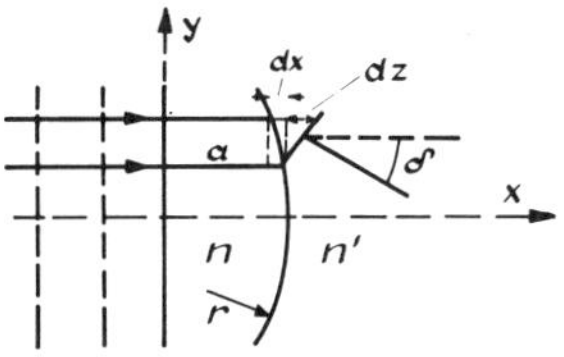

$$n(a-dx) + n'dx + n'dz = na$$

$$n'dx - ndx = -n'dz$$

$$dz = dx \frac{n-n'}{n'} \quad (I)$$

equation of the spherical surface: $x = \frac{y^2}{2r} + \frac{y^4}{8r^3} + \cdots$

$$dx = \frac{y}{r} dy + \frac{y^3}{2r^3} dy + \cdots$$

with (I): $\frac{dz}{dy} = \frac{n-n'}{n'} \left\{ \frac{y}{r} + \frac{y^3}{2r^3} + \cdots \right\} = \delta$

for $n'=1$: $\frac{d\delta}{dT} = \left\{ \frac{y}{r} + \frac{y^3}{2r^3} + \cdots \right\} \frac{dn}{dT} + (n-1) \left\{ -\frac{y}{r^2} \frac{dr}{dT} - \frac{3}{2} \frac{y^3}{r^4} \frac{dr}{dT} \right\}$

$$\frac{dn}{dT} = \beta^* \quad ; \quad \frac{dr}{dT} = r \alpha^*$$

$$\boxed{\frac{d\delta}{dT} = \frac{y}{r}(n-1)\left\{ \frac{\beta^*}{n-1} - \alpha^* \right\} + \frac{1}{2}\left(\frac{y}{r}\right)^3 (n-1)\left\{ \frac{\beta}{n-1} - 3\alpha^* \right\} \approx \frac{y}{r}(n-1)\gamma^*}$$

FIGURE 2

Such a glass pair is, for instance, the SK 5/SF 5. From this glass pair, an achromat with an $f$-number 13 and a focal length of 4.5 m has been computed. Figure 5 shows the optical setup and the chromatic aberration in the axis. With regard to its temperature behavior, this achromat should be compared with an achromat made from nonathermal lenses. Figure 6 shows the optical setup and the chromatic correction of this reference objective. As can be seen, the correction state of the two objectives is comparable.

Figure 7 illustrates the temperature dependency of the Gaussian focus for these two systems, namely, for a temperature range from 0 to 80 °C. The superiority of the athermal achromat is clearly noticeable.

For a normal achromat the relative change in the focal length is $3.3 \times 10^{-5}/°C$ within a medium temperature range, and $10^{-7}/°C$ for the athermal achromat. These examples were computed for a focal length of 4500 mm, for which the absolute focus change of a normal achromat amounts to 0.15 mm/°C, and that of the athermal achromat to $4.9 \times 10^{-4}$ mm/°C. Compared with the normal achromat, the focus change of the athermal achromat is smaller by a factor of 330.

A homogeneous temperature change of 1.25° will cause defocusing by one Rayleigh criterion (0.187 mm) for a normal achromat with a

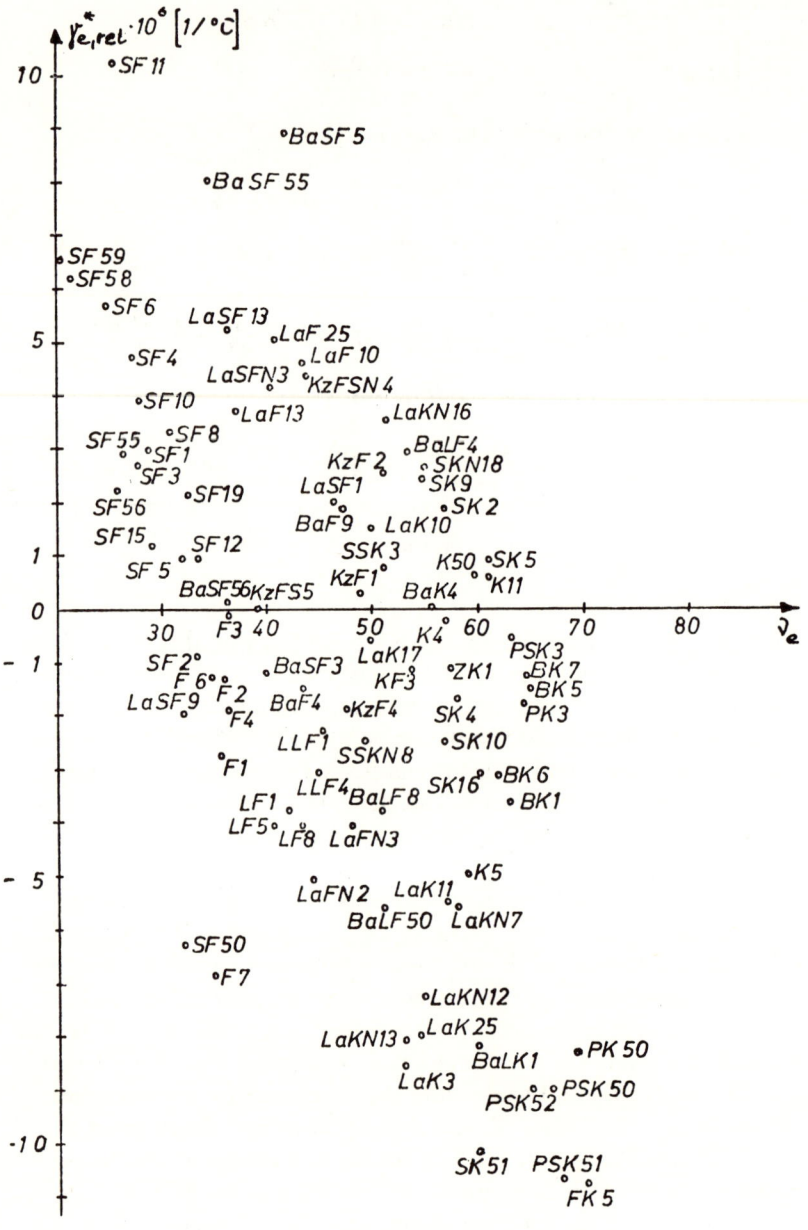

FIGURE 3 $\gamma_{rel}^*$-$\nu$ diagram.

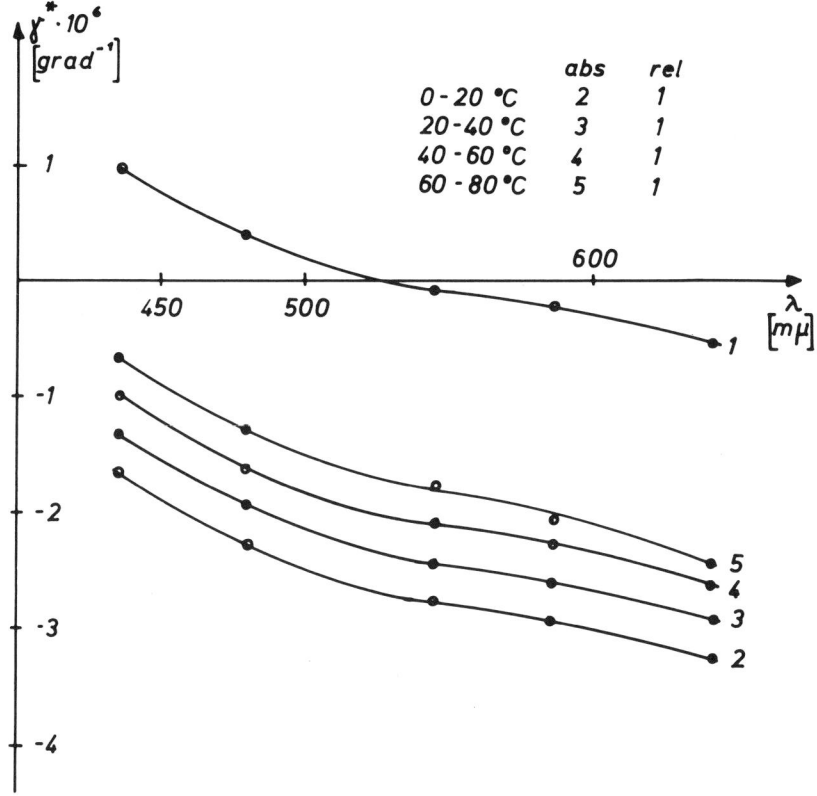

FIGURE 4 $\gamma^*$ as a function of the wavelength of the glass type SK 5.

focal ratio of 12.85. With an athermal achromat of the same focal ratio, a homogeneous temperature change of 50° would only cause defocusing by 1/8 Rayleigh criterion.

As a supplement to the above, Figure 8 shows the wavefront aberration computed with the aid of the formulas given by Herzberger [1957] for the normal and the athermal achromats as a function of the lens temperature. An analysis of the curves proves that the wavefront aberration generally corresponds to a pure defocusing, which is numerically identical to the values given in Figure 7. Here, too, a factor of 330 exists between the thermal defocusing of the normal achromat and the athermal achromat.

The change in the spherical aberration in comparison with the change in focus is negligibly small: trigonometric computation of the normal achromat yields for a temperature difference of 60° a change in the

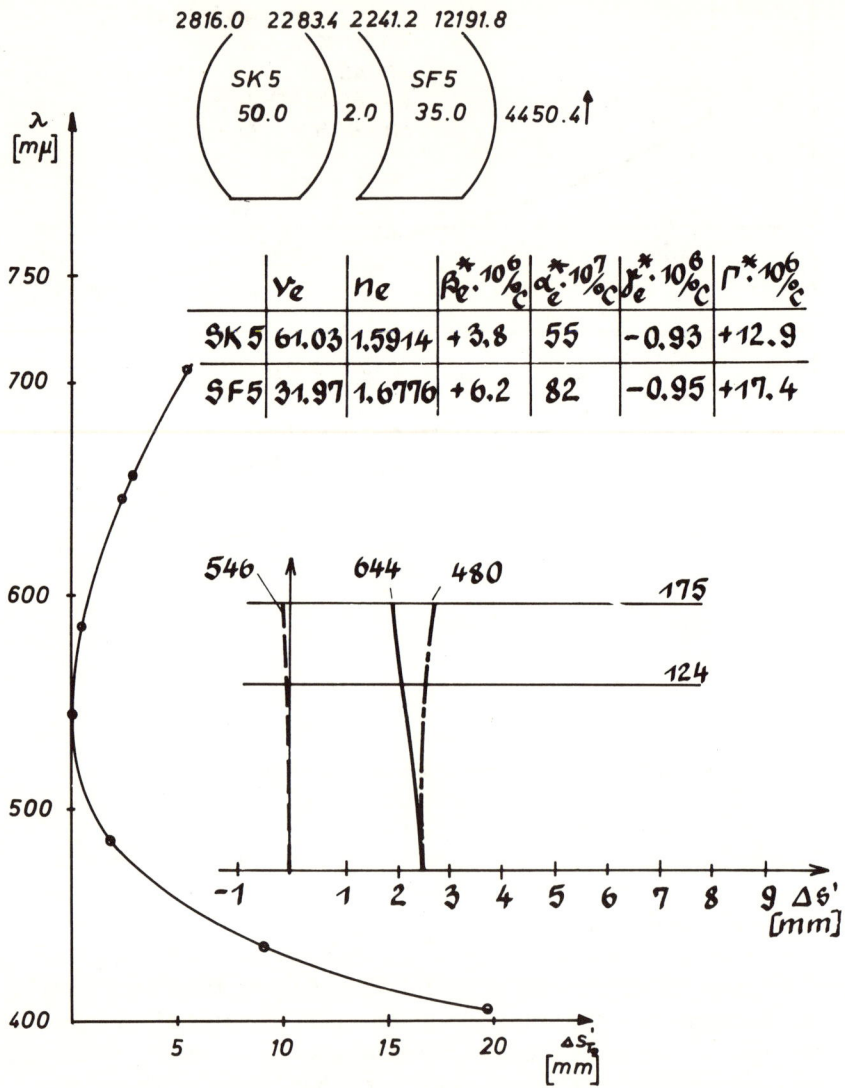

FIGURE 5 Achromat athermalized for homogeneous temperature variations.

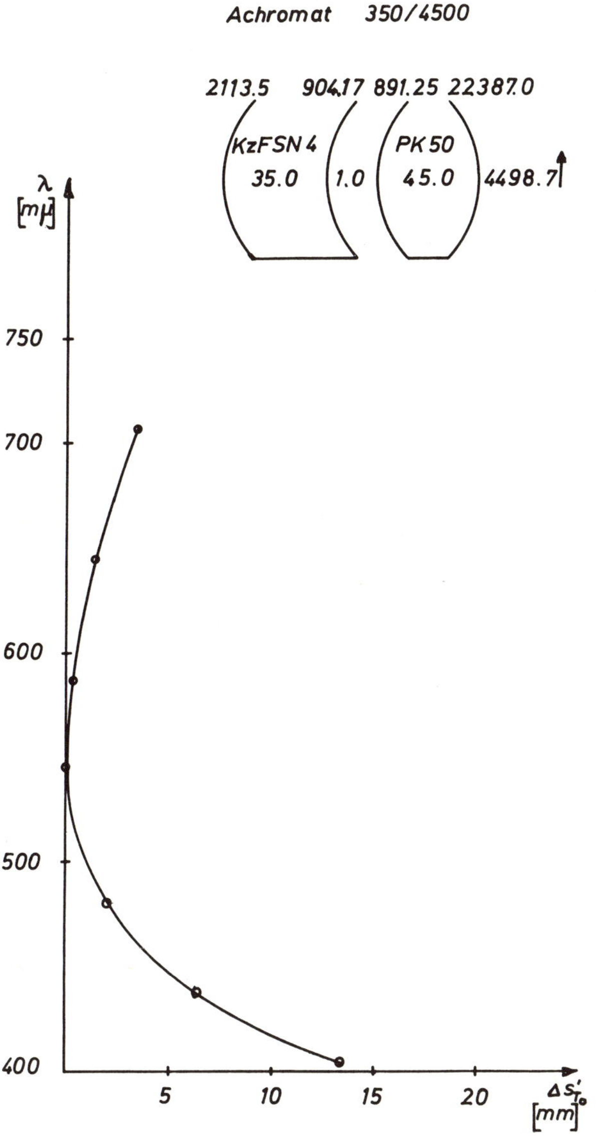

FIGURE 6  Nonathermalized achromat.

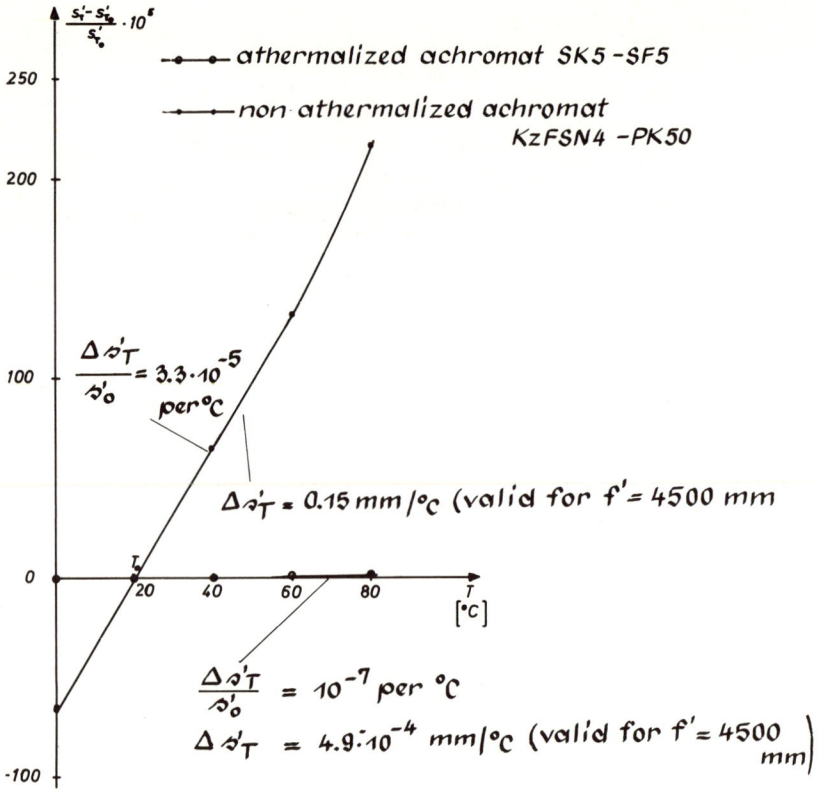

FIGURE 7 Variation of the Gaussian focus with homogeneous temperature variations.

spherical aberration of less than 1/10 Rayleigh criterion (1/40λ). The change is smaller by a factor of 10 for the athermal achromat.

Ketterer [1969] has examined the influence of homogeneous temperature changes on the aberration coefficients. The relationships are very complex and did not clearly establish a relationship between the aberration coefficients and $\gamma^*$. This study confirms, however, the statement that the influence of temperature on the changes of all geometrical optical image aberrations and/or corresponding wave aberrations is smaller than the influence of focus changes by some tenth powers.

As shown in Figure 4, the thermal glass constant $\gamma^*$ varies considerably as a function of the wavelength. Figure 9 shows this relationship for the glass type SF 5. (SK 5 and SF 5 are used in our athermal achromat.) According to the numerical computation, this wavelength dependency of $\gamma^*$ causes only a very small achromatic error. For the athermal

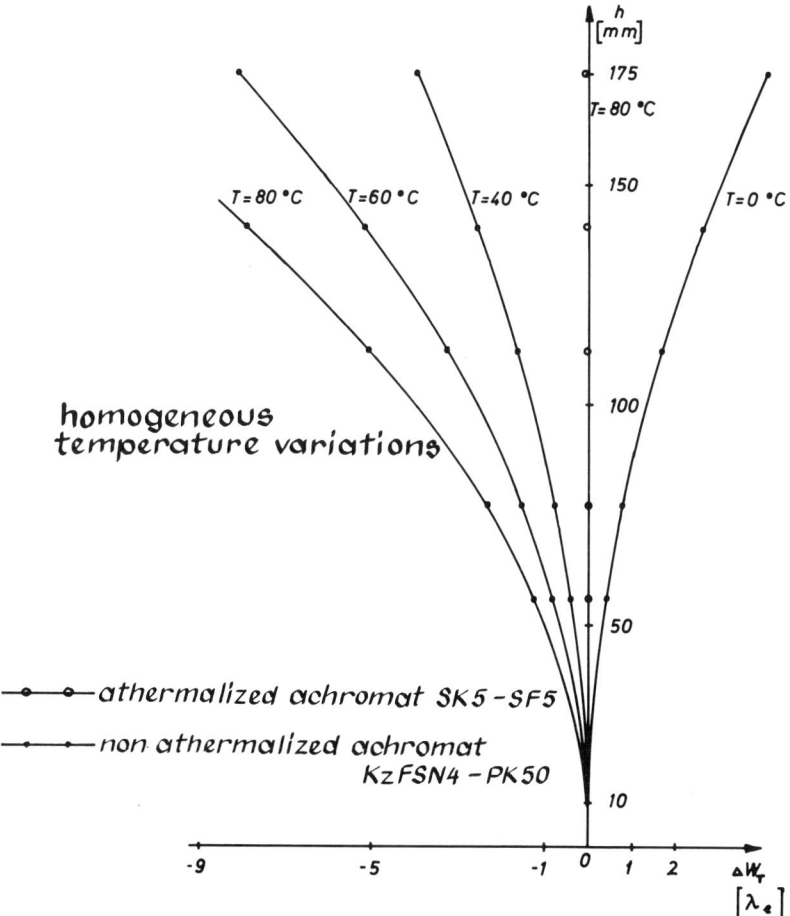

FIGURE 8  Variation of the wave aberration with temperature.

achromat, the chromatic longitudinal aberration between $C'$ and $F'$ is $6 \times 10^{-3}$ mm per degree temperature change. In other words, to achieve the tolerance limit given by the $\lambda/4$ criterion, the achromat ought to be heated up homogeneously by 60°.

As early as 1958, Volosov discussed the question of what extent an athermal achromat can be made from glasses whose $\gamma^*$ values differ from zero. In order to enter into further particulars, Figure 10 illustrates a somewhat more general theory of athermal achromats in the range of the Gaussian dioptrics. The equations are valid for a system of two thin lenses.

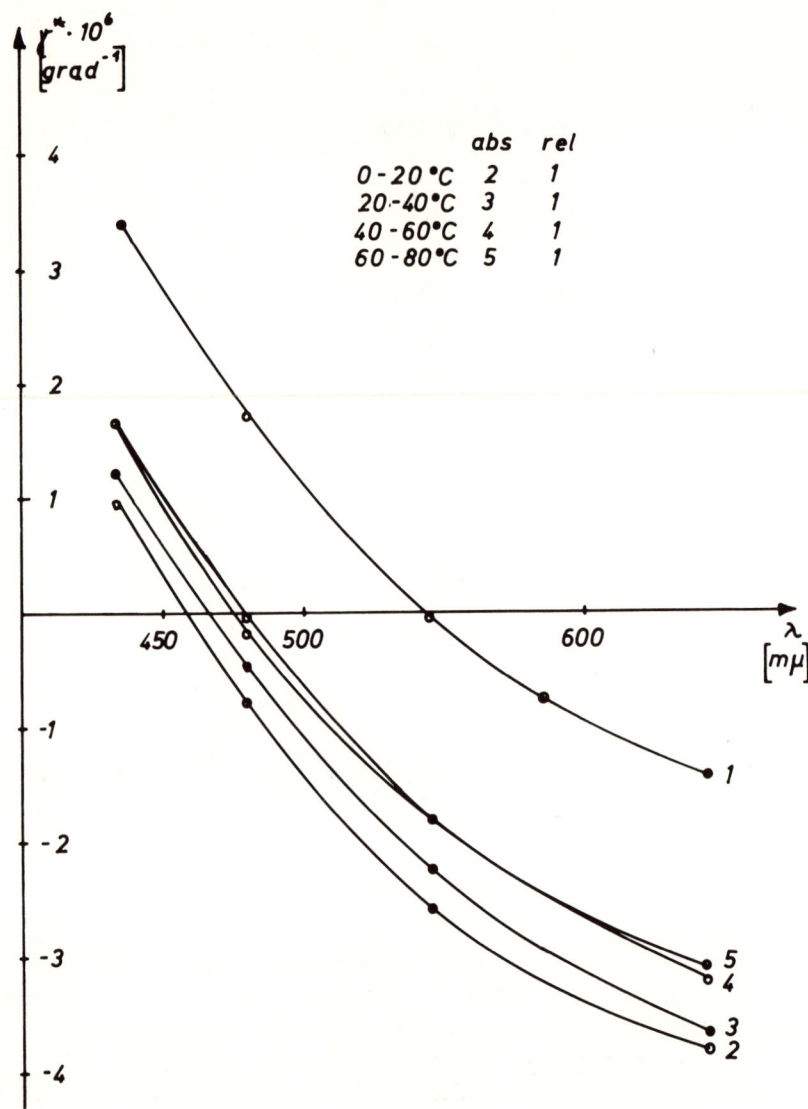

FIGURE 9  $\gamma^*$ as a function of the wavelength of the glass type SF 5.

$$\phi_1 = \phi \frac{\nu_1}{\nu_1 - \nu_2} \quad (1) \quad ; \quad \phi_2 = -\phi \frac{\nu_2}{\nu_1 - \nu_2} \quad (2)$$

$$\phi_1 \gamma_1^* + \phi_2 \gamma_2^* = -\phi \cdot \frac{1}{f'} \frac{da'}{dT}$$

$$\frac{da'}{dT} = a \alpha_g$$

$$\phi_1 \gamma_1^* + \phi_2 \gamma_2^* = -\phi \frac{a}{f'} \alpha_g = -\phi K$$

$$K = \frac{a}{f'} \alpha_g$$

**with (1) and (2) :**

$$\phi \left\{ \frac{\nu_1}{\nu_1 - \nu_2} \gamma_1^* - \frac{\nu_2}{\nu_1 - \nu_2} \gamma_2^* \right\} = -\phi K$$

$$K = \frac{\nu_1 \gamma_1^* - \nu_2 \gamma_2^*}{\nu_1 - \nu_2} \quad ; \quad \text{for } a = f' : \alpha_g = \frac{\nu_1 \gamma_1^* - \nu_2 \gamma_2^*}{\nu_1 - \nu_2}$$

**for $K = 0$ :**

$$\boxed{\nu_1 \gamma_1^* = \nu_2 \gamma_2^*}$$

FIGURE 10

If the system should be achromatic for two colors, Eqs. (1) and (2) for the partial refractive powers $\phi_1$ and $\phi_2$ must be satisfied. The third equation then gives the change of the Gaussian paraxial image plane. In general, the right side of this equation will not be made equal to zero; but it can be specified that the thermal change of the Gaussian paraxial image plane is, for instance, to correspond to the thermal expansion of the tube or part thereof. If this mechanical component is of a length $a$, and if its thermal expansion coefficient is $\alpha_g$, the equation in the fourth line will result. If for $\phi_1$ and $\phi_2$ the values from Eqs. (1) and (2) are introduced, the relationship obtained will correspond to the sixth and seventh line in which the nondimensional quantity $(a/f')\alpha_g = K$. If the thermal expansion of the housing should be disregarded, i.e., if $K = 0$, the athermalization condition achieved will be as follows:

$$\nu_1 \gamma_1^* = \nu_2 \gamma_2^* \text{ (also when } \gamma_1^* = \gamma_2^*\text{)}.$$

To give an idea of the consequences, the values of $\nu\gamma^*$ for some optical glasses are given in Figure 11 as a function of $\nu$. The relationship de-

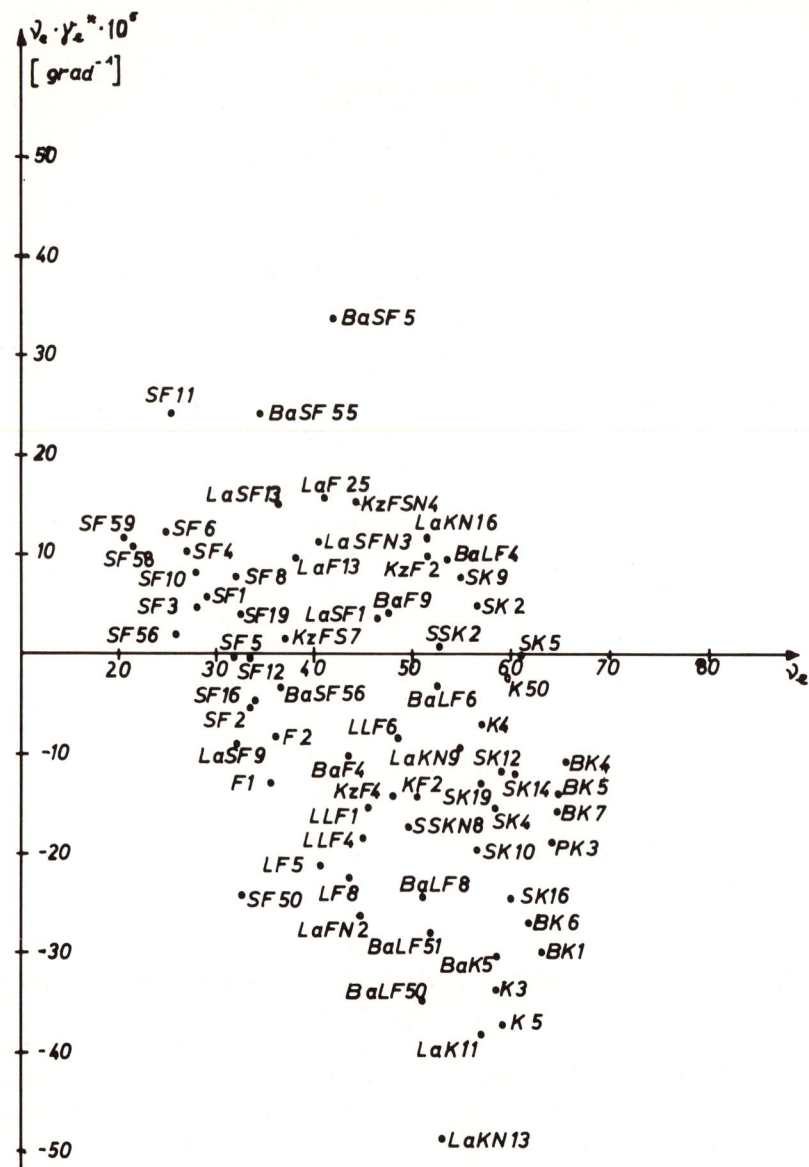

FIGURE 11 The $\nu\gamma^*-\nu$ diagram.

rived for $K = 0$ thus means that the Gaussian paraxial image plane of the achromat is independent of temperature if the glasses lie parallel to the abscissa in the graph. When taking into account the thermal expansion of the housing, the equations in the sixth and seventh line of Figure 10 apply. The right side of the equation in the sixth line means nothing but the gradient of the straight line connecting two glass pairs in Figure 11. If the tube length in particular is identical to the focal length, the auxiliary quantity $K$ is equal to the expansion coefficient $\alpha_g$ of the housing, and the inclination of this straight line is identical to the thermal expansion coefficient of the housing.

Since the quantity $\gamma^*$ is a function of the wavelength (as shown before) and therefore an exact athermalization can in general only be achieved for one single wavelength (as mentioned above) an approximate athermalization in a large spectral range can only be obtained if the glass pair is situated in the neighborhood of the abscissa of the graph in Figure 10.

In the following, we would like to deal with the focus changes caused by inhomogeneous temperature changes, especially in the event of radial temperature gradients, which are essentially larger than the effects mentioned before. These effects are of relevant importance for practical work. We have therefore abandoned the manufacture and experimental study of the achromat described here, which is insensitive to homogeneous temperature changes, and have carried out practical and experimental examinations of an achromat that is said to be insensitive to radial temperature gradients. It is with this achromat that we now deal.

## III. Temperature Function of Optical Elements with Temporarily Stationary Radial Temperature Gradients

In the following, we show that an optical lens system with a radial temperature gradient is a preferred model, which, no doubt, describes a whole array of practical applications.

As shown by Strähle, a radial temperature gradient occurs in a rotation–symmetrical optical element when the optical element is assumed to be in a mount of infinite extension and infinite thermal conductivity with a temperature deviating from the ambient temperature $T_u$.

In the stationary temperature equilibrium, the heat flow, which passes into the environmental medium through the lens surface by convection, equals the heat flow due to the thermal conductivity glass mount. This temperature balance results in an ordinary differential equation of the Bessel type, which is shown as Eq. (1) in Figure 12. This equation can-

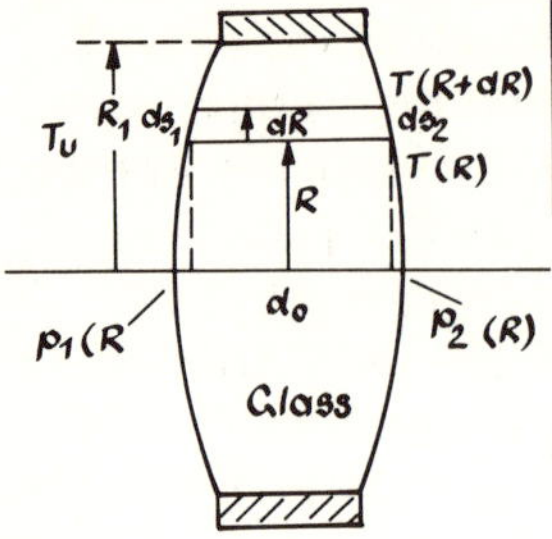

$$\frac{d^2T}{dR^2}+\left\{\frac{1}{d}\frac{dd}{dR}+\frac{1}{R}\right\}\frac{dT}{dR}-\frac{\alpha_T}{\lambda_T d}\left(\sqrt{1+\left(\frac{dp_1}{dR}\right)^2}+\sqrt{1+\left(\frac{dp_2}{dR}\right)^2}\right)T(R)$$
$$=-\frac{\alpha_T}{\lambda_T d}\left\{\sqrt{1+\left(\frac{dp_1}{dR}\right)^2}+\sqrt{1+\left(\frac{dp_2}{dR}\right)^2}\right\}T_U \quad (1)$$

$p_1(R)$; $p_2(R)$ : sagittae, $d(R)$ : thickness

$\lambda_T$ : thermal conductivity

$\alpha_T$ : heat transfer coefficient

$$\frac{d^2T}{dR^2}+\frac{1}{R}\frac{dT}{dR}-\frac{2\alpha_T}{\lambda_T d_0}T=-\frac{2\alpha_T T_U}{\lambda_T d_0} \quad (2)$$

$$T(R) = T_U + (T_0-T_U)\cdot I_0\sqrt{(-)\frac{2\alpha_T}{\lambda_T d_0}}\cdot R \quad (3)$$

$$T(R) \approx T_0 + (T_0-T_U)\left\{\frac{1}{8}\frac{\alpha_T}{\lambda_T d_0}R^2 + \frac{1}{256}\left(\frac{\alpha_T}{\lambda_T d_0}\right)^2 R^4 + \cdots\right\} \quad (4)$$

$$T(R) \approx T_0 + a_2 R^2 \quad (5)$$

$$a_2 = \frac{T_0-T_U}{8}\cdot\frac{\alpha_T}{\lambda_T d_0} = \frac{\Delta T(R_1)}{R_1^2} \quad (6)$$

FIGURE 12 Temperature function of an optical element with radial temperature gradient.

not be integrated as a whole. For several specific cases, Strähle has integrated it numerically according to the Runge-Kutta method.

The differential equation is considerably simplified when the lens is replaced by a plane-parallel plate. The differential quotients $dp_1(R)/dR$, $dp_2(R)/dR$, $dd(R)/dR$ become zero, and the result is Eq. (2) in Figure 12. This equation can be integrated as a whole. The result is shown in Eq. (3) of Figure 12. $T_o$ is the temperature of the optical axis in stationary equilibrium, which is generally different from the ambient temperature $T_u$. $I_0$ is the Bessel function of zero order. Equation (4) is obtained by expansion into a series of the Bessel function. Because of the good convergence of the series, it is possible to restrict with good approximation to the first term of the series. That means that under the given boundary conditions the temperature between axis and edge increases with the square for a plane-parallel plate, as shown in Eq. (5).

Coefficient $a_2$ results formally from the difference between axis and ambient temperatures, the heat-transfer coefficient $\alpha_T$, the thermal conductivity $\lambda_t$, and the center thickness $d_0$. As a rule, these parameters cannot be determined so that practical considerations are based on a given temperature difference between axis and lens edge $\Delta T(R_0)$, and $a_2$ is assumed to be given according to Eq. (6):

$$a_2 = \Delta T(R_1)/R_1^2.$$

To find out whether this relationship, which is approximately true for plane-parallel plates, is also applicable to higher values of $R$ and $a_2$ of the plane-parallel plates and to lenses as well, Strähle has integrated in some cases the differential Eq. (1) numerically according to the Runge-Kutta method and compared the result with approach (5).

The result is shown in Figure 13. Curves No. 3 refer to the plane-parallel plate, curves No. 4 to the positive lens of 100-mm focal length, and curves No. 2 to a negative lens of $-1000$-mm focal length. Curves No. 1 are the parabolas that osculate with the curve vertices. The curves are drawn for three different values of $T_o - T_u$. Up to values of $R$ of about 50 mm the numerically computed curves have a good correspondence with the parabolas of the approach, whereas in the ranges above this value the correspondence is less good. The deviations are, however, not so marked that for the following considerations it could not be assumed that the temperature between axis and edge follows a square function of the radius.

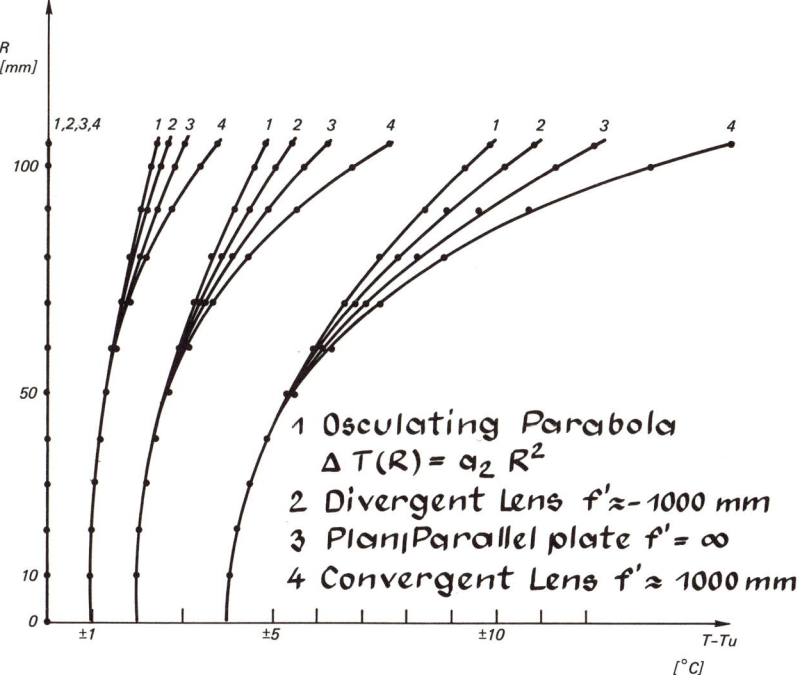

FIGURE 13 Theoretical temperature function of optical elements made of BK 7.

With thermoelements, the temperature distribution was measured on two plane-parallel plates with heating coils on the edges producing after about 60 min stationary temperature equilibrium, a constant temperature difference between axis and edge. The results are shown in Figures 14 and 15. The parabola osculating with the vertex of the measuring curve is also shown. Coincidence between this touching parabola and measured curve is even better than that of the preceding theoretical curves. Following these results, it may well be assumed for all following considerations that the temperature between axis and edge follows a square function.

## IV. Influence of Radial Temperature Gradients on the Image-Forming Properties of Lens Systems

For a judgment of the influence of radial temperature gradients on optical image-forming parameters, we have based our considerations on

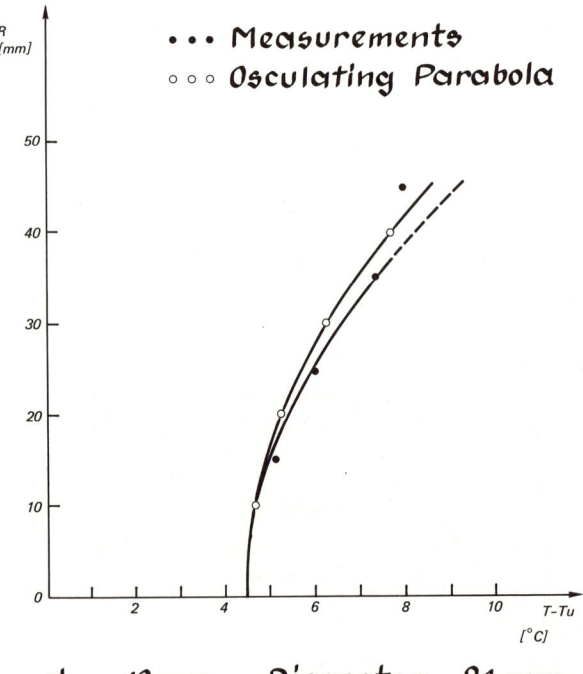

$d_o = 19\,mm$, Diameter $= 91\,mm$

FIGURE 14  Measured temperature function of an FK 50 plate.

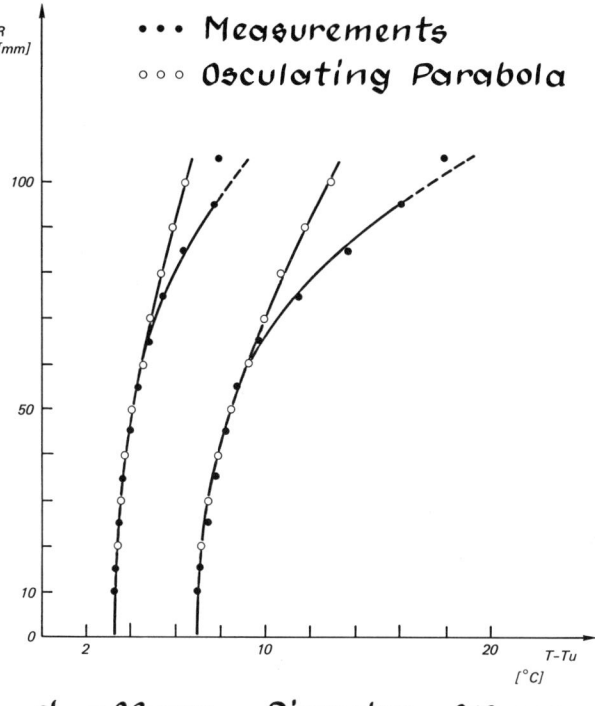

FIGURE 15 Measured temperature function of an Bk 7 plate.

the iconal function of a lens set up with negligence of changes in the ray direction. Equation (1) of Figure 16 shows this iconal function. Differentiation is made with respect to the radius in the pupil $R$ [see Eq. (2)]. The refractive index $n$, thickness $d$, and the lens radii $r_1$ and $r_2$ depend on temperature $T$ and thus on pupil radius $R$.

The first term in Eq. (2) of Figure 16 is of special importance, since it includes the changes of the iconal zero-order term $E_0$ as a function of the temperature. This important differential quotient $dE_0/dT$ is shown again separately in line 3 of the figure, and differentiation was made. $\Gamma^*$ was introduced for the form $[\beta^*/(n-1)] + \alpha^*$ [see Eq. (4)]. The differential quotient $dE_0/dT$ thus obtains the form, $dE_0/dT = d_0(n-1)\Gamma^*$ [see Eq. (3)], this being the same term that is obtained for the change in the optical path of a plane-parallel glass plate of thickness $d_0$. The abbreviation $\Gamma^*$ refers to the important glass constant mentioned above, which also governs changes in the optical path of a plane-parallel plate as a function of temperature.

$$E = -(n-1)d_0 + \frac{R^2}{2}(n-1)\left(\frac{1}{r_1}+\frac{1}{r_2}\right) + \frac{R^4}{8}(n-1)\left(\frac{1}{r_1^3}-\frac{1}{r_2^3}\right) + \cdots \quad (1)$$

$$\frac{dE}{dR} = -\frac{d[(n-1)d_0]}{dT}\cdot\frac{dT}{dR} + (n-1)\left(\frac{1}{r_1}+\frac{1}{r_2}\right)R + R^2 A\frac{dT}{dR} + R^3 B + R^4 C\frac{dT}{dR} + \cdots (2)$$

$$\frac{dE_0}{dT} = \frac{d[(n-1)d_0]}{dT} = -\left(d_0\frac{dn}{dT} + (n-1)\frac{dd_0}{dT}\right) = -d_0(n-1)\left(\frac{\beta^*}{n-1}+\alpha^*\right)$$

$$\boxed{\frac{dE_0}{dT} = d_0(n-1)\Gamma^*} \quad (3) \qquad \boxed{\Gamma^* = \frac{\beta^*}{n-1}+\alpha^*} \quad (4)$$

$$\frac{dE}{dR} = \phi_0 R - d_0(n-1)\Gamma^*\frac{dT}{dR} + R^2 A\frac{dT}{dR} + R^3 B + R^4 C\frac{dT}{dR} \quad (5)$$

with $T(R) \approx T_0 + a_2 R^2$ ; $\frac{dT}{dR} = 2a_2 R$ , we obtain:

$$\frac{dE}{dR} = \phi_0 R - d_0(n-1)\Gamma^* 2a_2 R + 2a_2 R^3 A + R^3 B + 2a_2 R^5 C + \cdots \quad (6)$$

$$E = \phi_0 \frac{R^2}{2} - d_0(n-1)\Gamma^* a_2 R^2 + \text{terms of } 4^{th} \text{ and higher order} (7)$$

$$\boxed{W_{20} = -d_0(n-1)\Gamma^* a_2 R^2}$$

FIGURE 16

Formally the coefficient of the second term in Eq. (2) is identical with the refractive power $\phi_0$ of a thin single lens in air. All other terms of Eq. (2) are of third and higher orders. In the integration to be made later they characterize spherical aberrations and their changes. The amount of the coefficients $A, B, C \ldots$ is of no interest. With the abbreviations mentioned before, Eq. (5) results from Eq. (2). Introduction of the temperature function $T = T_0 a_2 R^2$ derived in the previous section results in the expression of Eq. (6) for Eq. (5).

Integration with respect to $R$ results in iconal $E$ as a function of $R$ and the parabolic temperature coefficient $a_2$ [see Eq. (7)].

The first term of this equation corresponds to the iconal of an undisturbed spherical wave belonging to the Gaussian refractive power $\phi_0$. It is temperature-independent and characterizes the Gaussian image point at the initial temperature. The second term also corresponds to the iconal of an undisturbed spherical wave, which is, however, temperature-dependent due to the coefficient $a_2$. It characterizes the wave aberration $W_{20}$, which is the result of defocusing due to a radial temperature gradient. This term is shown in Eq. (8), and we shall deal with it in the following.

The terms of the third and higher orders in $R$ characterize the spherical wave aberration of the fourth and higher orders. We purposely abandon the exact reproduction of these terms, since the neglect with which the iconal function according to Eq. (1) was set up does not permit any reliable statements about these terms. Sliusarev [1959] and Strähle [1971] have estimated these terms. According to their estimates, the wave aberration of telescope lenses corresponding to these terms is at least smaller by a factor of 10 than the focus shift corresponding to the wave aberration $W_{20}$ according to Eq. (8). We shall prove this statement by numerical computations.

Within the range of wave aberrations of the second and fourth orders, the wave-aberration coefficients are additive. Assuming that for a system of thin lenses with small air spaces the temperature difference between axis and edge is the same for all lens elements, the wave aberrations of the individual lens elements according to Eq. (8) can be added up, since in a system of thin lenses all lens elements have the same axial radius of the pupil.

Figure 17 shows as Eqs. (1a) and (1b) the maximum aberration $W_{20\ max}$, that applies to a system of thin lenses. The known relationship between focal-point shift $\Delta f'$ and appertaining wave aberration according to Eq. (2) yields the amounts of defocusing given in Eq. (3), which apply to a system of thin lenses with equal radial temperature gradients of all lens elements, characterized either by the constant $a_2$ or by the real temperature difference $\Delta_T$ between axis and edge.

Figure 16 explains also why another glass constant causes more defocusing in the event of radial temperature gradients than in the event of homogeneous temperature changes. When the temperature changes are homogeneous, all terms multiplied by $dT/dR$ disappear in Eq. (5), which means there remains only the first term in Eq. (5), which is independent of $\Gamma^*$.

To determine the temperature influence of homogeneous temperature changes, the first term in Eq. (7) of Figure 16 or the second term in Eq. (1) of Figure 16 must be differentiated with respect to temperature, which yields $dE_1/dT = (R^2/2)\,\phi_0\gamma^*$ (see Figure 18), showing the dependence of defocusing with homogeneous temperature changes on the constant $\gamma^*$. The coefficient $A$ in Eqs. (2), (5), and (6) in Figure 16 is obtained as $A = \frac{1}{2}\phi_0\gamma^*$.

Figures 19 and 20 show the glass constant $\Gamma^*$ as a function of the Abbe number $\nu$ for a number of optical glass types taken from the catalog of Jenaer Glaswerk Schott & Gen., Mainz ($\Gamma^*$ and $\nu$ are for the mean wavelength $e$. In Figure 19, $\Gamma^*$ applies to optical elements in air characterized by $\Gamma^*_{e,\ rel}$.

In Figure 20, $\Gamma^*$ is given for elements in a vacuum, characterized by

$$W_{20}\,max = -\alpha_2 R_1^2 \sum_{\nu=1}^{k} (n_\nu - 1) d_{o\nu}\, \Gamma_\nu^* \qquad (1a)$$

$$= -\Delta T \sum_{\nu=1}^{k} (n_\nu - 1) d_{o\nu}\, \Gamma_\nu^* \qquad (1b)$$

$$\Delta f' = -\frac{2 W_{20}}{h_1^2} f'^2 \qquad (2)$$

$$\Delta f' = 2\alpha_2 f'^2 \sum_{\nu=1}^{k} (n_\nu - 1) d_\nu\, \Gamma_\nu^*$$

$$= 2\,\Delta T \frac{1}{\sin^2 u'} \sum_{\nu=1}^{k} (n_\nu - 1) d_\nu\, \Gamma_\nu^* \qquad (3)$$

$$= 8\,\Delta T \cdot k^2 \sum_{\nu=1}^{k} (n_\nu - 1) d_\nu\, \Gamma_\nu^*$$

FIGURE 17  Defocusing as a consequence of radial temperature gradients.

$\Gamma^*_{e,\,abs}$. The latter values are of special interest for space optics. The glass types listed so far in the Schott catalog are marked by solid dots. Only the glass types FK 50, FK 51, and FK 6 fulfill the requirement $\Gamma^* \approx 0$, whereas for all other known optical glass types $\Gamma^*$ differs considerably from zero. Considering the increasing importance of athermal optical systems, Schott & Gen. have developed new glass melts in order to achieve the smallest possible or even a negative value of $\Gamma^*$. These glass types are not yet commercially available but will soon be included in the official catalog. They are indicated by a circled dot in Figures 19 and 20. Their preliminary designations are PK 51, 3036, 3038, and 3136. Note that even negative values of $\Gamma^*_{rel}$ have been obtained (for

$$\frac{dE_1}{dT} = \frac{d\phi}{dT} \frac{R^2}{2}$$

$$= \frac{R^2}{2}\left\{\left(\frac{1}{r_1} + \frac{1}{r_2}\right)\frac{dn}{dT} + (n-1)\left(-\frac{1}{r_1^2}\frac{dr_1}{dT} + \frac{1}{r_2^2}\frac{dr_2}{dT}\right)\right\}$$

$$= \frac{R^2}{2}\left\{\frac{\phi_0}{n-1}\beta^* + \phi_0\alpha^*\right\}$$

$$\frac{\beta^*}{n-1} + \alpha^* = \gamma^*$$

$$\frac{dE_1}{dT} = \frac{d\phi_0}{dT}\cdot\frac{R^2}{2} = \frac{R^2}{2}\phi_0\gamma^*$$

$$\frac{d\phi_0}{dT} = \phi_0\gamma^*$$

FIGURE 18

the glass melts PK 51 and 3038). With regard to the above statements, this should have a favorable effect on the athermal behavior of a lens produced of these glass types, because the thermal wave aberration of the collective lens element could be compensated by that of the dispersing lens element.

Considering the present state of the glass technology, there are, according to our above statements, six possible optical glass types for the design of athermal achromats with radial temperature gradients. Their characteristic constants are listed in Figure 21, which shows that the most favorable results with regard to aberration correction and athermal behavior are to be expected of the combination FK 51 and 3038.

As you may gather from column 6 of this table, these six athermal glass types have a comparatively high linear expansion coefficient, between $100 \times 10^{-7}$ and $150 \times 10^{-7}$. Such high values are required for the compensation of the term $\beta^*/(n-1)$ in the expression for $\Gamma^*$. Because of the high expansion coefficient, thermal strains occur in the case of inhomogeneous heating up. These strains cause additional wave aberration and strain birefringence with a disturbing influence in the case of high-temperature gradients. These phenomena become noticeable only

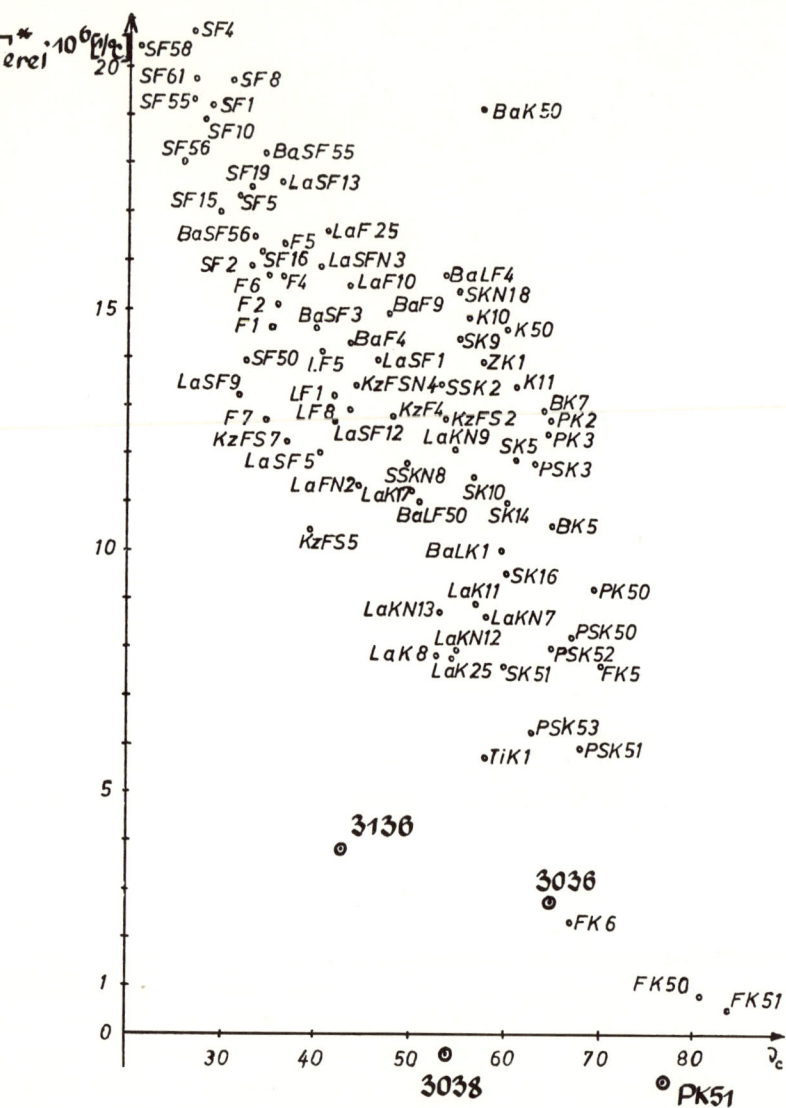

FIGURE 19 $\Gamma^*_{\text{rel}}-\nu$ diagram.

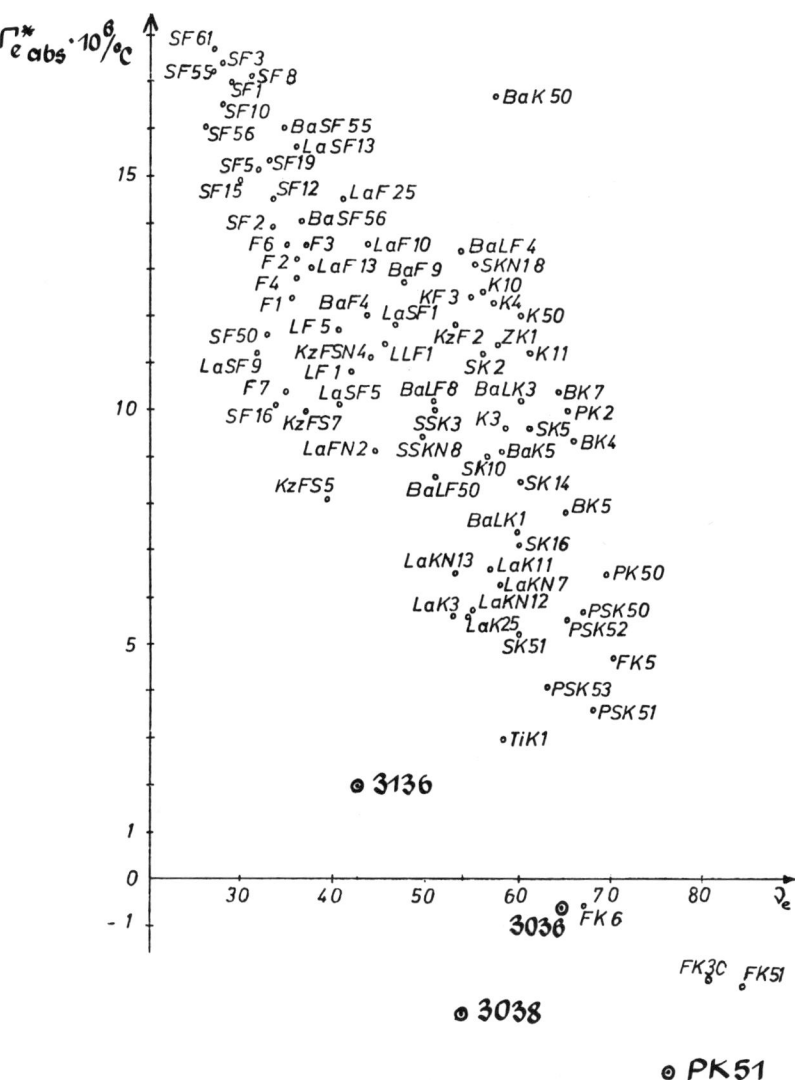

FIGURE 20  $\Gamma^*_{abs}$-$\nu$ diagram.

| Glass Type | $\nu_e$ | $n_e$ | $\beta^*_{rej}\cdot10^6$ [1/°c] | $\beta^*_{eabs}\cdot10^6$ [1/°c] | $\alpha^*\cdot10^7$ [1/°c] | $\Gamma^*_{e\,rej}\cdot10^6$ [1/°c] | $\Gamma^*_{e\,abs}\cdot10^6$ [1/°c] | $\gamma^*_{e\,rej}\cdot10^6$ [1/°c] |
|---|---|---|---|---|---|---|---|---|
| FK51 | 85 | 1.4876 | −6.5 | −7.8 | 136 | +0.27 | −2.4 | −26.9 |
| PK51 | 76 | 1.5340 | −8.4 | −9.9 | 147 | −1.0 | −4.26 | −30.4 |
| FK6 | 67.2 | 1.4479 | −4.0 | −5.3 | 112 | +2.3 | −0.63 | −20.1 |
| 3036* | 65 | 1.5833 | −5.4 | −7.0 | 113 | +1.87 | −0.69 | −20.6 |
| 3038* | 53.8 | 1.6320 | −8.0 | −9.5 | 121 | −0.60 | −2.98 | −24.7 |
| 3136* | 42.6 | 1.4479 | −3.6 | −5.0 | 103 | +4.4 | +2.08 | −16.3 |

\* new glass types, not yet scheduled

FIGURE 21 Characteristics of some athermal glasses for inhomogeneous temperature variation II.

when the temperature gradients achieve values that are not to be expected of optical image-forming systems. The high values of thermal expansion of the athermal glasses are nearly the same as those of steel. This is advantageous to avoid strains by the expansion of the mount.

Like athermal glass constant $\gamma^*$, which is decisive for homogeneous temperature changes, glass constant $\Gamma^*$ for radial temperature gradients discussed here, depends on the wavelength and on the temperature range concerned. Figure 22 proves this for glass type FK 50. We will see later that the thermochromatic aberration caused hereby will lie within acceptable limits.

## V. Numerical Results

To prove our theoretical statements and for a quantitative analysis of the changes of geometrical optical image aberrations to be expected with radial temperature gradients, we have computed an athermal achromat of 4500-mm focal length and 350-mm free aperture to be made of the glass types FK 51 and 3038, which are assumed to be the best for the purpose. By a ray-tracing program developed by Strähle, it was checked for its thermal behavior and compared with the achromat made of the

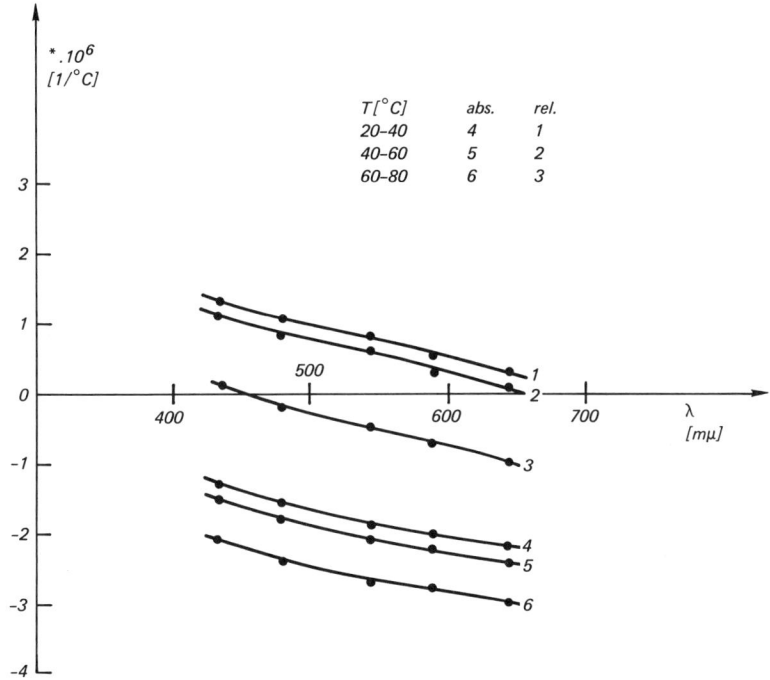

FIGURE 22  $\Gamma^*$ as a function of wavelength for FK 50.

glass types SK 5 and SF 5. The latter achromat was so designed as to be almost insensitive to *homogeneous* temperature changes.

We now discuss the ray-tracing program briefly. The change of the geometrical shape of the lens caused by thermal expansion is computed for a temperature distribution given by an even polynomial. Due to the thermal expansion, the spherical boundary surfaces of the lens are converted into aspheric surfaces, which can again be represented by an even polynomial with the distance of the axis as variable. The program determines the constants of the polynomial. The intersection point of a ray with the aspheric surface $P^*_0$ and the ray direction $\sigma_0'$ are then determined according to a method given by Herzberger (see Figure 23).

The stepwise ray tracing through the inhomogeneous medium is made according to a proposal of Montagnino [1968]. A step width $\Delta_s$ is given (here $\Delta_s = 1$ mm) and is used to compute the new ray coordinates according to Eq. (1) of Figure 23, where $\mathcal{R}_{so}$ is the ray curvature vector. The new ray direction is then given by Eq. (3). This results in the

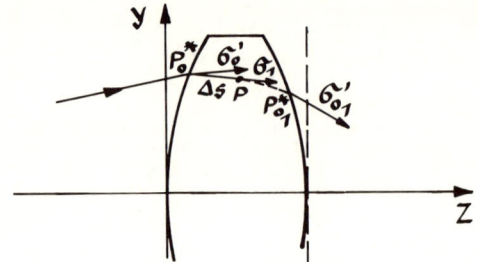

$$\mathscr{V}(\pmb{\nu}_0 + \Delta \pmb{\nu}) = \mathscr{V}(\pmb{\nu}_0) + \vec{\sigma}_{\pmb{\nu}_0} \Delta \pmb{\nu} + \frac{1}{2} \vec{\mathcal{R}}_{\pmb{\nu}_0} (\Delta \pmb{\nu})^2 \quad (1)$$

$$\vec{\mathcal{R}}_{\pmb{\nu}_0} = \frac{1}{n_{\pmb{\nu}_0}} \left\{ (\text{grad } n)_{\pmb{\nu}_0} - (\text{grad } n \vec{\sigma})_{\pmb{\nu}_0} \cdot \vec{\sigma}_{\pmb{\nu}_0} \right\} \quad (2)$$

$$\vec{\sigma}(\pmb{\nu}_0 + \Delta \pmb{\nu}) = \vec{\sigma}(\pmb{\nu}_0) + \vec{\mathcal{R}}_{\pmb{\nu}_0} \Delta \pmb{\nu} \quad (3)$$

FIGURE 23   Ray tracing through a nonhomogeneous lens system.

new orbital point $\mathscr{V}(s_0 + \Delta_s)$ and the new ray direction $\sigma(s_0 + \Delta_s)$. According to Herzberger, it can be used to compute the intersection point of the tangent with the orbital curve through the next aspheric surface, which is a control as to whether the ray tracing through the inhomogeneous medium is terminated or not.

The final result is the ray intersection point $P^*_{01}$ and the ray direction $\sigma_{01}$, which can be used to determine the direction of the refracted ray and to start the same procedure once again for the next inhomogeneous medium. Figure 24 shows the flow diagram.

Now the results: First for the achromat made of the glass types SK 5 and SF 5, which is insensitive to homogeneous temperature changes as was shown in Figure 5. Let us remind you that in the case of homogeneous temperature changes the focus change amounted to about 0.01 mm for a temperature difference of 20°, which was lower by a factor of 300 than that of an achromat of nonathermal glass types for homogeneous temperature changes.

Figure 25 shows the ray-tracing result for a radial temperature gradient of 1.53° between axis and edge, which corresponds to a value of $a_2 = 5 \times 10^{-5}$ °C/mm². It is seen that with fixed image plane (shown by curve 3), the result is a wave aberration of about 2λ. Refocusing by

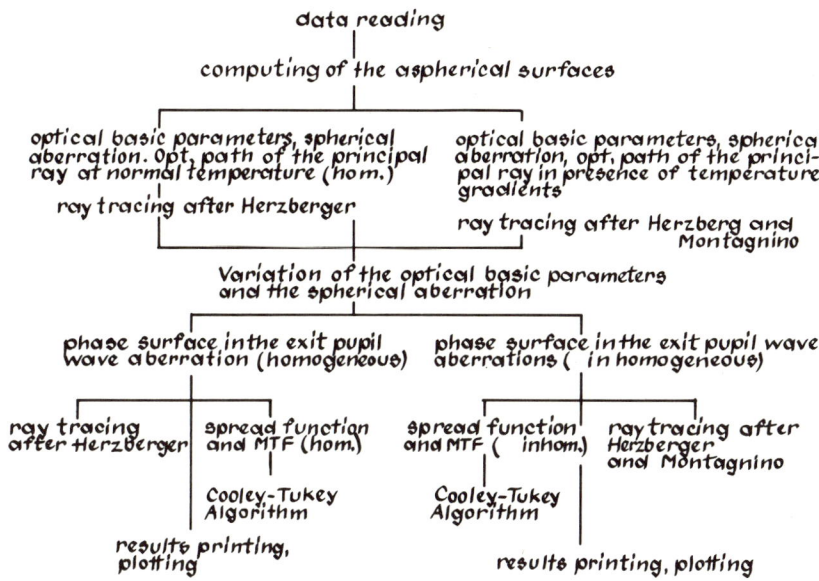

FIGURE 24  Flow diagram of ray tracing.

$\Delta_{s'}$ = 1.63 mm (curve 2) reduces this wave aberration to +0.5λ, which almost corresponds to the curve for normal temperature (curve 2). The difference between curves 1 and 2 is virtually the change in spherical aberration, corresponding to a value of about 0.023λ. This proves that defocusing caused by radial temperature gradients is higher than that caused by homogeneous temperature changes. Furthermore, the change in spherical aberration is smaller by a factor of about 10 than the change in thermal focus with radial temperature gradients.

Figure 26 shows the optical design and the state of correction of the athermal achromat made of glass types FK 51 and 3038. The table in the figure shows the glass properties that are of special interest in this instance.

Figure 27 shows the wave aberrations determined by the above-mentioned special ray-tracing program. For the ray tracing of this example, the value of the parabolic temperature constant $a_2$ was selected twice as large as that of the achromat SK 5 and SF 5. It is $10^{-4}$ °C/mm$^2$, which corresponds to a temperature difference of 3.05 °C between axis and edge. You may gather that the wave aberration represented by curve 3, which is true for a fixed image plane, is the same as that represented by curve 2 valid for the best image plane, namely,

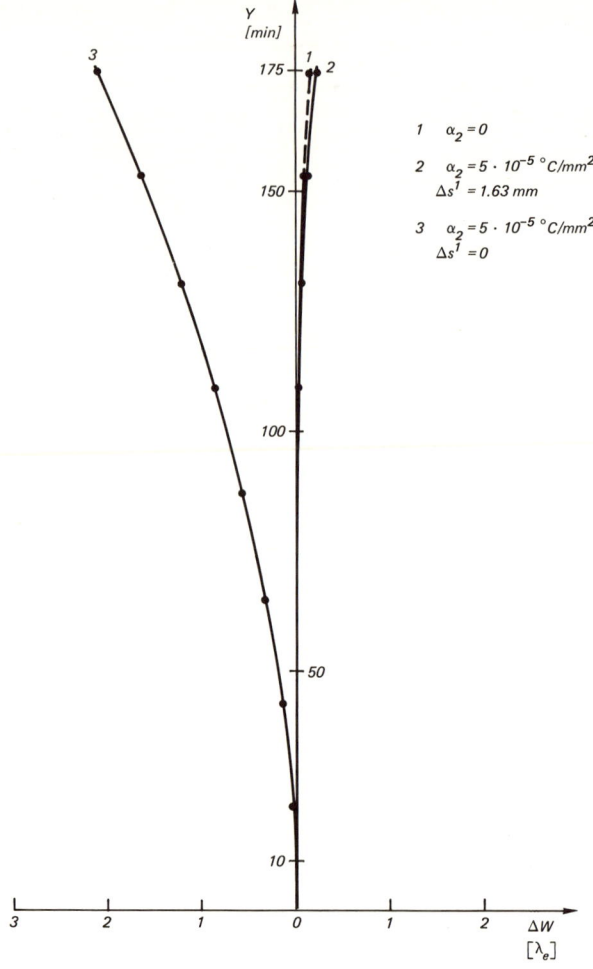

FIGURE 25  Achromat 350/4500 SK 5-SF 5 wave aberration in axis.

$0.10\lambda$. The wave aberration at normal temperature (curve 1) is $0.23\lambda$. The difference between the two curves of $0.13\lambda$ corresponds to the variation of the spherical aberration caused by a temperature gradient.

The defocusing between the fixed-image plane and the best focus has a calculated value of 0.003 mm. That value corresponds to the thermal defocusing, which in this example is better by a factor of 500 than that of the achromat making use of SK 5 and SF 5, which was

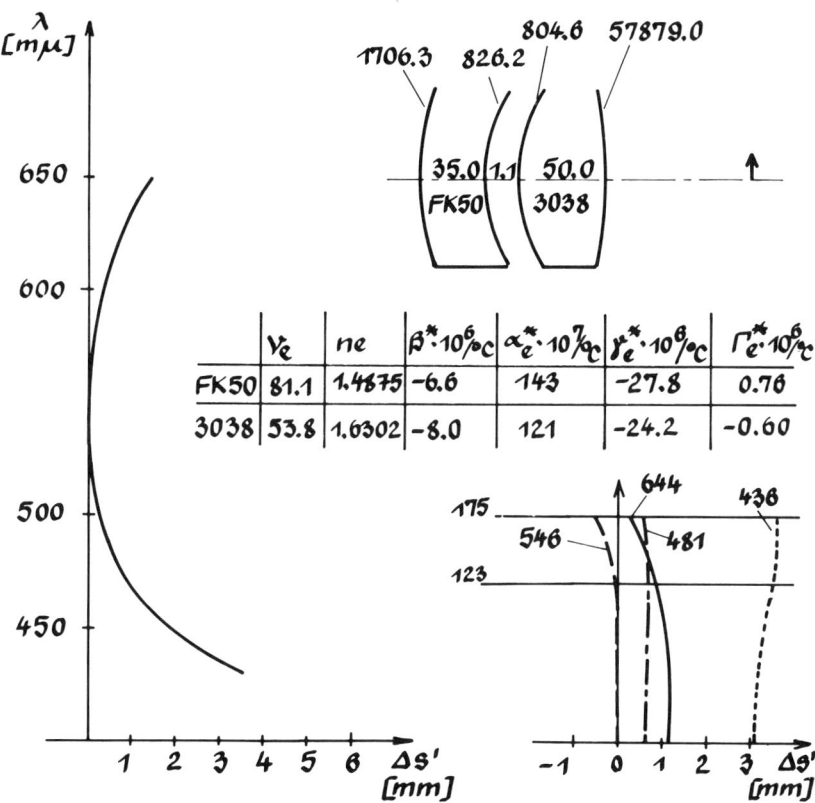

FIGURE 26 Achromat 350/4500 insensitive to radial temperature gradients.

optimized for homogeneous temperature changes ($\Delta_{s'}$ = 1.63 mm). The difference of the maximum wave aberration between normal temperature and the influence of the temperature gradient of $0.13\lambda$, which is likely to correspond to the thermal change of the spherical aberration, is sufficiently small for practical purposes.

We would like to mention that the computation of the thermal focus change according to the analytical formulas mentioned above yielded the following results:

for the lens made of SK 5-SF 5, 1.66 mm;
for the athermal achromat, 0.033 mm;

values that exactly correspond to the numerically computed ones.

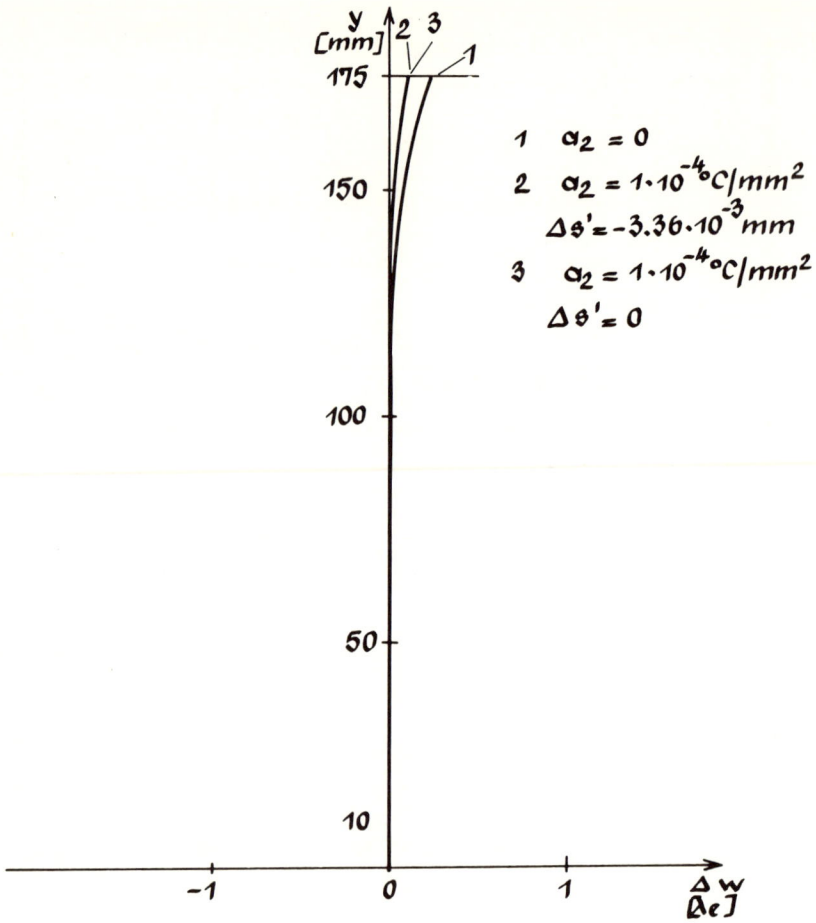

FIGURE 27  Wave aberration of the achromat insensitive to radial temperature gradients.

## VI. Studies Regarding Practical Application

After termination of our theoretical studies, Jenaer Glaswerk Schott & Gen., Mainz, can now make available a number of glass types guaranteeing small values of the thermal glass constant $\Gamma^*$, which were listed in Figure 21. This has enabled us to compute several telescope lenses and start their manufacture. All prototypes have diameters of 150 mm and focal lengths of 2250 mm, which means a focal ratio of 1 : 15. Figure 28 shows the lenses that we have computed. The list also includes several

| Type | Glasses | | $v_1$ | $v_2$ | $v_3$ | $\Delta v$ | $\cdot 10^6/°C$ $\Gamma_1^*$ | $\Gamma_2^*$ | $\Gamma_3^*$ | $\cdot 10^7/°C$ $\alpha_1^*$ | $\alpha_2^*$ | $\alpha_3^*$ | $\Delta f'$ theor. | $\Delta f'$ num. | $\Omega_o$ | Gauss. Error | Zone Err. | |
|---|---|---|---|---|---|---|---|---|---|---|---|---|---|---|---|---|---|---|
| AL1 | 3138 | 3038 | 42.6 | 65.0 | | 22.4 | +4.4 | +1.9 | | 103 | 113 | | +0.067 | +0.076 | 2.88 | +0.39 | 0.00 | a |
| AL2 | 3038 | FK51 | 65.0 | 84.0 | | 19.0 | +1.9 | +0.27 | | 113 | 136 | | +0.019 | 0,028 | 0.94 | +0.80 | +0.82 | th |
| AL3 | 3138 | FK51 | 42.3 | 84.0 | | 41.7 | +4.4 | +0.27 | | 103 | 136 | | +0.044 | +0.052 | 1.91 | +0.23 | 0.00 | erm |
| AL4 | 3038 | PK51 | 65.0 | 76.0 | | 11.0 | +1.9 | -1.00 | | 113 | 147 | | +0.006 | 0.013 | 0.43 | -0.01 | 0.00 | a |
| AL5 | 3138 | PK51 | 42.3 | 65.0 | | 22.7 | +4.4 | -1.00 | | 103 | 147 | | +0.030 | +0.038 | 2.00 | +0.29 | -0.01 | l |
| AL6 | 3038 | FK51 | 53.8 | 84.0 | | 30.2 | -0.8 | +0.27 | | 121 | 136 | | -0.003 | -0.001 | 1.39 | +0.35 | +0.05 | |
| E | BK7 | F2 | 64.0 | 36.1 | | 27.9 | +12.8 | 15.1 | | 71 | 87 | | +0.26 | +0.25 | 3.24 | +0.37 | 0.00 | non- |
| AS | KZF2 | BK7 | 51.5 | 36.1 | | 15.4 | +4.5 | 12.1 | | 60 | 87 | | +0.26 | +0.27 | 2.25 | +1.44 | +0.34 | athermal |
| F | BK12 | SF4 | SF11 | 64.1 | 27.4 | 25.6 | 12.1 | 20.7 | 23.8 | 71 | 80 | 81 | +0.78 | +0.76 | 0.02 | -0.03 | -0.03 | |

FIGURE 28   Some characteristics of athermal and nonathermal objectives ($D = 150$ mm, $a^2 = 10^{-4}$, $T = 0.563$ °C, $f' = 2250$ mm).

conventional telescope lenses that have been manufactured by Carl Zeiss for years. We intend to compare them with some of the prototypes made of athermal glass with regard to their temperature behavior. Besides the characteristic glass constants, the table lists the thermal focus change to be expected for a constant $a_2$ of $10^4$ (°C/mm$^2$) (corresponding to a temperature difference of 0.563° between center and edge).

The newly designed athermal objectives are characterized as AL 1 ... AL 6. Together with a rough scheme, they are listed in the first column of Figure 28. The objectives E, AS, and F are well-known conventional objective types of Carl Zeiss.

$\Delta f'$ (theor.) corresponds to the values that were determined analytically according to our paraxial approximation formula, whereas the value $\Delta f'$ (num.) corresponds to the results of ray tracing according to the ray-tracing program that we have explained in an earlier section.

The $\Omega$-value is a measure for the quality of the correction of the secondary spectrum. That value is defined by the relation

$$\Omega = (s'_g - s'_{C'}) - [(\nu_1 \delta_2 - \nu_2 \delta_C)/(\nu_1 - \nu_2)] (s'_{F'} - s'_{C'}),$$

where $\delta$ is given by

$$\delta = (n_g - n_{C'})/(n_{F'} - n_{C'}).$$

The zone error $Z$ is defined here as $Z = \Delta s'(h_1/\sqrt{2}) - \frac{1}{2}\Delta a'(h_1)$.
AL 6 is the most favorable combination; that is, a combination of the new glass melt 3036 and a collective lens made of FK 51. The thermal focal-length change lies well within the Rayleigh tolerance criterion. With regard to thermal focal-length change, the lens AL 4 is much better still. However, due to the small $\nu$ difference, the zonal error is considerable and can only be compensated by deformation of one of the outer lens surfaces.

The comparison with the conventional lenses proves that the most favorable combination (e.g., AL 4) is better by a factor of 100 with regard to the heavy-flint apochromat F with its excellent chromatic correction. Even with regard to the well-known AS lens, it is better by a factor of 60. The AL 1, which is available as a prototype and which regarding its thermal behavior is not yet optimal, is still better by a factor of 10 with regard to the apochromat F and by a factor of 5 with regard to the AS lens.

We intend to produce prototypes of the most important new combinations and to compare them experimentally with conventional lenses. So far we have only realized the AL 1 and the AL 2 achromat. Further combinations can be expected in the coming years.

Figure 29 shows the corrections of the athermal lenses compared with conventional lenses at normal temperature.

For the athermal lenses AL 1 and AL 2 and the conventional objectives AS, E, and F, the thermal defocusing was calculated numerically as a function of the temperature gradient. The results are shown in Figure 30, in which it is seen that the thermal defocusing is quite proportional to the thermal constant $a_2$ and, respectively, to the temperature difference between axis and edge. The graphs also show the great differences in the thermal behavior of the four objectives and the superiority of the AL 2 type. The thermal defocusing of the AL 6 type is so small that it could not be reproduced on the scale of Figure 30. The values marked as o o o correspond to the analytical calculated values. They do not deviate considerably from the numerical values.

The wavelength dependency of some lenses is shown in Figure 31. That dependency is small enough for practical purposes.

Prototypes of an AL 1 type and an AL 2 type with a diameter of 150 mm and a focal length of 2250 mm were manufactured at Carl Zeiss. We have measured their thermal behavior in comparison with conventional

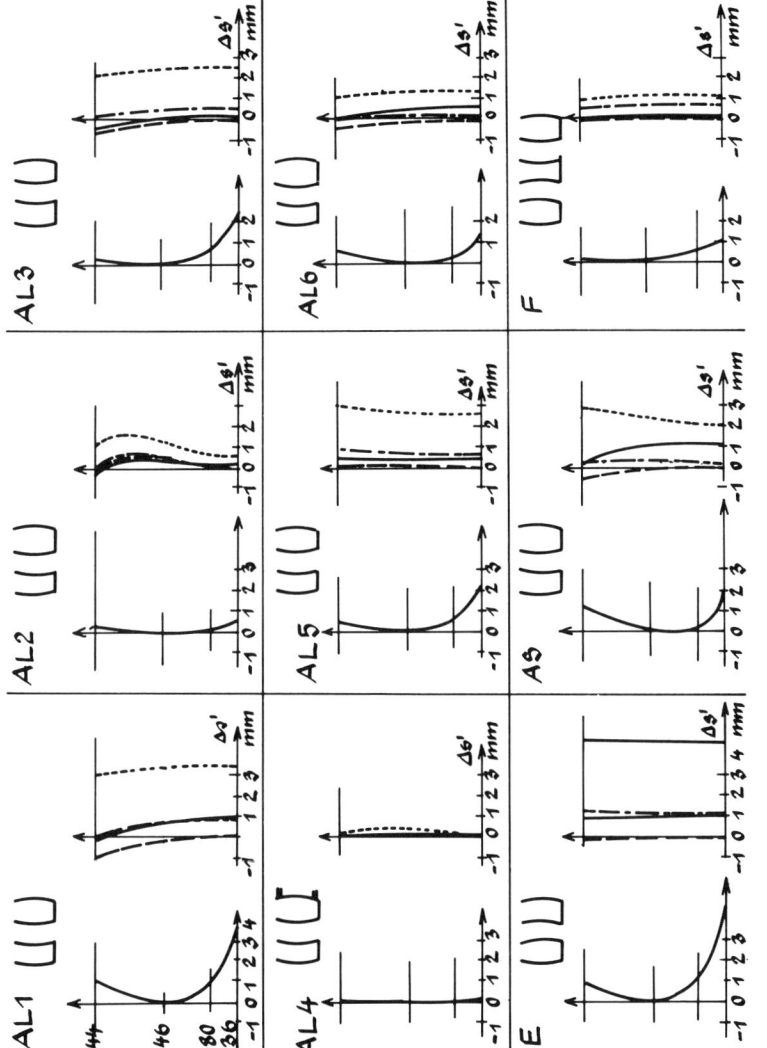

FIGURE 29 State of correction of some athermal and nonathermal objectives ($D = 150$ mm, $f' = 2250$ mm).

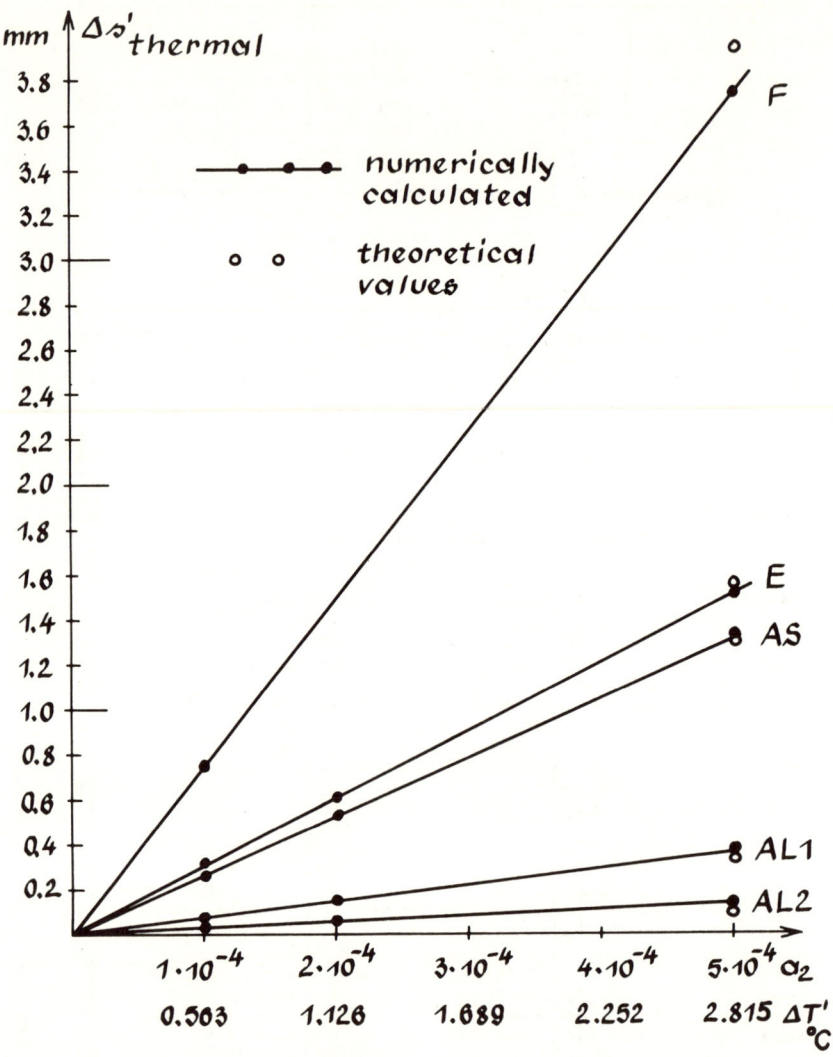

FIGURE 30 The thermal defocusing as a function of the temperature gradient for some objectives.

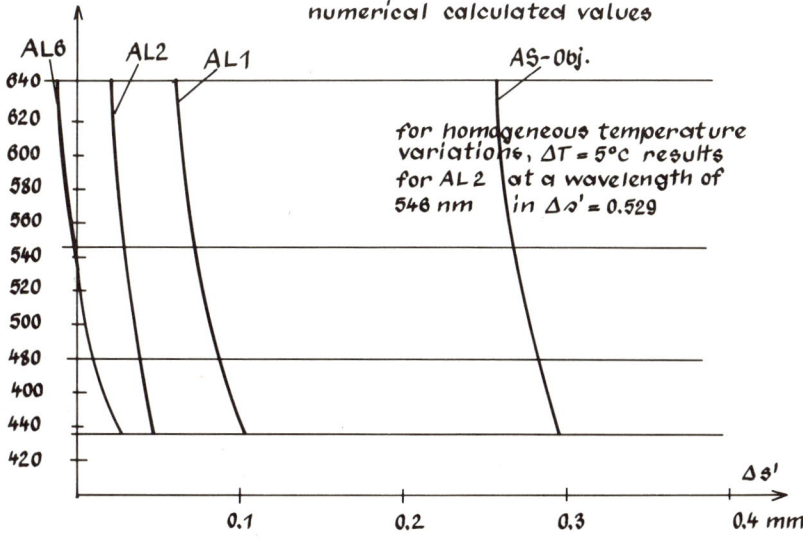

FIGURE 31 The thermal defocusing as a function of the wavelength for some objectives.

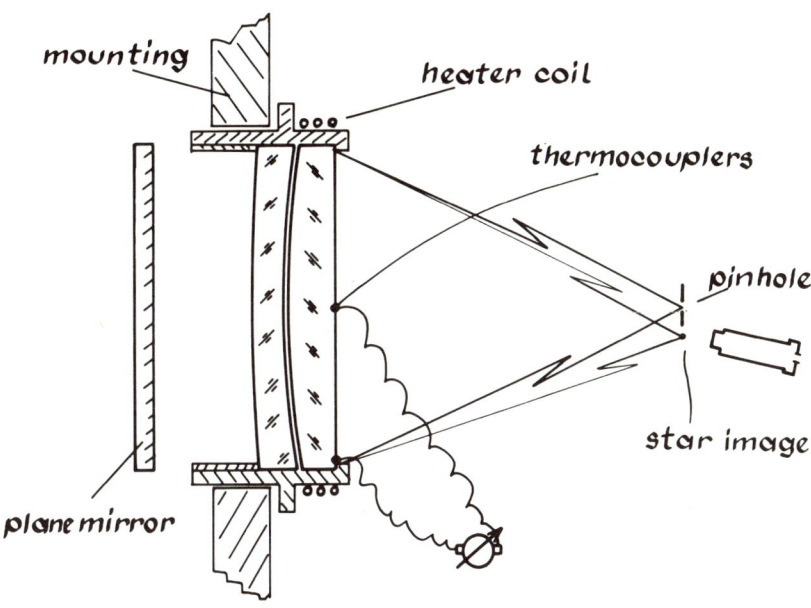

FIGURE 32 Measuring device.

objectives of the same aperture and the same focal length, namely, an AS type and an F type. Figure 32 shows the measuring device. The temperature difference between axis and edge was measured by thermocouples. The results are shown in Figure 33. The tendency of the measured curves corresponds to the predictions made by theory. The superiority of the AL 1 and AL 2 type to the other lenses is especially

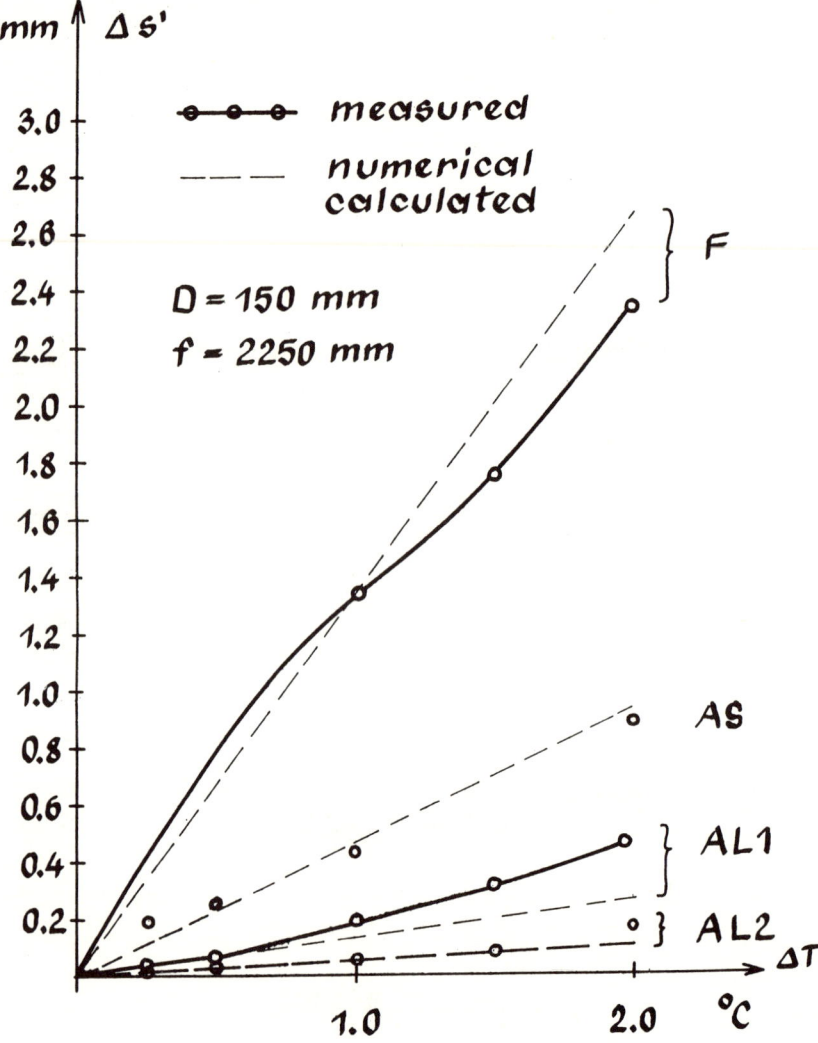

FIGURE 33 Measurements of the thermal defocusing of some objectives as a function of the temperature difference between axis and edge.

evident. The differences between measured and calculated values are probably caused by the fact that the temperature distribution has no exact rotational symmetry and that a longitudinal temperature gradient exists.

## VII. Conclusion

It could be shown that it is possible to design telescope objectives quite insensitive to inhomogeneous temperature variation. With new melted optical glasses, such objectives were produced and their thermal behavior was measured, and it was found that the thermal defocusing of the objective was much smaller than that of conventional objectives in accordance with the theory.

Our thanks are due to our colleagues Messrs. Kantor and Walter for their assistance with the calculations and for making the drawings. The theoretical work and the manufacturing of the prototype have been sponsored by the Bundesminister für Bildung und Wissenschaft within the Technology Program.

## References

Herzberger, M., *J. Opt. Soc. Am. 47*, 736 (1957).
Ketterer, G., Thesis, University of Stuttgart (1969).
Montagnino, L., *J. Opt. Soc. Am. 58*, 1667 (1968).
Perry, J. W., *Proc. Phys. Soc.* (London) *55*, 257 (1943).
Sliusarev, G. G., *Opt. Spectrosc. 6*, 134 (1959).
Strähle, F., Thesis, University of Stuttgart (1969).
Strähle, F., *Mitt. Astron. Ges. 30*, 85 (1971).
Volosov, D. S., *Opt. Spectrosc. 4*, 663, 772; *5*, 191 (1958).

JOHN D. MANGUS

# SPACE OPTICS—A REVIEW OF TELESCOPE DESIGNS AND TECHNOLOGY FOR THE SPACE-SCIENCE DISCIPLINES

## Introduction

The telescope designs to be discussed are those that have been designed for flight on earth-orbital spacecraft. There is an enormous variety of optical systems that could be reviewed, but the author has chosen to select a series of designs that span the largest spectral and observational range possible. Another purpose of this selection of designs, as well as the brief overview of criteria and formulas used in evaluating these telescope designs, is to acquaint the reader with one approach in the field of space-optics design that has worked quite well.

## Flight Designs

Space-science optical instrumental designs have relied primarily on the use of single- and two-mirror telescopes for optical systems to collect high-frequency spatial-resolution information over a broad spectral region. To

---

The author is at Goddard Space Flight Center, Greenbelt, Maryland 20771.

demonstrate this fact, a series of telescope designs developed for spaceflight applications are listed in Table 1. Note that the Cassegrain type of geometry is common to all of these designs, which cover the spectral region from 0.6 to 12600 nm. Specific surface geometry combinations range from the true Cassegrain (PEP) to a two-mirror general aspheric design (GEP). The large aperture combined with the optical imaging and folding properties of the concave primary mirror followed by a convex (concave for x-ray telescopes) confocal secondary mirror make this design particularly attractive. The versatility in application of this design approach is evident from Tables 1 and 2 by comparing the variety of observational applications and spacecraft to which they have been applied. Spacecraft orbital characteristics (Table 3) range from near-earth polar and equatorial orbits to synchronous orbits. Spacecraft platforms vary from spin-stabilized at synchronous orbit to three-axis stabilization in a near-earth equatorial orbit. Tolerances are listed (tilt, decentering, and axial displacement), which specify the accuracy to which the secondary mirror focus must be maintained relative to the primary-mirror principal focus. A variety of approaches have been taken to maintain the secondary position tolerances, which range from optically contacting to thermal-compensating structures.

Field-stop data have also been included for each of the telescope designs listed in Table 1. The field-stop size, except for the Goddard x-ray telescope [Mangus and Underwood, 1969] and the Large Space Telescope, limits the spatial resolution of the instruments. In some cases, the field stop is actually the entrance slit to a spectrometer (Goddard spectroheliograph, Princeton Experimental Package, Goddard Experimental Package) and as such also limits the spectral resolving power. The resolution for these instruments is listed in terms of the highest spectral resolution (smallest entrance slit) and the angle subtended by the entrance slit. Mechanisms were incorporated in these instruments to vary the size of the field stop so that there is flexibility in selecting spatial, spectral, and throughput properties depending on the nature of the object to be observed. The field-stop size listed for the multispectral scanner and visible infrared spectral scanner radiometer is actually the end size of one optical fiber pipe in an array of fibers located at the telescope focus. This instantaneous geometric field of view (IGFOV) of the telescope as determined by the fiber size is extended to a large full-field coverage by including a flat scanning mirror forward of the telescope. The motion of this mirror varies from a rapid oscillating rotation movement (20 rad/sec) in the case of the multispectral scanner to a step-function rotation (one 0.2 mrad step per 1.67 sec) in the case of the visible infrared spectral scanner radiometer.

TABLE 1  Spaceflight Telescope Designs

| | GXRT | GSH | PEP | GEP | MSS | VISSR | LST |
|---|---|---|---|---|---|---|---|
| GEOMETRY | | | | | | | |
| INSTRUMENT | GXRT | GSH | PEP | GEP | MSS | VISSR | LST |
| SPACECRAFT | ATM | OSO-7 | OAO-C | OAO-B | ERTS | SMS | LST |
| TELESCOPE TYPE | WI | WII | C | MRC | RC | RC | RC |
| APPLICATION | SOLAR PHYSICS | SOLAR PHYSICS | ASTRONOMY | ASTRONOMY | EARTH RESOURCES | METEOROLOGICAL | ASTRONOMY |
| WAVELENGTH RANGE | 6Å to 33Å | 170Å to 400Å | 1000Å to 3300Å | 1200Å to 4000Å | 5000Å to 10,000Å | 5400Å to 7000Å | 1000Å to 3μ |
| FIELD STOP | 35 min | 8 sec | 0.3 sec | 11 sec × 5 min | 18.7 sec | 4.3 sec × 5.2 sec | 6 min |
| FULL FIELD | 35 min | 30 min | 8 min | 6 min × 5 min | 11.6 deg | 20 deg | 30 min |
| APERTURE DIAMETER | 24.7 cm | 4.75 cm | 82.2 cm | 96.5 cm | 22.9 cm | 40.6 cm | 300 cm |
| AREA | 14.8 cm$^2$ | 11 cm$^2$ | 3,116 cm$^2$ | 6,200 cm$^2$ | 330 cm$^2$ | 1,090 cm$^2$ | 6.4×10$^6$ cm$^2$ |
| TELESCOPE FOCAL RATIO | 7.7 | 18 | 19.8 | 5 | 3.6 | 7.17 | 12 |
| PRIMARY FOCAL RATIO | 15.63 | 1.025 | 3.4 | 1.65 | 1.80 | 2.30 | 3.12 |
| SECONDARY MAGNIFICATION | 0.49 | 17.6 | 5.82 | 3.03 | 2.0 | 3.12 | 3.85 |
| RESOLUTION | 2 to 5 sec | 1Å @ 8 sec | .01Å @ 3.5 sec | 2Å @ 11 sec | 79 m | 0.8 km   3.2 km | 0.01 sec |
| MIRROR MATERIALS | | | | | | | |
| PRIMARY | FUSED SILICA | ULE | FUSED SILICA (EGG CRATE) | BERYLLIUM + E-Ni | FUSED SILICA | BERYLLIUM + E-Ni | Cer-Vit/ULE |
| SECONDARY | FUSED SILICA | ULE | FUSED SILICA | FUSED SILICA | FUSED SILICA + E-Ni | BERYLLIUM + E-Ni | — |
| FLAT | | | | | INVAR RODS | BERYLLIUM + E-Ni | |
| INTERVERTEX CONTROL | OPTICAL CONTACT | BERYLLIUM SPACER | FUSED SILICA RODS | ALUMINUM & TITANIUM STRUCTURE | INVAR RODS | BERYLLIUM STRUCTURE | TITANIUM TRUSS STRUCTURE |
| TILT | ±1.0 sec | ±4 sec | ±7.5 min | ±5 min | ±5 sec | | ±2.0 sec |
| DECENTER | ±0.012 mm | ±0.004 mm | ±0.9 mm | ±0.03 mm | ±0.015 mm | ±0.051 mm | ±0.025 mm |
| AXIAL | ±0.075 mm | ±0.008 mm | ±0.004 mm | ±0.064 mm | ±0.006 mm | ±0.045 mm | ±0.075 mm |

TABLE 2  Definitions of Instruments, Spacecraft, Telescopes, and Materials

| INSTRUMENTS | | TELESCOPES | |
|---|---|---|---|
| GXRT | - GODDARD X-RAY TELESCOPE | W I | - WOLTER TYPE I TELESCOPE |
| GSH | - GODDARD SPECTROHELIOGRAPH | W II | - WOLTER TYPE II TELESCOPE |
| PEP | - PRINCETON EXPERIMENTAL PACKAGE | C | - CASSEGRAIN |
| GEP | - GODDARD EXPERIMENTAL PACKAGE | MRC | - A QUASI HYPERBOLIC PRIMARY AND TRUE HYPERBOLIC SECONDARY |
| MSS | - MULTISPECTRAL SCANNER | | |
| VISSR | - VISIBLE, INFRARED SPECTRAL SCANNER RADIOMETER | RC | - RITCHEY-CRETIEN |
| LST | - LARGE SPACE TELESCOPE | MATERIALS | |
| | | E-Nickel | - ELECTROLYSIS NICKEL DEPOSIT ON THE ORDER OF 0.15mm TO 0.25mm THICK |
| SPACECRAFT | | | |
| ATM | - APOLLO TELESCOPE MOUNT-SKYLAB | | |
| OSO | - ORBITING SOLAR OBSERVATORY | Cer-Vit | - AN ULTRA LOW EXPANSION POLYCRYSTALLINE CERAMIC GLASS COMPOSED OF LITHIUM ALUMINA SILICATE MANUFACTURED BY OWENS-ILLINOIS CORP. |
| OAO | - ORBITING ASTRONOMICAL OBSERVATORY | | |
| ERTS | - EARTH RESOURCE TECHNOLOGY SATELLITE | | |
| OAO-C | - IS NOW CALLED COPERNICUS | | |
| | | ULE | - AN ULTRA LOW EXPANSION TITANIUM SILICATE GLASS MATERIAL MANUFACTURED BY CORNING GLASS. |

## Glancing Incidence versus Near-Normal Incidence

The most significant configuration difference between the designs listed in Table 1 is that of the transition from near-normal to grazing-incidence geometries. This geometry change is caused by the reflectance properties [Cox et al., 1972; Bradford et al., 1969; Osantowski, 1974] of materials as illustrated in Figures 1 and 2. Figure 1 shows that the reflectance of state-of-the-art thin-film vacuum-deposition coatings drops rapidly below 100 nm. Single reflection near normal-incidence designs (such as an off-axis parabola combined with a grazing-incidence spectrometer) can be efficiently used down to 20 nm. In Figure 2, near-normal incidence systems still have an adequate effective reflectance down to 100 nm if the

TABLE 3  Spacecraft Orbital Parameters and Stabilization Modes

| SPACECRAFT | ORBIT | STABILIZATION |
|---|---|---|
| ATM | 378 km @ 50° | 2 AXIS + COARSE ROLL |
| OSO | 483 km @ 33° | 2 AXIS RASTER ON SPIN STABILIZED PLATFORM |
| OAO | 628 km @ 35° | 3 AXIS STABILIZATION |
| ERTS | 805 km @ POLAR | 3 AXIS STABILIZATION |
| SMS | 36,210 km @ SYNC-ORBIT | SPIN STABILIZED |

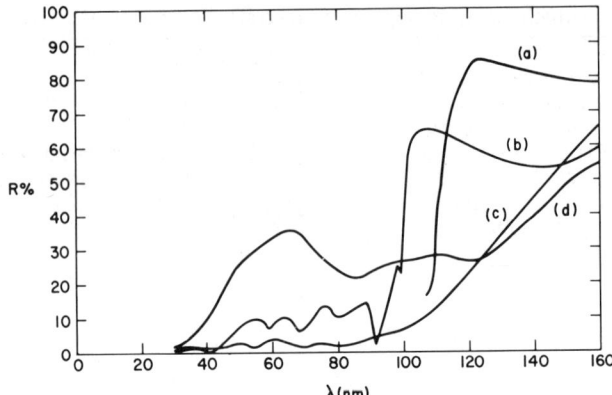

FIGURE 1  Percent reflectance of thin films at 15° angle of incidence versus wavelength for one reflection off of (a) Al + 25-nm MgF$_2$; (b) Al + 14-nm LiF; (c) aged aluminum; (d) 16-nm rhenium.

lithium fluoride–aluminum coating is used. There has been some hesitance to use this particular coating because of degradation in reflectance due to the hydroscopic properties of lithium fluoride. However, if care is exercised on humidity control in prelaunch operations, some very encouraging results have been realized, such as on the Princeton Experimental Package, which used this coating.

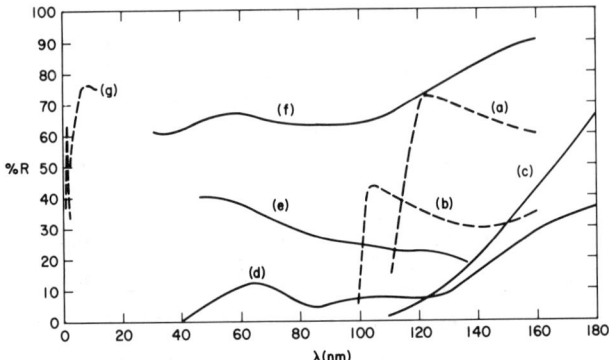

FIGURE 2  Percent reflectance versus wavelength for two reflections off of (a) Al + 25-nm MgF$_2$ at 15° angle of incidence; (b) Al + 14-nm LiF at 15° angle of incidence; (c) aged aluminum thin film at 15° angle of incidence; (d) 16-nm rhenium film; (e) ULE-grade fused silica at 80° angle of incidence; (f) aged aluminum thin film on glass at 85° angle of incidence; (g) quartz at 89.1° angle of incidence.

Below 100 nm, the gain in reflectance at glancing angles for two-mirror systems begins to emerge. For very short wavelengths of 0.2 to 15 nm, the use of glancing incidence is mandatory unless alternate imaging techniques are used, such as Fresnel zone plates or single and multiple pinhole arrays. The reflectance of quartz [Underwood and Muney, 1967] is shown in Figure 2, for this spectral region; however, beryllium overcoated with an optically polished electrolysis-deposited nickel has also been used very effectively [Giaconni, 1971]. The transition from near-normal incidence to glancing incidence is not straightforward but a matter of tradeoffs between effective throughput, spatial resolution, and alignment sensitivity. Of particular interest in glancing-incidence optical designs is the problem of matching the spatial resolution and throughput to the instrumentation located behind the field stop. Generally the obscuration ratio, $\epsilon$, associated with glancing-incidence telescopes is quite large. For x-ray and extreme-ultraviolet telescopes, $\epsilon$ is usually greater than 0.95 and 0.8, respectively, while near-normal-incidence designs have obscuration ratios of less than 0.3. The significance of the obscuration effect can be evaluated by considering the data in Figure 3. Note that for a fixed-aperture diameter the integrated energy in the diffraction pattern increases the field-stop size quite rapidly when $\epsilon > 0.8$. When compared to an unobscured aperture, the size of the field stop (spectrometer entrance slit, for example) must be increased by a factor of 6 ($\epsilon = 0.8$) to capture 80% of the energy in a diffraction-limited image or by a factor of 12 to capture 90% of the energy.

Another interesting property of glancing telescopes is the relationship

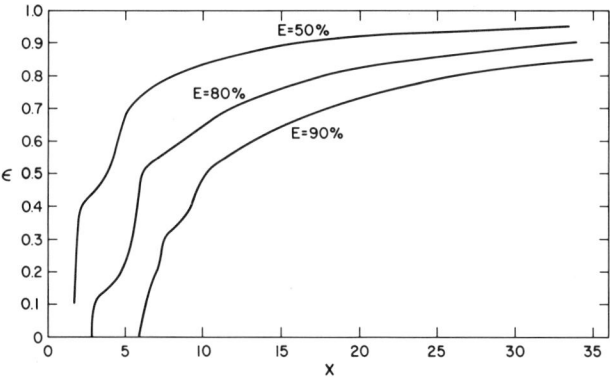

FIGURE 3 Percent energy $E$ contained in the diffraction pattern as a function obscuration ratio $\epsilon$ versus spot radius for $E$ = 90%, 80%, and 50%.

between the glancing angle and secondary mirror magnification (Figure 4). It is well known that an important factor in attempting to maintain the alignment of optical systems is the magnification of the secondary mirror. Typically, secondary-mirror magnifications for near-normal-incidence systems are constrained to be less than 3 or 4. In the design of extreme ultraviolet, Wolter Type II, telescopes [Wolter, 1952; Mangus, 1970] this is not practicable, and secondary mirror magnifications are generally quite large ($M > 10$). This effect of alignment sensitivity is offset somewhat by imaging properties due to the large obscuration ratio discussed above. In Figure 5, for example, the depth of focus of an $F_t/9$ glancing telescope with $\epsilon = 0.85$ is the same as an unobscured system used at $F_t/18$. This particular alignment sensitivity is equivalent for both telescopes, since the tolerance on intervex separation is directly proportional to the depth of the diffraction focus.

The effect of obscuration on the modulation transfer function of these telescopes must also be considered. The degradation in the modulation transfer function of an x-ray telescope (Figure 6) clearly indicates

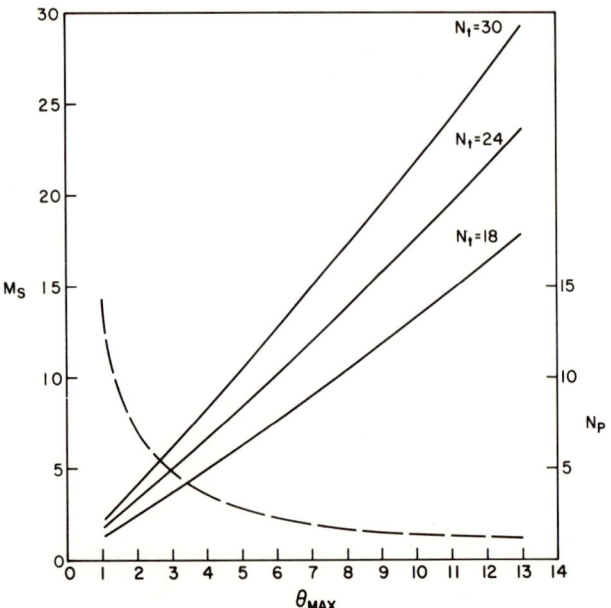

FIGURE 4 Type II telescope primary-mirror focal ratio $N_p$ versus maximum glancing angle and telescope focal ratio $N_t$ as a function of secondary-mirror magnification $M_s$ versus maximum glancing angle $\theta_{\max}$ for telescope focal ratios $N_t$ = 30, 24, and 18. $N_p$ a dashed line.

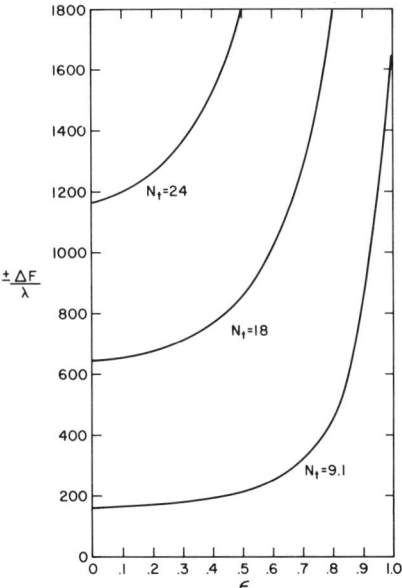

FIGURE 5 Type II telescope focal ratios $N_t$ as a function of depth of focus-to-wavelength ratio $\Delta F/\lambda$ versus obscuration ratio for $N_t =$ 24, 18, and 9.1

one reason why this telescope is not used at longer wavelengths. The problem of the obscured aperture in telescope design is also of concern in near-normal-incidence systems, even though $\epsilon < 0.8$, especially as space-optical-systems design performance must approach perfection.

The point to remember is that the transition from near-normal incidence telescopes to glancing-incidence geometries requires careful consideration of alignment tolerances; limits on field-stop size; instrumental requirements that follow the field stop, such as the focal ratio; spatial resolution requirements at the longer wavelengths; material selection; and, of course, the maximum glancing angle that is selected.

## Wave-Aberration Evaluation

As fore optical-system designs approach diffraction-limited performance, considerable emphasis has been placed on developing techniques to determine optical-system sensitivity to fabrication and alignment tolerances based on wave-aberration error budgets. The key to this analysis

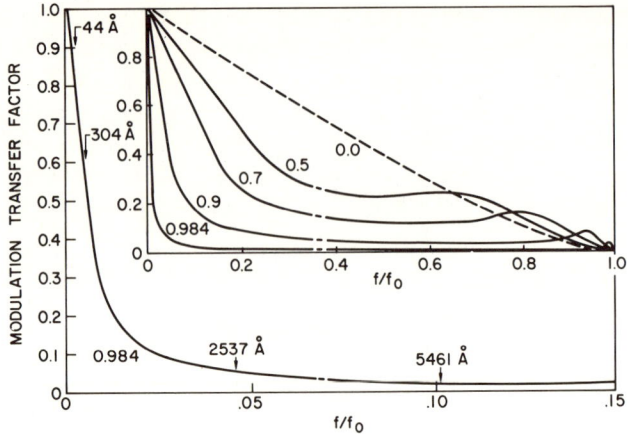

FIGURE 6 Modulation transfer function of a perfect imaging system with a highly obscured annular pupil. Inset: modulation transfer functions for other annular pupils. $f/f_0$ is the normalized spatial frequency. The curve for $\epsilon = 0.984$ is marked with arrows labeled with wavelengths at the points corresponding to 5 sec of arc spatial period for each particular wavelength.

has been the relationship of the peak intensity, $i(p)$, at the diffraction focus to the root-mean-square wave-aberration value, $\Delta\Phi$, of the aberration function $\Phi$.

This relationship [Born and Wolf, 1959] between the peak intensity and aberration function is given by

$$I(P) = \left(\frac{Aa^2}{\lambda R^2}\right) \left| \int_0^1 \int_0^{2\pi} \exp^i \left[k\Phi(Y,\rho,\theta) - v\rho\cos(\theta - \psi) - \frac{1}{2}u\rho^2\right]\rho d\rho d\theta \right|^2, \quad (1)$$

where $I(P)$ is the intensity at the Gaussian image. Since we are concerned about the effect of aberrations on a telescope performance as compared with what the performance would be in the absence of aberrations, it is more convenient to use the normalized intensity at the diffraction focus as a figure of merit. In this case, the maximum or peak intensity, called the Strehl intensity, is given by

$$i(P) = \frac{I(P)}{I_0} = \frac{1}{\pi^2} \left| \int_0^1 \int_0^{2\pi} \exp^i \left[k\Phi(Y,\rho,\theta) - v\rho\cos(\theta - \psi) - \frac{1}{2}u\rho^2\right]\rho d\rho d\theta \right|^2. \quad (2)$$

A much simpler version of this relationship may be derived, which is

$$i(P) \gtrsim [1 - (2\pi^2/\lambda^2)(\Delta\Phi)^2]^2, \quad (3)$$

where $\Delta\Phi$ is the root-mean-square departure of the system wavefront as compared to a reference sphere. This relationship, known as the Maréchal criterion, is regarded to define a well-corrected optical system when the intensity at the diffraction focus is greater than or equal to 0.8 ($\Delta\Phi = \lambda/14$). This approach provides a very powerful technique to control the final performance of an optical system if a method is devised whereby the error budget ($\Delta\lambda = \lambda/14$, for instance) can be apportioned between the designer, fabricator, and systems engineer.

## Design

The optical designer must continue to use the conventional geometrical design programs to arrive at optimum physical boundary conditions on the telescope geometry. Wave optical evaluation programs [Innes, 1971] may be used to determine sensitivity of the wave aberration, $\Delta\Phi$, to field angle and alignment (Figure 7). However, before proceeding with this conventional first step, the designer must be careful to consider that any obscuration placed in the system contributes to the aberrations

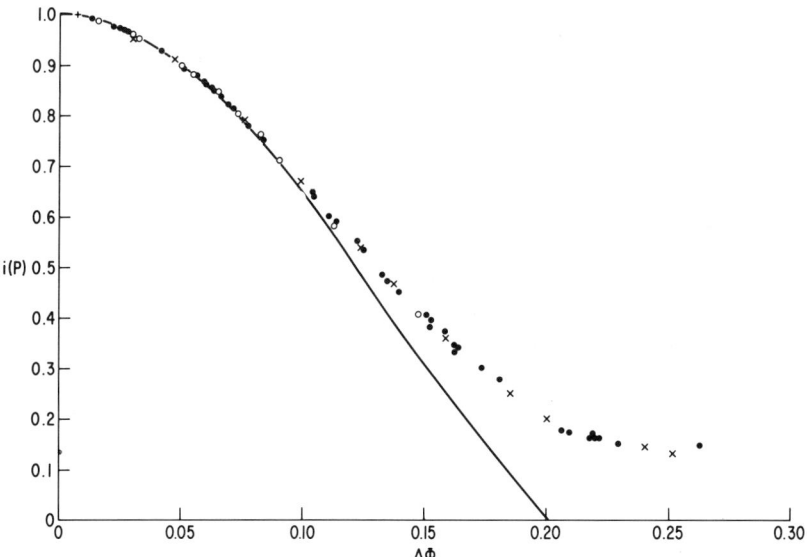

FIGURE 7 Peak intensity $i(P)$ in the diffraction focus versus the rms wave aberration $\Delta\Phi$; solid line is plot of Eq. (3); data points computed using wave-optical-analysis computer program for Ritchey-Crétien telescope ●, Type I telescopes +, Type II telescopes x, Cassegrain telescopes ○.

and, as we have seen above, affects the sensitivity to misalignments. The importance of this fact is illustrated by the example shown in Figure 8, where the Strehl intensity is plotted as a function of the obscuration ratio versus tolerable peak-to-peak wave aberrations, $\Delta\lambda$, to maintain the Strehl ratio at 0.8. In one case, the curves accompanied by an adjacent circle denote that the Strehl ratio is normalized to an unobscured aperture. For the other case, denoted by curves accompanied by a circle with a black dot in the center, the Strehl ratio is normalized to an obscured aperture. In this latter case, the obscuration is not treated as an aberration. In the first case, the peak waves of tolerable aberration rapidly decreases once $\epsilon$ exceeds 0.3. For the second case, this is obviously not true. As a matter of fact, the tolerance on third-order spherical aberration and coma actually increases. This behavior may be understood by considering that the wave aberration is a measure of the deviation of annular wavefront from a reference sphere. As $\epsilon$ approaches unity, any reference sphere can be fitted to a symmetrical "distortion" of the now zero-width annular wavefront.

## Fabrication

A common practice today is to check fabrication progress by calculating optical-path-difference maps over a mirror surface by means of interferometric measurements. The wavefront error contributions to degradation of the peak intensity at the diffraction focus is then computed and related back to mirror imperfections. It is important to know the nature, symmetry, and spatial frequency of these errors. For example, a low-frequency imperfection, which may in effect contribute to spherical aberration, has a different (lesser) impact on the overall error budget than a low-frequency nonsymmetrical error, of the same amplitude, that may for example contribute to astigmatism. A technique for evaluation of the location and contribution of imperfections in a mirror surface to loss of energy at the diffraction focus can be accomplished by relating the root-mean-square wave aberration to $i(p)$ to $\Phi$ in Eq. (1) by a polynomial expansion such as that given by Born and Wolf [1959]:

$$i(P) = 1 - \frac{2\pi^2}{\lambda^2} \frac{A^2 nm}{n+1}.$$

This approach requires that the interferometric wavefront display be digitized and fit to the polynomial expansion, which may be carried out as far as 25 terms. The critical figure errors may then be determined and corrected by recycling these data back into the fabrication process.

# John D. Mangus

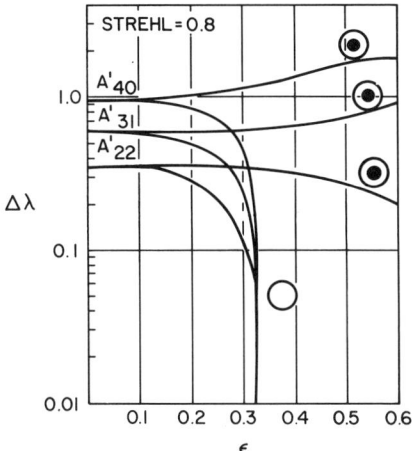

FIGURE 8  Obscuration ratio $\epsilon$ versus peak-to-peak tolerable wave aberration for Strehl ratio of 0.8. $A'_{40}$ = third-order spherical aberration, $A'_{31}$ = coma, and $A'_{22}$ = astigmatism.

## Systems Integration

The problems of systems design can become very complex when dealing with systems such as a 3-m-diameter aperture telescope. Dynamic as well as static errors must also be considered [Wetherell, 1973] but are beyond the scope of this paper. However, it is important to have some familiarity with those parameters that affect the sensitivity of two-mirror telescope systems to alignment errors.

Table 4 lists a series of equations that serve this purpose. Note that equations are also included to calculate the image quality of a system with no alignment errors. These equations have been derived from third-order theory and/or empirically. The equations for calculating the wave aberration of near-normal incidence Cassegrain type of geometries as a function of field angle and alignment errors (tilt, decenter and axial displacement of the secondary mirror) give excellent agreement with wave-optical computer analysis. These same equations indicate the correct trend of errors in the wave aberration for Wolter telescopes. However, computer wave-optical analyses show that the expression for $\Delta\Phi$ as a function of $\delta$ yields values for Type I and II telescopes that are low by a factor of 2 to 3. Also the equations for tilt and decenter sensitivity yield values for the rms wave aberration of Type II telescopes that are a factor of 10 to 20 less than those actually realized in practice. Equations have

TABLE 4  Imaging Sensitivity of Telescopes to Design and Alignment Parameters

| CASSEGRAIN TYPE | TYPE II | TYPE I |
|---|---|---|
| $\Delta\Phi_0 = \dfrac{\delta^2}{32(6)^{1/2}} \cdot \dfrac{D_P}{F_P + T} \cdot \dfrac{2m+1}{m} \cdot \dfrac{(1-\epsilon^2 + \epsilon^4)^{1/2}}{\lambda}$ | $\Delta\Phi_{II} > \Delta\Phi_0$ | $\Delta\Phi_I > \Delta\Phi_0$ |
| $\Delta\Phi_A = \dfrac{(1-\epsilon^2)}{16(3)^{1/2} N_T^2} \cdot \dfrac{\Delta IV}{\lambda}$ | $\Delta\Phi_{AII} = \dfrac{(1-\epsilon^2)}{16(3)^{1/2} N_T^2} \cdot \dfrac{\Delta IV}{\lambda}$ | $\Delta\Phi_{AI} = \dfrac{(1-\epsilon^2)}{16(3)^{1/2} N_T^2} \cdot \dfrac{\Delta IV}{\lambda}$ |
| $\Delta\Phi_D = 3.7\times 10^{-3} \dfrac{(1-m^2)}{m^2 N_P^3} \cdot \dfrac{\Delta D}{\lambda}$ | $\Delta\Phi_{DII} >> \Delta\Phi_D$ | |
| $\Delta\Phi_T = 1.07\times 10^{-6} \dfrac{(m-1)}{m^2} \cdot \dfrac{F_P + BFL}{N_P^3} \cdot \dfrac{\Delta\beta}{\lambda}$ | $\Delta\Phi_{TII} >> \Delta\Phi_T$ | |
| $\Delta C_0 = \delta \left[ \dfrac{3}{16} \dfrac{F_T}{N_T^2} + \left(\dfrac{3}{8} - \alpha\right)\left(\dfrac{m+1-K}{K}\right) \dfrac{F_T}{4 N_P^3} \right]$ | $\Delta C_{II} \approx \dfrac{3}{16} \dfrac{F_T}{N_T^2} \delta$ | $\Delta C_I = F_T (\text{TAN }\delta \text{ TAN }(\alpha + \delta) \text{ SIN }\alpha)$ |
| | | $\Delta\sigma_I = \dfrac{(\xi+1)}{10} \text{TAN}^2 \delta \cdot \dfrac{L_P}{Z_0} + 4\text{ TAN }\delta\text{ TAN}^2\gamma$ |

## RMS WAVE ABERRATIONS

$\Delta\Phi_0$  ON OPTIMUM FOCAL SURFACE AS A FUNCTION OF FIELD ANGLE, $\delta$
$\Delta\Phi_A$  DUE TO AXIAL DISPLACEMENT, $\Delta IV$, OF SECONDARY MIRROR
$\Delta\Phi_D$  DUE TO DECENTERING, $\Delta D$, OF SECONDARY MIRROR
$\Delta\Phi_T$  DUE TO TILTING, $\Delta\beta$, OF SECONDARY MIRROR ($\Delta\beta$ IN ARC MIN.)
m = SECONDARY MIRROR MAGNIFICATION
$\epsilon$ = OBSCURATION RATIO
$\lambda$ = WAVE LENGTH
$N_P$ = FOCAL RATIO OF PRIMARY MIRROR
$F_P$ = FOCAL LENGTH OF PRIMARY MIRROR
$D_P$ = DIAMETER OF PRIMARY MIRROR
$N_T$ = FOCAL RATIO OF TELESCOPE
BFL = BACK FOCAL LENGTH OF TELESCOPE
T = BFL/$D_P$
$F_T$ = FOCAL LENGTH OF TELESCOPE

$\Delta C$ = SIZE OF A COMATIC IMAGE AT A FIELD ANGLE $\delta$
$\delta$ = ANGLE BETWEEN THE OPTICAL AXIS AND INCIDENT RAY
$L_P$ = LENGTH OF PARABOLA
$Z_0$ = DISTANCE FROM AXIAL FOCUS TO PARABOLA-HYPERBOLA INTERSECTION PLANE
$r_0$ = DISTANCE FROM OPTICAL AXIS TO PARABOLA-HYPERBOLA INTERSECTION BOUNDARY
$\xi$ = RATIO OF TWO GRAZING ANGLES FOR AN AXIAL RAY STRIKING NEAR THE PARABOLA-HYPERBOLA INTERSECTION
$\sigma_I$ = RMS BLUR CIRCLE IN RADIANS
$\gamma$ = 1/4 $\text{TAN}^{-1}(r_0/Z_0)$
K = DISTANCE BETWEEN PRIMARY MIRROR FOCUS AND CASSEGRAIN FOCUS DIVIDED BY $F_P$
$\alpha$ = 0 FOR SPHERICAL PRIMARY MIRROR
$\alpha$ = 3/8 FOR PARABOLA PRIMARY MIRROR

been given [Van Speybroeck and Chase, 1972; Mangus, 1970; Wolter, 1971; Bowen, 1967] in Table 4 that are based on the sensitivity of the comatic image size, $\Delta C$, to field angle as a function of telescope design properties. These latter equations are of importance especially in Type I telescope design, since the problems of surface-scattering as well as fabrication precision override the considerations of diffraction-limited imaging.

**Large-Telescope Technology**

The development of a 3-m-diameter aperture telescope, the performance of which is to be limited by diffraction effects, is a key technology effort to advance space astronomy observational capabilities above 100 nm. Likewise, the development of up to 1-m-diameter aperture x-ray telescopes is key in advancing high-spatial resolution, high-energy astronomy capabilities.

The results of a design study for a Large Space Telescope (LST) that was undertaken by Goddard Space Flight Center (GSFC) are listed in Table 1. Construction of a 3-m-diameter lightweight Cer-Vit material technology mirror blank to conform to the geometry in Figure 9 was initiated by GSFC in June 1972. The final optical figure goal [Tschunko, 1971] on this blank is $\lambda/50$ rms. It is emphasized that the above LST data and mirror blank configuration reflect only preliminary study. Also, the mirror technology effort has recently been transferred from GSFC to Marshall Space Flight Center, as has the LST project responsibility, and therefore one may expect to see mirror blank configuration and telescope design parameter changes as the project progresses. For example, a 3-m flight blank may be made miniscus and have a focal ratio of 2.2 rather than concave-plano and $N_p/3.12$ for the purpose of weight savings and minimization of $\epsilon$. In addition to the LST, large-aperture infrared and solar (photoheliograph) telescopes are also of importance in advancing progress in infrared astronomy and solar physics. Arrays of nested Baez and concentric x-ray telescopes [Van Speybroeck et al., 1971] have also been proposed for the purpose of increasing the effective throughput of Type I telescopes.

It is interesting to note that the only addition to the design of large-aperture telescopes from those listed in Table 1 is a Gregorian geometry selected for the solar photoheliograph. This is a minor departure and further emphasizes the versatility of two-mirror fore optical designs. Perhaps the most significant change from the geometries that have been proposed is that of the Baez type of telescopes for x-ray astronomy.

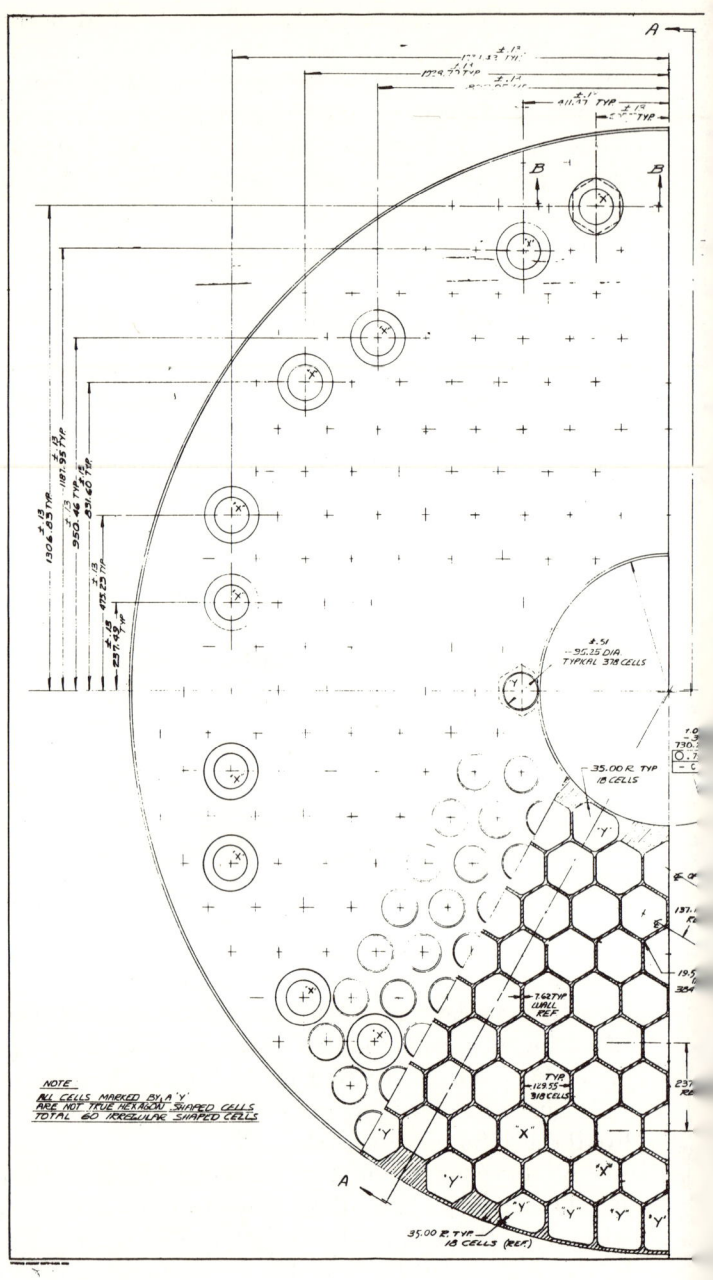

FIGURE 9 Lightweight design configuration for a 3-m-diameter large telescope

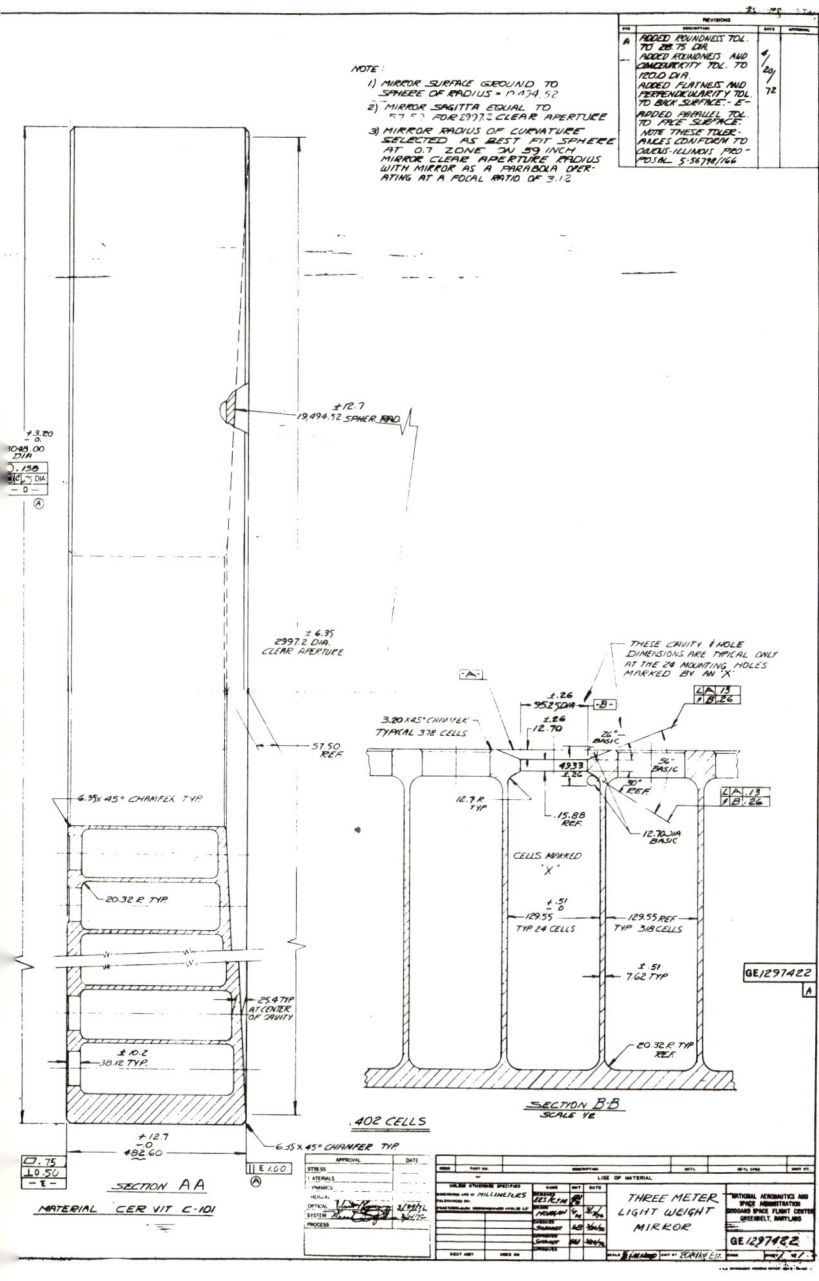

chnology mirror blank constructed of Cer-Vit 101 material.

## Conclusions

The optical industry will face very challenging space-optical-system fabrication problems, which span from the near infrared to the x-ray region.

Two-mirror fore optical telescope designs have proven to be very versatile in space-optical applications from astronomy to earth resources. Future programs indicate that this trend in telescope design will continue into the next generation of large space optical telescopes.

## References

Born, M., and E. Wolf, *Principles of Optics*, Secs. 9.1, 9.2, 9.3, Pergamon, New York (1959).
Bowen, I. S., *Ann. Rev. Astron. Astrophys. 5*, 45 (1967).
Bradford, A. P., G. Hass, J. F. Osantowski, and A. R. Toft, *Appl. Opt. 8*, 1183 (1969).
Cox, J. T., G. Hass, J. B. Ramsey, and W. R. Hunter, *J. Opt. Soc. Am. 62*, 781 (1972).
Giaconni, R., New Techniques in Space Astronomy, IAU, p. 104 (1971).
Innes, D. J., *J. Opt. Soc. Am. 61*, 694A (1971).
Mangus, J. D., and J. H. Underwood, *Appl. Opt. 8*, 95 (1969).
Mangus, J. D., *Appl. Opt. 9*, 1019 (1970).
Osantowski, J. F., *J. Opt. Soc. Am.*, to be published (1974).
Tschunko, H. F. A., *Appl. Opt. 10*, 2423 (1971).
Underwood, J. H., and W. S. Muney, *Solar Phys. 1*, 129 (1967).
Van Speybroeck, L. P., and R. C. Chase, *Appl. Opt. 11*, 440 (1972).
Van Speybroeck, L. P., R. C. Chase, and T. F. Zehnpfennig, *Appl. Opt. 10*, 945 (1971).
Wetherell, W. B., this volume, p. 55 (1973).
Wolter, H., *Ann. Phys. 10*, 94, 286 (1952).
Wolter, H., *Opt. Acta 18*, 425 (1971).

# TERRESTRIAL ENGINEERING OF SPACE-OPTICAL ELEMENTS

This discussion will review some recent efforts at Itek Corporation with particular emphasis on the mechanical and structural engineering problems associated with the design, fabrication, and testing of optical elements to be used in space. Let us note first that the use of extraterrestrial optical systems not only makes available some spectral bands that cannot be observed from the earth's surface but also makes it worthwhile to consider exploitation of the full resolving capability of very-large-aperture systems. The efforts described thus revolve around large, lightweight mirror elements.

We are presently conducting under the National Aeronautics and Space Administration's sponsorship a broad range of study tasks to support our progress toward a national space observatory. Much of this effort is focused on the Large Space Telescope, a 3-m-aperture orbiting astronomical telescope with a goal of diffraction-limited performance from the near infrared to a short wavelength in the neighborhood of 1800 Å. Our artist's concept of this telescope is shown in Figure 1. Here it is apparent that the 3-m-diam., 0.5-m-thick, lightweight, low-expansion primary mirror is a major element of the system.

---

The author is at Itek Corporation, Lexington, Massachusetts 02173.

FIGURE 1  Artist's concept of a 3-m space telescope.

Itek's interest in the design and fabrication of lightweight mirrors dates very nearly from the organization of the corporation in 1957. Some early examples are shown in Figure 2. Here we see two types of construction in aluminum: a machined Pyrex blank, a bonded structure of plates and tubes made of low-expansion ceramic material, a solid beryllium mirror, and one of the earlier fused silica eggcrate constructions made by Corning Glass Works. More recent efforts have focused on lightweight construction in low-thermal-expansion materials such as fully welded structures of ultra-low-expansion titanium silicate as fabricated by Corning Glass Works and machined lightweight elements of Cer-Vit glass–ceramic fabricated by Owens-Illinois Corporation. More detailed discussion on these materials is beyond the scope of this paper.

In retrospect, if not in actuality, our progress to our present status of knowledge has followed a logical pattern. The first, and perhaps the

FIGURE 2   Lightweight mirror design concepts.

most important, breakthrough in the technology of making large mirror elements was the application of the laser to interferometric optical testing. The coherence properties of the laser allow the design of interferometers with optical pathlength differences between test and reference beams of many meters. It became possible to design an interferometer for testing a 3-m element using a reference optical element no more than 5 cm in diameter, and thus one could package all elements of the interferometer, except the element to be tested, in a single assembly. An example is the Itek laser unequal-path interferometer shown in Figure 3.

The use of this instrument accentuated the need for controlling the ambient environment throughout the full length of the optical path being used. We undertook to build a variety of testing facilities, such as the vertical tank shown in Figure 4. An elevating mechanism in the tank is integrated with the fabrication mount of, for example, the Cer-Vit mirror shown here. The tank is 2.4 m in diameter and approximately 12 m long and is fitted with a laser interferometer at the upper end. In quick turn-around testing, the tank acts as an effective heat sink and shroud to minimize atmospheric effects. For the ultimate in testing accuracy, it can be evacuated to the moderately low pressures needed to eliminate all refractive effects of the atmosphere. A somewhat larger vertical tank shown in Figure 5 has been designed for ready access in the testing of complete camera systems. The lower end of this tank, which extends some 6 m below the floor level indicated here, is fitted with a 1.25-m (50-in.)-diameter Cassegrain-type collimating lens system. Of

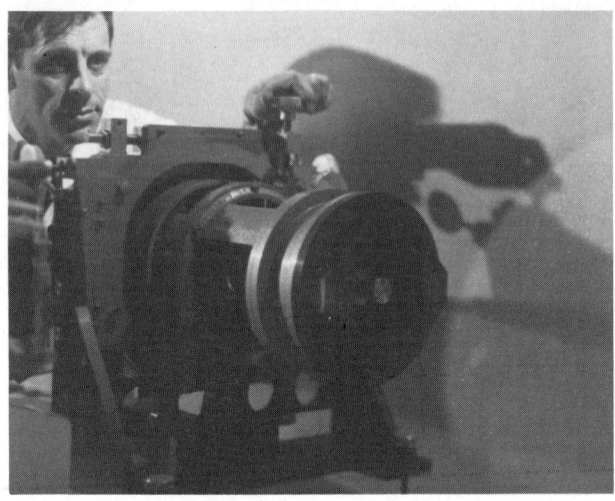

FIGURE 3  The Itek laser unequal-path interferometer.

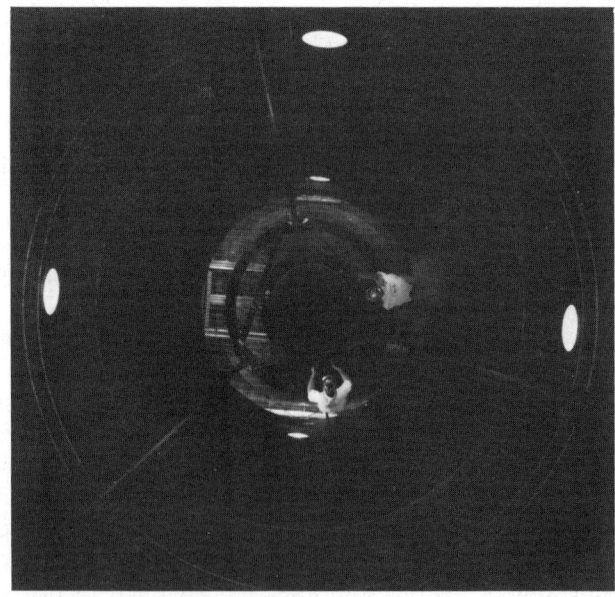

FIGURE 4  Downward view in a vertical test tank.

FIGURE 5  Side access vertical test tank.

somewhat older vintage, and indicative of our concern for better understanding of the effects of our gravitational environment on optical elements, is the tank shown in Figure 6. This 1.8-m-diameter, 10.6-m-long tank was fitted with an adsorption pumping system to provide rapid pump-down with little vibration. It may be tilted during vacuum operation from the horizontal to the vertical position or held at any position in between.

FIGURE 6  Tilting test tank.

Concurrent with these test methods and test facility developments, we applied and are continuing to apply much effort to improving the technology of grinding and polishing of the elements themselves. A major accomplishment here has been the development of Itek's Computer Assisted Optical Surfacing (CAOS) technology. The use of laser interferometry, modern optical scanning techniques, digital computation, and the latest in digital-control machine elements provides a considerable extension of an optician's skill in the working of surfaces, particularly large surfaces. The path traced out by a light attached to a working lap on our small CAOS machine as it is driven by computer-derived magnetic-tape instructions is shown in Figure 7. Here we see that the machine is being directed to produce some concentration of work toward the edge of the piece being worked. The degree of success achieved in this development spurred the construction of the machine shown in Figure 8. This machine will accommodate a 3-m mirror as presently assembled. Mirror diameters of 5 m or more would not require major modifications or redesign of this machine.

With all of these new and powerful tools available, it would seem that we are now well prepared to make mirrors of almost any size and configuration. We should first, however, conduct a mental experiment. First assume that there exists a perfectly surfaced mirror floating in a zero-$g$ environment. We then transport this mirror to a horizontal position resting on three support points in our normal gravitational environ-

FIGURE 7   1-m CAOS machine.

FIGURE 8  3-m CAOS machine.

ment. With appropriate optics we can then measure any departures of the surface from its former optical perfection. A contour map of these changes would closely resemble the interference pattern shown in Figure 9. This illustrates the need for careful design of both the mirror structure and the supporting system used during testing operations at 1 $g$ so that we may confidently predict the surface quality that will exist in a zero-$g$ environment. The first step in providing a solution will normally be to design the mirror structure for maximum stiffness using the materials and manufacturing processes that are most compatible with the totality of operational requirements that may be placed on the system. We may then turn attention to the supporting system and determine the extent to which it is necessary to approach zero-$g$ simulation by using a large number of support reactions. The closer we can approach the condition in which each differential element of the mirror mass is individually supported, the closer we will come to eliminating any effects produced by the gravitational acceleration environment, and the more nearly our support arrangement will simulate the true zero-$g$ condition.

At Itek, we have designed, constructed, and evaluated several approaches to mounts intended to minimize gravitational deflections. The first example here is the mirror mount for a 1.2-m aperture ground-based tracking telescope required to operate at elevation angles ranging from $-5°$ to $+95°$ and moderate tracking rates. The telescope itself is shown in Figure 10. The primary mirror is an $f/2.5$ paraboloid

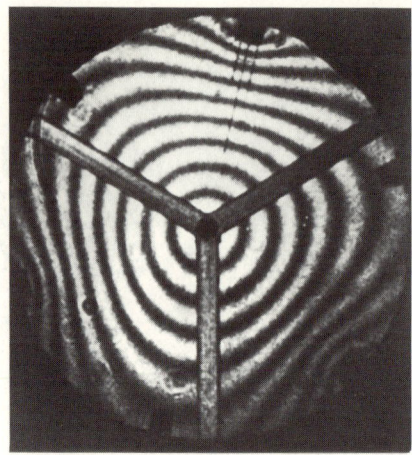

FIGURE 9 Mirror deflection contours, three-point support.

fabricated on a 1.25-m-diameter 0.2-m thick Cer-Vit blank. A front view of the mount is shown in Figure 11. The mirror is supported axially on 27 servo-controlled pneumatic actuators, providing equal force at each reaction point, while axial positioning and tilt are corrected using signals from the three displacement transducers shown. Radial support was provided by a pair of mercury-filled elastomer tubes. The pneumatic actuators are of a rolling diaphragm type. To provide

FIGURE 10  1.2-m tracking telescope.

FIGURE 11  1.2-m mirror mount, front view.

stable operation of the servo valves, it is necessary to continuously bleed gas through each of the actuators. Some appreciation of the complexity of this design may be obtained from Figure 12, which shows the aft end of the mount assembly. This mount was not used for fabrication and testing operations but was primarily intended to compensate for the operational variations in the gravitational vector.

A second example of our efforts is a mechanical 18-point support system shown in Figure 13. At each of the 18 support points, a loading pad was bonded to the mirror surface through a 3-mm-thick layer of silicone elastomer. All the mechanical connections were designed to take loads in any direction while maintaining a reasonable approximation to a statically determinate reaction system for distributing the reaction load equally. This mount was intended to support the mirror through all manufacturing, testing, and coating operations. The design proved adequate for supporting the mirror in all attitudes as illustrated in Figures 14 and 15. Again, these photographs are indicative of the complexity of construction and assembly using this design approach.

Our CAOS technology uses laps that are smaller than 40% of the diameter of the work piece, and finishing operations with CAOS are markedly less sensitive to blank deflections than are conventional optical finishing operations. Additionally, the requirement for launch survival generally

FIGURE 12   1.2-m mirror mount, rear view.

FIGURE 13   1.1-m mechanical, self-equilibrating support.

FIGURE 14  Mechanical support, horizontal attitude.

FIGURE 15  Mechanical support, inverted attitude.

ensures that the mirror blanks we design have sufficient mechanical integrity to withstand any sequence of reasonably well-engineered handling operations.

Our final example is the mount of Figure 16, for use in both fabrication and testing operations. The dual function contributes to minimizing deflections during fabrication operations, but the principal reason for its choice is to reduce the turn-around time during the normal iterative cycles of polishing, testing, polishing, testing, etc. Twenty-seven support points were used for a 1.8-m mirror. We have again chosen fluid actuators, but this time we are using a liquid and have no need for servo control of axial or tilt positioning. The actuators are of a rolling diaphragm type, all interconnected by tubing, with shutoff valves located to permit small groups of actuators to be isolated from their neighbors. In the design stages for this mount, it was felt that some departure from the self-leveling, self-balancing mount provided by fully open connections would be necessary during the varying load cycles of fabrication operations. In actual operation, however, it has been found that all valves may be left open. The piping restrictions are sufficient to prevent any significant undesirable motion of the mirror blank as the polishing lap moves.

As one of the initial steps toward the realization of our space observatory, we are presently undertaking fabrication studies of a 1.8-m-diam, lightweight mirror using this mount design.

FIGURE 16 Liquid piston support, 1.8-m mirror.

These efforts and concurrent progress in other areas have led us to the conclusion that, for space applications, a critical requirement for good zero-g simulation exists only for testing operations. It remains necessary, but less critical, to use a fabrication support arrangement that will reduce the blank deflections during grinding and polishing.

K. R. MASER and K. SOOSAAR

# ELASTIC ANALYSIS OF LARGE SPACEBOUND MIRRORS

## Introduction

The current effort to produce a 3-m orbiting telescope with diffraction-limited performance at visual wavelengths has imposed considerable demands on existing mechanical analysis and design capabilities for primary mirrors. Such reflectors experience a variety of environments from the time that they are fabricated until the time they are put in operation. Consequently, their design for structural integrity and ultimate optical performance requires an ability to analyze their behavior with considerable accuracy through all such environments.

A number of separate loading conditions may be isolated. The mirror is first polished, under the pressure of a polishing tool, on a given support configuration in a 1-$g$ environment. It will subsequently be tested, possibly on a different set of supports, for general optical performance. It is then launched, experiencing accelerations up to 10 $g$ and thermal changes, on a launch support structure differing from the one on which it was polished and tested. Finally, it must be operational in zero-$g$ environment and subject to additional thermal fluctuations. These environmental con-

---

K. R. Maser is at Foster-Miller Associates, Inc., Waltham, Massachusetts; K. Soosaar is at the C.S. Draper Lab., Inc., Cambridge, Massachusetts.

ditions define several analysis and design problems. One must be able to design the support configuration for polishing such that the zero-*g* mirror figure will be diffraction-limited. This implies, then, an analysis technique whose accuracy and reliability is commensurate with the ability to make such a prediction. Such a technique must also be able to analyze the performance of the mirror under thermal fluctuations expected in operation. In addition, one must assure that the mirror will survive the dynamic and thermal effects of launch.

Until recently, the theory of plates, shells, and finite-difference approximations to the equations of elasticity provided the only analytical tools for mirror analyses. These methods have severe limitations both within themselves and when applied to current concepts of primary mirrors for Large Space Telescopes. The finite-element method, developed extensively over the past ten years, has provided a numerical technique that can meet these requirements for mirror analysis and design.

This paper briefly describes the finite-element method and its application to mirrors and presents some numerical results that have been obtained from the authors' experiences in the analysis and design of primary mirrors for space applications. Particular emphasis is given to deep-slab mirrors with both honeycombed and solid cores.

While the development and application of the analysis techniques has been primarily aimed toward space implementation, the application to earthbound telescopes follows quite directly.

## The Finite-Element Method

The finite-element method is an analysis technique through which a continuum with an infinite number of degrees of freedom can be approximated by an assemblage of subregions (elements), each with a specified, but now finite, number of unknown degrees of freedom. The problem is solved by a variational principle that minimizes the potential energy functional within a class of functions defined over each subregion. While the primary application has been to elasticity problems, the approach is generally valid for all continua.

Under static loadings, the following systems matrix equations can be obtained to relate the forces and deformations in the elastic continuum:

$$kU = R, \qquad (1)$$

where

$$R = P + Q + S + B$$

and

   $k$ = stiffness matrix for the structural system,
   $U$ = deformation field of structure,
   $P$ = concentrated forces,
   $Q$ = thermal loadings,
   $S$ = surface forces,
   $B$ = body forces.

If $R$ is time-dependent, then the problem is dynamic and must include inertial and damping forces as well.

$$M\ddot{U} + C\dot{U} + kU = R(t), \qquad (2)$$

where

   $M$ = mass matrix,
   $C$ = damping matrix.

Equation (1) is usually solved by inverting the stiffness matrix ($k$) and thus obtaining the deformation field ($U$) for any arbitrary imposed loading vector $R$. Equation (2) is generally solved indirectly through eigenvalue or propagation means. In either case, the analyst can specify completely arbitrary structure geometries, material properties, boundary conditions, and loads.

Those interested in further study of this field are referred to the considerable literature in finite-element theory that has been generated over the past decade. These include a number of texts [Zienkiewicz, 1971; Desai and Abel, 1972; Przemieniecki, 1968] and untold quantities of scientific papers.

Finite-element methods have been incorporated into many highly sophisticated structural analysis systems such as NASTRAN, STRUDL II, SAMIS, and STAR, and one of those can be usually found at a large computing center. The mirror studies performed by the authors presented in this paper and elsewhere [Soosaar, 1971; Maser and Soosaar, 1972a, 1972b] have been computed primarily on the STRUDL II system. The application of one of these systems to the analyses of problems is quite straightforward. Consider the circular mirror in Figure 1. The mirror can be subdivided into as many of the triangular elements as the analyst considers desirable. With fewer elements, the accuracy is lower, while with larger numbers, the computer running costs become significant. The analysis system deals with the problem in the following manner:

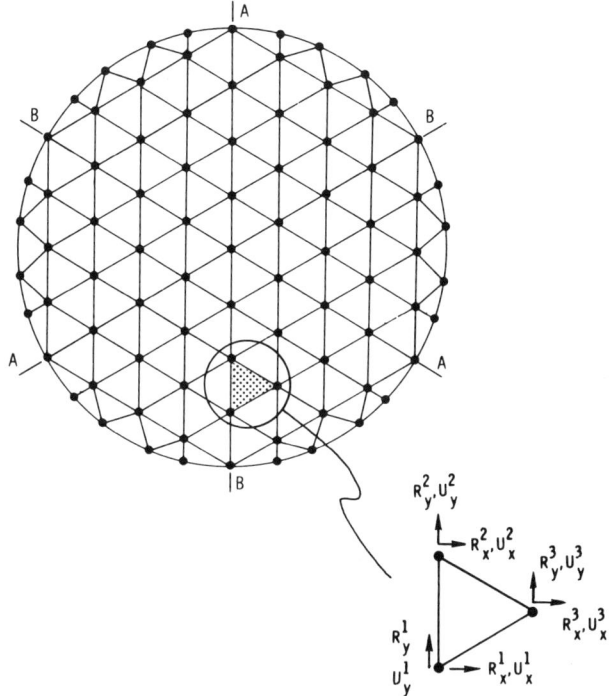

FIGURE 1 Finite-element subdivision of mirror.

1. The element subdivision is specified by the analyst.
2. The system computes the response of each element to applied loadings.
3. The system assembles the response of the total structure from the behavior of the individual elements.
4. The system solves for the displacement field using Eq. (1) or (2).

To make the displacement data meaningful to the optical designer, a best-fit reference surface can be generated from the obtained data and rms optical path differences and fringe patterns obtained therefrom.

The unit element in Figure 1 can represent many types of structural behavior, and it is the analyst's task to define the proper one for the physical case at hand. Figure 2 gives an overview of the various elements encountered in mirror studies. These range from simple plate bending and plane stress to full three-dimensional capability. A particularly useful element of the last category is the isoparametric solid element developed initially by Irons and Zienkiewicz [see Zienkiewicz, 1971] and is

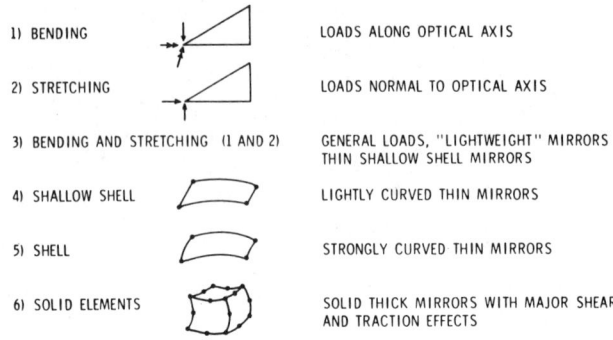

FIGURE 2  Finite-element types for optical-mirror applications.

implemented in most systems today. It consists of bricklike elements with arbitrarily curved sides and surfaces that are highly suitable for representing the optical surfaces of solid mirrors.

The following sections of this paper will consider various applications of the finite-element method to different types of primary mirror configuration and related support problems.

## Lightweight Mirrors

The stiffness-to-weight characteristics of mirror structures are often improved by a "lightweighting" technique. This approach generally acts to minimize the material at the neutral surface of the mirror, where it contributes relatively little to the bending stiffness. The total mirror weight then decreases more rapidly than the bending stiffness, and flexural displacements are reduced. The shear deflections, however, will increase and sometimes even offset the improved bending behavior. As the mirror weight has been reduced, however, the net stiffness-to-weight ratio generally improves.

Until recently, the lightweight structure type has been quite difficult to solve analytically. Some approaches can be made through orthotropic plate theory, but these can become very complex with nonorthogonal grids, circular mirror structures, and arbitrary support configurations. Finite-element techniques make the deformational analysis of lightweight mirrors relatively straightforward. As most of the components are plates, or can be idealized as plates, the generalized plate element can be used for both the ribs and the cover plates. If only a part of the mirror blank has been lightweighted, then the solid three-dimensional elements can be used wherever necessary in conjunction with the plate elements.

A number of candidate lightweight structures are analyzed here to study the sensitivity of the mirror behavior to parameter changes in the mirror lightweighting properties. The initial configurations are based on those given by Simmons [1969] for 64-in.-diameter lightweighted mirrors. To simplify the analysis, the fillets and the backplate holes are ignored, although for a detailed study they should be included. As the supports are continuous along the edge and the loading comes from gravity along the optical axis, symmetry assumptions can be used, and 30° sectors are analyzed for the triangular and hexagonal configurations and a 45° sector for the square-cavity case (Figure 3). A single element is used for the top and bottom plates for each triangle and rectangle, while each of the hexagons must be subdivided into two quadrilaterals. Single rectangle elements are used for the ribs between the intersection nodes. If concentrated loadings and stress effects are desired, it is advisable to subdivide these elements further.

Table 1 shows a comparison of the relative maximum deflections of the various configurations, assuming the material properties of Cer-Vit. The triangular configuration deflects a shade less than the hexagonal, although its weight is considerably higher. The weights, too, have been calculated assuming no fillets, backplate holes, or the additional lightweighting holes found in the junctions of triangular-core mirrors.

These results can now be compared with a mirror 12 in. thick and solid. It is evident here that lightweighting does not necessarily reduce the total mirror deflections in comparison to a solid of the same outside dimensions. The total weight of the mirror is, however, substantially reduced, and improved dynamic behavior should be expected.

Table 2 shows maximum deflection results for a triangular core mirror for a number of parameter tests. Keeping all other properties constant, the web and/or flange thicknesses were doubled or halved; then the total depth of the mirror was changed, keeping the webs and flange thicknesses constant, and finally the 1.5-in. backplate was removed from the mirror.

It is recognized here that definite constraints must be imposed on the sizes of the flange and plate members. Making the web or flange plates too thin can affect strength and handling problems. If the flange plate becomes too thin relative to the web stiffness and cell size, then the likelihood of the cell imprinting on the mirror figure during the polishing process becomes higher.

Additional later studies have shown that there is much less sensitivity to cell configuration than the results of Table 1 would indicate. The cell size in the cases presented is a fairly large fraction of the total mirror diameter, and inhomogeneities of cell shape occur near the boundaries with certain impact on the total deformational performance. With a cell size small relative to the mirror diameter, these localized effects are

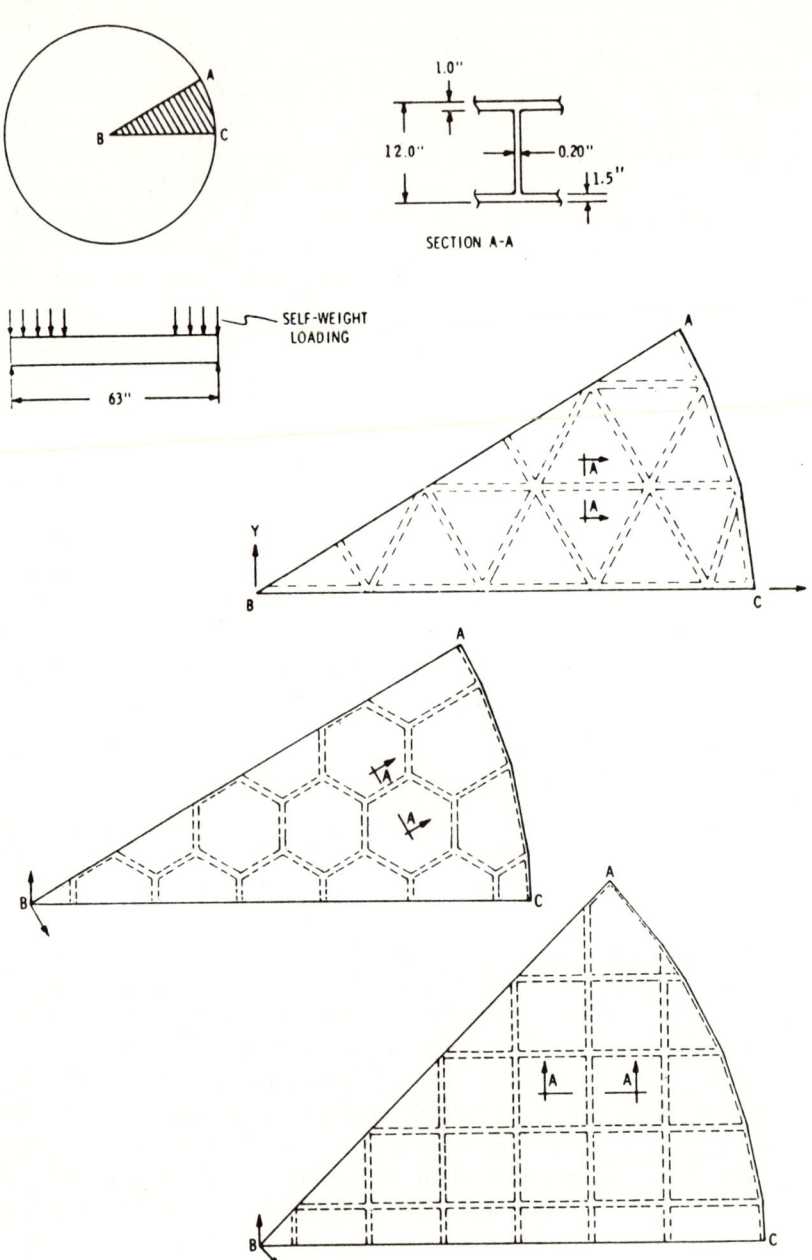

FIGURE 3 Lightweight-mirror core configurations.

TABLE 1  Lightweight and Solid Mirrors[a]

| Mirror Type | Deflection ($\times 10^{-6}$ in.) | Weight (lb) |
|---|---|---|
| Solid | 44 | 3450 |
| Lightweight, triangular | 49 | 1040 |
| Lightweight, square | 60 | 975 |
| Lightweight, hexagonal | 50 | 945 |

[a] Mirror diameter: 64 in.; thickness: 12 in.; loading: 1 g; material: Cer-Vit.

smoothed out, and all cell configurations perform approximately in a similar manner.

## Thick Solid Mirrors

The solid glass mirror is more traditional an approach than the lightweighted mirror, but its exact analysis is considerably more complex. Closed-form solutions are few and usually with somewhat restrictive boundary conditions, and thus nearly always, a numerical approach must be sought. Even then, a completely general three-dimensional approach is necessary, as the methods utilizing a bending with a shear-correction approach fall apart with discrete support cases. The approach taken by the authors involves the use of the 3-D isoparametric solids discussed earlier.

An example of the influence of solid elements on the results of an

TABLE 2  Variations on Lightweight, Triangular Dimensions and Loads as in Table 1

| Variation | Deflection ($\times 10^{-6}$ in.) | Weight (lb) |
|---|---|---|
| Original | 49 | 1040 |
| 0.5 × web | 65 | 865 |
| 2.0 × web | 44 | 1295 |
| 0.5 × flanges | 44 | 650 |
| 2.0 × flanges | 48 | 1730 |
| 0.5 × (web, flanges) | 50 | 505 |
| 2.0 × (web, flanges) | 48 | 2020 |
| 0.5 × total depth | 125 | 830 |
| 2.0 × total depth | 28 | 1370 |
| 1.5-in. backplate removed | 78 | 575 |

analysis can be seen in Figures 4 and 5. A 120-in. telescope mirror, 20 in. thick and supported on three points at the circumference, was analyzed both by isoparametric quadratic solid, as well as by bending elements. The surface transverse deformation contours are plotted on the element layout in both cases. While the distribution of contours is approximately similar in both cases, the net central deflection of the solid model is more than twice that in the bending model. The explanation for this can be seen in Figure 6. In the solid-element model, a large part of the deformation occurs in the neighborhood of the support where the three-dimensional stress state is high. This can be seen in the exaggerated deformation sketch shown by the dotted lines. The bottom graphs compare the top and bottom surface deflections of the solid model with the neutral-plane deflections of the bending model. The mirror center deflection is considered here as the datum for comparisons. The bottom surface distorts quite strongly, although the top surface of the solid and the bending model appear to be close. By geometrical-optics criteria they are not, however, as the changes in slope differ significantly.

When the mirror is uniformly supported at the circumference, there is relatively good agreement (within 15%) between the solid and bending approaches, as can be seen in Figure 7.

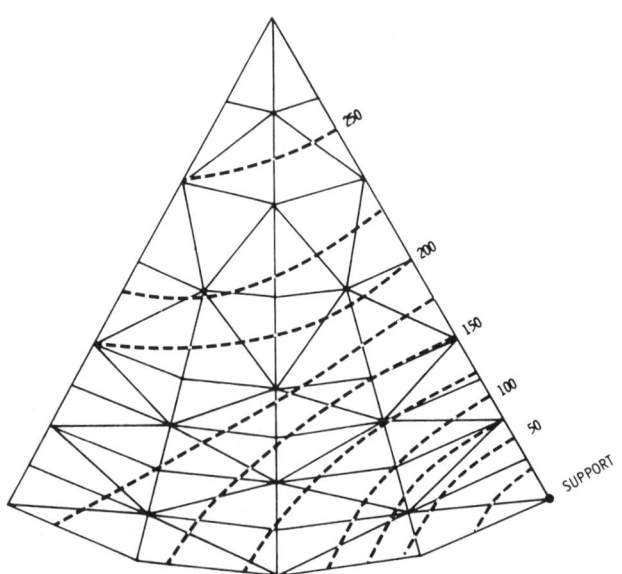

FIGURE 4  Top-surface transverse deflectors—bending mirror.

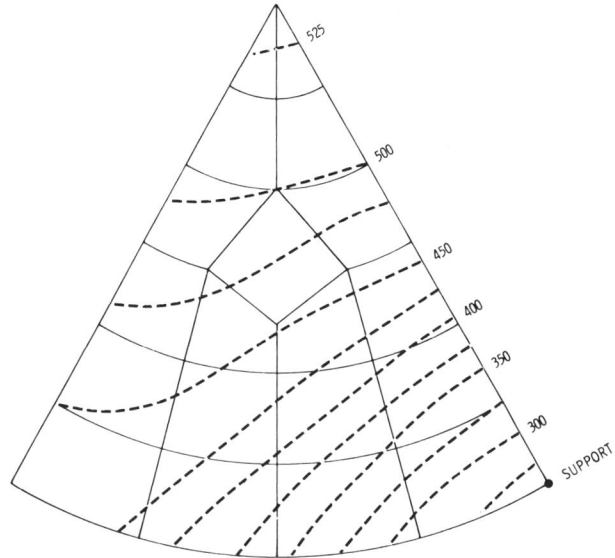

FIGURE 5 Top-surface transverse deflections—3-D mirror.

The implication is, of course, that discrete supports on a thick solid mirror cause local deformational effects, which are significant for the optical behavior and can only be detected via solid-element modeling.

## Solid Mirror on Real Supports

While the previous studies considered the mirror structures on idealized edge supports, the actual supports to be used for testing a large mirror for its manufactured figure will be quite different. A similar type of support may be used for launch purposes. Such a "real" support is depicted in Figure 8. It is a "wheel" supported at the rim to a solid foundation or to the spacecraft structure and is attached to the mirror by movable pads located on the spokes of that wheel. One objective of the study was to determine the location of the pads relative to the mirror such that there is a minimum impact on the optical figure. An element model was made of the solid mirror and the support structure. The support pads were then moved back and forth and deformations computed under a unit gravity load (Figure 9).

Two sets of pads were available, at different separations—16.2 and 18.7 in. The results are shown in Figure 10 in terms of the maximum

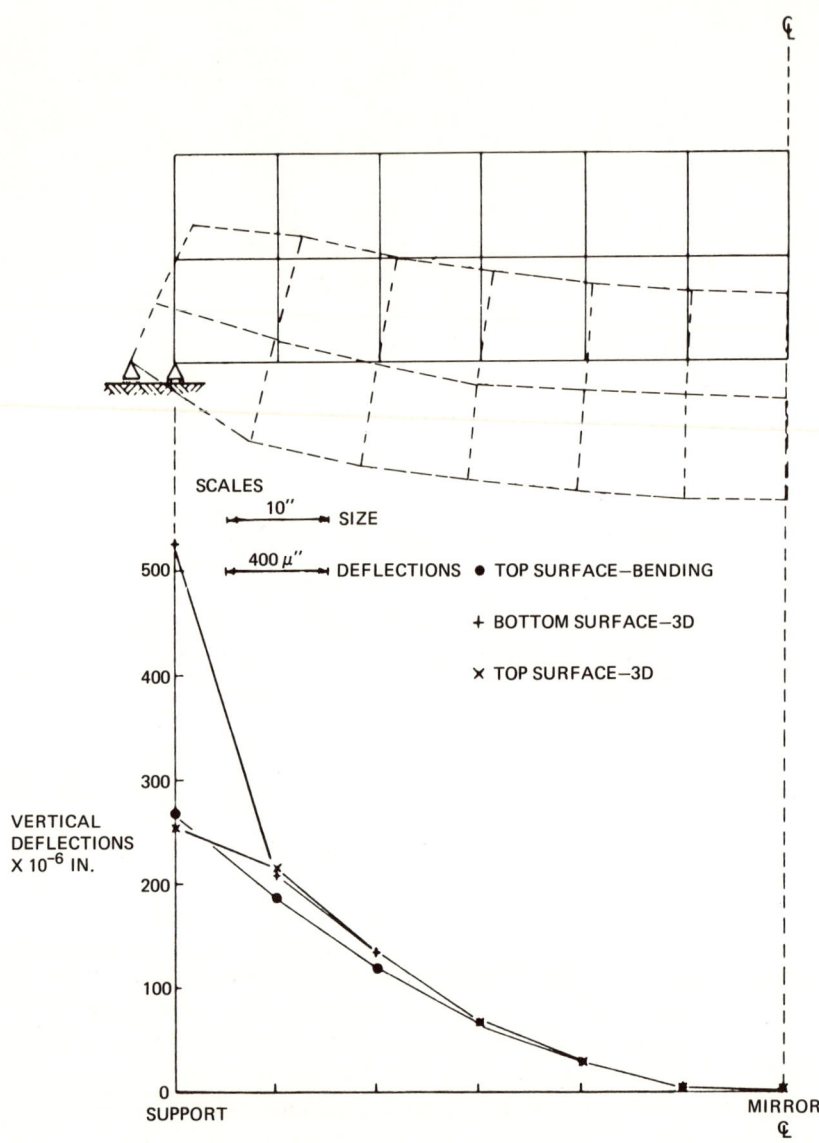

FIGURE 6  Solid versus bending—three-point supports.

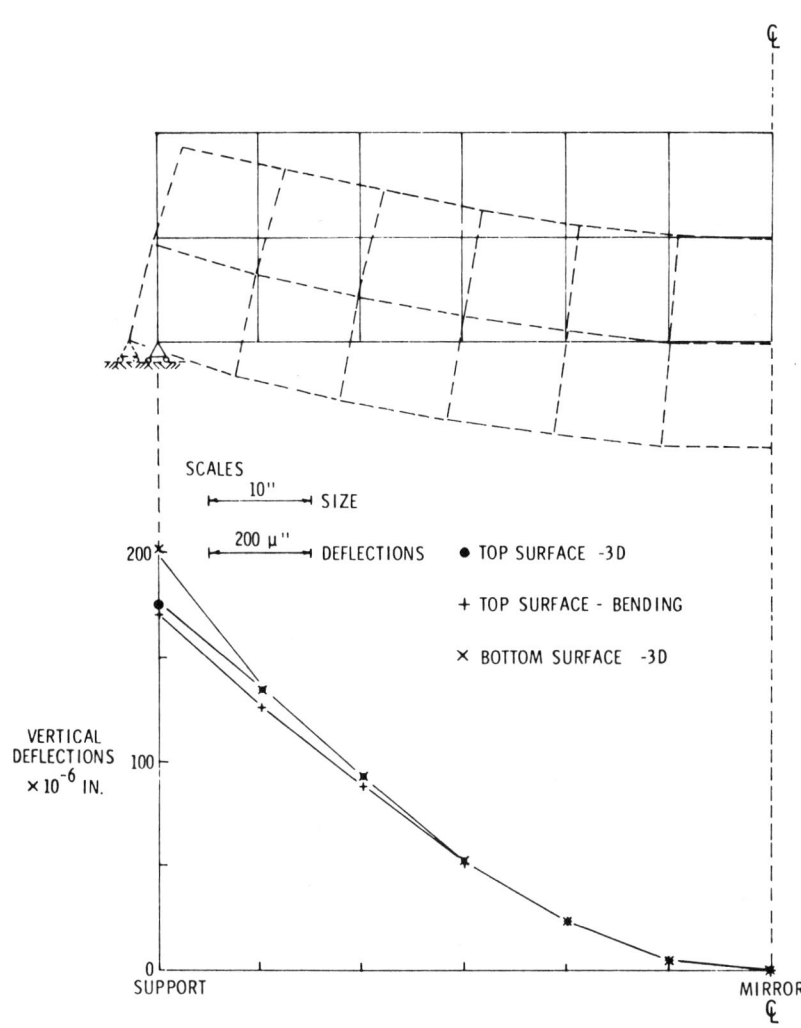

FIGURE 7  Solid versus bending—continuous supports.

FIGURE 8 Primary-mirror support structure.

optical-path difference and the rms of the optical path difference. Figure 11 shows typical contours of the deviations from the best-fit sphere to the deformed mirror surface.

**Sandwich Mirror**

An alternate approach toward minimizing the mirror weight is to use a sandwich of glass plates separated by a core of low-density foamed glass. The finite-element modeling uses plane-stress elements for the top and bottom plates, and three-dimensional solid elements are used for the core. Figure 12 shows a typical section and element layout for such a mirror.

In this case, the shear deformations dominate, and with a continuous rim support, the solid-element approach gives 2.5 times the bending deflections of an equivalent plate; with three-point supports, there is a factor of 5.5 between the solid and the bending. Although an equivalent plate bending element with transverse shear deformation would reduce the discrepancy somewhat, the three-dimensional effects in the neighborhood of the supports require the use of solid elements.

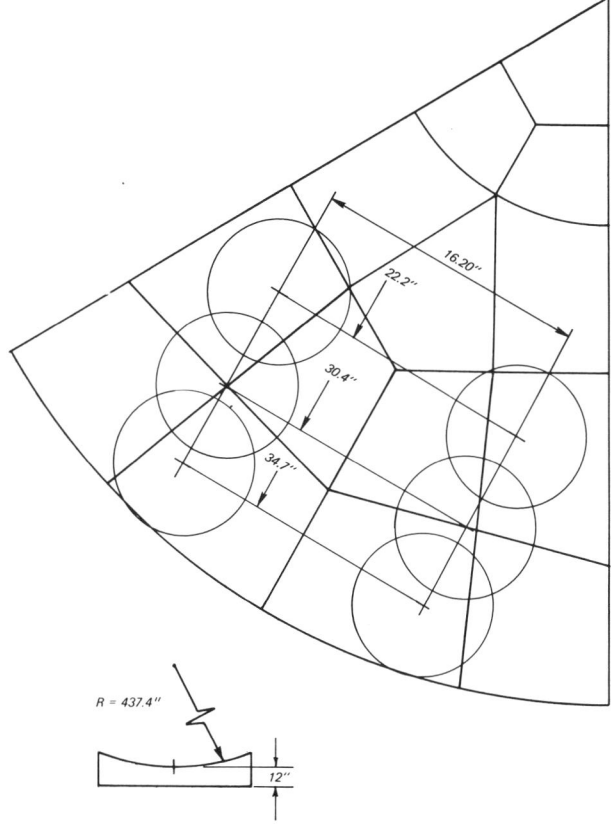

FIGURE 9 Positions of support pads.

## Cell Model Study

In the earlier section on honeycombed lightweighted mirrors, the overall deflections of the mirror were of primary concern. It has become apparent, however, that the local deflections within the individual cells may also have serious effects on the optical quality of the mirror. These local effects may be residual effects due to lightweighting and surface polishing, or they may be simply due to nonuniform deformations under gravity loading.

In order to study this problem in great detail, a cell model study was carried out. This was an attempt to trace, both analytically and experimentally, the stress-deformation history of the basic unit of a honeycombed mirror. The objective was to evaluate whether the grinding,

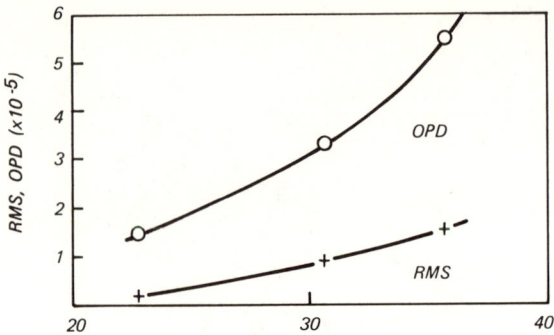

SUPPORT A - DISTANCE OF PAD AXIS FROM CENTER (in)

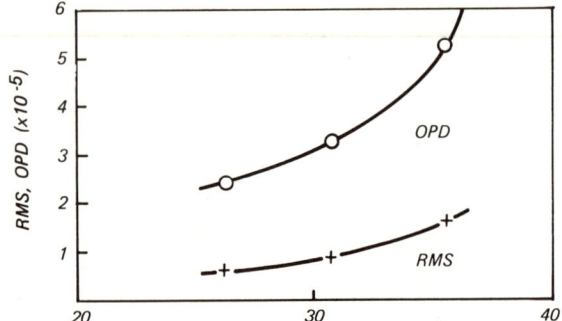

SUPPORT B - DISTANCE OF PAD AXIS FROM CENTER (in)

FIGURE 10  Mirror optical response to pad location.

polishing, and lightweighting processes, when performed in a 1-g field, could result in substantial mirror optical-surface distortions when the mirror was placed in a zero-g orbital environment.

The mirror slab (Figure 13) would first be studied in a solid configuration and the optical surface deformations determined for a number of support conditions under gravity and concentrated loads (the latter representing grinding and polishing tools). Once measurements had been made, and the analytic model adjusted if necessary, the mirror would be machined into the seven-hexagonal-cell configuration (Figure 14). The optical surface would be tested again for deformations, and by means of combining analytical and experimental data, it would be determined how much of the differences observed could be attributable to

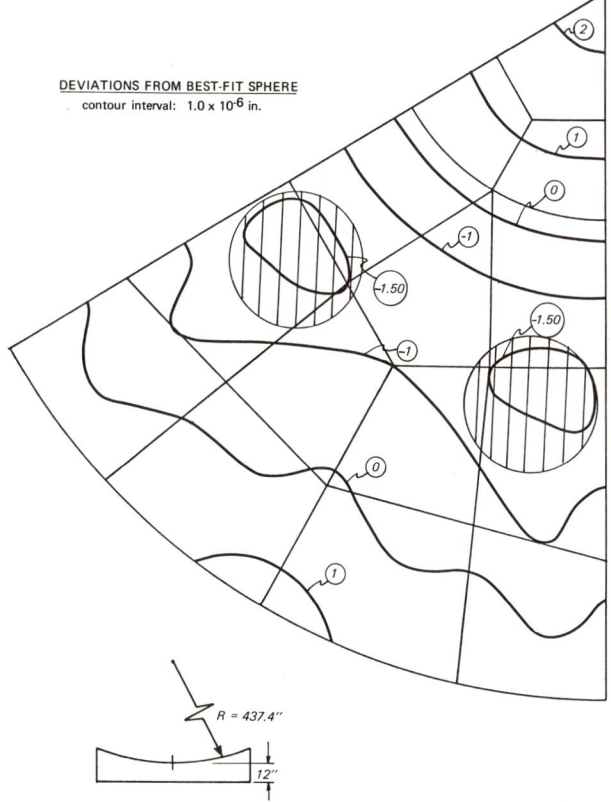

FIGURE 11 Contours of deviations from best-fit sphere.

changes in mass and stiffness and how much to the release of internal stress by machining. From these data, extrapolations might be made to the orbital performance of a full-size honeycombed mirror.

Figure 15 shows a typical layout of finite elements required to perform the cell-model analyses. Since the minimum number of supports is three at 120°, a 60° segment was required for the analyses. The elements are shaped as shown, so that the analyses of the lightweighted mirror blank can be performed by simply removing typical elements 12, 13, 22, 23, 32, and 33. Figure 16 shows typical optical-surface deflection results.

Experimental results required for the above study are forthcoming. The finite-element results have been checked with closed-form results

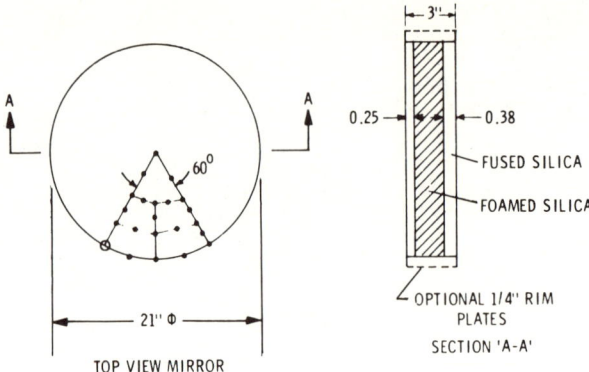

FIGURE 12  Sandwich-mirror analysis.

DISPLACEMENTS ON TOP SURFACE ALONG OPTICAL AXIS ($\times 10^{-6}$ IN)

| JOINT | CONTINUOUS SUPPORTS | | | 3 SUPPORTS | | |
|---|---|---|---|---|---|---|
|  | NO RIM | 0.25 RIM | PURE BENDING | NO RIM | 0.25 RIM | PURE BENDING |
| 1 | 3.7 | 3.4 | 1.5 | 13.9 | 9.6 | 2.4 |
| 2 | 3.4 | 3.2 |  | 13.5 | 9.2 |  |
| 3 | 2.9 | 2.7 |  | 12.6 | 8.3 |  |
| 4 | 2.4 | 2.1 |  | 11.5 | 7.3 |  |
| 5 | 1.8 | 1.5 |  | 10.1 | 5.9 |  |
| 6 | 1.0 | 0.8 |  | 7.6 | 3.8 |  |
| 7 | 0.7 | 0.1 | 0.0 | 6.8 | 2.0 | 1.6 |

for some highly simplified cases, and the deflection accuracy is within 5% at the $\lambda/200$ range. It is felt that, should the experimental procedure be able to measure deflections in this range accurately, the finite-element results are sufficiently accurate to complete the desired study.

The use of finite elements in this study led to some significant insights. One in particular was the sensitivity of the solid-element model to support conditions. Unlike the full mirror, where discrete supports can be modeled as mathematical points, the stubby cylinder cell model is extremely sensitive to the geometry and flexibility of discrete support pads. This fact, plus consideration of high strain gradients in the neighborhood of such supports, had to be considered in the analyses that were conducted.

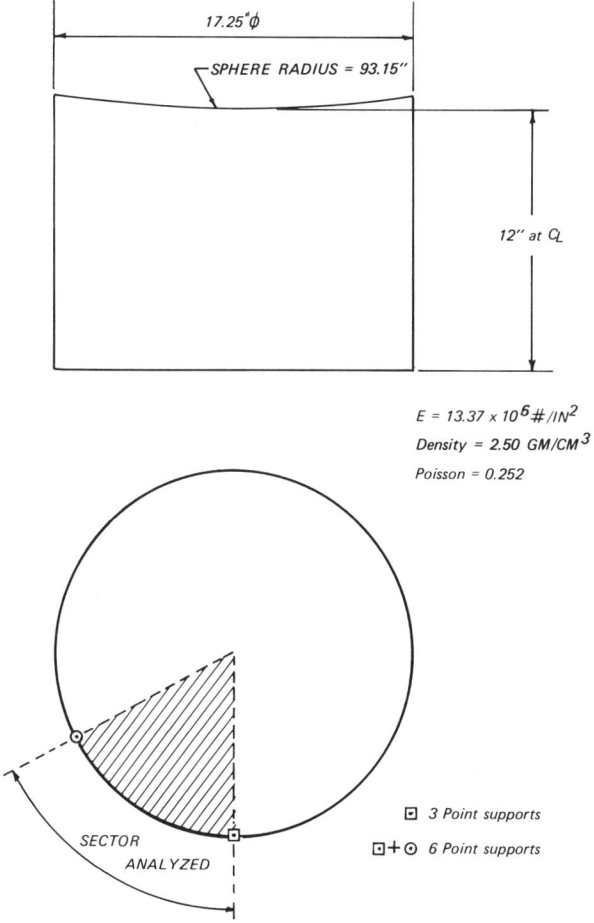

FIGURE 13 Cell model–prelightweighted.

## Mirror Clamping Study

While the majority of structural problems associated with the design of mirrors are those of deflection and stiffness, occasionally questions of strength are also encountered. A space-bound mirror must be held very tightly during the launch and docking procedures, yet the design of these attachments must not induce high stresses in the mirror itself.

A hexagonally lightweighted mirror, 120 in. in diameter, is supported on the wheel, hub, and pads type of structure described in Figure 8. A clamping device that will hold this mirror is depicted on Figure 17

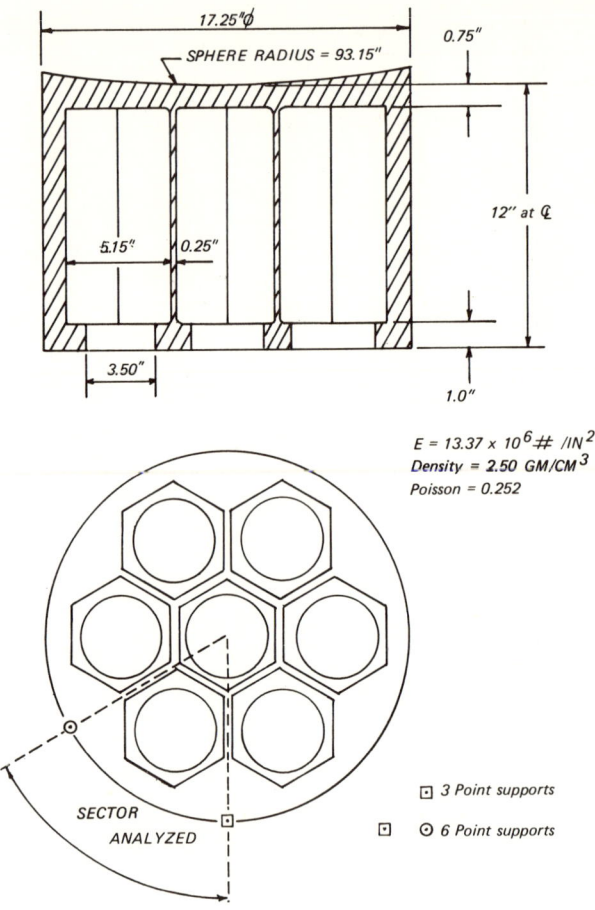

FIGURE 14 Cell model—lightweighted.

and applies tensile, compressive, and moment loads to the mirror. In order to determine the stresses in the mirror, a finite-element model of the clamping device and the mirror in the neighborhood of the clamp was assembled. This used solid elements throughout and modeled all the interacting elastic components.

Some of the results of this study are shown on Figure 18. After a number of design iterations, the fillet radius at the clamping area was increased from 0.25 to 0.50 in. to bring the tensile stresses in the mirror below 1500 psi under the maximum likely launch and docking loads. This form of information could not have been readily obtained by

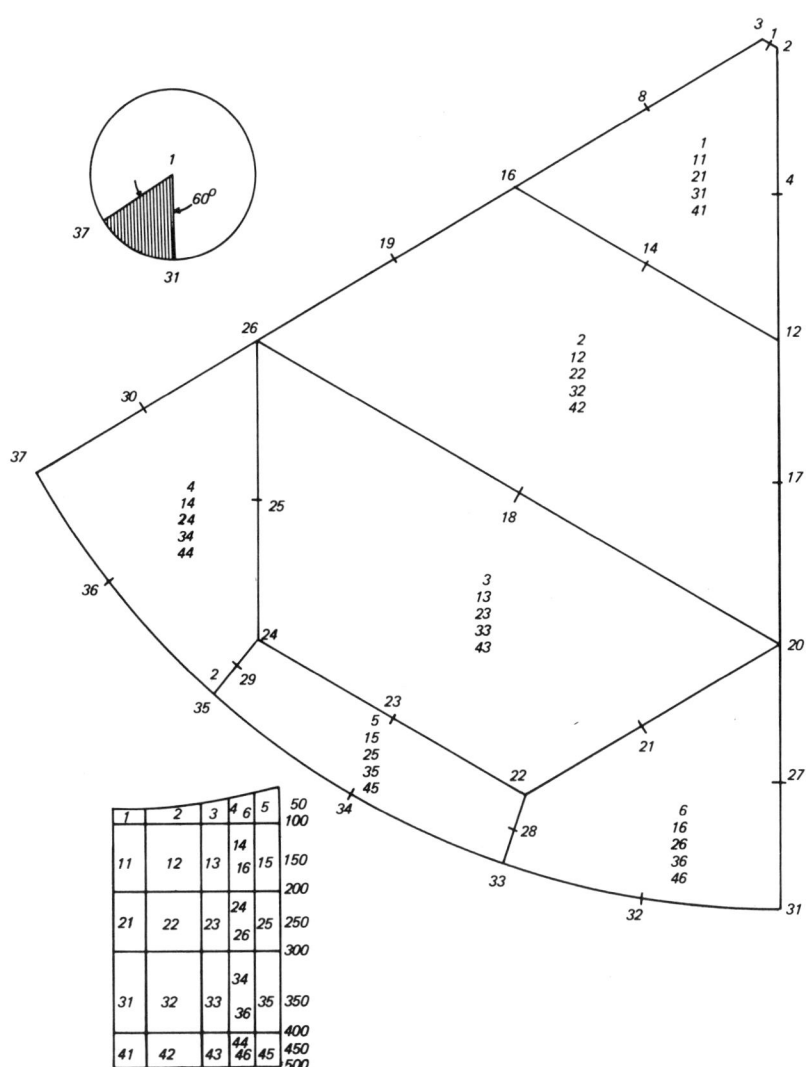

FIGURE 15  Cell model—element layout.

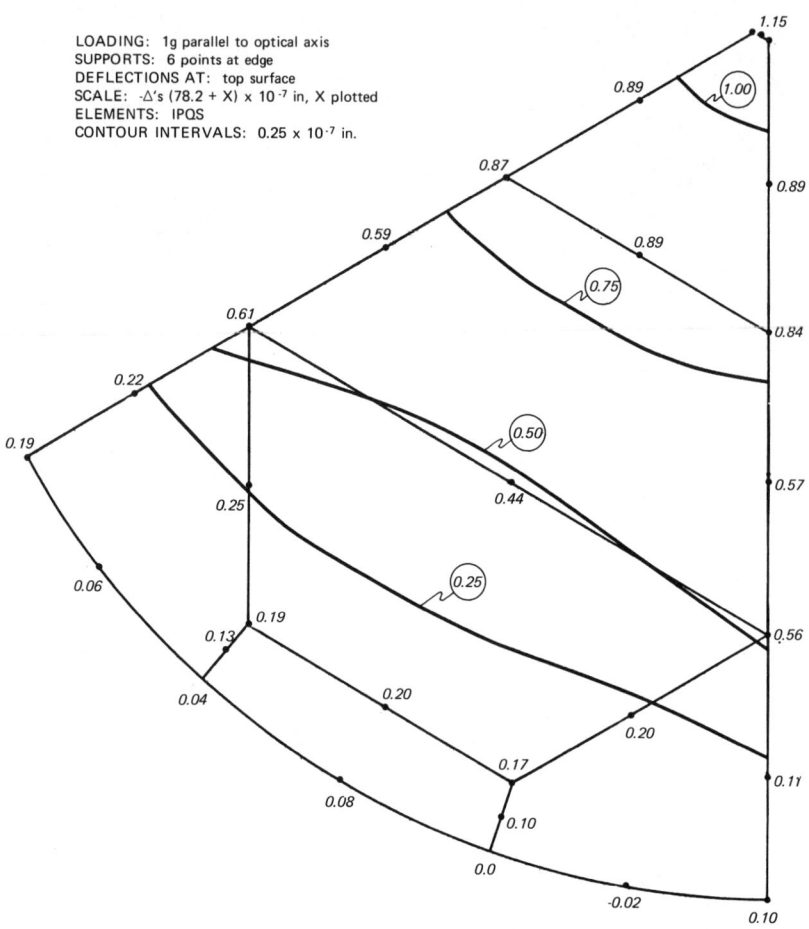

FIGURE 16  Cell model–typical deflection result.

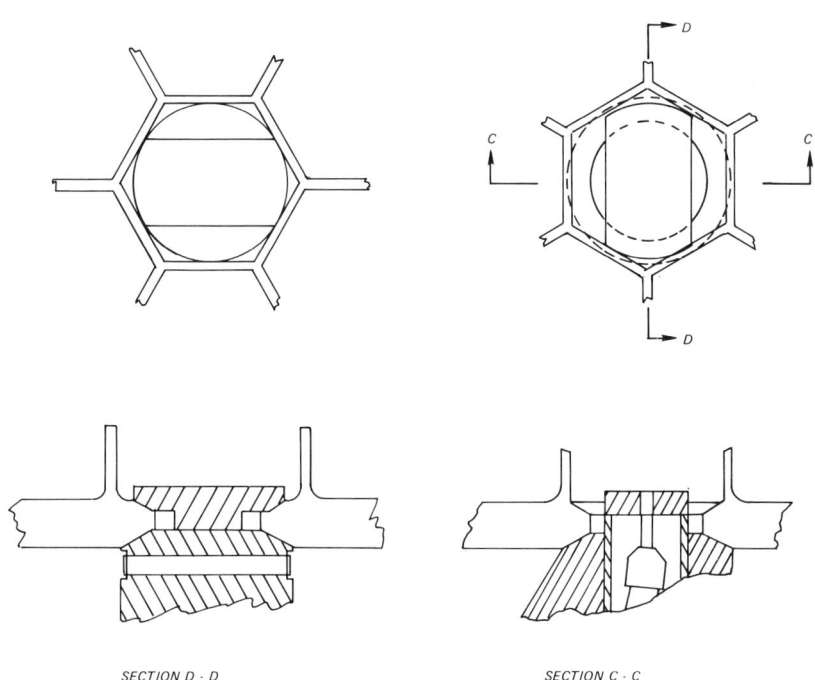

FIGURE 17 Mirror clamping mechanism.

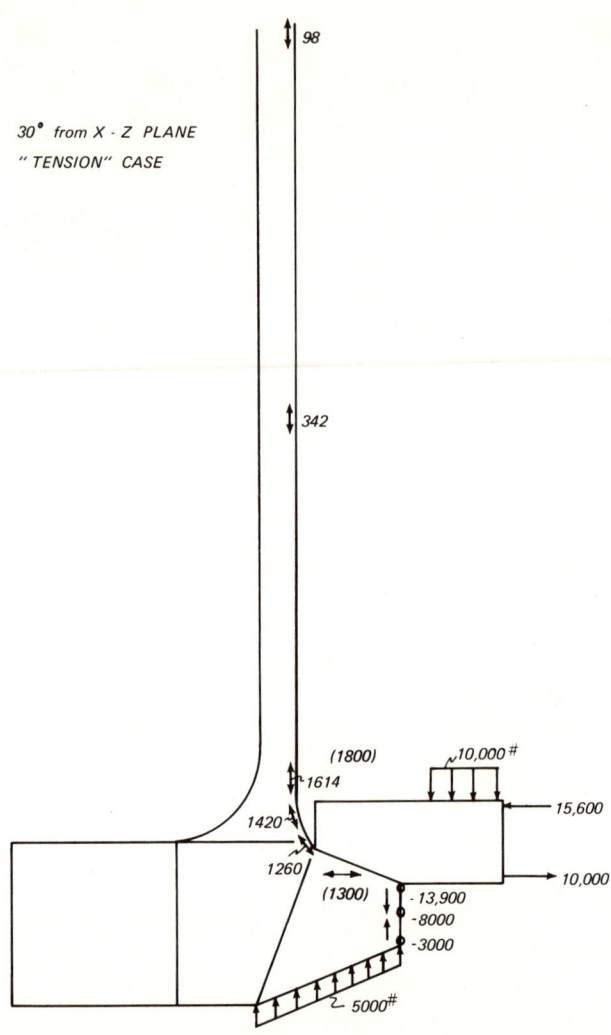

FIGURE 18  Mirror stresses under docking loads.

analytic means unless finite element methods were used and the costs of analysis were considerably less than an experimental study using strain gauge or photoelastic means.

## Conclusions

The techniques and results presented in this paper have demonstrated the application of the finite-element method of structural analysis to the analysis and design of primary mirrors and supporting structures for large telescopes, both orbital and earthbound. It has proved to be an effective tool for this problem, where the optical requirements demand an abnormally high degree of accuracy. The examples presented have illustrated the extreme versatility of this method, showing its application to honeycombed, sandwich, and solid slab mirrors, as well as to optimization of support structures. It is hoped that this analytical tool, which is available in its various forms at most large computing centers, will be used by many others for the design of our next generation of telescopes.

The authors wish to acknowledge the technical guidance and financial support provided by John Mangus of NASA Goddard Space Flight Center through Contract NAS-5-21542.

## References

Desai, C. S., and J. F. Abel, *Introduction to the Finite Element Method*, Van Nostrand Reinhold, New York (1972).
Maser, K., and K. Soosaar, "Analysis of Astronomical Mirrors by Finite Elements," *Proceedings of Conference on Finite Element Method in Civil Engineering*, McGill University, Montreal, Canada (1972a).
Maser, K., and K. Soosaar, "Structural Analysis of Large Telescope Mirrors and Supports," MIT-CSDL Rep. R-716 (1972b).
Przemieniecki, J. S., *Theory of Matrix Structural Analysis*, McGraw-Hill, New York (1968).
Simmons, G. A., "The Design of Lightweight CER-VIT Mirror Blanks," Optical Telescope Technology, Workshop at MSFC, NASA SP-233 (1969).
Soosaar, K., "Design of Optical Mirror Structures," MIT-CSDL Rep. R-673 (1971).
Zienkiewicz, O. C., *The Finite Element Method in Engineering Science*, McGraw-Hill, New York (1971).

CHANGHWI CHI, PRAVIN MEHTA, and
AARON OSTROFF

# Vibrational Modes of the Primary Mirror Structure in the Large Space Telescope System

## I. Introduction

In the field of astronomy and deep-space communications, the Large Space Telescope orbiting outside the earth's atmosphere provides a manifold improvement in the resolution of scientific data because it is minimally affected by the turbulence of the atmosphere, the scatter of sunlight, and the disturbances on the earth's surfaces that are both man-made and natural tremors.

For the Large Space Telescope, mirrors with diameters on the order of 3 m or larger are frequently considered for use as primary mirrors; however, the thickness of these mirrors should not be so excessive as to increase the overall payload weight. (A thinner mirror plate thickness is also desirable in situations where a feedback control system can be used to obtain the desired mirror surface contour.)

The primary and secondary mirror assemblies will form the space telescope structure; in general, the primary mirror will be an aspherical sur-

---

C. Chi and P. Mehta are at The Perkin-Elmer Corporation, Wilton, Connecticut; A. Ostroff is at the NASA Langley Research Center, Hampton Station, Virginia 23365.

face whose *f*-number is on the order of 10. The primary mirror with a large diameter results in increased image resolution and increased optical power of the received signal, thus increasing the signal-to-noise ratio. The higher signal-to-noise ratio, in turn, results in the improved telescope pointing accuracy.

In a primary mirror with such a large diameter, the resonant vibration of the mirror plate can become a significant factor in the design process and can no longer be neglected. The significance of the effect of the resonant vibration becomes apparent as we consider the requirement that rms error of the mirror surface be on the order of $\lambda/50$, where $\lambda$ is the wavelength of signal light. The telescope with a large primary mirror requires a correspondingly higher surface accuracy (or smaller rms surface error) in order to take advantage of its inherent ability to obtain higher image resolution.

The primary mirror of a space telescope is often mounted in a configuration where the primary mirror plate is supported at three points around the circumference, usually 120° apart and kinematically mounted. Such a three-point support configuration is capable of producing the small tilting movement in an arbitrary direction necessary to align the mirror. It is to be noted that the resonant vibrational mode shapes and the resonant frequencies depend on the manner in which the mirror plate is mounted.

The recently developed technology that controls the mirror surface by the feedback control requires the information of the resonant mode shapes and frequencies. In the simple proportionality feedback control system, information on the distribution of the resonant frequencies is needed when the stability, bandwidth, and control compensators of the closed-loop control system are to be optimized.

The modal control system [Creedon and Lindgren, 1970; Howell, 1974] is another approach and requires both the resonant mode shapes and frequencies. This technique controls and corrects the mirror surface error by reducing the error components in a set of predetermined (usually dominant) modes to zero; it is especially effective in situations where a small number of actuators are used. The need to investigate the vibrational modes and resonant frequencies thus arises.

In order to simplify the analysis, the behavior of a flat circular mirror plate will be analytically investigated instead of the mirror with small curvature, which is simply supported at three points around the circumference. This paper presents the highlights of the analytical solution for the flat-plate system and summarizes the characteristics of the resulting mode shapes and frequencies.

Subsequently, results will be presented that have been obtained by running the digital computer program (NASTRAN) to compute the resonant modes and frequencies of the flat-plate system and the curved-mirror surface systems.

It is to be noted that the mirror plate supported at three points along the circular boundary is a mixed boundary-value problem, which is usually complicated to solve and has an inherently high order of discontinuity due to the presence of the point loading at the support points (i.e., the pressure at the support points is an impulse function). In the presence of the high-order discontinuity, an effort was made to improve the accuracy and efficiency of the numerical computation. From this effort, the polar grid pattern and grid point numbering scheme have been devised, which are considered to be optimum in terms of computational accuracy and computational efficiency and most natural for the circular mirror geometry. Some of the representative results of the computer calculations for the flat plate and the plate with a curvature are presented and compared. Very often, the output of the computer calculation needs to be interpolated to find the values at the locations between the grid points, and a reliable interpolation scheme is therefore required.

A scheme, called the curvilinear bicubic spline fit interpolation, has been developed and is believed to be particularly suitable for interpolating the functions representing physical quantities in the circular geometry. A brief discussion of this interpolation scheme is presented in the latter portion of this paper.

## II. Analysis

A flat circular plate that is simply supported at three points on the circumference is shown in Figure 1. The vibrational behavior of this system has been considered previously [Chi, 1972]. We begin by presenting a brief summary of the mathematical derivation for the sake of completeness. The equation describing the vibration of a thin plate, with $e^{j\omega t}$ time dependence, is

$$\nabla^4 w - k^4 w = P/D, \qquad (1)$$

where

$w$ = the vertical deflection of the plate in $z$-direction,
$\nabla^4 \equiv \nabla^2 \nabla^2$ ($\nabla^2$ is Laplacian operator),
$P$ = the pressure in space coordinate

$$k^4 = (\mu/D)\,\omega^2, \qquad (2)$$

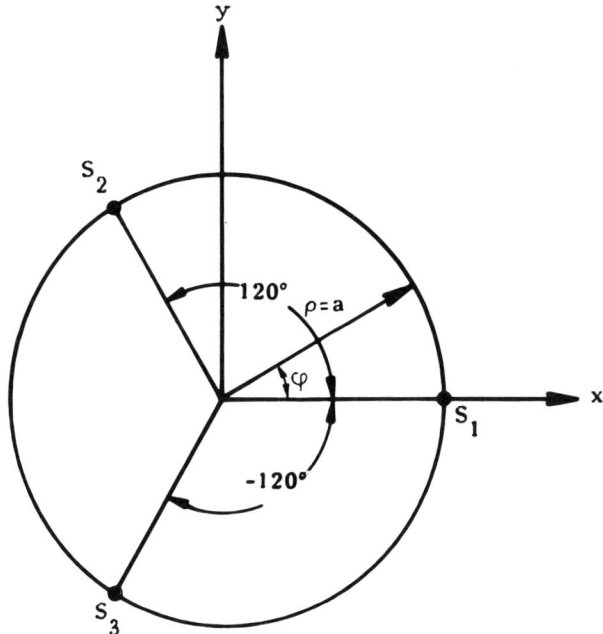

FIGURE 1  Thin flat circular plate with three simple support points $S_1, S_2, S_3$ on the circumference.

where

$\omega$ = angular frequency in radians per second,
$D = Eh^3/12(1-\nu^2)$,
$E$ = the modulus of elasticity,
$h$ = the thickness of the plate,
$\nu$ = Poisson's ratio,
$\mu$ = mass per unit area of the plate.

The system possesses the following boundary conditions:
1. The radial moment ($M_\rho$) along the circumference is zero:

$$M_\rho = -D\left[\frac{\partial^2 w}{\partial \rho^2} + \nu\left(\frac{1}{\rho}\frac{\partial w}{\partial \rho} - \frac{m^2}{\rho^2}w\right)\right]\bigg|_{\rho=a} = 0, \qquad (3)$$

where $\rho$ is the radial coordinate and $\varphi$ is the angular coordinate (see Figure 1).

2. The shear force ($V$) along the circumference is

$$V = \left( Q_\rho - \frac{1}{\rho} \frac{\partial M_{\rho\varphi}}{\partial \varphi} \right) \bigg|_{\rho = a} = 0 \text{ except at points } S_1, S_2, S_3 \qquad (4a)$$

(see Figure 1).

3. The displacement ($w$) is

$$w = 0 \text{ at the support points } S_1, S_2, S_3 \qquad (4b)$$

(see Figure 1).

4. The displacement at the center of the plate is finite and continuous, where (4c)

$$Q_\rho = -D \frac{\partial}{\partial \rho} \left( \frac{\partial^2 w}{\partial \rho^2} + \frac{1}{\rho} \frac{\partial w}{\partial \rho} + \frac{\partial^2 w}{\rho^2 \partial \varphi^2} \right),$$
$$M_{\rho\varphi} = (1 - \nu) D \left( \frac{1}{\rho} \frac{\partial^2 w}{\partial \rho \partial \varphi} - \frac{1}{\rho^2} \frac{\partial w}{\partial \varphi} \right). \qquad (5)$$

In the cylindrical coordinate system, the homogeneous solutions of Eq. (1) are

$$w = \sum_{m=0}^{\infty} A_m [J_m (k\rho) + B_m I_m(k\rho)] \cos m\varphi$$
$$+ \sum_{m=0}^{\infty} C_m [J_m (k\rho) + D_m I_m (k\rho)] \sin m\varphi, \qquad (6)$$

where $J_m(x)$ and $I_m(x)$ are the ordinary and the modified Bessel functions of the first kind, and $A_m$, $B_m$, $C_m$, and $D_m$ are constants.

Substituting Eq. (6) into Eq. (3), one obtains

$$B_m = D_m = \frac{\{[(1-m)/a] J_m (ka) + [(2m+1+\nu)/a] J_m (ka) - (m+1) J_{m+2} (ka)\}}{\{[(m-1)/a^2] I_m (ka) + [2m+1+\nu)/a] I_{m+1} (ka) + (m+1) I_{m+2} (ka)\}}. \qquad (7)$$

On the circumference of the plate, the shear force is zero except at the support points, and the shear forces exert the pressure at the three support points so as to make the displacement at the points zero. Thus, the shear force distribution is

$$P = \{P_1 \delta(\varphi) + P_2 \delta[\varphi - (2\pi/3)] + P_3 \delta[\varphi + (2\pi/3)]\} \bigg|_{\rho = a}, \qquad (8)$$

where

$\delta(\varphi)$ is the impulse function located at $\varphi = 0$,
$P_1, P_2, P_3$ are constants yet to be determined.

Equation (8) can be expressed in terms of its Fourier series components, and one obtains

$$P = \sum_{k=0}^{\infty} \left(\frac{1}{2\pi}\right) \left[P_1 + (P_2+P_3)\cos\left(\frac{2\pi}{3}k\right)\right] \cos k\varphi$$

$$+ \sum_{k=0}^{\infty} \left(\frac{1}{2\pi}\right) \left[(P_2-P_3)\sin\left(\frac{2\pi}{3}k\right)\right] \sin k\varphi. \quad (9)$$

Substituting Eq. (6), (7), and (5) into Eq. (4a) and (4b) one obtains

$$V = \sum_{m=0}^{\infty} F_m(ka)(A_m \cos m\varphi + C_m \sin m\varphi), \quad (10)$$

where

$$F_m(ka) = (D/a^2) \{ m [(ka)^2 + (1-\nu)m(m-1)] [J_m(ka)/a]$$
$$+ m[(ka)^2 - (1-\nu)m(m-1)] (B_m/a) I_m(ka)$$
$$- k[(ka)^2 + (1-\nu)m^2] J_{m+1}(ka)$$
$$- k[(ka)^2 - (1-\nu)m^2] B_m I_{m+1}(ka) \}. \quad (11)$$

Equating Eq. (9) to Eq. (10) one obtains

$$A_m = \frac{\{P_1 + 2P_e \cos[(2\pi/3)m]\}}{(2\pi) F_m(ka)},$$

$$C_m = \frac{2P_o \sin[(2\pi/3)m]}{(2\pi) F_m(ka)}, \quad (12)$$

where

$$P_e = \frac{1}{2}(P_2 + P_3), \quad \text{even-mode component};$$

$$P_o = \frac{1}{2}(P_2 - P_3), \quad \text{odd-mode component}.$$

Substituting Eq. (7) and (12) into Eq. (6), one obtains the deflection of the plate for a given set of forces at the support points. In particular, the deflections at the support points are expressed in the matrix form,

$$\begin{bmatrix} w^{(e)}(\rho=a, \varphi=0) \\ w^{(e)}\left(\rho=a, \varphi=\pm\dfrac{2\pi}{3}\right) \end{bmatrix} = \begin{bmatrix} \sum\limits_{m=0}^{\infty} \alpha_m(ka) & \sum\limits_{m=0}^{\infty} \alpha_m(ka)\cos\left(\dfrac{2\pi}{3}m\right) \\ \sum\limits_{m=0}^{\infty} \alpha_m(ka)\cos\left(\dfrac{2\pi}{3}m\right) & \sum\limits_{m=0}^{\infty} \alpha_m(ka)\cos^2\left(\dfrac{2\pi}{3}m\right) \end{bmatrix} \begin{bmatrix} P_1 \\ 2P_e \end{bmatrix}, \quad (13)$$

$$w^{(o)}\left(\rho=a, \varphi=\dfrac{2\pi}{3}\right) = \left[\sum_{m=0}^{\infty} \alpha_m(ka)\sin^2\left(\dfrac{2\pi}{3}m\right)\right](2P_o), \quad (14)$$

where

$$\alpha_m(ka) = \frac{J_m(ka) + B_m I_m(ka)}{(2\pi)F_m(ka)}. \quad (15)$$

In Eq. (13) and (14), the deflection and force distribution are decomposed into the symmetrical and asymmetrical components with respect to the $x$ axis, and they are called the even [$w^{(e)}$] and odd [$w^{(o)}$] modes, respectively.

In order to satisfy the boundary condition of Eq. (4b), the determinant of Eq. (13) and the coefficient of Eq. (14) are made zero; the resulting expressions are

$$\sum_{m=0}^{\infty} \alpha_m(ka) = 0 \quad (16a)$$
$$m = 0, 3, 6, 9 \ldots$$

and

$$\sum_{m=0}^{\infty} \alpha_m(ka) = 0. \quad (16b)$$
$$m = 1, 2, 4, 5, 7, 8 \ldots$$
$$m \neq 3, 6, 9 \ldots$$

The two sets of values of $k$ that are obtained from Eq. (16a) and (16b) are the eigenvalues of the plate system. It is to be noted that there is another set of values of $k$ that are obtained by letting

$$F_m(ka) = 0 \quad [F_m(ka) \text{ is defined in Eq. 11}]. \quad (16c)$$

It is physically meaningful and convenient to group the resonant modes into four types. The type I modes are those whose eigenvalues are obtainable from Eq. (16a); their mode shapes are even and given by

$$W_{Ii}^{(e)} = \sum_{m=0}^{\infty} \left\{ \frac{1 + G_i \cos\,[(2\pi/3)m]}{F_m(k_i a)} \right\} [J_m(k_i\rho) + B_m I_m(k_i\rho)] \cos m\varphi, \quad (17)$$

and it is a useful quantity in the investigation of the modal control technique for the space telescope mirror-control system.

The type II modes are those from Eq. (16b); their mode shapes are even and given by

$$w_{IIi}^{(e)} = \sum_{m=0}^{\infty} \left\{ \frac{1 + G_i \cos\,[(2\pi/3)m]}{F_m(k_i a)} \right\} [J_m(k_i\rho) + B_m I_m(k_i\rho)] \cos m\varphi, \quad (18)$$

where

$$G_i = \left(\frac{2P_c}{P_1}\right) = \frac{\sum_{l=0}^{\infty} \alpha_l(k_i a)}{\sum_{l=0}^{\infty} \{\alpha_l(k_i a) \cos\,[(2\pi/3)l]\}}. \quad (19)$$

The type III modes are those from Eq. (16b); their mode shapes are odd and given by,

$$w_{IIIi}^{(o)} = \sum_{m=0}^{\infty} \frac{\sin\,[(2\pi/3)m]}{F_m(k_i a)} [J_m(k_i\rho) + B_m I_m(k_i\rho)] \sin m\varphi. \quad (20)$$

The type IV modes are those from Eq. (16c); their mode shapes are odd and given by

$$w_{IVi,\,m}^{(o)} = [J_m(k_i\rho) + D_m I_m(k_i\rho)] \sin m\varphi, \quad (21)$$

where

$$m = 3, 6, 9, 12. \ldots$$

One finds that types I, II, and III exert force on the support points, while type IV does not. Consequently, types I, II, and III require the support points in order for them to exist. Type IV is comprised of the group of modes that are also members of vibrational modes of the free plate without support points; consequently, type IV does not require the support points in order for it to exist. As mentioned earlier, types I and

II are even with respect to the $x$ axis, while type III is odd. Type I, furthermore, possesses 120° rotational symmetry. (120° rotational symmetry means that, when the mode pattern is rotated ±120°, the pattern is unchanged.)

The mode shapes are shown in Figures 2 and 3 with Poisson's ratio ($\nu$) equal to 0.25. It is apparent that the low-order modes are those of types I, II, and III, which are characteristic to the plate with the particular support arrangement, while the modes of type IV appear as high-order modes. The figure that shows the lines and points of zero deflection for each vibrational mode is useful in identifying the particular mode in actual mirror vibration. The pattern of the reaction forces by the support points can be deduced from the mode shape near the support points. The important properties of each type of mode are listed in Table 1.

The chart in Figure 3 shows the inverse of the eigenvalues, $(k')^{-4}$, of each mode, which is calculated from the analysis, where $k' = ka$. The values are normalized so that the value for the first mode is 1. The chart shows how fast $(k')^{-4}$ decreases as the order of the mode increases.

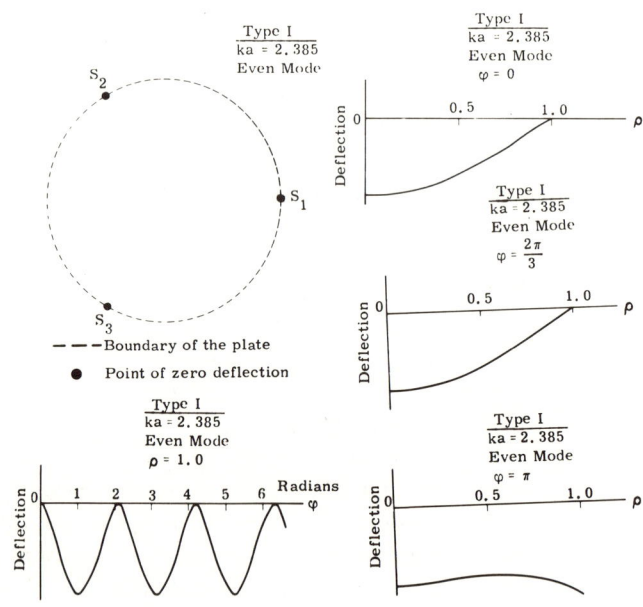

FIGURE 2 Vibrational mode shapes and eigenvalues normalized for unit radius of a flat plate simply supported at three points.

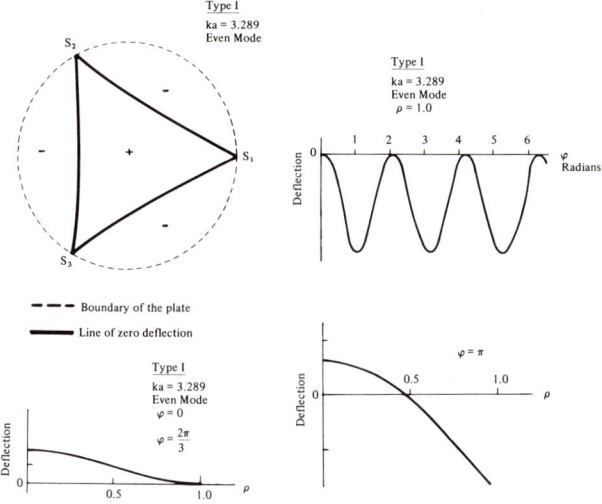

FIGURE 2 Continued

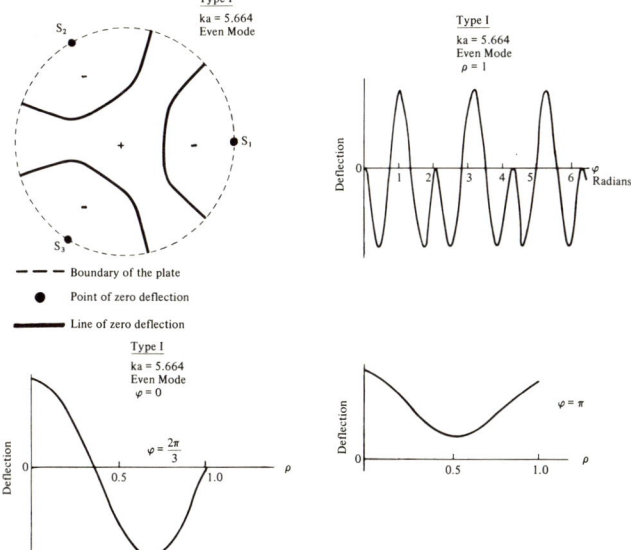

FIGURE 2 Continued

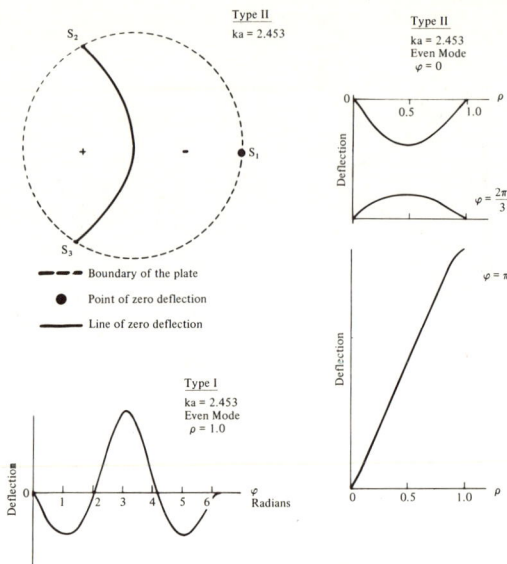

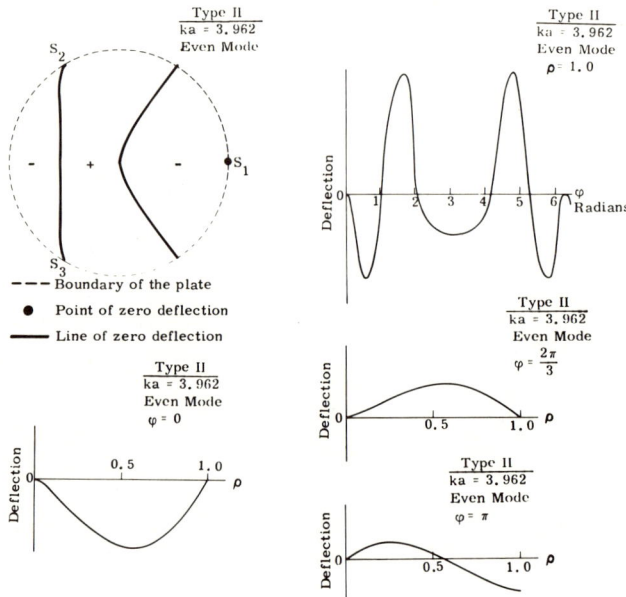

FIGURE 2 *Continued*

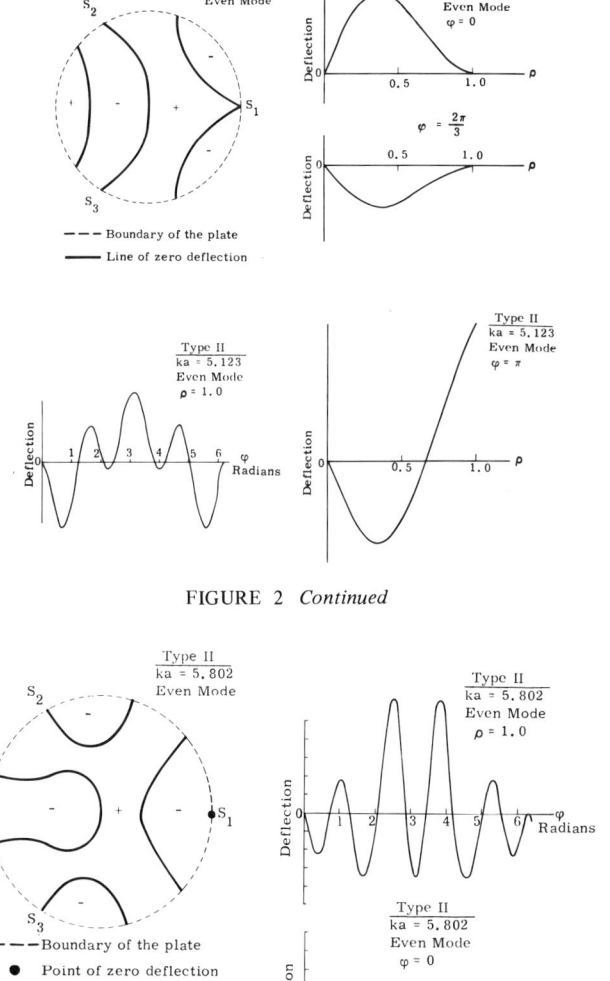

FIGURE 2 Continued

FIGURE 2 Continued

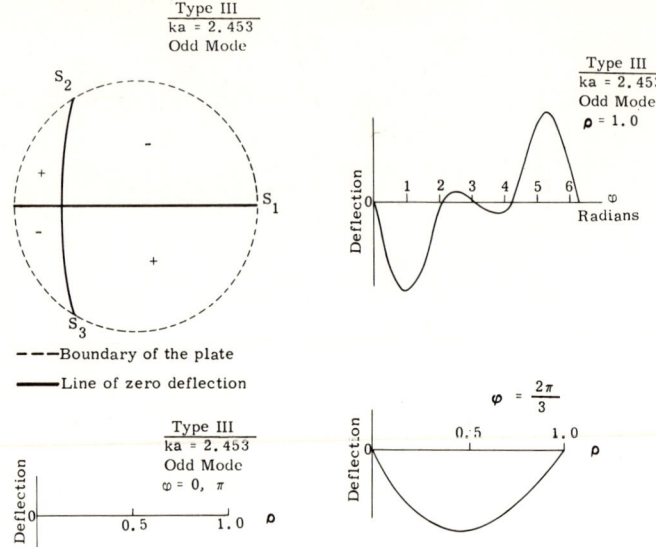

FIGURE 2 *Continued*

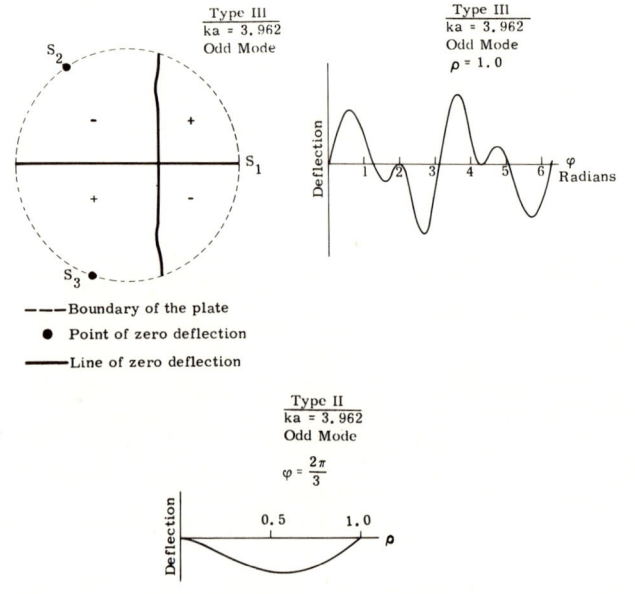

FIGURE 2 *Continued*

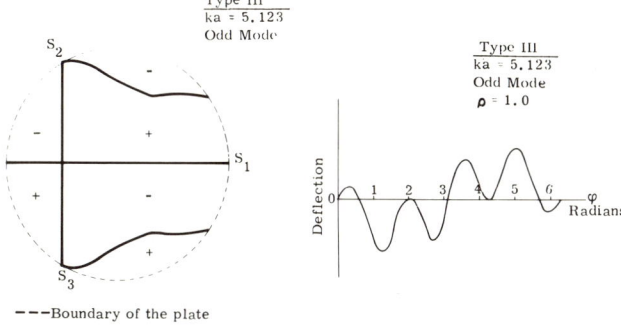

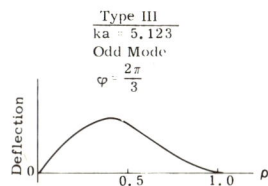

FIGURE 2  *Continued*

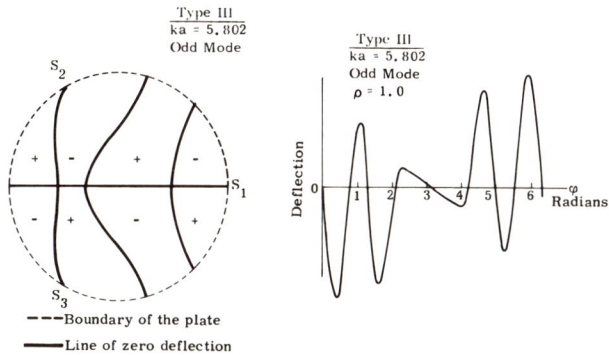

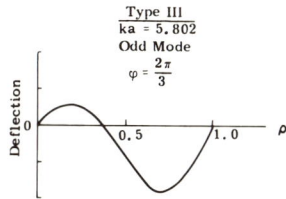

FIGURE 2  *Continued*

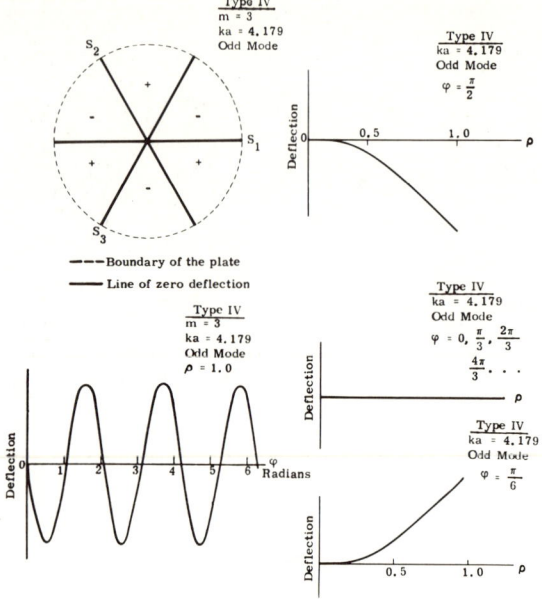

FIGURE 2 *Continued*

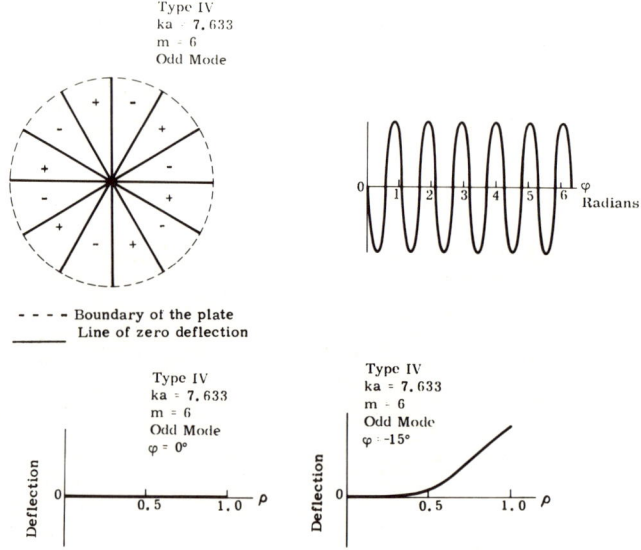

FIGURE 2 *Continued*

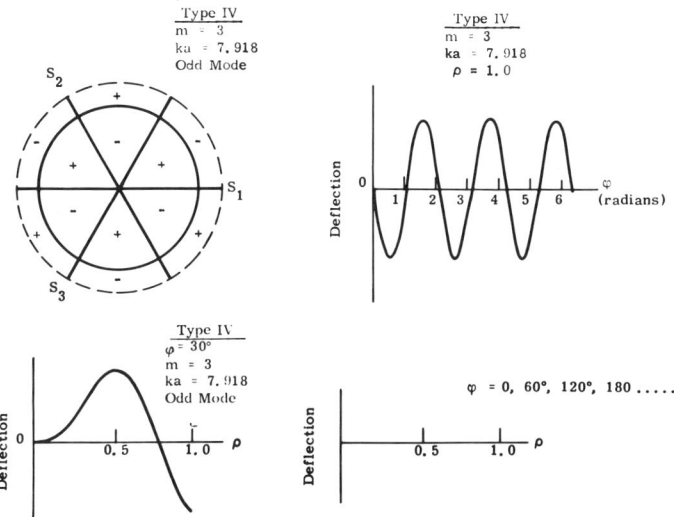

FIGURE 2 *Continued*

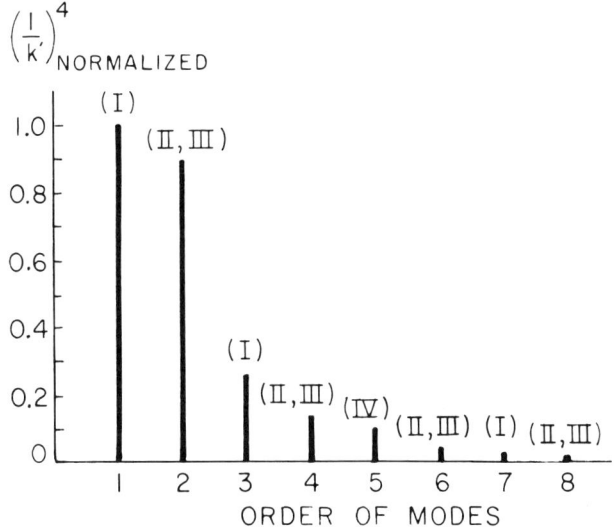

FIGURE 3 Distribution of $[1/(k')^4]$ for a flat plate ($\nu = 0.25$), where $k'$ is the normalized eigenvalue obtained by theoretical calculation. The value of $[1/(k')^4]$ is normalized so that it is 1 for the first mode. (Roman numerals in parentheses indicate the type of mode.)

TABLE 1  Properties of Mode Types I, II, III, and IV

*Type I Mode*
Mode shape has 120° symmetry and is symmetrical with respect to $x$ axis
All three support points exert the same force simultaneously
Three support points are required in order to excite this type of mode

*Type II Mode*
Mode shape is symmetrical with respect to $x$ axis
The support points exert force symmetrically with respect to $x$ axis
This is one of the degenerate modes (the frequency is the same as that of type III)
Three support points are required in order to excite this type of mode

*Type III Mode*
Mode shape is antisymmetrical with respect to $x$ axis
The support points exert force antisymmetrically with respect to $x$ axis
This is one of the degenerate modes (the frequency is the same as that of type II)
Only two support points ($S_2$, $S_3$) are required in order to excite this type of mode

*Type IV Mode*
Mode shape is antisymmetrical with respect to $x$ axis
The support points exert no force
This is a group of modes that are also free-plate vibrational modes
No support points are required in order to excite this type of mode

It is to be noted that $(k')^{-4}$ is a measure of how much each mode is excited for a unit disturbance in the particular mode, and it is a useful quantity in the investigation of the modal control technique for the space telescope mirror-control system.

Using the formula of Eq. (16c), we have calculated the eigenvalues of a free plate; they are shown in Table 2. Note that the eigenvalues of Leissa and our calculations agree quite well when the eigenvalues are large so that the condition $\lambda \gg m$ is satisfied.*

The asymptotic value given by Leissa in Eq. (2.17) on p. 10 [Leissa, 1969] is

$$k' \cong \pi/2(m + 2S).$$

For instance, the eigenvalues of $m = 1$, $S = 10, 11, 12$, and $m = 3$, $S = 10, 11, 12$ are very close to the values given by the asymptotic equation above.

---

* The eigenvalues of the free plate tabulated in Leissa's monograph [Leissa, 1969], p. 11, Table 2.6, are not exact but *approximate values*. These approximations are good when $\lambda \gg \kappa$ but not very accurate when the eigenvalue is small [see Eq. (2.15) and (2.16) on p. 10 of Leissa's monograph].

TABLE 2 Comparison of Eigenvalues ($k' = ka$), Values of Leissa's Table, and Values of Equation (16c)

| | Values of $k'$ | | | |
|---|---|---|---|---|
| | $m = 1$ | | $m = 3$ | |
| Order $S$ | Leissa's Values | Values of Eq. (16c) | Leissa's Values | Values of Eq. (16c) |
| 0 | | 1.917 | 3.571 | 4.179 |
| 1 | 4.518 | 5.048 | 7.291 | 7.918 |
| 2 | 7.729 | 8.256 | 10.600 | 11.216 |
| 3 | 10.903 | 11.435 | 13.821 | 14.452 |
| 4 | 14.024 | 14.600 | 17.015 | 17.656 |
| 5 | 17.218 | 17.757 | 20.203 | 20.844 |
| 6 | 20.368 | 20.907 | 23.363 | 24.015 |
| 7 | 23.516 | 24.059 | 26.526 | 27.183 |
| 8 | 26.663 | 27.207 | 29.685 | 30.346 |
| 9 | 29.809 | 30.354 | 32.841 | 33.504 |
| 10 | | 33.498 | | 36.659 |
| 11 | | 36.645 | | 39.813 |
| 12 | | 39.789 | | 42.965 |

## III. Computer Solutions

### A. GRID PATTERN DESIGN AND GRID-POINT NUMBERING SYSTEM

The presence of the strong discontinuity in the boundary condition on the circumference of the circular plate and its effect on the accuracy of the final results of the eigenvalues and eigenfunctions require that careful consideration be given to the selection of the grid pattern. The triangular grid pattern, which is frequently used and is shown in Figure 4, does not provide the regular pattern along the circumference. The design effort to produce the appropriate grid pattern resulted in the polar grid pattern shown in Figure 5. The polar grid pattern is most natural and compatible to the circular geometry of the mirror structure.

Radial spacing between two adjacent grid circles has been so chosen as to make the aspect ratio of the quadrilateral elements small, resulting in a nearly square element. This is desirable for maintaining accuracy of analysis.

Let $\bar{r}_n$ be the radius of the $n$th grid circle as shown in Figure 6. In terms of the outer diameter ($D_1$) and inner diameter ($D_2$) of the mirror plate, $\bar{r}_n$ is given by

$$\bar{r}_n = (\bar{c}_1 D_1 + \bar{c}_2 D_2)/2, \tag{22}$$

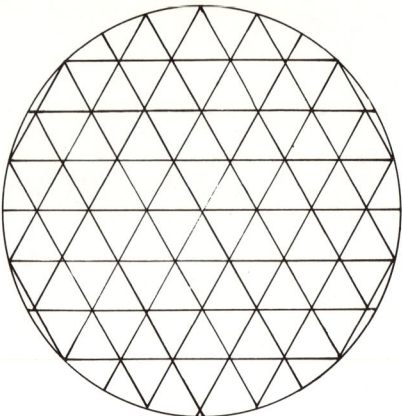

FIGURE 4 Triangular grid pattern.

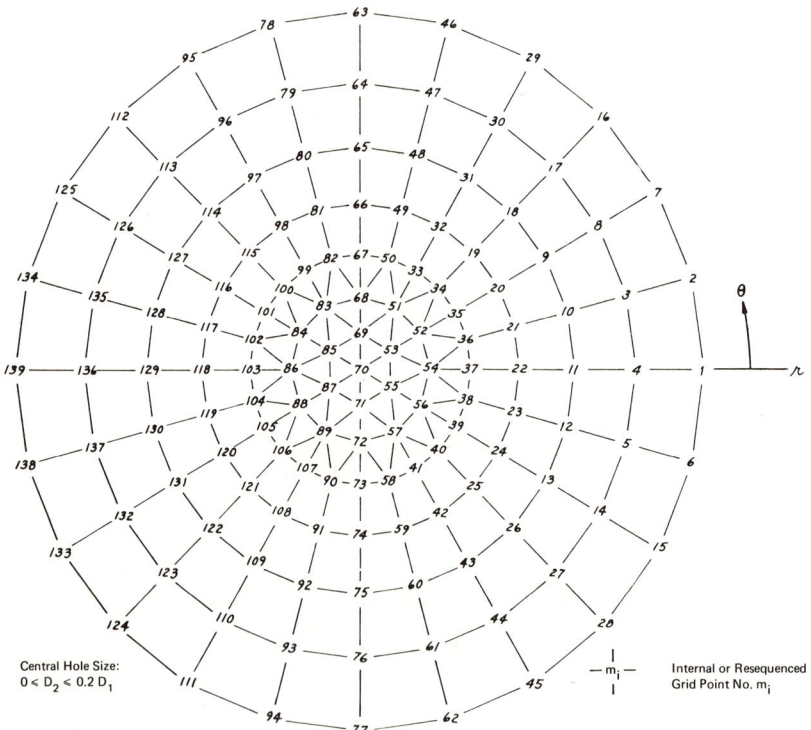

FIGURE 5 Proposed grid pattern and grid-point numbering system.

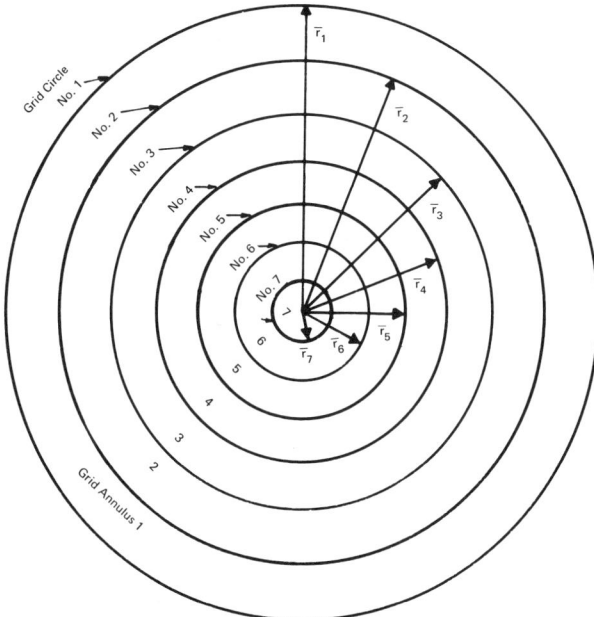

FIGURE 6 Mirror grid circles and grid annuli ($\bar{r}_8 = 0$ for the center, considered as grid circle No. 8).

where $\bar{c}_1$ and $\bar{c}_2$ are constants defined in Table 3. These constants have been so chosen that the resulting radial spacing between two successive grid circles keeps the aspect ratio of the plate elements in the neighborhood of unity. In other words, the deviations of the quadrilateral and triangular plate elements from the respective square and equilateral shapes are small, as mentioned earlier.

In the central region of the mirror, the plate elements are approximately equilateral triangles. If the radial rays were extended to the center of the mirror, the plate elements in this region would be quadrilaterals of high aspect ratio and nearly wedge-shaped triangles, and it is desirable to avoid such shapes.

The grid point numbering system of Figure 5 creates a stiffness matrix with a narrow band and no active columns as used in NASTRAN. In general, the computation time is less when the chosen grid-point sequence results in a narrow bandwidth and a small number of active columns. As a rule of thumb, the semibandwidth of a stiffness matrix is proportional to the maximum difference between any two connected grid-point sequence numbers. (The semiband is defined as the maximum number of

TABLE 3 Constants $C_1$ and $C_2$ of Equation (22)

| $n$ | $\bar{r}_n$ | Pattern 1 When $D_2 = 0$ (No Central Hole) (7 Annuli) | | Pattern 2 $0 < D_2 \leq 0.15D_1$ (Central Hole Present) (6 Annuli) | | Pattern 3 $0.15D_1 < D_2 \leq 0.2D_1$ (Central Hole Present) (5 Annuli) | |
|---|---|---|---|---|---|---|---|
| | | $\bar{c}_1$ | $\bar{c}_2$ | $\bar{c}_1$ | $\bar{c}_2$ | $\bar{c}_1$ | $\bar{c}_2$ |
| 1 | $\bar{r}_1$ | 1.00 | 0 | 1.000 | 0 | 1.00 | 0 |
| 2 | $\bar{r}_2$ | 0.80 | 0 | 0.781 | 0.219 | 0.76 | 0.24 |
| 3 | $\bar{r}_3$ | 0.62 | 0 | 0.586 | 0.414 | 0.54 | 0.46 |
| 4 | $\bar{r}_4$ | 0.46 | 0 | 0.412 | 0.588 | 0.34 | 0.66 |
| 5 | $\bar{r}_5$ | 0.32 | 0 | 0.257 | 0.743 | 0.16 | 0.84 |
| 6 | $\bar{r}_6$ | 0.20 | 0 | 0.120 | 0.880 | 0 | 1.00 |
| 7 | $\bar{r}_7$ | 0.10 | 0 | 0 | 1.000 | – | – |
| 8 | $\bar{r}_8$ | 0 | 0 | – | – | – | – |

columns included from the diagonal term in any row to the most remote term inside the band. Columns of a matrix containing nonzero terms outside the band are referred to as "active columns.") For our sequence, this maximum difference is 18.

The comparison of the normalized eigenvalues ($k'$) obtained from the analysis and NASTRAN computation is shown in Table 4. (It is to be noted that the eigenvalues of the same mode shapes should be compared.) By normalized eigenvalues, we mean the eigenvalues of the plate having the radius equal to 1 ($k' = ka$). Due to the relationship of Eq. (2), the percentage difference between $k'_{\text{theory}}$ and $k'_{\text{NASTRAN}}$ in Table 4 is smaller than the percentage difference between the corresponding frequencies. The values of $k'$ are indeed the appropriate quantity to compare because the solutions of the resonance equations produce $k'$ values, and the frequencies are the secondary quantity obtained from $k'$. The comparison of NASTRAN computation results for a flat plate having a 30-in. diameter and a 0.5-in. thickness, where $\nu = 0.2$, and that for a spherical plate having the similar dimensions with a 178-in. radius of curvature is shown in Table 5. We note that the eigenvalues of the spherical plate are larger than those of the flat plate in the lower modes; the differences between the two cases diminish as the order of modes increases. This trend is expected when one notes that as the order of the mode increases, the corresponding spatial period decreases; within the region of the spatial period of the higher order modes, the spherical plate is nearly flat.

The comparison of the mode shapes between the flat plate and the spherical plate described above shows that the nodal lines and the mode

TABLE 4  Normalized Eigenvalues of the Flat Circular Plate Simply Supported at Three Points on the Circumference ($\nu = 0.25$)

| Mode Number | Type I Mode (120° Rotational Symmetry Mode) | | Type II Mode (Conjugate Mode–Even) | | Type III Mode (Conjugate Mode–Odd) | | Type IV Mode (Member of Free-Plate Modes) | |
|---|---|---|---|---|---|---|---|---|
| | $k'$ theory | $k'$ NASTRAN | $k'$ theory | $k'$ NASTRAN | $k'$ theory | $k'$ NASTRAN | $k'$ theory | $k'$ NASTRAN |
| 1 | 2.385 | 1.86 | 2.453 | 1.87 | 2.453 | 1.875 | 4.179 | 3.54 |
| 2 | 3.289 | 3.11 | 3.962 | 3.61 | 3.962 | 3.613 | 7.918 | |
| 3 | 5.664 | | 5.123 | 4.48 | 5.123 | 4.485 | | |

TABLE 5  Comparison of the Computer-Calculated Values of the Eigenvalues for the Flat Plate and the Shallow Spherical Plate with Three Simple Supports along the Circumference

| Mode Number | Type I Mode (120° Rotational Symmetry Mode) | | Type II Mode (Conjugate Mode–Even) | | Type III Mode (Conjugate Mode–Odd) | | Type IV Mode (Member of Free-Plate Modes) | |
|---|---|---|---|---|---|---|---|---|
| | Flat Plate | Spherical Shell | Flat Plate | Spherical Shell | Flat Plate | Spherical Shell | Flat Plate | Spherical Shell |
| | $k'$ NASTRAN | $k'$ NASTRAN | $k'$ NASTRAN | $k'$ NASTRAN | $k'$ NASTRAN | $k'$ NASTRAN | $k'$ NASTRAN | $k'$ NASTRAN |
| 1 | 1.86 | 1.97 | 1.87 | 1.875 | 1.875 | 1.876 | 3.54 | 3.55 |
| 2 | 3.11 | 3.49 | 3.61 | 3.665 | 3.613 | 3.665 | | |
| 3 | | | 4.48 | 4.62 | 4.485 | 4.62 | | |

shapes are the same for all practical purposes. Sometimes, though not often, some mode shapes, especially the nodal lines, of the conjugate modes (types II and III) appear a little distorted. This occurs mostly because the computer outputs the mode shapes that are linear combinations of the two conjugate modes having the same eigenvalue. Experience with various mode shapes indicates that the mode shapes are not strongly dependent on the exactness of the eigenvalues.

B. CURVILINEAR BICUBIC SPLINE FIT INTERPOLATION SCHEME

This section gives a brief summary of the curvilinear bicubic spline fit interpolation scheme and some typical examples that illustrate the performance of the new scheme. The detailed analytical development will be presented in another paper [Chi, 1973].

The computer solution produces the amplitude of the mode shapes at the grid points. A reliable interpolation technique is then required to find the values between the grid points, and the bicubic spline fit interpolation technique is used for this purpose. When the system geometry is circular, as the primary mirror is, the currently available bicubic spline fit in a rectangular coordinate system [DeBoor, 1962; Vogl, 1971] is not conveniently adaptable for the circular boundary region. This fact is especially critical when the system has highly discontinuous boundary conditions such as the support point around the rim of the circular mirror. Furthermore, the choice of the polar grid pattern of Figure 5 over the pattern of Figure 4 requires that a new interpolation scheme be developed.

The bicubic spline fit interpolation technique, which is becoming increasingly popular, is adequate in many instances of engineering design; it precisely interpolates functions up to the third degree in each coordinate axis and is sufficient to accommodate the physical surface ($\psi$) that is expressible by the double Laplacian equation

$$\nabla^2 (\nabla^2 \psi) = 0, \qquad (23)$$

provided that the region is subdivided into sufficiently small patch areas.

We have extended the currently available bicubic spline fit scheme in the rectangular coordinate system to the scheme suitable for the polar coordinate system. In doing so, we have added a feature that has physical significance. It is to be noted that physical laws are based on the length as the basic unit (all the differential equations describing the physical phenomena are differentiated in terms of the length).

Therefore, we have taken the radial length ($r$) and the arc length ($\overline{r\theta}$) as the basis of the interpolation.

There are two approaches to take when developing the interpolation function in the polar coordinate system: one is to express the interpolation function in terms of ($\bar{r}$) and ($\bar{\theta}$), as shown in Figure 7(a), and the other is in terms of ($\bar{r}$) and ($\overline{r\theta}$), as shown in Figure 7(b). In the first approach, the formulation essentially requires the replacement of $x$, $y$ variables in the rectangular coordinate system by $r$, $\theta$ variables, respectively, in the polar coordinate system. In most applications involving the physical system, however, the latter formulation is more meaningful and implicitly takes the patch shape to be that shown in Figure 7(b) instead of that in Figure 7(a). The latter approach is used in the present scheme.

The region of the polar patch is defined by

$$R_{ij}: r_i \leqslant r < r_{i+1} \text{ and } \theta_j \leqslant \theta < \theta_{j+1},$$

where (from Figure 8),

$$r_i = \alpha, \quad r_{i+1} = \beta,$$
$$\theta_j = \eta, \quad \theta_{j+1} = \zeta$$

(24)

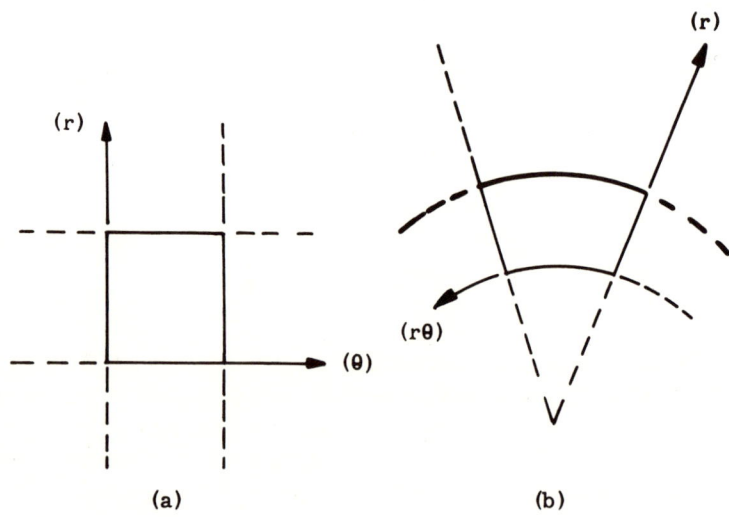

FIGURE 7 (a) Shape of the patch when rectilinear coordinate ($r\theta$) is used. (b) Shape of the patch when rectilinear coordinate ($r$, $r\theta$) is used.

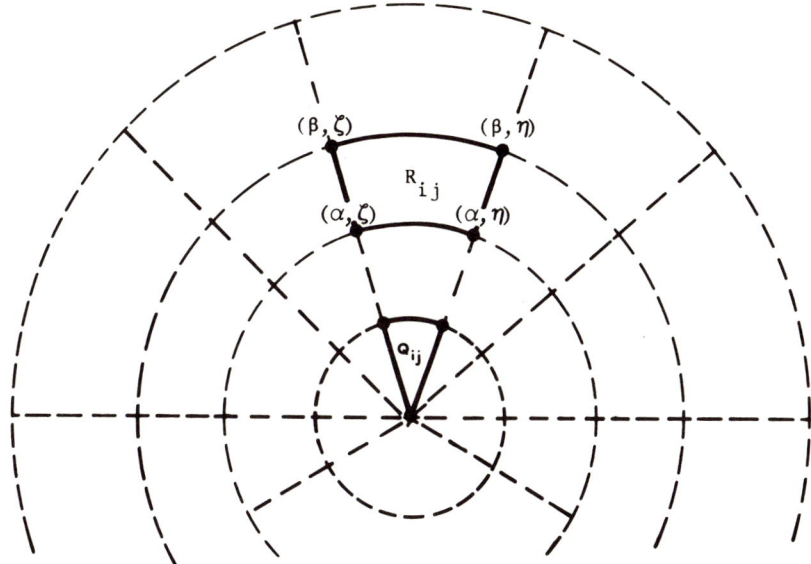

FIGURE 8  Designation of patch regions $R_{ij}$ and $Q_{ij}$.

and within each patch is defined the curvilinear bicubic polynomial,

$$T(r,\theta) = \sum_{m=0}^{3} \sum_{n=0}^{3} C_{mn} (r-\alpha)^m (r\theta - \alpha\eta)^n, \tag{25}$$

where $\alpha, \eta$ are constants.

The 16 members of $C_{mn}$ of Eq. (25) are to be found in each patch by imposing the requirements that $\partial^2 T/\partial r^2$ and $(\partial/r\partial\theta)(\partial T/r\partial\theta)$ are continuous at the four grid points of the patch, $R_{ij}$ (or three grid points of the wedge-shaped patch, $Q_{ij}$), shown in Figure 8.

Imposition of the above conditions results in three sets of recursion formulas:

$$\left. \begin{array}{l} b_{j-1} P_{i,j-1} + a_j P_{ij} + b_j P_{i,j+1} = x_{i,j}, \quad j = 0,1,2,\ldots N \\ f_{i-1} q_{i-1,j} + e_i q_{i,j} + f_i q_{i+1,j} = y_{i,j}, \quad i = 0,1,2,\ldots M \\ b_{j-1} s_{i,j-1} + a_j s_{i,j} + b_j s_{i,j+1} = z_{i,j}, \quad j = 0,1,2,\ldots N \end{array} \right\}, \tag{26}$$

where (see Figure 9)

$$P_{ij} = \frac{\partial T(r_i,\theta_j)}{\partial(r\theta)} \qquad s_{ij} = \frac{\partial^2 T(r_i,\theta_j)}{\partial r \partial(r\theta)}$$

$$q_{ij} = \frac{\partial T(r_i,\theta_j)}{\partial r} \qquad T_{ij} = T(r_i,\theta_j)$$

$$a_j = 2\left[\frac{k_{j-1} + k_j}{(k_{j-1})(k_j)}\right]$$

$$b_j = \frac{1}{k_j}$$

$$e_i = 2\left[\frac{h_{i-1} + h_i}{(h_{i-1})(h_i)}\right]$$

$$f_i = \frac{1}{h_i}$$

$$x_{i,j} = \left(\frac{3}{r_i}\right)\left[\frac{1}{(k_j)^2}(T_{i,j+1} - T_{i,j}) + \frac{1}{(k_{j-1})^2}(T_{i,j} - T_{i,j-1})\right]$$

$$y_{i,j} = \frac{3}{(h_i)^2}(T_{i+1,j} - T_{i,j}) + \frac{3}{(h_{i-1})^2}(T_{i,j} - T_{i-1,j})$$

$$z_{i,j} = \left(\frac{3}{r_i}\right)\left[\frac{1}{(k_j)^2}(q_{i,j+1} - q_{i,j}) + \frac{1}{(k_{j-1})^2}(q_{i,j} - q_{i,j-1})\right]$$

$$- \frac{3}{(r_i)^2}\left[\frac{1}{(k_j)^2}[T_{i,j+1} - T_{i,j}] + \frac{1}{(k_{j-1})^2}(T_{i,j} - T_{i,j-1})\right]$$

$$\left.\begin{matrix}\\\\\\\\\\\\\\\\\\\\\\\\\\\end{matrix}\right\} \cdot (27)$$

The matrix of coefficients in each set of the simultaneous equations is a symmetrical matrix; $p_{i,j}$, $q_{i,j}$, and $s_{i,j}$, obtained from Eq. (26), are used to find the coefficients $C_{mn}$ in Eq. (25).

In general, the recursion formulas of Eq. (26) require the boundary conditions at both ends of the line over which the interpolation is performed. These boundary conditions are either the derivative at the end grid points or the amplitude at an extra grid point outside the line over which the interpolation is performed. Usually, these boundary conditions are not known, and arbitrary (or some reasonable) boundary conditions are then imposed in order to proceed with the interpolation. Consequently, these arbitrarily imposed boundary conditions are the most important source of error in the interpolated values, and this error is most pronounced in the region close to the boundaries.

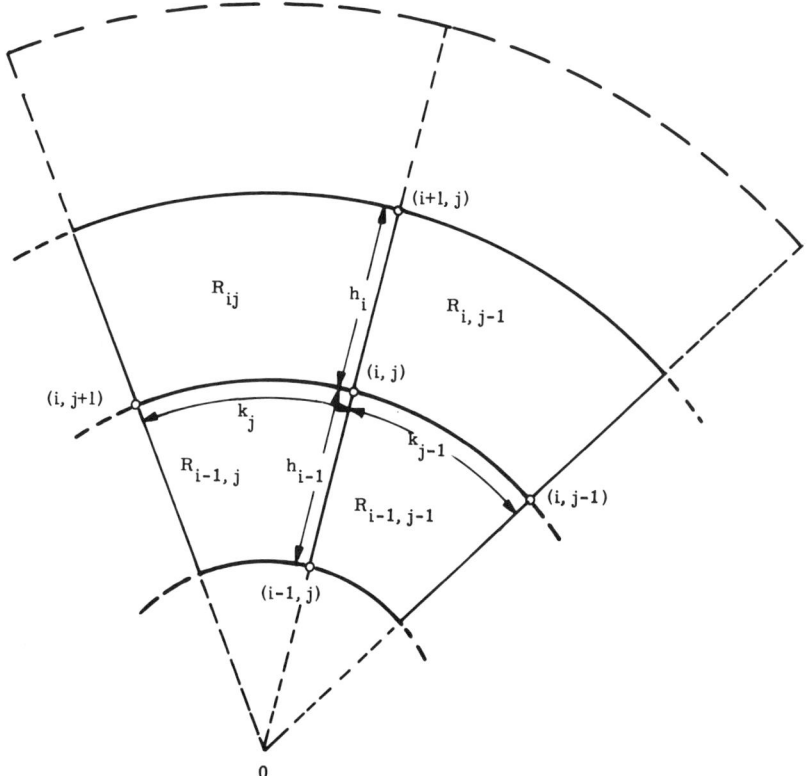

FIGURE 9 Definition of the grid point ($ij$), $h_i$, and $k_j$.

In the polar coordinate system, the boundary conditions are needed in $\theta$ direction because, as shown in Figure 10 (top), the boundary condition for the endpoint $N$ is the amplitude at the grid point 0, and the boundary condition for the endpoint 0 is the amplitude at the grid point $N$. The fact that arbitrary boundary conditions need not be made is an advantage when performing the interpolation in the polar coordinate system.

When the interpolation is performed in $r$ direction, boundary conditions must be specified at the endpoints $N$ and $M$, as shown in Figure 10 (bottom). The performance of the curvilinear bicubic spline is illustrated by performing interpolations for two test functions: one is a hemisphere (radially symmetrical function), and the other is a displaced hemisphere (asymmetrical function).

The isometric drawings of Figures 11 and 12 show a hemisphere and

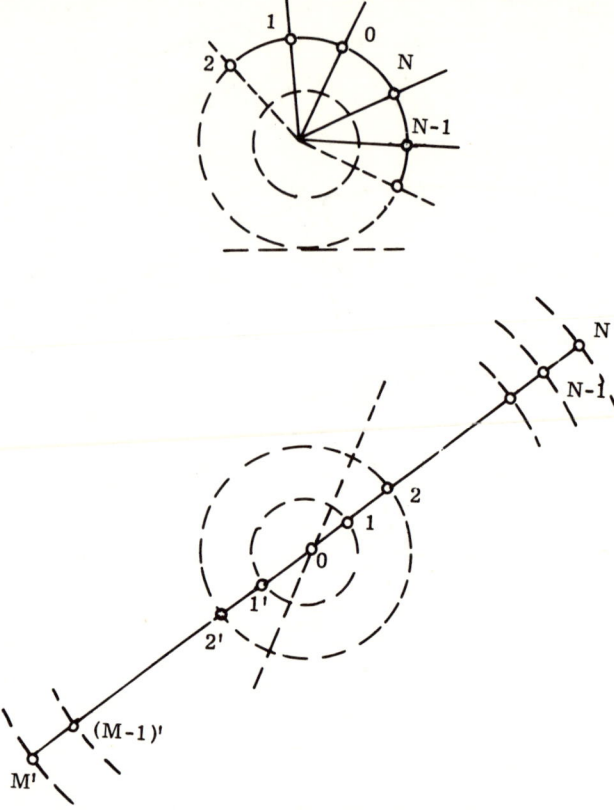

FIGURE 10 (Top) Boundary conditions for the recurrence formula in $r\theta$ direction. (Bottom) Boundary conditions for the recurrence formula in $r$ direction.

a displaced hemisphere, respectively, which are based on the interpolated values between the polar grid points. The functions representing the hemispheres were sampled at the grid points of the polar grid pattern shown in Figure 5. Specifically, 5 equally spaced rings (and the center point) and 12 equally spaced diameter lines were used. The isometric plotter scans the functions of the hemispheres in the rectangular coordinate system and requires the interpolated values along the straight lines in the $x$-$y$ coordinate system. These drawings serve to illustrate the working of the interpolation on the qualitative basis.

This work was supported by the National Aeronautics and Space Administration under contracts NAS1-9759 and NAS1-11205.

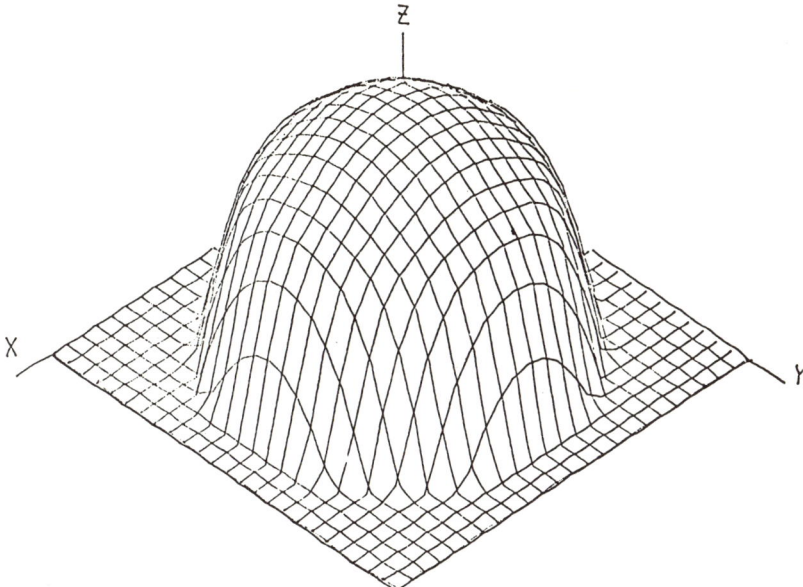

FIGURE 11  Isometric drawing of a hemisphere based on the curvilinear bicubic spline fit interpolated values.

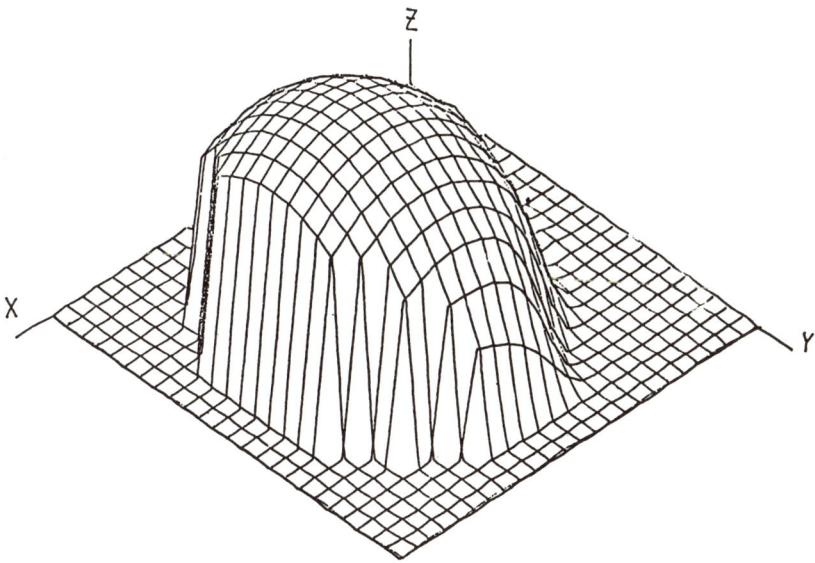

FIGURE 12  Isometric drawing of a hemisphere that is displaced 40% of the radius on the $x$ direction.

## References

Chi, C., "Curvilinear Bicubic Spline Fit Interpolation Scheme," *Op. Acta* (1973).
Chi, C., *Am. Inst. Aeronaut. Astronaut. J. 10,* 142 (1972).
Creedon, J. F., and A. G. Lindgren, *Automatica 6,* 643 (1970).
DeBoor, C., *J. Math. Phys. 41,* 212 (1962).
Howell, W. E., this volume, p. 239 (1974).
Leissa, A. W., *Vibration of Plates,* U.S. Govt. Printing Office, Washington, D.C. (1969).
Vogl, T. P., A. K. Rigler, and B. R. Canty, *Appl. Opt. 10,* 2513 (1971).

W. E. HOWELL

# RECENT ADVANCES IN OPTICAL CONTROL FOR LARGE SPACE TELESCOPES

## Introduction

If a large space telescope is to be as effective as possible and is to yield as much scientific information as possible it must have optics that perform at their theoretical maximum throughout their lifetime. We may take two approaches to this requirement. First, we may fabricate a mirror and anticipate variations in the figure caused by a change from 1 $g$ to zero $g$ and from thermal loading caused by changes in equipment and usage and changing orientation relative to the sun. As a second approach, we may recognize these as somewhat random, unpredictable effects and attempt to regulate the surface of the mirror to the desired diffraction-limited quality. It is this second approach that has been pursued at Langley Research Center and through various contractual efforts with industry and that is discussed here.

To implement this technique requires that the deviation of the mirror figure from the desired shape be measured, control signals generated, and forces applied to the mirror to align it physically to the desired shape. This is depicted in Figure 1 for one concept. A laser interferom-

---

The author is at the NASA Langley Research Center, Hampton, Virginia 23365.

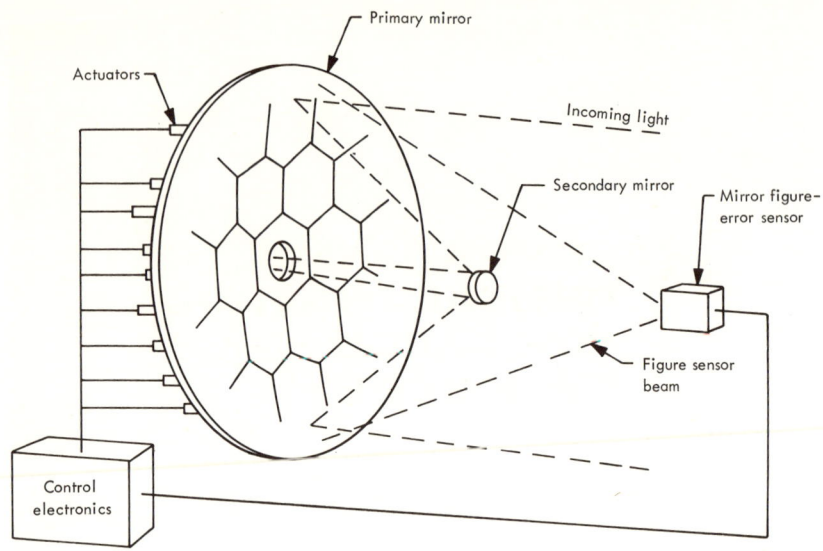

FIGURE 1 Schematic representation of a system for control of the optical surface of a telescope primary mirror.

eter is placed at the center of curvature of the primary mirror. From this vantage point it can measure the deviation of the mirror from the desired shape and relay these signals to the control electronics, which processes these raw data. The control electronics generates and relays the proper commands to a set of actuators on the rear of the mirror to bring it to the desired shape. The mirror itself may take one of two forms. It may be a segmented mirror, as depicted in Figure 1, in which each segment must be properly aligned with the others to effectively produce one large mirror, or, under an alternate arrangement, the mirror would be a single piece and the actuators would stress the mirror to the desired shape. Both techniques have been implemented and evaluated.

The four major areas of the system are, therefore, the mirror and its support, the control laws, the figure sensor, and, finally, the actuators. Each of these areas will be discussed starting with the alternate mirror concepts.

## Experimental Results

### SEGMENTED MIRROR

The experimental segmented mirror is shown in Figure 2. It began as a 56-cm (22-in.)-diameter blank, 10.2 cm (4 in.) thick. It was ground and

FIGURE 2 The 56-cm-diameter segmented mirror with actuators and white-light interferometers.

polished to an $f/3.5$ sphere, which was then cut into three pie-shaped segments as shown in the photograph. Each segment is controlled in two axes of tilt and one of focus by three actuators, one of which is visible in the upper left-hand corner. The devices on the edge of the front surface are white-light interferometers to resolve the integral wavelength ambiguity inherent in the laser interferometer. For a detailed description of its operation, see Robertson *et al.* [1966] and Robertson [1967].

Several techniques were used to evaluate the performance of this actively controlled system; however, the photographs of Figure 3 give a good indication of its performance. For this test, an illuminated 0.0005-cm (0.0002-in.) pinhole was placed just off the optical axis of the mirror and the return image examined. During the first part of the test, the segments were intentionally misaligned, which resulted in the three images shown on the left-hand side of the figure. Note the elliptical

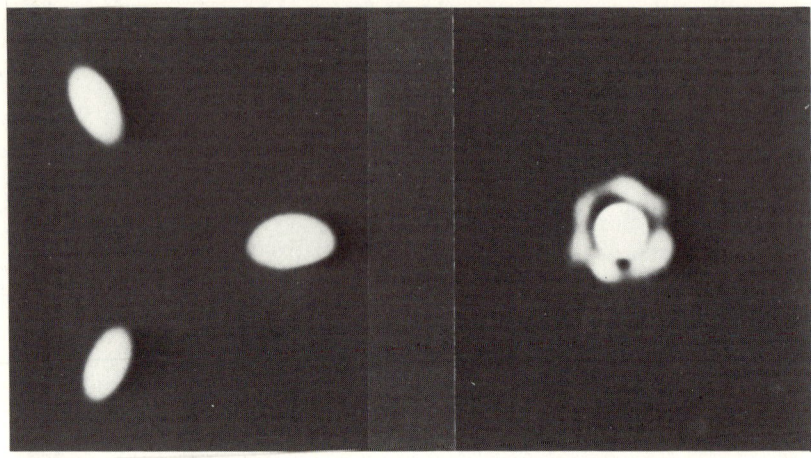

FIGURE 3 Images of point source located at center of curvature of segmented mirror. Left segments misaligned, right segments aligned.

shape of the images that result from the diffraction characteristics of the pie-shaped elements. During the second part of the test, the segments were aligned, which resulted in the characteristic Airy disk on the right. Note that some anomalies can still be seen in the first bright ring. Both of these photographs are reproduced at the same plate scale. From this and other data in Robertson [1966, 1967] it has been concluded that the rms error on this mirror was less than $\lambda/50$ ($\lambda = 0.6328$ $\mu$m) during the closed-loop operation of the active control system.

### FLEXIBLE MIRROR

The mirror used to investigate the second or flexible mirror approach is shown in Figure 4. This is a 76-cm (30-in.) diameter $f/3$, spherical mirror, which has had its thickness reduced (after figuring) to 1.27 cm (1/2 in.). There are three support points for this mirror located symmetrically around the edge and so arranged that they do not overconstrain the mirror. Fifty-eight actuators are evenly spaced over the rear of mirror on 9.46-cm (3.75-in.) centers. Each of these actuators is capable of exerting ±400 g (1 lb) of force on the mirror. The actuators are supported by the backing plate, which also provides the reaction force.

The results of this approach are summarized in Figure 5. In the upper left-hand side of the figure is an interferogram of the mirror taken

FIGURE 4  Flexible active-optics mirror.

shortly after the system was set up with no constraints applied by the actuators. Just below it is an image of a point source located, as before, at the center of curvature of the mirror. The interferogram indicates an rms error of approximately $\lambda/2$ ($\lambda = 0.6328$ $\mu$m). On the right-hand side of the figure are the corresponding photographs taken after the mirror was stressed into shape. Some dark areas can be distinguished in the interferogram, especially on the right side, indicating remaining errors; however, the image of the pinhole is of very high quality and greatly superior to the original image. The fine-line structure that may be visible in both interferograms is due to optics internal to the interferometer and is not a part of the error of the mirror. The final mirror error was less than $\lambda/50$. A complete description of this system and its characteristics is contained in Robertson [1970].

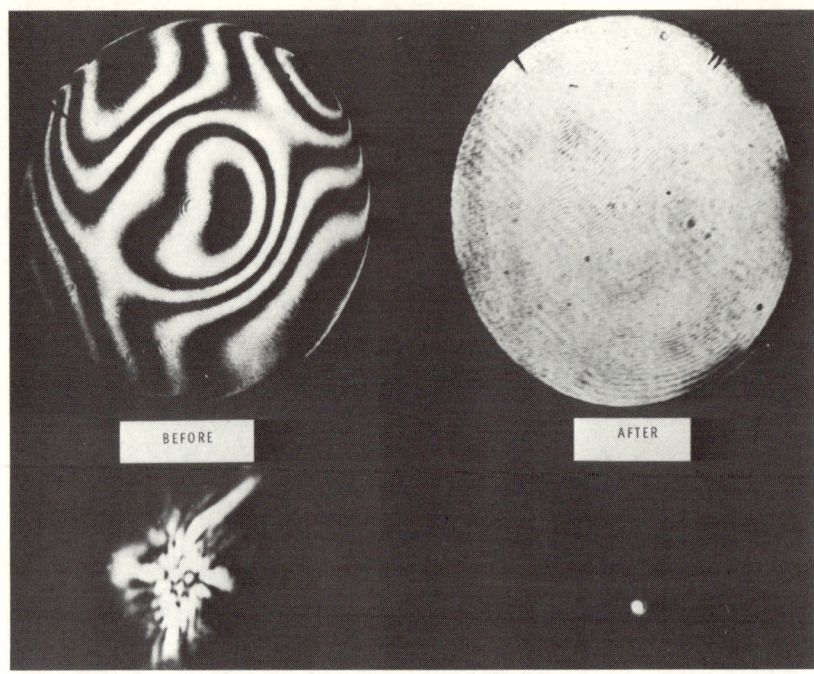

FIGURE 5 Flexible mirror interferograms and images. Left, as originally installed; right, after alignment by the control system.

## Theoretical Considerations for Flexible Mirror

MIRROR-CONTROL-SYSTEM CONCEPT

At this point the basic feasibility of either approach has been shown; however, there are some rather obvious questions that need answering. First is the question of tradeoff between the segmented-mirror approach and the single-piece concept. The segmented approach has the advantage that very large mirrors can potentially be orbited, since the segments could be stacked for launch and the telescope assembled in orbit. A disadvantage is that all segments are off-axis aspherics, which must be matched very precisely. The single-piece concept, however, need not be a radical departure from present mirror design and can probably attenuate errors to an even greater extent than the segmented concept. For these reasons, the major emphasis at Langley Research Center has been toward analyzing and developing the latter approach. The most critical questions concerning this latter approach are listed below:

1. What are the minimum number of actuators required to reach a given level of figure accuracy?
2. Where should these actuators be placed for maximum effect?
3. Under what conditions can we guarantee dynamic stability for the highly coupled, interacting system?

To answer these questions, a mathematical representation of the mirror and control system was needed. The development of this model is covered in detail in Creedon and Lindgren [1970]; however, the idea that was followed can be understood most easily by considering the two-dimensional example for a beam. First, the need to analyze and account for the dynamics of the problem led naturally to the representation of the beam in terms of its vibrational modes. For the simply supported beam in Figure 6 these are of the form

$$y_n = q_n \sin(n\pi/2)x; \quad n = 1, 2, 3, \ldots, \tag{1}$$

where

$y_n$ = deflection at point $x$ on the beam due to the $n$th mode,
$q_n$ = mode coefficient for the $n$th mode,
$n$ = mode number,
$L$ = beam length.

If it is assumed that this beam is disturbed by a uniformly distributed load, it will deflect as indicated in Figure 6 (top). This deflection may be

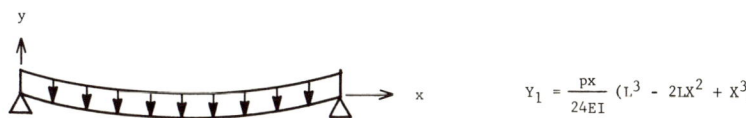

Deflection of a Beam Under a Disturbance Load of p kg/m

$$Y_1 = \frac{px}{24EI}(L^3 - 2Lx^2 + x^3)$$

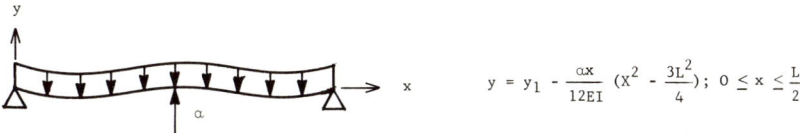

Deflection of a Uniformly Loaded Beam With Concentrated Corrective Load $\alpha$

$$y = y_1 - \frac{\alpha x}{12EI}\left(x^2 - \frac{3L^2}{4}\right); \quad 0 \leq x \leq \frac{L}{2}$$

FIGURE 6 Effect of correcting a beam disturbance by a concentrated load.

expressed in terms of the mode shapes of Eq. (1), and the first six coefficients of this series are listed in Table 1 under the heading $q^N$. To counter this disturbance, an actuator, which pushes up with a force $\alpha$, is placed at the center of the beam. The result of this operation is shown in Figure 6 (bottom), and the modal coefficients for the point load are tabulated in Table 1 under the heading $A^N$.

Now comes a critical point in the design: the selection of a control law, or how hard should you push with the force $\alpha$? At least three cases present themselves as reasonable alternatives:

*Case I*: Push until the center of the beam has zero deflection. This requires a force $\alpha = 0.625$ pL and results in 97% of the rms error being removed.

*Case II*: Push until mode one is zero. (This is referred to as modal control.) This requires a force $\alpha = 0.637$ pL and results in 98.5% of the rms error being removed.

*Case III*: Apply sufficient force to the beam to minimize the rms deflection. This is referred to as an optimal control law and requires a force $\alpha = 0.636$ pL approximately. The rms error is again reduced by 98.5%.

For this example the second case (modal control) comes within 1% of the optimal and, as will be shown later, provides considerable insight into problems of actuator placement and the number of actuators required.

Since it is not generally possible to describe the modes of a given mirror by elementary functions, it is necessary to use numerical procedures to obtain estimates of the mode shapes and eigenvalues. This results in a finite number of modes being used to represent the mirror. These can be obtained through computer programs such as SAMIS or NASTRAN

TABLE 1  Modal Coefficients for Beam

| Mode Number | $q^N$ Modal Coefficients for Disturbance (Uniform) Load | $a^N$ Modal Coefficients for Corrective (Point) Load |
|---|---|---|
| 1 | $0.6535 \times 10^{-2}$ pL | $-0.1026 \times 10^{-1} \alpha$ |
| 2 | — | — |
| 3 | $0.2689 \times 10^{-4}$ pL | $0.1267 \times 10^{-3} \alpha$ |
| 4 | — | — |
| 5 | $0.2091 \times 10^{-5}$ pL | $-0.1643 \times 10^{-4} \alpha$ |
| 6 | — | — |

[see Melosh and Christiansen, 1966; MacNeal, 1970]. They may also be obtained experimentally [Robertson, 1972]. The numerical results presented in this paper are based on data from SAMIS. When obtaining these estimates it is necessary to discretize the structural model of the mirror. The model used for this analysis is shown in Figure 7 and discussed at length in Creedon and Robertson [1969]. From this structural model, numerical estimates of the eigenvectors and eigenvalues were obtained for use in the control-system evaluation. The principal restriction of the discrete model is that it restricts the designer's freedom to the extent that actuator effects can only be evaluated at the node points of this model. For this model, therefore, there are 58 possible locations at which actuators can be placed.

MIRROR-CONTROL-SYSTEM DESCRIPTION

Figure 8 shows a block diagram of the mirror, figure error sensor, and actuators as they appear in a finite modal representation. The mirror itself is represented by the five blocks (matrices) labeled $H^N$ and $H^R$, $\lambda^N$ and $\lambda^R$, and $U^M$. The superscripts $N$, $R$, and $M$ have the relationship

$$N + R = M, \tag{2}$$

where

$N$ = number of controlled modes (numerically equal to the number of actuators),
$M$ = total number of modes used to model the mirror,
$R$ = remaining modes,
$U^M$ = matrix of eigenvectors.

In the physical world, the actuator forces ($\alpha^N$) are translated directly into mirror figure displacements ($W$); in the mathematical model, the $N$ forces are transformed by the $H^N$ and $H^R$ matrices into a set of force coefficients ($a^N$ and $a^R$), respectively, in the modal domain. These force coefficients are then transformed by the $\lambda$ matrices into the modal coefficients of displacement. These coefficients, generated by the control system, are summed with the coefficients representing the error that previously existed on the mirror, and the result, denoted by $C^N$ and $C^R$ is transformed by $U^M$ into the final displacement, $W$, according to the relationship

$$[U^M] \begin{bmatrix} C^N \\ C^R \end{bmatrix} = W^M. \tag{3}$$

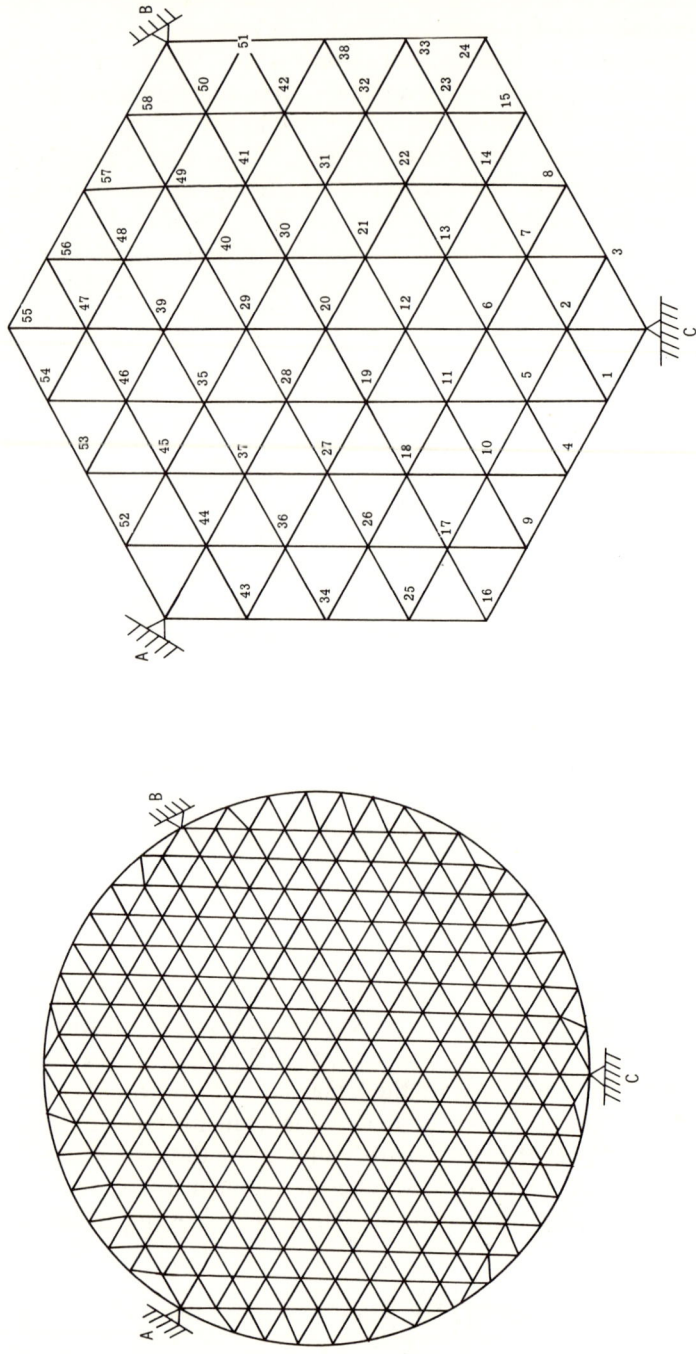

FIGURE 7 (a) Grid breakup used to analyze the structural response of the 76-cm thin mirror. Point A is constrained in $y$, $z$; B in $z$; C in $x$, $y$, $z$. (b) Grid arrangement and numbering used for control-system analysis.

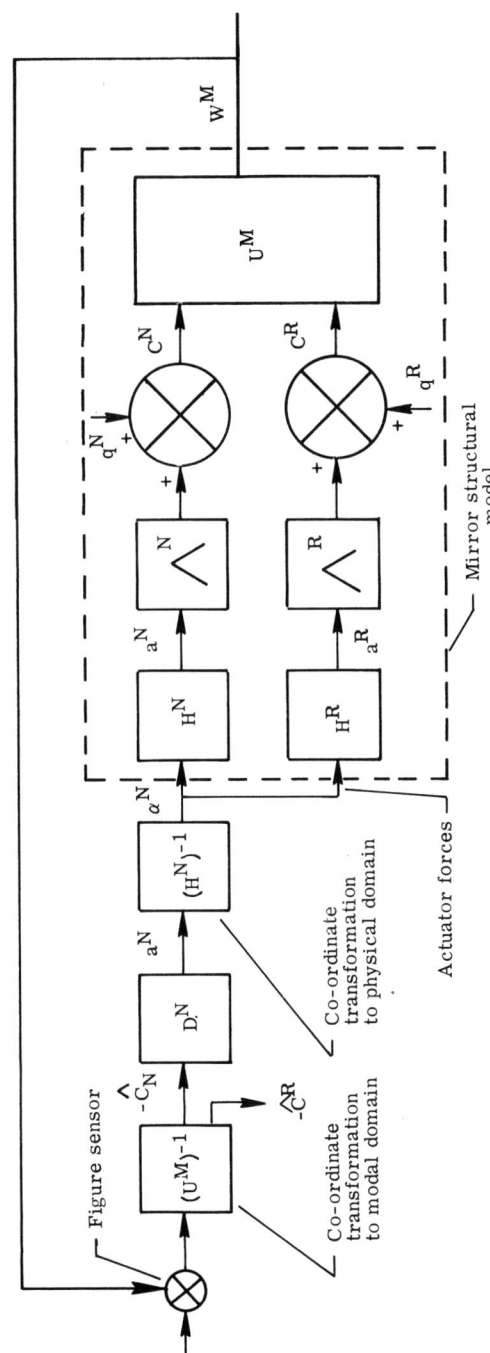

FIGURE 8 Block diagram of mirror and control system.

To combine this model into a control system requires a sensor to measure $W$. The output of this sensor is then transformed into the modal coordinate system by the proper transformation, which is to operate on its output by $[U^M]^{-1}$. Since $N$ actuators can control only $N$ modes, the subset of the $N$ selected modes to be controlled is usually all that is generated. In general, one would normally control only the first, or lowest, $N$ modes. The lowest mode is that eigenvector associated with the lowest eigenvalue, or, in general, the eigenvectors are ordered according to a hierarchy established by the eigenvalues.

The $N$ modes to be controlled are then fed through the dynamic compensation $D^N$ in which the proper gains and dynamic compensation are applied to each mode independently. If a type-one control system is used, as will be specified later, then each diagonal element of $D^N$ corresponding to one channel of the decoupled controller will contain an integration. By making the control system a type-one system, the error in the first $N$ modes is driven to zero. That is

$$C^N = 0; \quad i = 1, 2, 3, \ldots N. \tag{4}$$

This means that the final error in the mirror is made up of the algebraic sum of the externally induced error in the higher-order (uncontrolled) modes, and the error generated in these modes by the control system.

The output of $D^N$, denoted $a^N$, is in the modal domain. In fact, this output is a set of modal coefficients that describe the desired force patterns to be distributed on the mirror. To change these to discrete forces, which is the way they must be applied to the mirror, the values of $a^N$ must be transformed by multiplying by $[H^N]^{-1}$, which also accounts for the effect of the physical mechanism through which the actuator applies a load to the mirror. This completes the description of the control system. For a more complete and rigorous discussion see Creedon and Lindgren [1970].

ACTUATOR PLACEMENT CRITERION

In the example of the beam, it was fairly obvious that if you could use only one actuator then it should be placed at the center of the beam. It is also fairly obvious that if you have a complicated structure and associated mounting system, such as an astronomical mirror, and you are given several actuators that their placement for best results is anything but obvious; however, it is critical.

To fully optimize the actuator locations would require *a priori* knowledge of the mirror error; however, changes in this error can cause significant changes in the performance of the mirror. Since changes in the mirror error are expected during operation of the telescope, this deterministic approach is not suitable. An alternate approach is to place the actuators in such a manner that they can counter the error in the first $N$ modes but require that they generate minimum error in the higher-order, uncontrolled, modes. This is referred to as an uncorrelated treatment of the errors as opposed to the previous deterministic approach.

As shown in Creedon and Lindgren [1970], equations (40b) through (43b), the expected performance, $J$, may be determined by evaluating the expression

$$J = \sum_{i=1}^{N} \phi_i^2 \sigma q_i^2 + J_{R_0},\qquad(5)$$

where

$J$ = final mean-square expected error,

$J_{R_0}$ = mean-square error in the uncontrolled modes = $\sum_{i=N+1}^{m} \sigma q_i^2$,

$\sigma q_i^2$ = variance of the error in the $i$th mode,

$\phi_i^2$ = the diagonal elements of the matrix $\psi$ transpose $\psi$

$$\psi = \lambda^R H^R [\lambda^N H^N]^{-1}.$$

Here, of course, a finite number of modes, $M$, is assumed as opposed to an infinite number in Creedon and Lindgren [1970]. One would select that set of actuator locations that minimizes $J$; however, this semiempirical approach should be checked as many ways as possible to ensure that none of the assumptions leading to it have been violated. The two major assumptions are that the structural model is sufficiently fine and that the errors are truly uncorrelated.

Based on numerous examples, it has been found that locating the actuators near the node lines of higher-order modes tends to minimize the control-system-generated error. That is, if four actuators are to be used, they should be located on or near the node lines of modes 5, 6, 7, and so forth. For instance, in the example of the beam, placing the actuator in the center of the beam precludes generating any error in the even-numbered modes.

The technique of modal control was applied to the mirror of Figure 4. The results of this application are fully discussed in Robertson [1972].

### NUMBER OF ACTUATORS REQUIRED

Under the assumption that it is desirable to place actuators on the mirror in a pattern that generates minimum error in the uncontrolled modes, an estimate of the number of actuators required for a given final error can be obtained by answering the following question. If all the error in the first $N$ modes is removed by the control system and no additional error is generated, ($\phi_i^2 = 0; i = 1, \ldots N$) how good will the performance be? This estimate can be obtained by examining a plot of the rms error remaining in the higher-order modes versus $N$. Since the $\sqrt{J}$, Eq. (5), is the expected rms error, we get

$$J_1 = \sqrt{J} = \sqrt{J_{R_0}} = \sqrt{\sum_{i=N+1}^{M} \sigma q_i^2}; \quad N = 0, 1, 2, \ldots M, \tag{6}$$

where

$J_1$ = expected rms error of the mirror under the assumption of uncorrelated errors.

This plot for the error of the mirror in Figure 4 is shown in Figure 9. For instance, if four actuators were used ($N = 4$), an error of $0.433\lambda$ would be estimated as compared with the original error of $0.64\lambda$ rms. Of course, it remains to demonstrate that there exist actuator locations that meet these requirements.

### EXAMPLES OF ACTUATOR PLACEMENTS

For the mirror of Figure 4 and its structural model of Figure 7, numerous actuator arrangements were evaluated based on the previous theoretical considerations. For 1, 2, 3, and 4 actuators, all possible combinations of actuator patterns were tried. Those patterns that minimize Eq. (5) for each set of actuators are shown in Figure 10. The $\sigma q_i^2$ used in Eq. (5) were the $q_i^2$ of the mirror in Figure 4. An actuator arrangement for seven actuators is also shown; however, the search pattern was restricted to only 18 possible locations. These locations were chosen to be near the nodes of modes 8, 9, and 10.

Using the original error on the mirror, as seen in the left-hand side of Figure 5, the final errors were calculated. These errors are shown beside each actuator pattern. The top number of the trio represents the error predicted by Figure 9, the center number represents the calculated final error under the modal control law for the particular initial error of

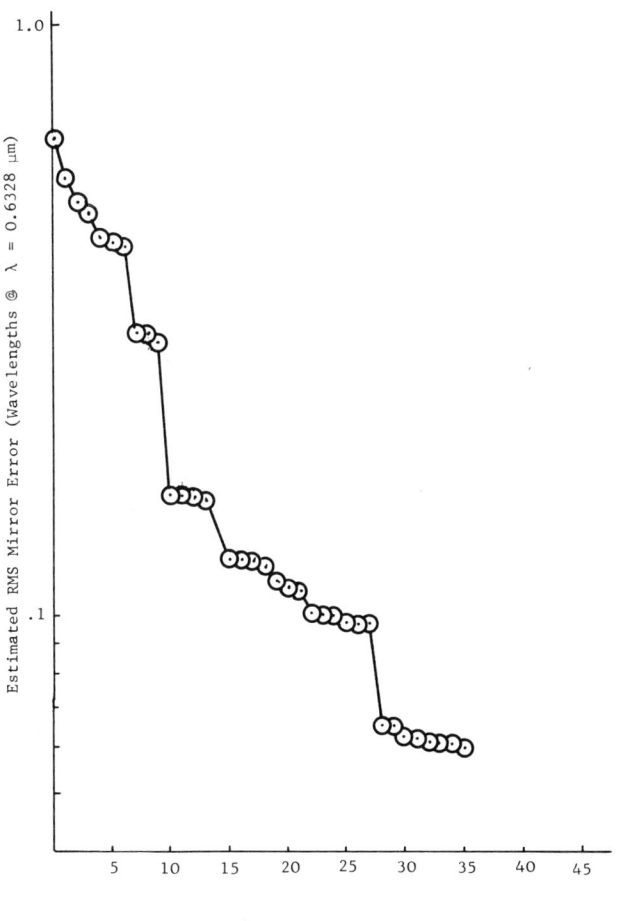

FIGURE 9 Estimated rms figure error (wavelengths) after $N$ actuators, assuming uncorrelated errors.

Figure 5, and the last number represents the error under the optimal control law. The units are rms wavelengths at 0.6328 $\mu$m and the original error was 0.64 wavelength rms.

These examples indicate that Figure 10 provides a good engineering estimate of the reduction in error a given number of actuators will achieve for a given error distribution. The seven-actuator case also shows that it is possible to find suitable locations without resorting to exhaustive searches as was done for 1, 2, 3, and 4 actuators.

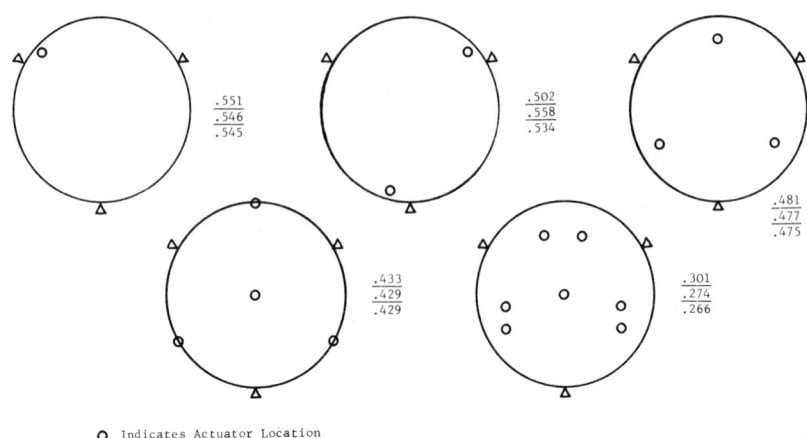

○ Indicates Actuator Location
△ Indicates Support Point

FIGURE 10  Actuator locations that generate minimum expected error in high-order modes. Top number indicates error predicted by Figure 10 (rms wavelengths), center number indicated calculated error under modal control law, last number indicates error under optimal control law.

Another point that Figure 9 shows is that it is desirable to concentrate most of the mirror errors in the lower-order modes so that the control system can have maximum effect. The most probable way for errors to get into the higher-order modes is during the figuring process when the mirror is worked with small laps.

SUMMARY OF DESIGN PROCEDURE

The suggested design procedure for an actively controlled mirror can be outlined as follows:

1. Select a candidate mirror and its mounting arrangement.
2. Structurally model the mirror and its mount for one of the available numerical programs and determine its eigenvectors and eigenvalues.
3. Estimate the variance of the expected errors in each mode ($\sigma q_i^2$) and plot $\sum_{i=N+1}^{M} \sigma q_i^2$ versus $N$ to use in estimating the number of actuators required (similar to Figure 9).
4. Select sets of appropriate trial actuator locations and evaluate Eq. (5), retaining for further analysis those sets of actuator patterns that yield minimal expected error.

## Figure Sensor Concepts

The figure error sensor, which determines the deviation of the mirror figure from the desired shape, was originally an automatic laser interferometer located at the center of curvature of the primary mirror. This device is described in Robertson *et al.* [1966]; a later version to work with aspheric primaries is discussed in Crane [1969]. A second concept, which allows the laser interferometer to be placed behind the primary mirror, is being developed and is described in Erickson [1970, 1972]. These two concepts are shown in Figure 11. The upper schematic shows the center-of-curvature sensor in which a collimated laser beam is split into a reference path that is modulated and an alternate path that is directed toward the primary mirror through a confocal lens. This latter beam is reflected by the primary mirror and returned to the confocal lens, which now recollimates the beam. The beam splitter recombines the two beams to form an oscillating interference pattern on the sensor (image dissector).

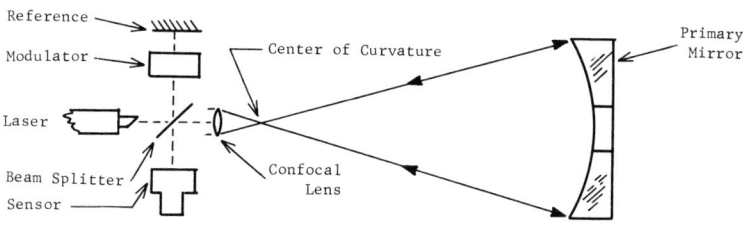

Center-of-Curvature Figure Error Sensor

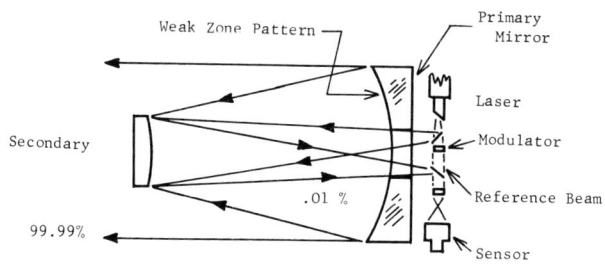

Zone Plate Figure Sensor

FIGURE 11 Conceptual diagrams of laser interferometer figure error sensors.

In the second method, a slightly different optical arrangement is used to direct the beam out of the central hole of the primary mirror. Almost all of this light is reflected out the telescope; however, a very light zone plate is impressed on the primary mirror and some of the energy is diffracted back to the optics behind the primary mirror, where it can be used to again form an interference pattern depicting the mirror errors. A third technique, which is presently receiving attention, is the use of the star image itself as a measurement source. Various tests of stellar-image quality are being considered as potential replacements for the previous techniques. Table 2 summarizes the above three methods as to type, basic measurement, their advantages and disadvantages.

### Actuators

The final part of the active control system is the actuator that positions the mirror—in the segmented case—or stresses it—in the flexible case. These two concepts require two different types of actuator. With segments, it is desirable to *position* the elements to within $\lambda/50$, while the flexible mirror concept requires a *force*. Two separate approaches to a space qualifiable position actuator were tried [LaFiandra, 1969; Fair-

TABLE 2 Figure Error Sensor Comparisons

| Type | Laser Interferometer at Center of Curvature | Laser Interferometer at Focal Plane | Stellar Source |
|---|---|---|---|
| Basic measurement | Primary mirror only | Primary + twice secondary | Primary + secondary |
| Major advantages | Well developed | Can probably test secondary directly | Tests complete system |
| | Easily interpreted | | |
| | | Tests complete system | Probably best long-term solution |
| Major problems | Requires long structure | Requires primary to be engraved | Undeveloped |
| | Secondary obscures primary | Only partially developed | Interacts with cont. system |
| | Must test aspheric wavefront | | Usable only on bright stars |
| | Cannot see final error | | |

child Hiller, 1968, 1969]; however, they met with only limited success. In both approaches, the actuators had the characteristic that when commanded to move in one direction they moved in the opposite direction first, then in the desired direction. While not intolerable, this is undesirable. There has been no attempt to develop a space-qualifiable force actuator. The differential spring actuators used in the laboratory experiment are covered in Robertson *et al.* [1966] and Robertson [1970] and are significantly different from the piezoelectric stepping concepts cited in LaFiandra [1969] and Fairchild Hiller [1968, 1969]; however, they are bulky, use considerable power, and are not ideal candidates for space use. This author feels that the lack of a good, sound concept for a space-qualifiable actuator is presently the weak point in an extremely sound and desirable system.

## Conclusions

The basic components and technology needed to analyze and actively control telescope mirrors has been developed and demonstrated. Refinement of components and techniques is required to realize the full potential of a working system; however, it is possible today to build large telescope mirrors of diffraction-limited quality and guarantee that quality throughout the operating environment by using active control.

## References

Crane, R., Jr., Advanced Figure Error Sensor, Final Rep. NASA Contr. NAS12-681, The Perkin-Elmer Corp., Norwalk, Conn. (Sept. 1969).
Creedon, J. F., and A. J. Lindgren, *Automatica* 6, 643 (1970).
Creedon, J. F., and H. J. Robertson, *IEEE Trans. Aerospace Electron. Syst.* AES-5, 287 (1969).
Erickson, K. E., Investigation Monitoring and Control of Large Telescope Performance, NASA CR-111811, Keuffel & Esser Co. (Sept. 1970).
Erickson, K. E., Fabrication and Evaluation of a Weak Zone Plate for Monitoring Performance of Large Orbiting Telescopes, NASA CR-112080, Keuffel & Esser Co. (July 1972).
Fairchild Hiller, Development of a Micro-Inch Actuator, NASA CR-66626 (1968).
Fairchild Hiller, Development of Micro-Inch Actuator, NASA CR-66795 (Apr. 1969).
LaFiandra, C., Research and Development Study of a Peristaltic Action Microinch Actuator, NASA CR-1658, The Perkin-Elmer Corp., Norwalk, Conn. (Sept. 1969).
MacNeal, R. H., ed., *The NASTRAN Theoretical Manual*, NASA SP-221 (Sept. 1970).

Melosh, R. H., and H. N. Christiansen, Structural Analysis and Matrix Interpretive System (SAMIS) Program, Tech. Rep., Jet Propulsion Lab., Calif. Inst. of Technol., Pasadena, Calif., NASA TM 33-311 (Nov. 1966).

Robertson, H. J., Active Optical Systems for Spaceborne Telescope, Final Rep., Vol. II, NASA CR-66489 (Dec. 1967).

Robertson, H. J., Development of an Active Optics Concept Using a Thin Deformable Mirror, NASA CR-1593 (Aug. 1970).

Robertson, H. J., Evaluation of the Thin Deformable Active Optics Mirror Concept, NASA CR-2073 (June 1972).

Robertson, H. J., R. Crane, and H. S. Hemstreet, Active Optical System for Spaceborne Telescopes, Final Rep., NASA CR-66297 (Oct. 1966).

FRED STEPUTIS and HAL NYLANDER

# LARGE SPACE TELESCOPE SUPPORT SYSTEMS MODULE AND ORBITAL OPERATIONS

## Introduction

Orbital support and operation of the Large Space Telescope (LST) is an appropriate subject for development in the next phase of the U.S. space program, which will be dominated by the use of the Space Shuttle. The LST will be one of the Space Shuttle's prime payloads and will take advantage of Shuttle capabilities for launch, on-orbit checkout, and periodic maintenance. Design studies have been conducted for a Support Systems Module (SSM) for the LST. Such a module will have application to a variety of future astronomy programs; however, the concept discussed herein is specifically for support of the LST.

The LST is a 3-m diffraction-limited Ritchey-Chrétien Cassegrain optical system with a primary focal ratio of $f/2.2$ and a system focal ratio of $f/12$. Instrumentation consists of imagers and spectrometers; spectral coverage is from 900 to 20000 Å. Observations will be made of stellar objects, planets, and possibly comets or other targets of opportunity. Operational life of the LST will be 15 years, with a launch by the Space

---

The authors are at the Martin Marietta Corporation, Denver Division, Denver, Colorado 80201.

Shuttle. It will be maintained by the Space Shuttle crew, on scheduled and nonscheduled flights. Compared with previous orbiting astronomy vehicles, the LST is larger; has more critical requirements for pointing, stability, thermal control, and data management; and has a longer operational life. To support these requirements, more complex systems are required than for any previous instrument. The operational goal of the LST is to provide maximum observations, using existing technology to a large degree.

### Support Systems Module Technology

Several NASA agencies and many contractors (representing a variety of technologies) are currently involved in LST programs. Because concept development and feasibility assessment are the major areas of progress to date, there are many different concepts of the LST and its required support systems. The concepts presented herein are for a module to support the Itek Corporation's 3-m diffraction-limited telescope, with an instrument package designed by Kollsman Instrument Corporation. The support functions are mostly automated, because ground contact is intermittent. Astronomers will be able to transmit operational programs to the module and receive some return data in near-real time. These data will permit updating of instrument adjustment and observational programs, thereby satisfying the prime requirement of controlling and managing the data on a continuous basis.

### Orbit Selection

A number of important orbital parameters were considered. A 330-naut mile altitude with a $28.5°$ inclination was selected based on Shuttle maintenance convenience, orbit lifetime, and viewing considerations.

### Configuration Selection

Configurations were studied based on various types of maintenance operation. Two configurations for on-orbit maintenance (pressurized access for the astronaut and unpressurized access using manipulators and teleoperators) and a configuration for ground maintenance only were considered. The estimated weights for the different configurations varied only slightly. The pressurized maintenance configuration was selected; additional studies will be required to determine the optimum configuration.

MAINTENANCE

The selected configuration permits maintenance by the Space Shuttle crew. The pressure shell provides meteoroid protection without weight or cost penalty, compared with alternative schemes. This concept provides efficient on-orbit maintenance, conducted at the subassembly replacement level in a shirtsleeve environment.

POINTING AND STABILITY

Target acquisition, pointing, and stability of the LST are critical considerations. The requirement for fine control at low frequency may be accomplished by articulating the secondary mirror of the telescope, if the basic vehicle does not achieve sufficient pointing performance. The attitude fine-error signal for both module and mirror would be furnished by the telescope focal-field fine-star trackers. The SSM uses a coarse sun-sensing system for initial orientation and the star trackers within the SSM, which permit acquisition of the telescope guide stars. Attitude control is accomplished by a momentum exchange system involving control-moment gyros.

THERMAL CONTROL

Two of the requirements almost dictate that thermal control be passive: the dynamics of pumps and fluid flow appear to be intolerable to telescope stability, and, for a 5-year lifetime, reliability of active systems is not good. A passive system offers options that will result in overall system compatibility and performance goals. Heat sources will have radiation paths to the outside. Auxiliary temperature control will be provided by louvers and heaters located in the module's surface, to satisfy temperature stability. A requirement for random orientation of the SSM means that heat rejection must occur, even on the module's sun side. This will be accomplished by optical solar reflector (OSR) coatings.

DATA MANAGEMENT AND COMMUNICATIONS

Because of the long on-orbit life, photographic film could not be retrieved in a timely fashion or safely stored for the operational time period. Therefore, the entire output of the scientific instruments will be digitized. Direct readout of the data stored on the image tube is considered possible when ground stations are within range. Commands and observation programs will be stored in an airborne computer.

### Scientific Objectives and Design Considerations

The LST will bring a versatile complement of state-of-the-art spaceborne instruments to bear on the study of astronomical and astrophysical problems.

Scientific objectives will require full use of LST capabilities and advantages, such as high spatial resolution due to elimination of atmospheric disturbance; broad unimpeded, high and low spectral resolution coverage free from atmospheric constraints; high temporal resolution; extended observing times and full sky coverage; and extended (15-yr) operational lifetime.

The mission elements that must be optimized to accomplish LST objectives successfully include the Optical Telescope Assembly (OTA), Scientific Instruments (SI), SSM, and mission design and operations.

Some LST design characteristics that must be established, consistent with the scientific objectives, are data handling, processing, storage, transmission, and management requirements; slewing rates, pointing and guiding accuracies, and offset search and guiding needs; optical, thermal, and mechanical stability; power and thermal considerations for spacecraft attitude constraints; and type and frequency of optical tests, calibrations, and observational controls.

Other considerations for LST operation include assessing orbital environmental effects, evaluating possible LST-limited performance, specifying reliability requirements, and assessing thermal effects on LST optical performance.

### Optical Telescope Assembly Objectives

The burden of the LST design effort centers on considerations of the Optical Telescope Assembly (OTA) subsystem. Primary scientific objectives require optimum OTA performance.

Optical design of a large-aperture, high-throughput system must provide diffraction-limited performance for a broad wavelength range, with a field of view sufficiently large to include adequate guide stars. Mirrors and mirror surfaces must be precisely figured and monitored and provisions made for thermal control. Alignment of the system must be critically maintained under the difficult thermal and mechanical environment. Mirror coatings must have high reflective efficiency from far uv to near ir. The projected 15-yr lifetime imposes extremely severe requirements on optical degradation of mirror surfaces.

Stability of the OTA subsystem must be ensured under normal LST operations. Protective elements must be adequate to ensure proper mechanical, thermal, and radiative conditions for OTA components. Adequate sensors must be integral parts of the OTA subsystem to test and monitor mirror figure and alignment to provide control of the optical surfaces and mirror alignments.

## Scientific Instruments

The Scientific Instruments (SI) package is an assembly of the instrumentation required to acquire data. The broad scope of the LST objectives and the diverse radiative characteristics of the astronomical sources impose severe requirements on SI capabilities.

The nature, extent, and types of observational requirements include high-resolution and wide-field multispectral imaging, broadband spectroscopy at high or low spectral resolution depending on source brightness, high temporal resolution imagery and spectroscopy on bright sources, and extended exposures for imagery and spectroscopy of very faint sources.

These requirements dictate the use of cameras with several image tubes of various cathode sensitivities. Spectral resolution and wavelength ranges are provided by high-resolution echelle and faint-object spectrographs using gratings, image tubes, and a Fourier spectrometer.

Auxiliary equipment necessary to ensure design performance of the LST subsystems includes guide star trackers to provide pointing and guiding accuracies that do not degrade performance of the OTA, test sensors to detect and maintain focus and figure tolerances within specified limits, calibration instruments to test achieved spatial and spectral resolution, and artificial calibration sources to control intensity and wavelength measurements.

## Performance of Scientific Instruments

The performance of the instruments in the Scientific Instruments assembly is given in Tables 1 and 2. Other instruments in the package include guide star trackers, focus sensor, and figure sensor.

The compactness of the LST design is obtained by using a reflector

TABLE 1  Performance of Camera Systems

| Instrument Cameras | Spectral Range (nm) | Field of View | Resolution, Angular (× Airy Disk Diam) |
|---|---|---|---|
| $f/12$ | 115–550 | 4.7 × 4.7 min of arc | 5.6 |
| $f/30$–slit jaw | 115–550 | 13 × 13 sec of arc/slit | 2.5 |
| $f/96$–Range I | 115–340 | 35 × 35 sec of arc | 1.5 |
| $f/96$–Range II | 160–550 | 35 × 35 sec of arc | 1.5 |
| $f/96$–Range III | 450–1100 | 35 × 35 sec of arc | 1.5 |

and six fold-mirrors to divert energy from the main OTA beam radially to instruments of the SI. Five fold-mirrors are located ahead of the reflector; one fold-mirror is located aft of the reflector and receives energy through an off-center opening in the reflector. Energy for the uv spectrograph passes to the principal focus plane through a slit in the reflector. Location of the data beams is given in Table 3. Figure 1 shows the fold-mirror location in the $f/12$ image plane.

## Instrument Locations in Scientific Instruments Radial Compartment

Instruments must be accessible for replacement or minor repair. This is accomplished by dividing the instruments, electronics, and associated drive mechanisms into self-contained modules with a method of dismounting for servicing and replacement. The instruments in the SI radial compartment (see Figure 2) are serviced through a primary mechanical interface with the central tube. Each camera is benchmarked for location to a flat reference surface, and a quick lock/disconnect feature is used to secure replaced instruments in proper alignment without need to refocus.

TABLE 2  Performance of Spectrograph and Interferometer

| Spectrographs | Spectral Range (nm) | Entrance Slot (sec of arc) | Spectral $\lambda/\Delta\lambda$ |
|---|---|---|---|
| High-resolution (echelle) I | 115–180 | 0.05–0.2 × 2 | $4.6 \times 10^4$ |
| High-resolution (echelle) II | 180–350 | 0.05–0.2 × 2 | $3.0 \times 10^4$ |
| Faint object I | 115–160–220 | 0.05–0.1 × 10 | $1 \times 10^3$ |
| Faint object II | 220–352–660 | 0.05–0.1 × 10 | $1 \times 10^3$ |
| Faint object III | 660–1000 | 0.05–0.1 × 10 | $1 \times 10^3$ |
| Mid-ir interferometer | 1–3.5 $\mu$m | 2.5–5.0 diam | TBD |

TABLE 3  Positions of Instruments

| To Instrument | Axial Position (in.) aft of Primary Mirror | Radial Position (deg) |
|---|---|---|
| Spectrograph, 0.22–0.66 μm | Sta-56.4 | 225 |
| Spectrograph, 0.66–1.0 μm | Sta-56.4 | 135 |
| Mid-ir spectrometer | Sta-56.4 | 315 |
| Focus sensor | Sta-56.4 | 0 |
| Figure sensor | Sta-56.4 | 45 |
| $f/12$ camera assembly | Sta-71.5 | 180 |
| $f/96$ camera assembly | Sta-66.6 | 0 |

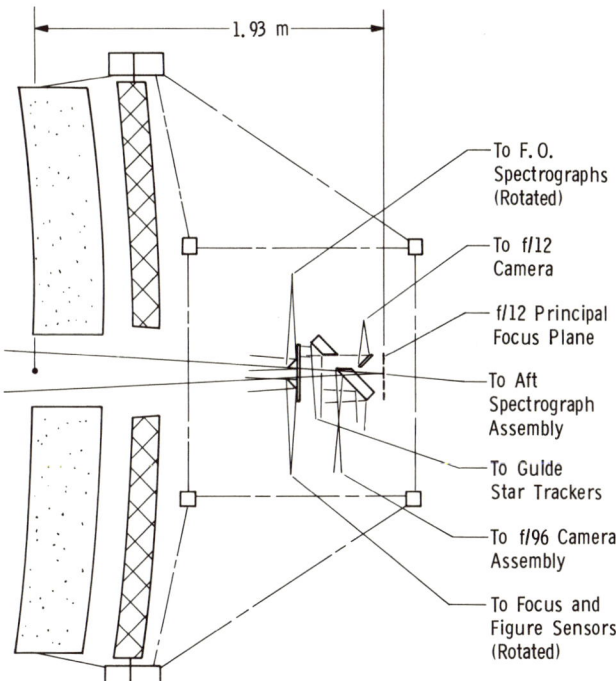

FIGURE 1  Schematic diagram of the fold-mirror location of the $f/12$ image plane.

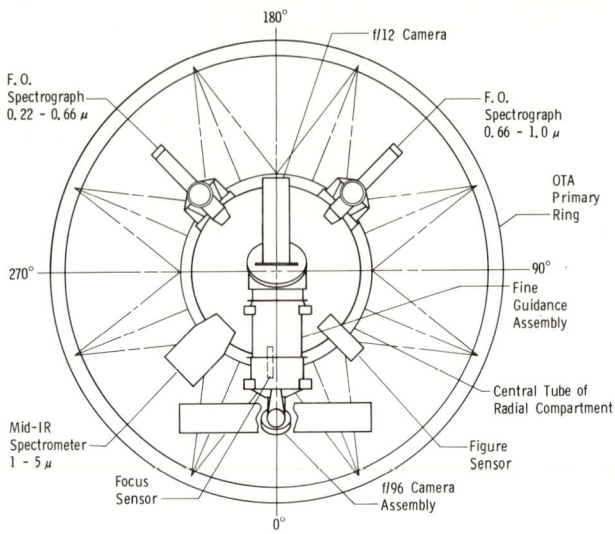

FIGURE 2 Instrument locations in SIP radial compartment.

All instruments in the illustration are shown mounted in their respective angular positions to receive radial data beams from fold-mirrors at Sta-56.4 in. The $f/12$ camera assembly is positioned aft of the reflector to receive energy from the fold-mirror at Sta-71.5 in. The guide star tracker assembly and the $f/96$ camera assembly receive radial beams directly from the reflector; energy for the $f/96$ camera assembly passes through a central opening in the guide star tracker.

Because several instruments, such as the $f/96$ camera assembly, will protrude from the envelope of the truss, protection is needed to prevent accidental disturbance of these instruments.

## Fold-Mirrors in the Image Plane—Radial

The optics deliver an image at the focal plane, which is 24 min of arc in diameter. A 10 min of arc diameter central portion of the focal plane is near-diffraction-limited and contains the fold-mirrors to receive light for the high-resolution scientific instruments (see Figure 3). The fold-mirrors are in fixed positions; the entire telescope is pointed to bring the target onto the image plane at the desired sensor. The 10 min of arc area is reserved for the $f/96$ camera, which receives the highest resolution data. The $f/12$ camera receives lower-resolution data within the 16

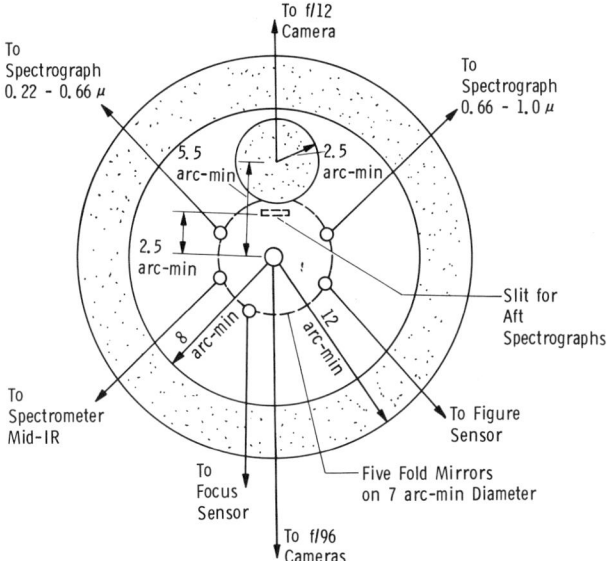

FIGURE 3  Fold-mirror locations in the $f/12$ image plane—radial.

min of arc field, which is the total field reserved for the scientific instrumentation. The outer field (16–24 min of arc diameter) contains the guide star sensors.

## Performance Considerations

Orbit altitude has a dominant influence on launch-vehicle payload capability, orbital decay rate, and trapped-particle radiation environment. Lower orbits improve payload capability and reduce undesirable radiation environment effects; higher orbits reduce orbit decay rates.

Ground-station contact time for a set of stations with given geographic locations is a function of orbit inclination, orbit precession, and orbit altitude. Increasing orbit altitude increases station view times. LST data and communication subsystem storage capacity and data rates must be compatible with orbit-constrained station contact time.

Target visibility refers to an object that can be seen without violating viewing constraints such as occultation by the sun, earth, or moon. Viewing too close to these bodies provides an additional constraint because of stray light entering the telescope. Uncertainties in the earth's atmosphere, airglow extent, and earth shine effects resulted in some initial arbitrary constraints, later revised as a result of stray-light suppression studies and sun-shade modification.

Viewing time (see Figure 4) is determined by how long an object can be viewed without interruption by earth, moon, or sun. For faint objects, viewing time is limited to observations conducted while the LST is in the earth's shadow. Additional limitations are imposed on viewing time by spacecraft maneuver rates, settling times, and the requirement that data from one observation be transmitted before proceeding to the next.

## Viewing Considerations

### VEHICLE

Looking close to the sun, moon, or earth allows unwanted stray light to enter the telescope. A truncated sun shade, with its 45° apex oriented toward the dominant stray-light source is designed to limit stray light. Target viewing is thus constrained to a sector between line of sight (LOS), 45° above LOS, and 135° below LOS. This sector places the sun behind and either above or below the LST.

With an orbital period of 97 min, efficient telescope use requires maneuvering to a new object in a relatively short portion of orbit time if more than one target is to be viewed. Momentum and torque available from control moment gyros (CMG's) determine maneuvering rates. Current design concepts permit two 90° maneuvers in 23 percent of orbit time without undue weight and power needs.

### SYSTEM

Specific spectrograph slit orientation may be needed to obtain polarization effects. Design options are the following: rotate the instrument, rotate the vehicle and decrease solar panel power output, schedule targets when vehicle can be rotated without excessive power loss, or add another degree of freedom to solar panel or sun-shield orientation. A combination of the second and third options appears operationally feasible.

Faint objects may be visible only from earth's shadow with little stray light. Viewing may be within ±5° of earth's limb, constrained by sensible atmosphere and airglow. Bright objects might be viewed from the light side, but earth shine may require sun-shade orientation for LOS between +50° and −95° to avoid internal sun-shade reflections.

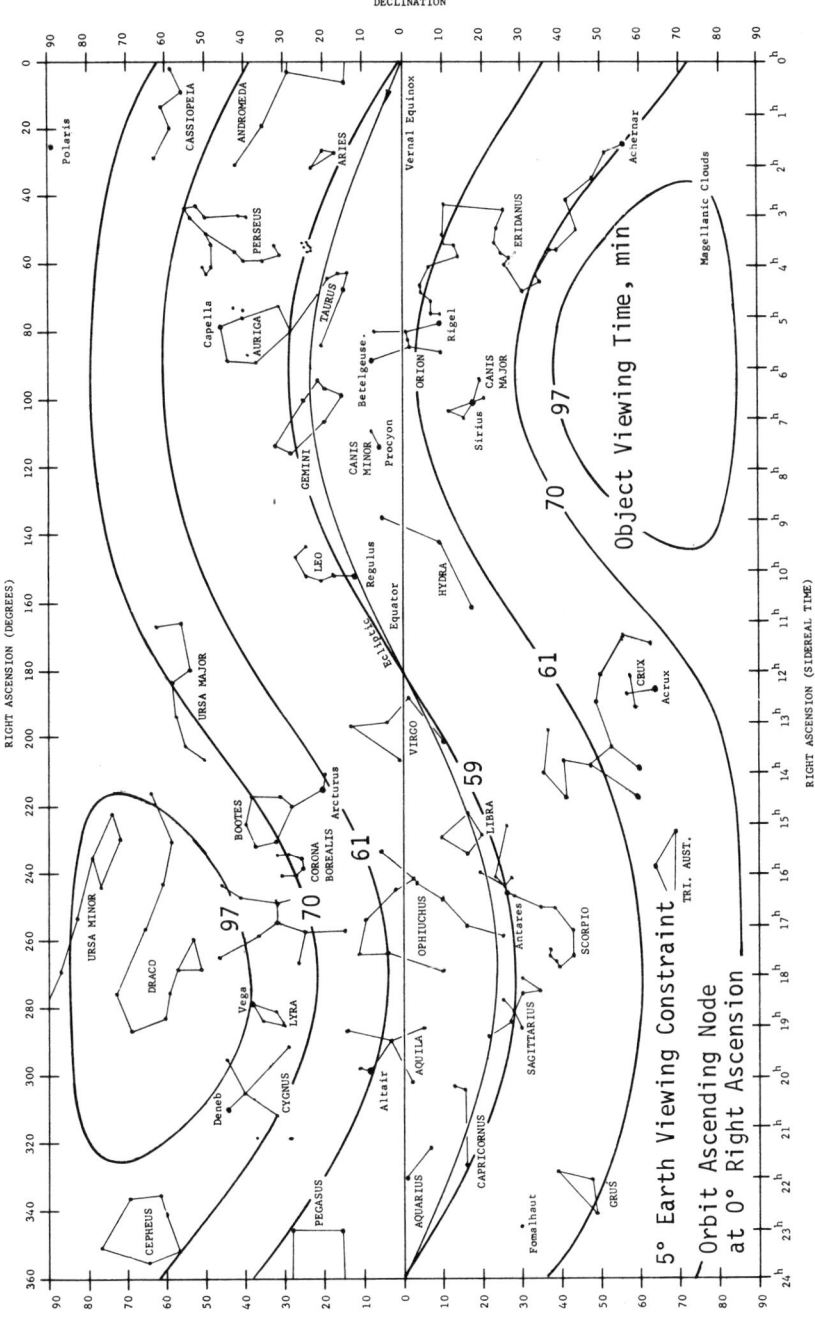

FIGURE 4 Stellar viewing time (unconstrained).

270   SPACE SYSTEMS

Ability for light side viewing adds the operational flexibility of viewing multiple targets in a single orbit if object locations permit time for LST maneuvering, settling, and viewing a secondary light-side target before reorienting for dark-side viewing.

## Constrained Viewing Time

Target viewing time, measured in degrees of orbit travel in each direction from the target longitude, is a function of the viewing constraint relative to the earth's limb and the target declination measured from the orbit plane. Considering stray light, the viewing constraints vary with LST sun-shade orientation and brightness of the earth limb closest to the LOS. The viewing constraint near the terminator is uncertain. The effect of such constraints on viewing time is illustrated for the condition where the 45° truncated sun shade is used with an earth-limb viewing constraint of ±5°. Both the stellar object and sun lie in the orbit plane (see Figure 5).

## Stellar Viewing Time (Unconstrained)

Viewing-time contours on a stellar background are shown in Figure 6. The contours are based on the base-line LST orbit (330 naut mile circular; 28.5° inclination). To prevent viewing through the earth's atmosphere, the LST must not look within 5° of the earth. No lighting constraints are considered. Note that for targets near the orbit poles, continuous viewing is possible during the 97-min orbit period. Also, viewing time does not change appreciably with target declination until the target approaches the orbit pole. These contours are for one revolu-

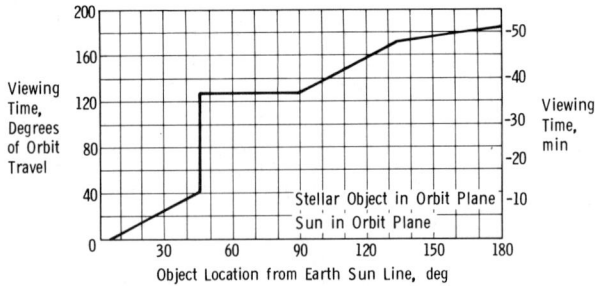

FIGURE 5  Constrained viewing time.

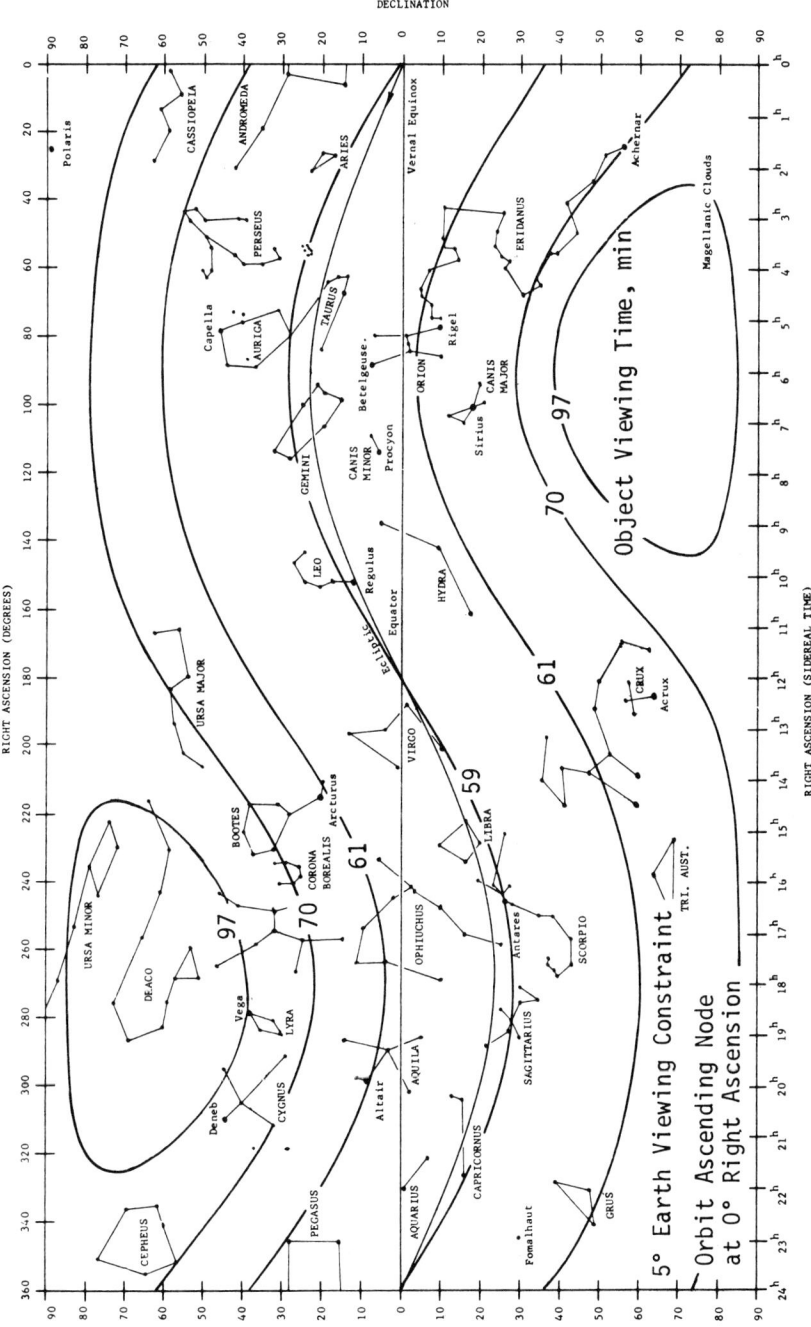

FIGURE 6 Stellar viewing time (unconstrained).

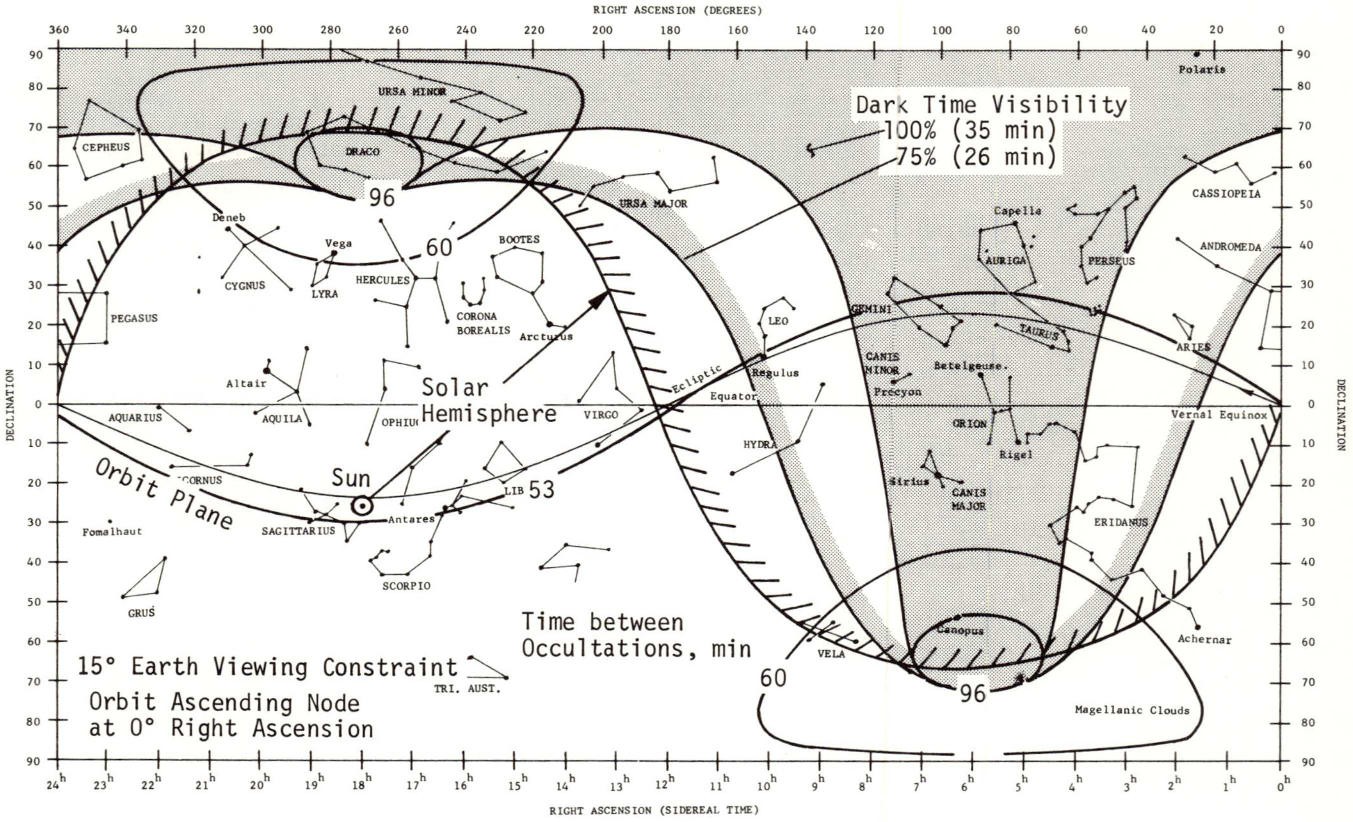

FIGURE 7 Typical viewing opportunities at $T + 0$ (December 21).

*Fred Steputis and Hal Nylander*

tion with the ascending node at 0° right ascension; the ascending node moves at about 6.36° per day. Thus, about 56.6 days will be required for one complete cycle. During this period the time contours will move around the celestial sphere. Note that targets lying above 30° declination and below −30° declination will at some time lie in the visible region for the full 97-min orbit. Objects near the equator will never be visible for more than about 60 min during one orbit.

### Typical Viewing Opportunities at $T + 0$ (December 21)

Target visibility limits for shadow viewing are illustrated in Figure 7 for the LST orbit. Stellar viewing while in the earth's shadow is currently considered to be the prime viewing time because of the absence of stray light and the fact that there are no viewing constraints that depend on sun for power generation. These data were generated for a constraint of not viewing within 15° of the limb of the earth, the sun's position on December 21, and the orbit ascending node at 0° right ascension. The contours shown differ from the previous chart because the earth-viewing constraint has been increased from 5° of limb to 15°. The solar hemisphere is shown, along with areas of the celestial sphere visible during dark time. Viewing of faint targets must be scheduled when they are in the shadow viewing region.

Two sets of contours are shown. One set shows the total time, during one orbit, that an object is not occulted by the earth. The times range from 96 min (at the orbital poles) to 53 min (at the equator). The other set shows objects that are visible while the spacecraft is in earth shadow. The objects in the shaded area are visible 100 percent of the time (35 min) that the spacecraft is in earth shadow. An additional contour is shown for 75 percent (26 min) of shadow viewing time.

### Typical Viewing Opportunities at $T + 2$ Weeks

Constrained target viewing opportunities are shown in Figure 8 for 2 weeks after $T + 0$. The orbit ascending node is now at about 90° right ascension, and the sun has moved. Different portions of the celestial sphere are now available for 100 percent shadow viewing. Targets near the celestial equator fall into the 100 percent shadow viewing region once a year for about 3 months at a time. Targets near the celestial poles fall into this region approximately once every 9 weeks for a few days at a time. These conditions illustrate the need for careful target

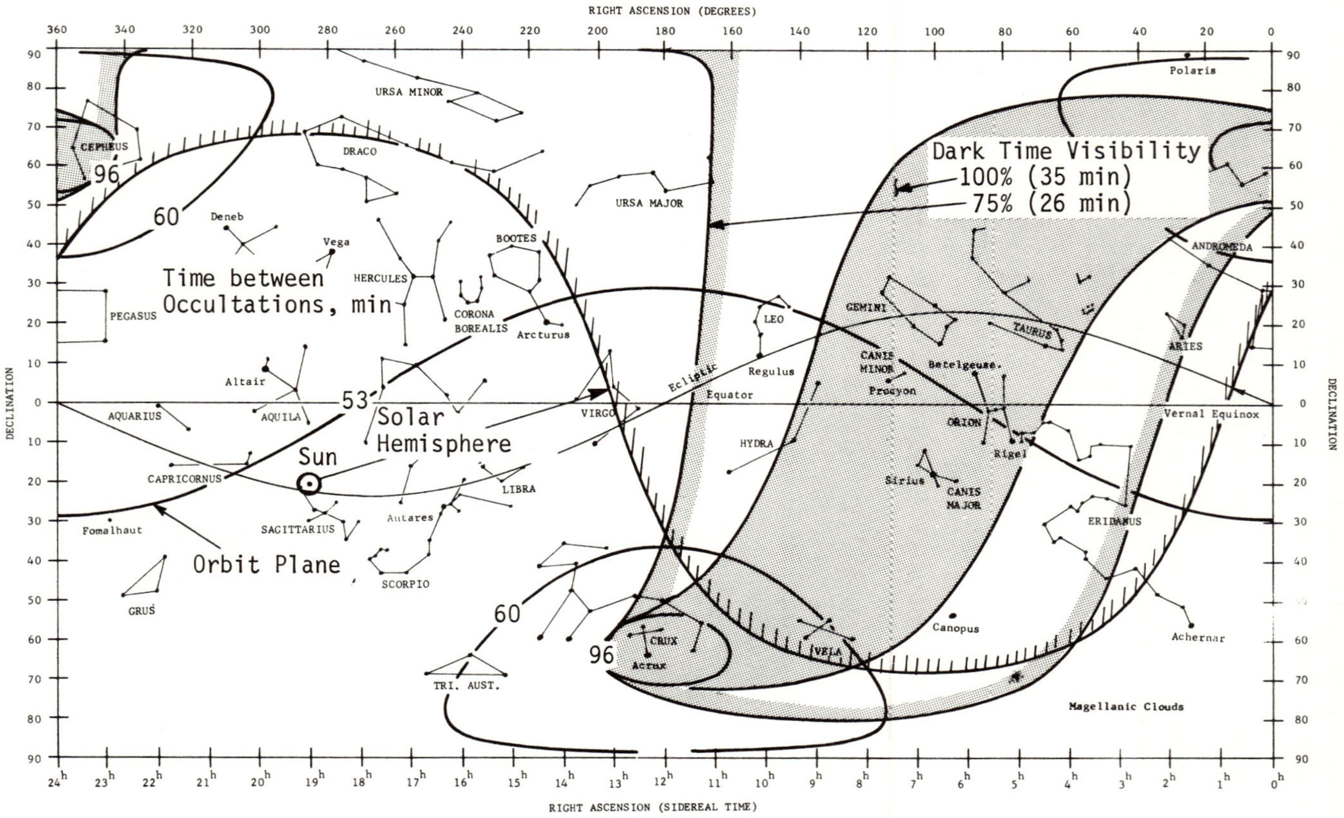

FIGURE 8 Typical viewing opportunities at $T + 2$ weeks.

selection, and long-range scheduling, to optimize use of prime-time viewing (especially for faint targets).

## Two-Object Viewing Example

Time available for secondary viewing is determined by maneuvering rates and settling times and light-side earth viewing constraints between LOS, earth, and sun.

Dark-side viewing of faint sources requires long exposure, which involves reacquisition on several successive orbits. To accomplish reacquisition, the fine star sensors must remain fixed on the target. It would be advantageous also to take data from bright sources while on the sun side of the orbit, which would require an additional set of fine star sensors.

Figure 9 shows that only 7.5 min of dark-time viewing is available before excessive earth shine enters the telescope (using the 95° angle between earth tangent and telescope LOS as the restriction). The example shown is only one of many that are possible; however, it is typical because of the difficulty encountered in positioning orbital plane, sun, earth, and object so that earth shine does not severely restrict operation of the telescope.

## Program Illustration for Two Targets per Orbit

Programming of two-target-per-orbit viewing is shown in Figure 10. The complexity of mission planning is illustrated. Maneuvers necessary to

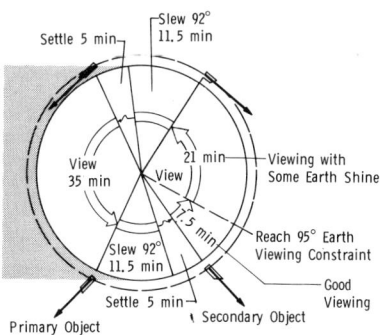

FIGURE 9  Two-object viewing example.

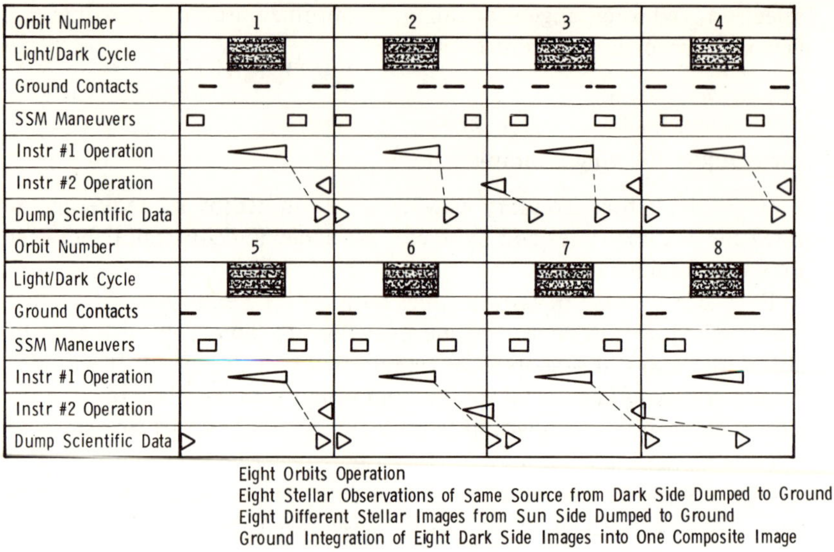

| Orbit Number | 1 | 2 | 3 | 4 |
|---|---|---|---|---|
| Light/Dark Cycle | ▓▓ | ▓▓ | ▓▓ | ▓▓ |
| Ground Contacts | — — | —— | — —— | — — |
| SSM Maneuvers | ▫   ▫ ▫ | ▫ ▫ ▫ | ▫ ▫ | ▫ ▫ |
| Instr #1 Operation | ◁ | ◁ | ◁ | ◁ |
| Instr #2 Operation | ◁ | ◁ | ◁ | ◁ |
| Dump Scientific Data | ▷▷ | ▷ | ▷ ▷ | ▷ ▷ |
| Orbit Number | 5 | 6 | 7 | 8 |
| Light/Dark Cycle | ▓▓ | ▓▓ | ▓▓ | ▓▓ |
| Ground Contacts | — — | —— | —— | — — |
| SSM Maneuvers | ▫ ▫ | ▫ ▫ | ▫ ▫ | ▫ |
| Instr #1 Operation | ◁ | ◁ | ◁ | ◁ |
| Instr #2 Operation | ◁ | ◁ | ◁ | |
| Dump Scientific Data | ▷ ▷▷ | ▷▷ | ▷ | ▷ |

Eight Orbits Operation
Eight Stellar Observations of Same Source from Dark Side Dumped to Ground
Eight Different Stellar Images from Sun Side Dumped to Ground
Ground Integration of Eight Dark Side Images into One Composite Image

FIGURE 10 Sequence of two-target-per-orbit viewing technique.

acquire additional targets are constrained by SSM slewing rates. The use of two different instruments (one for dark side and one for light side) and the need to reacquire a dark-side target add to programming complexity. Scientific data dumps must be coordinated with image data buildup and irregular station contact intervals. These constraints do not include restrictions relative to sun-shade orientations with respect to earth, sun, and object.

TRAPPED-PARTICLE RADIATION

This radiation may cause real-time loss of some data stored on the image tube. Accumulated dosage may permanently damage electronic components. Extrapolation of Skylab data indicates that these effects will not influence the concepts selected for instrument operation.

MAGNETIC FIELDS

Magnetic-field variations in the SI can cause unwanted electron-beam deflection during image tube readout. Internal LST fields are small compared with the earth's field. The worst variation in the earth's field is

TABLE 4  Orbital Environment

| | | |
|---|---|---|
| Trapped radiation (inside spacecraft) | | |
| Proton (max fluence) | Dose | 2000 protons/cm²/sec |
| Orbit dosage (average) | | 1 rad/day |
| Magnetic fields | | |
| Earth field | | 0.23 to 0.35 G |
| Spacecraft | | Less than earth field |
| External disturbances | | |
| Gravity gradient | | Cyclic and secular components |
| Aerodynamic (drag) | | Small |
| Micrometeoroid flux | | |
| Mass (particles) | | $10^{-12}$ to $10^{-5}$ g |
| Contamination | | |
| Clearing period | | |
| 1-μm particles | | 6 min |
| 50-μm particles | | 5 h |

the change in flux vector if maneuvering during readout. Magnetic shielding to avoid maneuvering limits needs to be further examined.

EXTERNAL DISTURBANCES

Gravity gradient torques produce components of momentum. The cyclic component is a factor in determining the size of CMG's; the secular component determines the size of the magnetic desaturation system. Aerodynamic disturbances are negligible at the selected orbital altitude (330 naut miles).

MICROMETEOROID FLUX

Using the NASA meteoroid model and analysis indicates that if all expected hits were of largest size and produced a pit 10 times bigger than the meteoroid's cross-sectional area, only 0.03 percent of the telescope's reflecting area would be damaged.

CONTAMINATION

Orbit insertion and orbital maintenance create the possibility of sensitive surfaces being contaminated by outgassing and other contaminants.

Skylab studies, scaled to LST orbit, show clearing in 6 min for 1-μm particles and 5 h for 50-μm particles. Orbital regression and atmospheric rotation prevent buildup on later orbits. The orbital environmental parameters are given in Table 4.

## Spacecraft Maneuvers—Design Considerations

### ARRAY ORIENTATION

The solar panel configuration uses active control to point the solar panels toward the sun. The attitude-control subsystem causes the spacecraft to roll about the LOS to the target so that the solar array axis is perpendicular to the sun line. The single axis drive then completes orientation of the array for full exposure to the sun. The left side of Figure 11 shows that the power-generating efficiency is a function of the telescope observed object displacement from the sun line for any roll angle away from the optimum position. Thus, it is possible to roll the spacecraft about targets that are close (within 60°) to the sun for short periods (one or two orbits) and not seriously degrade the power status. Such roll maneuvers are required for positioning spectrometer slits or polarimeters.

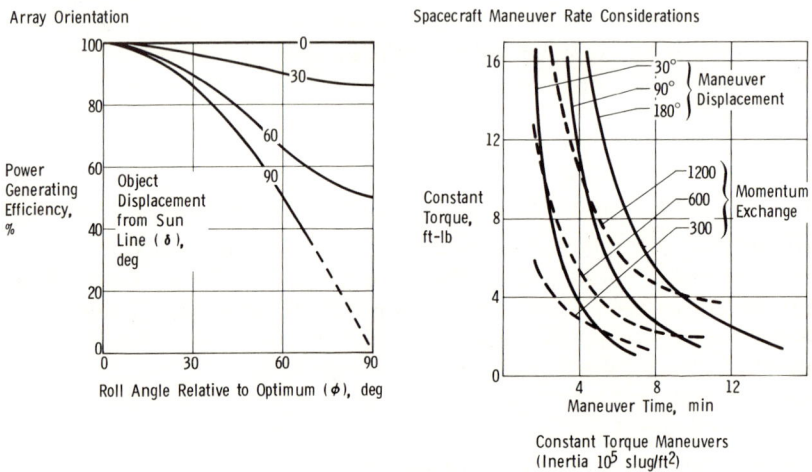

FIGURE 11 Design factors.

SPACECRAFT MANEUVER-RATE CONSIDERATIONS

The requirements for maneuver rates represent an effort to achieve a high utilization of the LST. The Manned Space Flight Center has established a minimum rate of 60°/40 min and a design goal of 90°/5 min, depending on actuator selection. In general, CMG's are more applicable to the relatively high rate requirements than are reaction wheels. To achieve a time-optimum maneuver with a constant torque-producing actuator, the vehicle is allowed to accelerate for one half of the desired displacement, then the process is reversed to remove the acquired rate at the time the desired displacement is achieved. The right side of Figure 11 shows the constant torque maneuvers for various displacements. Also shown is the momentum exchange required to achieve the minimum time maneuvers. As an example, to maneuver the stated inertia through an angle of 30° in 5 min would require about 2.0 ft-lb of constant torque and a momentum exchange of 300 ft-lb-sec. To achieve a 90° maneuver in the same time would require a momentum exchange of approximately 1200 ft-lb-sec. This increase in momentum exchange represents a weight and power penalty that must be assessed before final design selection is made.

## Electrical Power System

The electrical power system consists of solar array power sources; nickel–cadmium batteries for energy storage; and charging, conditioning, and distribution equipment (see Figure 12). The system is designed to be completely self-dependent during operation, although a man interface capability is provided for either earth command or maintenance purposes. Redundant energy storage and conditioning assemblies, feeders, and load centers provide for extended periods of continuous performance, even if some portions of the system become inactive during the course of the mission.

Power is collected on redundant buses from regulators that maintain a voltage range between 28 and 30 V. Redundant feeders carry the power to load centers, from which the loads of the scientific instruments and the supporting subsystems are supplied. A real-time command decoder or a stored command programmer performs switching. A command circuit provides signals to a decoder and driver acting on each load switch.

The two extendable foldout solar arrays, totaling 500 sq ft in area,

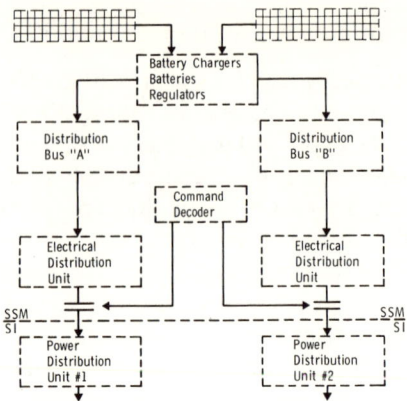

FIGURE 12  Electrical power system.

are located on the cylindrical structure of the spacecraft, diametrically opposite each other. The array extension drives, single-axis orientation drives, and slip-ring assemblies are located within the pressurizable portion of the spacecraft to allow shirtsleeve access for maintenance. The solar arrays require replacement at 5 years; no pressurized access is provided. Each array is made up of 44 panels with an active solar-cell area of 250 sq ft. This total cell area provides the average power requirement of 1335 W, including a 20 percent growth factor. To account for ultraviolet, charged particle, and thermal cycling damage, a degradation allowance of 30 percent has been made for solar-cell output during the 5 years. For the reference altitude of 330 naut miles the orbit period is 97 min. During the course of the mission, the sun eclipse time will vary from 25 to 36.6 min; the system is sized for the maximum eclipse time.

When in operation, the solar panels are oriented normal to the sun line. This is accomplished by rolling the spacecraft so that the solar array axis is perpendicular to the sun line. The single-axis drive then completes orientation of the array for full exposure to the sun.

The electrical load values (see Table 5) were determined by using data developed for other subsystems on power needs and duty cycles of each piece of equipment. The values in the table are based on an analysis of individual loads and their duty cycles, with the telescope and scientific instrumentation in use. The average electrical load for design purposes is 1335 W, after a growth factor of 20 percent is applied.

The solar arrays supply the raw electrical power for load requirements and recharging of the batteries during the illuminated portion of each orbit. During the eclipsed portion, loads are supplied by the nickel–cad-

TABLE 5  Electrical Power System

| Item | Electrical Loads | |
|---|---|---|
|  | Avg Watts | Peak Watts |
| Telescope and instruments | 750 | 950 |
| Communications and data management | 90 | 210 |
| Pointing, navigation, and control | 250 | 350 |
| Thermal control (SSM) | 20 | 150 |
| TOTAL | 1110 | 1660 |

mium batteries. Batteries are packaged (with a charger and a regulator) in a modular arrangement. Eight modules are connected to the bus at one time; four modules are maintained on standby. In the event of failure, a module is removed from the bus and disconnected from its solar-cell group by a solar group selector through command. A standby module is then connected into the circuit and controlled by command until it is operating normally.

The battery consists of 24, four-electrode, hermetically sealed cells connected in series. It has a rated capacity of 20 A-h. In addition to the positive and negative power electrodes, the battery has a third electrode used for charge control and a fourth electrode used for oxygen combination. The third electrode control concept was selected to avoid the conventional 5 to 25 percent overcharge (and the associated heat) applied to cycled batteries. In the selected spacecraft, the battery is passively cooled. Moderate temperature operation is essential to achieve the 3-yr life goal. When operating in the nominal orbit, 16,320 cycles of operation will be accumulated on the battery in 3 years. The battery temperature will vary from 40 to 70 °F (4.4 to 21 °C) depending on spacecraft attitude. Current test data show that for a life of the anticipated number of cycles the permitted depth of discharge can range from 18.5 to 23 percent. For the expected spacecraft load, including the 20 percent growth factor, battery discharge will range from 14.7 to 21.6 percent, which is compatible for the 3-yr life goal.

## Spacecraft Pointing Requirements

The telescope has a resolving power of 0.04 sec of arc, and its optical axis must be pointed to the target star to within ±0.1 sec of arc. It is essential not to degrade telescope performance by an inadequate space-

TABLE 6  Pointing Performance

|  | Low-Frequency Stability (sec of arc) (0–2 Hz) | High-Frequency Stability (sec of arc) (> 2 Hz) | Aiming Accuracy (sec of arc) |
|---|---|---|---|
| *Spacecraft* | | | |
| Pitch/yaw (min perf)[a] | 0.1 | 0.005 | 0.1 |
| (best perf)[a] | 0.005 | 0.005 | 0.1 |
| Roll | 1.0 | 0.1 | 1.0 |
| Target acquisition FOV | | | 30.0 (3σ) |
| *Optics* | | | |
| Pitch/yaw | 0.005 | 0.005 | 0.1 |
| Roll | 1.0 | 0.1 | 1.0 |

[a] Minimum performance depends on articulation of secondary mirror for fine stabilization, while best performance does not.

craft pointing and stabilizing system. To realize the full benefits of the LST, the optical axis of the telescope must be aimed to within ±0.1 sec of arc of the target LOS (Table 6). Furthermore, to realize the full resolution capability of the LST, it is necessary to maintain image position constant within ±0.005 sec of arc. Ultimately, it is desirable to achieve pointing and stability by the spacecraft control system alone, without recourse to control of the secondary mirror.

Factors such as disturbance torques, fine stabilization errors, vibration, and thermal effects combine to make up the total performance objective of the spacecraft pointing system. The pointing system objective is ±0.005 sec of arc pointing stability. Increasingly stringent requirements for pointing accuracies and LOS stability have resulted as the size of telescopes has increased. The objectives of other programs in the 1972 to 1977 period range from 0.1 to 0.05 sec of arc in 1977. By taking full advantage of these related technological developments, and by extreme care in the design of the LST spacecraft, the LST pointing objectives appear possible.

**Pointing System Description**

Key hardware characteristics needed to meet the LST accuracy and reliability goals include high-precision sensors with high signal-to-noise ratio and precision actuators. These actuators must provide the necessary torques for pointing stability against all external disturbances, without contaminating the telescope optics and without resonating with structural frequencies.

The control system attenuates the solar panel and spacecraft elastic modes and, at the same time, maintains a rigid body frequency in the vicinity of 1 to 2 Hz. The operational logic permits the spacecraft first to acquire reference stars with relatively coarse accuracy, then to progress sequentially to finer accuracy. Redundancy in the sensing and acquisition hardware permits continued operation after failures.

The compatibility of the above configuration with such characteristics as body bending, solar panel bending, sensor and actuator noise, and sensor dynamics is essential.

To date, analytical and laboratory hardware tests have indicated that the LST minimum performance goals can be achieved. Additional hardware development and analytical studies are being performed by NASA and industry to establish the performance for the LST fine-pointing system.

Figure 13 shows the relation between the sensors, computer, actuators, actuator desaturation system, and main telescope. The sun sensor and Canopus tracker establish a reference set of coordinates as soon after injection as possible. Angular errors from these two reference points are sent to the computer, which converts the angular errors to

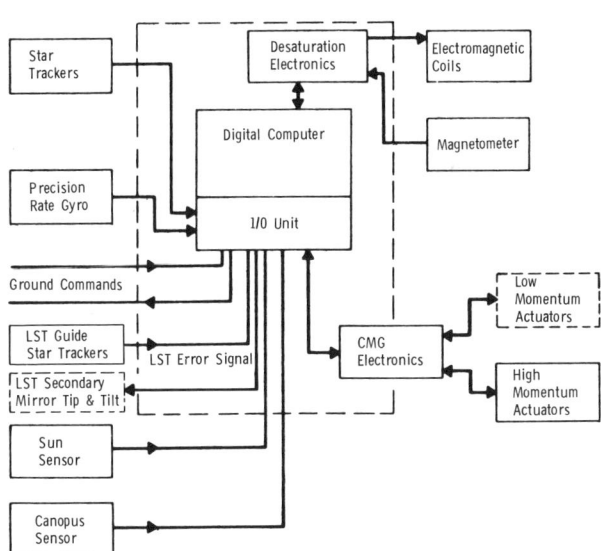

FIGURE 13  Overall pointing system.

signals to be sent to the torquers of the precision gyro system. These signals torque the gyro reference axes until they coincide with the axes established by the sun/Canopus trackers. The computer processes attitude and attitude rate signals in accordance with the CMG control laws that are designed to avoid CMG saturation and singularity areas. Power amplifiers then transmit the signals to the CMG's. Six SG-CMG's are used at each momentum level for actuators. However, performance can be maintained with up to two failures at either level. The high-momentum actuators are used to provide continuous desaturation to the low-momentum actuators and to provide maneuvering capability for the vehicle. The low-momentum actuators are for primary short-term control and provide the response necessary to meet the stability requirements.

**Data Management**

Science instrument commands are initiated by the Ground Control Center, transmitted to the LST, received and demodulated by the SSM, and either transmitted directly to the remote decoder unit (real-time commands) or stored in the command programmer memory for delayed execution (stored program commands) (see Figure 14). The remote decoder unit decodes the command word and transfers the proper logic control signals to the instrument over one of several separate interface lines. Additional synchronization, timing, and SSM operating mode signals are also similarly provided to allow all instruments and SSM service subsystems to operate in a compatible manner.

All science, instrument diagnostic, and SSM subsystem engineering data are transmitted to ground over a single RF link, modulated by a single time-multiplexed PCM encoded data channel. PCM format control originates within the SSM, and telemetry gate signals in time-multiplexed binary form are transferred to the applicable remote acquisition unit (RAU) over a single multiplex data-control bus. The RAU demultiplexes and decodes the control words and sends the appropriate telemetry gate-control signal (together with a data transfer clock signal) over one of several interface lines to the instruments' data-conditioning circuits. The addressed science or instrument diagnostic data have already been conditioned, converted to an eight-bit binary word, and stored in an appropriate buffer register. The input gate control and transfer clock allow the selected binary word to be transferred back to the RAU.

The RAU combines these data with other data from nearby sources and sends the time-shared data back to central SSM data storage and/

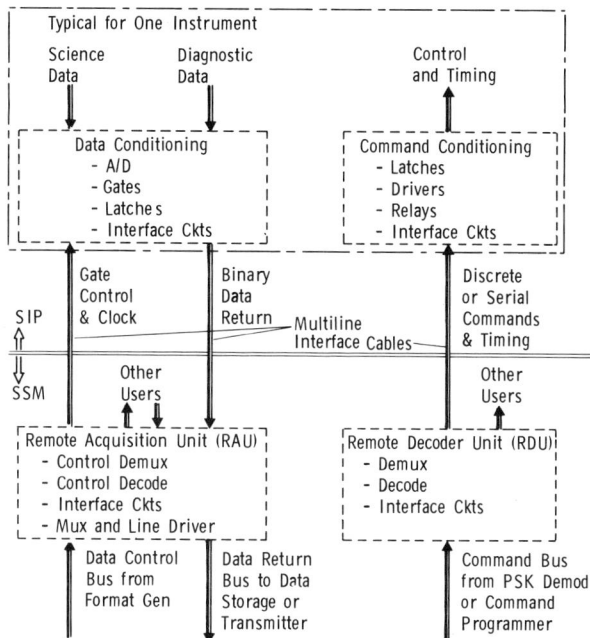

FIGURE 14 Data management.

or to the telemetry downlink modulation circuits. The presently preferred downlink mode requires data storage by the high-resolution image tubes; these data are then time-multiplex interleaved with real-time engineering data and stored data (readout of combined instrument and SSM data sampled during the time between ground stations).

## Data Management Parameters

The quantity of data generated by each of the science instruments varies between 2 million and 100 million bits per image (Table 7). The presently preferred technique is to use the storage inherent in the image tubes. However, the practicality of incorporating reliable mass-data-storage devices remains a consideration.

The transmission times listed are based on a present Manned Space Flight Network (MSFN) ground-station-receiver limit of 0.5 megabit per second. A normal mode of operation for high-resolution investigations will allocate the optical system to a single instrument, such as the $f/96$-1 camera, at any given time. The 4-min readout will be accommodated by

TABLE 7  Data Management Parameters

|  | Bits Generated per Image |
|---|---|
| Camera, $f/96$-1 | $120 \times 10^6$ |
| Camera, $f/96$-2 | $59 \times 10^6$ |
| Camera, $f/96$-3 | $9 \times 10^6$ |
| Camera, $f/12$ | $120 \times 10^6$ |
| Camera, slit jaw | $6 \times 10^6$ |
| Spectrograph, 0.11–0.22 $\mu$m | $1.8 \times 10^6$ |
| Spectrograph, 0.22–0.66 $\mu$m | $3.6 \times 10^6$ |
| Spectrograph, 0.66–1.0 $\mu$m | $1.8 \times 10^6$ |
| Echelle, 0.11–0.18 $\mu$m | $18 \times 10^6$ |
| Echelle, 0.18–0.35 $\mu$m | $18 \times 10^6$ |

a single ground-station pass (maximum time of 8 min above the 5° elevation limit). Relatively inexpensive modifications could be made for a higher data transfer rate if such rates are found to be necessary.

## Thermal System

The thermal system design incorporates passive thermal control methods, where possible, for improved reliability and long life and to minimize dynamic disturbances. Supplemental control, accomplished by heaters and louvers, is used only where required to provide temperature control for components with narrow temperature tolerances.

The passive system accomplishes thermal control by using OSR surfaces externally on the meteoroid shield. The high emittance of OSR provides the basic capability for dissipating internally generated heat to space. The extremely low solar absorptance, and flight-proven stability, of OSR's results in low sensitivity to variations in the external environment.

Subsystem components are designed to conduct and radiate heat directly to the pressure shell, which acts as a heat-sink surface. The pressure shell has sufficient radiating area to dissipate the heat to the OSR (meteoroid shield), which subsequently radiates the heat to space. Nonmetallic standoffs isolate the meteoroid shield from the pressure shell. Bimetallic controlled louvers between the pressure shell and meteoroid shield are used to control the amount of heat exchange between the two surfaces. Positioning of the louver blades controls the radiation interchange between the two surfaces. Through this control, the component temperature limits can be maintained in the cold external environment without supplemental electrical heating (except during inoperative periods).

The subsystem components are grouped in six thermally isolated compartments around the periphery of the pressure shell, which permits individual temperature control of each component group. Two compartments around the pressure shell periphery contain no equipment. The telescope scientific instruments that are contained within the SSM use variable-conductance heat pipes to control temperature and to transport their thermal energy to the two nonequipment sectors.

The exteriors of the OTA meteoroid shield and sun shield use low $\alpha/\epsilon$ coatings to minimize the influence of the orbital thermal environment. Multilayer insulation blankets isolate the OTA truss from temperature transients.

During pressurized manned maintenance, electrical wall heaters (powered by the Shuttle) are used to maintain touch-temperature limits and prevent wall-moisture condensation by compensating for wall heat leaks.

Steady-state analyses of the thermal design show that it is able to meet the objectives of controlling the equipment within the design temperature limits and of providing for growth in the system heat load. Table 8 illustrates the effectiveness of the OSR surface in rejecting system thermal energy in the hot extreme environment. In general, the minimum heat-rejection capability exceeds the design-assumed heat load by a factor of 2. The effectiveness of louvers in conserving the system thermal energy in the cold extreme environments has been analyzed. At

TABLE 8  Thermal System Considerations

| Equipment | Location Relative to Solar Arrays | Platform Temperature (°F) | Minimum Heat Rejection Capability (W) |
|---|---|---|---|
| Batteries | Two Bays at 45 deg | 70 | 540 |
| Communications Data Management GN&C | Two Bays at 0 deg | 90 | 830 |
|  | Two Bays at 90 deg | 90 | 600 |
| Scientific Instruments | Two Bays at 45 deg | 70 | 950 |

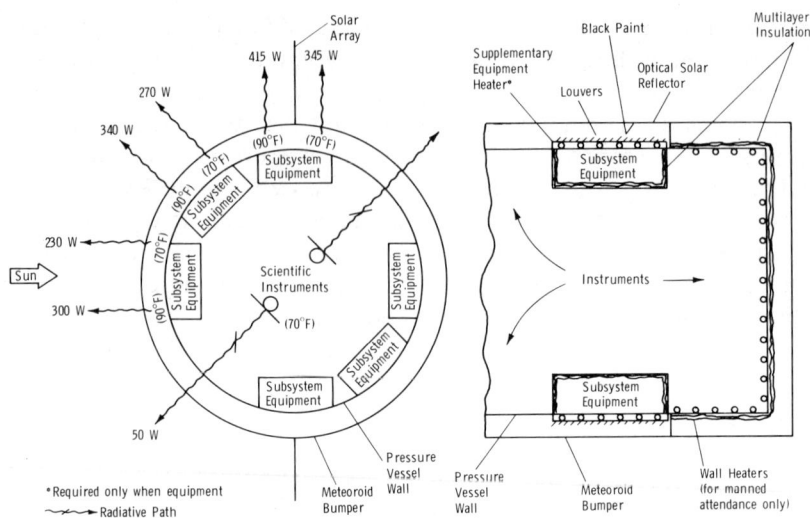

FIGURE 15  Thermal system considerations.

the 40 °F minimum subsystem design limit temperature, the heat dissipation from the six subsystem bays may be regulated down to 520 W, which is below the total subsystem heat load. The design is thus relatively insensitive to spacecraft orientation and orbital altitude and inclination.

The hot-case orientation was that the spacecraft be in sunlight for 100 percent of the orbit time, broadside to the sun, with the solar array axis perpendicular to the solar vector. The heat rejection capability values for the 70 and 90 °F heat sink temperatures are obtainable by

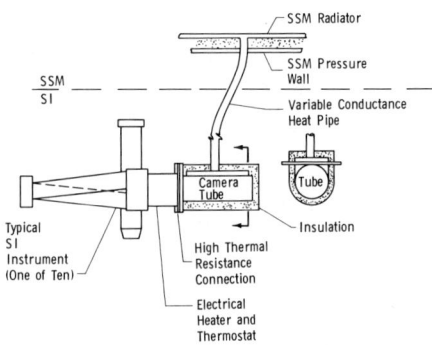

FIGURE 16  Detail of thermal system.

keeping the louvers fully open and by coating the inboard surfaces of the meteoroid shield with a high-emittance coating. Since the capability exceeds the requirements, this radiative coupling is then selected to balance actual internal power and to balance absorbed external energy against energy emitted for individual equipment bays. Thus, the coating on the inboard surface of the meteoroid bumper is tailored at relatively low cost to the upper temperature limit and power dissipation requirements of the equipment for specific bay locations. This enables the packaging arrangement to be optimized for maintenance and mass distribution requirements rather than for thermal requirements.

**Thermal System Considerations**

The thermal interface of the LST with the SSM is designed with the following major considerations (see Figures 15 and 16 and Table 8):

1. Transfer heat from SI instruments to the SSM for dissipation to space. The approach is to use heat pipes, as illustrated. This concept is based on a possible requirement to operate the camera tubes at $-50\,°C$, with up to ten cameras in the instrument compartment. Another approach to achieve the $-50\,°C$ operating temperature is to use Peltier coolers. This is a requirement only for the target of the SIT tube and is practical if the tube design permits cooling of the target only. If the entire tube must be cooled, then the heat pipe method will be used.

2. Minimize transfer of heat from SSM walls and equipment to SI and OTA equipment and structure. The approach is to use low-emissivity surfaces supplemented with insulation.

3. Use electrical power from the SSM to maintain LST structure and mirrors at desired temperatures and to make up heat loss to space. Present estimates are for an average power need of 350 to 450 W.

# Ultraviolet Instruments

R. J. SPEER

# ULTRAVIOLET SOLAR ECLIPSE SPECTROSCOPY FROM SPACE

## I. Introduction

The solar atmosphere may be defined as that fraction of the sun's mass ($\approx 10^{-10}$) that lies at greater heights than the optical edge (limb) of the visible disk. We recognize that most of our knowledge of the sun's outer layers, and its deeper interior, stem from observation of the electromagnetic spectrum accessible to us from this region. The corona forms the greater part of this atmosphere and is characterized by a low particle density ($\approx 10^8$ cm$^{-3}$) and an extremely high temperature ($T_e \approx 2 \times 10^6$ K).

Despite major advances in recent years, theories of the origin and energy balance of the corona remain hampered by the lack of uv and xuv data that combine very high spatial resolution with spectral resolution. An additional factor arises from the considerable experimental difficulties associated with measurements on the intrinsic coronal line-emission spectrum at visible and uv wavelengths. Ideally, such measurements are performed at times of total solar eclipse, and even on these rare occasions

---

The author was at the Institute for Astronomy, University of Hawaii, when this work was done, while on leave of absence from the Department of Physics, Imperial College, London S.W. 7.

the coronal continuum flux masks the weaker lines in the visible and near uv.

Visible data are obtained either at ground-based expeditions to eclipse sites or by the use of coronagraphs outside of eclipse at a limited number of high-altitude observing stations. A third, and relatively recent innovation, is the externally occulted white-light coronagraph, successfully flown, for example, on Skylab.

We note that the total solar eclipse situation improves on its ground-based optical counterpart—the coronagraph—by a factor lying between $10^2$ and $10^4$, depending on the actual observing conditions. Thus highly prized data are obtained during eclipse totality, and it is under these circumstances that new spectra and structure have recently been observed in the visible down to a level of $10^{-10}$ disk brightness.

Considerably greater experimental risks are involved before complementary data can be obtained in the uv. We list these difficulties briefly as (1) the requirement that the uv spectrographic system be above the effective absorbing fraction of the earth's atmosphere (say 100 km); (2) movement of the payload *into or out of* an eclipse shadow (umbra); (3) alignment of the optical axis to the *eclipsed* sun; and (4) recovery of photographic spectra.

Although several attempts have been made using rockets launched from temporary eclipse sites [1958 eclipse: Chubb *et al.*, 1961; Friedman, 1963; 1966 eclipse: Blamont and Malique, 1969], the North American eclipse of March 7, 1970, provided a major opportunity for such experiments, because of the passage of an eclipse shadow over the established NASA launch-range facility at Wallops Island, Virginia. Two Aerobee rockets were launched into the approaching umbra. The payloads consisted of intensity-calibrated uv grating spectrographs. Inertial pointing in three axes provided the necessary optical alignment to the eclipsed sun. These experiments recorded stigmatic spectra of the solar chromosphere and corona. Near second contact, these spectra are analogous to the visible eclipse "flash spectrum."

## II. Summary Perspective

The availability and sophistication of modern rocket and satellite platforms has led to a steady advance in our knowledge of the solar spectrum below the short-wavelength transmission limit of the terrestrial atmosphere at $\lambda \approx 3000$ Å. Even though this region accounts for only about 5% of the total solar photon flux, the spectrum exhibits a remarkable range of phenomena, and their study and interpretation have contributed greatly to our understanding of the structure of the sun's atmosphere. Probably the most significant observation is that a fundamental change occurs in the spectrum at $\lambda \approx 1600$ Å. At this wavelength, the spectrum is observed to change from absorption to emission. This changeover is due to a reversal in temperature gradient in the outer atmosphere of the sun. Satellite data accumulated in the last few years show that the temperature increases monotonically outward from $T_e \approx 4300$ K to $T_e \approx 2 \times 10^6$ K over a very narrow region. Thus an orbiting uv spectrograph, capable of observing say from 1600 Å to the strong Fe XVI emission line* at $\lambda = 361$ Å ($T_e \approx 3 \times 10^6$ K) is effectively probing the entire three-dimensional structure of the solar atmosphere from the temperature minimum to the coronal temperature maximum. Interpretation of uv data of this type, together with visible and infrared observations, appears to be leading to a common description of the average physical conditions in the solar atmosphere for the first time [Noyes, 1971].

Figure 1 shows the general run of temperature $T_e$ and total density $n_H$ as a function of height in the solar atmosphere, reproduced from the recent review by Noyes [1971]. The transition-zone chromosphere/corona is characterized by an extremely steep temperature gradient and illustrates the importance of uv observations. The approximate locations are marked for the emission from several abundant ions mentioned in the text.

The solar disk has a mean apparent angular diameter of 1927 sec of arc with 1 sec of arc equivalent to 720 km at the sun. Inspection of Figure 1 shows that the upward change in temperature to the corona occurs over a region well below the instrumental resolution of current uv experiments. Figure 1, however, has been derived from data obtained by Orbiting Solar Observatories 4 and 6 having uv spectroheliographic element resolutions of about 1 min of arc and 0.5 min of arc, respectively [Goldberg et al., 1968; Reeves and Parkinson, 1970].

Observations with the highest uv resolution have been conducted by the Naval Research Laboratory [Tousey, 1971], and spatial resolutions of 3 to 5 sec of arc have been achieved.

The total solar eclipse has two important and well-known optical properties that complement and extend these instrumental techniques for the study of the solar chromosphere and corona.

A. OBSERVATIONS EMPHASIZING THE INTRINSIC CORONAL SPECTRUM

The first advantage arises from the highly efficient suppression of the photospheric continuum from the disk, which masks the intrinsic

---

* Fifteen times ionized iron. We refer to the spectrum of the sixteenth electron.

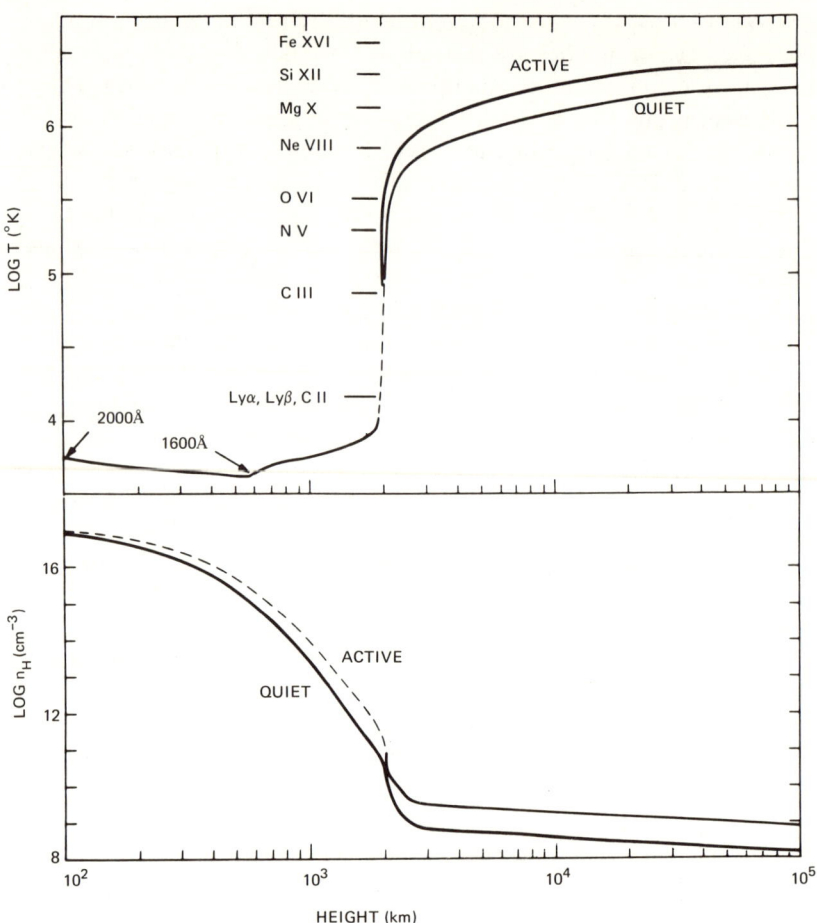

FIGURE 1 *Top:* The variation of temperature with height for quiet and active regions of the solar atmosphere. *Bottom:* Contiguous variation of total hydrogen density [Noyes, 1971].

coronal flux level by a factor of a million or more outside of eclipse. This suppression is so effective that if the eclipse combines with conditions of low sky radiance and good technique, detailed coronal structure may be observed from just above the limb out to five solar radii at $3 \times 10^{-10}$ disk intensity. Figure 2 shows the corona recorded by Newkirk and Lacey [1970] under such ideal conditions at the total eclipse of March 7, 1970. The visible radiation recorded here has three components: a Thomson-scattered photospheric continuum flux from the electron population (the *K* corona); a characteristic emission-line

FIGURE 2 Coronal photograph in unpolarized white light obtained by the 1970 High Altitude Observatory eclipse expedition to San Carlos Yautepec, Mexico [Newkirk and Lacey, 1970]. The effective wavelength is 6400 Å. A radially symmetric neutral-density filter has been used, permitting coronal structure to be traced out to $4.5 R_\odot$ ($3 \times 10^6$ km).

spectrum, consisting of forbidden (magnetic dipole) transitions in configurations of $p$ electrons in highly stripped ions; and a second scattered continuum component from the dust in the outer corona (the $F$ corona).

The advantages offered by suppression of the disk flux at total eclipse extend to all wavelengths, showing significant continuum emission, i.e., approximately 0.1 μm to 100 μm; however, it is only very recently that the technical possibility has existed to recover coronal eclipse spectra outside of the visible. The infrared has been investigated with Fourier transform spectrometers in high-flying aircraft in eclipse totality [Olsen et al., 1971]. In the uv, use has been made of developments in three-axis inertial rocket payload stabilization.

B. OBSERVATIONS EMPHASIZING SPATIAL RESOLUTION

Reference to Figure 1 indicates the intriguingly narrow height range for the transition zone chromosphere/corona, with theoretically deduced gradients reaching values as high as 5000 K/km. Observational evidence remains to be obtained for the precise location and structure of this region. In principle, observations at eclipse can tie down the heights of emission to a second of arc or less as a consequence of this exceptionally small angular range over which the transition region ions exist. A stationary ground-based spectrograph will record an apparent angular rate of the moon's limb across the sun's limb of $\approx 0.47$ sec of arc per sec of time, an impressively small rate that is exploited routinely in visible eclipse data. In the uv, however, a rocket platform will observe an apparent angular rate ranging somewhere between 0.2 and 5 sec of arc per sec depending on the detailed relationship between the ballistic motion of the vehicle and the moving eclipse shadow. Although there is good evidence that the transition zone contours the rough spicular structure of the chromosphere, uv stigmatic eclipse spectra can yield the relative volume of transition-zone material at each height and locate with precision the height at which emission commences. (Typically 1 sec of arc in movement of the lunar limb is equivalent to 1.7 km of translational movement of the rocket spectrograph in the rest frame of the umbra. A rocket can be located with a considerably greater precision than this.) Two experiments have been performed recently to exploit this situation at eclipse [Brueckner *et al.*, 1970; Speer *et al.*, 1970] and each obtained new uv data.

### III. The 1970 Eclipse

The event on March 7, 1970, was remarkable for the level of effort and quality of results that were recorded, and four extensive collections of data are now available: the *National Science Foundation 1970 Eclipse Bulletin; Nature,* Eclipse Issue, Vol. 226, No. 5242 (1970); *Solar Physics,* Vol. 21, Proceedings of IAU Symposium on the 1970 Solar Eclipse; and the *Journal of Atmospheric and Terrestrial Physics,* Vol. 34, 1970 Eclipse Issue.

The prime ground-based observing site was Mexico, where Newkirk and Lacey recorded the data shown in Figure 2. The eclipse track then passed along the eastern seaboard of the United States, reaching Virginia and the Wallops Island launch facility by 18.40 UT. The situation of this time was recorded from space by Applications Technology Satellite III, and the corresponding data frame from the spin-scan cloud camera is shown in Figure 3. The North American coastline is clearly seen, with the umbra/penumbra in the upper center of the picture. By coincidence, this frame recording the lunar shadow transit was obtained only minutes prior to the corresponding uv data described in the following paragraphs when a three-axis stabilized package entered into the umbra at a height of 143 km.

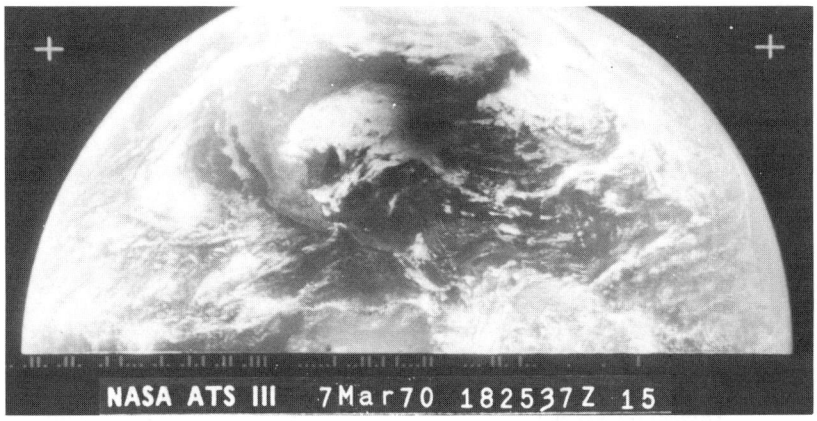

FIGURE 3 March 7, 1970, eclipse shadow transit relayed from 35,770 km and 85.12°W, 0.50°S by the multicolor spin-scan cloud camera on Applications Technology Satellite III [Suomi *et al.,* 1968]. This picture was obtained 14 min prior to the uv eclipse data described in Section IV when a three-axis stabilized package entered the umbra at a height of 143 km.

This is shown schematically in Figure 4. The view is from an elevated point in the U.S. mainland. The umbra, seen passing from right to left, is approximately 120 km in diameter and is to be intersected at the payload apogee. In the eclipse frame of reference, shown in Figure 5, the experimental package descends through the shadow almost radially as a consequence of their relative motions. The ballistic flight of this package was exceptional in that an over-performance of the rocket caused the flight plan (b, the most probable path) to be exceeded. The actual flight path (c) came very close to the optically ideal path (a).

The principal results from the two uv eclipse attempts on this day are to be found in Brueckner *et al.* [1970], Speer *et al.* [1970], Gabriel *et al.* [1971], Jones *et al.* [1971], Jordan [1971], and Gabriel [1971].

## IV. The 1970 Eclipse Data

Aerobee 4-312 US successfully carried two intensity-calibrated uv Wadsworth spectrographs through the March 7 eclipse shadow. The payload was prepared as a joint international effort between the Astrophysics Research Unit, Culham Laboratory; the Centre for Research in Experimental Space Science, University of York, Toronto; Harvard College Observatory; and Imperial College, London University.

Table 1 should be used in conjunction with Figure 5 to visualize the relationship between the recorded data and the flight path. A geometri-

FIGURE 4 Conceptual view of the passage of the uv payload launched by NASA Aerobee 4.317 US through the eclipse umbra. In this view, the eclipse shadow is seen moving from right to left off the eastern seaboard of the United States at a surface speed of 1500 mph, resulting in a rocket launch window of approximately 30 sec.

cal reduction relates the position of a payload in the umbra to the apparent instantaneous height of the moon's limb projected onto the sun along their common line of centers. The most important quantities are the absolute values of this height and its first rate of change. Fifty frames of uv data were recorded, and only the most important of these are tabulated. A cross-section schematic of the one-element optical system is shown in Figure 6, and light from the eclipsed sun enters from the top left. The short-wavelength spectrograph characteristics only are

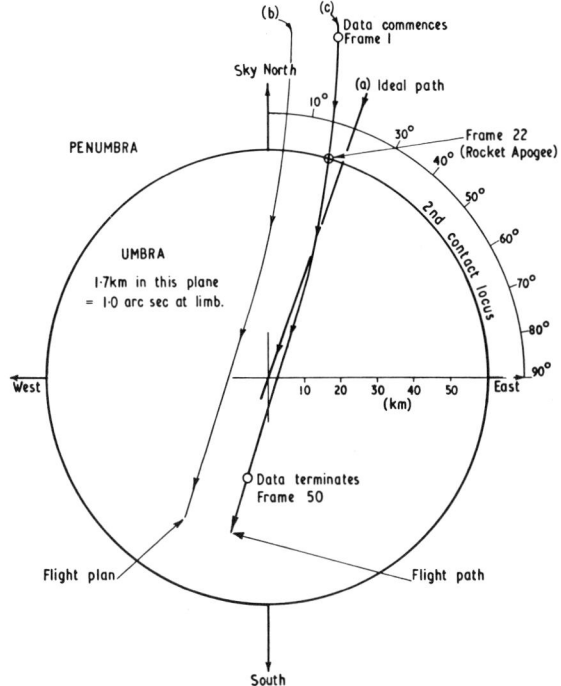

FIGURE 5 The passage of the uv spectrographic payload through the eclipse umbra. The plane shown is orthogonal to the line of central eclipse (0, 0), and a geometrical reduction shows that 1-km translation in this plane results in 1.7 sec of arc of lunar motion, when projected onto the sun's limb. Due to vehicle over-performance the payload trajectory (c) exceeded the flight plan (b) and approached radially to within 1.8 km of the line of central eclipse (equal midtotality).

listed in Table 2, as a system failure in the second spectrograph prevented the acquisition of data in the 2000–3000 Å range. Continuum-dominated spectra begin to appear above the scattered light level by the twelfth frame in the exposure sequence (Table 1). At this point, 11 sec of arc of the photospheric disk remain to be occulted by the moon at an apparent rate of 0.2 sec of arc per sec of flight time. Four key frames in the sequence are reproduced and discussed at length in Gabriel et al. [1971]. They show the characteristic rapid change in the spectrum as the payload crosses the eclipse shadow (second contact). By exposure 30, the spectrum is dominated by many circular coronal images showing spatially resolved variation in surface brightness according to solar limb position angle. We limit ourselves here to a brief description of this

TABLE 1  Summary of Data

| Exposure Sequence No. | Flight Time at Mid-exposure (sec) | Exposure Time (sec) | Height of Moon's Limb above Sun's at Line of Centers ($H_0$) (km) |
|---|---|---|---|
| 10 | 154.3 | 1.0 | − 8,810 |
| 11 | 157.8 | 0.2 | − 8,310 |
| 12 | 161.5 | 1.0 | − 7,740 |
| 13 | 165.0 | 0.2 | − 7,170 |
| 14 | 168.6 | 1.0 | − 6,540 |
| 15 | 172.2 | 0.2 | − 5,870 |
| 16 | 175.8 | 1.0 | − 5,160 |
| 17 | 179.4 | 0.2 | − 4,410 |
| 18 | 183.0 | 1.0 | − 3,620 |
| 19 | 186.5 | 0.2 | − 2,820 |
| 20 | 190.0 | 1.0 | − 1,990 |
| 21 | 193.7 | 0.2 | − 1,090 |
| 22 | 197.3 | 1.0 | −   180 |
| 23 | 200.8 | 0.2 | +   750 |
| 24 | 204.4 | 1.0 | + 1,750 |
| 25 | 207.9 | 0.2 | + 2,760 |
| 26 | 211.5 | 1.0 | + 3,830 |
| 27 | 215.1 | 0.2 | + 4,930 |
| 28 | 219.0 | 0.4 | + 6,160 |
| 29 | 222.3 | 0.2 | + 7,230 |
| 30 | 225.9 | 1.0 | + 8,440 |
| 31 | 229.4 | 0.2 | + 9,650 |
| 32 | 233.0 | 1.0 | +10,940 |
| 33 | 236.6 | 0.2 | +12,260 |
| 34 | 240.1 | 1.0 | +13,580 |
| 35 | (Missed) | − | − |
| 36 | 247.3 | 1.0 | +16,410 |
| 37 | 250.8 | 0.2 | +17,820 |
| 38 | 254.5 | 1.0 | +19,350 |
| 39 | 258.0 | 0.2 | +20,810 |
| 40 | 261.6 | 1.0 | +22,340 |
| 41 | (Missed) | − | − |
| 42 | 268.8 | 1.0 | +25,120 |
| 43 | (Missed) | − | − |
| 44 | 275.9 | 1.0 | +23,680[a] |
| 45 | 279.5 | 0.2 | +22,020[a] |
| 46 | 283.0 | 1.0 | +20,320[a] |
| 47 | 286.6 | 0.2 | +18,500[a] |
| 48 | 290.3 | 1.0 | +16,580[a] |
| 49 | 293.8 | 0.2 | +14,730[a] |

[a] Refers to third-contact limb.

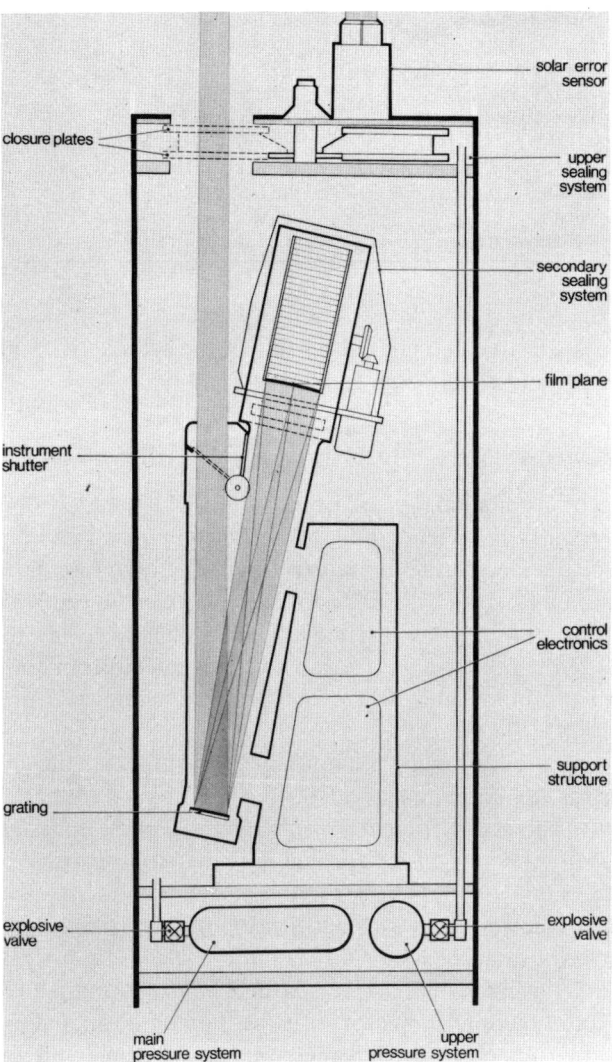

FIGURE 6 Schematic of uv payload instrumentation. The short-wavelength camera cycled and stored 50 frames of data during the period indicated on flight path (c) in Figure 5. Frame 22 of the sequence was close to uv eclipse second contact (photospheric extinction, $H_0 = 0$, Table 1) at the payload apogee height of 143 km.

TABLE 2  Spectrograph Characteristics, Wadsworth 1

| | |
|---|---|
| Wavelength range | 850–2220 Å |
| Focal distance | 50 cm |
| Spectroheliogram dispersion | 16.6 Å mm$^{-1}$ |
| Image diameter | 4.8 mm |
| Film resolution | 10 $\mu$m on film |
| | 0.17 Å in plane of dispersion |
| | 3000-km radial projection at sun |
| Grating | B&L 35-52-25-700 (Al + MgF$_2$) |
| Area | 15 cm$^2$ |
| Efficiency; blaze | 32% at 1500 Å ($n$ = +1) |

coronal data and to some related aspects of the solar activity observed in both the uv and visible data for March 7, 1970.

A. CORONAL LINES

Twenty-eight coronal images are observed, of which 17 have been identified by Jordan [1971]. Each line, of excitation potential $x$ eV, is characteristic of the electron temperature $T_e$ at which the calculable excitation function $g(T_e) = T_e^{-1/2} e^{-x/kT_e} \times$ (fractional ion abundance) has its maximum value [Pottasch, 1963]. The coronal lines may thus be ordered in temperature sequence and are reproduced from Jordan's paper in this form in Table 3. The isoelectronic sequence is also indicated.

In the case of collisionally excited optically thin lines the eclipse plate intensities for stigmatic images reflect the variation of $\int N_e^2 \, dz$ along the line of sight. This is a sufficiently good approximation in many cases to permit a structural analysis of coronal limb activity, provided some assumption is made about this structure in the $z$, or line-of-sight, direction. Figure 7 (lower) shows a small section of one uv limb spectroheliogram (exposure 26, 18:39:47 UT) in the spectral range 1330–1470 Å and conveniently illustrates both the data format and the varying excitation conditions for three western equatorial active regions. Reference to Table 4 shows that these images are associated with a range of excitation temperatures from $2 \times 10^4$ K to $1.7 \times 10^6$ K.

It is apparent from Table 3 that these newly observed coronal radiations are again from configurations $2p^n$ and $3p^n$. The lines are forbidden transitions within the ground term and in this respect complement and extend the visible and infrared spectra of the corona. At shorter wavelengths (xuv), the coronal spectrum is dominated by allowed transitions in coronal ions and has been extensively reviewed (see, for example,

TABLE 3  Newly Observed Coronal Lines, 1190-2185 Å

| $T_e(\times 10^6$ K) | Ion | Sequence | Transition | Obs. λ |
|---|---|---|---|---|
| 0.7 | Mg VII | C $2p^2$ | $^3P_1 - ^1S_0$ | 1190.2 |
| 0.7 | Si VII | O $2p^4$ | $^3P_2 - ^1D_2$ | 2147.4 |
| 0.8 | Si VIII | N $2p^3$ | $^4S_{3/2} - ^2D_{3/2}$ | 1446.0 |
| 1.2 | Si IX | C $2p^2$ | $^3P_2 - ^1D_2$ | 1715.3 |
| 1.2 | Si IX | C $2p^2$ | $^3P_1 - ^1D_2$ | 1985.0 |
| 1.2 | Si IX | C $2p^2$ | $^3P_2 - ^1D_2$ | 2149.5 |
| 1.3 | Fe XI | S $3p^4$ | $^3P_1 - ^1S_0$ | 1467.0 |
| 1.6 | Fe XII | P $3p^3$ | $^4S_{3/2} - ^2P_{3/2}$ | 1242.2 |
| 1.6 | Fe XII | P $3p^3$ | $^4S_{3/2} - ^2P_{1/2}$ | 1349.6 |
| 1.6 | Fe XII | P $3p^3$ | $^4S_{3/2} - ^2D_{5/2}$ | 2169.7 |
| 2.0 | O VII | He $1s^2$ | $^3S_1 - ^3P_2$ | 1624.0 |
| 2.0 | S XI | C $2p^2$ | $^3P_1 - ^1D_2$ | 1614.6 |
| 2.0 | S XI | C $2p^2$ | $^3P_2 - ^1D_2$ | 1826.0 |
| 2.0 | Ni XIII | S $3p^4$ | $^3P_2 - ^1D_2$ | 2126.0 |
| 2.0 | Ni XIV | P $3p^3$ | $^4S_{3/2} - ^2D_{5/2}$ | 1866.9 |
| 2.2 | Ni XIV | P $3p^3$ | $^4S_{3/2} - ^2D_{3/2}$ | 2185.1 |
| 2.5 | Ni XV | Si $3p^2$ | $^3P_1 - ^1D_2$ | 2085.7 |

Tousey [1967]). At wavelengths longer than 2200 Å, the coronal spectrum remains unknown until we reach the first line observed with ground-based instruments at λ3021 Å. Recovery of this missing spectrum remains one of the main objectives of future space experiments at eclipse.

B. LYMAN-α

Radiation at λ1216 Å is observed to dominate the coronal emission throughout all the uv eclipse data (Figure 7, top). This hydrogen Lyman-α radiation appears in the corona due to the resonance scattering of chromospheric Lyman-α, which is occulted by the moon in these data. The number of neutral hydrogen atoms ($\approx 1$ in $10^7$) remaining at coronal temperatures ($T \approx 2 \times 10^6$ K) is sufficient to produce the observed intensity. This problem has been considered in detail by Gabriel [1971].

C. THE PROMINENCES

Figure 8 is a schematic drawn from original 6563 Å H-α data recorded on eclipse day with the 10-cm coronagraph operated by the Institute for Astronomy, University of Hawaii. The heliocentric rotational coordinates and the corresponding direction of wavelength dispersion on the uv eclipse plates are shown. A loop structure, forming part of a

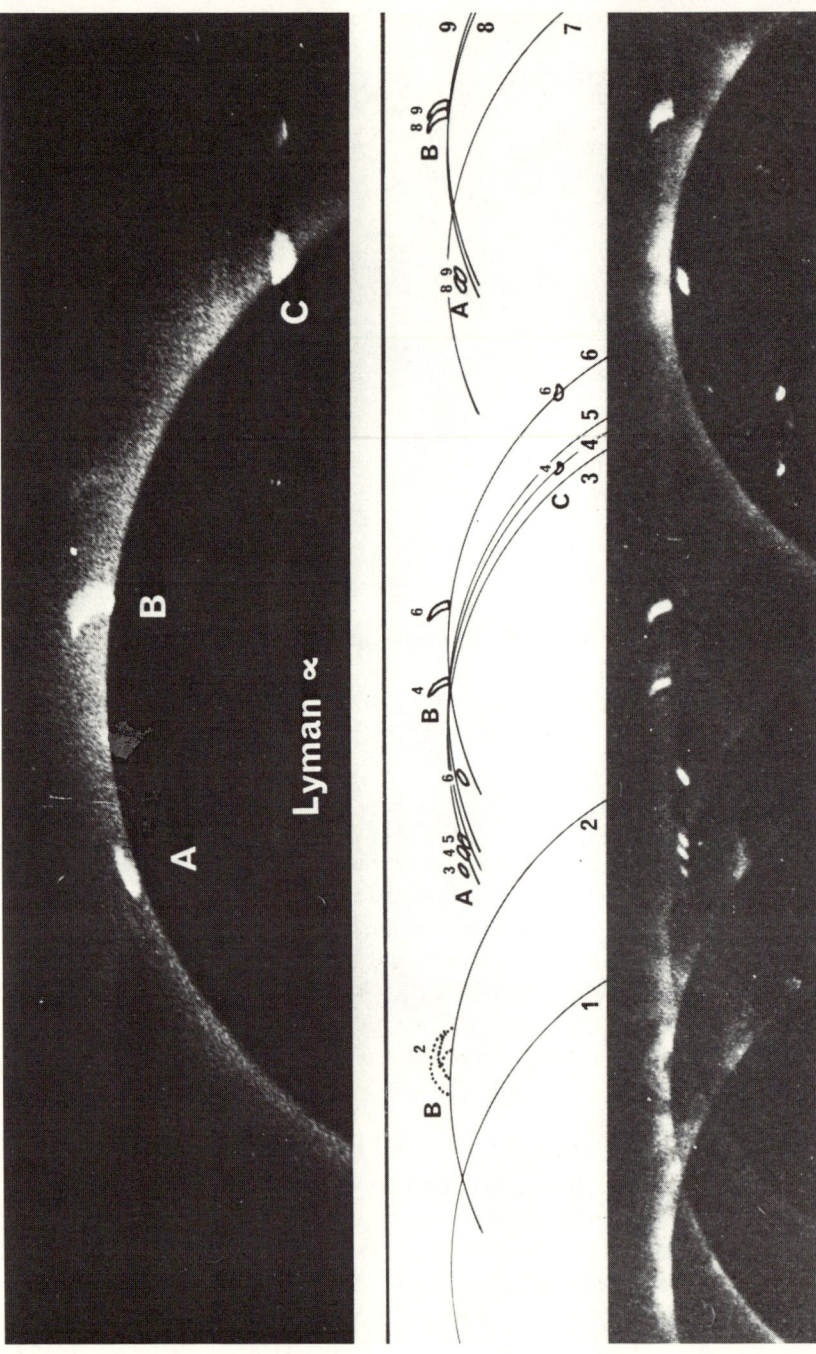

FIGURE 7 Eclipse spectroheliograms of the active limb regions A, B, and C. Upper section: Lyman-α 1216 Å; lower section: 1330–1470 Å. Features can be identified using the central key and Table 4.

TABLE 4  Key to Limb Spectroheliograms, Figure 7

| Temperature (K) | Ion | Spectro-heliogram | Wavelength (Å) | Transition | Prominence Image | | |
|---|---|---|---|---|---|---|---|
| | | | | | A | B | C |
| $2 \times 10^4$ | C II | 8, 9 | 1335.8 1334.6 | $^2P$–$^2D$ | $A_{8,9}$ | $B_{8,9}$ | Off figure |
| $6.5 \times 10^4$ | Si IV | 4, 6 | 1402.8 1393.8 | $^2S$–$^2P$ | $A_{4,6}$ | $B_{4,6}$ | $C_{4,6}$ |
| $1.2 \times 10^5$ | O IV | 3, 5 | 1404.8 1401.1 | $^2P$–$^4P$ | $A_{3,5}$ | Weak | Weak |
| $9.3 \times 10^5$ | Si VIII | 2 | 1446.0 | $^4S$–$^2D$ | Absent | Loop structure | Absent |
| $1.5 \times 10^6$ | Fe XI | 1 | 1467.0 | $^3P$–$^1S$ | Absent | Diffuse coronal activity | Absent |
| $1.7 \times 10^6$ | Fe XII | 7 | 1349.6 | $^4S$–$^2P$ | Absent | Intense coronal activity | Absent |

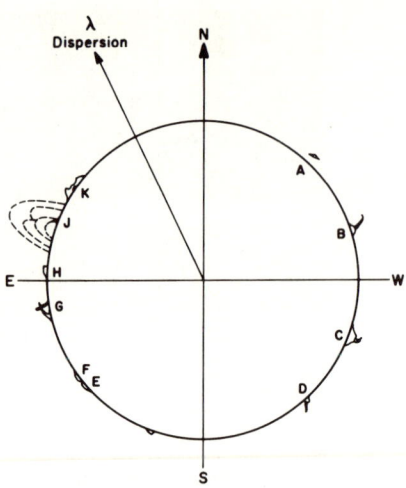

FIGURE 8 Solar limb schematic drawn from H-α coronagraph frame recorded at the Mees Solar Observatory, Haleakala, Maui. This data frame, timed at 18:40:24 UT on eclipse day, can be placed as 23 sec after the uv eclipse frame listed as exposure sequence No. 30 in Table 1. This schematic should be used in identifying the features shown in Figures 7, 9, 10, 13, and 14.

large coronal condensation, is present in the uv data alone, and we include it also in Figure 8, overlying the small H-α prominence J.

The structure that can be derived from the surface brightness of these objects in the uv is best visualized by arranging corresponding sections of different limb images in their temperature sequence with identical vertical registration.

Emission lines at the upper limit of the transition zone (Figure 1) are particularly sensitive in this respect, and Figure 9 chooses Si IV, O VI, and Si VIII to emphasize the rapidity of these changes with temperature (exposure 30, 1 sec 18:40:01 UT).

A systematic variation with temperature is observed for the loop system above J and is shown in Figure 10. This loop extends for 150,000 km into the corona (Si VIII) and surrounds the hotter core (Fe XII, S XI). The low-lying chromospheric emission at the limb within this loop (see, for example, in Si IV, Figure 9, and H-α, Figure 12, column J) probably represents quite different material along the line of sight.

Figures 11 and 12 are arranged to show the structural development with temperature for six of the nine active regions recorded in the uv data. Figure 11 shows principally chromospheric lines. Figure 12 continues the sequence through transition zone and coronal temperatures.

These arrays clearly show the very pronounced relative differences in the physical conditions of the transition zone prominence/corona. A comparison between A.H.K. and B.G.J. for Si VIII and Fe XI in Figure 12 shows three examples of an association between low emission at upper transition-zone temperatures and the quiescent type of

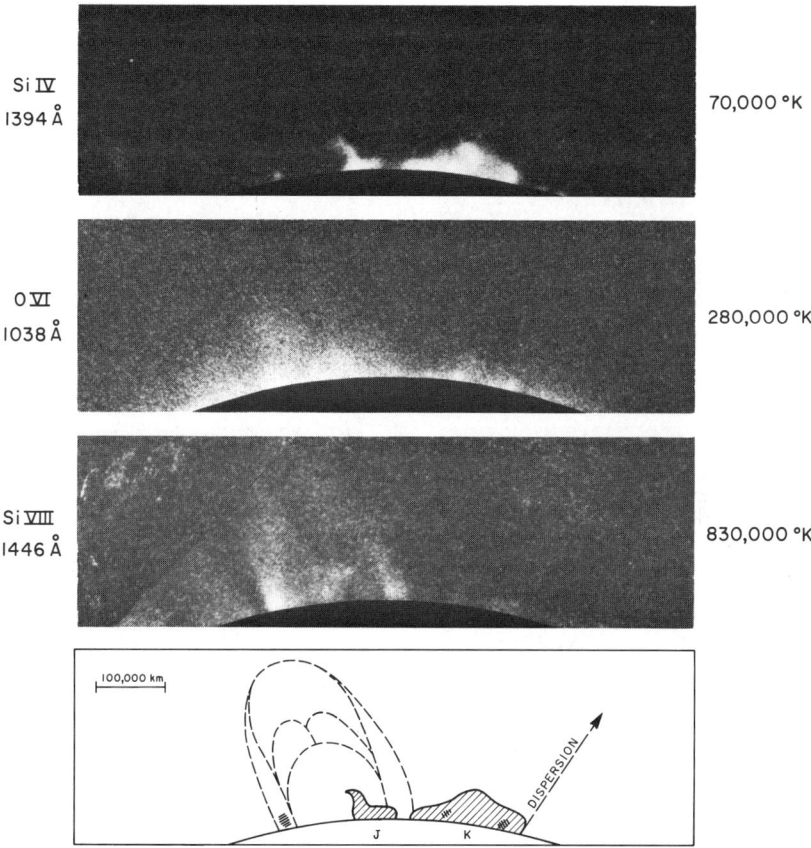

FIGURE 9 Eclipse spectroheliograms of the active limb regions J and K in Figure 8, recorded in Si IV, O VI, and Si VIII radiation with identical vertical registration. Unwanted images arising from partially overlapping spectra have been suppressed, and the position of the lunar limb has also been emphasized. The upper and lower images have been selected to show the rapid changes in emission that occur for $\Delta \log_{10} T \approx \pm 0.5$ about the upper limit of the steep temperature gradient shown in Figure 1. This region is characterized by the radiation of the lithiumlike ion O VI at $\log_{10} T_e = 5.4$, where emission arises partly from the transition zone and partly from the corona.

prominence where horizontal magnetic fields are known to predominate.

An important question concerns the evolutionary stage for these features observed in the uv. The uv data sequence is too transient in nature to have recorded physical changes in the solar atmosphere. However, by examining H-α records we may place these objects in some form of evolutionary sequence. One result of such analyses is that the feature A formed

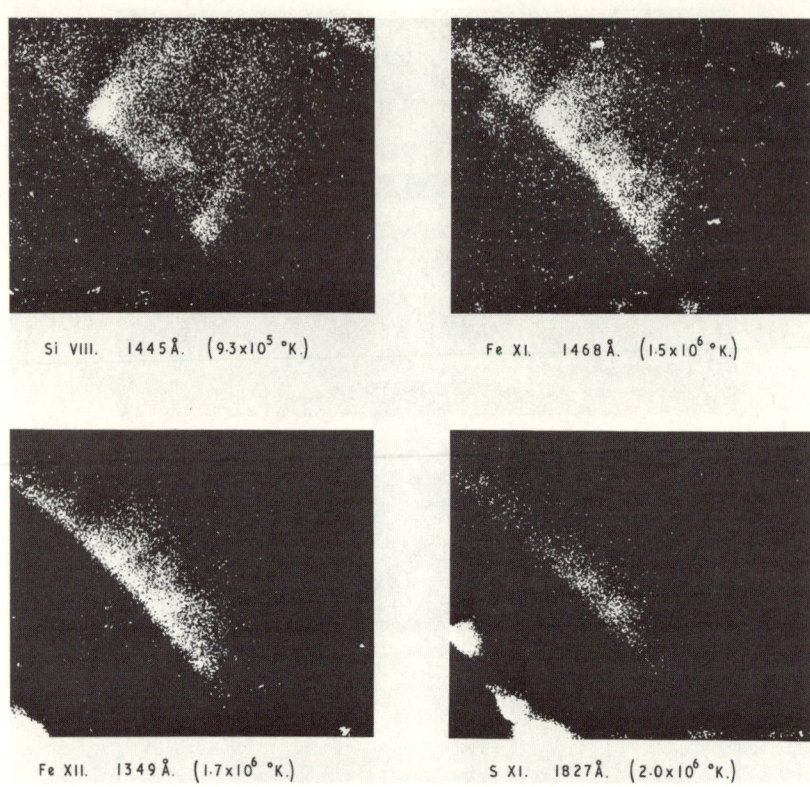

FIGURE 10  The Loop condensation of Figure 8 shown in four coronal lines in ascending coronal temperature. This temperature dependence of emission has been used in the identification of new uv coronal lines recorded at the March 7, 1970, eclipse [Jordan, 1971].

rapidly by condensation from the corona 7 min prior to the uv records (Figure 13). This simple object comes close to an idealized prominence model in which some unseen thermal instability triggers the condensation of coronal material, condensation proceeding until (in this case) gravitational forces cause an imbalance with the local magnetic field (last frame).

The evolution of prominence B is similar, although more complex phenomena can be seen in the sequence in Figure 14.

## V. Future Possibilities

Opportunities to extend and complement eclipse data of the type described above are rare. Table 5 lists the eclipses and transits that exist for

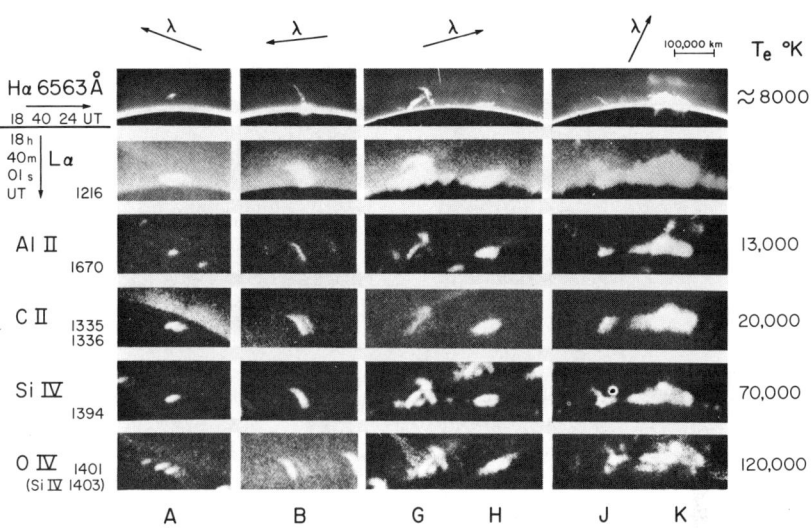

FIGURE 11  Chromospheric eclipse limb spectroheliograms arranged in ascending emission temperature and compared (top) to H-α prominences A, B, G, H, J, and K (Figure 8). Vertical registration is preserved, and an arrow indicates the differing directions of wavelength dispersion for each column. O IV and Si IV may be identified using the central key in Figure 7 and Table 4. Sequence continuation, Figure 12.

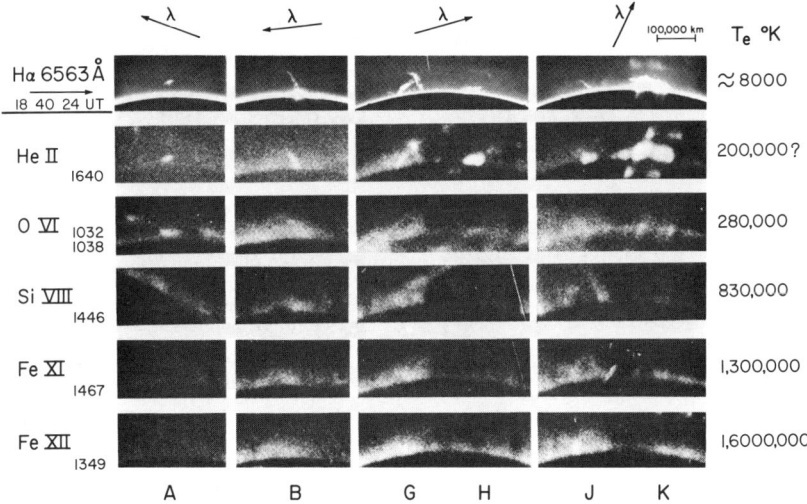

FIGURE 12  Eclipse limb spectroheliograms from transition-zone and coronal ions arranged in ascending emission temperature for $\log_{10} T_e > 5.5$. The temperature, and corresponding position for the He II 1640 Å images, may not be appropriate for all the regions shown. A temperature of $2 \times 10^5$ K has been estimated, by appearance alone, from the activity at B and G. All data in Figures 11 and 12 are from exposure No. 30 (Table 1) recorded 23 sec prior to the corresponding Haleakala H-α coronagraph data reproduced at the top of each column.

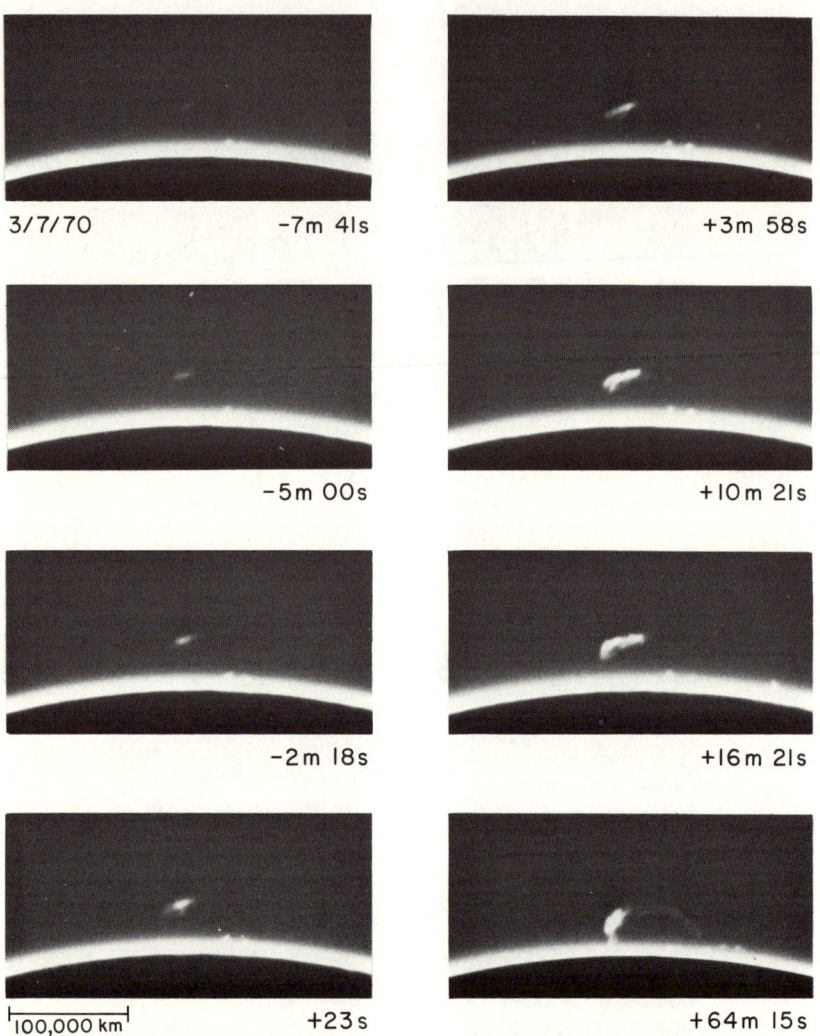

FIGURE 13 Time development of the H-α condensation A (Figures 7 and 8). This time sequence is referred to 18:40:01 UT for uv eclipse exposure sequence No. 30 (Table 1), at which point $t = 0$ in the sequence shown above. The data in Figures 13 and 14 were recorded by the 10-cm coronagraph at the Mees Solar Observatory, Haleakala. Filter $\Delta\lambda = \pm 3.5$ Å at H-α 6563 Å.

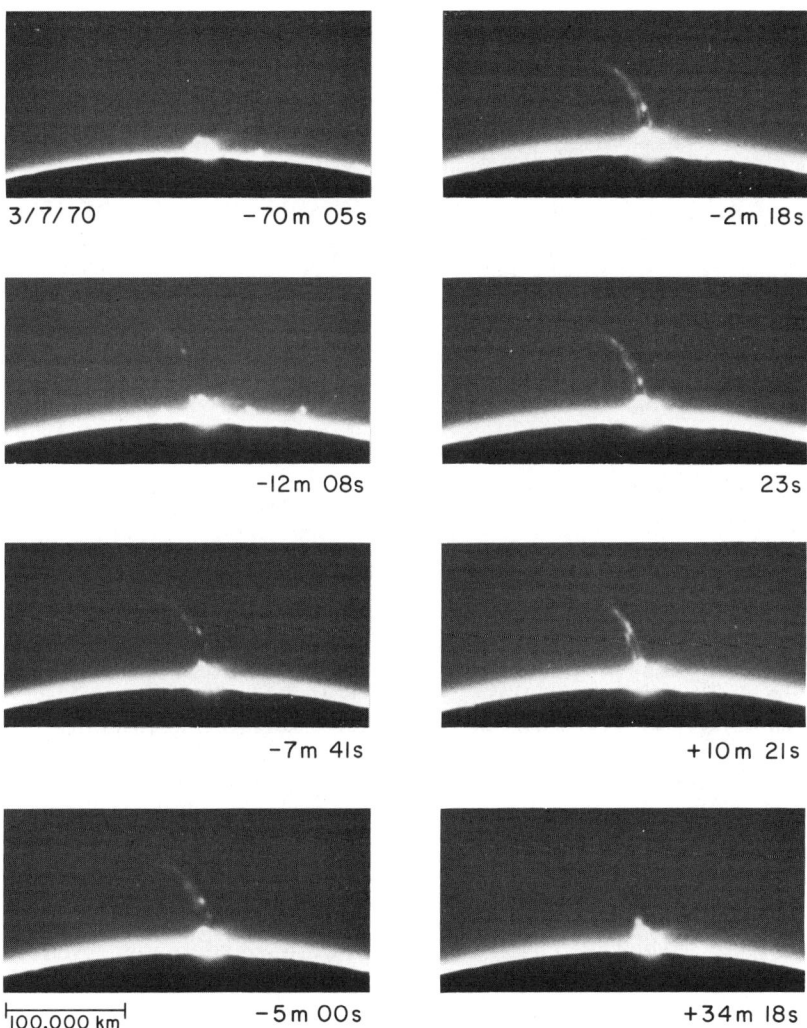

FIGURE 14 Time development of the prominence feature B in H-α; see Figures 7 and 8 and text.

TABLE 5  Four-Body Eclipse Possibilities

| Location of Rocket or Satellite | Site | Object | Eclipse or Transit |
|---|---|---|---|
| In earth shadow, between earth and moon, during a lunar eclipse | Earth | Moon | Lunar eclipse |
|  | Satellite | Earth | Solar eclipse |
|  | Satellite | Moon | Lunar eclipse |
| In earth shadow, between earth and moon's distance, during a solar eclipse | Earth | Moon | Solar eclipse |
|  | Moon | Earth | Lunar shadow transit of earth |
|  | Satellite | Moon | Lunar eclipse by earth |
|  | Satellite | Earth | Solar eclipse |
| In moon's shadow, between earth and moon, during solar eclipse | Rocket or satellite | Earth | Lunar shadow transit of earth (Figure 3) |
|  | Rocket or satellite | Moon | Solar eclipse (Figure 2) |

the earth–moon–sun system and a fourth body, namely, a rocket payload or satellite. The list is restricted to those situations favoring new solar or terrestrial data; however, we exclude the important case of observations of the gegenshein, zodiacal light, and corona from a *lunar-based* observatory as beyond the scope of this article. Moore and Schilling [1966] list the complete range of configurations.

Inspection of Figures 15 and 16 [Meeus *et al.*, 1966] shows that forthcoming opportunities of the 1970 type are limited to those listed in Table 6. Experiments are currently in preparation for the great eclipse event of June 30, 1973.

Figure 17 shows estimates for electron-scattered continuum levels ($K$ corona) to be expected at uv wavelengths down to 2000 Å, using disk continuum intensities published by Bonnet [1968]. Slitless eclipse spectra recorded with sufficient sensitivity will show these background levels unless they are masked by some unavoidable instrumental brightness. The units are those of photon brightness normalized to a central disk value of $10^{18}$ $h\nu$ cm$^{-2}$ sec$^{-1}$ sr$^{-1}$ Å$^{-1}$ at 5000 Å. Typical sky-brightness values are also shown and refer to ground-based observations in the visible (1, 2). The advantage of the externally occulted white-light

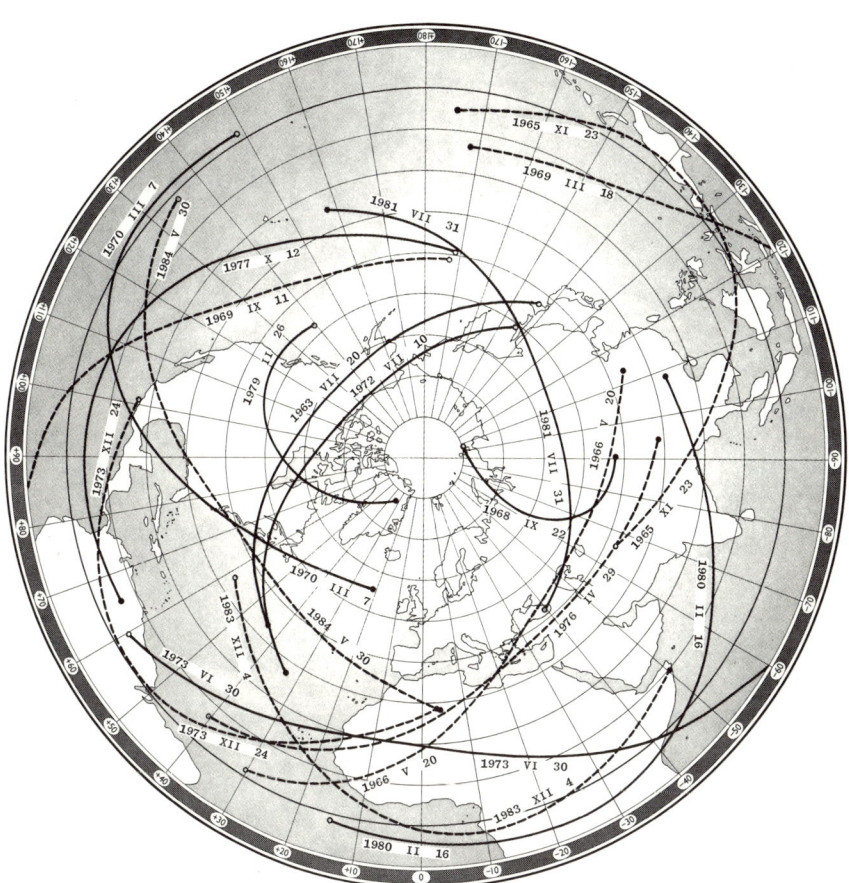

FIGURE 15  Northern hemisphere chart of future eclipse tracks up to 1984. Latitudes are marked in 10° intervals [Meeus *et al.*, 1966].

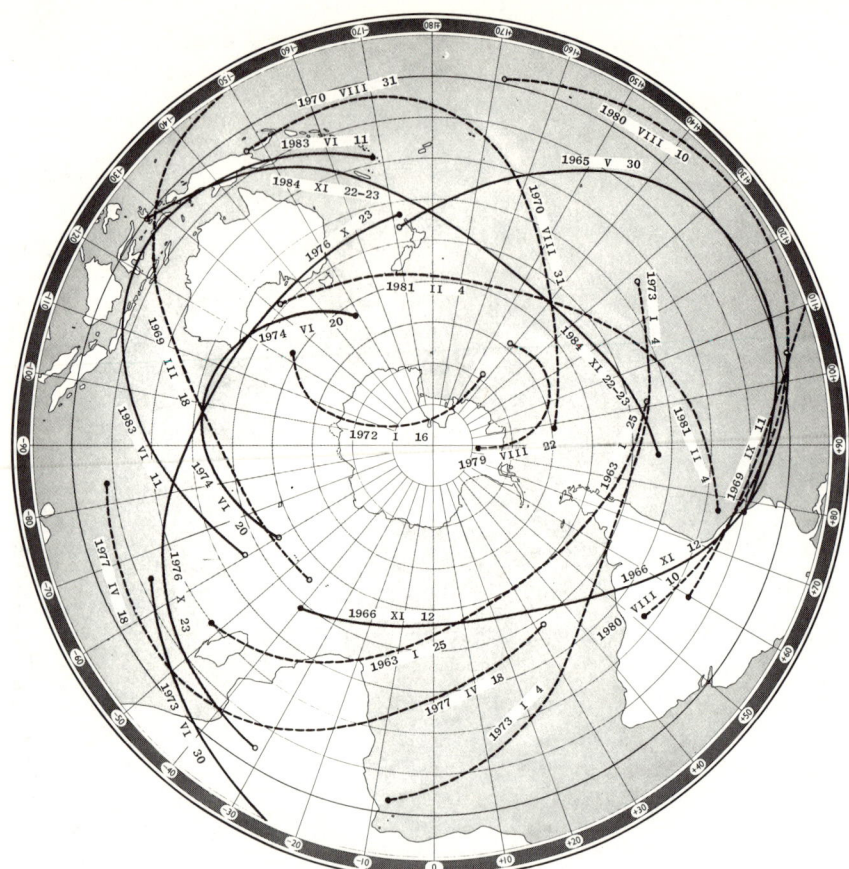

FIGURE 16  Southern hemisphere chart of future total eclipse tracks up to 1984. Open circles mark the beginning of the eclipse track at dawn; closed circles mark the termination of the eclipse at sunset [Meeus *et al.*, 1966].

TABLE 6  Forthcoming Total Eclipses

| Year | Date | Totality, Max. Duration | | Occurrence |
|---|---|---|---|---|
| | | min | sec | |
| 1974 | June 20 | 5 | 08 | S.W. Australia |
| 1976 | October 23 | 4 | 46 | S.E. Australia |
| 1977 | October 12 | 2 | 37 | Pacific |
| 1979 | February 26 | 2 | 48 | Canada |
| 1980 | February 16 | 4 | 08 | Africa, India |

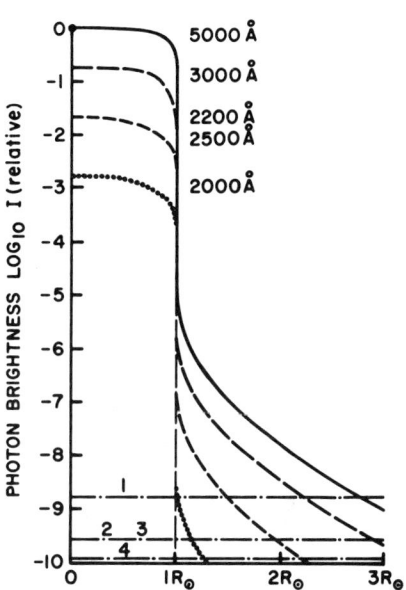

FIGURE 17 Estimates of electron-scattered continuum levels ($K$ corona) observable at eclipse for four uv wavelengths. The units are relative to a central disk photon brightness $I_o$ of $10^{18}$ $h\nu$ cm$^{-2}$ sec$^{-1}$ sr$^{-1}$ Å$^{-1}$. Short exposure and photographic background level limited the effective low signal of the 1970 rocket eclipse data on this scale to between $\log_{10} I$ (relative) = $-7$ and $-8$. Legends: (1) Van der Hulst [1953] sky brightness at midtotality. (2) Blackwell and Petford [1954], eclipse 30,000 ft, 6300 Å. (3) Newkirk and Lacey [1970], eclipse sky brightness 6400 Å. (4) Bohlin et al. [1971], instrumental brightness, white-light rocket coronagraph.

rocket or satelliteborne coronagraph is also evident (4). Short exposure and photographic background limited the effective low signal limit of the March 7, 1970, uv data to between $\log_{10} I$ (relative) = $-7$ and $-8$ on this scale, permitting the observation of a weak coronal continuum between 2000 and 2200 Å [Orrall and Speer, 1973].

The writer wishes to thank the following people for the use of material reproduced here: Marie McCabe (H-α coronagraph data, Haleakala Observatory); G. Newkirk (HAO white-light corona); Jean Meeus (eclipse charts); H. L. Galloway (NASA ATS-III).

The helpful advice and stimulus of Frank Orrall have been greatly appreciated.

Travel grants from the Royal Society and the International Astronomical Union are gratefully acknowledged.

## References

Blackwell, D. E., and A. D. Petford, *Mon. Not. R. Astron. Soc. 131*, 383 (1966).
Blamont, J. E., and C. Malique, *Astron. Astrophys. 3*, 135 (1969).
Bonnet, R., *Ann. Astrophys. 31*, 597 (1968).
Brueckner, G. E., J. F. Bartoe, K. R. Nicholas, and R. Tousey, *Nature 226*, 1132 (1970).
Chubb, T. A., H. Friedman, R. W. Kreplin, R. L. Blake, and A. E. Unzicker, *Mem. Soc. R. Sci. Liege 4*, 228 (1961).
Friedman, H., *Annu. Rev. Astron. Astrophys. 1*, 59 (1963).

Gabriel, A. H., W. R. S. Garton, L. Goldberg, T. J. L. Jones, C. Jordan, F. J. Morgan, R. W. Nicholls, W. J. Parkinson, H. J. B. Paxton, E. M. Reeves, C. B. Shenton, R. J. Speer, and R. Wilson, *Astrophys. J. 169*, 595 (1971).

Gabriel, A., *Solar Phys. 21*, 392 (1971).

Goldberg, L., R. W. Noyes, W. H. Parkinson, E. M. Reeves, and G. L. Withbroe, *Science 162*, 95 (1968).

Jones, T. L. J., W. H. Parkinson, R. J. Speer, and C. Yang, *Solar Phys. 21*, 372 (1971).

Jordan, C., *Solar Phys. 21*, 381 (1971).

Meeus, J., C. C. Grosjean, and W. Vanderleen, *Canon of Solar Eclipses*, Pergamon, New York (1966).

Moore, R. C., and G. F. Schilling, *J. Astronaut. Sci. 13*, 7 (1966).

Newkirk, G., and L. Lacey, *Nature 226*, 1098 (1970).

Noyes, R. W., *Annu. Rev. Astron. Astrophys. 9*, 209 (1971).

Olsen, K. H., C. R. Anderson, and J. N. Stewart, *Solar Phys. 21*, 360 (1971).

Orrall, F., and R. J. Speer, *Solar Phys. 29*, 41 (1973).

Pottasch, S. R., *Astrophys. J. 137*, 945 (1963).

Reeves, E. M., and W. H. Parkinson, *Astrophys. J. Suppl. 21*, 1 (1970).

Speer, R. J., W. R. S. Garton, L. Goldberg, W. H. Parkinson, E. M. Reeves, J. F. Morgan, R. W. Nicholls, T. L. J. Jones, H. J. B. Paxton, D. B. Shenton, and R. Wilson, *Nature 226*, 249 (1970).

Suomi, V. E., R. J. Parent, G. Warnecke, and W. S. Sunderlin, *Bull. Am. Meteorol. Soc. 49*, 75 (1968).

Tousey, R., in *Beam Foil Spectroscopy*, Vol. 2, p. 485, Gordon and Breach, New York (1967).

Tousey, R., *Philos. Trans. R. Soc. A270*, 59 (1971).

Van de Hulst, H. C., in *The Sun*, University of Chicago Press, Chicago, Ill. (1953).

B. BATES, C. D. McKEITH, G. R. COURTS, and J. K. CONWAY

# Astronomical Observations in the Middle Ultraviolet with High Spectral Resolution

## Introduction

In many of the spectroscopic astronomical observations to date the spectral resolving power employed has of necessity been less than desirable, particularly in the ultraviolet region of the spectrum. The reasons for this are straightforward: the radiant flux from astronomical sources is intrinsically very weak, and the resolving power of a given spectroscopic instrument can in general only be increased at the expense of its luminosity or light-gathering power. The situation is less severe in the visible region of the spectrum, where spectrometers can be combined with large ground-based telescopes. The difficulty in fulfilling the necessary requirements for astronomical observations with high spectral resolution increases progressively toward the shorter wavelengths, where the majority of the astrophysically important spectral lines of neutral and ionized atoms are to be found. For example, of considerable importance in the study of the chromospheres of the sun and of late-type stars are the resonance lines of Mg II at $\lambda 279.5$ and $280.2$ nm. In the solar spectrum

---

The authors are in the Department of Pure and Applied Physics, The Queen's University of Belfast, Belfast BT7 1NN, Northern Ireland.

these lines are evident as broad absorption features in the photospheric continuum with self-reversed emission components in their center (Figure 1). The shape and strength of the emission components are highly sensitive to large- and small-scale chromospheric activity, while the broader absorption contours yield information on the more general conditions in the atmosphere [Bates *et al.*, 1969, 1971a; Lemaire 1969, 1970, 1971]. As such, the Mg II lines are analogous to the H and K lines of Ca II in the visible region of the spectrum but are much more prominent in emission and absorption and can thus be studied in greater detail and to fainter stellar magnitudes [Bates and McKeith, 1972].

Resolving powers of the order of $10^5$ are required to fully resolve the detail of the emission features, although somewhat lower values are adequate to study the wider absorption profiles. The best reported Mg II satellite observations of stellar sources are those of the Utrecht experi-

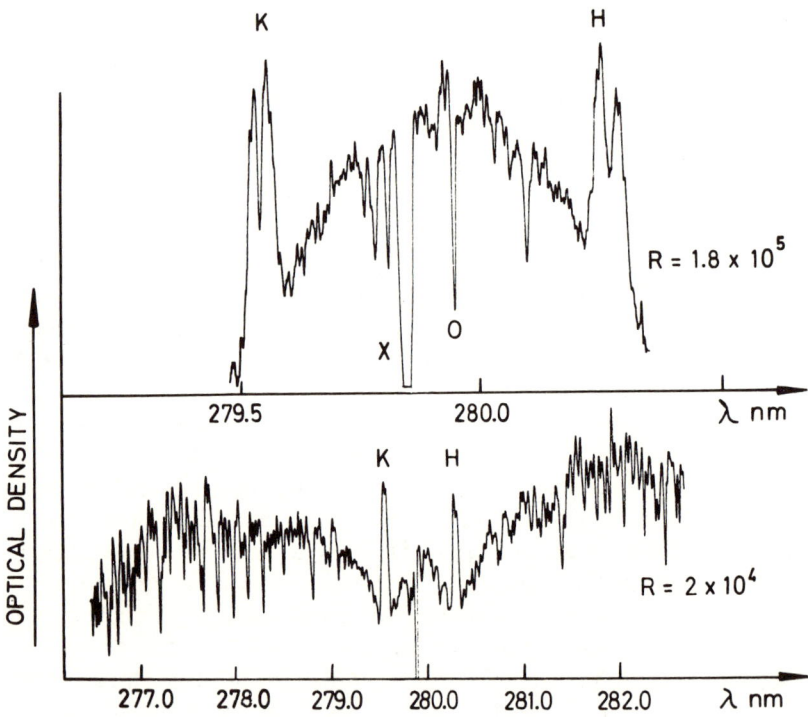

FIGURE 1 Microdensitometer traces of a solar Mg II Fabry-Perot interferogram (see Figure 10). The lower trace is obtained from the echelle dispersion alone, and the upper trace from the combined Fabry-Perot and echelle dispersion. (X and O are wavelength and position markers, respectively.)

ment on the ESRO TD-1A [Hoekstra et al., 1972] (Figure 2). With a resolution of 0.18 nm ($R = 1.45 \times 10^3$) the Mg II absorption features are well resolved, although the instrumental width is evidently of the order of the stellar line widths. Recently, stellar spectra in the same wavelength region have been obtained with a balloonborne spectrometer [Kondo et al., 1972] in which the resolution was between 0.025 and 0.05 nm. Here the resolving power was sufficient to detect the emission components in the spectra of several stars and to resolve the self-reversal in the exceptionally strong and broad features of the supergiant α Ori. The need for more general higher resolution ($R > 10^5$) observations is still outstanding.

Again, in theoretical model stellar atmosphere calculations that en-

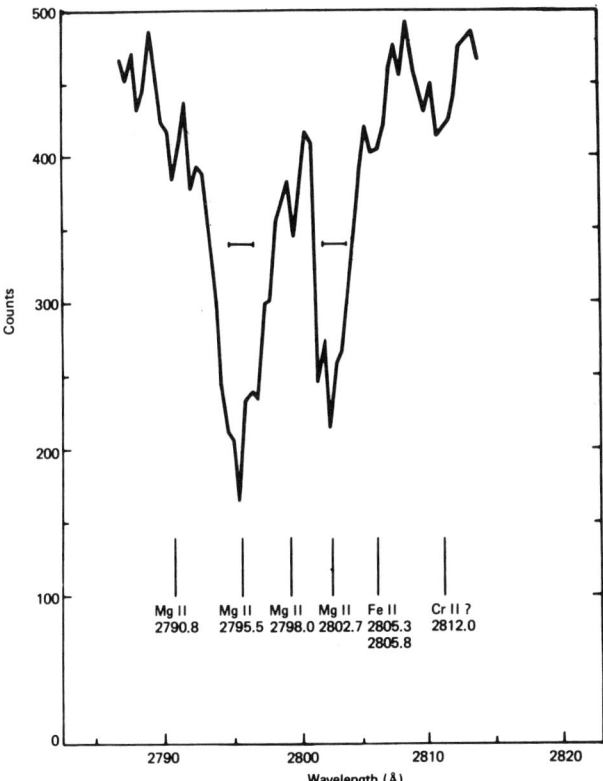

FIGURE 2 Mg II absorption line spectrum of β Aurigae (A2V) obtained by the ESRO TD 1A satellite [Hoekstra et al., 1972]. The horizontal bars indicate the half-width of the instrumental profile.

deavor to predict the absolute spectral energy distribution of the radiation emitted by stars, the number and strength of the many absorption lines included in the computation strongly influence the form of the emergent intensity. It is necessary to obtain high-resolution spectra of each stellar type in the ultraviolet, where this heavy line blanketing occurs, for comparison with the computed spectra. Progress in this direction is again demonstrated by the TD-1A data (Figure 3), where the decrease in flux observed at 253 nm in the earlier Orbiting Astronomical Observatory spectra ($\Delta\lambda = 2.5$ nm) is clearly seen with the higher resolution to be due to the strong influence of the many Fe II lines. The theoretical spectral distribution is also shown for comparison [Maran et al., 1968]. Additional observations with much higher spectral resolution are required in order to establish the collective importance of fine-scale absorption features in the spectra.

A pressing need for ultraviolet astronomical observations with spectral resolving powers in excess of $10^5$ is in the study of the interstellar medium. Here the individual clouds of intervening interstellar gas produce

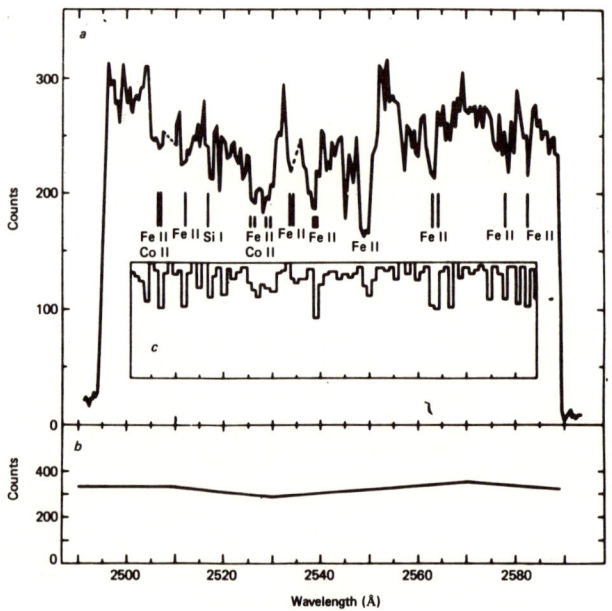

FIGURE 3 Spectrum of $\beta$ Aurigae (A2V) in the region 249 to 259 nm: (a) as observed by TD 1A; (b) as observed by OAO-2; (c) as predicted by LTE theory. [Hoekstra et al., 1972].

separate narrow absorption lines in the continuum spectra of hot early-type stars of class O and B. In many cases, as seen in the ground-based data, these lines may be as narrow as 0.006 nm (3 km/sec) [Hobbs, 1969; Marschall and Hobbs, 1972]. With several interstellar clouds in the line of sight to a background star and with moderate spectral resolution, the principal asset in the spectrum is the equivalent width of the combined absorption features. Much of the information on the motion of the individual clouds is masked by instrumental broadening, while also the analysis of the number density of the absorbing species involves certain assumptions about the prevailing velocity distribution. With high-resolution observations ($R > 10^5$) the form of the distribution becomes directly available, and information on the physical conditions within the clouds can be obtained from the detailed individual profiles. Observations of elements in more than one state of ionization are particularly informative, and the near ultraviolet lines of Mg, Fe, and Si will contribute significantly in this respect. Resolving powers of the order of $6 \times 10^5$ have been demonstrated to be warranted in such studies of the interstellar medium by Hobbs using an elaborate triple interferometer system (PEPSIOS) in the visible lines of Na and Ca.

As already indicated, space vehicles are required for astronomical observations in the ultraviolet because of the attenuation of the earth's atmosphere. To examine the complete ultraviolet spectrum longward of 91.2 nm requires observational altitudes in excess of 100 km, accessible only by rockets and satellites, each having its separate merits. A useful part of the middle-ultraviolet astronomical spectrum between 200 and 300 nm, however, including the important Mg I and II lines, is open to observation from balloons of $5 \times 10^5$ m$^3$ or larger, at altitudes of around 40 km. Scientific ballooning is particularly attractive for uv astronomy, as it offers the possibility of long flight periods, instrument operation by ground command, large payload capacities, short preparation and turn-around times for flights, and relatively low costs.

In ground-based astronomy using large optical telescopes, the flux in a spectral resolution element of a given spectroscopic instrument is ultimately limited by the atmospheric seeing conditions. In observations from space vehicles, however, the determining aspect is the accuracy and stability with which a telescope can be pointed at the object under investigation. The degree to which the pointing capability of the stabilized platform is effective in determining the spectral resolution of the observation is dependent almost entirely on the angular dispersion of the spectrometer or spectrograph. Two types of disperser well suited to space astronomy are the high blaze-angle grating, in particular the echelle grating, and the Fabry-Perot interferometer [Bradley, 1968].

## Grating Instruments

Astronomical spectrometers or spectrographs employing plane diffraction gratings may be either slit or slitless systems [Wilson, 1965; Rense, 1966]. The former are used in combination with flux-collecting telescopes. So too are the latter on occasions, as well as in the objective grating configuration. In those systems incorporating entrance slits, the spectral resolution is determined solely by the slit width for a given ruled grating, while the flux in a spectral resolution element is critically dependent on the absolute stellar acquisition (second of arc pointing is required). In slitless systems, the absolute pointing of the platform becomes only of secondary importance (typically of minute of arc accuracy), but the spectral resolution is ultimately dependent on the pointing noise for a given instrument.

For a plane grating used in autocollimation, the input angular dispersion is

$$d\alpha/d\lambda = 2 \tan \theta / \lambda,$$

and the angular width of a resolution element is

$$d\alpha = 2 \tan \theta / R$$

($\theta$ = the autocollimation blaze angle), demonstrating that the higher the blaze angle the greater the angular width of the spectral resolution element for a given resolving power.

When used in combination with the telescope, the pointing performance ($d\mu$) required to match the angular width ($d\alpha$) of the spectral element becomes

$$d\mu = (D_c/D_T)(2 \tan \theta / R),$$

where $D_c$ and $D_T$ are the spectrometer collimator and telescope aperture, respectively. For a 2160 lines/mm grating, used at a blaze angle of approximately $32°$ in the second order at wavelength 250 nm, then with $D_c$ = 100 mm and $D_T$ = 400 mm, the pointing required to match the resolution element for $R = 10^5$ is of the order of $\pm 0.3$ sec of arc. It is emphasized that this is only a telescope of moderate aperture, that the resolving power is only moderately high, and that the blaze angle is approaching the maximum for this type of grating. Thus, even if such exacting pointing is achieved, this system is already substantially stretched, and while the size of the grating could probably be trebled, allowing a corresponding increase in the telescope size, the cost and weight of the systems are increased considerably.

One method of easing the requirements on the platform pointing system is that of using an echelle grating with a blaze angle of the order of 63° ($\tan^{-1} 2$). The pointing for the above example is now relaxed to a slightly less demanding value of the order of ±1 sec of arc. The smaller free spectral range of the echelle, however (e.g., $\approx$10 nm at $\lambda$250 nm for a 300 lines/mm grating), requires a cross-dispersion system for order sorting, as also does the plane grating used in the second or third order. While in the latter case a simple small-angle prism will suffice, the echelle system generally requires a second grating to provide adequate separation of the echelle cycles. This adds complexity to the optical system, as well as becoming expensive for the larger-aperture systems. Nevertheless, it offers the overriding advantage of accommodating a wide spectral range in the form of many spectral cycles in a small two-dimensional array. The spectrum may then be photographed directly or in combination with an image intensifier, or integrated on an SEC target and read out by a television camera tube.

An example of current astronomical crossed-dispersion systems is the uv stellar echelle payload of the Astrophysics Research Unit, Culham Laboratory (Figure 4) [Burton *et al.*, 1971]. A stellar image is produced by a 350-mm aperture $f/12$ Cassegrain telescope at the spectrograph entrance aperture, which is located at the vertex of an inclined conical mirror. Reflected light from the periphery of the entrance aperture is used to generate an error signal, and the final acquisition and rms noise is expected to be of the order of 2.5 sec of arc. The crossed-dispersion system enables the entire wavelength range from 120 to 300 nm to be recorded simultaneously with a spectral resolution between 0.007 and 0.027 nm, respectively. The compact nature of the array is easily accommodated on the salicylate-coated window of the image tube and subsequently photographed on 35-mm film with an estimated gain in speed of 10 over direct photography (Figure 5). A similar system is under consideration for the proposed International Ultraviolet Explorer satellite, whose projected launch data is in 1976. Here the image tube is supplemented by an SEC vidicon camera and the spectral information telemetered to ground receiving stations.

One disadvantage of such simultaneous wide spectral coverage is that the near-ultraviolet end of the spectrum is often recorded with insufficient resolution. Thus for some studies the 200- to 300-nm region must be given separate consideration; and, being intermediate between the

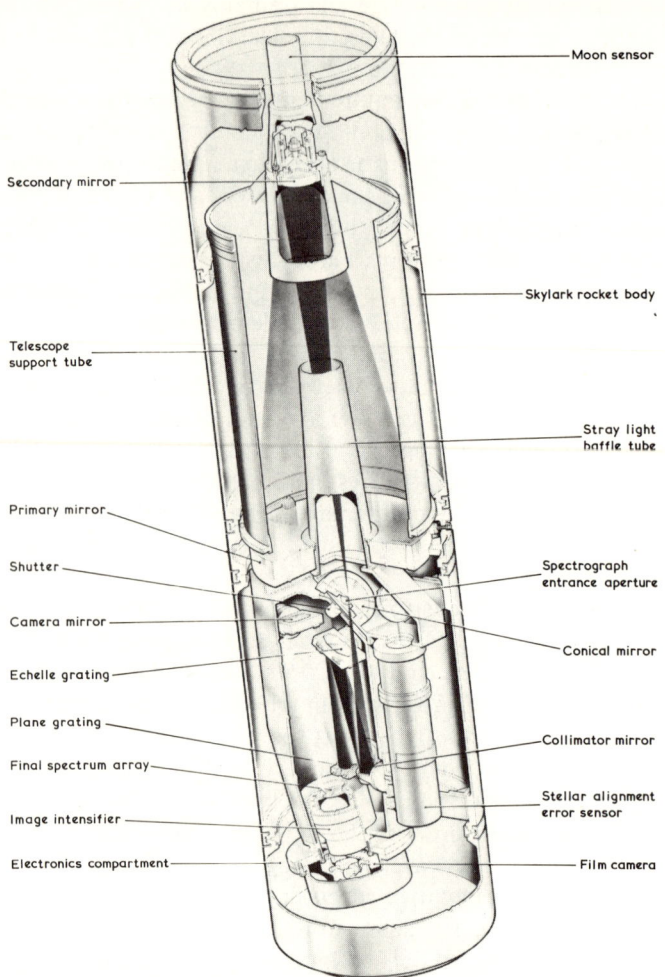

FIGURE 4 Ultraviolet stellar echelle payload (Astrophysics Research Unit, Culham Laboratory, U.K.).

atmospheric extinction and the quartz cutoff, it occupies a special position in that advantage can be taken of established ground-based spectroscopic techniques applied to space-vehicle observations.

Plane gratings may also be used with some advantage in the objective grating configuration. Here the relative aperture ratio of the telescope to the disperser or collimating mirror, $D_c/D_T$, disappears so that for a given resolving power the demands on the pointing requirements of the sta-

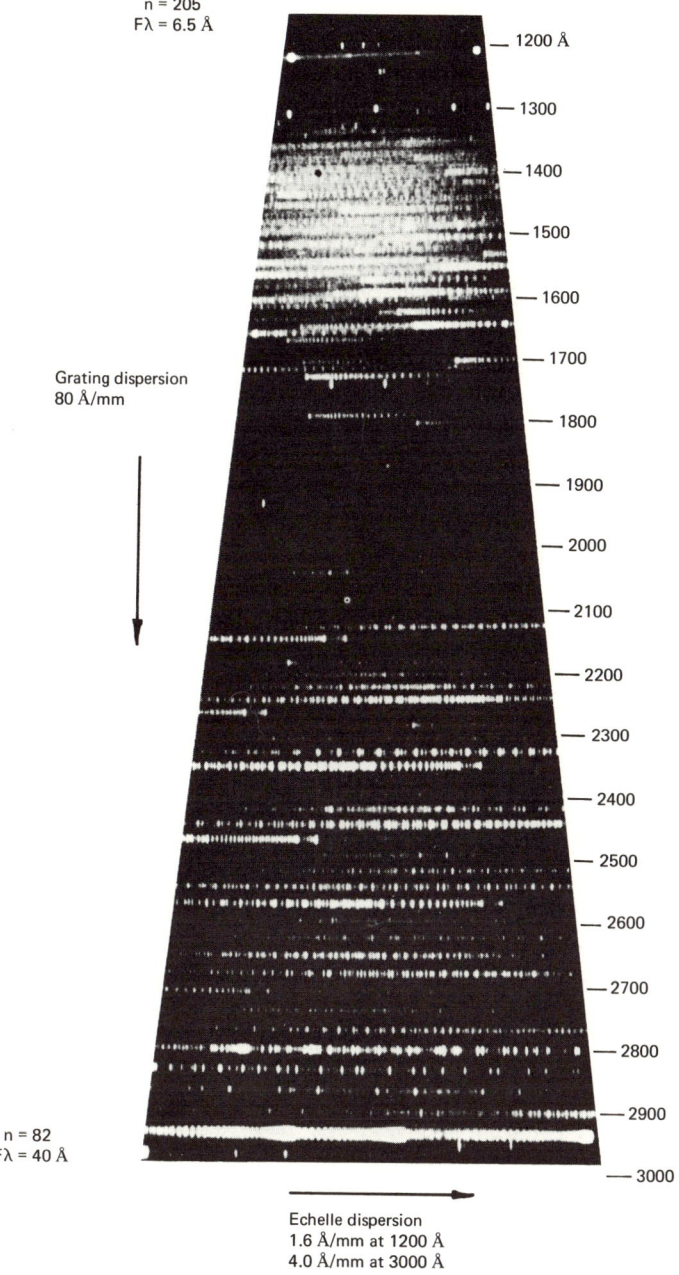

FIGURE 5 Echellogram of a microwave air discharge of the spectral range 1200–3000 Å (Culham Laboratory).

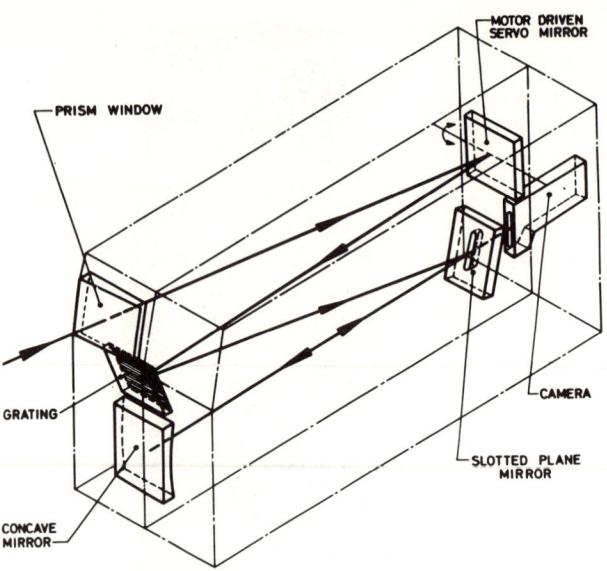

FIGURE 6  Balloonborne objective grating spectrograph (Queen's University of Belfast/University College of London).

bilized platform are relaxed by the same factor, but only at the expense of collector area, which is now the normal area of the grating. Although the maximum size of this is currently 208 × 208 mm$^2$, these are exceptional gratings and a more normal flight system would be typically 128 × 128 mm$^2$.

One disadvantage of large-objective-grating systems is that in general a large inertial mass is involved in the stabilization, which is therefore more difficult than in the case of fine stabilization of a telescope secondary mirror. For the smaller gratings, secondary stabilization becomes practical as in the balloonborne objective instrument prepared by the Queen's University of Belfast and University College, London, for an initial preliminary sky survey in the Mg I and II stellar and interstellar program.* The optical configuration of the spectrograph is shown in Figure 6. The 2160 lines/mm (102 × 128 mm$^2$) grating is used in the second order, close to the blaze angle of 32°41′, and the output angular dispersion is ≈6 × 10$^{-3}$ rad/nm. To achieve a desired minimum spectral resolution of 0.01 nm for the first flight, image stabilization to better than ±6 sec of arc is required in the direction of the grating dispersion. This is

*This experiment was successfully flown on October 4, 1972.

achieved by servo-control of a plane beam-folding mirror illuminating the grating. The whole payload is pointed in azimuth and elevation to approximately ±1 min of arc. The first, second, and third orders of the grating are separated by a quartz prism. The optical configuration finally adopted is particularly attractive for its relative simplicity and for its excellent image-forming properties. Laboratory tests on the spectrograph alone indicate a spectral resolution of the order 0.005 nm ($R \approx 6 \times 10^4$).

In summary, plane diffraction gratings offer the possibility of recording high-resolution spectra over a wide spectral range. It is, however, not practical to do so while maintaining sufficient spectral resolution at the longer wavelengths for line-profile studies. Furthermore, even with echelle gratings, for only moderate-size telescopes and average resolving powers, the demands on the platform pointing systems are in the region of ±1 sec of arc. In order to relieve the situation, very large gratings are required. These are expensive, particularly if associated with an additional order-sorting grating. Also the increasing weight and difficulty of maintaining the figure of such large gratings in support become critical with respect to the restricted capacities of space vehicles.

## Some Developments in Fabry-Perot Instruments for Ultraviolet Astronomy

It was first pointed out by Jacquinot [1954] that the luminosity of Fabry-Perot spectrometers could be many times greater than those using conventional high and low blaze gratings for equal dispersing areas and for the same resolving power. The considerable advantages to be gained in their application to ground-based astronomy have been discussed in detail [e.g., Vaughan, 1967; Meaburn, 1970]. Their greater luminosity arises directly from the higher angular dispersion of the interferometer, but of special relevance to space application is the simultaneous relaxation in the demands on the platform stabilization. The immediate disadvantage of the single-etalon system is its small free spectral range, thereby requiring some form of predispersing system.

As is well known, for an extended source a circular fringe pattern is produced in which the angular dispersion is simply

$$d\phi/d\lambda = 1/\lambda \tan \phi,$$

where $\phi$ is the angular distance from the center of the pattern. If a diaphragm in the form of a circular hole is located on axis ($\phi \to 0$) of angular radius equal to the half intensity width of the central fringe of zero internal radius, those wavelengths corresponding to the spectral resolution of the interferometer will be transmitted and others rejected. This angular diameter is

$$\Delta\phi = (8/R)^{\frac{1}{2}}.$$

For a given resolving power, this corresponds to the angular width of the spectral resolution element and, hence, to the pointing requirement of the space platform.

The immediate advantage is seen in the example of $R = 10^5$, where $\Delta\phi$ equals the large value of $9 \times 10^{-3}$ rad (i.e., $\approx \frac{1}{2}°$). In combination with a telescope, the requirement is correspondingly greater, e.g., for $D_T = 0.5$ m and $D_{FP} = 25$ mm the pointing necessary to maintain the stellar image within the analyzing diaphragm is of the order of ±45 sec of arc. Thus for telescopes of moderate size and moderate resolving powers, only coarse pointing is required. Alternatively, if a stabilization accuracy of ±10 sec of arc was available, then a small-diameter interferometer (~25 mm) could accept all the flux from a 2.25-m telescope for a resolving power of $10^5$. Thus unlike the grating instruments, even small interferometers are well beyond being platform pointing limited. Moreover, the large angular width of the analyzing diaphragm considerably relaxes the tolerances on the quality of the stellar image, permitting the use of large-aperture telescopes with only moderately figured optics.

To carry out a program of Fabry-Perot astronomical spectroscopy in the ultraviolet from space vehicles requires the following:

1. Good optical quality interferometer plates and high-reflectance, low-absorption coatings for high etalon finesses with good transmission characteristics;
2. A suitable predispersing system for isolating a single-interferometer free spectral range;
3. A means of scanning the interferometer in wavelength for photometric spectrometers;
4. A method of synchronously scanning the predisperser and interferometer for examination of a large spectral range;
5. A measure of instrument reliability and robustness and the capacity to be operated by remote control.

With recent advances in optical polishing techniques and thin-film deposition technology, interferometers with a finesse between 20 and 40 over the wavelength region 200–300 nm are now feasible [Bradley *et al.*, 1967].

For a series of experiments (Queen's University/Culham Laboratory) to measure the point-to-point variation of the Mg II profiles on the solar disk with high spatial resolution and with a resolving power of $2 \times 10^5$, permanently adjusted air-gap interferometers were developed to be used classically with photographic detection in rocketborne and balloonborne spectrographs [Bates et al., 1966]. The robustness and stability of alignment of the optically contacted, high-finesse (approximately 30 at λ280 nm) interferometers is demonstrated by the fact that all interferometers flown in the two rocket and two balloon flights were recovered intact with their preflight performance [Bates et al., 1969, 1971a].

The layout of the solar balloon instrument is shown in Figure 7. The interferometer spectrograph and control electronics were enclosed in a thermally insulated cylindrical envelope. The payload was mounted on a Ball Bros. Solar Pointer platform, which orientated the package to within a few minutes of arc of the solar center. Fine stabilization of the solar image to ±5 sec of arc on the spectrograph slit was achieved through

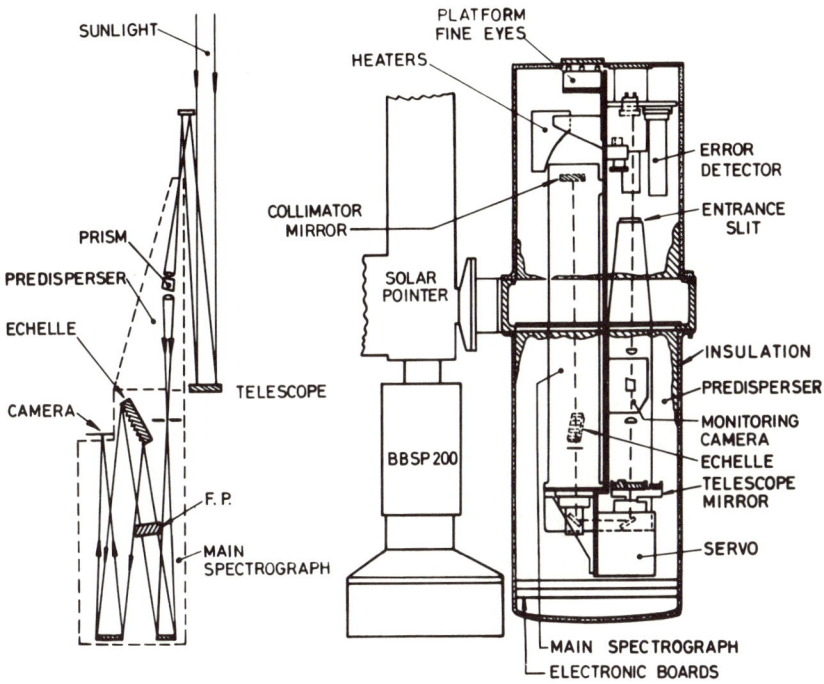

FIGURE 7  Solar interferometer spectrograph (Queen's University of Belfast/Astrophysics Research Unit, Culham).

servo-controlling the main telescope mirror. The high-dispersion unit comprised an echelle grating spectrograph, whose dispersion was crossed with that of a Fabry-Perot interferometer located in the parallel beam. In effect, the echelle grating separated adjacent free spectral ranges of the interferometer so that for a continuum spectrum a series of heterochromatic channels were formed in the film plane (Figure 8). For an 8° angle of incidence onto the interferometer, the dispersion along the channels was essentially linear, and the spectral resolution obtained was approximately 1.6 pm.

There are, however, difficulties associated with this classical photographic arrangement, since for the extended source there is a spatial or position variation as well as a spectral variation along any heterochromatic channel. In the static interferometer configuration, as in the rocket flights, it was not possible to separate the spatial and spectral variation, i.e., to derive the Mg II profiles for areas of the order of the spatial resolution of the instrument. With the longer time available during the balloon flights, the spectrum was scanned by tilting the interferometer in angular steps corresponding to the spectral resolution element and making an exposure at each step. By doing so some 15 times, 30 intensity points could be obtained for the construction of profiles at each spatial point.

While this method of angle scanning was successful, a more direct method of obtaining the profiles is that using photoelectric detection.

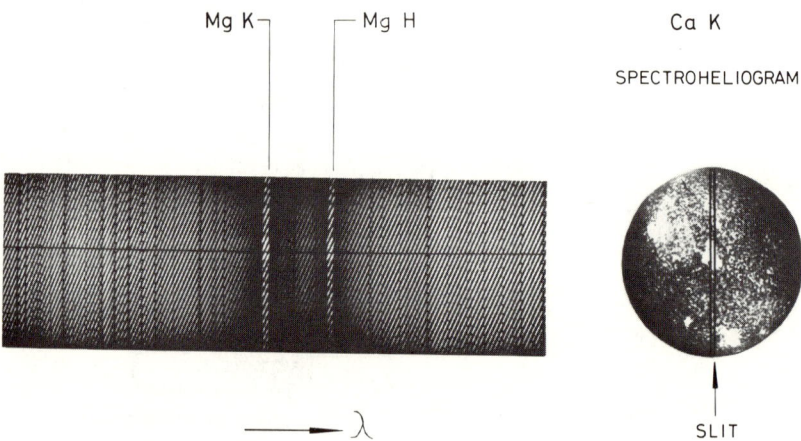

FIGURE 8 Solar Mg II interferogram recorded during a balloon flight, August 1971 (Queen's University of Belfast/Astrophysics Research Unit, Culham/Air Force Cambridge Research Laboratory, Bedford, Massachusetts).

Moreover, while the angle scanning technique was satisfactory for the small ($\approx 200$ sec of arc) angular free spectral range in this application, this was only possible because of the high angle of incidence onto the interferometer. The angular extent of the free spectral range and the degree of nonlinearity in the scan is substantially increased near normal incidence.

The alternative methods of wavelength scanning involve either a variation of the refractive index of the interferometer gap (i.e., pressure scanning) or physically altering the plate separation. Although the method of pressure scanning over several atmospheres is widely employed in ground-based observations, it has its obvious drawbacks for space applications. Plate-separation scanning has been effected in many ways with varying degrees of success, but the most common current technique is by means of piezoelectric transducers because of their near linear extension with voltage.

In view of the particular requirement for a high-finesse instrument that is compact and robust, stable in adjustment, and simple to use in remote-controlled operation, the scanning interferometer design was based on that of the previous air-gap interferometer. The constructed prototype interferometer is shown in Figure 9. In this, the spacers are cylindrical columns comprising two silica disks, a barium titanate disk, and a lead zirconate–titanate disk. Differential voltages are applied to the barium titanate disks for plate parallelism, and the same linear ramp voltage to the other disks provides the spectral scan [Bates et al., 1971b].

Tests in the visible region revealed a scanning defects finesse of $\sim 65$ and a wavelength scale linear with applied voltage to within 2% over a free spectral range. Placed in a small thermally stabilized enclosure, the interferometer was found to be extremely stable in adjustment and

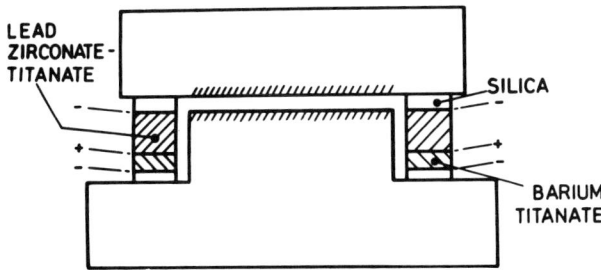

FIGURE 9  Scanning Fabry-Perot interferometer with piezoelectric transducers.

highly insensitive to shock and vibration. The same interferometer has been used for Ca II ($\lambda 396$ nm) and for Mg II ($\lambda 280$ nm) wavelengths, and the recorded finesse was 26 and 18, respectively. Since in the latter case the plates had been separated, coated, and reassembled several times, the results are encouraging; and with a new system a useful recorded finesse in excess of 20 at $\lambda 280$ nm seems perfectly feasible. The only drawback with this prototype instrument is the high voltage required for the piezoelectric disks ($\approx 800$ V per free spectral range at 280 nm). Current developments in this area involve the use of ceramic tubes and disk stacks for the column spacers.

To combine a scanning Fabry-Perot interferometer directly with a telescope requires some means of isolating the free spectral range of interest. Typically, for a resolving power of $10^5$ and a finesse of around 20 at $\lambda 280$ nm, a blocking filter of 0.06-nm bandpass is in order. The development of conventional interference filters in the middle ultraviolet, however, with spectral bandpass less than 0.5 nm, still remains a problem. On the other hand, with a narrow air gap and sufficiently large free spectral range, the Fabry-Perot interferometer is a high-performance filter in which the central transmission maximum is selected and all others rejected. This aspect of the interferometer is again well known and has been used successfully in the visible region for solar monochromatic photography [Ramsay *et al.*, 1970; Kononovich and Shcheglov, 1968]. Similar solar applications in the ultraviolet would also be important, particularly in the lines of Mg II.

The general features of the permanently adjusted interferometer again lend themselves well to this problem for a plate separation less than 100 $\mu$m, provided that a sufficiently high degree of plate parallelism is attained. Figure 10 shows schematically the construction of a prototype narrow-band ultraviolet filter. The lower plate is in two parts, which are optically contacted together. The spacer ring has three protruding contacting areas on each side, which are polished until the air gap is parallel to better than $\lambda/100$ over the full 30-mm clear aperture. At $\lambda 280$ nm, the full half-intensity width of the spectral bandpass was 0.06 nm for a free spectral range of 1.6 nm, and the transmission was approximately 40% [Bates *et al.*, 1973]. Developments in tunable narrow-gap filters incorporating piezoelectric ceramics are under way, and it should then be possible to extend the examinable range of the wider-gap interferometer by synchronously scanning the combination.

An alternative method of extending the spectral range is to synchronously scan the interferometer in combination with a grating monochromator. This offers the possibility of the simultaneous sampling of more than one spectral element by, for example, the use of channel

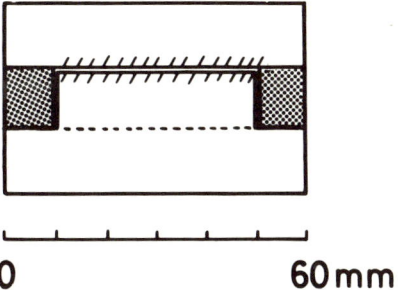

FIGURE 10 Narrow-band ultraviolet Fabry-Perot filter.

multipliers or a fiber-optic array. Such systems have been devised in various configurations [Chabbal and Jacquinot, 1955; Hirschberg and Kadesch, 1958; Jaffe *et al.*, 1955] and commonly employ pressure scanning of the interferometer with simultaneous pressure scanning of the monochromator.

For our future balloon program, we have investigated the feasibility of coupling the present piezoelectrically scanned interferometer to a suitable scanning monochromator. In this initial study, a recovery rocket echelle spectrograph was modified for the purpose (Figure 11). The interferometer is driven in a sawtooth fashion through a free spectral range at the same wavelength scan rate as the monochromator. The spectral width of the entrance and exit slits is approximately one half of the interferometer free spectral range for good spectral purity. The use of an echelle grating with its high angular dispersion permits a large physical width of the slits, which ideally should be equal to the width of the central fringe of the interferometer pattern [Wyller and Fay, 1972]. In this case, for a well-matched system the high throughput and rotational symmetry of the interferometer system should not be altered except by the efficiency of the echelle grating and associated optics. The overall spectral resolution of the grating spectrometer alone is, however, increased by a factor equal to the interferometer finesse.

In operation, the beam-folding mirror is driven by a precision stepper-motor drive in angular steps of 1/40 of the width of the exit slit. Synchronism has been achieved over wavelength intervals of 1 nm; but with the present system, beyond this, the cumulative effects of the changing free spectral range of the interferometer and second-order nonlinear terms in the grating dispersion lead to a gradual loss in coincidence. Spectra of a laboratory water vapor discharge in emission and absorption are shown in Figure 12, in which the high spectral resolution ($R = 10^5$) of the com-

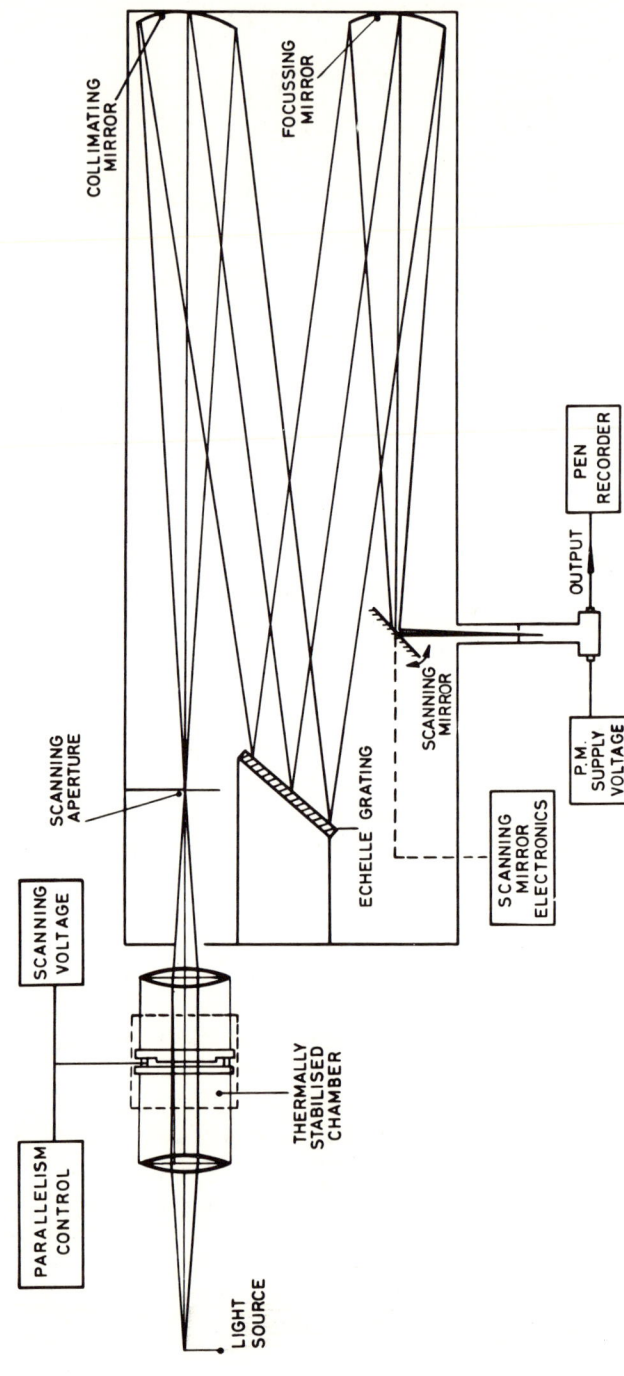

FIGURE 11  Scanning Fabry-Perot/echelle spectrometer.

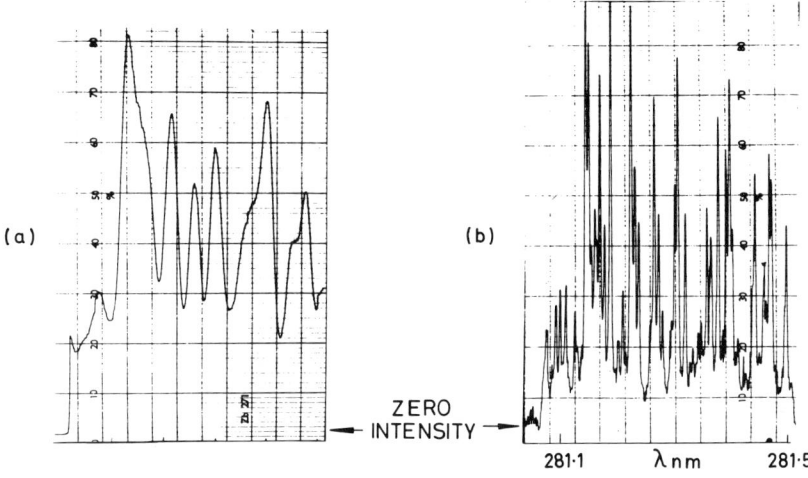

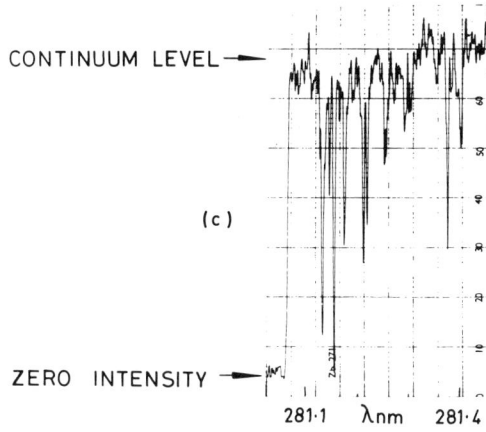

FIGURE 12 Spectra of an OH microwave discharge (a) in emission with the echelle dispersion alone, (b) in emission with the combined Fabry-Perot and echelle dispersion, (c) in absorption with the combined dispersions.

bined system and the excellent stray-light characteristics of the spectrometer are evident.

This prototype instrument goes some way toward demonstrating the capacity of this interferometer system in fulfilling the need for a highly luminous, high-resolution, robust, and compact spectrometer with the ability to examine an extended wavelength region. With multichannel detection, the simultaneous information content in an observation could be further increased over a wavelength range of perhaps 6 nm. This and other extensions of the present systems are currently being investigated for the program of Fabry-Perot astronomical spectroscopy.

We wish to acknowledge the general assistance of all our colleagues at The Queen's University of Belfast. The flight experiments described were carried out in collaboration with the Astrophysics Research Unit Culham Laboratory, U.K., the University College of London, and the Air Force Cambridge Research Laboratories, Bedford, Massachusetts. In particular we are grateful to D. J. Bradley, Queen's University, for his continued interest in this work, which was supported by the Science Research Council. One of us (J.K.C.) gratefully acknowledges receipt of a Ministry of Education for Northern Ireland Postgraduate Studentship.

## References

Bates, B., and C. D. McKeith, *Contemp. Phys. 13,* 225 (1972).
Bates, B., D. J. Bradley, T. Kohno, and H. W. Yates, *J. Sci. Instrum. 34,* 476 (1966).
Bates, B., D. J. Bradley, C. D. McKeith, N. E. McKeith, W. M. Burton, H. J. B. Paxton, D. B. Shenton, and R. Wilson, *Nature 224,* 161 (1969).
Bates, B., D. J. Bradley, D. A. McBride, C. D. McKeith, N. E. McKeith, W. M. Burton, H. J. B. Paxton, D. B. Shenton, and R. Wilson, *Philos. Trans. R. Soc. London A270,* 47 (1971a).
Bates, B., J. K. Conway, G. R. Courts, C. D. McKeith, and N. E. McKeith, *J. Phys. E, Sci. Instrum. 4,* 899 (1971b).
Bates, B., J. K. Conway, C. D. McKeith, and H. W. Yates, *Appl. Opt. 12,* 140 (1973).
Bradley, D. J., *Opt. Acta 15,* 431 (1968).
Bradley, D. J., B. Bates, C. O. L. Juulman, and T. Kohno, *J. Phys. 28,* C2-280 (1967).
Burton, W. M., N. K. Reay, D. B. Shenton, and R. Wilson, in *New Techniques in Space Astronomy,* IAU Symp. 41, p. 304, F. Labuhn and R. Lust, eds., Reidel, Dordrecht, Holland (1971).
Chabbal, R., and P. Jacquinot, *Nuovo Cim. 2,* 661 (1955).
Hirschberg, J. G., and R. R. Kadesch, *J. Opt. Soc. Am. 48,* 177 (1958).
Hobbs, L. M., *Astrophys. J. 157,* 135 (1969).
Hoekstra, R., K. van der Hucht, T. Kamperman, and H. Lamers, *Nature Phys. Sci. 236,* 121 (1972).
Jacquinot, P., *J. Opt. Soc. Am. 44,* 761 (1954).
Jaffe, J. H., D. H. Rank, and T. A. Wiggins, *J. Opt. Soc. Am. 45,* 636 (1955).
Kondo, Y., T. R. Giuli, J. K. Modisette, and A. E. Rydgren, *Astrophys. J. 176,* 153 (1972).
Kononovich, E. U., and P. V. Shcheglov, *Sov. Astron. AJ 12,* 235 (1968).
Lemaire, P. *Astrophys. Lett. 3,* 43 (1969).
Lemaire, P., in *Ultraviolet Stellar Spectra and Related Ground Based Observations,* IAU Symp. 36, p. 332, L. Houziaux and H. E. Butler, eds., Reidel, Dordrecht, Holland (1970).
Lemaire, P., in *New Techniques in Space Astronomy,* IAU Symp. 41, p. 231, F. Labuhn and R. Lust, eds., Reidel, Dordrecht, Holland (1971).

Maran, S. P., R. L. Kurucz, K. M. Strom, and S. E. Strom, *Astrophys. J. 153*, 147 (1968).
Marschall, L. A., and L. M. Hobbs, *Astrophys. J. 173*, 43 (1972).
Meaburn, J., *Astrophys. Space Sci. 9*, 206 (1970).
Ramsay, J. V., H. Kobler, and E. G. Mugridge, *Solar Phys. 12*, 492 (1970).
Rense, W. A., *Space Sci. Rev. 5*, 234 (1966).
Vaughan, A. H., Jr., *Annu. Rev. Astron. Astrophys. 5*, 139 (1967).
Wilson, R., in *Electromagnetic Radiation in Space*, p. 30, J. G. Emming, ed., Reidel, Dordrecht, Holland (1965).
Wyller, A. A., and T. Fay, *Appl. Opt. 11*, 1152 (1972).

DONALD F. HEATH and JAMES B. HEANY

# Observations on Degradation of Ultraviolet Systems on Nimbus Spacecraft

## Introduction

The launch of an ultraviolet radiation experiment aboard an earth-orbiting satellite places it into a very hostile environment, which is especially severe if small changes in uv radiation levels are to be observed over periods of months or years. This is critical if one is looking for changes in the solar–terrestrial radiation system that may be indicative of long-term climatological changes. Also, in the coming years one will be faced with assessing the nature of the problem associated with the pollution of the stratosphere and its relationship to man and his environment. This work is concerned with observations on the nature of degradation in uv systems that were designed for nominal one-year operational lifetimes on Nimbus 3 and 4 for the investigation of the uv solar–terrestrial radiation system.

Some possible sources of degradation that one must consider are the trapped particle radiation belts, particularly in the vicinity of the

---

The authors are in the Laboratory for Meteorology and Earth Sciences and the Engineering Physics Division, respectively, Goddard Space Flight Center, Greenbelt, Maryland 20771.

South Atlantic Anomaly; outgassing from the spacecraft; solar radiation that produces both thermal effects and uv-induced photochemical changes; and micrometeorite impacts on optical surfaces.

The observations on the space degradation of uv systems that are described in this paper were derived from one experiment on Nimbus 3 and two on Nimbus 4. The Nimbus 3 and 4 spacecraft, earth-oriented, were put into circular, 10° retrograde (near polar), sun-synchronous orbits at an altitude of 1100 km in April 1969 and April 1970, respectively. The spacecraft traverse the ascending node of the orbit near local noon. The location of the experiments is such that they are subject to a changing angle of solar illumination from about 45° from the daylight side of the terminator through the passage into satellite night, which occurs about 30° past the terminator.

The Monitor of Ultraviolet Solar Energy (MUSE) experiment was flown on Nimbus 3 and 4. Its objective was to investigate the magnitude and types of variability of the solar irradiance in relatively broad spectral bands in the 1200 to 3000 Å region, which is important for meteorology. The MUSE experiment on Nimbus 3 was operated continuously from launch in April 1969 until the spacecraft was deactivated in January 1972. This experiment on Nimbus 4 has operated continuously since launch in April 1970.

The BUV, Backscatter Ultraviolet, experiment on Nimbus 4 consists of a double monochromator of 0.25-m focal length, which is basically a tandem Ebert-Fastie type. The BUV measures the earth radiance and solar irradiance at 12 wavelengths (10 Å bandpass) from 2550 to 3400 Å and at 3800 Å with a 50 Å bandpass photometer channel. This experiment completed 12,000 orbits of operation in September 1972.

The MUSE and BUV experiments were the sources for the observations that are discussed in this work. The principal objectives of this work were to determine (a) the principal source of degradation in uv space optical systems and (b) the temporal character of the system degradation. A knowledge of these should help in the design of future experiments and in the analyses of the observations to separate slowly varying solar–terrestrial radiations from system changes.

## Instruments

A MUSE sensor package, shown in Figure 1, is located in the Nimbus sensory ring 180° from the velocity vector. The five broadband photometer channels have a nominal 90° field of view and are illuminated at near-normal incidence over the northern terminator. A digital solar as-

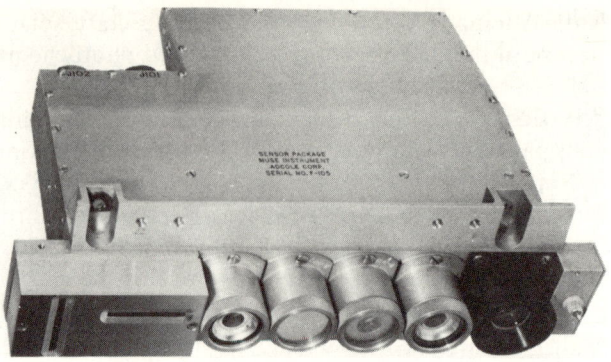

FIGURE 1  MUSE experiment sensor package with five sensors and digital solar aspect sensor at the top.

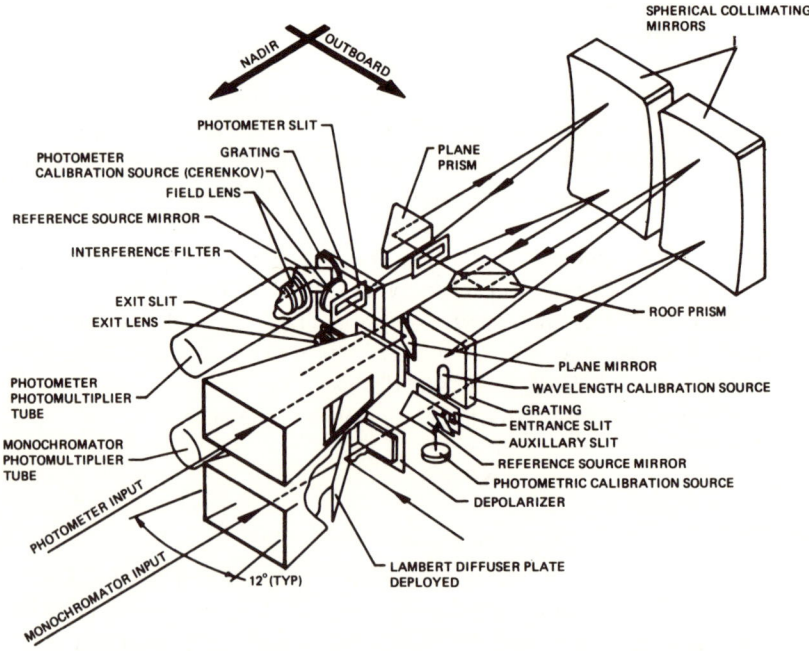

FIGURE 2  Optical component diagram of Nimbus 4, BUV experiment.

pect sensor located in the upper section of the sensor package provides the angle of solar illumination of the sensors at 0.7° increments. The passband of a photometer channel is determined by the short-wavelength cutoff of uv transmitting materials and the long-wavelength rejection of different "solar blind" photocathodes.

A functional diagram of the BUV experiment is shown in Figure 2. The optical path of the earth radiance from a 12° field of view traverses an $Al_2O_3$ particle radiation shield, a double-Lyot calcite depolarizer, two Ebert mirrors, the first coated with a multilayer coating of Al and $MgF_2$ (peaked at 2500 Å) and the second with Al, two Al-coated replica gratings, high-purity fused-silica transfer prisms, and field lenses in the path through the double monochromatic to the $Al_2O_3$ window of the photomultiplier. At the northern terminator, a diffuser plate of ground Al and overcoated with pure vacuum-deposited Al is deployed for measurement of the solar irradiance. The diffuser plate is illuminated at an angle of 55° to the normal when the solar vector lies in the plane of the optic axis and the diffuser plate normal. The diffuser plate is exposed at all times to space and solar radiation in the vicinity of the northern terminator. When the plate is in the stowed position, light baffles restrict the instrument field to 12° of the terrestrially scattered solar radiation.

More complete descriptions of the instruments may be found in the Nimbus 3 and 4 *User Guides* [1969, 1970].

## Observations

### DEGRADATION IN RESPONSE

Prior to the fabrication of the MUSE experiment, it was believed that the most likely source of degradation in a uv optical system operating in the planned Nimbus orbit would be the high-energy trapped-particle radiation environment. High-energy electrons in the MeV energy range can degrade uv optical materials through the formation of color centers, which shift the short-wavelength transmission limit to longer wavelengths. The color centers can be formed by either direct ionization produced as the electron loses energy in its passage through the material or by the bremsstrahlung produced by the collision between the electron and a surface. The resulting high-energy photons subsequently produce ionization in their passage through the optical materials.

An estimate of the anticipated charged particle flux that Nimbus 3 would encounter in one year was $10^{13}$ $e^-/cm^2$ at an equivalent energy of 1 MeV, which gives an energy deposition of about $2 \times 10^7$ ergs $cm^{-2}$

yr$^{-1}$ as an upper limit. A series of investigations on the effect of 1- to 2-MeV electrons on uv transmitting materials [Heath and Sacher, 1966] and uv photocathodes [Heath and McElaney, 1968] at $10^{14}$ e$^-$/cm$^2$ indicated that the effects on MUSE and BUV experiments should be negligible at the end of one year in space. Corresponding degradation studies with protons indicated that their contribution to degradation in the Nimbus orbit would be trivial [Heath, McElaney, and Sacher, unpublished work].

Some recent results on the effect of increasing doses of high-energy electron irradiation on the transmittances of LiF and MgF$_2$ are shown in Figure 3. The transmittance is proportional to the logarithm of the dose [Heath and Fedor, to be published]. The multiple straight-line segments may indicate possible saturation effects, although admittedly there are too few data points.

The actual signal decay curves for two of the MUSE sensors on Nimbus 3 are shown in Figure 4. The sensor with the MgF$_2$ window and W photocathode responds principally to H Lyman-α and exhibited an initial decay period ($e^{-1}$) of 77 days for the period between day 100 and 230 in 1969. The secondary period characteristic of the signal after day 230 is 284 days. The other sensor consisted of an MgF$_2$ outer particle shield and vacuum photodiode using a semitransparent CuI photocathode deposited on a CaF$_2$ window. Since both sensors use MgF$_2$ as the element exposed to space, the most likely source of degradation is H Lyman-α, which is absorbed in the CaF$_2$ window. Considering that

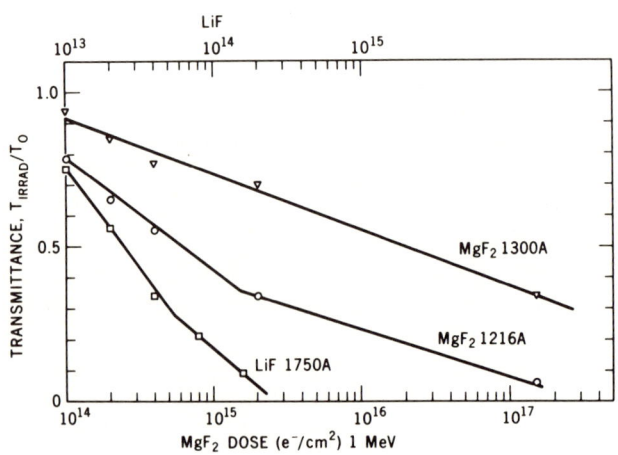

FIGURE 3  Change in transmissions in LiF and MgF$_2$ with increasing doses of 1-MeV electrons.

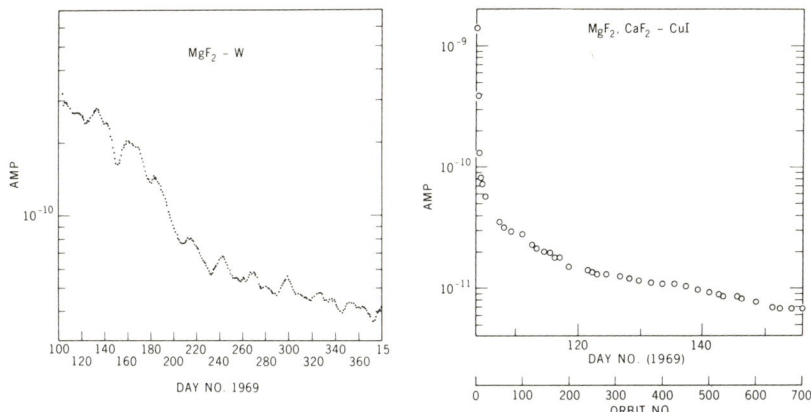

FIGURE 4 Probable radiative induced change in transmittance of $CaF_2$ induced by solar H Lyman-α and satellite-space environment-induced changes in the $MgF_2$-W sensor, which responds principally to solar H Lyman-α.

the sensor is illuminated for about 20 min for each 107-min orbit, the transmittance of the outer $MgF_2$ element, and the solar irradiance at H Lyman-α, the rate of Lyman-α energy absorption is about $1 \times 10^7$ ergs cm$^{-2}$ yr$^{-1}$. The observed initial decay period is 0.12 day. The secondary period is 14.4 days, and the tertiary period is 43.7 days. For both of these sensors the logarithm of the signal is a linear function of time with negative slope over particular time intervals.

A composite of 30-day signal averages of sensors common to Nimbus 3 and 4, which have been corrected for the annual variation of earth-sun distance, is shown in Figure 5. The signals have been averaged over 30 days to smooth the 27-day variations, which are associated with the solar rotational period. In general, the logarithm of the sensor signal is a linear function of time with negative slope over specific time intervals, and furthermore the slopes are a decreasing function with increasing time. This characteristic that can be associated with saturation phenomena has been used to infer times when actual changes in the solar irradiance may occur.

For example, this has led to the conclusion that there was an overall increase in the irradiance at Lyman-α that peaked in the spring of 1969 and 1971. This is indicated in sensors A. There is also an indication from sensor B on Nimbus 3 of a significant change in the solar irradiance at 1750 Å toward the end of 1969.

A summary of the sensor degradation observed in the MUSE experiments is given in Table 1. It should be noted that on-board electronic

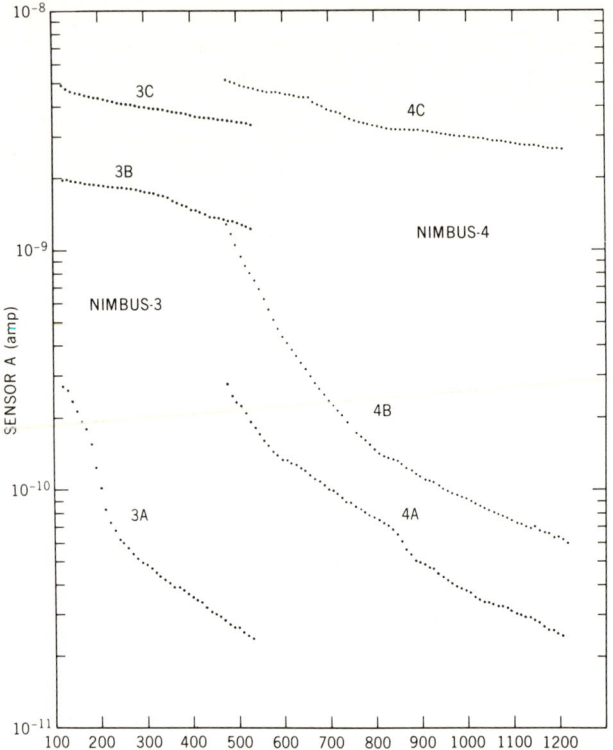

FIGURE 5 Summary of MUSE 30-day signal averages showing change with time for the three sensors common to Nimbus 3 and 4. Signals have been corrected for changing earth–sun distance.

TABLE 1 Summary of Long-Term Signal Changes in MUSE Sensors on Nimbus 3 and 4

| Sensor signal ratio | 1216 Å | 1750 Å | 2100 Å | 2800 Å | 2900 Å |
|---|---|---|---|---|---|
| *Nimbus 3* |  |  |  |  |  |
| Orbit 11158/3 |  |  |  |  |  |
| Apr. 1969–July 1971 | 0.017 | 0.40 | – | – | 0.52 |
| *Nimbus 4* |  |  |  |  |  |
| Orbit 12111/5 |  |  |  |  |  |
| Apr. 1970–Sept. 1972 | 0.056 | 0.034 | 2.59 | 0.95 | 0.434 |

calibrations and the fact that all sensors are switched into a common electrometer make it very unlikely that the signal decrease with time can be attributed to an electronic malfunction. On Nimbus 4, sensor A, which responds to Lyman-$\alpha$, experienced a degradation that was one third that observed on Nimbus 3; however, sensor B (1750 Å) degraded by a factor of 12 on Nimbus 4. For sensor C (2900 Å) the results were comparable. The factor of 2.6 increase of signal of the 2100 Å sensor is attributed to the increase in leakage in the sidebands of the interference filter. This same effect probably would not be observed with the 2800 Å sensor, since it most likely would occur at a wavelength where the photocathode quantum efficiency was sufficiently low to make its contribution negligible.

The BUV experiment observations at 12 wavelengths (2550–3400 Å) with the double monochromator and at 3800 Å with the filter photometer are of two types. These are direct measurements of earth radiance and measurements of the solar irradiance at the northern terminator from a ground aluminum diffuser plate that had been overcoated with vapor-deposited aluminum. An $Al_2O_3$ high-energy particle-radiation shield was placed at the entrance slit of the double monochromator. In addition, extensive laboratory testing of the diffuser plate was done in vacuum and subjected to uv, high-energy electrons and protons. There were no indications of any degradation having occurred.

The observations of the solar irradiance, which had been normalized to orbit 32 when the BUV was turned on, are shown in Figure 6. The time interval spans 1.8 years, and it is apparent that the degradation is an increasing function of decreasing wavelength.

The decrease of the solar irradiance signal at 2557 Å as a function of time is shown in Figure 7. As observed with the MUSE sensors, the logarithm of the signal is a linear function of time, which is characterized by a negative slope that decreases in magnitude with increasing time intervals.

Over this same time interval, one has measurements of the equatorial terrestrial radiance, which should exhibit good long-term stability if the photochemistry of ozone remains constant in the equatorial upper stratosphere. The BUV equatorial observations, uncorrected for the changing earth–sun distance, are shown in Figure 8. The equatorial terrestrial radiance below 2976 Å is strongly influenced by clouds and surface albedo. One easily can see a gradual decrease with time of the apparent radiance and also the detector gain that was determined by independent means. The long-term changes in the apparent equatorial radiances are given in Table 2 for those wavelengths that originate principally above the cloud levels. Also given are values for the correspond-

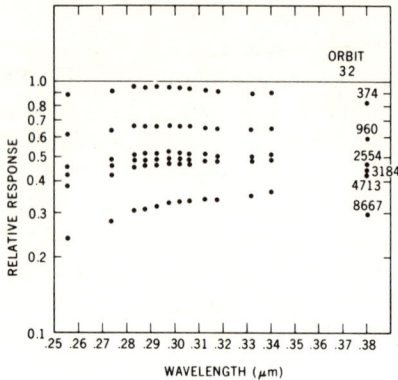

FIGURE 6 Apparent change in solar irradiance from Nimbus 4, BUV observations of solar-illuminated diffuser plate at northern termination. Signals are normalized to the experiment turn-on orbit no. 32.

ing solar irradiances, which were recorded with the diffuser plate. Knowing that the total ozone has been shown to have no correlation with the 11-year sunspot cycle and that the earth radiance that is most strongly influenced by the total amount of ozone is in the 2900–3000 Å range, one may conclude that in the vicinity of 3000 Å the long-term change in signal may be explained by a slowly decreasing gain of the detector with time. On the other hand, the equatorial radiance at 2557 Å is indicative of a decreasing amount of upper stratospheric ozone with time

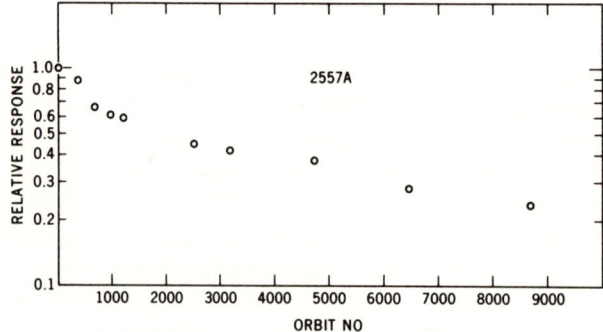

FIGURE 7 Relative change in BUV experiment 2557 Å signal from solar-illuminated diffuser plate with Nimbus 4 orbit number.

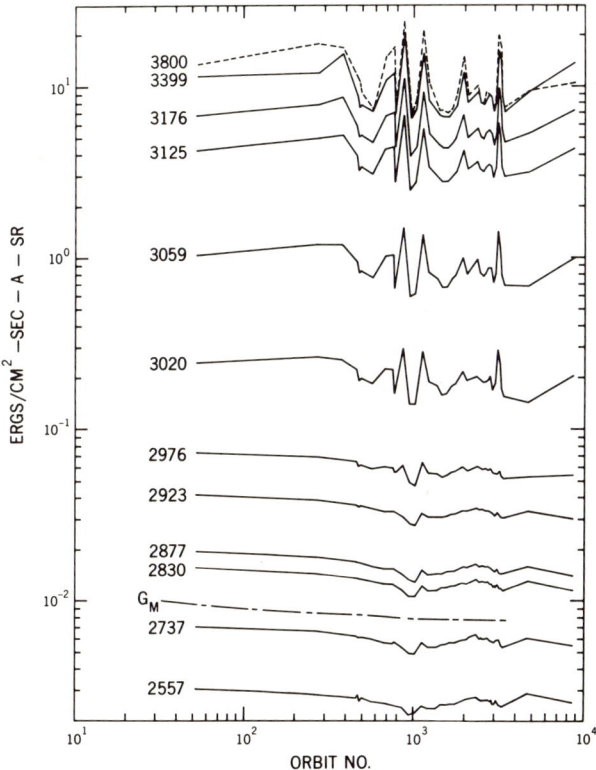

FIGURE 8 Summary of equatorial radiances from the vicinity of 200°W longitude based on prelaunch absolute calibration of BUV experiment. Also shown is normalized detector gain curve ($G_m$) derived from simultaneous pulse counting and current measurements.

TABLE 2 Summary of Long-Term Changes in BUV Experiment Observations of Equatorial Atmospheric Radiances that Originate above the Cloud Levels and Corresponding Solar Irradiances Observed off the Diffuser Plate

| Ratio | 2557 A | 2737 A | 2830 A | 2877 A | 2923 A | 2976 A | Gain |
|---|---|---|---|---|---|---|---|
| *Equatorial Radiance* Orbit 8667/32 | | | | | | | |
| Apr. 1970–Jan. 1972 | 0.95 | 0.85 | 0.83 | 0.82 | 0.82 | 0.73 | 0.76 |
| *Solar Irradiance (Diffuser Plate)* | 0.24 | 0.28 | 0.31 | 0.31 | 0.32 | 0.33 | |

past solar maximum. The decrease of ozone in the upper stratosphere is associated with an increase in atmospheric radiance.

To a first approximation, one may conclude that the monochromator signals are decreasing at a rate that may be explained by the changing gain in the photomultiplier. At the same time, it appears that a very real decrease in the diffuse reflectance of the diffuser plate is being observed.

CHANGES IN ANGULAR RESPONSE

The changing angle of solar illumination of the Nimbus spacecraft in its sun-synchronous, local noon orbit provides additional data for the investigation of the observed degradation. The angular scan for the three MUSE sensors for orbits 101 and 10,502, which spans 2.2 years, is shown in Figure 9. Sensor A has an $MgF_2$ window and an opaque tungsten photocathode, whereas sensors B and C have identical outer $Al_2O_3$ particle-radiation shields and semitransparent photocathodes deposited on $Al_2O_2$. The median sensor responses are at 1216, 1750, and 2950 Å, respectively. The ordinate is a normalized sensor current divided by the cosine of the angle of solar illumination to the sensor normal. The post terminator angular response is given only for sensor C, since it exhibited the largest asymmetry in angular response. The asymmetry is in the form of an azimuthal variation in angular response for a constant angle of incidence. The combination of a semitransparent photocathode, the cathode vapor-deposition technique, and the diode mechanical design all combine to produce a small secondary cathode on a wall.

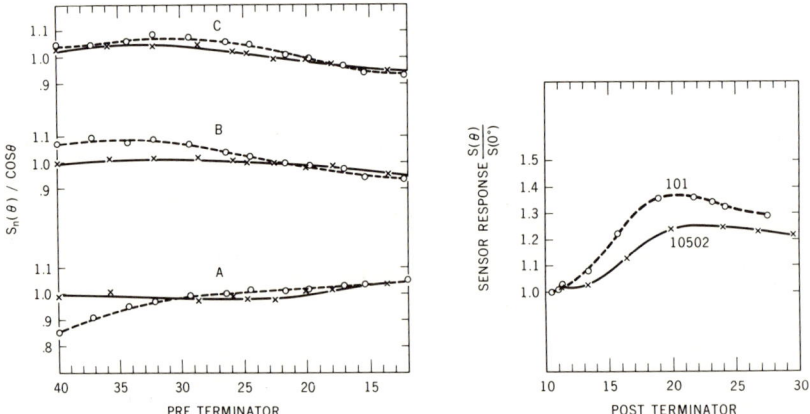

FIGURE 9 Change in angular response of MUSE sensors on Nimbus 3 for orbits 101 and 10,502. The angle of solar illuminations of the sensor normal at the terminator is +3° for these orbits. Ordinate is the observed sensor current/terminator sensor current × cos θ.

The anode ring has two side holes; however, only one was used in the cathode processing. This resulted in a secondary cathode being deposited onto the opposite wall, which is electrically connected to the semitransparent cathode on the front window. This effect was quite significant in the post terminator response for sensor C. The significant feature illustrated in Figure 9 is that the sensors are tending toward a cosine response with increasing time.

Changes have also been observed in the angular response of the diffuser plate in the BUV experiment with the passage of time. The angular response of the diffuser plate was calibrated in the instrument prior to launch, which is shown as a straight line in Figure 10 for 3398 Å. The deviations from the preflight calibration are shown for orbits 492 and 8908, where the diffuser plate was deployed for the entire orbit. The x's labeled "L" represent the case of an ideal Lambertian diffuser. There is a definite indication that the ground aluminum diffuser plate is losing some of its specular component that was present in its prelaunch calibration and is tending toward a Lambertian response.

## Discussion

There is considerable evidence from both flight and laboratory experiments to support the contention that outgassing from spacecraft materials is a primary cause of degraded performance in satelliteborne optical instrumentation. Cothran et al. [1971] have reviewed laboratory and

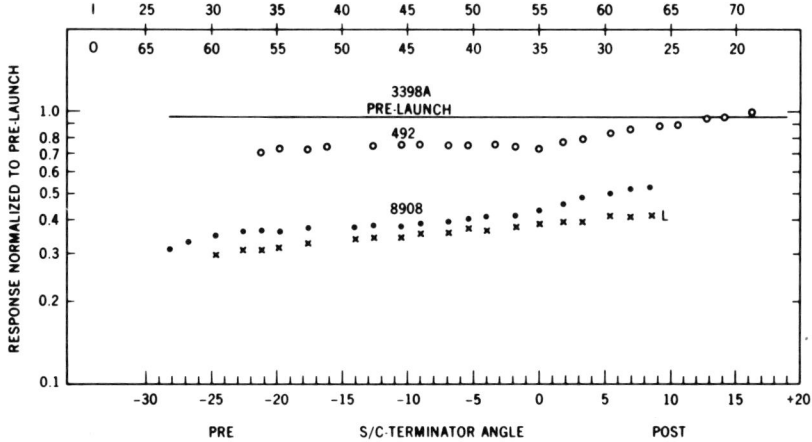

FIGURE 10 Change in angular characteristics of the diffuse reflectance of the BUV diffuser plate with time in orbit. The x's represent a Lambertian diffuser.

flight test data pertaining to this problem and have documented a number of interesting cases. Reduction in transmittance of some Gemini and Apollo windows, for example, was observed by the astronauts on board and was later attributed to the outgassing of an RTV silicone elastomer used for a window sealing. McKeown and Corbin [1970] used quartz crystal microbalances aboard OGO-6 (whose orbital environment is comparable with that of Nimbus 3 and 4) to demonstrate that solar panels baking out in the sun outgassed onto the spacecraft. Their estimates of contaminant desorption activation energy identified epoxies and vacuum oils as likely suspects.

Flight instrumentation need not view solar panels to become contaminated since the opportunities for self-contamination are numerous. An investigation of the particle dynamics associated with the return and deposition of outgassed products onto critical surfaces has shown that spacecraft self-contamination is quite possible, either in flight or during vacuum chamber testing [Scialdone, 1972].

Condensates may form in a variety of ways and at different rates depending on such factors as varying substrate temperatures, molecular weights, ambient pressure, time in sunlight, and sputtering loss caused by upper atmospheric neutral impacts. They have been observed to condense as small droplets, which Shapiro and Hanyok [1968] and Shapiro [1970] in their tests found to be 1–5 $\mu$m in diameter. These droplets eventually evaporate in the vacuum of the test chambers. However, Hass and Hunter [1970] have shown that irradiation of oil-contaminated mirrors with uv, electrons, or protons is sufficient to make the contaminant residue permanent.

Most common contaminants, such as high-molecular-weight organic materials, are quite transparent prior to photolysis, and their presence on an optical window may not be detected by an untrained observer or in a routine transmittance measurement. The chemical decomposition induced by exposure to energetic photons, electrons, and protons alters this situation by creating products that absorb strongly, especially in the short wavelength regions in which the MUSE sensors operate. Even a few monolayers of a strongly absorbing surface film is sufficient to drop the transmittance of a window at Lyman-$\alpha$ by a factor of 10 or more.

The wavelength-dependent nature of optical surface degradation in space was clearly demonstrated by the ATS-3 reflectometer experiment [Heaney, 1970]. Highly specular mirror surfaces that experienced a drastic reduction in reflectance in the 300–400-nm spectral region showed little or no loss for wavelengths longer than 650 nm. The loss of specular reflectance, although large at first, tended to saturate with time over a 2-year period. These observations are consistent with the MUSE data and suggest a common mechanism.

Evaporated aluminum surfaces flown on the ATS-3 reflectometer suffered a loss of reflectance that was linear with time and did not exhibit the same tendency toward saturation as those mirror surfaces with dielectric overlayers. Since only the specular component of reflectance was monitored, it is not possible to say with certainty that the mirrors were being roughened. The data do support this interpretation, however, and are quite consistent with the change in angular reflectance exhibited by the BUV scatter plate.

Another possible source of degradation is through surface cratering by micrometeorite impacts. The estimated particle flux for the 1100-km circular Nimbus orbit [O. Berg, private communication, 1972] for $2\pi$ sr at 90° to the spacecraft velocity vector is $\Phi = 2 \times 10^{-8}$ impact cm$^{-2}$ sec$^{-1}$. The assumption of an average spherical particle mass of $10^{-13}$ g and a typical density of 1 g/cm$^3$ yields a particle radius of 0.3 µm. The crater $r_c = 2r$. The damaged area per impact, $A_D = \pi r_c^2 \approx 1.1 \times 10^{-8}$ cm$^2$. The number of impacts cm$^{-2}$ yr$^{-1}$ $\approx 0.6$ cm$^{-2}$ yr$^{-1}$, which leads to a damaged area of $0.7 \times 10^{-8}$ cm$^2$, which is completely negligible in relation to the typical MUSE sensor apertures of $\approx 0.4$ cm$^2$.

From experimental tests in the laboratory and uv observations on the Nimbus 3 and 4 spacecraft with the MUSE and BUV experiments, the following have been observed:

1. High-energy electrons at MeV energies produce a change in transmittance ($T$), where

$$T \propto -\log \text{Dose } (e^-/\text{cm}^2).$$

2. All forms of degradation can be represented by straight lines that show the effect of saturation through decreasing slope with increasing degradation.

3. Transmittance loss from solar H Lyman-$\alpha$ with time ($t$)

$$T \propto e^{-\alpha t} \alpha(t_1)/\alpha(t_2) = 120, \quad t_2 > t_1.$$

4. Nimbus 4, BUV diffuse reflectivity ($R$)

$$R \propto e^{-\alpha t}, \quad \alpha(t_1)/\alpha(t_2) = 4.4.$$

5. Nimbus 4, MUSE (1216 Å)

$$T \propto e^{-\alpha t}, \quad \alpha(t_1)/\alpha(t_2) = 1.8.$$

6. Nimbus 3, MUSE (1216 Å)

$$T \propto e^{-\alpha t}, \quad \alpha(t_1)/\alpha(t_2) = 4.2.$$

7. Both the MUSE sensors and the BUV diffuser plate tend toward a cosine response with increasing time in orbit.

8. The surfaces that were exposed directly to uv solar radiation experienced the greatest degradation.

It appears that the most likely source of uv degradation may be attributed to the deposition of $\mu$m size droplets resulting from spacecraft outgassing and the subsequent formation of permanent residues under the action of the uv solar radiation.

## References

Cothran, C. A., M. McCargo, and S. A. Greenberg, "A Survey of Contamination of Spacecraft Surfaces," AIAA Paper No. 71-457 presented at the 6th Thermophysics Conference, Tullahoma, Tenn. (Apr. 1971).

Hass, G., and W. R. Hunter, *Appl. Opt. 9*, 2101 (1970).

Heaney, J. B., "Results from the ATS-3 Reflectometer Experiment," *Thermophysics: Applications to Thermal Design of Spacecraft*, PIAA, Vol. 23, pp. 249–274, J. Bevans, ed., Academic Press, New York (1970).

Heath, D. F., and J. H. McElaney, *Appl. Opt. 7*, 2049 (1968).

Heath, D. F., and P. A. Sacher, *Appl. Opt. 5*, 937 (1966).

McKeown, D., and W. E. Corbin, "Space Measurements of the Contamination of Surfaces by OGO-6 Outgassing and Their Cleaning by Sputtering and Desorption," pp. 113–127 in NBS Special Publ. 336, *Space Simulation*, J. C. Richmond, ed. (Oct. 1970).

Nimbus 3 *User's Guide*, National Space Science Data Center, Goddard Space Flight Center, Greenbelt, Md. (Mar. 1969).

Nimbus 4 *User's Guide*, National Space Science Data Center, Goddard Space Flight Center, Greenbelt, Md. (Mar. 1970).

Scialdone, J. J., "Predicting Spacecraft Self-Contamination in Space and in a Test Chamber," pp. 349–360, NASA SP-298, *Space Simulation* (1972).

Shapiro, H., Visual Appearance of Polymeric Contamination, NASA Optical Tech. Note, TN D-5839 (1970).

Shapiro, H., and J. Hanyok, *Vacuum 18*, 587 (1968).

F. L. ROESLER, R. A. KRUGER, and
L. W. ANDERSON

# AN ALL-REFLECTION INTERFEROMETER WITH POSSIBLE APPLICATIONS IN THE VACUUM ULTRAVIOLET

## I. Introduction

This paper reports studies on an all-reflection interferometer suitable for development as a Fourier transform spectrometer. Some aspects of this device have already been described [Kruger *et al.*, 1972, 1973]. The unique feature of this interferometer is the use of a combination of diffraction gratings to replace the beam splitter used in a conventional Michelson interferometer. The motivation for development of the concept of the all-reflection interferometer was provided by recognition, first, of the difficulties of obtaining large beam splitters of suitable transmittance and optical quality below about 1700 Å and, second, the potential advantage of the large étendue of an interference spectrometer for the study of faint, extended ultraviolet sources either in space or in the laboratory.

It must be acknowledged at the outset that the avoidance of transmitting elements does not solve all the problems. Reflectivities, which may be 90% or more above about 1000 Å, drop off quickly for shorter wavelengths to not more than about 30% at present. Furthermore, grat-

---

The authors are in the Department of Physics, University of Wisconsin, Madison, Wisconsin 53706.

ing efficiencies in the uv are not all one might hope for. However, it seems some improvement in short-wavelength reflectivities and grating technology can be expected. Furthermore, for special applications it appears possible to use an interferometer design involving only three reflections, or no more than for most conventional grating instruments.

While the optical principles to be described evolved from the desire to do ultraviolet interferometry, it should be recognized that they are applicable in any spectral region where suitable reflectivities are obtainable. In fact, it is planned to test an interferometer based on the concepts to be described here for use in the far infrared, where technological problems with transmitting beam splitters have so far prevented the successful operation of a conventional liquid helium-cooled Fourier transform spectrometer. The components of the all-reflection beam splitter to be described here already have demonstrated reliability in a liquid helium-cooled rocketborne grating spectrometer.

## II. The Reflection Grating as a Beam Splitter

Figure 1 shows the simplest form of a grating beam splitter. In this simple form the idea is quite old [Barus, 1911], and interferometers similar to the one shown schematically in Figure 1 have been previously studied to a limited extent [Munnerlyn, 1968; Davis, 1940; Connes, 1959]. Monochromatic light from the source region S (the upper half of the elliptical central Haidinger fringe) is collimated by mirror M and is normally incident on the plane reflection grating G. The grating diffracts the light into symmetric beams that go to plane mirrors $M_1$ and $M_2$, respectively. These mirrors reflect the light antiparallel to the incident light, and the two beams recombine at G and are focused by M onto the exit region of the central fringe. (The Haidinger fringes are not circular. It can be shown that the fringes are elliptical with eccentricity $\zeta = \cos\theta$, where $\theta$ is the grating angle of diffraction.) If the path difference is changed by translating mirror $M_1$, modulation of the exit intensity is produced. However, because of the dispersion of the grating G, the system is not achromatic, and modulation is produced only for a spectral range equal to the resolution limit of the grating. This configuration is therefore impractical as a broadband Fourier transform spectrometer.

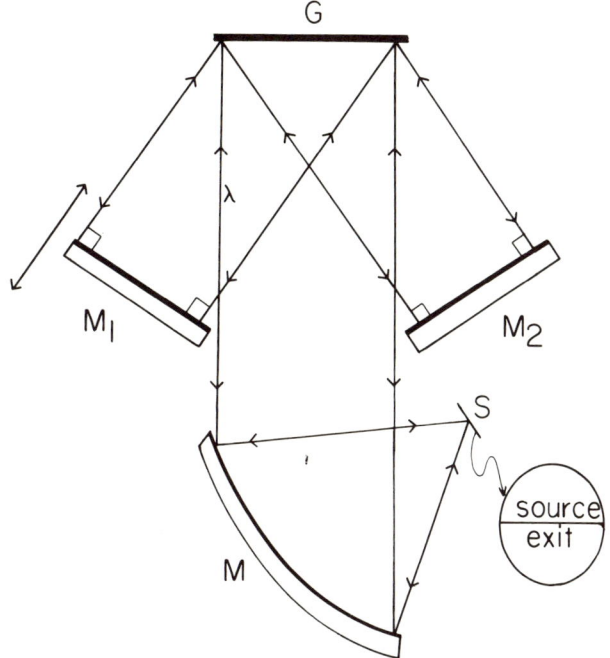

FIGURE 1 Schematic diagram of a simple interferometer using a grating G to split a beam of light into two beams. Light goes from the source to the concave mirror M and then to the grating G. Mirrors $M_1$ and $M_2$ reflect the two beams back to the grating where they are recombined. Haidinger fringes are observed in the image of the entrance aperture.

## III. A Reflection Grating Beam Splitter for Fourier Transform Spectrometry

The essential problem to be overcome is that of the dispersion of the grating. This can be accomplished by using an additional grating in each arm of the interferometer to undo the effects of dispersion produced by the primary grating. In Figure 2, if gratings $G_1$, $G_2$, and $G_3$ are parallel, and have identical groove spacings, then light of wavelengths $\lambda_1$ and $\lambda_0$, for example, leave $G_1$ at different angles but leave $G_2$ or $G_3$ parallel. Mirrors $M_1$ and $M_2$ return the beams through the instrument to the exit position of the aperture. Displacement of mirror $M_1$ (or $M_2$) produces modulation at all wavelengths. Furthermore, if the distances $S_1$ and $S_2$ between $G_1$ and $G_2$ or $G_3$, respectively, are equal, zero path difference occurs simultaneously for all wavelengths when the distances $L_1$ and

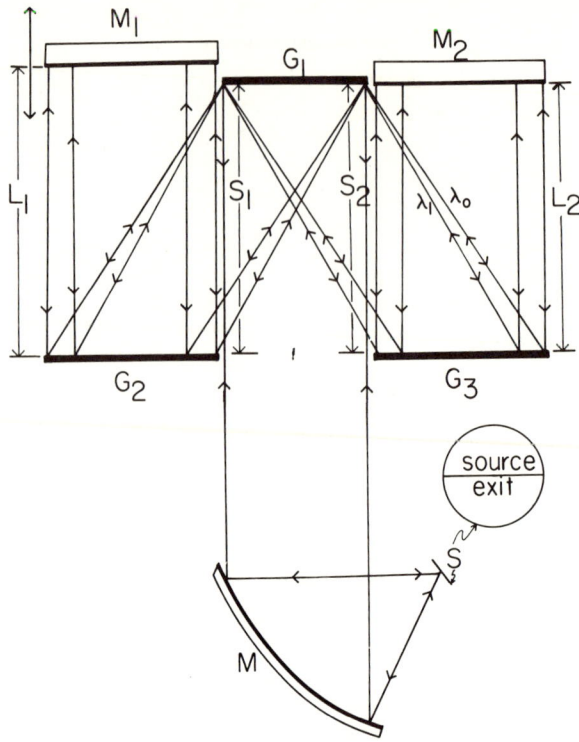

FIGURE 2 All-reflection interferometer for use as a Fourier transform spectrometer. $G_1$, $G_2$, and $G_3$ are diffraction gratings with identical groove spacings, and $M_1$ and $M_2$ are plane front surface mirrors. The mirror M is a concave mirror. The path difference $\Delta = 2(L_1 - L_2)$ for normal incidence rays is the same for the different wavelengths $\lambda_0$ and $\lambda_1$.

$L_2$ as shown in the figure are equal. Thus this system can be used as a broadband Fourier transform spectrometer in the usual manner. If $S_1 \neq S_2$, it is obvious that zero path difference cannot occur simultaneously for all wavelengths, since the distance the light travels between $G_1$ and $G_2$ or $G_3$ is wavelength-dependent. The situation is similar to that in a conventional Michelson interferometer with an improperly adjusted compensating plate.

It remains to be shown that the fringe shape (hence the étendue) of this interferometer is the same as for a conventional Michelson interferometer. To aid in the analysis, consider Figure 3, which is an unfolded representation of the interferometer shown in Figure 2. The

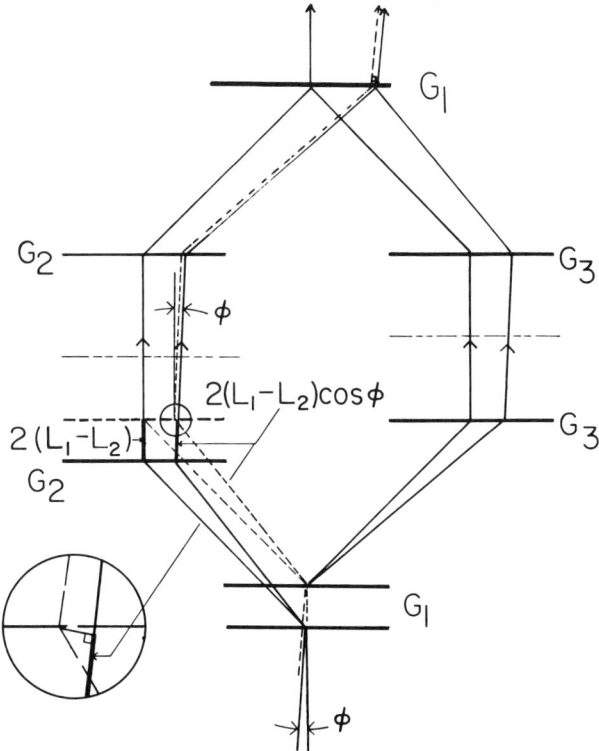

FIGURE 3 The interferometer of Figure 2 unfolded and with gratings and grating images replaced by transmission gratings. Bold lines indicate gratings or grating images. If the path difference for normal incidence rays between arms in the interferometer is $\Delta = 2(L_1 - L_2)$, then the path difference for nonnormal incidence rays is $\Delta = 2(L_1 - L_2) \cos \phi$, where $\phi$ is the angle of incidence. The inset is an enlargement of the circled portion in the figure.

images of the gratings in the two mirrors are shown, and the reflection gratings are replaced by transmission gratings. The grating $G_1$ is shown separately for each arm (hence doubled) at the bottom of the figure, since it must be equidistant from $G_2$ and $G_3$ even when $L_1 \neq L_2$. For any small angle of incidence $\phi$, the ray path in one beam between the bottom representations of $G_1$, $G_2$, and $G_3$ cancels with the opposite ray path in the upper representation of $G_1$, $G_2$, and $G_3$, so that the only remaining path difference is that due to the difference between $L_1$ and $L_2$. With the aid of the inset in the figure, the path difference $\Delta$ is seen to be $\Delta = 2(L_1 - L_2) \cos \phi$ as for a conventional interferometer. Thus the fringe shape and behavior of this instrument is the same as in the Michelson interferometer.

In tests of this interferometer, a holographic grating has been used for the primary grating, and ordinary blazed gratings have been used for the secondary gratings. The holographic grating splits the incident light into plus and minus first orders with very nearly equal intensity. The blaze angle of the secondary gratings is chosen to send the light efficiently to mirrors $M_1$ and $M_2$. Present holographic gratings are very satisfactory for the primary grating in the visible region of the spectrum, but efficiencies in the ultraviolet fall below those of conventional ruled gratings. A conventional ruled grating would need a symmetrically blazed ruling and would be difficult and expensive to produce with diffraction-limited quality in the ultraviolet. There seems to be no fundamental reason why efficient ultraviolet holographic gratings should not eventually be produced.

In the near future, an interferometer of this basic design will be tested in the far infrared using conventional ruled gratings, and a careful comparison will be made of its efficiency with that of a conventional Fourier transform spectrometer. In the far infrared, the production of the symmetrical groove appears relatively easy. For the ultraviolet, an alternative design to be discussed below appears more attractive.

## IV. Alternate Design

For ultraviolet applications, work has concentrated on the new design of the all-reflection interferometer shown in Figure 4 [Kruger *et al.*, 1973]. This design requires two fewer reflections, which can result in a large efficiency gain in spectral regions where only low reflectivities are obtainable. Light incident on grating $G_1$ is split symmetrically into two first-order diffracted beams. Gratings $G_2$ and $G_3$ have the same groove spacing as $G_1$ but are ordinary blazed gratings designed to throw light in second order directly back to $G_1$. The interferometer is scanned by displacing one of the secondary gratings as indicated in the figure. Zero path difference occurs for all wavelengths when $S_1$ and $S_2$ as shown in the figure are equal. It is evident from the figure that when the interferometer is scanned, the path difference is wavelength-dependent. For a displacement of one of the gratings by an amount $\Delta/2$ as shown, the path difference is $\delta = \Delta \cos \theta$. The angle $\theta$ depends on the wavelength according to the grating equation, giving $\cos \theta = (1 - \lambda^2/d^2)^{1/2}$.

This property is readily incorporated into the usual equations for Fourier transform spectroscopy. For a spectral distribution $S(\sigma)$, one

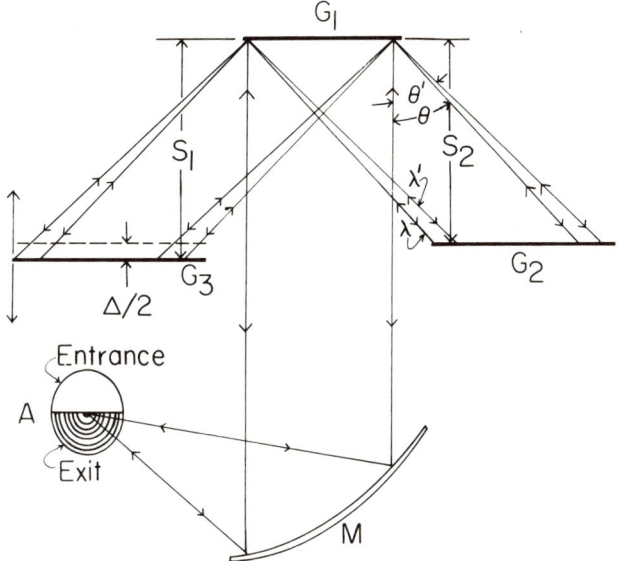

FIGURE 4 Arrangement of the second design for the all-reflection interferometer. The gratings $G_2$ and $G_3$ return the light directly to $G_1$. Scanning is achieved by displacement of $G_3$. The path difference in this infterferometer is wavelength-dependent, as shown by the rays drawn for wavelengths $\lambda$ and $\lambda'$.

has

$$I(\Delta) = \int S(\sigma) \cos (2\pi\Delta\sigma \cos \theta) d\sigma.$$

Defining a pseudo-wavenumber $\sigma' = \sigma \cos \theta = \sigma(1 - \sigma^{-2}d^{-2})^{1/2}$, this becomes

$$I(\Delta) = \int S(\sqrt{\sigma'^2 + 1/d^2})(d\sigma/d\sigma') \cos 2\pi\Delta\sigma' \, d\sigma'$$

$$= \int Z(\sigma') \cos 2\pi\Delta\sigma' \, d\sigma',$$

where $Z(\sigma')$ is the pseudo-spectrum given by

$$Z(\sigma') = S(\sigma')(d\sigma/d\sigma').$$

The inversion is done in the usual way to obtain the pseudo-spectrum $Z(\sigma')$, which is converted to the real spectrum using the defining equations.

Figure 5 shows the layout of the all-reflection interferometer that has been used to obtain simple interferograms to test the principles presented here. Grating $G_1$ is a 1200 groove/mm holographic grating, and gratings $G_2$ and $G_3$ are conventional blazed gratings with 1200 grooves/mm. Baffle B has been added in order to separate zero-order from first-order reflections. This is not essential and was done only to obtain a quick test of the concept. Grating $G_3$ was mounted on a movable slide, and its motion was calibrated using the auxiliary interferometer and laser source shown in the figure. The output of the auxiliary interferometer was made to activate the event marker of a recorder, while the output of the all-reflection interferometer was being recorded. Figure 6 shows the interferogram and calibration record for visible white light. Figure 7 shows the record for the green and yellow lines of mercury. Figure 8 shows the inversion of Figure 7. Only 15 points were carried through by hand to check the principle. The pseudo-spectrum $Z(\sigma')$ is shown at the top. In the middle it is shown scaled in amplitude, and at the bottom it is shown adjusted to true wavenumber. The two peaks coincide with the expected green and yellow lines of mercury.

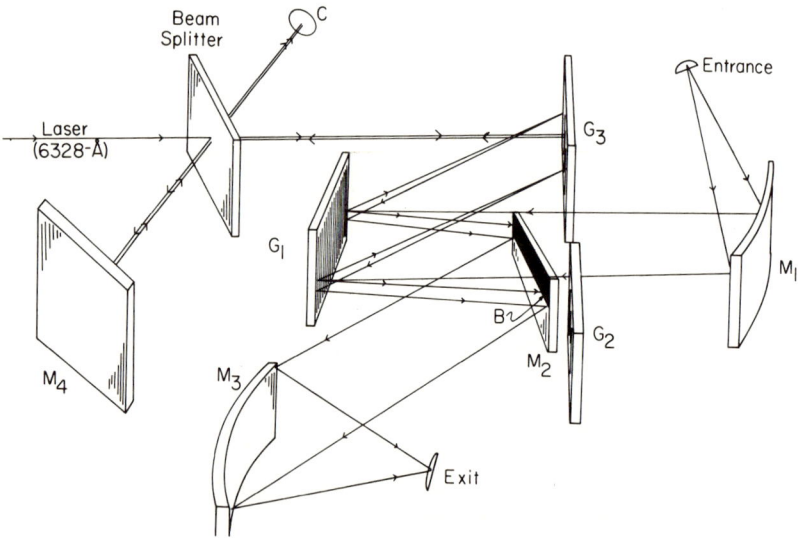

FIGURE 5  Actual test setup of the interferometer of Figure 4. The mirrors $M_2$ and $M_3$ as well as the absorbing baffle B have been added for convenience and are not necessary. Their function is to separate the zero-order reflection of the grating $G_1$ from the recombined interfering beams. The grating $G_1$ is a 48-mm, circular, 1200 groove/mm holographic grating. The motion of the grating $G_3$ is monitored by using $G_3$ in one arm of a Michelson interferometer as shown. The output C of the Michelson is made to operate the event marker of the recorder.

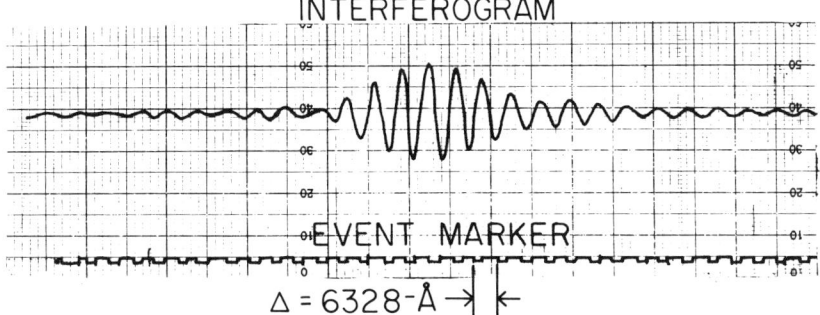

FIGURE 6  Interferogram taken with the interferometer shown in Figure 5. A tungsten light was used as the source. Also shown is the event marker recording (below), which indicates the path-length difference $\Delta$ normal to the grating $G_3$ in intervals of 6328 Å.

As a final test, Figure 9 shows the results in the near ultraviolet using the interferometer adjusted for the 2537 Å line of mercury. By comparing the calibration mark spacing with the interferogram fringe spacing it can be varified that the interferogram corresponds to the pseudo-wavelength expected for the 2537 Å line of mercury.

## V. All-Reflection Interferometer with Focusing Grating

A brief test was made to check the possibility of reducing the total number of reflections to three by using a focusing primary grating. With the source at the focus of the grating, collimated beams are sent to the

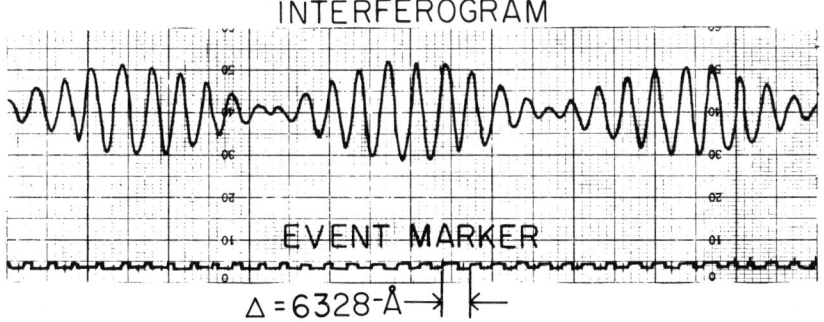

FIGURE 7  A portion of the interferogram obtained while using the visible radiation near 5500 Å from a hot Hg lamp as the source.

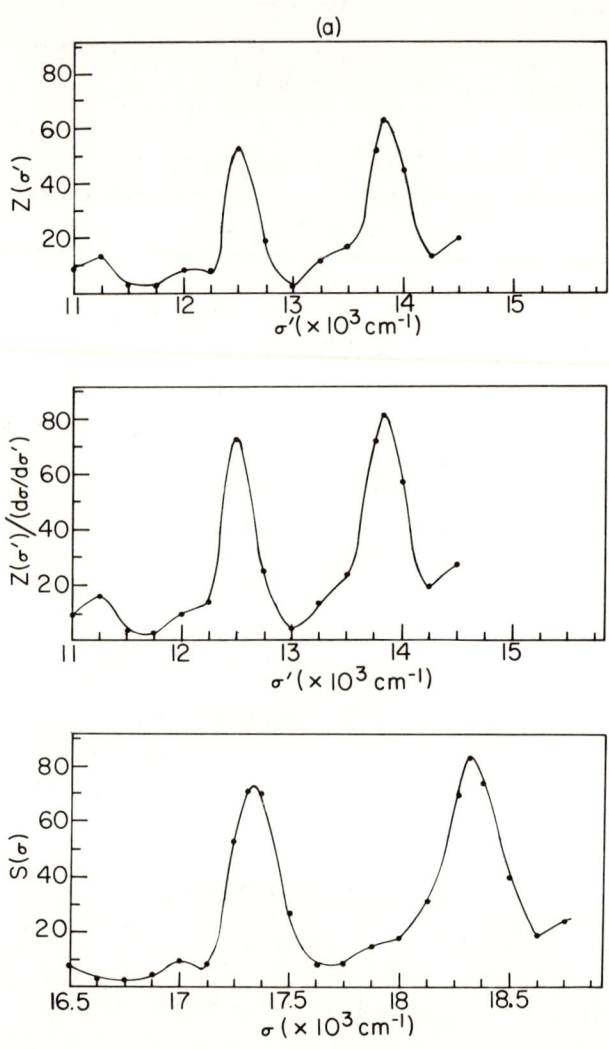

FIGURE 8 (a) The pseudo-spectrum $Z(\sigma')$ as a function of pseudo-wavenumber $\sigma'$. $Z(\sigma')$ was obtained by Fourier inversion of the interferogram shown in Figure 7. (b) The weighted pseudo-spectrum $Z(\sigma')/(d\sigma/d\sigma')$. (c) The actual spectrum $S(\sigma)$ showing the 5461 Å line and the unresolved 5770 Å, 5790 Å yellow doublet. The resolution is $\sim$60.

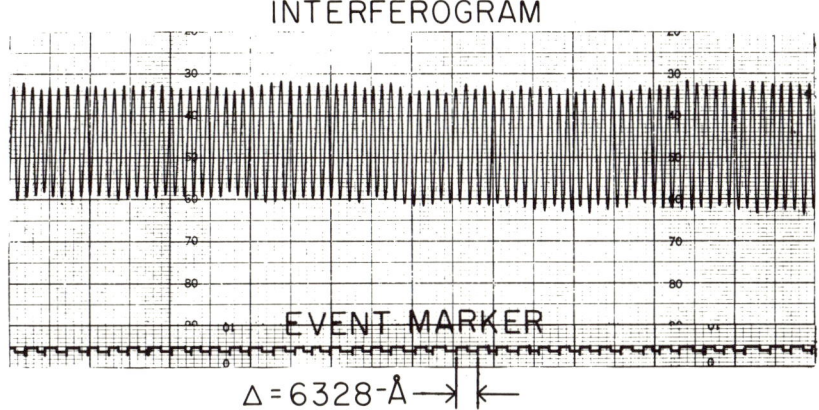

FIGURE 9  Interferogram of the 2537 Å emission from a cold Hg lamp.

secondary gratings and return through the system to the focal plane. Because of the rapid change of focal length of the spherical grating with wavelength, fringes could be observed only over a narrow wavelength range. The aberrations encountered when using conventional ruled concave gratings present a further problem. However, because of the importance of reducing the number of reflections in spectral regions where reflection coefficients are small, it is planned to pursue this configuration further. If satisfactory performance could be achieved over even very small spectral ranges, it would be possible to obtain selected line profiles.

## VI. Conclusion

Two types of all-reflection interferometer for use in Fourier transform spectroscopy have been demonstrated to be feasible. One of the major remaining unanswered questions is that of the overall efficiency of the systems presented. While some approximate theoretical considerations of the efficiency have been made [Kruger et al., 1972, 1973], such considerations are expected to be of only limited value. It is hoped in the near future to make some experimental tests of the efficiency of this device in the far infrared and in the vacuum ultraviolet.

This work has been supported in its entirety by the Research Committee of the Graduate School of the University of Wisconsin with funds provided by the Wisconsin Alumni Research Foundation.

Figures 1 through 3 are from Kruger *et al.* [1972], and Figures 4 through 9 are from Kruger *et al.* [1973]; these figures are reproduced with permission.

## References

Barus, C., *The Interferometry of Reversed and Non-Reversed Spectra*, Carnegie Institution of Washington Publ. 149, Part I (1911).
Connes, P., *Rev. Opt. 38*, 198 (1959).
Davis, J. R., An Interferometer without Essential Transmission, Thesis, U. of Rochester, Rochester, N.Y. (1940).
Kruger, R. A., L. W. Anderson, and F. L. Roesler, *J. Opt. Soc. Am. 62*, 938 (1972); *Appl. Opt. 12*, 533 (1973).
Munnerlyn, C. R., *Appl. Opt. 8*, 827 (1968).

GORDON N. STEELE

# Development and Fabrication of Large-Area Extreme-Ultraviolet Filters for the Apollo Telescope Mount

## Introduction

Since 1961 [Tousey et al., 1964], ultrathin metal foils of aluminum have been successfully used as filters in ballistic space probes that detect and measure the solar extreme ultraviolet (xuv) radiation. More sophisticated requirements are imposed on aluminum foil filters, however, in their application to the astronaut operated instruments of the Apollo Telescope Mount (ATM) [Winter, 1969]. Two of these instruments, the S082A xuv spectroheliograph and the xuv monitor require filters that not only provide perfection over large areas but will withstand the acoustic and vibration environment of the Saturn V booster launch. The spectroheliograph requires a filter providing an uninterrupted transmitting area 4.1 cm × 24.4 cm. This is about ten times larger than any previously launched filter. The position of the spectroheliograph filter near the focal plane of the instrument requires a freedom from pinholes, which has been the major problem in its fabrication. The xuv monitor requires an entrance filter 10.4 cm × 10.4 cm and another intermediate filter 4.6 cm in diameter.

---

The author is at Sigmatron, Inc., Santa Barbara, California 93111.

The methods used to make the filters are reported in detail here and are an adaptation of the techniques developed at the Naval Research Laboratory (NRL) for making thin-foil xuv filters for laboratory and rocket instruments. The NRL process will be reviewed below in connection with discussion of the process development.

## Methods

### FACTORS CONCERNING FILTER QUALITY

The unique character of metal foil filters for the xuv and the special requirements for the ATM instruments dictated a program of development concurrent with the supply of filters to serve in the overall ATM instrument development. Thus, certain specifications imposed at the outset were ultimately relaxed, whereas others became more stringent.

The filters must withstand thermal cycling between $-26\,°C$ and $+52\,°C$, a 3-min exposure to 126-dB acoustic noise, and a peak acceleration of $3.0\,g$ at 50–140 Hz of sine-wave-type vibration. Optically, the transmittance at 304 Å must be not less than 35% and at 584 Å not less than 15%. Visible and near-ultraviolet transmittance must be $10^{-7}/cm^2$ or less. Particulate contamination must meet Class 100 Cleanliness Standards, and all materials must meet stringent requirements with regard to optical contamination by outgassing. The latter requirement is very restrictive with regard to the selection of organic materials [NASA, 1972].

The visible and near-ultraviolet transmittance of thin films has two components: that which is not absorbed in the film and transmittance due to pinholes. If necessary, the transmitted component can be reduced simply by increasing the filter thickness as long as the minimum transmittances at 304 Å and 584 Å are maintained. Pinhole transmittance, on the other hand, is much more difficult to correct and, in fact, was the most enduring problem encountered in this development. Whereas visual inspection may indicate that filters are pinhole-free, the active pinhole transmittance may be $10^{-6}/cm^2$. In solar spectroscopy the visible and near-ultraviolet fluxes are approximately $10^7$ times more intense than the xuv flux, therefore pinhole transmittance of $10^{-7}/cm^2$ or better is required to prevent fogging in photographic recording.

It has been assumed that factors affecting the quality of vacuum-deposited aluminum mirrors for the xuv are also the factors affecting the quality of aluminum foil filters [Hunter *et al.*, 1965]. Six factors known to affect the quality of the mirror are rate of deposition, vacuum

*Gordon N. Steele*

pressure, substrate temperature during deposition, thickness of the deposited film, purity of the aluminum, and aging conditions [Hass, 1955, 1956; Hass *et al.*, 1957]. Accordingly, deposition rates should be greater than 300 Å/sec (high compared with normal rates in other applications) and at a chamber pressure of not more than $10^{-5}$ Torr. The substrate temperature during deposition of mirrors should be less than 50 °C to avoid crystal growth that would cause roughening of the film. Although a purity of 99.99% gives best results for aluminum mirrors, purity of aluminum for xuv filters is less critical.

PROCESS DEVELOPMENT

The greatest problem in producing aluminum foil filters for use in rockets and satellites is that of eliminating pinholes. Pinholes may be caused by faults on the substrate, by contamination of the substrate surface, or during subsequent processing. Therefore, methods for fabricating filters must be designed with this goal always in mind. In the NRL process as developed by Angel [Hunter *et al.*, 1965], a collodion layer is stripped from the surface of a well-cleaned, polished-glass substrate in vacuum. This layer removes from the surface the last traces of particulate contamination. The collodion layer is not easily stripped *in vacuo* because of the strong electrostatic forces generated by this step. Therefore, it is found necessary to do it in the presence of a glow discharge that allows the electrostatic charge to dissipate. With continued pumping, a pressure of $10^{-5}-10^{-6}$ Torr is achieved, and a film of the organic dye, fuchsin, which will be the soluble layer, is evaporated from a tungsten boat onto the substrate. Heavy tungsten filaments, located 50 cm below the substrate, evaporate aluminum at a deposition rate of 1000 Å/sec. Fast-acting shutters, controlled manually, are opened and closed at the proper times so that the film is just opaque to the light from the tungsten filaments. A special glass tank is used to dissolve the water-soluble fuchsin and float foils as large as 100 cm$^2$. Foils are inspected visually by using illumination from beneath the tank while they are floating on the water surface, and the best areas are "cut" from the whole with a sharp blade driven by a hand-held vibrator. These pieces are floated to a separate part of the tank and picked up on 80 mesh electroformed nickel screen cemented over a metal frame. In drying, the foils tend to form a bond to the screen. A technique of precoating the screen with a lacquer and then exposing the dry filter to the lacquer thinner vapor produces a more reliable bond.

Although the results of the NRL process were acceptable for high-quality rocket filters, there was reason to believe that the process could not yield the same quality in filters as large as the 4.1 cm × 24.4 cm filter required for the spectroheliograph. Its main weaknesses lie in the possible damage to such delicate films by the strong forces of surface tension and long exposure to water, which was suspected of generating pinholes. Thus, a new technique was developed, which incorporated parts of the NRL process and comprised the following steps:

1. Thoroughly chemically and ultrasonically clean the glass substrate.
2. Coat the substrate with a parlodion layer that was stripped from the substrate just prior to pump-down.
3. Coat the substrate with a vacuum-deposited film of the organic dye, fluorescein.
4. Vacuum deposit the aluminum filter film.
5. Cement a fine mesh nickel screen to the filter film while it is on the substrate.
6. Release the film and screen from the substrate through dissolution of the fluorescein by immersion in acetone.
7. Cement selected areas of the foil-screen filter stock to filter frames.

This process is treated in detail below.

### SUBSTRATE MATERIAL AND CLEANING

The two most promising substrate materials were polished plate glass, which had been finished with a submerged super final polish, and microsheet, a thin glass sheet with a natural fire-polished surface. Although polished plate glass is more convenient to use, it did have surface blemishes that were revealed by comparison of repetitive pinhole patterns in the foils produced from a particular substrate. Codes 7059 and 0211 microsheet, obtained from Corning, were found to be superior for the purpose. Of these, the Code 0211 soda lime glass, 0.75 mm thick and 5.0 cm × 27.4 cm has been adopted as the standard in this work, because the 7059 glass was slightly etched by the cleaning process. These substrates are normally used once and discarded.

Cleaning of substrates has undergone a continuous process of refining. The different aspects of this most difficult problem will be discussed below.

1. Substrates, received packed between sheets of lens paper, are inspected for gross defects, then chemically cleaned in chromic acid solution and rinsed with deionized water. Those with gross defects are discarded.

2. The substrates are then subjected to ultrasonic cleaning in a detergent solution, Micro (manufactured by International Products Corporation, Trenton, New Jersey), flushed with deionized water, and then cleaned again ultrasonically in filtered deionized water.

3. After ultrasonic cleaning, a thorough flushing is done using a dental Water Pik with filtered deionized water, followed by a free flush with high-purity conductivity water. The substrate is blown free of water and dried by a jet of dry nitrogen gas, after which it is coated with a parlodion lacquer. The composition of this lacquer has been varied considerably to develop the proper stripping characteristics. The composition offering the best qualities is as follows: strip parlodion, 10%; amyl acetate, 26%; ethyl acetate, 8%; acetone, 46%; octoil, 10%.

4. After all preparations for deposition have been completed, air within the bell jar is displaced with humidified helium at atmospheric pressure and the parlodion layer is stripped from the substrate and removed from the vacuum chamber. Care must be used during initial pumping to avoid stirring dust onto the clean substrate [Jorgenson and Wehner, 1963].

VACUUM DEPOSITIONS

The vacuum fixture employed for deposition of films in this work is shown in Figure 1. It is unique in two respects. First, it is built on a chamber that has a titanium getter pump in addition to a conventional oil-pumped system. Before deposition, a nude ionization gauge within the chamber indicates a pressure in the low $10^{-8}$ Torr range. During the aluminum deposition, the pressure is in the lower half of the $10^{-6}$ Torr range. Second, depositions are made by moving vapor sources with specially shaped apertures past the substrate, exposing it to vapor much like a focal-plane shutter exposes photographic film. This is achieved by raising and lowering a carriage, to which several vapor sources are attached, with a motor-driven lead screw. One pass takes 10 sec. The vapor sources for fluorescein and silicon monoxide are laboratory-built furnace-type, calibrated to produce a known deposition rate. The aluminum vapor source is composed of six tungsten filaments, which function by wicking molten aluminum from a reservoir to an evaporation zone. These sources are shown in Figure 2.

The orientation of the multiple-source array makes possible a coating consisting of up to three layers, deposited in one pass of the array, for example, coatings consisting of SiO, Al, and SiO.

Ejection of microscopic droplets of molten aluminum, believed to be due to violent outgassing of the melt in the evaporation zone, was elimi-

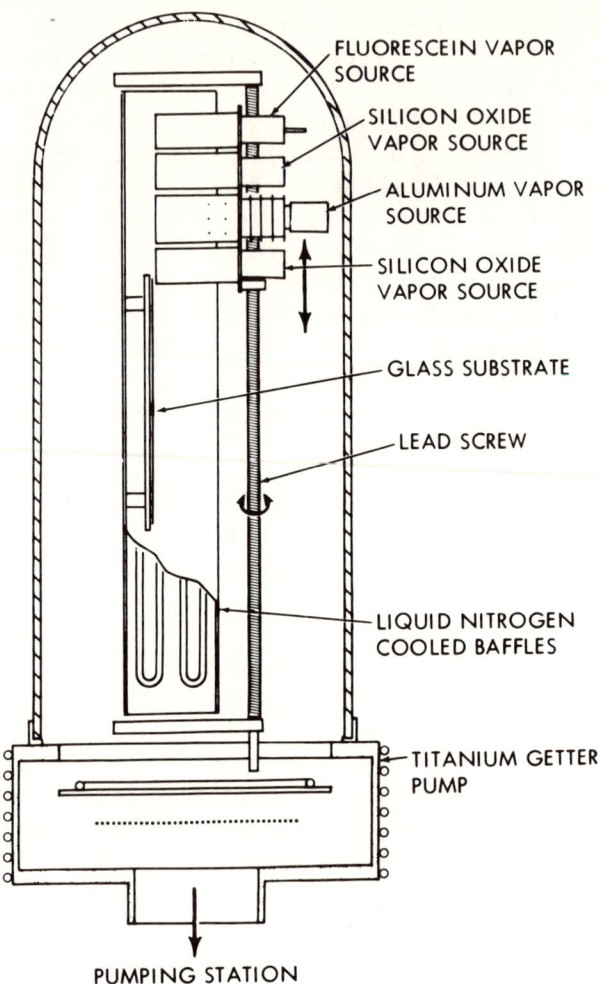

FIGURE 1  Fixture used for vacuum deposition of metal foils.

nated by the chemical gettering action of a small piece of tantalum wire attached to the filament.

A rate-of-deposition quartz crystal microbalance was used to maintain the aluminum evaporation rate through a servo system that controlled the filament power. Vapor from the sources is confined by the interior of a box, the end of which has an opening about 2.5 cm wide in the direction of motion and as wide as the substrate. These boxes also serve to prevent cross-contamination between sources and restrict the

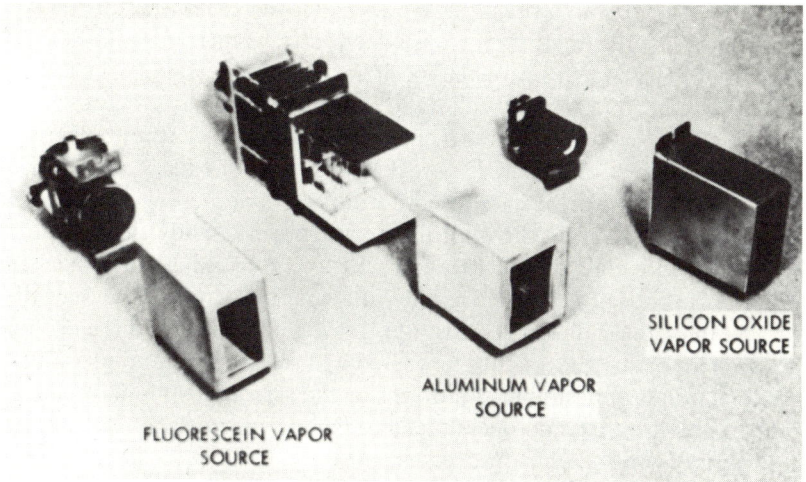

FIGURE 2  Vapor sources used for vacuum deposition of metal foils.

parts that must be cleaned after each run. The aluminum source-to-substrate distance in this case is 7 cm, the deposition rate averages 1500 Å/sec, and the source power is about 500 W. The fluorescein film is deposited on the substrate at a comparatively low rate, about 20 Å/sec, to avoid thermal decomposition. This is done in a separate slow pass to produce a film about 800 Å thick.

The substrate is mounted between two liquid nitrogen-cooled panels, which provide some cryopumping close to the deposition area. Clearance between the substrate and the vapor boxes is about 2.0 mm. Each film thickness is determined by multiple-beam interferometry as measured on an optical flat monitor coated as the sources pass off the end of the substrate. Proximity of the cooled panels chills the substrate prior to deposition an unknown amount by radiant losses. The temperature rise in the substrate due to heat from the vapor sources during a pass has been calculated to be not more than 15 °C, thus the substrate temperature is below the 50 °C limit.

SCREEN SUPPORT

A significant departure of this process from preceding ones lies in attaching the support screen to the film prior to its release from the substrate. The screen used (Buckbee-Mears Company, St. Paul, Minnesota; Dynamics Research Corporation, Wilmington, Massachusetts) is a nickel mesh of 28 lines/cm (70 lines/in.), electroformed on a photoprocessed substrate developed from a ruled master and removed from the substrate by peeling. The resultant screen has a frontal transmittance of 80%, and the more or less oval wires of the mesh are 0.018 mm (0.0007 in.) thick and 0.033 mm (0.0013 in.) wide. The quality of this material is excellent for selected areas up to about 40 cm$^2$, but larger areas without electroforming imperfections, such as "breaks" and "nickel balls," have been difficult to obtain. Breaks are discontinuities where there was no electroplating at some point in the photoformed pattern. Usually these appear as rounded wire ends butted together with no intervening space. Nickel balls are spurious growths on the side of a wire that are thicker than the wire. Both of these defects contribute to the pinholes in a foil, and thus the screen must be thoroughly inspected and selected to minimize this problem. Other undesirable conditions of the screen include wrinkles from the peeling step and occasional contamination with threads of photoresist from the photoprocessed substrate.

The working size for screen is 7.6 cm × 30.5 cm of which the central 5 cm × 27.4 cm is required to be of high quality. The screen is obtained in sheets 30.5 cm$^2$ so there is some latitude for selection. Selected screen of working size is annealed at 500 °C while hung under slight tension. Annealing softens the screen, reduces the severity of the wrinkles, burns off the photoresist, and reveals breaks more prominently by separating the butted ends. The most critical inspection is made after annealing, and the best screen is matched with the best films. Poor screens are used to make small filters.

Epoxy cement in dilute solution of chloroform is applied to the screen, mounted on a processing frame, by spraying to give an equivalent coating on the wire of about 1 $\mu$m thick. The cement thickness is monitored by a quartz crystal microbalance behind the screen during the spraying. This quantity of cement is just enough to form a filet between the wire and the aluminum film, when they come in contact, so that it does not extend beyond the frontal shadow of the wire. Contacting the prepared screen to the film on the substrate requires a technique that avoids spurious contact between the screen and the film. This is achieved by means of a cementing fixture, shown in Figure 3. The screen is suspended a few millimeters above the film on the substrate and pressed into contact by inflating a rubber diaphragm. In this process, the initial contact is central, spreading outward toward the edges in a smooth manner. To prevent any tugging of the diaphragm on the screen, an intervening layer of polyethylene film is provided. During this pressing step, the frame is allowed to travel somewhat with the screen by spring mounts to minimize any distortion at the edges.

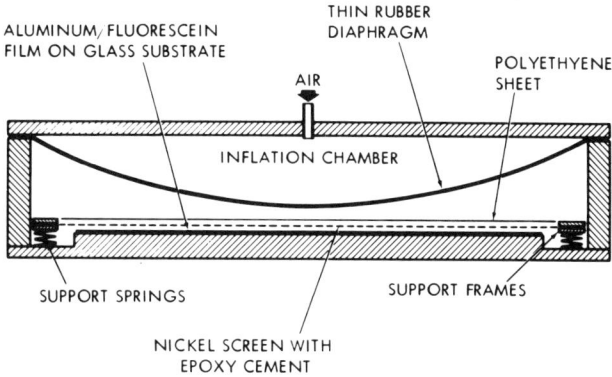

FIGURE 3 Schematic of device used for cementing fine screen to metal films.

FOIL RELEASE

The fluorescein deposited as the soluble layer is moderately soluble in acetone. Dissolution of this layer to release the screen and foil is achieved by immersion in this solvent over a long period of time. The cement employed to secure the screen to the frame dissolves quickly on initial immersion, minimizing any forces that might be transmitted through the film to the substrate and generate pinholes. To avoid gravitational stresses, the release is done with the substrate flat in a covered tray. Vibration during release was determined to be a source of crack-like pinholes that were eliminated by carrying out this step on a vibration-free platform. Release is completed in about 10 h. A released foil may be readily handled with clips that grab a centimeter or more of the selvage edge screen. It is important to rinse the foil with acetone and Freon and dry it in the absence of moisture, because condensation resulting from the evaporation of solvents is particularly damaging with respect to pinholes. Foils prepared in this manner can be stored on frames in a desiccated condition almost indefinitely without pinhole degradation.

FILTER FABRICATION

Filters with very low pinhole transmittance may be selected by a technique wherein a foil is strongly backlighted and the pinhole pattern photographically recorded. By this means, pinholes that may be too small to be detected visually are recorded on film to provide a map of their location and size. Such a record showing the areas selected for some rocket filters is presented in Figure 4. This photograph was obtained from a negative exposed for 4 min to bring out the details that would not be detectable by the unaided eye.

Foil stock must be cemented to frames, sometimes with accurate positioning to avoid poor areas but always smoothly and with regulated tension. For this purpose a hinged, trampolinelike fixture, shown in Figure 5, can be used. In this device, weak springs are hooked to the selvage edge screen to support the foil in a flat condition. Frames to which the foil is to be cemented are placed over the appropriate area of the accurately placed, full-sized pinhole photograph so that when the foil is lowered the frames are covered by the selected areas.

Epoxy cement applied to the top surface of the frames secures the foil material, and the excess foil is removed with a sharp blade. To avoid contaminating the filter with particulate matter during trimming, it is important to design the frame with a raised perimeter around the aperture. This provides a narrow strip for cementing and aids the trimming operation.

Small circular filters 4 cm in diameter, fabricated by the process described above, have successfully survived rocket flights. Noncircular filters such as the spectroheliograph filter and the xuv monitor main-aperture filter, shown in Figure 6, are less robust in that degradation varying from minor to complete destruction has resulted from exposure to the simulated environmental stress of the Saturn V booster rocket. Actually, the screen-foil structure is surprisingly strong and will support

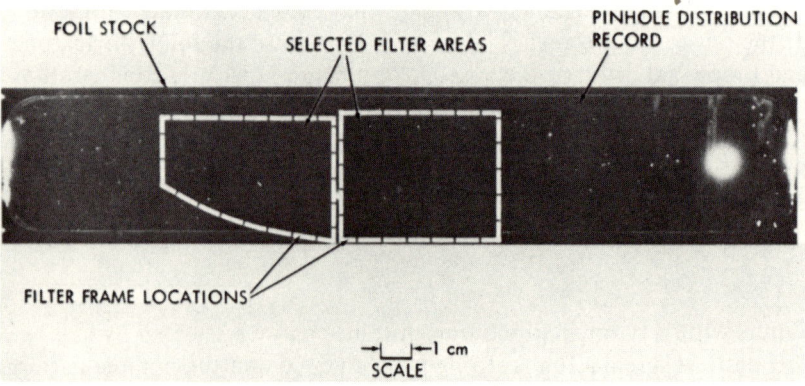

FIGURE 4  Photograph of pinhole distribution in a sheet of foil and the selection of areas for two Cal-Rock filters.

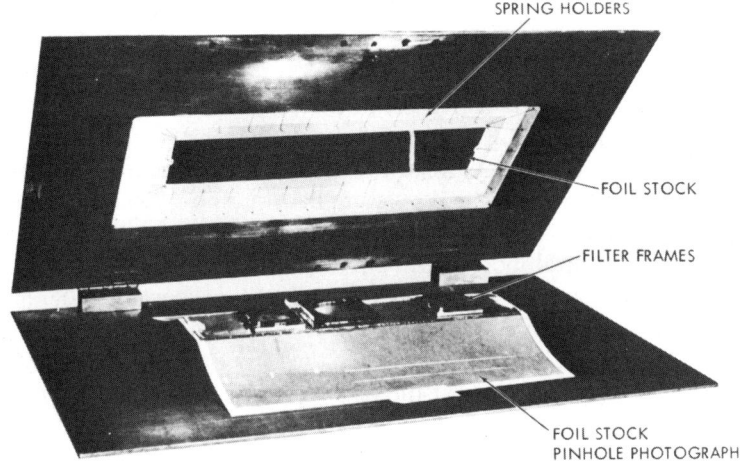

FIGURE 5  Trampoline-type device for cementing selected areas of a stock foil to filter frames.

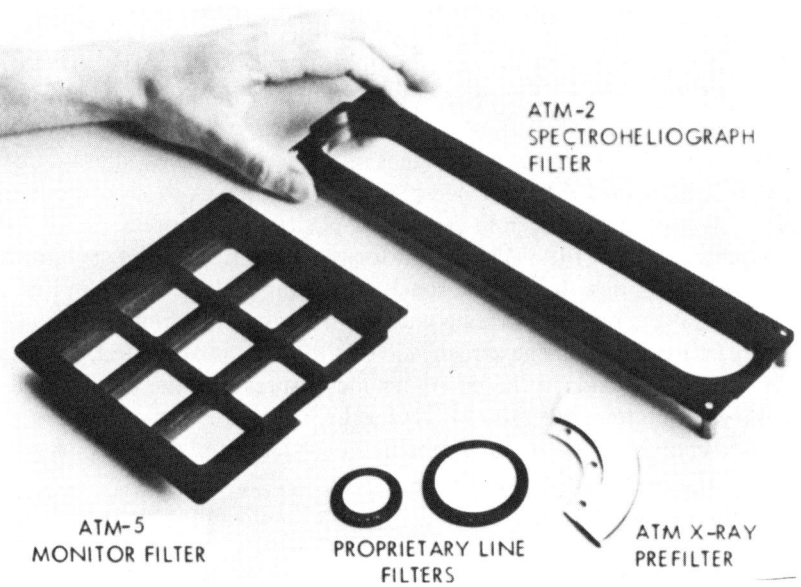

FIGURE 6  Large-area filters fabricated for the Apollo telescope instruments.

a pressure differential of 10 to 20 Torr without damage, but damage due to vibration occurs because of uneven loading and fatigue. In almost every case, large filters that are bolted hard to the mounting surface warp sufficiently to create slight puckers and wrinkles in the foil. Therefore, under vibration or acoustic stress some squares of the screen flex to a rhomboidal shape tearing the foil within the squares. To avoid this, large filters are best retained by compliant standoff mounts, such as polyurethane foam pads, or by cementing in place.

Failure of the support screen at the point of attachment to the frame was a serious problem in the early models of the spectroheliograph filter. During acoustic testing at peak levels of 130 dB, the flexure of the foil is of large amplitude, causing the screen wires to fatigue by the sharp bending that results at the frame attachment. A combination of two features has overcome this problem. The first, referred to as "margining," distributes the bending near the frame by stiffening the foil-screen structure. This is done with a thin layer of epoxy cement applied adjacent to the frame and extending into the aperture area about 2 mm. This epoxy is applied in a highly diluted form from the flattened end of a small tube. The second feature involves achieving tautness in the framed foil. The foil stock cannot be pulled as smooth and wrinkle-free as is desired while on the trampoline fixture because of the nonuniform tension around the perimeter. However, if the frame is pinched slightly prior to cementing the foil, when it is released the foil is pulled smooth. This creates an even tautness in the filter foil and restrains frame flexure during vibration testing so that no wrinkles or slack occur. Frames for filters of this type are compressed 0.05 mm/cm of aperture. The restoring force of the pinched frame should not be large enough to stretch the nickel mesh. Long frames are pinches in the short dimension only, but square frames should be pinched in both directions. A square filter and pinching tool may be seen in cementing position in Figure 5.

Each of the xuv filters developed for the ATM has certain structural features worth describing. The spectroheliograph filter, shown in Figure 7a, illustrates a one-piece frame with a narrow raised perimeter around the aperture to which the screen side of foil stock is cemented. The raised perimeter aids in the trimming and ensures that the cement will make good contact with the foil. It has been found necessary to use an opaque epoxy for the foil attachment since clear cement transmits a significant amount of light.

In the technique of margining, shown in the diagram, the screen side of the filter adjacent to the frame is coated with a very thin layer of epoxy cement. On the spectroheliograph filter, the coating is about 2 mm wide for ease of application. Actually, margining is equally effec-

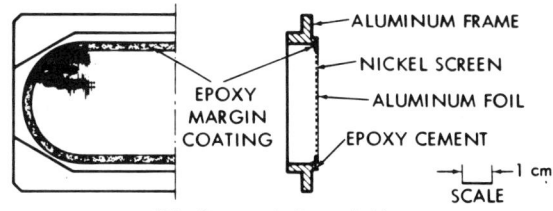

a. ATM-2 spectroheliograph filter.

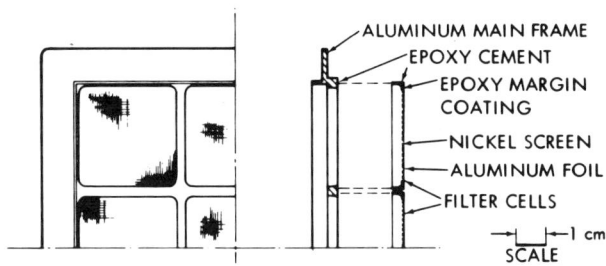

b. ATM-5 XUV monitor filter—main aperature.

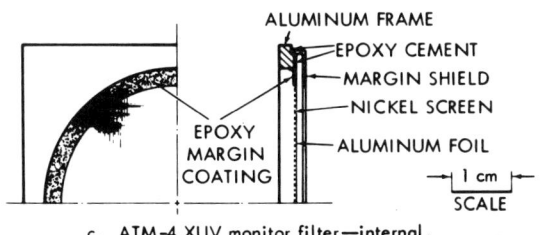

c. ATM-4 XUV monitor filter—internal.

FIGURE 7 Structural details of the xuv filters employed in instruments on the Apollo Telescope Mount.

tive when it is only 1 mm wide, but it is more difficult to apply. The margining material is, of course, opaque to the xuv, and, therefore, the optical aperture is smaller than the frame aperture by the width of the margining coating.

The main aperture monitor filter is shown in Figure 7b and comprises nine smaller filters (cells) cemented to one large main frame to avoid the problem of covering a 10 cm × 10 cm aperture with foil stock 5 cm wide. Each cell is 3.3 cm$^2$ with a frame section 0.5 mm thick and 3 mm high. The frames are pinched slightly during cementing to achieve the taut foil condition and are margined 1 mm wide. The width of the opaque grid of the main frame is 3 mm.

The internal monitor filter, shown in Figure 7c, has a margin shield, an additional feature considered to be important where low pinhole transmittance is required. The weakest part of foil filters, as has been pointed out, is adjacent to the frame. Although the margining technique prevents failure of the screen, it is only partially able to prevent pinholes in the foil at the line of attachment to the frame. The margining material may, of its own nature, etch pinholes in the foil. Therefore it is advisable to add the margin shield, machined to fit closely to the filter and shade the margined area. This is cemented in place after the filter has been otherwise completed. The spectroheliograph filter does not incorporate this feature, but it is provided elsewhere in the instrument. The main aperture filter for the monitor is provided with some margin shielding by the main frame, but shielding is much more effective when it is close to the filter, as with the intermediate monitor filter.

## Results

### XUV TRANSMITTANCE

The xuv transmittance for pure aluminum foil 800 Å thick on 80% transmitting screen is presented as a solid line in Figure 8. This curve has been derived from published data [Hunter *et al.*, 1965; Rustgi, 1968] by multiplying the published data by 0.8 to make it directly comparable with the filter data. The undulating character of this curve is typical of interference effects in transparent films. Datum points plotted for comparison include aluminum of 99.99% purity coated with a layer of silicon monoxide, 100 Å thick, on each side and high-purity aluminum alloyed with 1% silicon by weight, with and without the evaporated silicon monoxide layers. Silicon monoxide coatings were deposited in the same deposition pass with the aluminum. There is evidence that silicon monoxide has a smaller absorption coefficient than aluminum oxide [Sampson, 1967], hence depositing the SiO layers before the formation of aluminum oxide should result in a filter of higher transmittance.

Silicon, alloyed with aluminum for microstructural purposes to be described below, was selected for the unique match of xuv transmittance between it and aluminum [Hunter *et al.*, 1965].

Focusing attention on the open triangle datum points, one observes that a pure aluminum foil protected with silicon monoxide and more than one and one-half times thicker than the comparison foil has about the same or slightly better transmittance. This indicates that a silicon

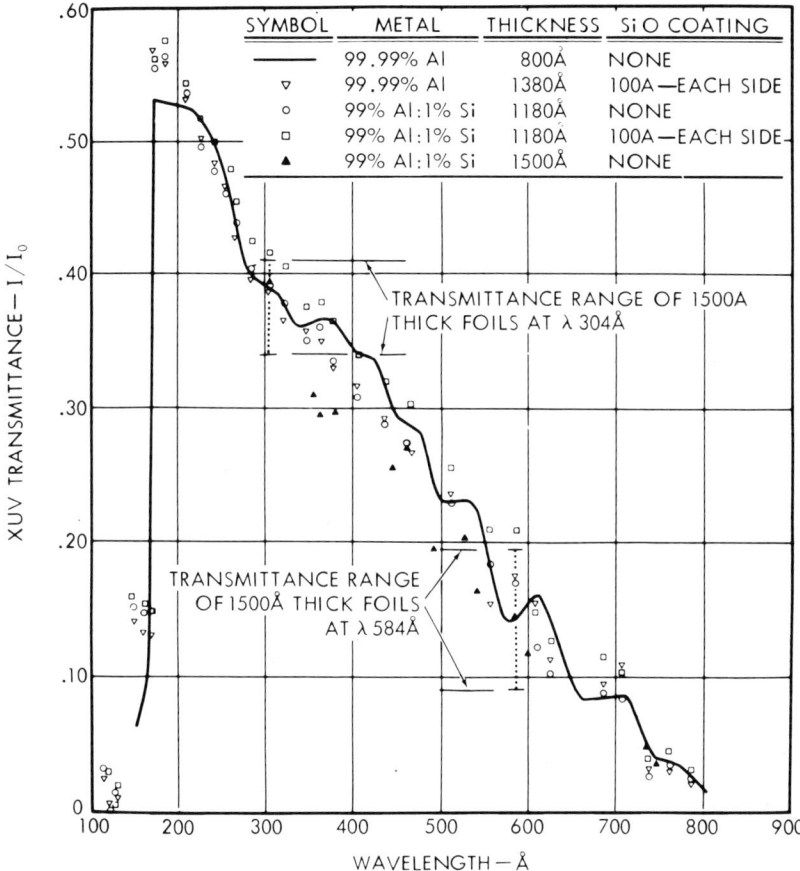

FIGURE 8 Transmittance of pure aluminum and aluminum–1% silicon alloy with and without silicon oxide coating on 80% transmitting screen. The solid line provided for reference is from Hunter *et al.* [1965], adjusted to compare with 80% transmitting screen support. All measurements are by W. Hunter, C. Finter, J. Young, and J. Kruly of NRL and are gratefully acknowledged.

monoxide coating 100 Å thick does indeed have a smaller absorptance than the aluminum oxide film that it replaces.

A more convincing demonstration of the role of silicon monoxide as a protective layer is shown in the datum points for the aluminum–silicon alloy. In a single-pass alloy film deposition, the silicon monoxide sources were operated such that both surfaces of a portion of the alloy film were coated with silicon monoxide. The xuv transmittance measurements obtained from filters prepared of the two areas are plotted as the open circles and squares. These points show consistently higher values for the case of the silicon monoxide protective coating.

A comparison of the xuv transmittances of the pure aluminum foil versus the aluminum–silicon alloy is also contained in this figure and indicates very little, if any, absorption as a result of the presence of the silicon. The 1500 Å thick foil, shown by the solid triangle datum points was determined to have transmittances at wavelengths of 304 Å and 584 Å, equivalent to the comparison of 800 Å thick pure aluminum foil, but it has smaller transmittances between these wavelengths. It is the measurements at 304 Å and 584 Å that have guided this work rather than transmittance measurements over the entire wavelength range from 800 Å to 170 Å.

Probably the most important factor affecting the variability of the xuv transmittance of aluminum foils lies in the terminal thickness of the aluminum oxide. Although the process described in this report minimizes exposure to moisture, the oxide thickness may have varied from the terminal thickness of 35–40 Å usually found on aluminum mirrors [Hass, 1955]. The range of variations in the transmittances at the wavelengths of 304 Å and 584 Å, determined from many foils about 1500 Å thick, are also shown in Figure 8. At 304 Å, transmittance varies from about 34% to 41% with a median value of 36 ± 0.02%. At 584 Å, the transmittance varies from 9% to 19% with a median value of 14 ± 0.02%. Although the cause of this variation is of unproven origin, it is suspected that it is the result of small differences in the terminal thickness of the aluminum oxide.

PINHOLE TRANSMITTANCE

Pinhole and white-light transmittances are measured in this laboratory with a photometer that has a dynamic range of $10^{-4}$ to $10^{-9}$. The area of measurement is 1 cm$^2$, and thus only small filters can be measured. Filters of larger area are evaluated photographically.

An example of what is achievable in pinhole transmission of large area filters, ruggedized to withstand the rigors of a Saturn V booster launch is shown in Figure 9. Figure 9a shows a spectroheliograph filter before it was subjected to a 24-h vacuum bake at 60 °C, sine and random vibration in three axes, and a 3-min exposure to 127-dB peak acoustic stress. Figure 9b, a photograph taken after the tests, indicates only a small increase in pinhole transmission, most of which is limited to the margin area. For comparison, standard photographs of pinholes nominally 1 and 5 $\mu$m in diameter are shown in Figure 9c, and a fairly uniform transmittance of $4 \times 10^{-7}$/cm$^2$ is shown in Figure 9d. From these re-

a. ATM-2 spectroheliograph filter before environmental testing.

b. ATM-2 spectroheliograph filter after environmental testing.

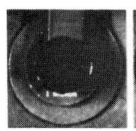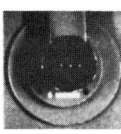  c. Multiple photographic images of single pinholes 1 and 5 microns in diameter.     d. Photographic image of foil with nearly uniform transmissive properties determined to be of $4 \times 10^{-7}$.

FIGURE 9  Photographic record of the white-light pinhole transmittance of an ATM-2 spectroheliograph filter before and after thermal, vibrational, and acoustic testing. Standard photographic techniques are used to provide visual comparison of areal and pinhole white-light leakage with measured standards.

sults, it is evident that the filter has a transmittance in the low $10^{-7}/\text{cm}^2$ range and, between the pinholes, in the $10^{-9}/\text{cm}^2$ range. The pinholes are about 5 μm in diameter or smaller.

Considering that a single large pinhole in a filter the size of a spectroheliograph filter may render it unusable, pinhole plugging has some attraction. Tests in which single squares of the mesh have been filled with an opaque epoxy cement demonstrate that it will pass the qualification tests required. These plugs are opaque to the xuv and would have to be limited to the noncentral areas of the filter to avoid loss of detail. Tests are in process at NRL and in this laboratory to determine the efficacy of this procedure.

It is of some interest to develop a semiquantitative relationship between filter size and area of foil to fabricate a filter. A sufficient amount of foil and numbers of filters have been produced in the course of this work to develop such a picture. Figure 10 shows the estimated yield from foil, of filters of various sizes that have been produced in quantities ranging from several hundred small test filters to about 15 spectroheliograph filters. The yield has been determined for foil 1500 ± 100 Å

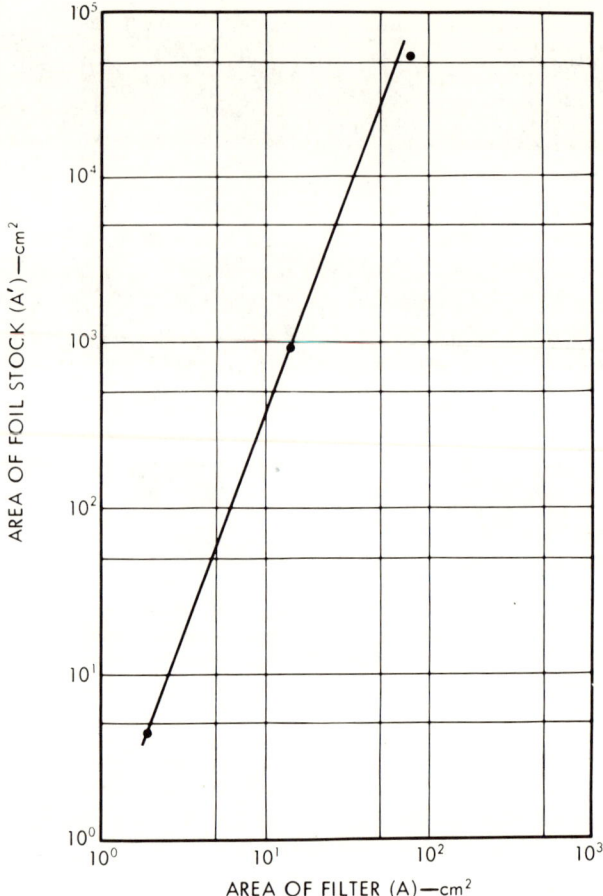

FIGURE 10 Area ($A'$) of foil stock 1500 ± 100 Å thick, required by this process to produce one filter of given area ($A$), wherein a filter is required to have a pinhole transmittance of $10^{-7}$ over an area of 1 cm².

thick of the aluminum–silicon alloy and with pinhole transmittances of $10^{-7}/\text{cm}^2$ or better over the filter aperture. Filters the size of the test filters, having an area of 1.9 cm², require about 5 cm² of foil stock— 2½ times the filter area. The main-aperture monitor filter cells, which have an area of 14 cm², require about 950 cm² of foil stock—about 68 times the filter area. A spectroheliograph filter with an area of 80 cm² requires about 55,000 cm² of foil stock per filter—about 690 times the filter area. These data indicate the desirability of making large filters from several smaller cells whenever practical, e.g., the main-aperture monitor filter.

## FOIL MICROSTRUCTURE

The crystallite boundaries of aluminum foils, nominally 1000 Å thick, may be examined directly when the foil is mounted on the stage of a transmission electron microscope. No crystal growth was ever observed in the initial or during prolonged exposure to the electron beam. Electron microscopy of NRL foils and of foils prepared using the present process showed that the NRL foils had a crystal structure that was noticeably smaller. The reason for this was never identified, but it did suggest that a finer structure was desirable from the standpoint of aging and strength. The effect of sandwiching aluminum foils with silicon monoxide and of alloying on the crystallite size is shown in Figure 11. Figures 11 a, b, and c are from a single foil in which the masking techniques described earlier were used to leave a strip of the pure aluminum foil bare, another strip coated on both sides with 50 Å of silicon monoxide, and, finally, a strip similarly coated with 100 Å. Figures 11 d, e, and f are correspondingly the same except that the aluminum–1% sili-

a. Pure Al—no SiO.    b. Pure Al—50A SiO on each side.    c. Pure Al—100A SiO on each side.

d. 99 Al:Si—no SiO.    e. 99 Al:Si—50A SiO on each side.    f. 99 Al:Si—100A SiO on each side.

FIGURE 11 Effect of alloying and coating with silicon monoxide on the microcrystalline structure of aluminum foil. Silicon monoxide of the indicated thicknesses was deposited immediately before and after the metal layer. Each photograph shows on area of 1 $\mu m^2$.

con alloy was used for the foil material. The areas shown are 1 $\mu m^2$. A study of the crystallite sizes shows that pure uncoated aluminum has a median grain size a little larger than 0.3 $\mu$m. This is not changed when coated with 50 Å of silicon monoxide, but it is reduced to a median size of 0.2 $\mu$m when coated with 100 Å of SiO. A study of the aluminum–silicon alloy shows that the uncoated foil has a median crystal size of 0.2 $\mu$m and that this is not affected by either thickness or the silicon monoxide coatings.

The role of the SiO coatings in reducing grain size is probably due to the energy of absorption being greater on the silicon monoxide than it is on fluorescein and works to increase nucleation density and decrease coalescence during film growth. Apparently, 50 Å of silicon monoxide is too thin or discontinuous to dominate the surface properties.

The addition of silicon to aluminum probably retarded the crystal growth during the coalescence of smaller crystals during film growth. The silicon monoxide coating did not seem able to further affect the coalescence.

### AGING EFFECTS IN ALUMINUM XUV FILTERS

By the time foils have been processed into filters, the protective aluminum oxide layer has formed, and repeated tests of xuv transmittance over a period of a few months show no significant change. This is not true, however, for pinhole transmittance. It has been observed that the pinhole transmittance increases with time at certain points, and this is attributed to breakdown of the protective oxide layer at these points.

Test filters of various materials were stored at room temperature in a controlled humidity environment and periodically removed to measure the pinhole transmittance. Anhydrous calcium sulfate was used to provide the very dry environment for the control filters. A strong sodium hydroxide solution provided a stabilized relative humidity (R.H.) for test filters stored at 20% R.H.; and at the other humidities, saturated solutions of sodium and potassium carbonates were used so that the R.H. could be chosen by the proper sodium-to-potassium ratio. Figure 12 shows the effect of humidity on pinhole transmittance at various humidity levels for test filters of foils of pure and alloyed aluminum with and without 100 Å of silicon monoxide on each side. For comparison, a similar test was conducted of an aluminum–1% silicon alloy film on glass.

At 20% R.H. and lower, no significant increase in pinhole transmittance has ever been observed for exposures up to 150 days. The 60–85% R.H. range is about equally damaging to foils. At 84% R.H. an aluminum–1%

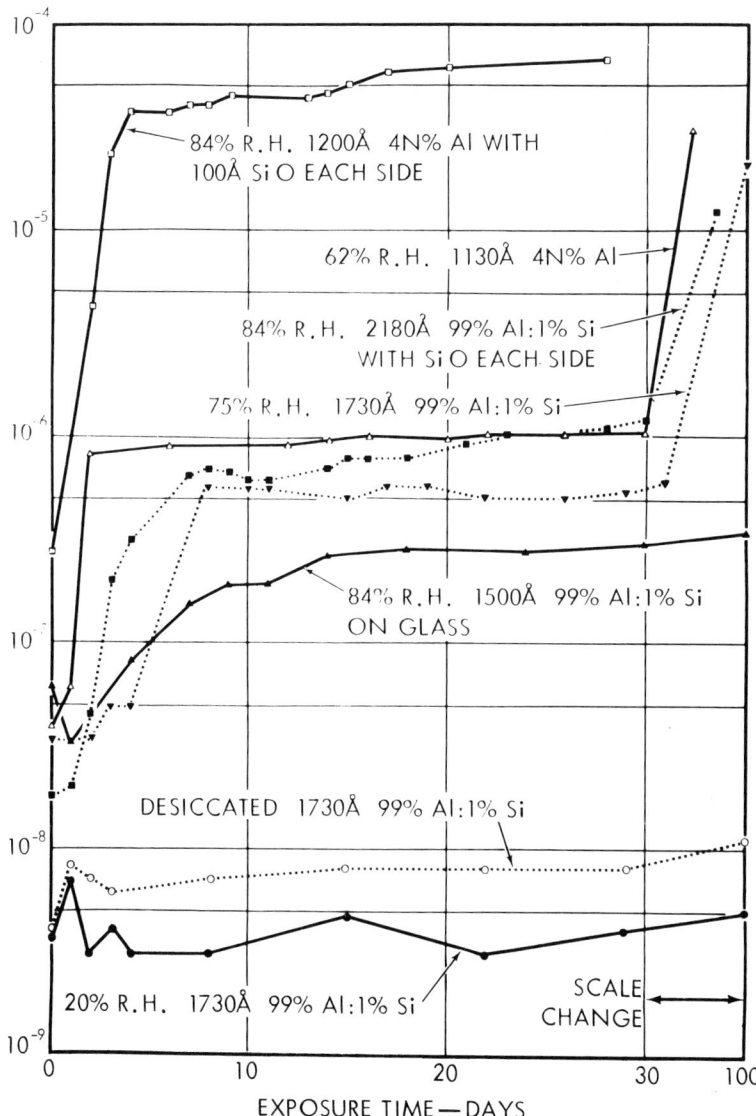

FIGURE 12 Effect of humidity on pinhole transmittance of foils of 99.99% pure aluminum and aluminum alloyed with 1% silicon with and without 100 Å of silicon oxide coating on both sides. Similar exposure of an aluminum–1% silicon alloy film tested *in situ* on the glass substrate is shown for comparison.

silicon alloy film on glass is degraded by moisture, but within a few weeks it tends to stabilize with an increase in transmittance to about $3 \times 10^{-7}/cm^2$. By comparison, a foil of the same composition, 1730 Å thick and exposed to 75% R.H., after making a similar initial change, did not stabilize, and its transmittance increased to the $10^{-5}/cm^2$ range in about 3 months.

Similar results were observed with the alloy coated with silicon monoxide and the pure aluminum without silicon monoxide coating. The results are considerably different, however, for silicon monoxide-coated pure aluminum. This foil material rapidly degraded in high humidity so that the pinhole transmittance was in the $10^{-5}/cm^2$ range within a few days. This effect is real and is observed to a lesser extent when the silicon monoxide coating is thinner. It is believed to be due to stresses generated within the chemically active silicon monoxide coatings as they absorb moisture. One frequently sees heavier films crack and craze after vacuum deposition when the vacuum system is vented to the atmosphere. Why this did not occur with the alloy is unknown, but these data led to the adoption of the 99% aluminum–1% silicon alloy for the ATM filters in the place of pure aluminum.

## Summary

It is possible to make high-quality, large-area xuv filters of either monolithic or cellular structure as large or larger than those reported here. Certain features of the process and filter design will increase the probability of success. These are (1) the area of foil stock from which the required area will be chosen should be a minimum of four times the required area; (2) margining and taut-foil features should be provided in the frame; (3) optical aperture definition should be provided by a margin shield; (4) filters that will be subjected to severe vibrational and acoustic stress should be mounted in the instrument in an unstressed manner to avoid any frame warpage; and (5) the foil thickness should not be less than about 1500 Å.

Filters having an area up to 80 cm$^2$ obtained from foils of 135-cm$^2$ area have been made successfully with white-light transmittances in the low $10^{-7}/cm^2$ range. This visible transmittance is almost entirely from pinholes a few micrometers in diameter. Excluding pinholes, the foil has a transmittance in the $10^{-9}/cm^2$ range.

Silicon alloyed with aluminum reduces the crystal size, which is believed to increase the strength of the foils. When used in a small amount, about 1% by weight, there is no noticeable change in xuv transmittance.

Silicon monoxide protective coatings are not of substantial aid in improving the xuv transmittance of aluminum foils and probably contribute to the pinhole degradation when exposed to normal ambient moisture levels. Corrosion of thin aluminum foils, to create pinholes, occurs above 60% R.H. increasing the pinhole transmittance to the $10^{-6}/cm^2$ range in a few days and to the $10^{-5}/cm^2$ range in a few months. No moisture corrosion is observed when foils are exposed to 20% or lower R.H.

The author would like to express particular appreciation to W. R. Hunter of the Naval Research Laboratory for the many helpful suggestions in the pursuit of this work. This work was supported by the Naval Research Laboratory under Contract No. N00014-67-C-0547.

## References

Hass, G., *J. Opt. Soc. Am. 45,* 945 (1955).
Hass, G., *J. Opt. Soc. Am. 46,* 1009 (1956).
Hass, G., W. R. Hunter, and R. Tousey, *J. Opt. Soc. Am. 47,* 1070 (1957).
Hunter, W. R., D. W. Angel, and R. Tousey, *Appl. Opt. 4,* 891 (1965).
Jorgensen, G. V., and G. K. Wehner, "Pinholes in Thin Films," p. 388 in *1963 Transactions of the Tenth National Vacuum Symposium,* Macmillan, New York (1963).
NASA Document "ATM Material Control for Contamination Due to Outgassing," MSFC 50M02442, Rev. W (1972).
Rustgi, O. P., "Study on a Technique for Detecting Photons in the 100–1000A Wavelength Region," p. 50, NASA Contractor Rep., NASA CR-1096 (June 1968).
Sampson, J. A. R., *Techniques of Vacuum Ultraviolet Spectroscopy,* Chap. 6 (Wiley, New York, 1967).
Tousey, R., J. D. Purcell, W. E. Austin, D. L. Garrett, and K. G. Widing, *Space Research,* Vol. 4, p. 703, North-Holland, Amsterdam (1964).
Winter, T. C., *Astronaut. Aeronaut. 7,* 64 (1969).

# INFRARED METHODS

C. B. ROUNDY and R. L. BYER

# Sensitive Pyroelectric Detectors

## I. Introduction

There has been significant interest recently in the use of various materials for low-NEP (noise equivalent power) pyroelectric detectors. Pyroelectric detectors have the advantages of room-temperature operation, broad spectral response, low power consumption, and simple construction, which make them ideal for space applications.

Putley [1970] has written a review article describing developments in pyroelectric detectors up to 1970, and Weiner and Beerman [1972] have compiled extensive data on more recent developments in pyroelectric detector materials as well as advances in field-effect transistors (FET's) to be used in the first amplifier stage. Some significant developments that have been described are a TGS detector [Beerman, 1971] of area 1.5 × 1.5 mm$^2$, which achieved an NEP of $1.63 \times 10^{-10}$ W/Hz$^{1/2}$ at 15 Hz; a polyvinylidene fluoride (PVF$_2$) detector [Glass et al., 1971] of area 0.02 cm$^2$ with an NEP of $5.7 \times 10^{-9}$ W/Hz$^{1/2}$ at 90 Hz; and an alanine-doped TGS detector [Lock, 1971] with an NEP of $2.5 \times 10^{-11}$ W/Hz$^{1/2}$ below 10 Hz with an area of 0.5 × 0.5 mm$^2$. We have designed and

---

The authors are at Stanford University, Palo Alto, California.

tested a LiTaO$_3$ pyroelectric detector with an area of 1 mm$^2$, which has an NEP of $1.3 \times 10^{-10}$ W/Hz$^{1/2}$ at 20 Hz. Although the alanine-doped TGS yields the most sensitive detectors, and PVF$_2$ has the advantages of potentially low-cost, large-area detectors [Phelan et al., 1971], LiTaO$_3$ has the advantages of being nonhydroscopic, relatively insensitive to ambient temperature, easy to polish to thin samples, and very uniform in quality giving a high yield of sensitive detector elements.

In this paper we describe the noise sources that limit the sensitivity of pyroelectric detectors, compare LiTaO$_3$ with other pyroelectric materials, and describe the construction and performance of the low-NEP LiTaO$_3$ detector.

## II. Noise Sources and Calculated NEP

There are a number of considerations in constructing low-NEP pyroelectric detectors. We have evaluated the noise contributions and compared various materials for the low-NEP detector application. The minimum NEP is limited by the root mean square of the noise contributions from five sources: Johnson noise in the dc load resistor, $V_J$; Johnson noise due to the ac dielectric loss of the detector, $V_D$; amplifier voltage noise, $e_N$; amplifier current noise, $I_N$; and noise from thermal fluctuations of the crystal, $V_T$. The equations for the noise from these sources are listed in Table 1. We have found that these noise sources have relative signifi-

TABLE 1  Noise Sources for Pyroelectric Detectors[a]

| Source | | Equation | Limit when $\omega R_L C \gg 1$ |
|---|---|---|---|
| $V_J$ | = | $\dfrac{\sqrt{4kTR_L}}{1+j\omega R_L(C_c+C_a)}$ | $\dfrac{\sqrt{4kT}}{\omega(C_c+C_a)\sqrt{R_L}}$ |
| $V_D$ | = | $\sqrt{4kT}\sqrt{D/\omega C_c}$ | $\sqrt{4kT}\sqrt{D/\omega C_c}$ |
| $V_e$ | = | $e_N$ | $e_N$ |
| $V_I$ | = | $\dfrac{I_N}{1+j\omega R_L(C_c+C_a)}$ | $\dfrac{I_N}{\omega(C_c+C_a)}$ [b] |
| $V_T$ | = | $\dfrac{\sqrt{4kT}\sqrt{TGp(T)R_L}}{\rho c_p a[1+j\omega R_L(C_c+C_a)]}$ | $\dfrac{\sqrt{4kT}\sqrt{TG}p(T)}{\rho c_p a \omega(C_c+C_a)}$ |

[a] Noise given for 1-Hz bandwidth.
[b] $I_N = (2eI_L)^{1/2}$, where $I_L$ = gate leakage current.

TABLE 2  NEP Equations for Pyroelectric Detectors

| Noise Source | | Including $C_a$ | Limit for $C_c \gg C_a$ |
|---|---|---|---|
| $NEP_J$ | = | $\dfrac{\rho c_p a \sqrt{4kT}}{p(T)\sqrt{R_L}}$ | $\dfrac{\rho c_p \sqrt{4kT}\sqrt{\sigma A a}}{p(T)}$ |
| $NEP_D$ | = | $\dfrac{\rho c_p a (C_c + C_a)\sqrt{4kT}\sqrt{D\omega/C_c}}{p(T)}$ | $\dfrac{\rho c_p \sqrt{4kT}\sqrt{D\omega\epsilon}\sqrt{Aa}}{p(T)}$ |
| $NEP_e$ | = | $\dfrac{\rho c_p a \omega (C_c + C_a) e_N}{p(T)}$ | $\dfrac{\rho c_p \omega \epsilon e_N A}{p(T)}$ |
| $NEP_I$ | = | $\dfrac{\rho c_p a I_N}{p(T)}$ | $\dfrac{\rho c_p a I_N}{p(T)}$ |
| $NEP_T$ | = | $\sqrt{4kT}\sqrt{TG}$ | $\sqrt{4kT}\sqrt{TG}$ |

cance depending on the detector material, the frequency of operation, the detector area and thickness, and the particular FET used for the first amplifier stage. Thus, a material figure of merit based on only one parameter such as the pyroelectric coefficient can be misleading.

In the high-frequency limit, where $f \gg 1/\tau_E$ and $f \gg 1/\tau_{TH}$, the responsivity of a pyroelectric detector is given by

$$V/W = \frac{p(T)}{\rho c_p a \omega (C_c + C_a)} \quad (1)$$

$$= \frac{p(T)}{\rho c_p \omega \epsilon A}, \quad C_c \gg C_a, \quad (2)$$

where $\tau_E$ and $\tau_{TH}$ are the electrical and thermal time constants, and $C_c$ and $C_a$ the crystal and amplifier capacitance. The resulting equations for NEP are listed in Table 2 for the five noise sources of Table 1. In these equations, $p(T)$ is the pyroelectric coefficient, $\rho$ the crystal density, $c_p$ the heat capacity, $D$ the dielectric loss or loss tangent [$D$ can be found from the measurement of ac conductivity by $D = \sigma(\omega)/\omega\epsilon$], $\sigma(\omega)$ the ac conductivity, $a$ the thickness between electrode surfaces, and $A$ the detector area. $I_N$ and $e_N$ are the current and voltage noise of the FET amplifier. For thermal conductance we assume, as a practical consideration, that $\tau_{TH} = 1$ sec, so $G = H = \rho c_p A a$; this gives $G \cong 6 \times 10^{-5}$, which is a factor of 5 larger than thermal radiation conductance given by $G_R = 8\eta\sigma_s A T^3$ with $\eta$ the emissivity and $\sigma_s$ the Stefan-Boltzmann constant. We have included the amplifier capacitance $C_a$ as well as the crystal capacitance $C_c$ because a typical FET amplifier with voltage gain in the first stage has an input capacitance of about 10 pF, which can equal or exceed the detector capacitance.

From the above equations we note that $NEP_J$, $NEP_I$, and $NEP_T$ are frequency-independent but that $NEP_e$ varies approximately as $\omega$ and $NEP_D$ varies as $\sqrt{\omega}$. We also note the dependence on crystal size: $NEP_I$ varies as $a$, $NEP_e$ as $A$, and both $NEP_J$ and $NEP_D$ vary as $\sqrt{Aa}$. Depending on which noise dominates, a change in frequency or crystal size can have a significant effect on the comparison of different materials. Table 3 lists the measured crystal parameters for six important pyroelectric materials [Byer and Roundy, 1972; Keve et al., 1971], and Table 4 summarizes the NEP of those materials for each of the noise sources described above. We have assumed an area $1 \times 1$ mm² and thickness 30 μm to conform to typical thicknesses to which TGS is polished.

Table 4 illustrates that depending on which noise source is considered, different materials would provide the lowest NEP. For example, $Sr_{0.73} Ba_{0.27} Nb_2 O_6$ would have an NEP of $7 \times 10^{-12}$ if only amplifier current noise were present, but loss tangent noise increases it to $6 \times 10^{-11}$ (at 1 Hz). With $LiTaO_3$ the loss tangent noise would dictate an NEP of $9.5 \times 10^{-12}$ (at 1 Hz), but current noise limits the minimum NEP to $1.64 \times 10^{-10}$.

Figures 1–3 show the relative noise contributions as a function of frequency for three representative materials, again for 1 mm × 1 mm × 30 μm detector elements. In Figure 1, for $LiTaO_3$ the amplifier current noise and voltage noise dominate over the entire frequency range. Figure 1 also shows that a different FET may be chosen to optimize the NEP for low- or high-frequency operation. Figure 2 for TGS shows a material-limited NEP, where amplifier noise is negligible and dielectric loss noise dominates over the entire frequency range. Figure 3, drawn for the new ATGS (alanine-doped TGS) with lower dielectric loss and dielectric constant, is still loss tangent noise limited at most frequencies of interest.

Figure 4 shows the rms NEP for the six materials listed in Table 3. Both $LiNbO_3$ and $LiTaO_3$ are FET current noise- and voltage noise-limited, while $Sr_x Ba_{1-x} Nb_2 O_6$, TGS, and ATGS are loss tangent noise-limited over most of the frequency range. A reduction in crystal thickness from 30 μm to 10 μm reduces the NEP for $LiTaO_3$ and $LiNbO_3$ by three at low frequencies, while reducing the NEP for SBN, TGS, and ATGS by $\sqrt{3}$.

Figure 4 shows that where low temperature, low power, and hydroscopic characteristics are satisfactory the new ATGS will provide

TABLE 3  Material Parameters

| | $Sr_{0.73}Ba_{0.27}Nb_2O_6$ | $Sr_{0.48}Ba_{0.52}Nb_2O_6$ | $LiTaO_3$ | $LiNbO_3$ | TGS | ATGS |
|---|---|---|---|---|---|---|
| $p(T) \times 10^{-6}$ C/cm² °C | 0.28 | 0.065 | 0.0176 | 0.0083 | 0.035 | 0.035 |
| $\epsilon_r$ | 8200. | 380. | 43. | 28. | 50. | 23 |
| $\rho$ g/cm³ | 5.2 | 5.2 | 7.45 | 4.64 | 1.69 | ≈1.69 |
| $c_p$ J/g | 0.4 | 0.4 | 0.43 | 0.635 | 0.97 | ≈0.97 |
| $D$ | 0.031 | 0.045 | 0.0002 | 0.001 | 0.03 | 0.0045 |
| $\sigma_{DC}\Omega^{-1}$ cm⁻¹ | $10^{-10}$ | $10^{-11}$ | $10^{-13}$ | $10^{-13}$ | $10^{-12}$ | ≈$10^{-13}$ |
| $C_c$ pF (1 × 1 × 0.03 mm³) | 2420. | 112. | 12.7 | 8.25 | 14.7 | 6.8 |

TABLE 4  Calculated NEP

| | $Sr_{0.73}Ba_{0.27}Nb_2O_6$ | $Sr_{0.48}Ba_{0.52}Nb_2O_6$ | $LiTaO_3$ | $LiNbO_3$ | TGS | ATGS |
|---|---|---|---|---|---|---|
| $V/W \times 10^3$ | $2.98/f$ | $14.6/f$ | $21.4/f$ | $16.0/f$ | $72.3/f$ | $137/f$ |
| $NEP_J \times 10^{-10}$ | 0.57 | 0.79 | 0.70 | 1.38 | 0.33 | 0.09 |
| $NEP_D \times 10^{-10}$ | $0.6\sqrt{f}$ | $0.7\sqrt{f}$ | $0.095\sqrt{f}$ | $0.35\sqrt{f}$ | $0.32\sqrt{f}$ | $0.097\sqrt{f}$ |
| $NEP_e{}^a \times 10^{-3}$ | $e_Nf/2.98$ | $e_Nf/14.6$ | $e_Nf/21.4$ | $e_Nf/16.0$ | $e_Nf/72.3$ | $e_Nf/137$ |
| $NEP_I \times 10^{-10}$ (E1600) | 0.07 | 0.29 | 1.64 | 3.24 | 0.42 | 0.42 |
| $NEP_T \times 10^{-10}$ (for $\tau_T = 1$ sec) | 0.17 | 0.17 | 0.17 | 0.17 | 0.17 | 0.17 |

[a] TI E1600 FET:

| $I_N$ in $A/Hz^{1/2}$ | $e_N$ in $nV/Hz^{1/2}$ | | | | |
|---|---|---|---|---|---|
| | 1 Hz | 10 Hz | 100 Hz | 1 kHz | 10 kHz |
| $3 \times 10^{-16}$ | 73 | 50 | 35 | 25 | 17 |

TI E8002 FET:

| $I_N$ in $A/Hz^{1/2}$ | $e_N$ in $nV/Hz^{1/2}$ | | | | |
|---|---|---|---|---|---|
| | 1 Hz | 10 Hz | 100 Hz | 1 kHz | 10 kHz |
| $6 \times 10^{-16}$ | 22 | 13 | 7.6 | 4.5 | 3.1 |

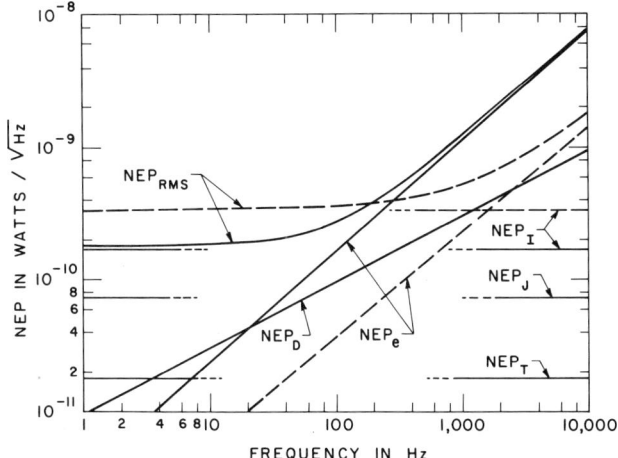

FIGURE 1 Calculated NEP for an $LiTaO_3$ pyroelectric detector using a TI E1600 FET, —— ; a TI E8002 FET, --- .

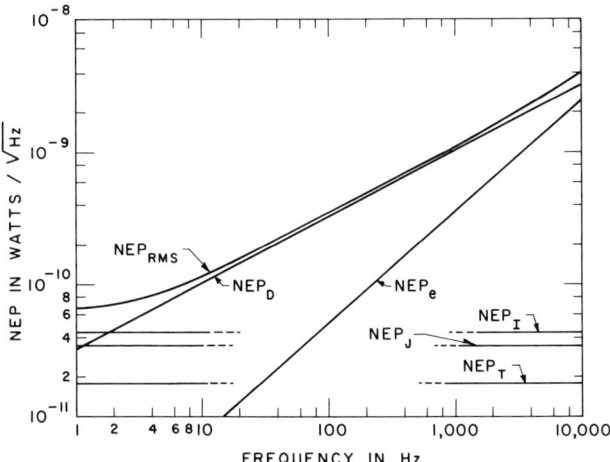

FIGURE 2 Calculated NEP for a TGS pyroelectric detector versus frequency using a TI E1600 FET.

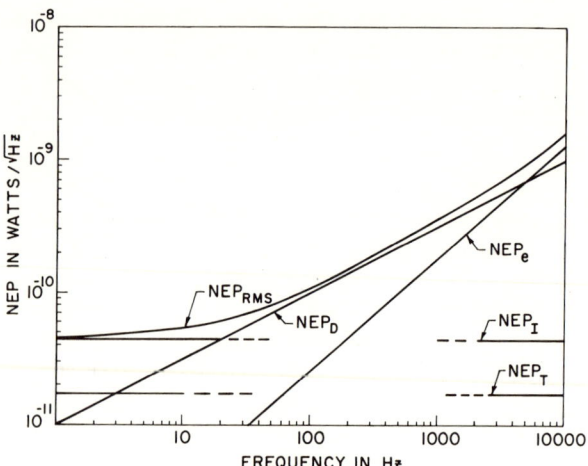

FIGURE 3 Calculated NEP for an ATGS pyroelectric detector using a TI E1600 FET.

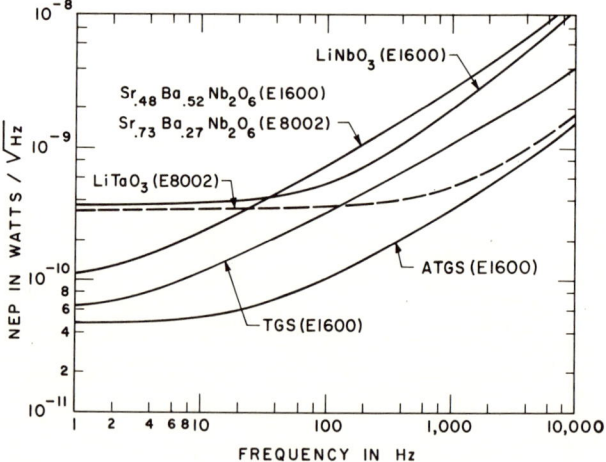

FIGURE 4 Calculated rms NEP for six materials versus frequency for the specified FET.

superior detectors at frequencies below 1 kHz. For nonhydroscopic crystals, SBN or LiTaO₃ appear reasonable, especially if polished to 10 μm or thinner, which is possible because of their favorable mechanical properties. It should be emphasized that the above comparison could be altered significantly for different size detectors as the various noise sources take precedence.

## III. Measured NEP

We have constructed LiTaO₃ detectors with 10-μm-thick elements. The calculated rms noise versus frequency is shown in Figure 5 with the measured noise shown as individual points. The calculated and measured responsivity, which shows a measured thermal break frequency of about 10 Hz, as well as the NEP, which is calculated by dividing the rms noise by the responsivity, are also shown in Figure 5. We see excellent agreement between predicted and measured noise and NEP in the 10- to 100-Hz region. Above 100 Hz, the measured NEP is better than the predicted value due to a lower FET voltage noise than the quoted typical value.

The detector element is a wafer of LiTaO₃ approximately 0.5 cm² with the optic axis perpendicular to the face and polished to 10 μm thickness. Using a vacuum pickup device and moderate care in handling we achieved over 50% yield of unbroken detector elements. The thin wafers were placed in a mask with 1-mm² holes and 1-mm spacings and

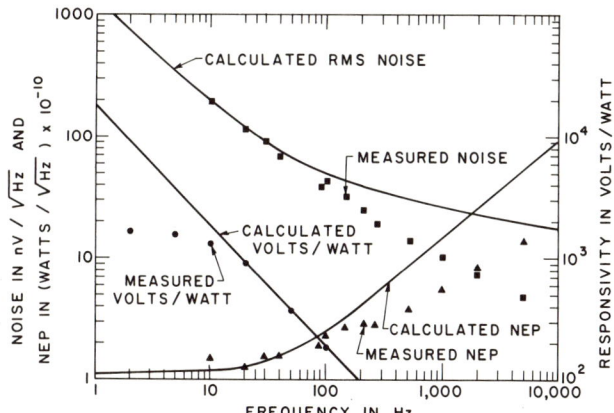

FIGURE 5 Calculated and measured noise, responsivity, and NEP for LiTaO₃ detector 1 × 1 × 0.01 mm³ and TI E1600 FET.

evaporated with a Cr–Au rear electrode, which had approximately 100% absorption in the visible and a thin, nearly transparent, front Cr electrode.

The spectral response was similar to the response of the coating shown in Figure 6, which was developed for very fast pyroelectric detectors [Roundy and Byer, 1972] and is nearly 50% absorbing from less than 0.3 $\mu$m to greater than 50 $\mu$m. We have used detectors with this coating to absorb radiation at the third and ninth harmonic of the 1.064-$\mu$m Nd:YAG laser at 3546 Å and 1182 Å, respectively. The detector has also detected x rays at the Cu K-$\alpha$ line at 1.5 Å.

After dicing the individual elements on a wire saw they were then mounted on T05 headers on two beads of conductive epoxy (spaced 1 mm apart), which made electrical contact for the rear electrode and provided an air gap in the center of the detector for thermal isolation. A 1-mil gold wire was attached to the front electrode with conductive epoxy. A sleeve was then placed over the T05 can for mechanical protection of the detector but still leaving it in open air. The detectors were then mounted inside an rf shielded metal box housing a low-noise FET input amplifier (courtesy of Molectron Corporation, Sunnyvale, California) with a gain of 1000. The first stage of the amplifier used a low-noise E1600 FET from Texas Instruments in England [Glass et al., 1971]. Figure 7 is a photograph of the detector element mounted on the T05 header.

The noise was measured on an HP 302A wave analyzer, with a 7-Hz

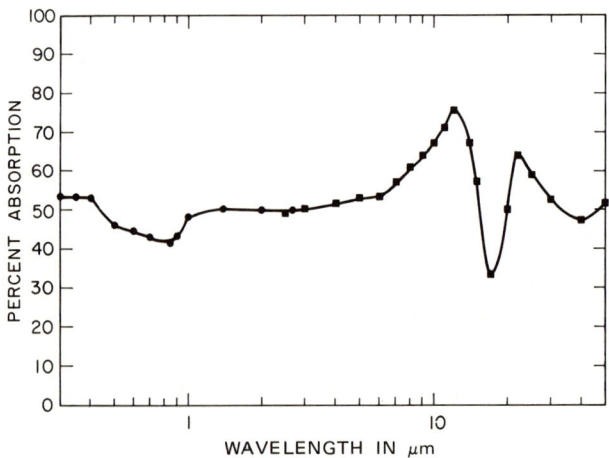

FIGURE 6 Absorption of LiTaO$_3$ versus wavelength when coated with fast thermal time-constant black.

FIGURE 7 Photograph of 1 × 1 × 0.01 mm³ detector mounted on T05 header.

bandwidth, with the signal output coupled to a slow time constant, true rms, voltmeter to average the rapid noise fluctuations. The meter reading was then divided by $1000\sqrt{7}$ to obtain the equivalent input noise.

The responsivity versus frequency was measured by chopping a 1-mW He–Ne laser, attenuated by 100 to $10^{-5}$ W, with a variable-speed chopper. The 6328 Å light was focused to the center of the crystal to ensure total absorption and to obtain the maximum thermal time constant. The NEP was then calculated by dividing the measured noise by the measured responsivity. In addition, we measured the NEP using a PAR lock-in amplifier and an He–Ne beam attenuated to $5 \times 10^{-10}$ W. The measured signal-to-noise ratio verified the above results. Three detectors were made and tested and were all within 10% in noise and responsivity.

In conclusion, it is clear that no one figure of merit can be used when comparing different pyroelectric crystals. For the most sensitive pyroelectric detector at frequencies below 1 Hz, TGS and more recently alanine-doped TGS, appears to be the best material until a significant reduction in FET current noise is made. However, for higher frequencies and for applications requiring higher-temperature operation, $LiTaO_3$, $LiNbO_3$, and SBN are all capable of low-NEP operation. We have constructed three $LiTaO_3$ 10-$\mu$m-thick detectors, and all have a measured NEP within 10% of the calculated value. Thus $LiTaO_3$ is a very useful pyroelectric material yielding detectors with an NEP near $10^{-10}$ W/Hz$^{1/2}$.

We wish to acknowledge Bob Griffin for polishing the $LiTaO_3$ to 10 $\mu$m and Forrest Futtere for help with vacuum evaporation technology.

### References

Beerman, H. P., *Ferroelectr. 2*, 123 (1971).
Byer, R. L., and C. B. Roundy, *IEEE Trans. Sonics Ultrasonics SU-19*, 333 (1972).
Glass, A. M., J. H. McFee, and J. G. Bergman, Jr., *J. Appl. Phys. 42*, 5219 (1971).
Keve, E. T., K. L. Bye, P. W. Whipps, and A. D. Annis, *Ferroelectr. 3*, 39 (1971).
Lock, P. J., *Appl. Phys. Lett. 19*, 390 (1971).
Phelan, R. J., Jr., R. J. Mahler, and A. R. Cook, *Appl. Phys. Lett. 19*, 337 (1971).
Putley, E. H., in *Semiconductors and Semimetals*, Vol. 5, p. 259, R. K. Willardson and A. C. Beer, eds., Academic Press, New York (1970).
Roundy, C. B., and R. L. Byer, *Appl. Phys. Lett, 21*, 512 (1972).
Weiner, W., and H. P. Beerman, Development of Improved Pyroelectric Detectors, Contr. NAS5-21655, NASA Goddard Space Flight Center, Greenbelt, Md. (1972).

JOHN A. DECKER, JR.

# Hadamard-Transform Instrumentation for Infrared Space Optics

## Introduction

The performance obtainable with infrared space optical systems is constrained—sometimes quite severely—by the limits placed by observing time, signal-to-noise ratio, spatial and spectral resolution, and the spacecraft environment. The latter is particularly limiting, as the necessary weight, power, reliability, and temperature constraints severely restrict the detector performance and the system's throughput and degree of complexity.

We will briefly discuss here a comparatively new technique—Hadamard-transform optics—that offers the designer of infrared space optical systems substantial help in dealing with these problems. Specifically, Hadamard-transform techniques provide spectral and/or spatial *multiplex* capability (that is, the ability to view a number of discrete spectral/spatial resolution elements simultaneously with a single detector) and *very high étendue* (optical throughput, say, in mm² steradians)—and hence an increase in signal-to-noise ratio of several orders of magnitude over conventional imaging and dispersive optical techniques. Hadamard-

---

The author is at Spectral Imaging, Inc., Concord, Massachusetts 01742.

transform optical systems use *binary* optical coding techniques in conjunction with "conventional" dispersive and/or imaging optics—they are *not* interferometers—and are hence well understood, of a very low degree of complexity, and highly reliable. The performance gains are sufficient, in some cases, to allow the use of uncooled (or radiation-cooled) infrared detectors in situations where cryogenically cooled detectors would ordinarily be required. Additionally, Hadamard-transform systems inherently have a very low output-signal dynamic range and hence require a substantially lower telemetry bit rate than other multiplex systems for most applications. Finally, the necessary computer decoding is extremely fast and easy: the fast Hadamard-transform algorithm [Nelson and Fredman, 1970] runs about ten times faster (per dimension) than the equivalent fast Fourier transform [Pratt *et al.*, 1969].

Hadamard-transform optics—that is, the use of *binary* orthogonal codes based on Hadamard matrices for optical multiplexing [Decker, 1972, and the references cited therein]—was first proposed over 15 years ago [Fellgett, 1958] but has only quite recently been developed as a practical technique and hence is very little known. It is, however, well described in the current literature [Decker, 1971, 1972; Decker and Harwit, 1968, 1969; Harwit, 1971, 1973; Harwit *et al.*, 1971; Phillips and Harwit, 1971; Sloane *et al.*, 1969]. We will therefore discuss here several approaches to the use of the technique in infrared space and airborne astronomical and earth-resources measurements. First, we will describe a quite simple airborne spectrometer that took part in a recent airborne expedition of Mars observations and then discuss an extension of this technique into the very-high-performance regime that is under current study—an example of the state of the art as it were. We will then discuss at some length the recently developed imaging spectrometer* [Harwit, 1971], describe the bench-test prototype of this instrument [Harwit, 1973], and close with a brief discussion of the application of this unique class of instrument to spacecraft observations.

### Performance of Infrared Optical Systems

Before discussing specific hardware systems, I feel it necessary to first answer the questions, "Why go to all this trouble? Why would one use

---

* The imaging spectrometer, in all its ramifications, is proprietary to Spectral Imaging, Inc., is covered by U.S. Patent 3,720,469, and should not be used without written permission.

multiplex optical systems in the first place?" (And, in fact, "Why is it that 99% of all infrared spectroscopists do *not* use multiplex methods?")

Let us answer the questions by taking the example of an infrared spectrometer. A figure of merit, approximately proportional to the signal-to-noise ratio to be expected from a given instrument making a given observation, may be defined by

$$E = RLN^{\frac{1}{2}}A_d^{-\frac{1}{2}} \langle \epsilon_\lambda D^*_\lambda \rangle_\lambda,$$

where $R$ is the spectral resolution $\lambda/\Delta\lambda$, $L$ is the étendue, $N$ the degree of multiplexing (the number of spectral resolution elements $\Delta\lambda$ viewed simultaneously, equal to the number of *exit* slots on an HTS code mask), $A_d$ is the infrared detector area, and the quantity in triangular brackets is an average over the spectrometer's wavelength band of the product of the net efficiencies of filters, gratings, beam splitters, lenses, etc., and the detector detectivity. If we now let $n$ be the number of *entrance* slots of a multislit–multiplex (or "two-ended") HTS grating spectrometer [Harwit et al., 1971], $l$ the ratio of focal length to slit height, $A_g$ the area of the grating, and $A_M$ the aperture of a Michelson interferometer (FTS spectrometer) of equivalent $R$ and $N$, then the performance ratios for an HTS instrument to a conventional scanning monochromator and a Michelson interferometer spectrometer are approximately, ignoring detector effects,

$$E_{HTS}/E_{m-\lambda} \approx \tfrac{1}{2}nN^{\frac{1}{2}}$$

and

$$E_{HTS}/E_M \approx (n/2\pi l)(A_g/A_M).$$

For a typical state-of-the-art HTS design, as would be used in spacecraft systems, with $N \approx 2000$, $n \approx 250$, $A_g \approx 10^4$ mm², these numbers can become quite substantial. These performance gains are also realized by spatial and spectral/spatial multiplexing systems [Harwit, 1971] and are quite often increased in practice by the HTS systems' more efficient use of their detectors, in particular their ability to take advantage of the variation of detectivity with wavelength for broad-band systems. As an example of what can be done with a comparatively simple system—and of what is available off the shelf *now*—our Model HTS-255-15 Analytical Spectrometer has an efficiency (as defined above) some *3500* times that of the conventional thermal-detector monochromators normally used for infrared spectrochemistry; this figure increases to about *50,000* if one uses liquid nitrogen-cooled detectors, instead of the standard room temperature detectors, in the HTS.

## Airborne "Mars Observation" HTS

The conventional answer to the question "Why doesn't everyone use them?" has been that they (multiplex spectrometers) are complex—and hence expensive and prone to reliability problems—and hard to use in the field. As a partial answer to this position, we would like now to describe a small, simple, rugged, light-weight Hadamard-transform spectrometer (or HTS) designed for infrared airborne astronomical measurements (Figure 1). This instrument, the Spectral Imaging, Inc., Model HTS-19-1, was designed and constructed for Cornell University and flew as one of the three primary experiments on board NASA's Convair 990 observatory aircraft during the 1971 Mars opposition observations.*

The question of the specific chemical composition of the Martian surface and its relationship to the "bright" and "dark" regions observed visually has been under study in the astronomy community for some years. Recently, the question of the presence of water of hydration in the surface rocks on Mars was raised [Pollack and Sagan, 1969]. As existing ground-based and spacecraft spectra were insufficient to provide a conclusive answer to this question, a group of astronomers at Cornell University and the NASA Ames Research Center proposed to observe the appropriate region of the Martian reflection spectrum from a high-altitude aircraft during the 1971 Mars opposition. The astronomical results will be presented elsewhere [Houck *et al.*, 1973]; we will limit ourselves here to a description of the design and use of the spectrometer.

A spectrometer for these observations has to meet certain basic requirements: a spectral coverage from 2.5 to 3.8 $\mu$m (4000 to 2630 cm$^{-1}$) at a resolution $\Delta\lambda \leq 0.05$ $\mu$m ($\Delta\nu \leq 35$ cm$^{-1}$). The major design problem was the extremely low incident flux and comparatively short observing time. The 30-cm telescopes on NASA's Convair 990 observatory aircraft could focus only some $10^{-10}$ W/$\mu$m into the spectrometer within the appropriate spectral interval. The short observing times available with aircraft observations, as compared with ground-based measurements, led to the requirement that the instrument must be able to achieve a signal-

---

* Supported by NASA through contract NGR 33-010-148 and Cornell University Purchase Order 1-78722.

FIGURE 1  Model HTS-19-1 Hadamard-transform airborne astronomical spectrometer.

to-noise ratio of at least 5 with an observing time per data point of 1 sec. In addition, of course, it had to be lightweight and rugged enough to operate successfully in the airborne environment; the primary problem area here was vibration resistance, principally during takeoff and landing.

A simplified optical schematic of the Model HTS-19-1 airborne HTS spectrometer is given as Figure 2. Because of severe cost and time restraints (the instrument had to be delivered within six weeks of receipt of the order), it was built around off-the-shelf commercially available optical components. The dispersive system was a 250-mm, $f/3.5$, Ebert spectrometer, fitted with 19-slot cyclic $S$-matrix HTS coding masks [Sloane et al., 1969] at both the entrance and exit focal planes. The resulting entrance aperture was 6.3 mm wide by 1.0 mm high; as the focused image of Mars was expected to be only 0.338 mm in diameter, the instrument was actually operated with all but one entrance slot blocked, in the "single-ended" mode. The exit coding mask was traversed by a stepper-motor-driven translation stage, whose control cir-

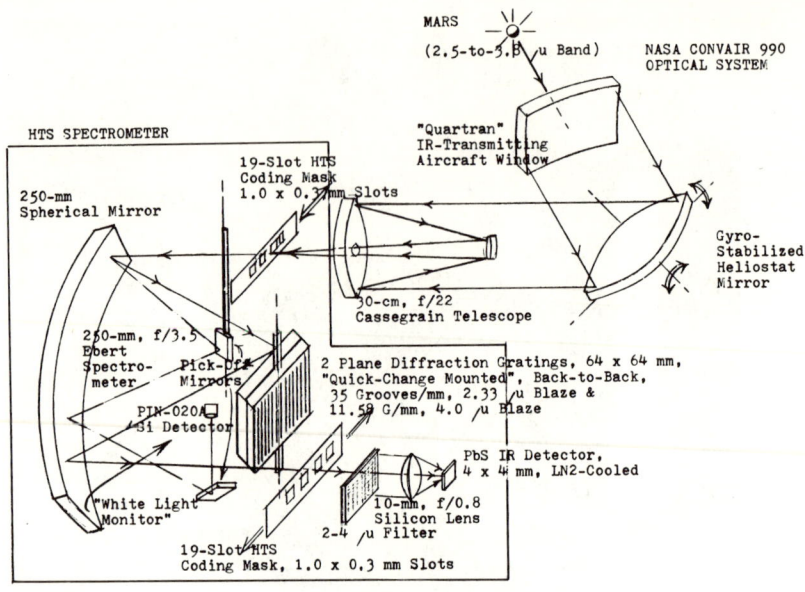

FIGURE 2 Optical schematic of the Model HTS-19-1 spectrometer and the NASA observatory aircraft optics.

cuits also provided the command and indexing pulses for the punched-paper-tape digital data-recording system used for the flight observations.

Two 64 mm × 64 mm Bausch & Lomb plane diffraction gratings were used, mounted in a back-to-back quick-change mounting. The primary grating was ruled with 11.58 grooves/mm and blazed for a wavelength of 4.0 μm. It simultaneously observed a wavelength range of 2.13 μm at a resolution $\Delta\lambda$ = 0.112 μm ($\Delta v$ = 78 cm$^{-1}$). The second grating, which was used for higher-resolution measurements of selected spectral regions, had a simultaneous wavelength span of about 0.7 μm and a resolution $\Delta\lambda$ = 0.037 μm ($\Delta v$ = 26 cm$^{-1}$); it was ruled with 35 grooves/mm and was blazed for a wavelength of 2.33 μm. A 2- to 4-μm bandpass filter preceded the 10-mm, $f$/0.8, silicon lens used to focus the radiation passing the exit mask onto the 4 mm × 4 mm liquid nitrogen-cooled lead sulfide (PbS) detector. The spectrometer was fitted with a white-light monitor, used to normalize the spectra with respect to changes in atmospheric transparency and aircraft window contamination. This consisted of a 1 cm² pick-off mirror, which intercepted a portion of the light incident on the grating and returned it to a focus on a low-noise silicon PIN photodiode. The electronic system was completely conventional and was, in fact, assembled from laboratory electronic components in daily use at the Cornell Astronomy Department and at Spectral Imaging, Inc.

The NASA observatory aircraft used for the 1971 Mars opposition observations, Convair 990 *Galileo,* is fitted with three 30-cm, $f/22$ Cassegrain telescopes, each equipped with a gyro-stabilized heliostat pointing mirror and an appropriate optical window through the aircraft skin; in our case, this window was of Quartran infrared-transmitting glass. Fine guiding was accomplished by correction signals applied to the heliostat gyro system; the guiding image was obtained by a beam splitter in the optical train. A stepped-segment rotating-mirror chopper was used (which had the effect of shifting the telescope field of view on and off the planet), and the chopped beam was focused onto the entrance slit by a 25-mm, $f/1.0$, Irtran II lens.

A series of observation flights were made during the 1971 Mars opposition and included two flights in which both Mars and the moon were observed, two flights observing Mars only, and two short calibration sun observations during the ferry flights from the observing station, Hickham Air Force Base, Hawaii, to the aircraft's normal base, NASA Ames Research Center, California. Each flight provided up to 3 h of observing time at altitudes from 11.9 to 12.5 km (39,000 to 41,000 ft). All in all, about 8 h of Mars data were collected and appropriately less data from the moon and sun. Spectral scanning times varied with the brightness of the source: most Mars spectra were taken at 190 sec/spectrum, while the lunar and solar spectra were taken at 19 sec/spectrum, which was about as fast as the paper-tape punch would operate reliably. To eliminate the effects of the spectral absorption of the earth's atmosphere, as well as spectral deficiencies in the aircraft window and the various filters, lenses, mirrors, etc., the spectra were reduced as *ratio* spectra: Mars/sun and Mars/moon. These spectra and their interpretation will be presented elsewhere [Houck *et al.,* 1973] ; we can note, however, that there is water of hydration on the Martian surface.

**State-of-the-Art Airborne HTS Spectrometer**

Figure 3 sketches the optical layout of a Hadamard-transform spectrometer under current study for high-resolution, high-sensitivity analysis of the chemical composition of the upper atmosphere from a high-altitude aircraft. The entire instrument would be cooled to 77 K by liquid nitrogen. It represents the current state of the art in airborne HTS spectrometers and is also typical of what could be achieved in high-resolution spacecraft systems. The only major change from the design sketched— other than constructional techniques, of course—would be the use of uncooled or radiation-cooled detectors in place of the liquid nitrogen-cooled detector shown on Figure 3.

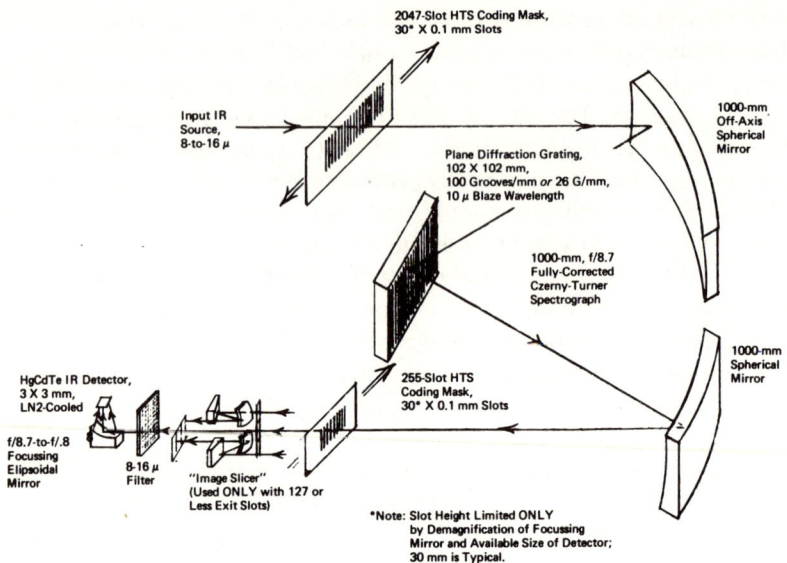

FIGURE 3  Optical schematic of a high-resolution, high-étendue airborne HTS under study for upper-atmosphere trace-constituent measurements.

This instrument would cover the 8- to 16-$\mu$m (1250 to 625 cm$^{-1}$) spectral region at an average resolution of about $\Delta v = 0.1$ cm$^{-1}$. The operator would have the choice of viewing the spectral range at full resolution in four segments (i.e., with four grating settings) or viewing the whole spectral range at once at reduced resolution (about 0.4 cm$^{-1}$). As currently seen, the system would use a 1000-mm, $f/8.7$, asymmetric Czerny-Turner dispersive system fitted with two 102 mm × 102 mm gratings—of 100 grooves/mm and 25 grooves/mm, both blazed for a wavelength of about 10 $\mu$m—in a quick-change mount. The spectral multiplexing "exit" mask would have 2047 slots, 0.095 mm wide × 24.3 mm high, and the étendue-increasing "entrance" mask would have 255 similar slots. Since the spectrometer would be viewing a uniform extended source, we would dispense with "dedispersion" [Decker, 1971] and, instead, operate the system in the "backwards" configuration (i.e., with the "exit" mask at the entrance to the spectrometer and the "entrance" mask at the exit, as in Figure 3; see Nelson and Fredman [1970]). A single 3 mm × 3 mm mercury–cadmium telluride (HgCdTe) detector, cooled to 77 K by liquid nitrogen, would be used, preceded by the appropriate bandpass filter and an $f/8.7$ to $f/0.8$ ellipsoidal focusing mirror. Other detectors (such as liquid helium-cooled silicon bolometers) are easily utilized, as would be the addition of dichroic order sorters and additional detectors to extend the simultaneously observed spectral range to higher orders (say, to cover the 4- to 16-μm band simultaneously—this is used in our "commercial" analytical spectrometer). Different numbers of "entrance" slots and/or image slicers [Strong and Stauffer, 1964] could be used if different fields of view were required.

This design, as sketched above, has an étendue of about 7.8 mm$^2$ sr and a noise equivalent spectral radiance of a bit less than $4 \times 10^{-7}$ W cm$^{-2}$ sr$^{-1}$ μm$^{-1}$. This is almost a full order of magnitude better than a 25-mm aperture, 0.1 cm$^{-1}$ resolution Michelson interferometer–spectrometer, which is considerably more complex, more difficult to operate in aircraft and spacecraft environments, and difficult to cool to the required cryogenic temperatures.

The point here, briefly, is that it is possible to obtain, by the use of Hadamard-transform optical techniques, very high total system performance while retaining the comparative simplicity, ruggedness, and economy of "conventional" dispersive spectrometers.

### Imaging Spectrometer*

We noted briefly above that it is possible to use the entrance mask in an HTS system to give the instrument a *spatial* resolution capability. The "conventional" multislit–multiplex HTS has a one-dimensional spatial capability (i.e., it can see strips across an extended source, corresponding to the discrete entrance slots on the entrance mask; see Phillips and Harwit [1971])—if the data are reduced in the appropriate manner. If one then replaces the multislit entrance mask with a two-dimensional binary-coded mask—of the type used in recent years for "pseudo-random imaging" [Girard, 1971; also Figure 4] —the resulting system has a full three-dimensional (two *spatial* dimensions and one *spectral* dimension) spectral imaging capability [Harwit, 1971]. That is, it either provides a *spectrum* at *each* spatial point of an image or, equivalently, it provides an *image* at *each* wavelength of the spectrum.

This system—the Hadamard-transform imaging spectrometer—is quite new and has yet to be used for any operational missions. It is, however, the *only* optical system available to record the full spectral/spatial information about a scene; this is especially evident in the infrared, where

---

*U.S. Patent 3,720,469: here, again, we must stress that the imaging spectrometer is proprietary and must not be used without the written permission of Spectral Imaging, Inc.

FIGURE 4 Hadamard-transform imaging mask for 255-element (17 horizontal × 15 vertical) imaging.

even the semiquantitative techniques of color photography are unavailable. As such, it has many potential applications in planetary astronomy and earth-resources monitoring, to name only two of many possible applications areas.

A feasibility-demonstration prototype imaging spectrometer has recently been completed and will be reported fully elsewhere [Harwit, 1973]. Since it illustrates the inherent simplicity of the technique—and the fact that it *works*—I will briefly describe it here. The prototype Hadamard-transform imaging spectrometer (or HADIS) views an image as 63 spatial resolution elements (9 horizontally and 7 vertically), and then analyzes *each* of these resolution elements in terms of 15 colors (spectral resolution elements). This is done, in the same manner as with conventional HTS spectrometry, by means of binary-coded masks at exit and entrance of the dispersive optical system; the total light passing both masks is recorded by a single photodetector for each of the positions of each mask—here, a total of 945 measurements per spatial/spectral "frame." In the prototype, the simplest possible optics were used: an ordinary light bulb as illumination, the light chopped by a rotating slotted disk, focused upon a color "slide" as "object," two-dimensionally modulated ("spatial" modulation) by a 63-element mask mounted on a microscope $x$-$y$ stage, dispersed by a prism, "spectrally" modulated by a 15-slot mask on a second single-axis micrometer stage, and focused upon a silicon photodiode for detection. The system, which is shown on Figure 5, is so simple it is almost literally a breadboard model.

Yet even this simple system works quite well. Figure 6 is a 15-wavelength "spectral image" of a color transparency of the Ring nebula in Lyre, which is shown in black and white in the next-to-bottom row, second-from-right "image." The original transparency is predominantly

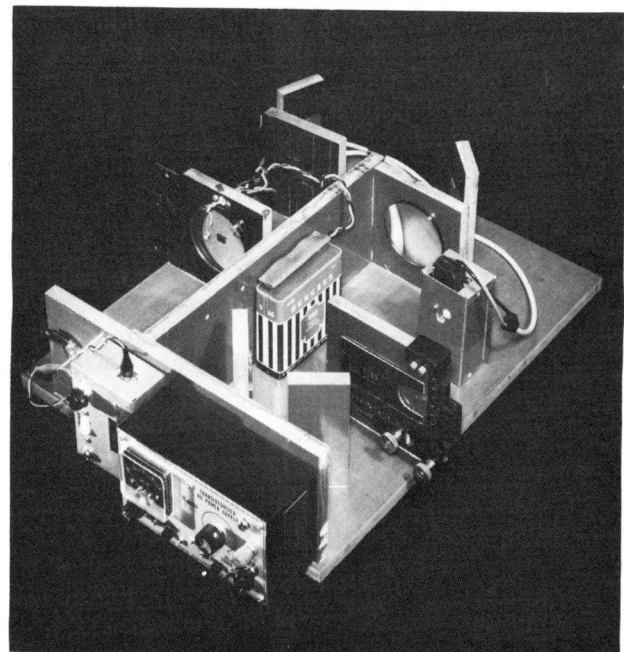

FIGURE 5 Breadboard feasibility-demonstration prototype Hadamard-transform imaging spectrometer (from Harwit [1973]).

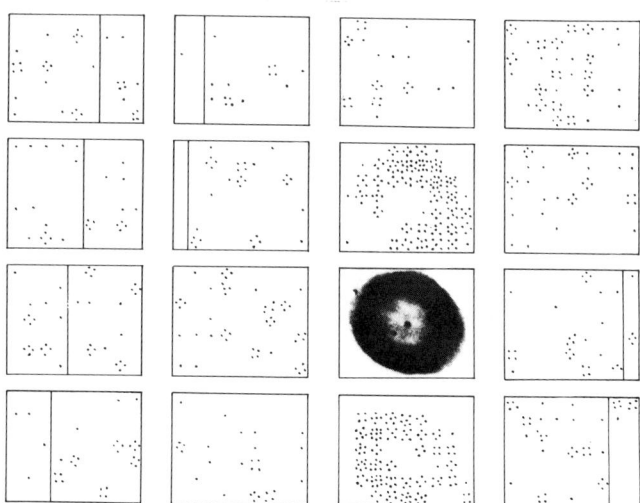

FIGURE 6 Fifteen-color "spectral image" of Ring nebula in Lyre (shown in black and white as the third "image" of the third row). The two colors principally visible in the original transparency, red and yellow, appear in the "images" above and below the photograph.

two-color, red and yellow, and these show prominently in the "images" above and below the photograph of the nebula; for a more detailed analysis, refer to Harwit [1973].

## HADIS Multispectral Scanner

Typical of the possible spacecraft applications of the Hadamard-transform imaging spectrometer is that of the multispectral scanner for earth-imaging applications, principally for weather satellites and earth-resources monitoring satellites, although other applications are obvious with some thought. Figure 7 sketches the optical layout of one specific—hardware—design proposed to NASA for a requirement for a visible/thermal-infrared seven-band multispectral scanner for an advanced earth-applications satellite.

The specifics of the design are not of particular interest, as the application is "far off optimum" as far as the technique itself is concerned (it has many too few spectral elements, compared with what is easily possible). Those features of interest—the use of an all-reflecting folded Schmidt telescope design, a conical HADIS mask tangent to the spherical focal surface of the Schmidt, light-pipe detector condensing optics, and filter wavelength discrimination—are noteworthy primarily because they illustrate that the full repertoire of "tricks" of the classical optical designer are applicable to this new technique.

What is noteworthy about this design, at least to us, is that—even for a far-from-optimum set of requirements—the HADIS system would operate at higher scan speed than a conventional line-scanner (because its lower dynamic range lowered the number of telemetered bits per scan), would impose a much lower inertial load on the spacecraft attitude-control system (because its mask is light and runs continuously, rather than oscillating), and would allow the use of noncryogenic infrared detectors (because of the performance gained by multiplexing). Additionally, the system is of low complexity—and therefore of high reliability and low cost—compared with both conventional line-scanning multispectral systems and interferometric spectrometers.

Again one sees—and to sum up the point of the paper—that Hadamard-transform optical techniques allow one to reap the orders-of-magnitude performance gains inherent in multiplex, high-étendue systems while retaining the conventional optics virtues of simplicity, economy, and flexibility.

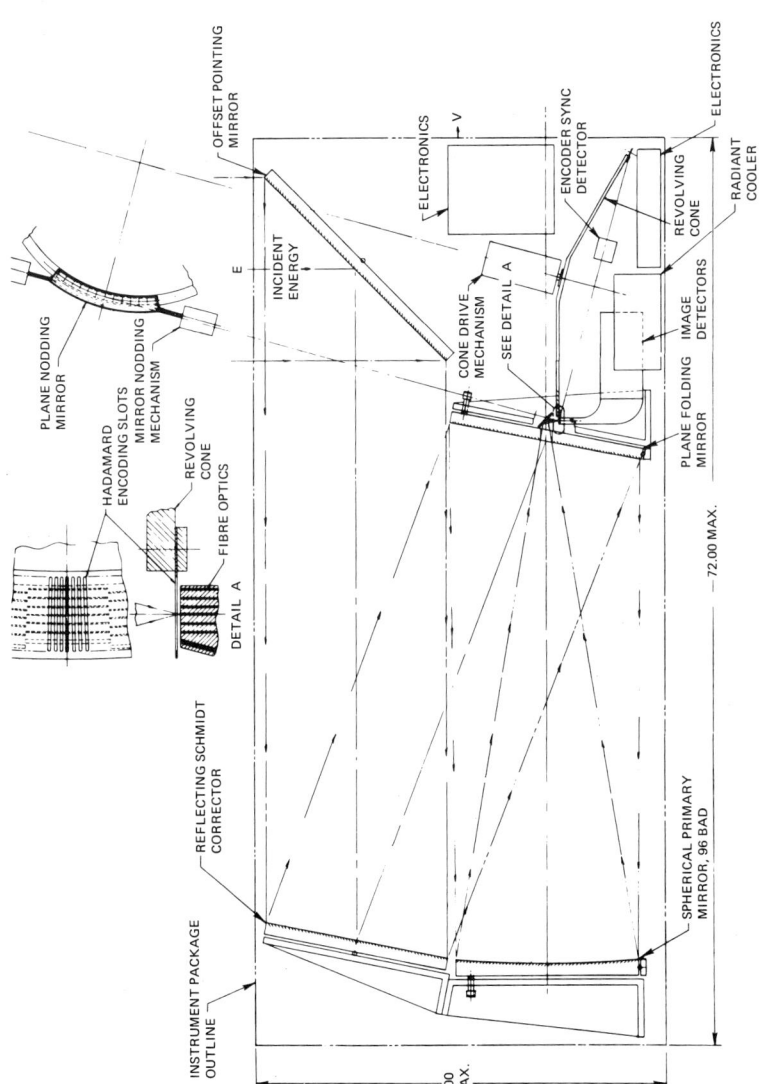

FIGURE 7 Optical layout of a proposed seven-band HADIS multispectral scanner for earth-resources applications.

## References

Decker, J. A., Jr., *Appl. Opt. 10,* 510, 1971 (1971).
Decker, J. A., Jr., *Anal. Chem. 44,* 127A (1972).
Decker, J. A., Jr., and M. Harwit, *Appl. Opt. 7,* 2205 (1968); *8,* 2252 (1969).
Fellgett, P., *J. Phys. Rad. 19,* 187 (1958).
Girard, A., *T.P. no. 972* (ONERA, 29, Ave. de la Division Leclerc, 92 Chatillon, France, 1971).
Harwit, M., *Appl. Opt. 10,* 1415 (1971).
Harwit, M., *Appl. Opt. 12,* 285 (1973).
Harwit, M., P. G. Phillips, T. Fine, and N. J. A. Sloane, *Appl. Opt. 10,* 1415 (1971).
Houck, J., J. B. Pollack, D. Schaack, C. Sagan, and J. A. Decker, Jr., *Icarus* (to be published, 1973).
Nelson, E. D., and M. L. Fredman, *J. Opt. Soc. Am. 60,* 1664 (1970).
Phillips, P. G., and M. Harwit, *Appl. Opt. 10,* 2780 (1971).
Pollack, J. B., and C. Sagan, *Space Sci. Rev. 9,* 243 (1969).
Pratt, W. K., J. Kane, and H. C. Andrews, *Proc. IEEE 57,* 58 (1969).
Sloane, N. J. A., T. Fine, P. G. Phillips, and M. Harwit, *Appl. Opt. 8,* 2103 (1969).
Strong, J., and F. Stauffer, *Appl. Opt. 3,* 761 (1964).

GEORGES GAUFFRE and MICHEL CHATANIER

# An Infrared Pneumatic Transducer with Capacitive Detection

## I. Introduction

For a long time pneumatic cells have figured among the best infrared detectors operating at room temperature. In two articles Golay [1947, 1949] explained the theoretical working principle and described a practical realization. Some time later, at ONERA, K. Luft designed a new type of pneumatic cell with capacitive measurement of the membrane motion.

Today, computer techniques permit one to design pneumatic cells precisely and to determine the best operating conditions. Progress in electronic design permits one to include a portion of the necessary electronics within the transducer housing. An instrument has been developed at ONERA called TRIAS (for TRansducteur Infrarouge pour Applications Spatiales). The instrument is an infrared transducer of high sensitivity, which will operate over a broad temperature range.

---

The authors are in the Office National d'Etudes et de Recherches Aérospatiales (ONERA) 92320 Chatillon, France.

## II. Operating Principle

The detector is made of an airtight enclosure, filled with a gas at reduced pressure and divided into two chambers by a deformable diaphragm (Figure 1). The infrared radiation to be measured penetrates into the first chamber through a window. In the middle of the chamber, opposite the window, is set a sensitive surface, or target. This target transforms the infrared incoming radiation into heat. It is a collodion membrane covered with a very thin metallic layer. Its main property is a high absorbing power, uniform within the whole infrared spectrum. Being in contact with the heated target, the surrounding gas heats up, expands, and causes the deformation of the diaphragm separating the two chambers. This diaphragm constitutes one of the plates of a capacitor. The deformation entails capacitance variations proportional to the incident infrared flux. The other plate, fixed, lies in the rear chamber. The variations of capacity are detected by the associated electronic unit. The front and rear chambers communicate through a small duct. The gas passing through this duct balances, with a large response time, the pressures within the two chambers. The receiver is thus insensitive to slow changes in the incoming radiation. The gas inside the cell is xenon at a pressure of 100 mbar. This choice results from a compromise between sensitivity and bandwidth.

The electronic unit processes the information provided by the detector in order to make its use easy. Thus the unit has three functions: to

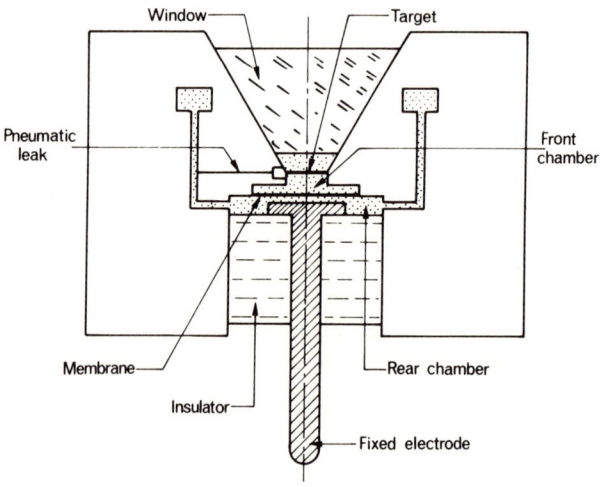

FIGURE 1  Infrared pneumatic cell.

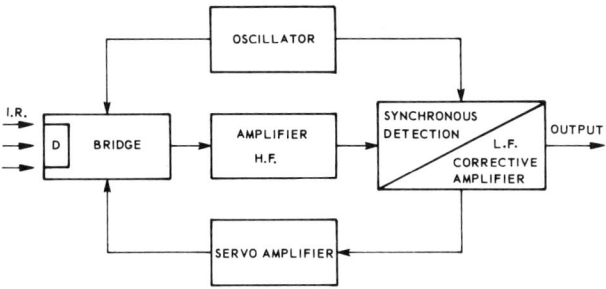

FIGURE 2  Diagram of electronics.

transform capacitance variations into voltage variations, to increase the signal level, and to correct the receiver defects by flattening the response curve and compensating the slow drifts.

The block diagram summing up these functions is given Figure 2. The cell is inserted in one of the branches of a measuring bridge. This bridge is supplied by a 455-kHz oscillator. The main qualities of this oscillator are an unusual high spectral purity and a good temperature stability [Rutman, 1972]. The rf amplifier is a very-low-noise unit. The rf high-level signal feeds the synchronous detector controlled by the oscillator. The low-frequency signal is processed in an equalizing amplifier circuit that has various functions: to eliminate the residual carrier frequency, to ensure a passband centered on the optimal detectivity frequency, and to offer a low output impedance. Last, a control loop feeds back the bridge with the very-low-frequency part of the output voltage spectrum. This voltage acts on the electrostatic force, attracts more or less the diaphragm, and maintains constant the mean value of the detector capacitance by stabilizing the interplate distance. This ensures the balance of the bridge and reduces the variation of the responsivity versus ambient temperature variations.

## III. Transducer Theoretical Performance

Some years ago, Jones explained how the responsivity, the noise, and the noise equivalent power (NEP) were the most useful parameters to describe the performance of radiation detectors [Jones, 1953]. The responsivity can be deduced from the previous functional diagram: incoming radiation, temperature rise of the target, expansion of the heated gas, and motion of the diaphragm. Each step is governed by a set of equations; the coefficients to be introduced in these equations are related to the design parameters: cell volume, nature and pressure of the gas, membrane elastic tension, etc. Some equations are difficult to solve, particularly the thermal balance one; and some numerical values of parameters, such as the flow rate of the duct between the two chambers, are not directly available. Thus the complete analytical solution is not accessible.

The problem may be solved to a good approximation by using an electrical analog somewhat similar to that proposed by Golay [1949]. This analog is suitable for computer calculations, with special software for use in conjunction with electrical networks. Figure 3 illustrates the equivalent circuit of the ONERA pneumatic cell. It is split into four parts: thermal, pneumatic, mechanical, and electrical. The various components have the following meaning:

$R_1$, resistance to radiation heat transfer between membrane and thermostat;

$C_1$, thermal reactance of the membrane;

$R_2, R_3, R_4, R_5, R_6$, semidistributed resistance to gas heat transfer between membrane and thermostat;

$C_2, C_3, C_4, C_5$, semidistributed thermal reactance of gas;

$C_2', C_3', C_4', C_5'$, semidistributed compliance of gas;

$C_7$, compliance of dead volume (part of gas whose temperature does not change);

$R_7$, resistance to gas flow in the leak duct;

$C_8$, compliance of the capacitor membrane;

$R_8$, resistance to membrane motion;

$\overline{e^2}$, analog of a noise voltage generator, translated into thermal units, representing the fluctuations of the electrostatic attractive forces due to the noise amplitude modulation of the oscillator waveform.

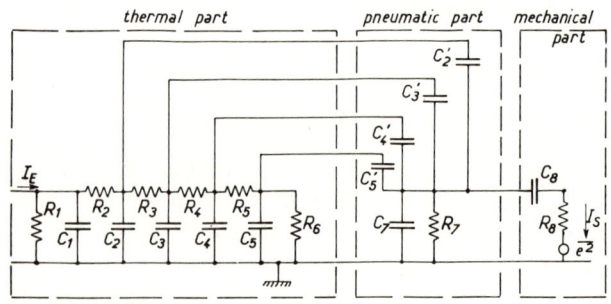

FIGURE 3 Analogical sketch.

In this scheme the noise due to the rf amplifier following the measuring bridge is omitted. The actual amplifier has such a low noise level that its contribution to the overall noise may be neglected in practice.

In order to facilitate further computation all the components are to be translated into a unique system: thermal, pneumatic, or mechanical.

As an example, in thermal representation of the network, the variables are

$\theta$, temperature deviation at a point with respect to absolute temperature of thermostat as "voltage,"

$(1/T)(dQ/dt)$, entropy flux flowing in a branch as "current."

The choice of such a set of variables, whose product has the dimension of a power, makes possible noise computation. The component values are translated with the aid of the following operations:

| components | thermal | multiplied by | pneumatic | multiplied by | mechanical |
|---|---|---|---|---|---|
| R | $R_T$ | $P^2/T^2$ | $R_P$ | $S^2$ | $R_M$ |
| C | $C_T$ | $T^2/P^2$ | $C_P$ | $1/S^2$ | $C_M$ |

where $T$ is the absolute temperature, $P$ the pressure, and $S$ the membrane surface.

As prescribed by Jones, the responsivity is the ratio of the rms value of the output variable $x$, divided by the rms thermal power $W$ in the incoming radiation, $R(\omega) = x/W$, with $x = I_s T/\omega PS$, where $I_s$ is the current in the output branch.

Callen and Welton [1951], studying the relation between fluctuations of a system and its dissipative properties, gave an extension of the Nyquist formula to a generalized impedance. In the actual circuit this formula takes the form:

$$\overline{(i_s^2)} = 4kTX/(1/C_8^2 \omega^2),$$

where $\overline{(i_s^2)}$ is the current fluctuation in the output branch and $X$ is the real part of the impedance seen at terminals of membrane compliance $(C_8)$.

If we recall that the output variable is related to the current flowing in the output branch, the expression of noise for the cell alone becomes

$$B(\omega) = \left[(T^2/P^2 S^2)\, 4kTXC_8^2\right]^{1/2}, \tag{1}$$

and the overall noise

$$B'(\omega) = \left[ (T^2/P^2S^2) 4kTXC_8^2 + \overline{e^2} \, C_8^2 \right]^{1/2}. \qquad (2)$$

The high spectral purity attained in the oscillator voltage waveform leads to very low sideband amplitudes: $-134$ dB and $-146$ dB at, respectively, 5 Hz and 50 Hz of the carrier frequency. This makes practically negligible the $\overline{e^2}$ term in the above formula. So physical measurement of the entire noise is very close to the $B(\omega)$ value.

Computation of responsivity and of noise versus angular frequency has been achieved with the aid of an ECAP program on a 360/50 IBM computer. The components had the following values:

$R_1 = 20 \times 10^6$    $R_5 = 0.7 \times 10^6$    $C_7 = 0.8 \times 10^{-9}$
$C_1 = 2 \times 10^{-9}$    $R_6 = 0.3 \times 10^6$    $R_7 = 5 \times 10^6$
$R_2 = 2.4 \times 10^6$    $C_2 = C_3 = C_4 = C_5 = 0.8 \times 10^{-9}$    $C_8 = 4 \times 10^{-9}$
$R_3 = 2 \times 10^6$    $C_2' = C_3' = C_4' = C_5' = 0.5 \times 10^{-9}$    $R_8 = 0.4 \times 10^6$
$R_4 = 1.2 \times 10^6$

Figure 4 presents the theoretical responsivity and noise curves versus frequency. Figure 5 presents the corresponding experimental curves. The shapes of the NEP curves presented on Figure 6 are comparable. The theoretical curve includes neither the reflection losses in the window nor the part of incoming energy that is not absorbed by the target.

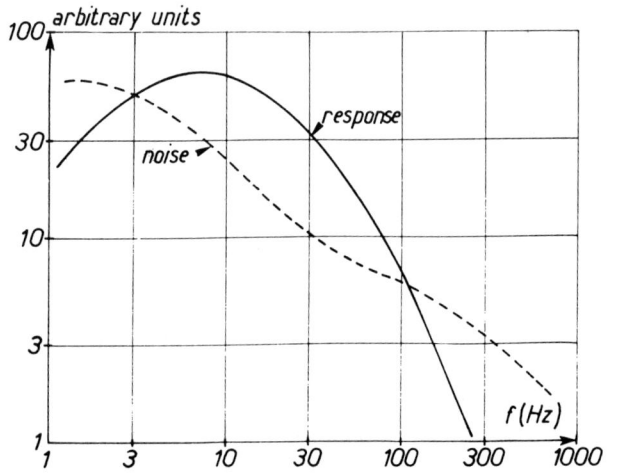

FIGURE 4 Theoretical response and noise spectrum.

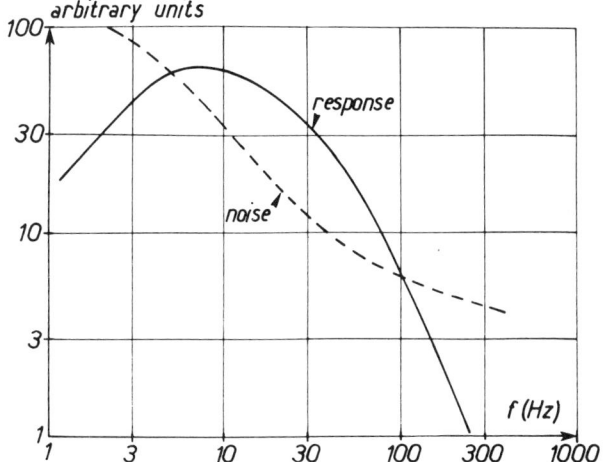

FIGURE 5 Experimental response and noise spectrum.

Agreement between the formulation exposed in Eq. (1) and experimental results confirms the hypothesis of several authors [Golay 1949; Smith *et al.*, 1968], who assumed that noise in a pneumatic cell was of fundamental thermodynamical origin.

As pointed out by Jones, the radiation exchange is the most fundamental limit in the performance of pure thermal detectors and allows

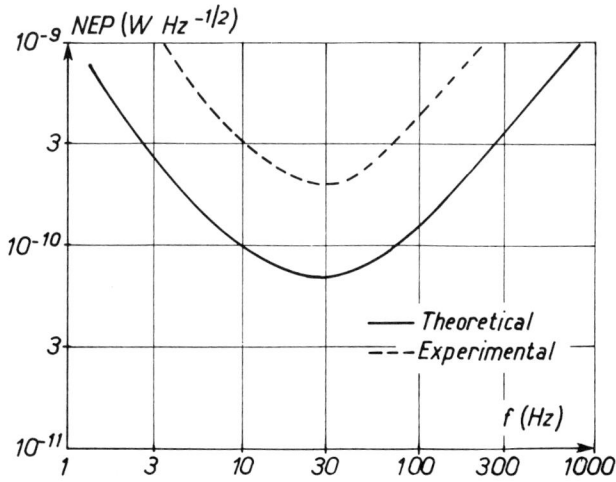

FIGURE 6 NEP, theoretical and experimental.

the concept of reference detectivity $D^*$. The value of $R_1$, representing that exchange, is so high that it has very little effect on the cell performance. This explains why pneumatic cells have a $D^*$ that slightly increases with target area, as was experimentally found.

## IV. Example of Design

The pneumatic cells described above have been especially built for field applications [Eddy et al., 1970; Girard and Lemaitre, 1970; Chatanier and Gauffre, 1972]. This achievement was obtained mainly through integrating the cell and the electronic unit as a single component. TRIAS is really a transducer: infrared power as input, volts as output. The major feature of this design is that it allows matching the cell and the electronic performance. The other ones are compensation of cell electrical capacity temperature drift, reduction of the sensitivity variations with respect to temperature variations, reduction of stray capacities, and correction of the cell response curve.

Another point of importance for operational systems is that the transducer presents a high reliability against mechanical shocks, accelerations, and vibrations.

These results are achieved through three types of precaution: first, the cell and the electronic unit are very compact; second, the chambers on either side of the capacitive membrane have slightly different centers of gravity in order to cancel the membrane inertia, so that axial accelerations do not induce motion of the membrane; third, the servo-loop reacts against low-frequency inertia solicitations of the membrane.

Figure 7 illustrates the effect of temperature on the transducer re-

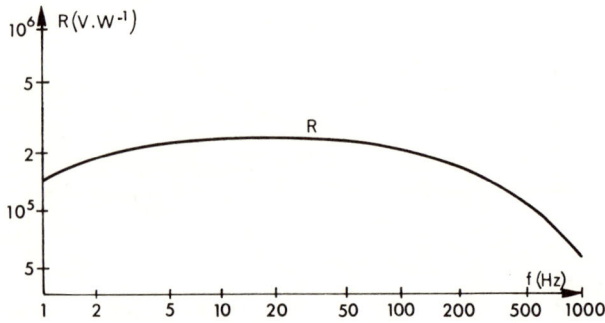

FIGURE 7   TRIAS responsivity versus frequency.

sponse. Such a small drift could be obtained only by a careful choice of materials, components, and assembly design in both the cell and the electronic unit.

The performance characteristics of the present TRIAS are

Circular target: 2-mm diameter,
Angle of acceptance: 60°,
Spectral range limited only by the window material,
Response: $3 \times 10^5$ V/W maintained with ±1 dB from $-20$ to $+40$ °C (Figure 8),
3-dB passband: 2 to 200 Hz,
NEP: $2 \times 10^{-10}$ W Hz$^{-1/2}$ at 30 Hz.

Technical data:

Shape: cylinder, 70-mm diameter, 90 mm long,
Weight: 400 g,
Supply voltage: ±15 V,
Power consumption: 0.25 W,
Output impedance: 50 Ω.

Withstanding to environment:

Pressure: from atmospheric pressure to zero,
Acceleration: 20 $G$, rms white noise from 20 to 2000 Hz,
Shock: 100 $G$ for 5 msec,
Temperature: $-20$ °C to $+40$ °C.

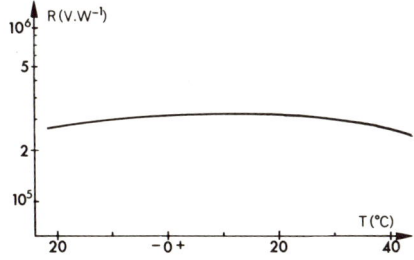

FIGURE 8 Responsivity versus temperature.

## V. Conclusion

In spite of the growing competition of cooled quantic cells, thermal detectors remain very useful in infrared technology and are of particular interest for field applications. In this family, TRIAS is a rugged unit that has the best detectivity within the frequency range of its bandwidth, 2 to 200 Hz.

The equivalent network technique for computing both signal response and noise spectrum, as described in this paper, may be useful for other applications: either to optimize the same kind of transducer in different operating conditions or to facilitate the comparative analysis of the performance of other kinds of transducers.

## References

Callen, H. B., and T. A. Welton, *Phys. Rev. 83,* 34 (1951).
Chatanier, M., and G. Gauffre, *J. Br. Interplanet. Soc. 25,* 13 (1972).
Eddy, J. A., R. H. Lee, P. J. Lena, and R. M. McQueen, *Appl. Opt. 9,* 439 (1970).
Girard A., and M. P. Lemaître, *Appl. Opt. 9,* 903 (1970).
Golay, M. J. E., *Rev. Sci. Instrum. 18,* 347 (1947).
Golay, M. J. E., *Rev. Sci. Instrum. 20,* 816 (1949).
Jones, R. C., *Advan. Electron. 5,* 1 (1953).
Rutman, J., Sur les générateurs de fréquence de très hautes performances, thesis, Paris University; also publ. 142, ONERA, Châtillon.
Smith, R. A., F. E. Jones, and R. P. Chasmar, *The Detection and Measurement of Infra-Red Radiation,* 2nd ed., Clarendon Press, Oxford (1968).

C. R. MUNNERLYN and J. W. BALLIETT

# A TUNABLE FILTER FOR CALIBRATING RADIOMETERS IN THE 11-19-MICROMETER SPECTRAL REGION

## Introduction

Radiometers are commonly used in space applications for the measurements of radiation emitted by the earth, moon, stars, or other planets. One application for a very-narrow-spectral-band, moderate-field-of-view radiometer is in the measurement of the atmospheric vertical temperature profile from satellite platforms. The Vertical Temperature Profile Radiometer (VTPR), recently developed by Barnes Engineering Company [Falbel and Zink, 1971] is the latest in a series of multispectral radiometers designed to measure the temperature of the earth's atmosphere at different altitudes by observing the amount of radiation received above the atmosphere within narrow spectral windows corresponding to the $CO_2$ absorption lines near 15 $\mu$m. The radiometer also contains filters for observing an $H_2O$ absorption band at 18.7 $\mu$m and an atmospheric window at 12 $\mu$m. A detailed theory of this technique of measurement is given by Kaplan [1959]. In order for the calculations to be made with sufficient accuracy, the radiometer must be calibrated accurately as to its spectral transmission function when its entire field of view is taken into account.

---

The authors are with Tropel, Inc., Fairport, New York 14450.

TABLE 1 Radiometer Filter Characteristics at 35 °C

| Filter No. | Center of Spectral Response Wavenumber (cm$^{-1}$) | Bandwidth and Tolerance | |
|---|---|---|---|
| | | 50% Response (cm$^{-1}$) | 10% Response (cm$^{-1}$) |
| 1 ($Q$-branch) | 668.5 ± 0.5 | 3.5 ± 0.5 | 10.5 ± 1.5 |
| 2 | 695 ± 1 | 10.0 ± 2.5 | 20 ± 5 |
| 3 | 725 ± 1 | $10.0^{+1.0}_{-2.0}$ | $20^{+2}_{-4}$ |
| 4 | 535 ± 1 | $10.0^{+1.0}_{-2.0}$ | $20^{+2}_{-4}$ |
| 5 | 835 ± 1 | $8.0^{+1.0}_{-2.0}$ | 16 ± 2 |
| 6 | $747^{+2}_{-1}$ | 10.0 ± 2.5 | 20 ± 5 |
| 7 | 708 ± 1 | 10.0 ± 2.5 | 20 ± 5 |
| 8 | $677^{+2}_{-1}$ | 10.0 ± 2.5 | 20 ± 5 |

The VTPR contains eight narrow-band dielectric interference filters. The characteristics of these filters are given in Table 1. Their center frequencies range from 535 to 835 cm$^{-1}$. The half-power bandwidth of the filters are from 3.5 to 10 cm$^{-1}$. The apparent bandwidth and center frequencies of the filter can be modified due to radiation from different angles within the $f/3$ focusing beam and from different field points. Manufacturing variations in the filters and other optical surfaces and in the detector can lead to an uncertainty in the final spectral characteristics of the instrument.

The best spectral calibration of the radiometer can be obtained after the instrument has been assembled by irradiating it with a tunable monochromatic field of radiation and observing the VTPR output signal. The purpose of this paper is to describe the Precision Spectral Calibrator (PSC) designed to transform a blackbody radiation source into a tunable monochromatic source that subtends a field of view greater than the field of the instrument to be tested. The design concepts employed can be used to modify the design for other instruments with different fields of view and different spectral bands.

### Design Concept

The VTPR has a Cassegrain-type reflective optical system as shown in Figure 1. It uses a pyroelectric detector to cover the 11- to 19-$\mu$m spectral band and has an aperture of 73.5 mm and a square field of view

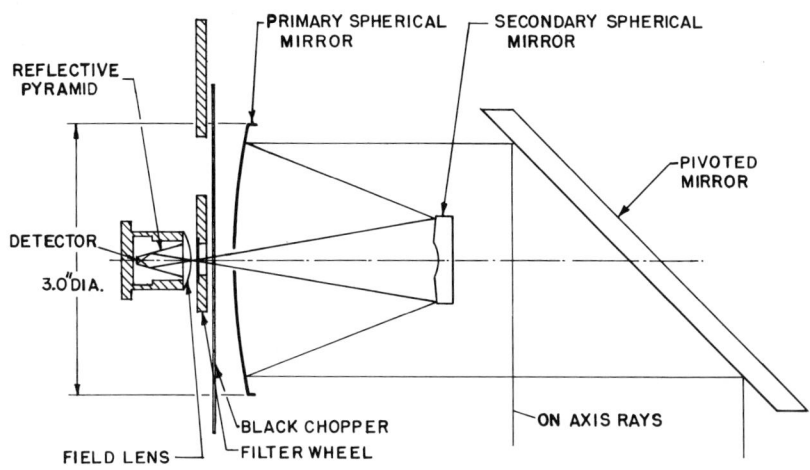

FIGURE 1 Optical schematic of the VTPR.

with a diagonal of 3.4°. Additional optical and scan parameters are given in Table 2. The eight narrow-band filters are mounted on a filter wheel in front of the detector and rotated at 120 rpm. To provide the appropriate spectral data, either a Fabry-Perot (F-P) or a Michelson interferometer could be used. A grating-type spectrometer is impractical because of its low étendue or energy throughput as compared with the interferometers. The filter bandwidths are narrow enough relative to the desired spectral resolution so that they fall well within the free spec-

TABLE 2 Optical and Scan Parameters of the VTPR

| | |
|---|---|
| *Optical Parameters* | |
| Primary-mirror diameter | 73.5 mm |
| Secondary-mirror diameter | 28.6 mm |
| Entrance aperture diameter ($D_0$) | 66 mm |
| Focal length of Cassegrain optics | 197 mm |
| Field of view (FOV) | 2.235° × 2.135° |
| Detector area ($A_d$) | 1.5 × 1.5 mm² |
| Effective focal length (efl) = $\sqrt{A_d}/\sqrt{FOV}$ | 38.5 mm |
| Effective system $f$/no. = efl/$D_0$ | 0.584 |
| | |
| *Scan Parameters* | |
| Cross-track scan angle | 63.4° |
| No. of steps per scan | 23 |
| Time per 23 steps | 11.5 sec |
| Scan retrace time | 1.0 sec |

tral range of an F-P interferometer. Therefore, the F-P interferometer offers a more direct solution to measuring the spectral bandwidths than the Fourier transform method required with a Michelson interferometer.

Neglecting absorption, the transmission function for an F-P interferometer as a function of wavenumber, $\sigma$, is

$$T\sigma = \left\{ 1 + [4R/(1-R)^2] \sin^2(2\pi d\sigma \cos\theta) \right\}^{-1}, \quad (1)$$

where $R$ is the mirror reflectivity, $d$ is the mirror spacing, and $\theta$ is the angle of incidence of the light on the parallel plates. This function repeats for values of $\sigma = m/2d \cos\theta$, where $m$ is an integer. A detailed analysis of the F-P interferometer is given by Born and Wolf [1964].

As the reflectivity of the F-P mirrors increases, the transmitted bandwidth decreases. The ratio of the instrumental bandwidth to the free spectral range is called the finesse and is given by

$$F = \sqrt{R}/(1-R). \quad (2)$$

Figure 2 shows a plot of a single free spectral range as a function of the finesse. The free spectral range for normal incidence, given in wavenumbers is

$$\Delta\sigma_{FSR} = 1/2d, \quad (3)$$

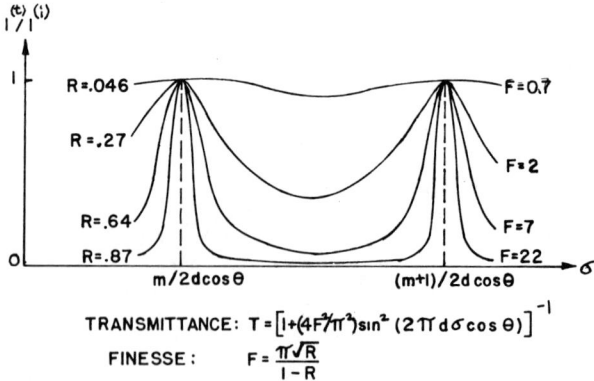

FIGURE 2 Transmission of a Fabry-Perot interferometer as a function of wavenumber for different values of finesse over one free spectral range.

so that the instrumental bandwidth is

$$\Delta\sigma_{BW} = 1/2dF. \qquad (4)$$

From Table 1, one sees that the filters fall within two classes: those with bandwidth $< 4$ cm$^{-1}$ and those with bandwidth between 8 and 10 cm$^{-1}$. The F-P interferometer should then be designed so that the spacing can be changed by at least a factor of 2. Also, the transmission in the mid-region of the FSR should be low to reduce the energy outside the main transmission peaks. For the high reflectivity, the midrange transmission is

$$T_{MR} = \pi^2/4F^2. \qquad (5)$$

For a finesse greater than 30, $T_{MR}$ is less than 0.3%. This value of finesse is easily achieved with a reflectivity of about 90%. By selecting a bandwidth about one fifth of the filter bandwidth, one has an FSR at least six times the bandwidth of the filter. The filter provides its own blocking of less than 0.1% outside a band about four times its bandwidth. Therefore, other orders of the F-P interferometer can be effectively blocked, and the output can be treated as a scan by a single spectral line about one fifth of the filter bandwidth. This arrangement is illustrated in Figure 3. A scanning function of this type is adequate to de-

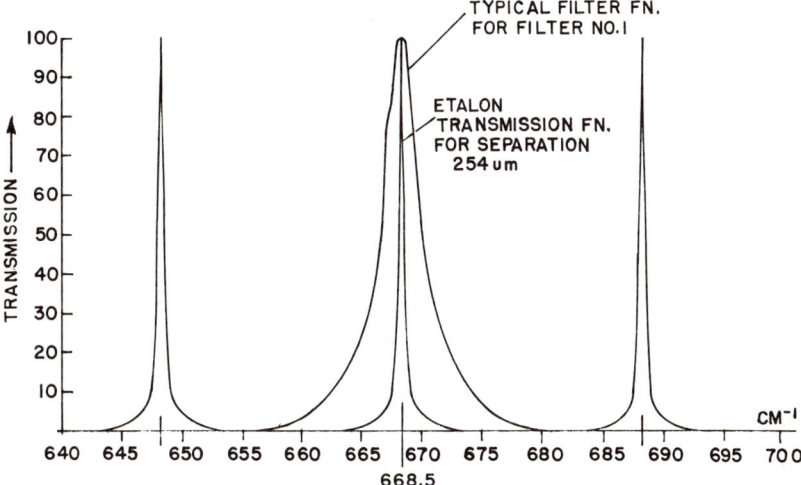

FIGURE 3 A comparison of the $Q$-branch filter transmission function with the Fabry-Perot transmission function including three orders.

fine the filter bandwidth to better than ±0.1 times its bandwidth. A more exact evaluation of this subject is given under the calibration section.

The discussion so far does not take into account the effect of the angular broadening of the F-P transmission function. For small angles, the shift in wavenumber of the peak transmission is

$$\Delta\sigma \simeq \sigma \sin^2 \theta/2. \qquad (6)$$

An angle of 2° shifts the wavenumber by 0.16 cm$^{-1}$ at 668.5 cm$^{-1}$. This contribution of bandwidth due to the field of view is excessive for the narrowest filter. The angle of the rays within the F-P can be reduced by imaging the entrance pupil of the VTPR onto the F-P interferometer at a greater than 1:1 magnification. Figure 4 shows the basic optical configuration of the Precision Spectral Calibrator (PSC). In order to obtain maximum instrumental transmission, the number of optical elements is held to a minimum by using one F-P plate as a collimator lens and the other as a focusing lens. This design minimizes the angle of the off-axis rays through the F-P for a given size interferometer.

Because of the other factors—flatness and field of view—it was necessary to raise the reflectivity of the mirrors to 93% at 668 cm$^{-1}$ to obtain a finesse of 30.

The specifications for the Precision Spectral Calibrator are given in Table 3.

The diameter of the field lens is 13 cm, and the clear aperture of the

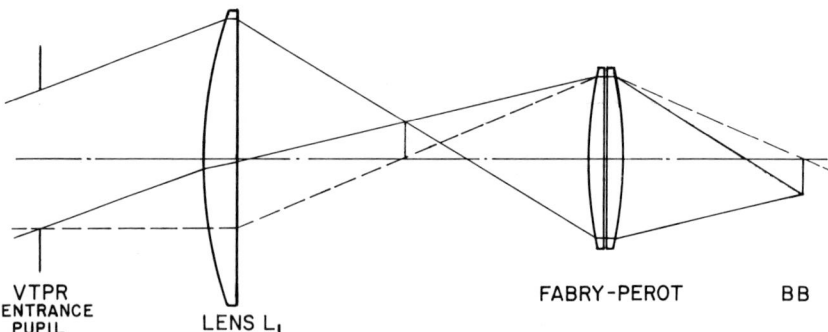

FIGURE 4 Optical layout schematic for a single channel. The Fabry-Perot lenses image the blackbody source into a narrow spectral band source at BB. Lens $L_1$ then collimates this source to fill the entrance pupil of the VTPR. Lens $L_1$ also images the VTPR entrance pupil onto the Fabry-Perot plane at the appropriate magnification.

TABLE 3  Precision Spectral Calibrator Specifications at 668.5 cm$^{-1}$

| | |
|---|---|
| Free spectral range | 20 cm$^{-1}$ |
| Integrated spectral half-width | 0.65 cm$^{-1}$ |
| Absolute PSC calibration | ±0.2 cm$^{-1}$ |
| Absolute calibration at calib. spacing | ±0.1 cm$^{-1}$ |
| Field of view 100% | 2.167° square |
| Field of view 10% | 2.392° square |
| Minimum peak transmission | 20% |

interferometer is 9.7 cm. The field lens is made of KBr, and the interferometer plates are made of Irtran IV. The diameter of the blackbody source was selected so that any ray traced from the entrance pupil of the VTPR through the required field angles would fall within the diameter of the blackbody source even when the chromatic aberration of the system was taken into account. The selected diameter of the blackbody was 3.17 cm.

A multiple-layer coating with adequate reflectivity and transmission could not be designed to cover the entire spectral bandwidth from 11 to 19 μm. Therefore, two channels were required: one covering 11 to 16 μm and one covering from 17 to 19 μm. Figure 5 shows an optical layout that permits a common blackbody and field lens to be used in both channels. The second interferometer is made from Irtran VI and has a clear aperture of 7.1 cm. The smaller size was due to the size limitation of the Irtran VI material, but was permitted due to the wider bandwidth associated with the filter being evaluated with this interferometer. An additional KBr lens is required in this channel to locate the image of the blackbody source properly. Two mechanically coupled mirrors are used to switch from one channel to the other. These same mirrors permit the introduction of a $CO_2$ laser beam for calibration of the spacing of the interferometer mirrors.

## Mechanical Design

The final tests of the VTPR are conducted in vacuum. The entire F-P system was designed to go into a 60-in. vacuum chamber, while the $CO_2$ laser and He–Ne laser were mounted external to the vacuum chamber. Invar differential screws with stepper-motor drives were used to move the parallel plates. Each step of the motors corresponded to 0.02 μm. The longer wavelengths at which tests were to be conducted permitted mechanical tolerances 20 to 30 times greater than required for visible

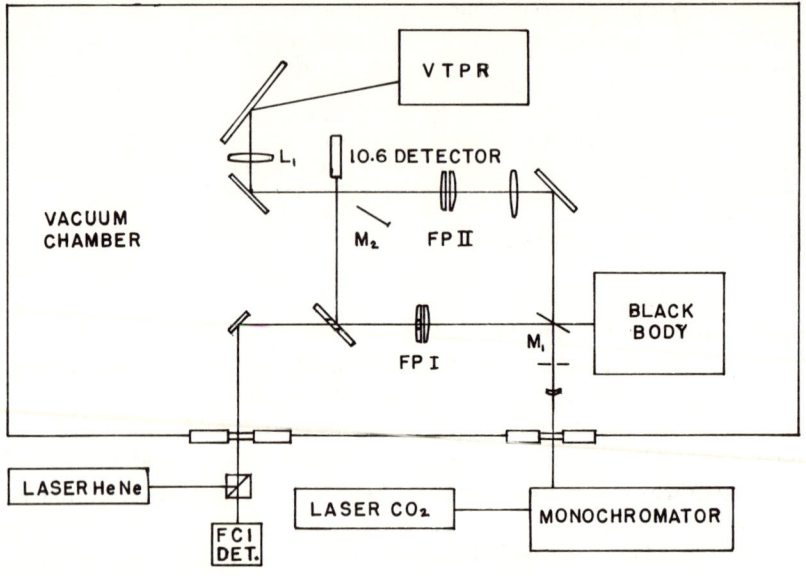

FIGURE 5 Optical layout of the PSC incorporating two Fabry-Perot interferometers. Mirrors $M_1$ and $M_2$ are remotely movable and determine the interferometer being used. The detector is used to receive the interference pattern from the $CO_2$ laser for calibration. The He-Ne laser is used in a fringe-counting interferometer to monitor the separation of the interferometer plates.

interferometers. The flatness of the Irtran plates was only $\lambda/5$ for visible light. The details of the F-P interferometer section are shown in Figure 6. Water cooling was used to reduce the heating effect of the stepper motors and the blackbody source. Invar parts were used to reduce the effects of any residual temperature variations within the vacuum chamber.

The two-sided mirror in Figure 4 used to switch from the blackbody source to the $CO_2$ laser was made by stretching 10-$\mu$m-thick Mylar film over a metal frame. The resulting mirror was flat to a few wavelengths in visible light and reduced the field-angle requirements for the blackbody source.

### Calibration Techniques

Since absolute spectral information is required, it is necessary to know accurately the separation of the interferometer plates at all times during the spectral scan. The scanning is accomplished by three stepper motors

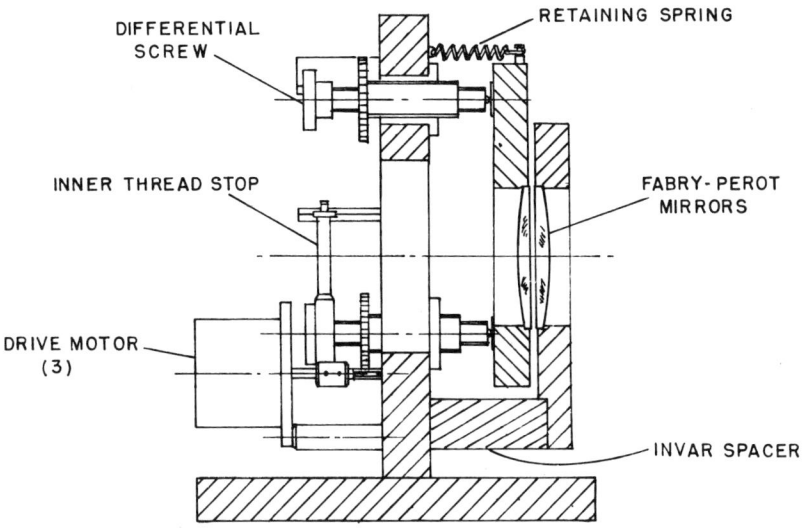

FIGURE 6  A cross section of the Fabry-Perot interferometer.

driving differential screws in synchronization. The plate parallelism can be controlled by stepping the motors independently. A $CO_2$ laser source is used as a primary reference wavelength for determining the plate separation.

The output of the laser passes through a monochrometer to separate the different transitions at which a $CO_2$ laser can radiate. The laser energy then passes through an on-axis aperture and through the interferometer. The transmitted beam is collected on a pyroelectric detector, which is connected to a chart recorder. As the plate separation is changed, the transmission function is plotted. The stepper-motor position is noted for each peak in transmission. The wavelength of the laser is changed, and the transmission function is replotted and positions of the stepper motor noted. By selecting two $CO_2$ laser wavelengths such as 10.590 and 10.274 $\mu$m, transmission peaks for both wavelengths will occur at a separation of 344.2 $\mu$m. The stepper motors can repeat a position of the plates to an accuracy of better than 0.1 $\mu$m. This repeatability is checked using a visible-light interferometer operating at 0.6328 $\mu$m. Once the plate separation at one transmission peak for the $CO_2$ laser is determined, the separation for the other peaks is known to be an integer number of half wavelengths difference. Therefore, each transmission peak becomes a reference mark at about 5-$\mu$m intervals. Table 4 illustrates how the absolute plate separation can be verified using two

TABLE 4  $CO_2$ Calibration Wavelengths

| $CO_2$ Line | Wavenumber (cm$^{-1}$) | m | d (μm) | Δd (μm) |
|---|---|---|---|---|
| P-20 | 944.195 | 64 | 338.9131 | 0.1434 |
| R-16 | 973.289 | 66 | 339.0565 | |
| P-20 | 944.195 | 65 | 344.2086 | 0.0149 |
| R-16 | 973.289 | 67 | 344.1937 | |

wavelengths of the $CO_2$ laser. By adjusting the spacing to correspond to a given $CO_2$ laser transition such as P-20, an order number can be found for a different laser transition that corresponds to the same spacing. Changing the spacing by one order number produces an easily detectable difference in the location of the peak.

## Bandwidth Measurements

The absolute spectral bandwidth of the PSC cannot be measured directly. However, its value can be determined within narrow limits by making measurements of the mirror reflectivity, flatness, alignment, and field of view of the instrument. A detailed analysis of the contribution of each factor is given by Steel [1967]. The modified finesse due to surface irregularities is

$$F = F_0 \left[1 - \overline{\sigma^2} F_0^2 (3+R)/2\pi^2\right], \qquad (7)$$

where $\overline{\sigma^2}$ is the mean-square surface deviation, $R$ is the reflectivity, and $F_0$ is the finesse in the absence of defects. For very small alignment errors, the alignment error is equivalent to a linear surface defect. For large angular tilts, the fringe pattern loses its symmetrical shape. Within the PSC, a $CO_2$ laser is used as a source for checking the parallelism of the plates by observing the symmetry of the fringes as the spacing is scanned. The transmitted spectral shape is also a measure of the instrumental bandwidth due to tilt and irregularity. However, the bandwidth must be corrected due to a different reflectivity at 10 μm.

The contribution to the bandwidth due to a finite field of view is obtained by convolving an Airy function with a finesse determined by the reflectivity and the surface irregularity over all angular points within the field of view. The result is that in addition to changing the finesse, the center frequency and line shape will also change. A theoretical trans-

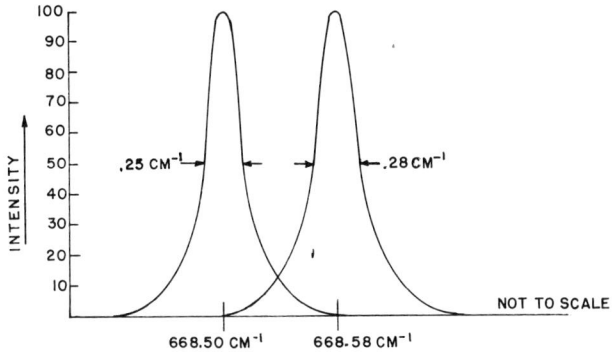

FIGURE 7  Change of bandwidth and center frequency of the scanning function due to the field angle (FSR $\sim 20$ cm$^{-1}$, $R = 96\%$).

mission function for the PSC is shown in Figure 7, which takes into account both the field of view and the surface deviation. This function is compared to the transmission function for reflectivity only.

## Measurement Procedure

The VTPR is tested by aligning it so that the output of the PSC fills its entire field of view and its entrance aperture. The output signal of the VTPR is then recorded for each of the eight channels as the spectral output of the PSC is scanned. The output data are the convolution of the spectral bandwidths of the PSC and the VTPR. If both the filter and the F-P transmission functions are considered to be Lorentzian, the convolved bandwidth is the sum of the bandwidths of the two functions. Data obtained by convolving the angle-broadened F-P function with a measured filter function gave an increased bandwidth of about one half the bandwidth of the F-P function. Figure 8 is a plot of the filter function. A Gaussian function of the same bandwidth would have produced an even smaller increase in bandwidth.

Each of the eight filters is scanned by only one F-P transmission line as the data are taken. In order to minimize the total required mechanical scan, a plate separation was selected that placed a transmission line near the center of each of the eight filters and also near the peak of an order for a $CO_2$ laser line. The optimum plate separation was found to be 254 $\mu$m for the seven filters in one channel.

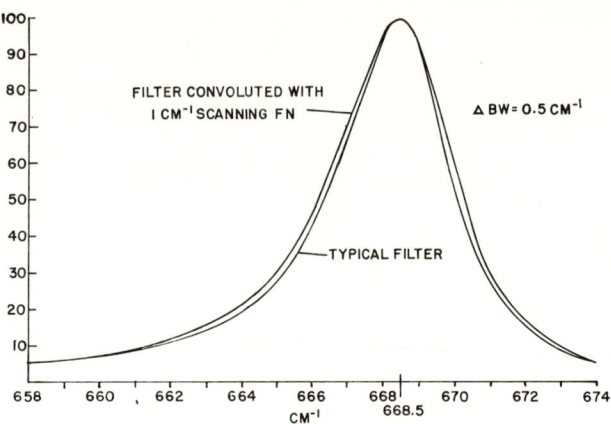

FIGURE 8  Bandwidth increase of output data due to bandwidth of scanning function.

## Total System Transmission

In order to have adequate signal-to-noise ratio, it is desirable to have a peak system transmission of at least 15 to 20%. This was, in turn, primarily limited by the interferometer transmission due to coating and substrate absorption. Irtran IV and Irtran VI were selected rather than a haloid material, which would have improved the transmission but would have been incapable of maintaining the required flatness on the plano side. The outside surfaces of the interferometer plates were given a radius to act as the collimator to avoid the necessity of another element in the 11- to 15-$\mu$m channel. Because of the unavailability of Irtran VI 10 cm in diameter, it was necessary to introduce a collimating element of KBr in that channel to maintain the magnification.

The effect of the high reflectance coatings on the peak transmission is given by

$$T_p = [T/(A + T)]^2, \quad (8)$$

where $A$ and $T$ are the absorption and transmission of the coating, respectively. It can be seen from this relationship that the peak transmission falls rapidly as the ratio of transmission to absorption decreases. Since the absorption is limited by the material and thickness of the coating, increasing the reflectivity tends to decrease the peak transmission.

Another problem that must be considered in selecting a high reflectance coating is the rate of change of reflectivity and absorption with frequency. If these values change appreciably over the width of a single filter under test, the radiometer output will be affected by the changing calibrator throughput in addition to the filter function being calibrated. This effect must be taken into account when interpreting the data.

## Conclusions

The design approach offers enough latitude to control the spectral bandwidth of the F-P interferometer when the aperture and field-of-view requirements are taken into account. However, a limiting factor in determining the maximum reflectivity was the absorption of the dielectric coatings. Attempts to increase the reflectivity much above 93% resulted in a transmission absorption ratio that was too small to produce adequate peak transmission. Several acceptable theoretical reflectivity curves were obtained for the required coatings, but in practice a number of attempts were required to produce coatings with the required spectral uniformity.

The authors wish to express their appreciation to Robert Astheimer of Barnes Engineering Company for originating the idea of testing the VTPR using an F-P interferometer system and for the support and encouragement of other Barnes, NOAA, and NASA personnel.

## References

Born, M., and E. Wolf, *Principles of Optics,* pp. 328-369, Macmillan, New York (1964).
Falbel, G., and D. Zink, *Proc. Inst. Electron. Radio Eng.,* p. 231 (1971).
Kaplan, L. D., *J. Opt. Soc. Am. 49,* 1004 (1959).
Steel, W. H., *Interferometry,* pp. 119-122, Cambridge U.P., New York (1967).

H. VAN DE STADT, TH. DE GRAAUW, J. C. SHELTON, and C. VETH

# NEAR-INFRARED HETERODYNE INTERFEROMETER FOR THE MEASUREMENT OF STELLAR DIAMETERS

## I. Introduction

Heterodyne detection at visible and infrared wavelengths is nowadays a well-established technique, which is used in spectroscopy to combine the advantages of high signal-to-noise ratio and extremely high resolution without cumbersome classical spectrographs [Cummins and Swinney, 1970] and in communications to combine the advantages of very-high-modulation bandwidth and high angular sensitivity, particularly useful in space applications. We will report here on another application, i.e., the measurement of brightness distributions of astronomical objects with interferometers.

Heterodyne detection is the result of interaction between two waves originating from a coherent radiation source (laser) and an incoherent radiation source, e.g., an astronomical object. The two waves combine on a photocell and produce a signal at the difference frequency containing all essential information about phase and amplitude of the incoherent radiator. The electrical signals produced by the mixing process of the coherent and incoherent light signals at two spatially separated telescopes

---

The authors are at The Astronomical Institute at Utrecht, The Netherlands.

can be used to exhibit the degree of coherence of the incoherent source and thus its angular dimensions.

## II. Astronomical Introduction

Recent sky surveys at different wavelengths show many more bright infrared objects than was expected some 10 years ago. Many of these objects have an excess in radiation in the infrared with respect to an extrapolation of the visible radiation if the source were to radiate like a blackbody. Most stellarlike objects show strong evidence of circumstellar dust clouds consisting of silicate or graphite or other molecular grains [see review by Neugebauer *et al.*, 1971]. We present here only a few items that are of interest for the present situation.

The infrared excess is not restricted to one stellar type. The excess has been observed in young, intermediate, and late-type stars. For late-type stars the origin of the ir radiating shells of dust clouds is the usual mass loss of highly evolved stars. Such clouds absorb the visible and uv light emitted by the central star, and the infrared excess consists of thermal reradiation of the cloud. The expected angular sizes of bright objects range up to 1 sec of arc. For young type stars, the possibility of observing the remains of the prestellar cloud is very attractive. The ir spectrum of these shells tends to peak at 10 $\mu$m and sometimes also at 20 $\mu$m. Angular sizes are uncertain but are estimated to be of the order of 0.01 sec of arc. Some rather cool stars show shells with ir spectral properties remarkably similar to protostars. Another class of galactic sources is formed by the H II regions. The observations of OH and $H_2O$ maser lines in very small point sources in those regions, together with infrared observations, make those regions very interesting. Recent ir observations by Wynn and Williams, give reasonable coincidence with the maser sources. High spatial resolution in the ir is needed to understand the physics of these regions, where stars are expected to be formed.

A number of unresolved point sources have been discovered in our own galactic center. Several external galaxies, especially bright Seyfert galaxies, appear to be strong ir radiators. For these objects, an explanation for the origin of their radiation is still difficult to give. We conclude that for an explanation of the origin of many ir sources it is of prime importance to know precisely the spatial extent of these objects, including their cross-sectional brightness distribution and possible asymmetry. Once the angular size of ir radiating objects is measured, the physical or linear size is known if the distance is available from other measurements. If the data are combined with the measured absolute monochromatic flux, the measured angular diameter gives direct values of the absolute fluxes and brightness temperatures of the particular object.

## III. Observational Techniques with High Spatial Resolution

It is interesting to know the angular size of many stellarlike objects at wavelengths between approximately 4 and 10 μm. With a single telescope, this, however, is often difficult to obtain because of the diffraction-limited resolution of a medium-sized telescope, which for a diameter of 2 m is 0.5 and 1.2 sec of arc at 4 and 10 μm, respectively, whereas an angular resolution of 0.1 sec of arc or less is required. In the visible, the resolution of the same telescope would be of the order of 0.06 sec of arc if the seeing were perfect, where seeing is defined as fluctuations in phase and amplitude of the transmitted wavefront due to atmospheric fluctuations in the refractive index of air.

Theories based on the assumption of homogeneous isotropic turbulence [see Strobehn, 1971; Fried, 1966, 1967] indicate that the so-called "seeing-disk" or "effective coherent size" [Fried and Yura, 1972] of a telescope is proportional to $(\lambda)^{6/5}$, where $\lambda$ means wavelength. Thus the angular resolution of telescopes often increases toward infrared wavelengths in spite of the increasing wavelength!

We give a short but necessarily incomplete survey of possible techniques for realizing higher angular resolutions than is possible with conventional telescopes [see also Goodman, 1970].

An elegant technique in the visible-wavelength region, proposed by Labeyrie [1970], is speckle interferometry, where a large number of stellar images are photographed in the focus of a large-aperture telescope with very high time resolution in order to "freeze" the speckle pattern formed by seeing. Superposition of the Fourier transforms of many of these patterns suppresses the Fourier components due to seeing, yielding a final picture with diffraction-limited resolution: 0.02 sec of arc has been achieved with a 5-m telescope [Gezari et al., 1972]. This remarkable technique can be used only for circularly symmetric objects and is apparently restricted to visible wavelengths because of sky background radiation in the infrared and also because of the required highly sensitive two-dimensional recording means that are not available for longer wavelengths.

Another not too complicated technique makes use of lunar occultations. This technique can be applied for all possible wavelengths from the radio region to the visible. Some recent infrared measurements [Toombs et al., 1972] show resolutions of the order of 0.02 sec of arc. This method seems to be well suited for infrared measurements, although the total number of observable objects as well as the actual number of possible observations are somewhat limited.

A special class of different observational techniques is that of stellar interferometry. This is based on the principle that an interferometer does not form directly a physical image of the observed object but instead measures the (complex second-order) degree of coherence of the received wavefront. Interferometers consist of two or more spatially separated telescopes or, alternatively, two or more diaphragms in the otherwise obscured aperture of a single telescope. The degree of coherence is related in a unique way to the intensity distribution over the radiating object [Born and Wolf, 1970].

The first example of a stellar interferometer is the well-known Michelson interferometer [Michelson and Pease, 1921]. Visual observations of the modulation of the fringe pattern using a base line of 15 m have yielded a resolution of 0.008 sec of arc [Hanbury Brown, 1968]. There are two major problems with Michelson interferometers. First, there is random motion of the interference pattern due to seeing. This difficulty can be circumvented [Currie, 1967; Rogstad, 1968] by applying modern electrical detection systems for the fringes, but to our knowledge no practical astronomical measurements have been published so far. The second and major difficulty is to equalize the two light paths, from the star to the points where the fringes are formed, to within the necessary precision of several tens of wavelengths. This problem is inherent to the broadband nature of the measurement, and the requirements can only be relaxed when much smaller wavelength bands are used, like in the two kinds of stellar interferometer to be mentioned later.

The techniques mentioned up to this point are broadband techniques. They yield a relatively high signal-to-noise ratio, but they are restricted in their application to relatively short base lines and, accordingly, to angular resolving powers. A Michelson interferometer might be applicable in the near infrared, but for longer wavelengths (i.e., above $\sim 5$ $\mu$m) increased background radiation will probably rule out its application.

A second class of stellar interferometer is formed by intensity interferometers. Here, correlations are measured between fluctuations in the photocurrents that are caused by focused stellar radiation at different telescopes [Hanbury Brown and Twiss, 1956]. These interferometers measure the square of the modulus of the degree of coherence with the important advantage of being insensitive to atmospheric phase fluctuations. This also implies that relatively low-quality flux collectors can be used as telescopes. The narrow-band nature of the measuring technique makes it possible for the time delay between the two paths of this interferometer to be many orders of magnitude greater than the delay in a Michelson interferometer, which makes it possible to realize very long base lines with a limited mechanical precision, e.g., the Narrabri interferometer has a base line of 188 m, yielding a resolution of $4 \times 10^{-4}$ sec of arc at a wavelength of 443 nm. While the small bandwidth is in a certain sense an advantage, it also means that the signal-to-noise ratio is relatively small and that large light collectors are required. The method is said not to be suited to the measurement of cool objects [Hanbury Brown, 1968]; toward the infrared some additional difficulties are appearing. First, the nature of the usual detectors (i.e., photomultipliers) is such that they cannot be used for wavelengths beyond 1 $\mu$m, while solid-state detectors, which can be used, should not be able to provide shot-noise-limited performance (however, it remains to be shown that this is also true for modern helium-cooled photoconductors); second, the increasing background radiation in the infrared should degrade the signal-to-noise ratio by contributing additional shot noise.

A third type of stellar interferometer is formed by heterodyne interferometers, resembling the intensity interferometer in their narrow-band performance but also resembling Michelson interferometers in that they measure the degree of coherence directly. Heterodyne interferometers appear to be most suitable for infrared measurements at long base lines, where all other kinds of interferometer seem to engender insurpassable problems. We will discuss this in more detail in the following sections; arguments for the latter statement will be given.

## IV. Signal-to-Noise Ratio of Heterodyne Detection Systems

The basic elements for building a heterodyne detector are arranged as follows. First, one needs a telescope to collect and focus the radiation from the object of interest. Next, a laser is needed with suitable wavelength, sufficiently intense for providing shot-noise-limited performance and with a reasonable stability and mode structure (i.e., a single transverse mode and longitudinal modes not too close together). Then a beam combiner is needed for superposition of the two beams and a detector (or two balanced detectors) for generating the difference frequency signal and suitable electronics consisting of a broadband radio-frequency (rf) amplifier, square-law detector, and some combination of chopper and synchronous (linear) detector or lock-in amplifier.

The signal-to-noise ratio (SNR) that can be achieved with this equipment has been published [Ross, 1966; Siegman, 1966], and we restrict ourselves here to quoting the complete expression in a form suitable for illustration of infrared measurements:

$$\text{SNR} = \frac{2i_L i_s R}{P_A + 2eRB(i_L + i_d + i_s^* + i_b^*) + 2R(i_L i_b + i_s i_b)}, \quad (1)$$

where $R$ is the equivalent electrical impedance of the amplifier circuit as seen by the photocells, $i$ is the electrical current at the input of the rf amplifier, while the suffixes $L$, $s$, $d$, and $b$ mean laser, signal, dark current, and background; a current due to radiation of an optical power $I$ obeys the relation

$$i = (\eta e/h\nu)I, \quad (2)$$

with $h\nu$ the photon energy, $\eta$ the detector quantum efficiency, and $e$ the charge of an electron; $I$ is the optical power over a bandwidth of $2B$ Hz, where $B$ is the electrical rf bandwidth. Where an asterisk (*) is added to $i$ and $I$, they should be integrated over the total optical bandwidth transmitted by the telescope optics and detected by the detector. Note that Eq. (2) is only exact for photoemitting diodes; for photoconductors, different formulas apply. $P_A$ is the amplifier noise power, also referred to its input. Equation (1) refers to an equipment that discriminates against background by using as chopping technique a switching between background and source.

Normally the laser power is so high that $i_L$ prevails over most other noise sources, so that Eq. (1) reduces to

$$\text{SNR} = \frac{\eta I_s}{h\nu B + \eta I_b} = \frac{2\eta N_s}{1 + 2\eta N_b}, \quad (3)$$

where $N$ stands for the total number of detected photons per hertz bandwidth and per second.

Expressions (1) and (3) assume an ideal detector, in the sense that both polarizations of the source are detected, while no loss occurs due to chopping techniques.

An even simpler formula can be derived, when the source is treated as a blackbody radiator of effective temperature $T_e$:

$$\text{SNR} = 2\eta (A/A_c) \delta_s [1/(1 + 2\eta \delta_b)], \quad (4)$$

where $\delta$ is the degeneration factor according to Planck's radiation formula

$$\delta = 1/[\exp(h\nu/kT) - 1] \quad (5)$$

and the suffixes $s$ and $b$ again mean source and background; $A$ is the area of the receiving aperture of the telescope, and $A_c$ is the so-called coherence area, quantified by $A_c = \lambda^2/\Omega$ if $\Omega$ is the spatial angle subtended by the source as seen by the telescope. For most astronomical objects $A_c \gg A$ and Eq. (4) can be applied; but for a few objects, like the sun, moon, and planets, $A_c \ll A$. In the last case, SNR grows to a maximum, given by

$$\text{SNR} = 2\eta\delta_s[1/(1 + 2\eta\delta_b)]. \quad (6)$$

A graph of $\delta$ as a function of $\lambda$ and $T$ is given in Figure 1. Assuming that background radiation originates mainly in the earth's atmosphere with a temperature of around 300 K, it is clear that background radiation becomes important only at wavelengths around 50 μm and longer. For shorter wavelengths, the term $2\eta\delta_b$ in Eqs. (4) and (6) can be omitted.

It is important to remark that in an interferometer the two telescopes collect background radiation from different parts of the sky and these contributions will in general not be correlated: background radiation in stellar interferometry is thus of even less significance than is suggested by Eqs. (4) and (6).

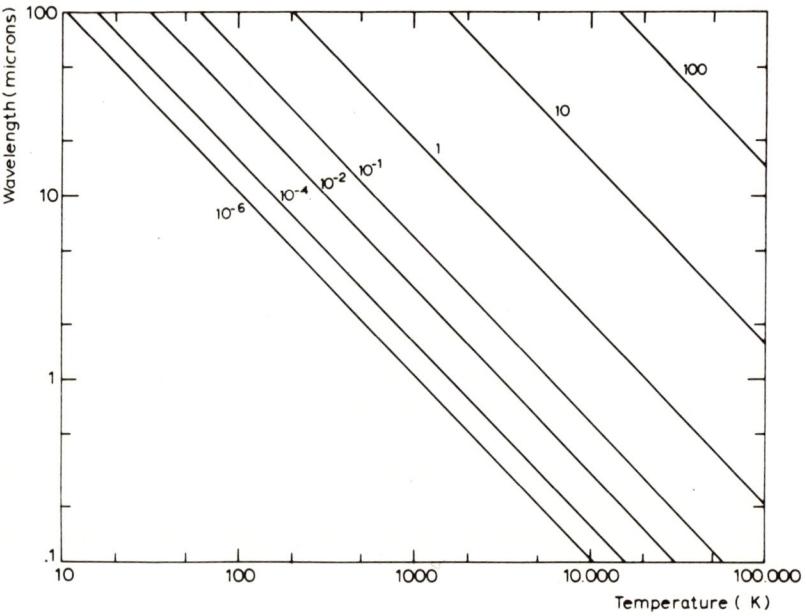

FIGURE 1 The degeneration factor $1/[\exp(h\nu/kT) - 1]$ according to Planck's radiation formula as a function of wavelength and temperature.

The actual SNR that can be realized in practice depends on the brightness temperature of the radiator, the degree of partial coherence $A/A_c$, and the quantum efficiency of the detector. The value according to Eqs. (4) or (6) is, finally, to be multiplied by $(Bt + 1)^{1/2}$ [Cummins and Swinney, 1970], where $B$ is the electrical bandwidth of the rf amplifier, and $t$ the integration time. Fast detectors, broadband amplifiers, high quantum efficiencies, long integration times, and big telescope apertures are clearly needed for good SNR's in astronomical applications, where many radiators yield small values for $\delta_s$ and $A/A_c$. Maximum usable apertures are limited by seeing; an indication of practical values has been given by Fried [1967].

Signal-to-noise ratios for a heterodyne interferometer consisting of two identical telescopes have been derived theoretically by Gamo [1961] and by Nieuwenhuijzen [1969] and were tested experimentally by van de Stadt [1970]. An electronic circuit that adds and squares the rf signals from the telescopes gives

$$\text{SNR} = (\eta I_s/h\nu B)\,[1 + |\gamma_s|\cos(\phi_L - \phi_s)], \qquad (7)$$

where $|\gamma_s|$ is the modulus of the degree of coherence of the stellar light and $\phi$ denotes the phase difference of the two waves falling on the detectors, while the subscripts $L$ and $s$ refer to the laser light and starlight, respectively. A multiplying circuit yields

$$\text{SNR} = (\eta I_s/h\nu B)\,|\gamma_s|\cos(\phi_L - \phi_s). \qquad (8)$$

Equations (7) and (8) are simplifications since they assume that laser light and starlight have identical intensities at the two telescopes, while detectors and amplifying circuits have exactly identical properties. A schematic diagram of such an interferometer is shown in Figure 2.

The phase factor $(\phi_L - \phi_s)$ is not constant when related to telescope distances of a few meters or more: the phase difference $\phi_L$ of the laser is determined by the optical path-length difference between the two telescopes and is thus very sensitive to small temperature variations in the air. This can be circumvented by transporting the laser beam in an evacuated tube. But random variations in $\phi_s$ due to atmospheric seeing are a much bigger problem. In the visible and the near infrared the frequency spectrum of the seeing may contain frequencies up to 100 Hz. A possible detection method is then to phase-modulate one laser beam or star beam at a frequency $f_m$, much higher than the highest seeing frequency (see Figure 2). The resulting modulation after correlating the rf signals can then be amplified in a narrow-band amplifier tuned at the frequency $f_m$ and with a

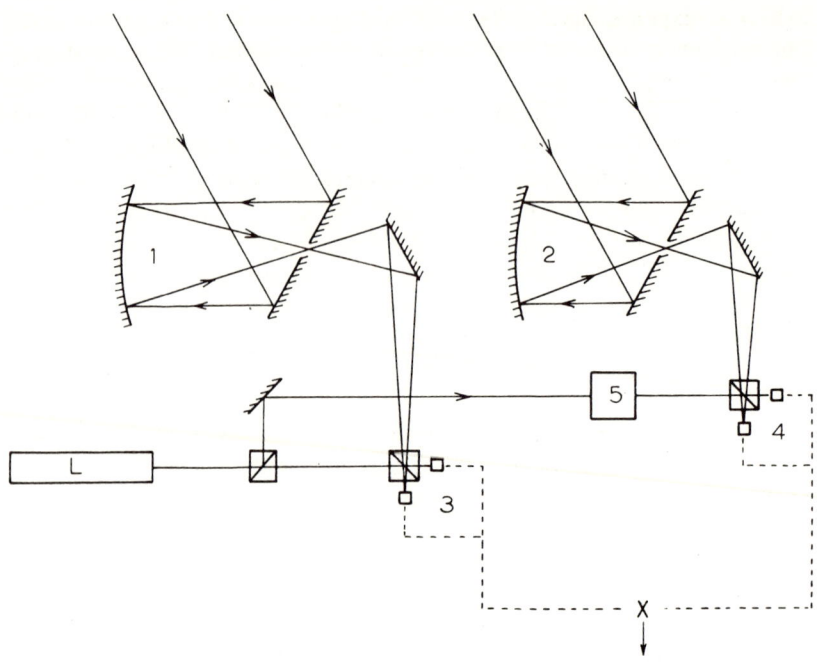

FIGURE 2 Schematic view of a heterodyne stellar interferometer. L, laser; 1 and 2, telescopes; 3 and 4, balanced detectors; 5, phase modulator.

bandwidth $B_m$. At the exit of this amplifier the SNR is equal to the expression (8) multiplied by the factor $(B/B_m + 1)^{1/2}$. This signal can then be detected with a (third) square-law detector followed by an integrater with time constant $t$, yielding finally

$$(\text{SNR})_{\text{H.I.}} = (\eta I_s/h\nu B)^2 \; |\gamma_s|^2 \; B \, [(t/B_m)^{1/2}] \tag{9}$$

if $B \gg B_m \gg 1/t$.

We give in Table 1 the signal-to-noise ratios for various wavelengths in the near infrared. A lower limit for $B_m$ is set by the seeing; the telescope aperture is set to 1 m², which is too optimistic for the shorter wavelengths; values for $I_s/B$ correspond to a number of bright sources [see Neugebauer et al., 1971; Low, 1968; Low and Johnson, 1964]; $\eta = 0.2$ was used. From this table we conclude that heterodyne interferometry is a useful technique for wavelengths of 3 µm and longer, but its applicability for wavelengths shorter than 3 µm is uncertain.

From Eq. (9) it follows that phase information is lost, as in intensity

TABLE 1  Expected Signal-to-Noise Ratios of a Heterodyne Interferometer and an Intensity Interferometer for Bright Sources in the Near Infrared and an Aperture of 1 m²

|  | $\lambda$ ($\mu$m) | | |
| --- | --- | --- | --- |
|  | 1.25 | 3.4 | 10.2 |
| $B$ (Hz) | $10^8$ | $10^8$ | $10^8$ |
| $t$ (sec) | $10^2$ | $10^2$ | $10^2$ |
| $B_n$ (Hz) | $10^2$ | 30 | 10 |
| $h\nu$ (J) | $1.8 \times 10^{-19}$ | $6 \times 10^{-20}$ | $2 \times 10^{-20}$ |
| $I_s/B$ (W/Hz)/m$^{-2}$ | $2 \times 10^{-22}$ | $10^{-22}$ | $4 \times 10^{-23}$ |
| $|\gamma_s|^2$ | 0.2 | 0.2 | 0.2 |
| (SNR)$_{H.I.}$ | 1 | 3.9 | 10 |
| (SNR)$_{II}$ | 4.5 | 6.7 | 8 |

interferometry. This seems to be an inevitable concession set by atmospheric seeing, but it seems nonetheless to be possible to recover an unambiguous brightness distribution [Bates, 1969]. It remains interesting to calculate (SNR)$_{II}$ for an intensity interferometer with shot-noise-limited performance:

$$(\text{SNR})_{II} = (\eta I_s/h\nu B) |\gamma_s|^2 \sqrt{Bt}. \tag{10}$$

Calculated values are also given in Table 1.

## V.  Heterodyne Detection at 633 nm

In this section we describe some experiments with a balanced optical detector using an He–Ne laser at 633 nm as local oscillator. The balanced optical detector consists of a combination of a beam combiner with semitransparent dielectric coating, two photocells, and an amplifier circuit. Advantages of this setup over a single detector photocell are the elimination of unwanted laser noise (e.g., beat frequencies from longitudinal modes) and a higher detection efficiency [see Oliver, 1961; Waite, 1965]. A diagram of the electrical circuit of the balanced mixer is given in Figure 3. We use silicon PIN diodes, type 4204, manufactured by Hewlett-Packard. They are connected in the reverse bias direction to dc voltages adjusted for minimum noise of the detector; at a wavelength of 633 nm the factor $\eta e/h\nu$ is equal to 0.35 mA/mW. Resistors of 100 $\Omega$ are inserted to measure the dc current during lining up of the photocells. Capacitors

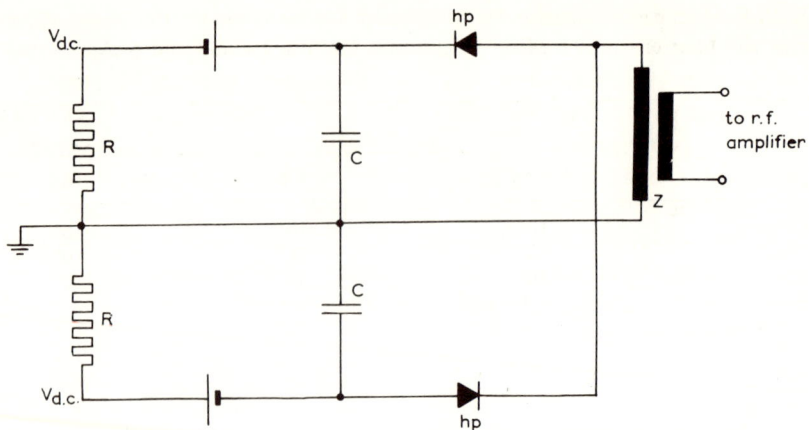

FIGURE 3 Diagram of electrical circuit of balanced optical detector for 633 nm; the diodes are Hewlett-Packard PIN photodiodes type 4204; $R = 100\ \Omega$; $C = 1$ nF; $Z = Z$-match transformer type 50200A.

of 1 nF provide a low-impedance bypass for the rf signals that were effectively added through an impedance stepdown transformer, for which we used a Z-match wide-band transformer type 50200A. The rf amplifier is a 300-MHz wide-band amplifier type CA 1003, manufactured by E & M laboratories; the amplifier noise figure is 2.5 dB. Using a Spectra-Physics model 120 laser, output 6.5 mW at 633 nm, the resulting ratio of laser shot noise to amplifier noise was measured to be 6.6:1.

Highest SNR's were realized using two balanced detectors detecting simultaneously the two perpendicular linear polarization directions of a thermal source (see Figure 4). One polarization is totally reflected by the polarizing prism, 4, mixed in beam combiner, 2, with half of the laser power and detected by a detector, 6. The other polarization is totally transmitted by the polarizing prism, subsequently rotated over 90° by a half-wave plate and finally detected in a detector, 7, with the other half of the laser power. The two independently detected signals can be added, and the result is a $\sqrt{2}$ increase in signal-to-noise ratio over that obtainable with a single balanced detector.

The described detector was calibrated using a tungsten ribbon lamp of known brightness temperature (2670 ± 15 K at 18.00 A). Measured SNR for an integration time of 10 sec was 4.0. In the experimental situation, the size of the detector was appreciably bigger than one coherence area, so that Eq. (6) applies with $\delta_b = 0$, while $\delta_s = 2 \times 10^{-4}$ for the quoted temperature. Assuming a quantum efficiency $\eta = 0.7$, a 70% transmission of the interference filter that was used in the experiment, and a factor

$\sqrt{2}$ loss due to chopping, one can calculate the effective rf bandwidth of the detector to be 115 MHz. It is concluded that the present detection system may be improved in the future by using better rf equipment.

Another experiment with a slightly different heterodyne detector at 633 nm was performed by pointing it to the sun. When the detector is pointed at the sun without any optics except one flat mirror, it has a receiving angle of 4 min of arc, corresponding to a coherence area of 0.2 mm$^2$, the sensitive area of the detector. The solar diameter is 30 min of arc, hence we have again $A_c \ll A$. Figure 5 shows the results of a scan over the solar disk. Integration time was 5 sec and the scanning speed was exactly the daily motion of the earth, because the solar light was reflected toward the detector by a fixed mirror. The degeneracy of the center of the solar disk at 633 nm is 2.56 $\times$ 10$^{-2}$ ($T = 6160$ K); and based on the ribbon-lamp experiment, an SNR of 80 was expected for the sun at an elevation angle of 60°. Actual SNR was 50, which is acceptable, because of a number of rather uncertain parameters, such as atmospheric transmission. In any case, this rather simple experiment using normally available components and a plain laser without any stabilization demonstrates clearly that earlier predictions about the feasibility of such ex-

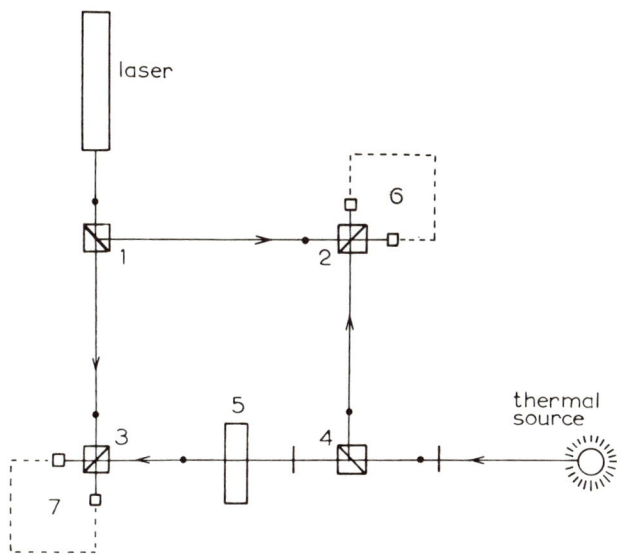

FIGURE 4 Setup for a heterodyne detector for two directions of polarization; 1, 2, and 3, semitransparent beam combiners; 4, polarization beam cube; 5, $\lambda/2$ plate under 45°; 6 and 7, balanced detectors.

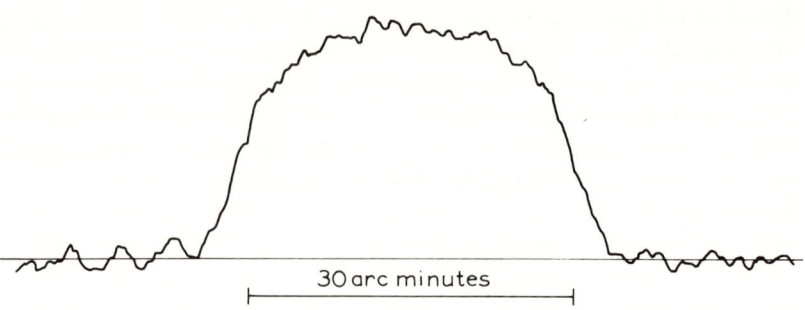

FIGURE 5  Scan over solar disk with a heterodyne system at 633 nm; integration time, 5 sec; angular resolution, 4 min of arc; spectral bandwidth, $4 \times 10^{-4}$ nm. No telescope was used.

periments [Siegman, 1966; Kopeika and Bordogna, 1970] were much too pessimistic.

At this point it is not difficult to predict a possible SNR for stars with heterodyne detectors at 633 nm. Suppose that we observe Sirius with a telescope at some favorable site, giving an effective coherent size of 30 cm, corresponding to a 0.5 sec of arc seeing disk. The brightness temperature of Sirius is 10380 K, and its measured angular size [Hanbury Brown et al., 1967] makes that $A/A_c = 1.4 \times 10^{-4}$. This means that even after an integration time of 1 h the SNR is not more than 2.2, which makes measurements in the visible practically impossible. Whereas the earliest measurements carried out by our group [Nieuwenhuijzen, 1970] were rather encouraging, more detailed measurements made at two different sites in Europe during the winter of 1971–1972 on several bright stars did not show evidence of a reliable signal. These measurements were not performed under optimum seeing conditions. But although the detectors have been improved since that time and additional improvement may be gained by rapidly tracking the wavefront tilt [Fried, 1972], we do not believe that useful astronomical measurements on stars with a visible heterodyne detector will be possible in the future.

We conclude this section by describing some results of a laboratory simulation experiment of a heterodyne interferometer. The schematic setup is given in Figure 6. We used the same Spectra-Physics model 120 laser, operating at 633 nm, as laser (L). An He–Ne gas-discharge tube (Spectra-Physics model 124 laser without front mirror) was used as thermal radiator (T). The brightness temperature of this gas-discharge tube was measured to be 9600 K, and this yielded a comfortably high SNR for this experiment. The balanced detectors, 1 and 2, simulate two telescopes both illuminated with laser light and light from the incoherent source.

A rotatable plane-parallel plate, P, serves to vary the phase factor $\phi_s$ as described by Eqs. (7) and (8); results have been published earlier [van de Stadt, 1970]. Large rotation angles of P will also decrease $|\gamma_s|^2$, because the incoherent source is virtually shifted sideways as seen by detector 1. The measured shift together with the known angular size of the source and the wavelength result in a calculated decrease in $|\gamma_s|^2$, which agrees with the measurements: in Figure 7 the detected heterodyne signal at detector 1 is shown to be nearly constant as a function of the rotational shift of P in units of millimeters of linear shift; also the square of the correlated heterodyne signals from detectors 1 and 2 is shown, from which the equivalent uniform disk of the gas discharge can be calculated to be 2.0 mm, while the internal diameter of the discharge tube is 2.2 mm.

Also indicated in Figure 6 are two possible optical delay lines $D_1$ and $D_2$, as well as an electrical delay line $D_3$. Allowed maximum delays are determined by

$$E_2 + E_3 < c/4B, \tag{11}$$

where $E_2$ and $E_3$ are the equivalent optical path differences introduced

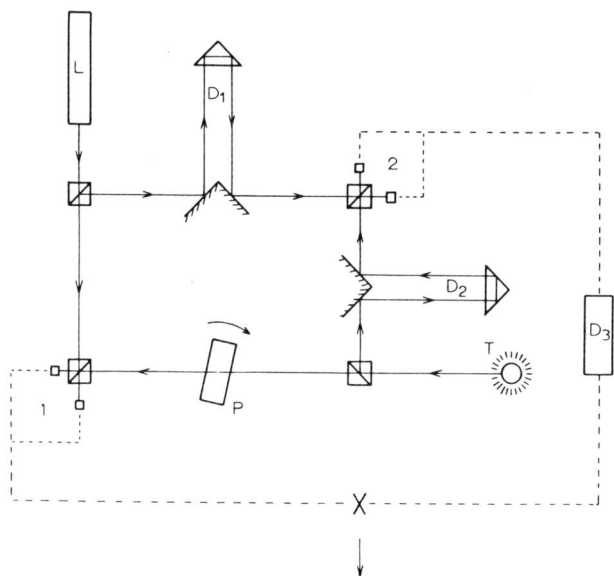

FIGURE 6 Simulation experiment of a heterodyne interferometer. L, laser; T, thermal light source; P, rotatable plane-parallel glass plate; 1 and 2, balanced detectors; $D_1$ and $D_2$, optical delay lines; $D_3$, electrical delay.

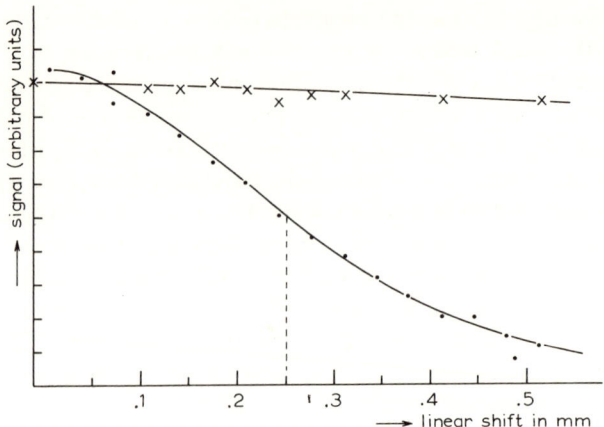

x = heterodyne signal at one detector
• = square of correlated heterodyne signals from two detectors

FIGURE 7 Correlated and uncorrelated signals from a laboratory simulated interferometer.

by $D_2$ and $D_3$, respectively, $c$ is the speed of light, and $B$ is the rf bandwidth. The factor 1/4 determines the point where the SNR is decreased to half of its maximum value. Experimental results were in good agreement with Eq. (11).

It is concluded that a heterodyne stellar interferometer for near-infrared wavelengths using two medium-sized telescopes and a laser as local oscillator is technically feasible and astronomically very interesting.

We want to thank Prof. H. G. van Bueren for stimulating this project. Three authors want to express their appreciation for the very profitable, stimulating, and pleasant period of nine months that Chris Shelton spent in Utrecht during 1972. Our gratitude goes to Ir. Bezemer of the physical laboratory in Utrecht for his careful calibration of the tungsten ribbon lamp. This project has been made possible by a grant from the Netherlands Organization for the Advancement of Pure Research (Z.W.O.).

### References

Bates, R. H. T., *Mon. Notices R. Astron. Soc. 142,* 413 (1969).
Born, M., and E. Wolf, *Principles of Optics,* Chap. X, Pergamon, New York (1970).
Cummins, H. Z., and H. L. Swinney, *Progress in Optics VIII,* North-Holland, Amsterdam, Holland (1970).
Currie, D. C., Woods Hole summer study, Vol. 2, p. 35 (1967).
Fried, D. L., *J. Opt. Soc. Am. 56,* 1372 (1966).
Fried, D. L., *Proc. IEEE 55,* 57 (1967).

Fried, D. L., *J. Opt. Soc. Am.* **62**, 729 (1972).
Fried, D. L., and H. T. Yura, *J. Opt. Soc. Am.* **62**, 600 (1972).
Gamo, H., *Advances in Quantum Electronics*, p. 252, Columbia U.P., New York (1961).
Gezari, D. Y., A. Labeyrie, and R. V. Stachnik, *Astrophys. J.* **173**, L1 (1972).
Goodman, J. W., *Progress in Optics VIII*, North-Holland, Amsterdam, Holland (1970).
Hanbury Brown, R., and R. Q. Twiss, *Nature* **178**, 1046 (1956).
Hanbury Brown, R., *Annu. Rev. Astron. Astrophys.* **6**, 13 (1968).
Hanbury Brown, R., J. Davis, L. R. Allen, and J. M. Rome, *Mon. Notices R. Astron. Soc.* **137**, 393 (1967).
Kopeika, N. S., and J. Bordogna, *Proc. IEEE* **58**, 1571 (1970).
Labeyrie, A., *Astron. Astrophy.* **6**, 85 (1970).
Low, F. J., and H. L. Johnson, *Astron. J.* **139**, 1130 (1964).
Low, F. J., *Highlights of Astronomy*, p. 136, Reidel, Dordrecht, Holland (1968).
Michelson, A. A., and F. G. Pease, *Astron. J.* **53**, 249 (1921).
Neugebauer, G., E. Becklin, and A. R. Myland, *Annu. Rev. Astron. Astrophys.* **9**, 67 (1971).
Nieuwenhuijzen, H., *Mon. Notices R. Astron. Soc.* **150**, 325 (1970).
Nieuwenhuijzen, H., *Bull. Astron. Inst. Neth.* **20**, 300 (1969).
Oliver, B. M., *Proc. IRE* **49**, 1960 (1961).
Rogstad, D. H., *Appl. Opt.* **7**, 585 (1968).
Ross, M., *Laser Receivers*, Wiley, New York (1966).
Siegman, A. E., *Appl. Opt.* **5**, 1588 (1966).
Strobehn, J. W., *Progress in Optics IX*, North-Holland, Amsterdam, Holland (1971).
Toombs, R. I., E. E. Becklin, J. A. Frogel, S. K. Law, F. C. Porter, and J. A. Westphal, *Astrophys. J.* **173**, L71 (1972).
van de Stadt, H., *Opt. Commun.* **2**, 153 (1970).
Waite, T., *Proc. IEEE* **53**, 334 (1965).

# Communications and Radiometry

R. J. D'ORAZIO and NICHOLAS GEORGE

# MATCHED-FILTER DETECTION OF MODE-LOCKED LASER SIGNALS

## I. Introduction

Multitone lasers of the mode-locked [Hargrove et al., 1964] and cavity-dumped [Steier, 1966] types emit their energy in short pulses. Sensitive detection of these emissions for point-to-point communications or echo-ranging systems can be accomplished by using appropriate filtering at the optical frequencies before detection and radio-frequency amplification. In the present work we describe our approach to matched filtering for these signals.

In the literature, related prior studies of laser detection include scanning Fabry-Perot cavities using a single passband of the passive cavity to analyze laser radiation [Fork et al., 1964], spatial filtering techniques [Kogelnik and Yariv, 1964], and various laser heterodyne techniques [Uhlhorn and Holshouser, 1970; Sonnenschein and Horrigan, 1971; Rudd, 1969].

Our optical receiver for mode-locked gas-laser signals consists of a

---

The authors were at the California Institute of Technology, Pasadena, California 91109, when this work was done; R. J. D'Orazio is now at the Bell Telephone Laboratories, Holmdel, New Jersey 07733.

passive laser cavity controlled in length and a photodetector with its associated electronics. The length of the passive Fabry-Perot cavity is chosen roughly equal to the cavity length of the transmitting laser but with provision for fine fractional wavelength control of its length. In addition to the selective filtering characteristics of the passive cavity (passbands of unity transmission matching the frequencies of the multimode laser), a readout of the vernier length control, peaking the output, provides for an extremely wide range of velocity measurements with either an active or a passive vehicle moving relative to the receiver.

## II. The Passive Cavity: A Matched-Filter for Mode-Locked Laser Radiation

Consider the passive cavity as shown in Figure 1, where $h_0$ is the cavity length, $M_1$ is the fixed cavity mirror, and $M_2$ is the movable cavity mirror. The amplitude transmission function $T(\omega)$ may be shown to be [Eq. (1) is a generalization, to include frequency variations of $t_1$, $t_2$, $r_1$, and $r_2$ of the well-known expression for cavity transmissivity in a Fabry-Perot, see e.g., Born and Wolf, 1970]

$$T(\omega) = L \frac{\exp[i\psi(\omega)]}{\{1 + P_2 \sin^2[\omega(h_0/c)]\}^{1/2}}, \qquad (1)$$

where

$$P_2 = \frac{4 r_1 r_2 \exp(-2\alpha h_0)}{[1 - r_1 r_2 \exp(-2\alpha h_0)]^2}, \qquad (2)$$

$$\psi(\omega) = -\frac{\omega}{c} h_0 - \tan^{-1}\left\{\frac{r_1 r_2 \sin[2\omega(h_0/c)]}{1 - r_1 r_2 \cos[2\omega(h_0/c)]}\right\}, \qquad (3)$$

and

$$L = \frac{t_1 t_2 \exp(-\alpha h_0)}{1 - r_1 r_2 \exp(-2\alpha h_0)}. \qquad (4)$$

$t_1$ and $t_2$ are the transmission functions for mirror 1 and mirror 2, respectively; $r_1$ (mirror 1 right-side incidence) and $r_2$ (mirror 2 left-side incidence) are the reflection coefficients, $\alpha$ is the cavity loss per unit length, and $c$ is the speed of light.

To study the passive cavity as a multitone filter we use the matched-filter criterion resulting from the optimization of the signal-to-noise ratio. So for a linear system with impulse response function $H_1(t)$ and input

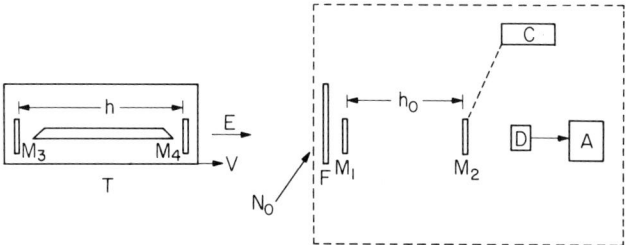

FIGURE 1 Passive cavity receiver. The components are T, laser transmitter; $M_3$, $M_4$, laser mirrors; $h$, laser cavity length; $E$, signal; $V$, velocity of laser relative to receiver; $N_0$, noise; F, coarse bandpass filter; $M_1$, $M_2$, passive cavity mirrors; $h_0$, passive cavity length; D, detector; A, detector electronics; C, mirror control.

$F_1(t)$ the total time-varying output is $G_1(t) = H_1(t) \otimes [F_1(t) + N_1(t)]$, where $N_1(t)$ is the additive input signal and $\otimes$ indicates convolution. The matched-filter criterion specifies that the amplitude transmission function $H(\omega)_m = AF^*(\omega)/S_n(\omega)$, where $H(\omega)_m$ and $F(\omega)$ are the Fourier transform of $H_1(t)$ and $F_1(t)$, respectively, $S_n(\omega)$ is the power spectral density of the additive input noise, the asterisk denotes the complex conjugate, and $A$ is any nonzero complex constant [Brown, 1963]. The subscript 1 will be used to denote time-varying signals. Hence we see that the signal for which the passive cavity is a matched filter is given by $F(\omega) = S_n(\omega) T^*(\omega)/A^*$. Since the signals we will be considering have a finite number of frequency peaks, we will approximate $T(\omega)$ expanding around the zeros, $\omega_p$, of $\sin^2(\omega h_0/c)$ for a finite number of peaks around $\omega_0$ so that from Eq. (1) dropping $\psi(\omega)$ we have

$$T(\omega)_A = L \left\{ 1/[1 + (2\Delta\omega/\Delta\omega_p)^2]^{1/2} \right\}, \quad (5)$$

where $\Delta\omega = \omega - \omega_p = \omega - (\omega_0 + p\omega_{c0})$, $\omega_{c0} = \pi c/h_0$ and $\Delta\omega_p = 2c/(h_0 \sqrt{P_2})$. We note that $\Delta\omega_p$ is the full width at half-power of each Lorentzian line shape function generated by Eq. (5). Generally we will assume white noise so that $S_n(\omega) = N_0$ is uniform over the frequencies of interest. Similarly we will assume that $t_1, t_2, r_1, r_2$ are constant over the frequencies of interest. Thus from Eq. (5) we see that the signal for which the passive cavity is a matched-filter is

$$F(\omega) = \frac{S_n(\omega) L}{A^*} \sum_{p=-N}^{N} \frac{1}{[1 + (2\Delta\omega/\Delta\omega_p)^2]^{1/2}}. \quad (6)$$

Now for convenience we will assume that the idealized electric-field amplitude at the output of an unmodulated mode-locked laser with $2N + 1$ modes is given by

$$E_1(t) = \sum_{p=-N}^{N} \exp\left[i(\omega_0 + p\omega_c)t\right] = \exp(i\omega_0 t) \frac{\sin\left[(2N+1)(\omega_c t/2)\right]}{\sin(\omega_c t/2)}, \quad (7)$$

where $\omega_0 = n\pi c/h$ is the center frequency of the laser, $n$ is some large integer, $\omega_c = \pi c/h$ is the free spectral range, and $h$ is the effective cavity length. Then the Fourier transform of Eq. (7) is

$$E(\omega) = 2\pi \sum_{p=-N}^{N} \delta[\omega - (\omega_0 + p\omega_c)], \quad (8)$$

where $\delta(\omega)$ is the Dirac delta function. Thus we see that $T(\omega)_A$ is a comb filter for $E(\omega)$ for $h = h_0$, i.e., $T(\omega)_A$ has passbands of unity transmission matching the frequencies of the multimode laser.

For an actual laser signal, writing $E(\omega)$ for a multitone laser with finite line width will yield an expression as an alternative to the monochromatic idealization of Eq. (8). We note that the spacing between the tones of a mode-locked laser are determined by free spectral range, $\omega_c = \pi c/h$, of the laser cavity. It is our contention that the passive cavity transmission function will control the line shape of the laser output if the gain $\alpha$ and the dispersion in the cavity are independent of frequency around a resonant peak. One may further consider this observation by noting that for He–Ne the width of the Lorentzian shaped hole that is burned into the Doppler-broadened gain profile at saturation is much broader than the mode width of the laser cavity. Thus for $\alpha < 0$ the frequency variation of $P$, Eq. (2), and $L$, Eq. (4), are negligible around the resonance.

In the literature, related prior studies of the laser line shape include lumped element LGC circuit models [Blaquiere, 1953; Gordon et al., 1955]. Freed and Haus [1965] used the solution of the nonlinear Van der Pol oscillator equation to describe the spectrum of the laser output. The basic result of using a lumped circuit model is that the line-shape function is Lorentzian [Grivet and Blaquiere, 1963]. The interesting point is that if one started with the transmission function for a cavity or a transmission line, the line shape would be controlled by equations similar to Eq. (1).

Note that both the lumped-circuit and passive-cavity approaches to the laser spectrum assume that the random-cavity mirror vibrations and effective cavity-length fluctuations are negligible. If one dropped these assumptions, the line shape could be considered Gaussian or some other line-shape function.

So, as the gain curve saturates, the right-half-plane poles of Eq. (1), for $s = i\omega$, will migrate to the $i\omega$ axis. We select the value of $\alpha$ that is an amount $\epsilon$ from the saturated pole so that Eq. (2) becomes

$$P_1 = 4 \exp(-2\epsilon h_0)/[1 - \exp(-2\epsilon h_0)]^2. \tag{9}$$

Then we may write the spectrum of the laser signal with finite line widths as

$$E(\omega) = \frac{t_1 t_2}{2(r_1 r_2)^{1/2}} (P_1)^{1/2} \frac{1}{[1 + P_1 \sin^2 \omega(h_0/c)]^{1/2}}. \tag{10}$$

We note that a good approximation to Eq. (10) may be obtained by expanding around the zeros $\omega_p$ of $\sin^2(h_0/\omega_c)$ for $2N + 1$ tones analogous to the approximation of Eq. (5) so that the alternative to Eq. (8) becomes

$$E(\omega) = \frac{t_1 t_2}{2(r_1 r_2)^{1/2}} (P_1)^{1/2} \sum_{p=-N}^{N} \frac{1}{[1 + (2\Delta\omega/\Delta\omega_l)^2]^{1/2}}, \tag{11}$$

where $\Delta\omega = \omega - \omega_p = \omega - (\omega_0 + p\omega_c)$, $\Delta\omega_l = 2c/(h\sqrt{P_1})$. Thus comparing Eq. (6) with Eq. (11) indicates that for $\Delta\omega_p = \Delta\omega_l$ and $h = h_0$ the passive cavity of Eq. (1) is a matched filter for the multitoned signal of Eq. (10).

## III. Signal-to-Noise Ratio

We compute the predetection time-varying signal-to-noise ratio at the output of the passive cavity as

$$\text{SNR}(t) = |T_1 \otimes E_1|^2 / R_{hn}(0), \tag{12}$$

where $E_1$ is the inverse Fourier transform of Eq. (10) and $T_1$ is the inverse Fourier transform of Eq. (1) and $R_{hn}(0)$ is the autocorrelation function evaluated at zero that is equivalent to the mean-squared value of the additive noise at the output of the passive cavity given by

$$R_{hn}(0) = \frac{1}{2\pi} \int_{-\infty}^{\infty} S_n(\omega) |T(\omega)|^2 \, d\omega. \tag{13}$$

The numerator of Eq. (12) may be expressed by the inverse Fourier transform as

$$\left| T_1 \otimes E_1 \right|^2 = \left| \frac{1}{2\pi} \int_{-\infty}^{\infty} T(\omega) E(\omega) e^{i\omega t} d\omega \right|^2 \tag{14}$$

and substituting Eq. (1) and Eq. (10) we may write the modulus-squared value of the signal portion of the output of a passive cavity as

$$\left| T_1 \otimes E_1 \right|^2 =$$

$$\left| \frac{L}{2\pi} \frac{t_1 t_2}{2(r_1 r_2)^{1/2}} (P_1)^{1/2} \int_{-\infty}^{\infty} \frac{\exp(i\omega t) d\omega}{[1+P_1 \sin^2 \omega(h_0/c)]^{1/2} [1+P_2 \sin^2 \omega(h_0/c)]^{1/2}} \right|^2. \tag{15}$$

Since the signal has only $q = 2N + 1$ modes, we may write Eq. (15), with the substitutions $\omega = \omega_0 \quad (c/h_0) x$, $t = h_0/c$, $\omega_0 = n\pi c/h_0$, where $n$ is some larger integer, $e^{-ix} = \cos x - i \sin x$, and noting that the sine integral over symmetric limits vanishes, as

$$\left| T_1 \otimes E_1 \right|^2 = \left( \frac{q}{2\pi} \right)^2 L^2 \left[ \frac{t_1 t_2}{2(r_1 r_2)^{1/2}} \right]^2 \left( \frac{2c}{h_0} \right)^2 K^2(m), \tag{16}$$

where for $1/P_2 > 1/P_1 > 0$, $K(m)$ is the complete elliptic integral of the second kind given by

$$K(m) = \int_0^{\pi/2} \frac{d\theta}{(1 - m^2 \sin^2 \theta)^{1/2}}$$

and

$$m = [(1/P_2) - (1/P_1)]^{1/2}/(1/P_2)^{1/2}.$$

The denominator of Eq. (12) may be expressed from Eq. (1) and Eq. (13) as

$$R_{hn}(0) = \frac{L^2}{2\pi} \int_{\frac{\pi c}{q \, h_0}} \frac{S_n(\omega) d\omega}{1 + P_2 \sin^2[\omega(h_0/c)]} = \frac{L^2 N_0}{2\pi} q \frac{\pi c}{h_0} \frac{1}{(1+P_2)^{1/2}}, \tag{17}$$

where as in the integration above $q$ is the number of free spectral ranges over which the integration is taken. Thus the peak SNR for the passive cavity taking the ratio of Eq. (16) to Eq. (17) is given by

$$\text{SNR}_p = \frac{4(q/2\pi) [t_1 t_2 / 2(r_1 r_2)^{1/2}]^2 (c/h_0)(1+P_2)^{1/2} K^2(m)}{\pi N_0}. \tag{18}$$

In a similar fashion, expressing $E \otimes E^*$, and $R_{hn}(0)$ for the matched filter to Eq. (10) and using the same assumptions as above, the peak SNR for the matched filter is

$$\text{SNR}_\varrho = \frac{4(q/2\pi)\,[t_1 t_2/2(r_1 r_2)^{1/2}]^2\,(c/h_0)\,(\tan^{-1}\sqrt{P_1})^2\,(1+P_1)^{1/2}}{\pi N_0}. \tag{19}$$

Thus the departure of the passive cavity from the matched filter is, from Eqs. (18) and (19)

$$\frac{\text{SNR}_P}{\text{SNR}_I} = \frac{(1+P_2)^{1/2}}{(1+P_1)^{1/2}} \frac{K^2(m)}{[\tan^{-1}(\sqrt{P_1})]^2}. \tag{20}$$

Since $P_1 \gg 1$, $\Delta\omega_p = 2c/\left(h_0\sqrt{P_2}\right)$, and $\Delta\omega_l = 2c/\left(h_0\sqrt{P_1}\right)$, then

$$\frac{\text{SNR}_P}{\text{SNR}_I} = \left(\frac{2}{\pi}\right)^2 \left(\frac{\Delta\omega_l}{\Delta\omega_p}\right) K^2(m). \tag{21}$$

Hence the SNR of the passive cavity approaches that of the matched filter when $\Delta\omega_l \cong \Delta\omega_p$, as may be seen in the plot of Eq. (21) in Figure 2.

The Fabry-Perot cavity is probably as close a physical realization to a matched filter for the multitoned laser as can be attained in a passive system. Even so, gain narrowing invariably results in $\Delta\omega_l < \Delta\omega_p$, thereby limiting the observed improvement in SNR from its optimal value [Yariv, 1967]. For high-gain lasers with cavities of low finesse, the receiver can be made closer to ideal, while greater departures are to be expected in the case of low gain. We note, too, that larger bandwidths, $\Delta\omega_p$, are called for with information-modulated lasers and cavity-dumped lasers where mode locking may not have been employed.

To obtain the predetection SNR improvement with use of the passive cavity we note that the SNR without the cavity is given by the ratio of the modulus-squared value of Eq. (7) to Eq. (17) with $P_2 = 0$. Thus the SNR improvement with use of the passive cavity with respect to no cavity is given by Eqs. (1), (7), (12), (14), and (17) as

$$\text{SNR}_{\text{with}}/\text{SNR}_{\text{without}} = (1+P_2)^{1/2}. \tag{22}$$

Since the signal portion of the time-varying output of the passive cavity for a mode-locked laser input is given by the real part of $T_1 \otimes E_1$, then

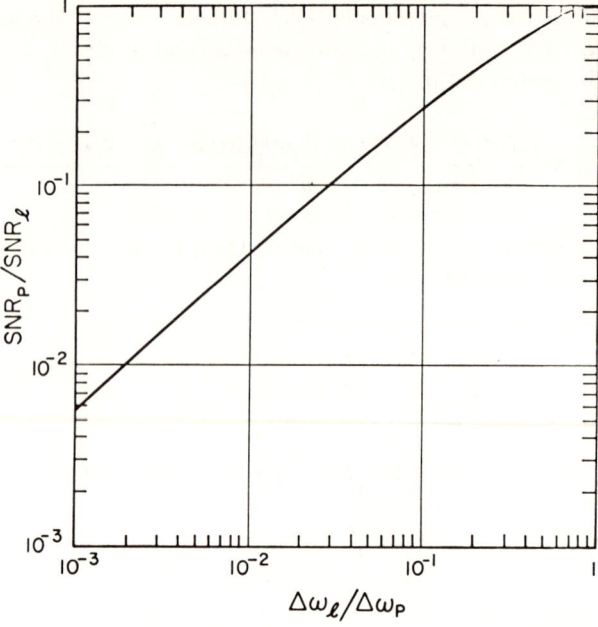

FIGURE 2 Departure from matched filter versus relative line widths.

from Eq. (1) and (8), dropping $\psi(\omega)$, we have

$$G_1(t) = \sum_{p=-N}^{N} \frac{\cos(a-b)}{(1+P_2 \sin^2 b)^{1/2}}, \qquad (23)$$

where $a = [\omega_0 + (\pi c/h)]t$ and $b = (h_0/c)[\omega_0 + (p\pi c/h)]$, then we see the peak SNR occurring for $h = h_0$. The SNR as a function of time and a function of relative cavity length $h_0/h$ is illustrated in Figure 3. The number of modes oscillating is nine with a peak SNR of 49 and a period of 8 nsec. The increment of relative cavity length in $\Delta h_0/h = 0.0001$.

## IV. Rise Time of the Passive Cavity

In consideration of the rise time, $\tau$, of the Fabry-Perot resonant cavity we start with the Laplace transform representation of the amplitude transmission function given by setting $s = i\omega$ in Eq. (1); we obtain

$$T(s) = e^{-as} T'(s) = \mathcal{T}_\alpha e^{-as}/(1 - \mathcal{R}_\alpha e^{-2as}), \qquad (24)$$

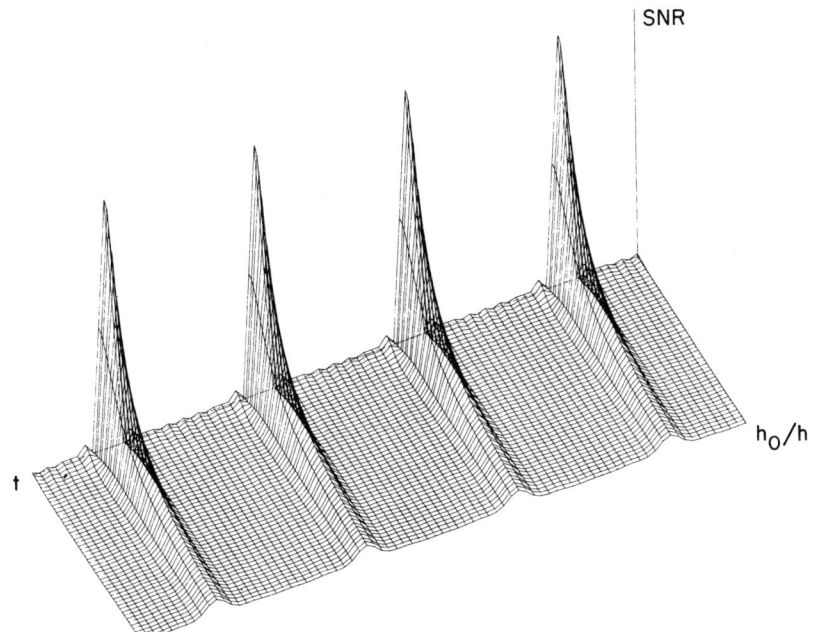

FIGURE 3 Signal-to-noise ratio as a function of time and relative cavity length.

where $\mathcal{T}_\alpha = t_1 t_2 \exp(-\alpha h_0)$, $\mathcal{R}_\alpha = r_1 r_2 \exp(-2\alpha h_0)$, $a = h_0/c$. If we expand $T'(s)$ in partial fractions and group the complex conjugate poles, we obtain

$$T'(S) = \frac{\mathcal{T}_\alpha}{a} \sum_{m=0}^{\infty} \frac{s - \sigma}{(s - \sigma)^2 + \omega_m^2}, \qquad (25)$$

where $\alpha = (1/2a) \ln \mathcal{R}_\alpha$ and $\omega_m = m\pi c/h_0$. Now the Laplace transform of the real part of Eq. (7) becomes

$$E(s) = \mathcal{L}[\text{Real } E_1(t)] = \sum_{p=-N}^{N} \frac{s}{s^2 + \omega_p^2}. \qquad (26)$$

So the signal output of the passive cavity for the idealized mode-locked laser input is given by

$$G(s) = e^{-as} G'(s) = e^{-as} E(s) T'(s), \qquad (27)$$

where we note that $e^{-as}$ for a nonnegative real constant corresponds to a time shift in the time domain. The inverse Laplace transform of a general term of $G'(s)$ is given by

$$\mathcal{L}^{-1}\left[\frac{s}{s^2+\omega_p^2}\cdot\frac{(s-\sigma)}{(s-\sigma)^2+\omega_m^2}\right]$$

$$=\left[\frac{1+\sigma^2}{(\sigma^2+\omega_m^2-\omega_p^2)^2+4\sigma^2\omega_p^2}\right]^{1/2}\left[\omega_p\sin(\omega_p t+\psi_1)+\omega_m\sin(\omega_m t+\psi_2)\right]. \tag{28}$$

Thus for $\omega_p = \omega_m \gg \sigma, \sigma \gg 1$ the fraction of the maximum steady-state output that is obtained in $\tau$ sec is given by Eq. (27) and (28) as

$$f = 1 - e^{\sigma(\tau-a)}u(\tau-a) = \{1 - \mathcal{R}_\alpha^{[(\tau-a)/2a]}u(\tau-a), \tag{29}$$

where $u(x) = \begin{cases} 1, & x > 0 \\ 0, & x < 0 \end{cases}.$

The $\omega_p \neq \omega_m$ terms are neglected in that the coefficients of the sine terms of Eq. (28) are on the order of $\ln \mathcal{R}_\alpha$, which are negligible for $\mathcal{R}_\alpha \approx 1$. Equation (29) is plotted in Figure 4 as a function of time and number of pulse-train bounces parameterized by $\mathcal{R}_\alpha = r_1 r_2 e^{-2\alpha h}$.

Thus the rise time of the passive cavity is given by

$$\tau = -\frac{1}{\sigma} = -2a\frac{1}{\ln \mathcal{R}_\alpha} \cong 2\frac{h_0}{c}\frac{\mathcal{R}_\alpha}{1-\mathcal{R}_\alpha}, \tag{30}$$

and from Eq. (2)

$$\tau = \frac{h_0}{c}\sqrt{P}\sqrt{\mathcal{R}_\alpha} = \frac{2\sqrt{\mathcal{R}_\alpha}}{\Delta\omega_p} = \frac{\sqrt{\mathcal{R}_\alpha}}{\pi\Delta\nu_p} \tag{31}$$

and thus the rise time–bandwidth product is

$$\tau\Delta\nu_p = \sqrt{\mathcal{R}_\alpha}/\pi, \tag{32}$$

where for $\mathcal{R}_\alpha = 0.991$, $\tau\Delta\nu_p = 0.3175$.

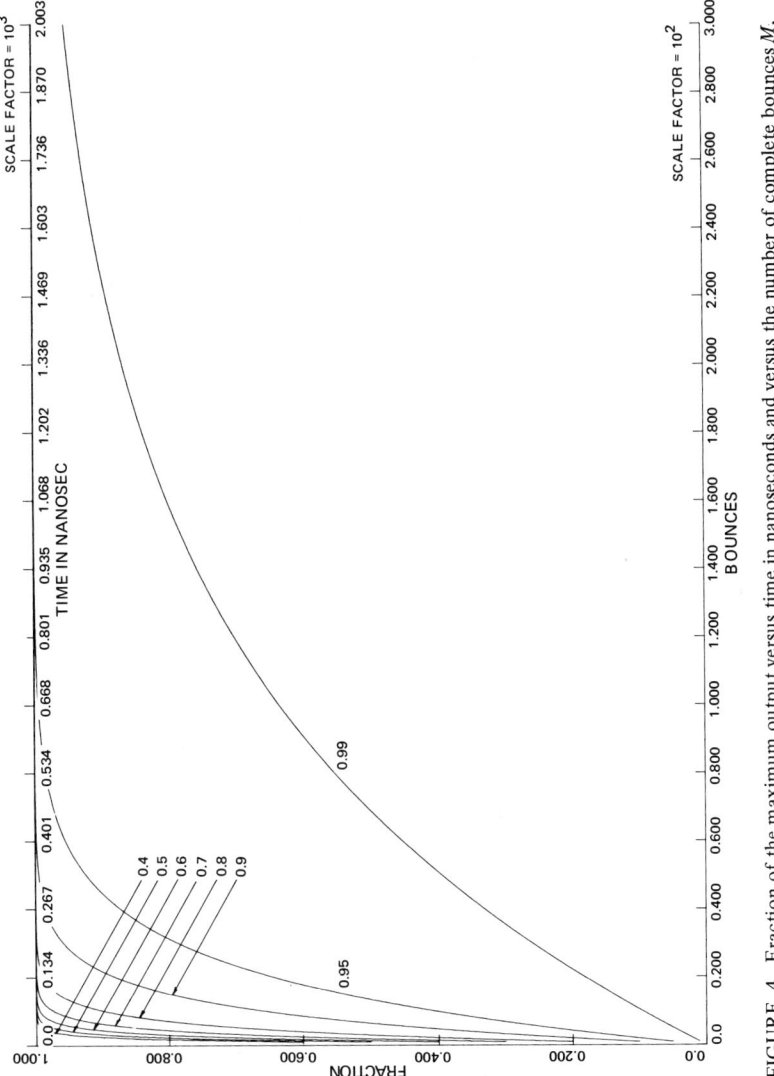

FIGURE 4 Fraction of the maximum output versus time in nanoseconds and versus the number of complete bounces $M$, parameterized for various $R_\alpha = r_1 r_2 e^{-2\alpha h}$.

## V. Doppler Measurements with the Passive Cavity

Suppose the mode-locked laser is moving toward our receiver with a velocity $v$ as indicated in Figure 1. For TEM waves [Papas, 1965], an emitted frequency $\omega'$ will be observed up-shifted to $\omega$ given by $\omega = \gamma(1 + v/c)\omega'$ in which $\gamma = [1 - (v/c)^2]^{1/2}$. Assuming normal incidence, by Eq. (7) the input signal, i.e., the Doppler-shifted electric field, is readily expressed as follows:

$$E_1(t) = \frac{\sin\{(2N+1)\omega_c\gamma[1+(v/c)](t/2)\}}{\sin\{\omega_c\gamma[1+(v/c)]t/2\}} \exp\{i\omega_0\gamma[1+(v/c)]t\} \quad (33)$$

Thus, in the case where there is relative motion, optimal detection of the mode-locked laser signal requires a receiver with a cavity length, $h_0$, given by

$$h_0 = \frac{\pi c}{\omega_c\gamma[1+(v/c)]} = \frac{h}{\gamma[1+(v/c)]}. \quad (34)$$

Similarly, if the mode-locked laser and the passive cavity were on a common platform, then the echo from a vehicle moving toward this platform with velocity $v$ would be shifted to $\omega = (1 + 2v/c)\omega'$, where we have set $\gamma = 1$. So by vernier adjustments (PZT-driven mirror) of the passive cavity we can read a large range of approach velocities, with a resolution independent of the velocity $v$, i.e., $\delta h/h \approx \delta v/c$ for $v/c \ll 1$. Thus with $\delta h/h = 3.3 \times 10^{-8}$ we find a resolution of $\delta v \cong 10$ m/sec.

## VI. Experiment

In this experiment, the SNR improvement by predetection filtering of the optical input by a passive Fabry-Perot resonant cavity was measured as a function of several parameters: relative cavity lengths and passive cavity finesse (Figure 1). Since the passive cavity is a good approximation to the matched filter for mode-locked lasers only when the cavity lengths are matched, the detector output was monitored for various relative cavity lengths. The mode-locked laser signals were obtained from a self-mode-locking He–Ne laser operating at 0.633 μm of length 1.2 m with an average power output of 2 mW. The change in length of the laser $h$ was provided by the motion of mirror $M_3$ on Teflon runners, while the fine fractional wavelength control (1 μm full scale) of the passive cavity length $h_0$ was provided by a PZT-driven mirror $M_2$. The passive cavity was fabricated with the mirrors on Invar rods to reduce thermal variations while the entire experiment was performed in a controlled acoustical environment that provided isolation (~70 dB) from external turbulence and mechanical fluctuations. The passive cavity was scanned at a 1-Hz rate to eliminate the remaining fluctuations. Thus continuous monitoring of the cavity output, while the length of the laser cavity was changed, allowed us to match the cavity lengths exactly by peaking the output. Note that all the modes of the laser are transmitted simultaneously in the matched condition. The power output was observed to go as approximately $(2N + 1)^2$, i.e., as the square of the number of modes oscillating.

To measure the SNR improvement, white noise $N_0$ from an ac-driven tungsten lamp at 3200 K was introduced axially into the system. The power from the noise source, passed through a coarse bandpass filter, F, at the input of the passive cavity, was 1.5 W in a 100 Å band around 0.633 μm. The thermal noise of the TIXL55 avalanche diode detector in a 500-MHz range was far above the shot noise value $(2eI_0 \Delta f)^{1/2}$ for operating currents of 0.1 μA; so through this discussion we will be considering the signal-to-unwanted signal ratio. Mirror $M_1$ is a standard Spectra-Physics flat laser reflector with reflectivity 0.991. Mirror $M_2$ is a standard Spectra-Physics laser reflector with radius of curvature 2 m and reflectivity 0.991. Both reflectivities were chosen to maximize the ideal finesse $F \equiv \pi\sqrt{P}/2$ by minimizing loss. In all cases the measured finesse (245 max.) was lower than the ideal (346) due to mirror-surface roughness and scattering from occlusions in the mirror multilayer. This factor as well as the finesse were measured by means of a Spectra-Physics model 119 single-mode laser, which provided the delta in frequency required to study the spectral response of the cavity.

With the cavity set for the largest Fresnel number (50), i.e., end apertures were limited by the mirror dimensions, the signal and noise were measured with $M_1$ and $M_2$ aligned to maximize the passive cavity finesse. The finesse was also measured and found to be 245. The mirrors of the passive cavity were then removed, and the signal and noise were measured again. The ratio of SNR$_{with}$ to SNR$_{without}$ was 156, and from Eq. (22) we see that the theoretical improvement using the experimentally determined finesse is $2(245)/\pi = 156$. The experiment was repeated for various values of finesse, and the results are summarized in Figure 5 along with a plot of Eq. (22). All measurements were made using density filters calibrated at 0.633 μm to avoid nonlinearities in the detector electronics.

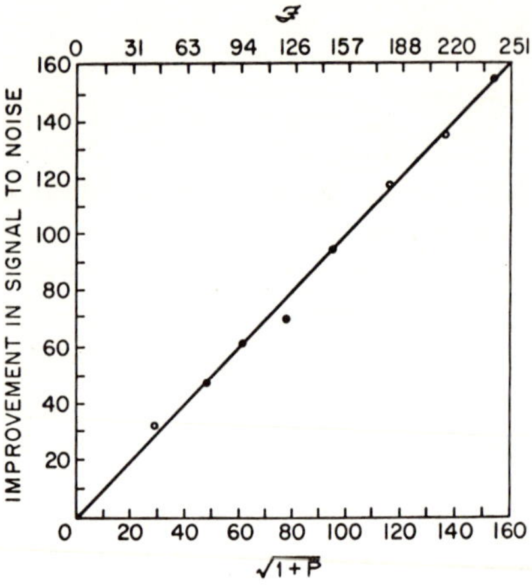

FIGURE 5 Signal-to-noise improvement with passive cavity.

## VII. Summary

In this work we have shown that the passive Fabry-Perot resonant cavity, which is equal in length to the laser cavity, is probably as close a physical realization to a matched filter for multitoned mode-locked gas laser signals as can be attained in a passive system. For the passive cavity in contrast to no cavity the SNR improves by the factor $(P_2 + 1)^{1/2}$, which is typically 100 to 200 for a cavity of good finesse. Also, the peak value of the temporally varying SNR improves as $(2N + 1)^2$, i.e., as the peak power of the mode-locked laser.

An alternative approach to the mode-locked laser line shape is presented in Eq. (10), along with the departure of the passive cavity from the matched filter for $\Delta\omega_l < \Delta\omega_p$ and $h \neq h_0$. We obtain an expression for the rise time that is also a function of $\Delta\omega_p$ and for cavities of high finesse independent of the number of modes detected.

Further improvement of the SNR above that obtained with the passive cavity may be accomplished with subsequent processing of the detector output, such as boxcar integration of time-sampled displays. In applications of the receiver to information-modulated multitone lasers, the effective bandwidth of the passive Fabry-Perot can be controlled by appropriate choices of $r_1$ and $r_2$ in Eq. (2).

R. J. D'Orazio would like to thank the Bell Telephone Laboratories for their doctoral support fellowship during the completion of this study.
The research reported in this paper was supported in part by the Air Force Office of Scientific Research.

## References

Blaquiere, A., *Ann. Radio Electron. 8*, 36 (1953).
Born, M., and E. Wolf, *Principles of Optics*, 4th ed., pp. 62, 327, Pergamon, New York (1970).
Brown, W. M., *Analysis of Linear Time-Invariant Systems*, p. 245, McGraw-Hill, New York (1963).
Fork, R. L., D. R. Herriott, and H. Kogelnik, *Appl. Opt. 3*, 1471 (1964).
Freed, C., and H. A. Haus, *Appl. Phys. Lett. 6*, 85 (1965).
Gordon, J. P., H. J. Ziegler, and C. H. Townes, *Phys. Rev. 99*, 1264 (1955).
Grivet, P., and A. Blaquiere, *Symposium on Optical Masers*, p. 69, Polytechnic Institute of Brooklyn, Brooklyn, N.Y. (1963).
Hargrove, L. E., R. L. Fork, and M. A. Pollack, *Appl. Phys. Lett. 5*, 4 (1964).
Kogelnik, H., and A. Yariv, *Proc. IEEE 52*, 165 (1964).
Papas, C. H., *Theory of Electromagnetic Wave Propagation*, p. 225, McGraw-Hill, New York (1965).
Rudd, M. J., *J. Phys. E, Sci. Instrum. 2*, 55 (1969).
Steier, W. H., *Proc. IEEE 54*, 1604 (1966).
Sonnenschein, C. M., and F. A. Horrigan, *Appl. Opt. 10*, 1600 (1971).
Uhlhorn, R. W., and D. F. Holshouser, *IEEE J. Quantum Electron. QE-6*, 775 (1970).
Yariv, A., *Quantum Electronics*, p. 409, Wiley, New York (1967).

W. N. PETERS

# SCINTILLATION AND POLARIZATION COMPENSATION TECHNIQUES FOR OPTICAL COMMUNICATIONS

## I. Introduction

The optical modulation technique of PCM/PL (pulse code modulation with polarized light) encodes binary information on an optical carrier by switching the state of polarization of the laser output between two orthogonal polarization states, typically with an electrooptic modulator. If these polarization states are portrayed as Stokes vectors [Peters and Arguello, 1967], the binary signal states are represented by *any* two antipodal points on the Poincaré sphere. Two commonly used techniques utilize the two orthogonal plane polarizations or the two circular polarizations of opposite sense to represent the binary signal states. This modulation technique is operationally unique to the visible portion of the electromagnetic spectrum since it is easily implemented and has been theoretically shown to be one of the most efficient methods for impressing information on the output of continuous lasers [Peters, 1965]. This paper discusses the interaction of the PCM/PL link with the

---

The author was at The Perkin-Elmer Corporation, Norwalk, Connecticut 06856, when this work was done. He is presently at General Research Corporation, Arlington, Virginia 22209.

transmission medium. However, many of the concepts developed in this paper can be applied to most of the coherent and incoherent optical modulation techniques presently being used. Topics discussed will include the effect of polarization errors and scintillation upon the performance of the link. The paper presents techniques for the compensation of induced polarization errors through the use of a polarization compensator (patent applied for) and techniques utilizing diversity to reduce the effects of scintillation. Three theorems are stated and rigorously proved that demonstrate that for the major class of polarization errors anticipated in the optical systems it is possible to detect the polarization errors (even if they are time-dependent) and introduce the polarization compensation required to regain optimum performance of the link. This compensation may be performed either with rotatable phase plates or electrooptic modulators. Assuming that the scintillation statistics are log normal and distributed with log amplitude variance $\sigma^2$, it is shown that the use of spatial-coherence diversity raises the system performance to a level nearly equivalent to that of a similar system of same radiated power operated on by a fading medium that has a log amplitude variance $\sigma^2/2$. The coherence requirements for the diversity are satisfied by the use of independent free-running lasers or through the use of a "decoherencer" network.

## II. Polarization "Noise"

The insertion of "noise" into a communication link usually is not only detrimental to the performance of the system under consideration but also changes the signal so that the original noise-free signal cannot be recovered. Common sources of noise are amplitude and phase perturbation of the carrier waveform. One dimension of an optical communication signal is polarization. Often the *a priori* knowledge of the signal polarization is useful for suppressing background. However, the interaction of the signal with a depolarizing medium is especially critical in a link using PCM/PL. Decoding of a PCM/PL signal is performed by a polarization discriminator (e.g., a Wollaston prism), which spatially separates the received optical signal so that one photodetector detects all the signal energy representing one binary state (say, the 0) and the complementary photodetector detects the other (1) binary state (Figure 1). As a polarization error is introduced into an initially optimum system, this division of energy between the photodetectors is not mutually exclusive, but rather the energy in a given signal state is divided between the two photodetectors. The exact quantitative measure of this division is determined by the magnitude of the polarization error. This cross talk between

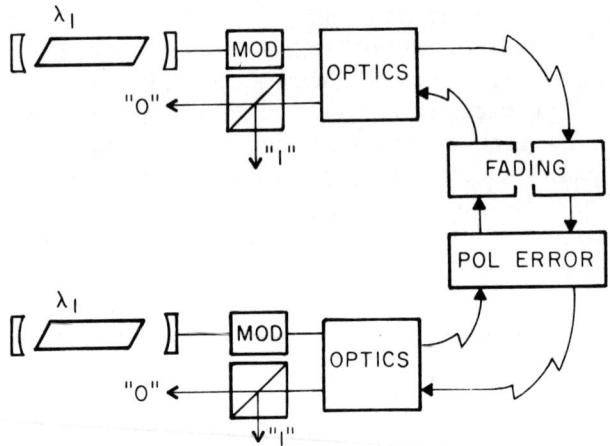

FIGURE 1 Optical schematic of PCM/PL optical communication system.

photodetectors inevitably leads to a reduced efficiency (higher bit error rate) for the communication channel. Unfortunately, many of the more complex optical communication links have a number of optical elements that introduce polarization errors into the communication channel. Phase errors of up to 15 degrees can be introduced by light being reflected from coated metal mirrors at nonnormal incidence. This error is especially bothersome if the orientation of these mirrors is continuously changing throughout the mission, as in the case of a ground telescope utilizing a coudé focus to track a low-orbit satellite. Improperly biased modulators and modulator drivers can also introduce polarization errors [Chen, 1970]. However, the depolarization expected from the atmosphere is extremely small [Saleh, 1967].

At first, it may appear that PCM/PL is an unacceptable modulation technique for such a mission because of these polarization errors. However, we should raise the following question: Is it possible to detect the magnitude of the polarization noise and somehow correct for the error to return the signal back to its optimum state? For most cases of induced phase polarization errors, we will show that it is possible to compensate *completely* for the phase polarization errors with the result that there is *no* degradation of the system performance.

## III. The Concept of Orthogonality

One of the basic precepts of information theory is that the information-carrying capacity of any channel is maximized when the allowed signal states are orthogonal. In Section I, we demonstrated that the PCM/PL system complies with this precept, since the two polarization states incident on the polarization discriminator are orthogonal linear polarizations.

One must not be overly restrictive in his consideration of the orthogonality of the PCM/PL signal states. For example, if we choose a given elliptical polarization of orientation and ellipticity such as ↻ , there is always another elliptical polarization ↺ , that is orthogonal. (It is interesting to note that these two orthogonal polarization states are always on a diameter, i.e., antipodal, in the Poincaré sphere representation.) Unfortunately, the polarization discriminator *by itself* cannot uniquely partition the energy of these two orthogonal (but elliptical) polarization states between the two photodetectors. Our purpose here is to present techniques that may be employed to utilize the inherent orthogonality of the optical signal states that usually (but not always) exist in order to optimize the performance of the PCM/PL link.

Reiterating, it is the orthogonality of the two polarization states representing the binary data states, rather than the polarization states *per se*, that is important in evaluating the performance of a link utilizing the modulation technique of PCM/PL.


## IV. Polarization Compensation Theorems

The preceding sections provided a basis on which we may present three important theorems. Proofs of these theorems are found in Appendix A.

1. The introduction of a spatially uniform arbitrary polarization phase error into the laser beam (either at the transmitter or receiver) does not alter the orthogonality of the PCM/PL optical signal states.

2. In general, at least two operations with birefringent elements are required to alter the polarization states of a set of two orthogonal signals of arbitrary polarization to a given pair of orthogonal polarization states as required by the system polarization discriminator.

3. The introduction of a selective polarization absorption will usually force the signals to be nonorthogonal. Thus, these types of polarization errors are only partially compensatable.

Since the expected polarization errors in most optical transceivers are predominantly of phase rather than amplitude, these theorems state that the PCM/PL link will suffer *no* degradation despite the source or magnitude of the phase polarization error if a two-element polarization compensator is in the optical path at either transceiver optical system (Figure 1). Only one compensator (consisting of two elements) is required to correct the composite system polarization error resulting from both the ground station and satellite (aircraft). Also, this single compensator automatically corrects for both the up and down links simultaneously if all the polarization errors occur in the common path. Exceptions to this are when the biases of the modulators are in error or when a system using two lasers of different wavelengths has birefringent errors that are wavelength-dependent. However, this type of error can be compensated at the cost of added complexity by introducing a similar bias "error" at another location in the system.

## V. Implementation of Polarization Compensation

There are two methods available for determining the magnitude of the polarization error. An operational system may use either technique or may even have the capability of using both techniques. The analytic technique utilizes diagnostic data from every source of polarization error in both transceivers. A computer program then calculates the composite polarization error. The alternate technique monitors the output of the difference amplifier following the two PCM/PL photodetectors. Since the correct amount of polarization compensation maximizes the signal from the difference amplifier and vice versa, it is possible to optimize the overall system performance merely by monitoring the pristine electronic PCM waveform.

Two of the many possible polarization compensators will be discussed in detail. Rather than presenting the compensation in strictly mathematical terms, the operation of the compensator will be described in terms of operations on the Poincaré sphere. As described by Shurcliff [1962] and O'Neill [1963], the Poincaré sphere is an abstract mapping of all possible polarization states onto the surface of a sphere. As demonstrated in Figure 2, all plane polarization states occur on the sphere's equator and the circular polarizations at the poles, the orientation and ellipticity of the remaining elliptical polarization states being determined by the longitude and latitude, respectively. Choosing an arbitrary elliptical polarization state of azimuth $\rho$ and retardance $\delta$ to represent the perturbed signal, note that a quarter-wave plate whose fast axis is at angle

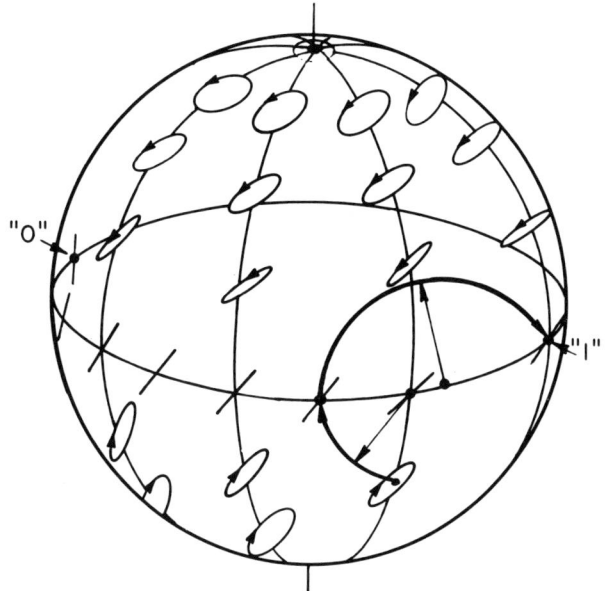

FIGURE 2 Poincaré sphere mapping of arbitrary polarization state to required state using a quarter-wave and a half-wave birefringent plate.

$\rho$ will generate a linear polarization as shown in Figure 2. This linear polarization can then be rotated to the required polarization state (1 in Figure 2) by orienting a half-wave plate with its fast axis at angle $(\rho + \delta)/2$. As discussed before, the 0 polarization state is also mapped to the required vertical linear polarization state as required.

An alternate compensator can be constructed from two electrooptic modulators. The first modulator has its fast axes parallel to one of the binary polarization states (say 1); the second modulator has its fast axis at ±45° to the same binary polarization state. Operating on the same arbitrary polarization state as discussed earlier, Figure 3 demonstrates that the first modulator maps the polarization state onto a great circle through the 0 and 1 polarization states, and the remaining modulator converts this elliptical polarization to the required plane polarization state.

## VI. Scintillation-Reduction Techniques

A commonly used statistical model for the fading statistics of an optical signal operated on by atmospheric turbulence assumes log normal statis-

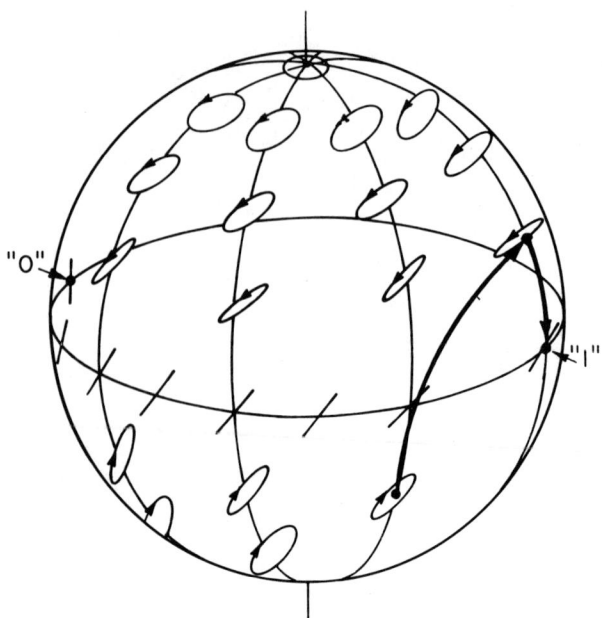

FIGURE 3  Poincaré sphere mapping of arbitrary polarization state to required state using two electrooptic modulators at orientations of 0° and 45°, respectively.

tics [Fried, 1966]. For ground-based receivers, the variance (i.e., depth of fade) of the distribution may be reduced by using larger apertures. This effect is referred to as aperture averaging and has been discussed by Fried [1967]. Log normal statistics of intensity fluctuations for a fading optical communications channel were used to determine the relative increase in the signal power required for the system to operate at an *a priori* error rate. The bit error-rate (BER) probability for the quantum-limited case is BER $(S) = \exp(-S)/2$. This function is integrated over the log normal probability density function to yield the average bit error rate of the scintillating signal.

$$\langle \text{BER}(S,\sigma^2) \rangle = \int_0^\infty dS \frac{\text{BER}(S)}{\sqrt{2\pi}\,\sigma_x 2S} \exp - \left[\ln(S/S_0)/2 + \sigma_x^2\right]^2 / 2\sigma_x^2.$$

This function was solved for several values of the log amplitude variance $(\sigma_x^2)$, and the results are plotted in Figure 4. The variance equal to zero corresponds to the no-fading case. For the no-fading case, 8.5 signal

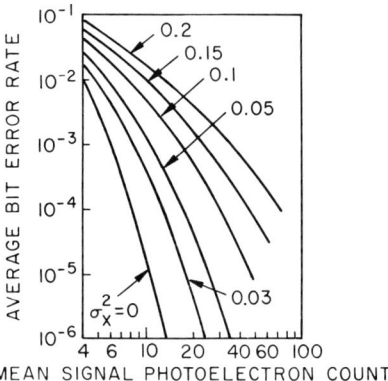

FIGURE 4 Mean signal count for a given bit error rate and fading of variance $\sigma_{\chi}^2$. The link is quantum noise limited and has no diversity.

counts are required for BER = 0.0001. Assuming a log amplitude variance of 0.03, in order to maintain ⟨BER⟩ = 0.0001 the signal must be increased to 13 counts per decision, which is equivalent to a 1.8-dB loss in link margin.

The performance of an optical link terminated at a receiver located in the atmosphere can be improved by increasing the size of the receiver aperture. Estimates of the reduction in the log amplitude fading as a function of the receiver diameter can be obtained from Fried [1967]. Unfortunately, the required aperture size of a receiver orbiting the earth and viewing a ground-based laser becomes very large (typically tens of meters for a synchronous satellite) if significant averaging of scintillation is required. Also, the use of large transmitting apertures on the ground does little to reduce the scintillation of the signal as viewed in space.

This paper proposes a diversity system containing two or more transmit apertures. Two requirements must be satisfied in such a diversity system. First, the fading signal statistics from the two transmit apertures as detected at the receiver must not be highly correlated—and preferably must be uncorrelated. This is accomplished by physically separating the transmit apertures by at least typically 8 to 12 in. The second requirement is that the signals should add together at the receiver in an incoherent manner. This requirement is automatically satisfied if each transmit aperture is illuminated by its own free-running laser. However, let us now consider the possibility of generating the mutually incoherent sources from a single laser. A method to satisfy this demand is obvious when we consider the self-coherence function of a multimode laser that has deep nulls, or zeros. In the case of a multimode laser, the coherence function is repetitive at a period related to the round-trip propagation time of the wavefront within the cavity. The width of the individual

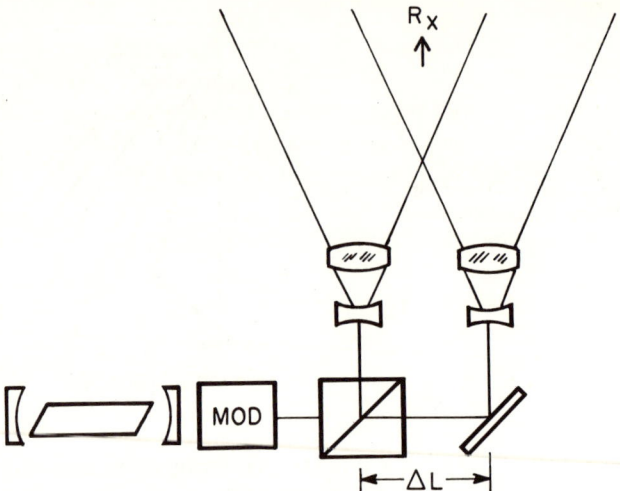

FIGURE 5 Optical schematic of transmitter with diversity of order 2 and laser "decoherencer" network.

coherence "spike" linearly decreases as the number of longitudinal modes is increased. A good example of a laser with this property is the Nd:YAG laser. The cw-operated laser has a tendency to operate with a line width of typically $10^{-5}$ $\mu$m. For the specific case being described, the emission line width increases to typically $10^{-4}$ $\mu$m because the mode-locking function tends to force cavity resonances of marginal gain to oscillate. This in turn forces the width of the coherence spike to be typically less than 1 nsec.

The purpose of this discussion becomes obvious when we consider a laser "decoherencer" that permits a ground station to transmit $n$ mutually incoherent beams through the transceiver aperture with all the beams originating from a single laser. Such a system is automatically implemented when a time delay $\tau = \Delta L/C$ sufficient to suitably reduce the mutual coherence is inserted in each beam relative to the remaining beams (Figure 5).

## VII. Analysis of Diversity System

The performance of a quantum-noise-limited PCM/PL link can be easily calculated for the case of statistically independent fading and incoherent summation of the two beams at the receiver by noting that an error in this limiting case occurs only when zero events are detected by the receiver from both received beams. Thus, using the notation of the previous equation, the average bit error rate for a link with diversity of order 2 under the constraint of constant total radiated power is

$$\langle \text{BER}(S, \sigma_x^2, 2) \rangle = [\langle \text{BER}(S/2, \sigma_x^2) \rangle]^2.$$

These results are plotted in Figure 6.

By comparing Figures 4 and 6 and extrapolating the results to order $N$, a useful approximation can be deduced for estimating the performance of a system with a diversity of order $N$:

$$\sigma_x^2 \Big|_{\text{Diversity of order } N} \cong \sigma_x^2 / N.$$

Thus, an estimate of the quantum-limited performance of a PCM/PL system utilizing a diversity of order $N$ can be deduced by reducing the data of Figure 4. For example, an atmospheric condition that generates a scintillation of say $\sigma_x^2 = 0.2$ for a single channel would require nearly 80 counts per bit on the average to support a link at a bit error rate of 0.0001. A transmitter with a diversity of order 4 would permit the link to appear to be operating through an atmosphere with $\sigma_x^2 = 0.05$. Thus, for the same 0.0001 bit error rate the average signal count would be reduced to less than 20. If radiated power is at a premium, the added complexity of multiple apertures to yield diversity may well be a cost-effective alternative.

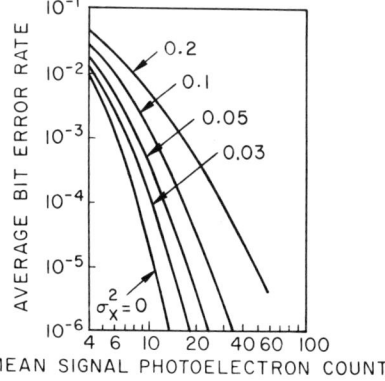

FIGURE 6 Mean signal count for link with a diversity of order 2.

## VIII. Application of Concepts to Other Modulation Techniques

Both the concept of polarization compensation and transmitter diversity are readily applied to all the other commonly used optical modulation techniques. The polarization compensator may be used in any system that is polarization sensitive. Two major applications are systems that use polarization to discriminate against background (e.g., PCM/On–Off [Curren and Ross, 1965] and PPM [Peters, 1964; Karp and Gagliardi, 1968]) and systems that utilize coherent detection of the received optical energy.

Noting that the quantum-limited performances of both PCM/PL and PPM are identical [Peters and Arguello, 1967; Peters, 1964], the curves and accompanying theory of both Figures 4 and 6 can be readily applied to a PPM system. Since the performance of a PCM/On–Off link is critically affected by the threshold implementation, the performance gain realized by such a system with the use of diversity requires further study.

The author credits the significant inputs of C. McIntyre for the theorem proofs and S. Gowrinathan and R. E. Hufnagel, who made helpful comments in the study of the atmospheric modeling.

## Appendix A

PROOF–THEOREM I

*The introduction of a spatially uniform arbitrary polarization phase error into the laser beam (either at the transmitter or receiver) does not alter the orthogonality of the PCM/PL optical signals states.*

First, define two orthogonal polarization signals $S_0$ and $S_1$ in terms of Jones matrices [Shurcliff, 1962]:

$$S_0 = \begin{bmatrix} A_x e^{i\epsilon_x} \\ 0 \end{bmatrix}, \quad S_1 = \begin{bmatrix} 0 \\ A_y e^{i\epsilon_y} \end{bmatrix}, \quad (A1)$$

where

$\epsilon_x$ = phase of E-vector in $x$-direction,
$A_x$ = amplitude of E-vector in $x$-direction.

A sufficient condition for orthogonality is that the Hermitian product is zero.

$$S_0^\dagger S_1 = \begin{bmatrix} A_x e^{-i\epsilon_x} & 0 \end{bmatrix} \begin{bmatrix} 0 \\ A_y e^{i\epsilon_y} \end{bmatrix} = 0, \quad (A2)$$

where

$S_0^\dagger$ = Hermitian conjugate of $S_0$.

Each of these signals will now be operated upon by a matrix $P$, which represents the lumped optical system birefringence of arbitrary orientation and phase,

$$P = \begin{bmatrix} \bar{C}_1^{\,2} e^{i\delta/2} + \bar{S}_1^{\,2} e^{-i\delta/2} & \bar{C}_1 \bar{S}_1 \, 2i \sin \delta/2 \\ \bar{C}_1 \bar{S}_1 \, 2i \sin \delta/2 & \bar{C}_1^{\,2} e^{-i\delta/2} + \bar{S}_1^{\,2} e^{i\delta/2} \end{bmatrix}, \quad (A3)$$

where

$\bar{C}_1 = \cos \rho,$
$\bar{S}_1 = \sin \rho,$
$\delta$ = phase delay between fast/slow axes of lumped system birefringence,
$\rho$ = azimuth of birefringent element representing lumped system birefringence.

Theorem I is proved if we can prove that the vectors $PS_0$ and $PS_1$ are orthogonal. Straightforward matrix multiplication of the following equation yields a result that is identical to zero for all $\delta, \rho$:

$$(PS_0)^\dagger (PS_1) = 0. \quad (A4)$$

PROOF–THEOREM II

*In general, at least two operations with birefringent elements are required to alter the polarization states of a set of two orthogonal signals of arbitrary polarization to a given pair of orthogonal polarization states as required by the system polarization discriminator.*

Following the procedure outlined in the main text accompanying Figures 2 and 3, we will use a geometric "proof" utilizing the Poincaré sphere for this theorem. As discussed by Shurcliff the locus of polarization states obtainable from an arbitrary polarization state by a single birefringent element can be represented by a line on the Poincaré sphere. Likewise, a line of the sphere through the desired linear polarization state can be generated by the *second* birefringent element. If these two lines intersect for the set of all allowable arbitrary polarization states, an acceptable polarization compensator has been devised. A good example where this criterion is not met is with two half-wave plates. The locus of points for this case is such that the two lines do not intersect, except in a trivial case of polarization states where $\delta = 0$.

One may argue that the polarization compensator may consist of a single device such as a Soleil compensator. However, we still require two *operations* (translation and rotation) to introduce the required compensation.

PROOF—THEOREM III

*The introduction of a selective polarization absorption will, in general, force the signals to be nonorthogonal. These types of polarization errors are thus only partially compensatable.*

The matrix operator for a selective absorber is

$$D = \begin{bmatrix} e^{-\alpha_1} \cos^2 \theta + e^{-\alpha_2} \sin^2 \theta & (e^{-\alpha_1} - e^{-\alpha_2}) \sin \theta \cos \theta \\ (e^{-\alpha_1} - e^{-\alpha_2}) \sin \theta \cos \theta & e^{-\alpha_1} \sin^2 \theta + e^{-\alpha_2} \cos^2 \theta \end{bmatrix}, \quad (A5)$$

where

$\theta$ = angular orientation of absorber,
$\alpha_1, \alpha_2$ = absorption coefficients for $\theta$ and $\theta + \pi/2$ directions.

Following an identical (although algebraically tedious) procedure as used in the proof of Theorem I, we generate the Hermitian scalar product.

$$(DS_0)^\dagger (DS_1) = A_x A_y e^{i(\epsilon_x - \epsilon_y)} \sin \theta \cos \theta (e^{-2\alpha_1} - e^{-2\alpha_2}). \quad (A6)$$

In general this is unequal to zero; thus proving Theorem III. Note that they are orthogonal if $\theta = 0, \pi/2$ (i.e., absorber axes aligned to signal state axes) or $\alpha_1 = \alpha_2$ (or equivalently, the loss is not polarization-sensitive).

## References

Chen, F., *Proc. IEEE 58*, 1440 (1970).
Curran, T. F., and M. Ross, *Proc. IEEE 53*, 1770 (1965).
Fried, D. L., *J. Opt. Soc. Am. 56*, 1667 (1966).
Fried, D. L., *J. Opt. Soc. Am. 57*, 169 (1967).
Karp, S., and R. M. Gagliardi, M-ary Poisson Detection and Optical Communications, NASA Tech. Note, NASA TN D-4623 (June 1968).
O'Neill, E. L., *Introduction to Statistical Optics*, Addison-Wesley, Reading, Mass. (1963).
Peters, W. N., "PPM Optical Communication System," in *Proc. of NATCOMX*, pp. 94–105, New York (Oct. 1964).
Peters, W. N., *Proc. NEC 21*, 467 (1965).
Peters, W. N., and R. J. Arguello, *IEEE J. Quantum Electron. QE-3*, 332 (1967).
Saleh, A. A. M., *IEEE J. Quantum Electron. QE-3*, 540 (1967).
Shurcliff, W. A., *Polarized Light*, Harvard U.P., Cambridge, Mass. (1962).

A. CONSORTINI, P. PANDOLFINI, L. RONCHI,
and R. VANNI

# IMAGE DETERIORATION DUE TO ATMOSPHERIC TURBULENCE

## I. Introduction

The propagation of light in the turbulent atmosphere is mostly treated by assuming the turbulence to be described by the so-called two-thirds law, according to which the structure function of, say, the refractive index is given by

$$\mathcal{D}_n(P_1, P_2) = C_n^2 r^{2/3} \quad \text{for } 0 \leqslant r \leqslant \infty, \tag{1}$$

where $r = |P_1 - P_2|$ and $C_n^2$ is denoted the refractive-index structure constant.

According to Kolmogorov's theory, the use of Eq. (1) should be limited to values of $r$ much larger than the so-called inner scale of turbulence and much smaller than the outer scale of turbulence [Tatarski, 1967]. Its use for $0 \leqslant r \leqslant \infty$ has been discussed in a number of papers [Consortini and Ronchi, 1969; Lutomirski and Yura, 1971; Consortini et al., 1972]. In particular, it has been noted that it cannot be justified

---

The authors are at the Istituto di Ricerca sulle Onde Elettromagnetiche, of CNR, Firenze, Italy.

from a physical point of view, by invoking the processes with stationary increment, but only from a mathematical point of view [Consortini *et al.*, 1971]. As a matter of fact, it simplifies the mathematical treatment of many problems of propagation; however, it seems necessary to check, problem by problem, the degree of approximation of the results or their limits of validity.

In the present paper, we consider the problem of practical interest of the quality of an atmospherically degraded image. In one case, concerned with vertical upward propagation, occurring, for example, when the earth's surface is imaged by a perfect diffraction-limited objective lens placed on a space vehicle, the use of Eq. (1) turns out to yield sufficiently accurate results. In another case, concerned with horizontal propagation, the use of a model of turbulence that describes the "saturation" of $\mathcal{D}_n(P_1, P_2)$ when $r > L_0$ yields results that may differ both quantitatively and qualitatively from those obtained by the use of Eq. (1), depending on the strength of turbulence and on the length of the path from the object to the lens.

The problem is treated along the line used by Fried [1966a, 1966b] described in Sec. II. In Sec. III, we treat the vertical upward propagation, and in Sec. IV the horizontal propagation. The results are discussed in Sec. V.

## II. Statement of the Problem

Following Fried [1966a, 1966b], the quality of an atmospherically degraded image will be described by the quantity $\mathcal{R}$ defined as

$$\mathcal{R} = 2\pi \int \langle \tau(f) \rangle f df, \qquad (2)$$

where $f$ denotes a spatial frequency and $\langle \tau(f) \rangle$ the modulation transfer function of the optical system formed by the lens and the turbulent medium. For $\langle \tau(f) \rangle$ we can write [Fried, 1966a]

$$\langle \tau(f) \rangle = \tau_0(f) \exp\left[-\frac{1}{2} \mathcal{D}_w(\lambda F f)\right], \qquad (3)$$

where $\tau_0(f)$ denotes the modulation transfer function of the lens and is given by

$$\tau_0(f) = \frac{2}{\pi}\left[\arccos \frac{\lambda F f}{D} - \frac{\lambda F f}{D}\sqrt{1 - \left(\frac{\lambda F f}{D}\right)^2}\right] \quad \text{for } f < D/\lambda F,$$

$$\tau_0(f) = 0 \qquad \text{for } f > D/\lambda F. \qquad (4)$$

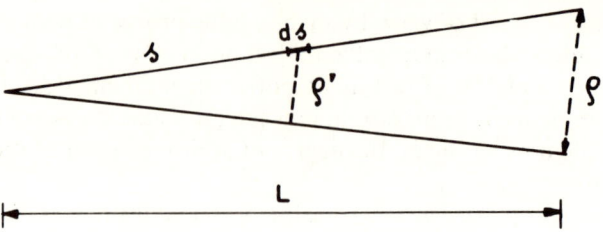

FIGURE 1 Geometry of the wave path.

Moreover, $\lambda$ denotes the wavelength, $F$ the focal length of the lens, and $\mathcal{D}_w(\rho)$ the so-called wave-structure function in the plane of the lens, corresponding to a point source in the plane of the object. In both cases under consideration $\mathcal{D}_w(\rho)$ is the wave-structure function of a spherical wave.

As is well known, the wave-structure function describes the phase and amplitude fluctuations at two points separated by $\rho$, due to the turbulence.

The wave-structure function must be evaluated at least in the Rytov approximation, since the involved path lengths are rather large. In the Rytov approximation the wave-structure function of a spherical wave is given by the same expression as the phase-structure function of a spherical wave in the geometrical optics approximation [Carlson and Ishimaru, 1969]. Accordingly,

$$\mathcal{D}_w(\rho) = \int_0^L \mathcal{D}_S(\rho')ds, \qquad (5)$$

where $\mathcal{D}_S(\rho')ds$ denotes the phase-structure function of a plane wave after a path of length $ds$ in the turbulence, and $\rho'$ is the distance, after a length $s$ from the source, of two straight rays that on the plane of the lens are spaced by $\rho$ (Figure 1). Accordingly,

$$\rho' = \rho s/L. \qquad (6)$$

For the model (1) of turbulence we have

$$\mathcal{D}_S(\rho')ds = 2.91 k^2 C_n^2(s) \rho'^{5/3} ds,$$

and therefore

$$\mathcal{D}_w(\rho) = 2.91 k^2 \rho^{5/3} \int_0^L C_n^2(s) \left(\frac{s}{L}\right)^{5/3} ds, \qquad (7)$$

which expression has been used by several authors.

For our analysis, we will use the von Kármán model of turbulence, according to which the refractive-index structure function in a homogeneous and isotropic turbulence is given by

$$\mathcal{D}_n(P_1, P_2) = 1.05 C_n^2 \kappa_0^{-2/3} [1 - 0.59(\kappa_0 r)^{1/3} K_{1/3}(\kappa_0 r)], \qquad (8)$$

where $\kappa_0$ is of the order of $L_0^{-1}$ and $K_\nu$ denotes a modified Bessel function of order $\nu$ [Abramowitz and Stegun, 1965]. The expression (8) of $\mathcal{D}_n$ takes into account the behavior of $\mathcal{D}_n$ for distances $r$ larger than $L_0$. For distances smaller than the inner scale $\lambda_0$, one has [Tatarski, 1967]

$$\mathcal{D}_n(P_1, P_2) = C_n^2 \lambda_0^{-4/3} r^2, \qquad (9)$$

where $C_n$ and $\lambda_0$ are constant in the case of a homogeneous and isotropic turbulence.

## III. Vertical Upward Propagation

Let us first examine the case of the vertical upward propagation. We assume, as usual, the turbulence to be stratified parallel to the earth's surface, so that $s = z$, $C_n(s) = C_n(z)$, $\kappa_0 = \kappa_0(z)$, $\lambda_0 = \lambda_0(z)$. By using Eq. (8), we have

$$\mathcal{D}_S(\rho') ds = 1.56 C_n^2(z) k^2 \kappa_0^{-5/3}(z) [1 - 0.994(\kappa_0 \rho')^{5/6} K_{5/6}(\kappa_0 \rho')] dz, \qquad (10)$$

and therefore, with Eqs. (5) and (6),

$$\mathcal{D}_w(\rho) = 1.56 k^2 \int_0^L C_n^2(z) \kappa_0^{-5/3} \left[1 - 0.994\left(\kappa_0 \rho \frac{z}{L}\right)^{5/6} K_{5/6}\left(\kappa_0 \rho \frac{z}{L}\right)\right] dz. \qquad (11)$$

Clearly, the evaluation of $\mathcal{D}_w(\rho)$ requires one to know $\kappa_0(z)$ and $C_n(z)$. If we assume [Tatarski, 1967]

$$L_0 = \kappa z, \qquad (12)$$

where $\kappa \sim 0.4$ is referred to as the von Kármán constant, Eq. (11) takes the form

$$\mathcal{D}_w(\rho) = 1.56k^2 \left[ 1 - 0.994 \left( \frac{\rho}{\kappa L} \right)^{5/6} K_{5/6}\left( \frac{\rho}{\kappa L} \right) \right] \kappa^{5/3} \int_0^L C_n^2(z) \, z^{5/3} \, dz. \quad (13)$$

In practice, since for the evaluation of $\mathcal{R}$ as given by Eq. (2) the range of $\rho$ of interest is $0 \leq \rho \leq D$, we have

$$\rho \ll \kappa L, \quad (14)$$

at least for path lengths $L$ larger than $\sim 10$ m. Accordingly, by disregarding the cases of scarce practical interest where (14) is not verified, the expression (13) of $\mathcal{D}_w(\rho)$ can be replaced by its asymptotic expression holding for $\rho \to 0$, which coincides with Eq. (7).

This analysis, which yields the conclusion that the model (1) is sufficient for the investigation of the images in looking downward through the atmosphere, does not include the case for which $\rho'$ remains much smaller than the inner scale of turbulence along that portion of the path where the turbulence is not negligible. This is the case of an optical system placed on a space vehicle to observe the earth's surface. It appears from the literature [Hufnagel and Stanley, 1964; Hufnagel, 1966; Minott, 1972] that the turbulence is negligible at altitudes larger than $\sim 100$ km. Thus only the portion of the integration path in Eq. (7) with $s = z < 100$ km contributes to $\mathcal{D}_w(\rho)$. In the case of space vehicles, one can assume $D/L \leq \sim 10^{-6}$, which holds even for the big telescope of Copernicus ($D = 80$ cm, $L \sim 800$ km) though it does not observe the earth's surface. If we compare $\rho'(D) = zD/L$ with $\lambda_0(z)$ (Figure 2), as given by Zimmerman [1966], we note that for $D/L < 10^{-6}$, $\rho'(D)$ is much smaller than $\lambda_0$ almost all over the path. Consequently, the expression of $\mathcal{D}_S(\rho')$ to be used in Eq. (5) is that holding for $\rho' < \lambda_0$, rather than Eq. (10), namely [Tatarski, 1967],

$$\mathcal{D}_S(\rho') dz \simeq 3.28 k^2 C_n^2(z) \lambda_0^{-1/3}(z) \rho'^2 dz. \quad (15)$$

Upon introduction of (15) into Eq. (5), we obtain

$$\mathcal{D}_w(\rho) = (k^2/L^2) \alpha^2 \rho^2, \quad (16)$$

where

$$\alpha^2 = 3.28 \int_0^L C_n^2(z) z^2 \lambda_0^{-1/3}(z) dz, \quad (17)$$

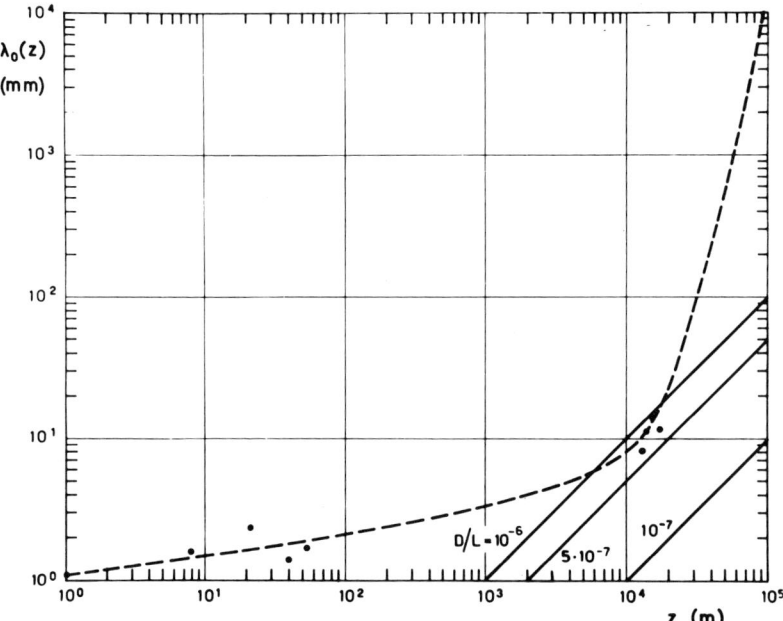

FIGURE 2 Approximate trend of $\lambda_0(z)$ versus $z$. The dots represent experimental values taken by Zimmerman [1966].

and therefore, by using Eqs. (3) and (2),

$$\mathcal{R} = \frac{\pi}{4}\left(\frac{D_0}{\lambda F}\right)^2 \left\{1 - \exp\left(-\frac{D^2}{D_0^2}\right)\left[I_0\left(2\frac{D^2}{D_0^2}\right) + I_1\left(2\frac{D^2}{D_0^2}\right)\right]\right\}. \quad (18)$$

Here $I_0$ and $I_1$ denote the modified Bessel functions of orders 0 and 1, respectively, and

$$D_0 = 1.56\,(L/k\alpha). \quad (19)$$

Note that for $D \to \infty$, $\mathcal{R} \to \mathcal{R}_{max}$, where

$$\mathcal{R}_{max} = (\pi/4)(D_0/\lambda F)^2.$$

Figure 3 shows $\mathcal{R}/\mathcal{R}_{max}$ plotted versus $D/D_0$ (solid line). It appears that $\mathcal{R}/\mathcal{R}_{max}$ has the same trend as the analogous quantity investigated by Fried [1966a, 1966b], and that $D_0$ plays the same role as the quan-

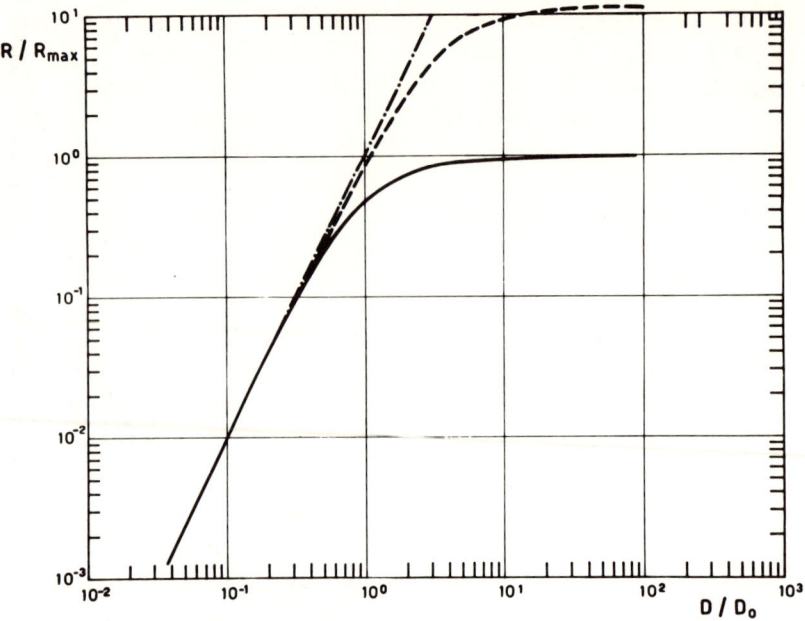

FIGURE 3  $R/R_{max}$ plotted versus $D/D_0$ (solid line). Dashed line represents Fried's results for $(D_0/R_0)^2 \sim 11$. The straight line represents the no-turbulence case.

tity $R_0$ defined by Fried, as

$$R_0 = 1.67L \left[ k^2 \int_0^L C_n^2(z) \, z^{5/3} \, dz \right]^{-3/5}. \tag{20}$$

For a comparison with the results obtainable by using Eq. (7), we have evaluated $D_0$ and $R_0$ by using for $C_n(z)$ the data represented by the dashed line in Figure 4, taken from Minott [1972], and for $\lambda_0(z)$ the data represented by the dashed line in Figure 2. The data for $\lambda_0(z)$ are rather arbitrary; however, they are expected to be sufficient for our purposes, since $D_0$ depends on $\lambda_0^{-1/6}$. By using such data, we obtain (in mks units)

$$D_0 = 123(L/k), \quad R_0 = 1.03 \times 10^4 Lk^{-6/5}. \tag{21}$$

In Figure 3, the dashed line reports the results that would be obtained by using Eq. (7), with $k = 10^7$ m$^{-1}$. The numerical values for drawing it are taken from Table I of Fried's paper [1966a]. While the solid line is independent of the particular set of values taken for $C_n^2(z)$ and $\lambda_0(z)$, the dashed line depends on it, through the ratio $D_0/R_0$. It is easily seen that the ratio between the saturation value of the dashed line and that of the solid line is $(R_0/D_0)^2$. In the particular case of the data of Figure 4, this ratio $(R_0/D_0)^2 \simeq 10$. However, the factor $\sim 10$ has no practical interest, since our results are applicable when $D/L < 10^{-6}$. By virtue of the first expression (21), we have, with the data of Figures 2 and 4,

$$\frac{D}{D_0} \simeq \frac{D}{L} k \times 10^{-2}.$$

Accordingly, for $k = 10^7$ m$^{-1}$, $D/D_0$ must be smaller than $10^{-1}$, for the applicability of our results. Figure 3 clearly shows that for $D/D_0 < 10^{-1}$ the dashed and solid lines cannot practically be distinguished. However, we cannot avoid noting that for $D/D_0 < 10^{-1}$ the effect of

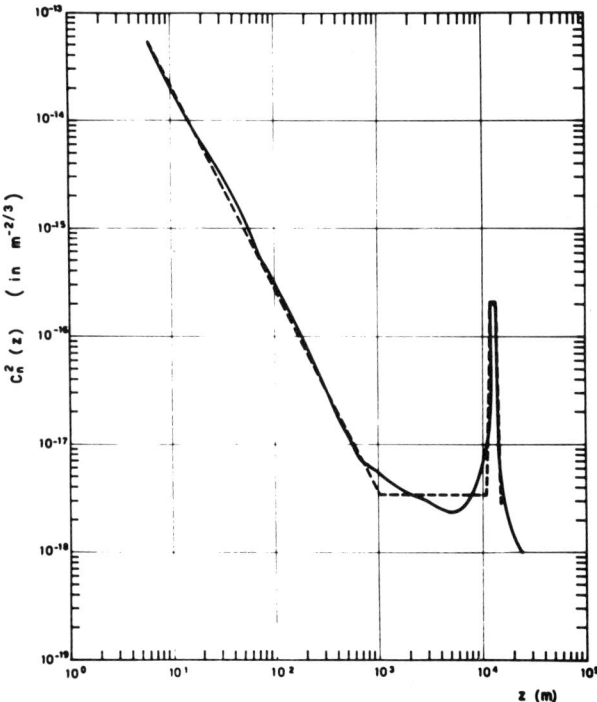

FIGURE 4 Approximate trend of $C_n^2(z)$ versus $z$. The solid line is taken from Minott [1972].

the turbulence evaluated either by the two-thirds law or by the von Kármán model is completely negligible.

## IV. Horizontal Propagation

In the case of horizontal propagation, $C_n^2$ and $\kappa_0$ are constant along the integration path and

$$\mathcal{D}_w(\rho) = A \frac{1}{\kappa_0 \rho} \int_0^{\kappa_0 \rho} [1 - 0.994 \xi^{5/6} K_{5/6}(\xi)] \, d\xi, \qquad (22)$$

with

$$A = 1.56 k^2 C_n^2 \kappa_0^{-5/3} L. \qquad (23)$$

By inserting Eq. (22) into Eq. (10) and then into Eq. (8) we have

$$\mathcal{R} = 4 \left(\frac{D}{\lambda F}\right)^2 \int_0^1 \exp\left\{ -\frac{A}{2} \frac{1}{\kappa_0 Du} \int_0^{\kappa_0 Du} [1 - 0.994 \, \xi^{5/6} K_{5/6}(\xi)] \, d\xi \right\}$$

$$\times [\arccos u - u(1-u^2)^{1/2}] \, u \, du. \qquad (24)$$

Since

$$\lim_{\kappa_0 \rho \to \infty} D_w(\rho) = A$$

we have that, for $\kappa_0 D \to \infty$, $\mathcal{R}$ increases proportionally to $D^2$, while according to the two-thirds law it would reach saturation.

The asymptotic trend of $\mathcal{R}$ turns out to be

$$\mathcal{R}_{\text{asym}} = \frac{\pi}{4} \left(\frac{D}{\lambda F}\right)^2 \exp\left(-\frac{A}{2}\right). \qquad (25)$$

This in principle implies that the atmosphere does not put an upper limit to the resolution, and that an increase of the diameter $D$ over a certain value corresponds to an improvement of the quality of the image. This conclusion has been confirmed by the computation of $\mathcal{R}$ as a function of $\kappa_0 D$; however, in many cases, it has a purely theoretical interest as may be seen from Figure 5.

Figure 5 shows $\mathcal{R}$ plotted versus $\kappa_0 D$ for several values of $A$. It ap-

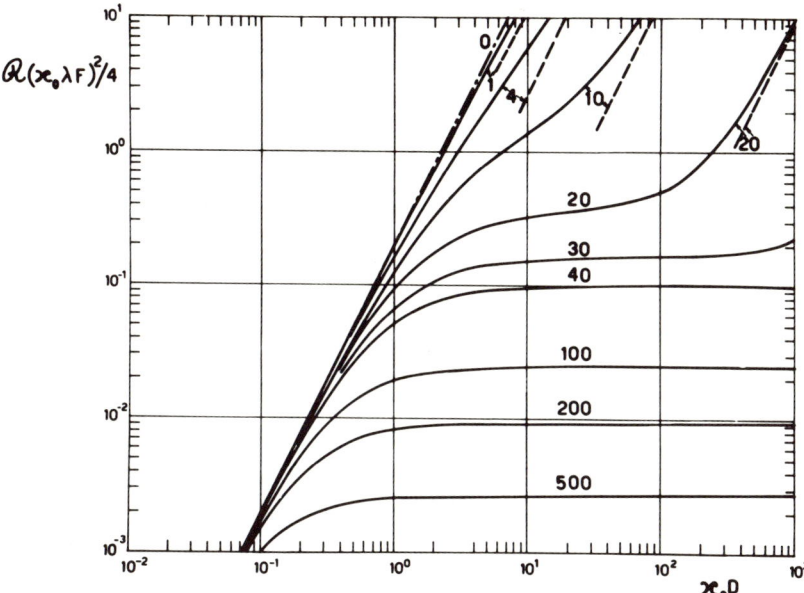

FIGURE 5  Horizontal propagation. $[\mathcal{R}(\kappa_0 \lambda F)^2/4]$ plotted versus $\kappa_0 D$ for several values of $A$. Dashed lines represent asymptotic trends.

pears that for $A < \sim 20$ the curves increase monotonically with $\kappa_0 D$ and tend to their asymptote (dashed line). On the contrary, for $A > \sim 20$, the curves present a flat horizontal region that looks like a saturation. After that region, they increase and tend to their asymptote, but this happens at so large values of $D$ that it has no practical interest.

The "saturation" value for $A > 20$ does not coincide with the value resulting from the two-thirds law but tends to it when $A$ increases (Figure 6). Figure 7 shows the comparison between the results derived from the von Kármán model and those derived from the two-thirds law [Fried, 1966a], for three values of $A$.

## V.  Discussion of the Results

The results described in the present paper indicate that the two-thirds law (1) brings sufficiently correct results as regards the quality of the atmospherically degraded images obtained by looking downward through the atmosphere. This holds both when the objective lens is im-

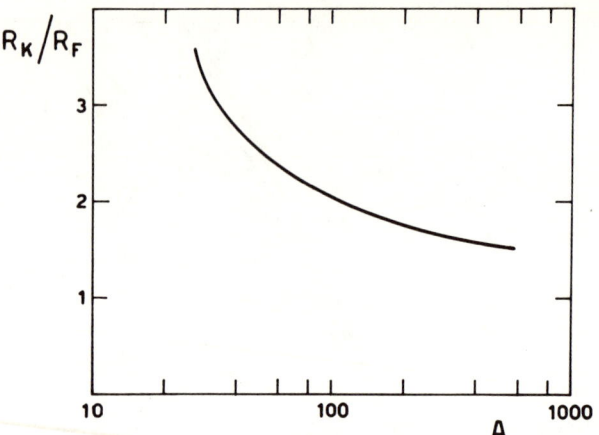

FIGURE 6 Ratio of the "saturation" values of the curves of Figure 5 for $A \geq 30$ to the corresponding saturation values obtained from Fried's theory, plotted versus $A$.

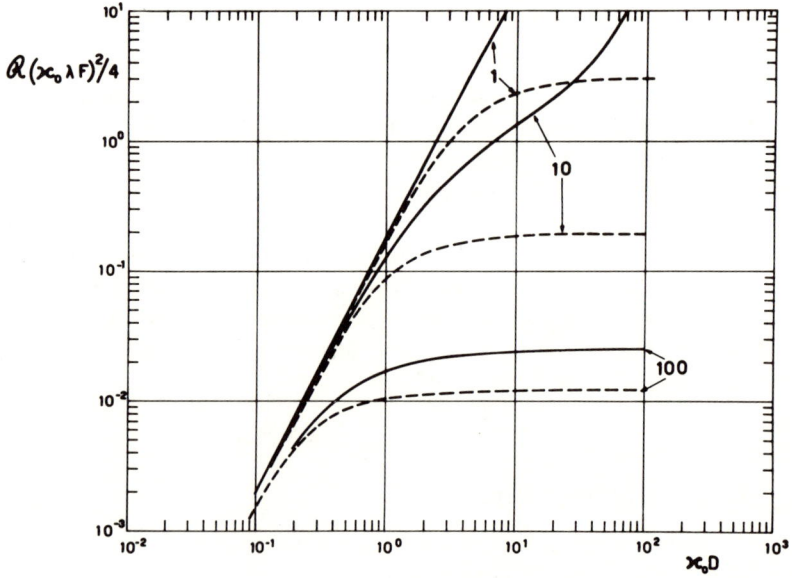

FIGURE 7 Comparison of $[\mathscr{R} (\kappa_0 \lambda F)^2 /4]$ as given by the von Kármán model (solid lines) and as given by the two-thirds law (dashed lines), for three values of $A$.

bedded in the atmosphere, because of the fact that

$$D \ll \kappa L,$$

and when the objective lens is placed on a space vehicle. In the latter case, however, due to the fact that

$$D \ll D_0$$

the effect of turbulence appears to be completely negligible: the quality of the images is only diffraction-controlled.

In the case of horizontal propagation, the use of the von Kármán model yields results that differ from the two-thirds law qualitatively and quantitatively when the parameter $A$ given by Eq. (23) is smaller than $\sim 20$. When $A > \sim 20$ the von Kármán model and the two-thirds law yield analogous results; however, the question arises if $A > 20$ corresponds to a situation in which the Rytov approximation is no longer appliable [Tatarski, 1967; Strohbehn, 1971].

## References

Abramowitz, M., and I. A. Stegun, *Handbook of Mathematical Functions*, p. 375, Dover, New York (1965).
Carlson, F. P., and A. Ishimaru, *J. Opt. Soc. Am. 59*, 319 (1969).
Consortini, A., and L. Ronchi, *Lett. Nuovo Cim. 2*, 683 (1969).
Consortini, A., G. Fidanzati, A. Mariani, and L. Ronchi, *Lett. Nuovo Cim. 2*, 259 (1971).
Consortini, A., G. Fidanzati, A. Mariani, and L. Ronchi, *Appl. Opt. 11*, 1229 (1972).
Fried, D. L., *J. Opt. Soc. Am. 56*, 1372 (1966a).
Fried, D. L., *J. Opt. Soc. Am. 56*, 1380 (1966b).
Hufnagel, R. E., in *Restoration of Atmospherically Degraded Images*, Vol. 2, Appendix 3, p. 15, National Academy of Sciences–National Research Council, Washington, D.C. (1966).
Hufnagel, R. E., and N. R. Stanley, *J. Opt. Soc. Am. 54*, 52 (1964).
Lutomirski, R. F., and H. T. Yura, *J. Opt. Soc. Am. 61*, 482 (1971).
Minott, P. O., *J. Opt. Soc. Am. 62*, 885 (1972).
Strohbehn, J. W., in *Progress in Optics*, Vol. IX, p. 75, North-Holland, Amsterdam, Holland (1971).
Tatarski, V. I., *Wave Propagation in a Turbulent Atmosphere*, Nauka Press, Moscow (1967).
Zimmerman, S. P., *J. Geophys. Res. 71*, 2439 (1966).

RICHARD C. WILLSON

# Absolute Radiometry and the Solar Constant

## I. Introduction

The definition of an absolute radiation scale, based on fundamental physical principles, can be effected by standard detectors or by standard sources of radiation. Standard sources, usually termed blackbody sources, are cavity radiators operated at high temperature. The specific intensity of such a source is determined from an accurate knowledge of its temperature, radiating area, and emittance. Calibration of secondary radiometers can be carried out in the laboratory by exposing them to the irradiance of a standard source. Measurements made in remote locations by the radiometer may then be reported on the absolute radiation scale as defined by the standard source.

The standard source method has some serious disadvantages. The calibration of secondary radiometers is an added experimental step with associated indeterminacies regarding the absolute temperature of the source and the radiative transfer between source and radiometer. At the temperatures accessible for a well-controlled source, the total irradiance at a reasonable working distance is small, requiring assumption of, or

---

The author is at the Jet Propulsion Laboratory, Pasadena, California 91103.

further experimentation to establish, the linearity of response for radiometers designed to measure radiant energy at solar irradiance levels.

The definition of the absolute radiation scale by standard detectors obviates many of the standard source-associated problems. It is this approach that has been pursued at the Jet Propulsion Laboratory in making absolute measurements of solar radiation.

A standard detector defines the absolute radiation scale through an accurate knowledge of its instrumental optical, mechanical, and electrical parameters. With these quantities known, the interaction of the radiometer with irradiant fluxes can be accurately predicted from theory.

Standard detectors are calorimeters in which the heating effect of an unknown irradiance on a detector is compared with the heating effect of a known electrical power. The electrical power is dissipated in a heater placed in intimate thermal contact with the detector or is dissipated in the body of the detector itself. An accurate knowledge of (1) the effective area through which the detector receives the irradiance, (2) the detector absorptance for the radiation to be measured, and (3) the electrical heating power facilitates accurate irradiance measurements on an absolute basis. The Angstrom pyrheliometer (1895) and the Smithsonian water flow pyrheliometer (1913) are well-known examples of early developments in standard detector technology. Some features of these instruments have recently been combined with more modern instrumentation methods. The resulting instruments, described here, are a family of absolute cavity radiometers, standard detectors of high accuracy.

## II. The Active Cavity Radiometer

A series of cavity radiometers has been developed at the Jet Propulsion Laboratory (JPL) for the accurate measurement of irradiance in absolute units [Kendall and Berdahl, 1970; Willson, 1969, 1971a, 1971b, 1972, 1973]. A group of instruments, functionally described as active cavity radiometers, have evolved from the SACRAD and PACRAD radiometers developed at JPL. While the SACRAD and PACRAD are primarily laboratory instruments, the active cavity radiometers (ACR's) are designed for automatic, remote operation in any environment and have potential usefulness in astrophysical, meteorological, and engineering solar-radiation measurement programs [Willson, 1969, 1971a, 1971b, 1972, 1973]. ACR's have been developed that can make irradiance measurements from low-level infrared sources up to 30 solar constants (4 W/cm$^2$) with small absolute uncertainty.

The essential physical features of the ACR are shown in Figure 1. The

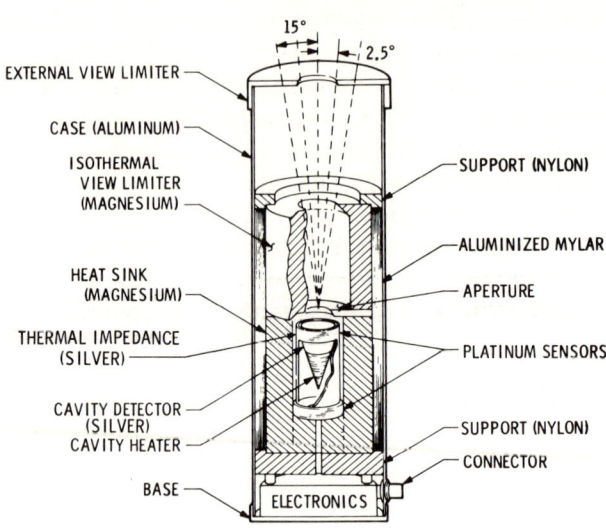

FIGURE 1  JPL active cavity radiometer type III.

ACR's conical cavity detector is connected by a low thermal impedance to a heat sink. The heat sink is insulated from the external environment to minimize the rate of change of its temperature when subjected to fluctuating thermal environments.

The interior surface of the conical cavity detector is thinly coated with a surface material whose effective absorptance for solar radiant flux is high ($\alpha_s = 0.95 \pm 0.01$). The cavity's 2-cm-diameter base is partially shaded by a round aperture that both defines the detector area for the radiometer and enhances the effective cavity absorptance. The aperture can be shaped precisely and its size determined with a high degree of accuracy. Enhancement of the effective cavity absorptance is achieved by confining the incident irradiance to the lower portion of the conical detector, near its apex. A large fraction of the radiation not initially absorbed is scattered to the higher walls of the cavity and absorbed in a second interaction. The magnitude of the enhancement depends on the solid angular subtendance of the incident beam. It can be seen, however, that with a high surface absorptance of 0.95 secondary and higher-order interactions of the scattered incident irradiance rapidly attenuate the amount of flux eventually scattered out of the cavity's aperture.

The cavity absorptance has been analytically shown to be 0.996 ± 0.001, effecting a tenfold decrease in the uncertainty of the cavity surface absorptance. The field of view for the cavity is isothermally confined to 30° by the heat sink or to 5° with the addition of an external view limiter.

The dissipation of a fixed amount of power in the cavity will produce a constant temperature drop across the thermal impedance. This drop, transduced by resistance temperature sensors, can be calibrated as a direct measure of the irradiant power or used to actively control an electronic servo system as in the active cavity radiometer. Circuitry housed in the base of the radiometer automatically maintains constant cavity power dissipation by controlling a dc voltage supplied to a fixed-resistance heater on the cavity. A schematic drawing of this circuit is shown in Figure 2.

The ACR operates in a differential mode. The radiant source is chopped at low frequency and the cavity heating power monitored in each phase. The irradiance is then determined from these two electrical power measurements, along with the instrumental constant that is the reciprocal of the cavity area–absorptance product.

## III. Operation of the Active Cavity Radiometer

The active cavity radiometers are electrical substitution calorimeters. Using the platinum windings as sensors, an electronic circuit maintains a constant temperature drop across the thermal impedance connecting the cavity detector to the heat sink. This is accomplished by dissipating dc electrical power in the cavity heater winding. The ACR is operated in a differential mode by chopping the radiant flux to be measured at a slow rate. In the observation phase, the thermal impedance temperature drop is maintained by the combined inputs of incident radiant power and dc electrical power to the cavity detector. In the reference phase, all the power required for maintenance of the temperature drop is provided by electrical heating. By careful design, the only significant differences be-

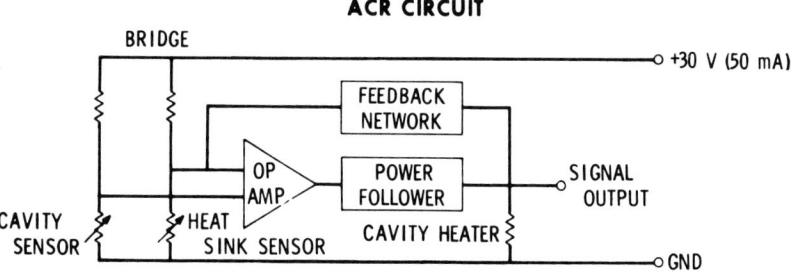

FIGURE 2  Active cavity radiometer electronics.

tween the two phases of measurement are the presence or absence of the radiant flux to be measured and the electrical heating power. The difference in electrical power required in the two phases is directly proportional to the radiant flux. A constant of proportionality relates the absorption of radiant flux by the cavity to the dissipation of electrical power. In simplified form, the equation describing ACR operation is

$$H = K(P_r - P_o), \qquad (1)$$

where

$H$ = measured irradiance,
$K$ = standard detector constant,
$P_r, P_o$ = reference and observation phase electrical powers,
$K = (\alpha A)^{-1}$,
$\alpha$ = effective cavity absorptance,
$A$ = effective cavity area.

## IV. Accuracy of the Active Cavity Radiometer

Equation (1) is an approximation to the general ACR working equation, which has been rigorously developed in the literature [Willson, 1971a, 1973]. To the extent that this approximation departs from the complete working equation, errors are introduced into the measurements. A detailed analysis of these errors is presented in Willson [1973] and is summarized here in Table 1.

TABLE 1  Uncertainties of Type III Active Cavity Radiometer Measurements Relative to the Absolute Radiation Scale

| Radiometer | Irradiance Range (W/cm²) | Absolute Uncertainty (mW/cm²) |
|---|---|---|
| Low-irradiance ACR | 0 → 0.010 | 0.030 → 0.030 |
| One-solar-constant ACR[a] | 0 → 0.150 | 0.104 → 0.316 |
| Ten-solar-constant ACR | 0 → 1.50 | 0.916 → 1.790 |
| Twenty-solar-constant ACR[a] | 0 → 3.00 | 1.515 → 3.467 |
| Thirty-solar-constant ACR | 0 → 4.20 | 2.216 → 4.963 |

[a] The uncertainties of the one- and twenty-solar-constant ACR's as functions of irradiance are shown in Figures 3 and 4, respectively.

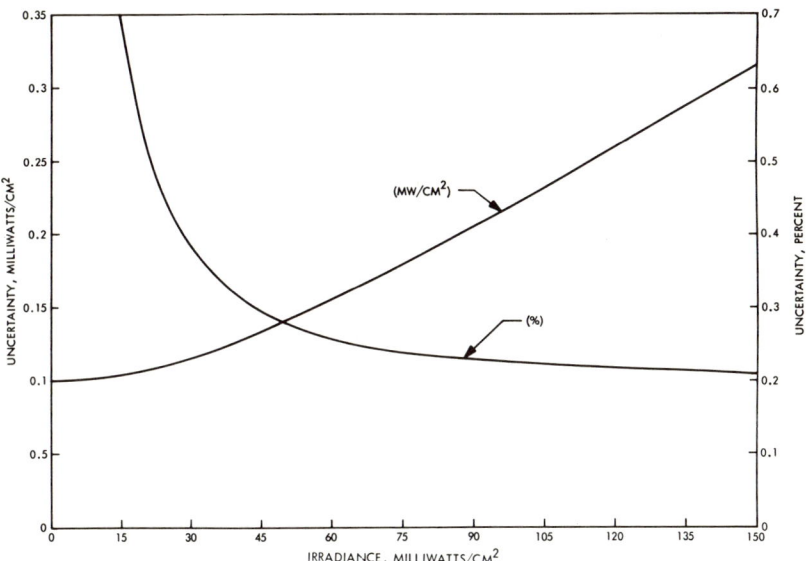

FIGURE 3  Uncertainty of the one-solar constant active cavity radiometer (ARC1) as a function of irradiance.

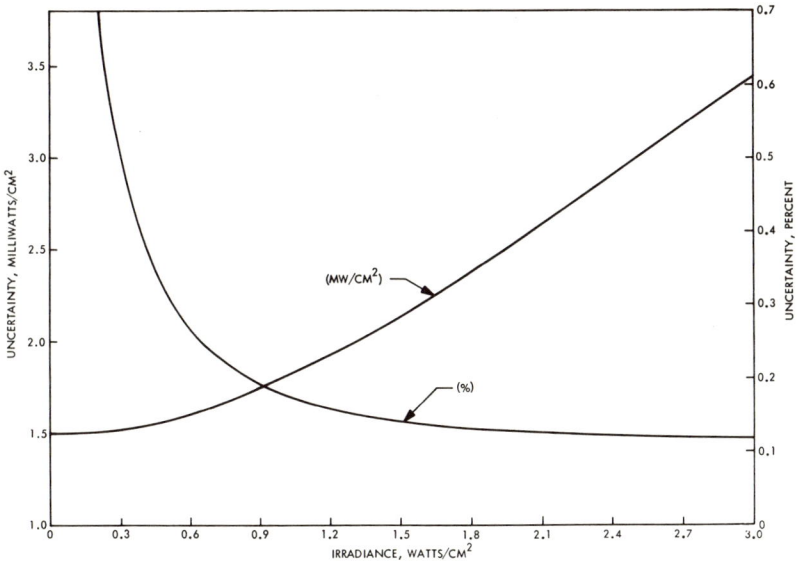

FIGURE 4  Uncertainty of the twenty-solar-constant active cavity radiometer (ACR 20) as a function of irradiance.

## V. Comparisons of Radiation Scales

The International Pyrheliometric Scale (IPS) is a radiation measurement scale defined by the International Radiation Commission in 1956. It was adopted by the World Meteorological Organization in 1957 as the scale of reference for the International Geophysical Year. The IPS was a compromise between the two prominent radiation scales in use at the time: the Angstrom Scale and the Smithsonian Scale. These scales were defined by standard detectors developed around 1900: the Angstrom pyrheliometer developed in Sweden and the Abbot water flow pyrheliometer developed at the Smithsonian Institution in the United States. At the time the IPS was defined, intercomparisons of these two scales were considered to have yielded a systematic difference of 3.5%, with measurements on the Smithsonian Scale exceeding those of the Angstrom Scale. The IPS was defined to be 1.5% above the Angstrom Scale and 2.0 percent below the Smithsonian Scale.

A series of comparisons between the absolute radiation scale, as defined by the active cavity radiometer, and the International Pyrheliometric Scale, as defined by Eppley Angstrom pyrheliometers were carried out between 1968 and 1970. The site of these tests was the California Institute of Technology's Jet Propulsion Laboratory Observatory at Table Mountain, California (60 miles northeast of Pasadena, California, at an elevation of 2.25 km).

The format of the intercomparisons was the synchronous measurement of direct solar irradiance by both Angstrom pyrheliometers and active cavity radiometers. The results of these tests, summarized in Table 2, demonstrate a systematic difference between the measurements by the two types of instrument that is significant relative to the uncertainty with which the active cavity radiometer defines the absolute radiation scale. The weighted-average result yielded a 2.2% difference with measurements on the absolute scale exceeding those of the IPS. The uncertainty of the ACR-defined absolute radiation scale was less than ±0.3%.

A similar result was reported from the third international comparison of pyrheliometers held at Davos, Switzerland, in September 1970. Measurements of solar irradiance on the absolute radiation scale, defined by the JPL's PACRAD radiometer and an absolute cavity radiometer developed by the U.S. National Bureau of Standards, were observed to exceed those of the Davos IPS-defined Angstrom pyrheliometer No. 210 by 1.8 ± 0.4%.

Instrumental field of view can be an important factor in the intercomparison of radiometers when significant atmospheric aerosol scattering is present. The ACR has a circular detector geometry and is used

TABLE 2  Intercomparison of JPL Active Cavity Radiometers and the International Pyrheliometric Scale

| Test Date (Mo/day/yr) | Average Value of Scale Difference (%) | Standard Deviation of Scale Difference ±(%) | Absolute Uncertainty of Scale Difference ±(%) | Weight[a] |
|---|---|---|---|---|
| 5/10/68 | 2.0 | 0.13 | <0.5 | 355 |
| 5/11/68 | 2.4 | 0.09 | ↓ | 370 |
| 9/23/68 | 2.1 | 0.07 |  | 612 |
| 9/24/68 | 2.3 | 0.06 |  | 4444 |
| 9/25/68 | 2.1 | 0.04 |  | 12500 |
| 4/23/69 | 2.4 | 0.19 |  | 138 |
| 9/22/69 | 2.3 | 0.21 |  | 91 |
| 9/23/69 | 2.3 | 0.05 |  | 12400 |
| 8/25/70 | 2.1 | 0.25 | ↓ | 128 |
| Weighted averages | 2.21 | ±0.09 | ±0.5 |  |

[a]Equal to the number of independent measurement periods in each test divided by the square of the standard deviation of the scale difference.

with a 5° circular field of view. The Angstrom pyrheliometer has a rectangular detecting surface and a 4.4° × 10.6° rectangular field of view. Several circular apertures providing fields of view ranging from 5 to 20° were tried with the ACR's to evaluate the experimental sensitivity to this parameter. On the days during which data were taken, the differences observed between measurements made within this range did not exceed the ACR experimental uncertainties (±0.3%) indicative of a "turbidity parameter" [Schöne, 1966] of less than $m\beta = 0.05$. The effective circular field of view for the Angstrom pyrheliometer, under these conditions, would be 5 to 6°. It does not appear that the effects of circumsolar aerosol scattering could produce a significant fraction of the observed difference between the IPS and the absolute scale. According to our results, measurements on the Angstrom Scale, the International Pyrheliometric Scale, and the Smithsonian Scale are 3.7, 2.2, and 0.2% low, respectively, relative to the absolute scale.

## VI. Measurement of the Solar Constant

Total solar irradiance has been measured by ACR's in two high-altitude balloon flights. Two ACR-Type II's [Willson, 1969, 1971a] measured the solar irradiance at an altitude of 25 km in 1968. The solar constant was derived from these measurements by correcting them for extinction due to the remaining earth's atmosphere above 25 km and for the earth–sun distance at the time of measurement. The average value of the 1968 measurement was a solar constant value of $H_o = 137.0$ mW/cm$^2$ (1.964 cal cm$^{-2}$ min$^{-1}$) [Willson, 1971a] with an uncertainty of less than ±2%. The more accurate ACR III made measurements in 1969 at an altitude of 36 km, producing a solar constant value of $H_o = 136.6 \pm 0.7$ mW/cm$^2$ (1.958 ± 0.010 cal cm$^{-2}$ min$^{-1}$) [Willson, 1971b].

## References

Kendall, J. M., and C. M. Berdahl, *Appl. Opt. 9*, 1082 (1970).

Schöne, W., "Results of Pyrheliometer Comparisons Made in Potsdam with Regard to Recent Investigations on the Influence of Circumsolar Sky Radiation," paper given at Precision in Pyrheliometric Measurements Seminar, Brussels (1966).

Willson, R. C., Experimental and Theoretical Comparisons of the JPL Active Cavity Radiometric Scale and the International Pyrheliometric Scale, Tech. Rep. 32-1365, Jet Propulsion Lab., Pasadena, Calif (1969).

Willson, R. C., *J. Geophys. Res. 76*, 4325 (1971a).

Willson, R. C., New Radiometric Techniques, International Solar Energy Soc., Goddard Space Flight Center (May 1971b).

Willson, R. C., *Nature 239*, 208 (1972).

Willson, R. C., *Appl. Opt. 12*, 810 (1973).

J. L. KOHL and W. H. PARKINSON

# Absolute intensity calibration of a high-resolution rocket spectrometer

## Introduction

The rocket-spectrometer project at Harvard is part of a space- and ground-based observational program in solar physics. Our two previous rocket flights (September 24, 1968, and September 11, 1969) were used to record the spectrum of the quiet sun between 140 and 185 nm with 6.0-pm resolution [Parkinson and Reeves, 1969]. The absolute intensity calibration of these two instruments, based on a Reeder thermopile detector, was reported elsewhere [Parkinson and Reeves, 1968].

Our most recent rocket flight was successfully launched to an altitude of 238 km on an Aerobee 170 sounding rocket from White Sands Missile Range on July 27, 1972, and it is the calibration of this rocket-spectrometer system that will be described here. During the flight we measured the absolute intensity of the quiet sun from the C I (2 uv) multiplet at 165.72 nm to the C IV (1 uv) multiplet at 154.91 nm. The wavelength resolution is defined by the instrumental profile, which has a full width at half-maximum efficiency of 2.9 pm.

Because of the complexity of the solar spectrum in the region of the

---

The authors are at the Harvard College Observatory, Cambridge, Massachusetts 02138.

temperature minimum (155–165 nm), a wavelength resolution greater than 6.0 pm is useful in resolving and identifying emission lines and in determining their absolute intensities. Increased resolution also permits a more accurate determination of the continuum level between lines. In addition, high resolving power provides a more precise measurement of the "true" solar line profiles, which we can analyze using spectral-line-formation theory.

**Instrument**

Figure 1 is a schematic diagram of the telescope–spectrometer system. The telescope utilizes an off-axis parabolic mirror of 93.0-cm focal length as the main collector with two additional plane-folding mirrors. In all cases, the surfaces are coated with aluminum and overcoated with magnesium fluoride.

The Ebert instrument has a 75-cm focal length with a 3600 line/mm plane grating blazed at 130.0 nm in first order. This results in a reciprocal dispersion of 0.35 nm/mm.

The detector is an EMR 641-G solar blind photomultiplier with a lithium fluoride side window and a sapphire filter used to discard Lyman-$\alpha$ radiation. We increased the resolution of the spectrometer to $2.9 \pm 0.1$ pm for this flight by reducing the width of the straight entrance and exit slits to 6 $\mu$m.

We measured the instrumental line profile at several wavelengths using a microwave-driven $^{198}$Hg single-isotope lamp. We took particular care to determine the wings of the instrumental profile because, as was recently pointed out by Griffin [1969], the profile wings influence the measured intensities of spectral lines. The wings of the instrumental profile can remove significant amounts of the central intensity of a spectral line and deposit it in the wings of the line.

In the case of emission-line intensities, this effect produces measured values that are too low, and the effect also causes measured values of the continuum intensity between lines to be too high.

Our most recent observations with this spectrometer have established that several spectral lines of the quiet sun spectrum between 165.0 and 155.0 nm display central absorption features. The degree to which these spectral lines are centrally absorbed was predicted by Vernazza (1972, personal communication), using a non-LTE solar-atmosphere calculation. The wings of the instrumental profile tend to fill in these absorption features of the spectrum with radiation that falls well outside the ab-

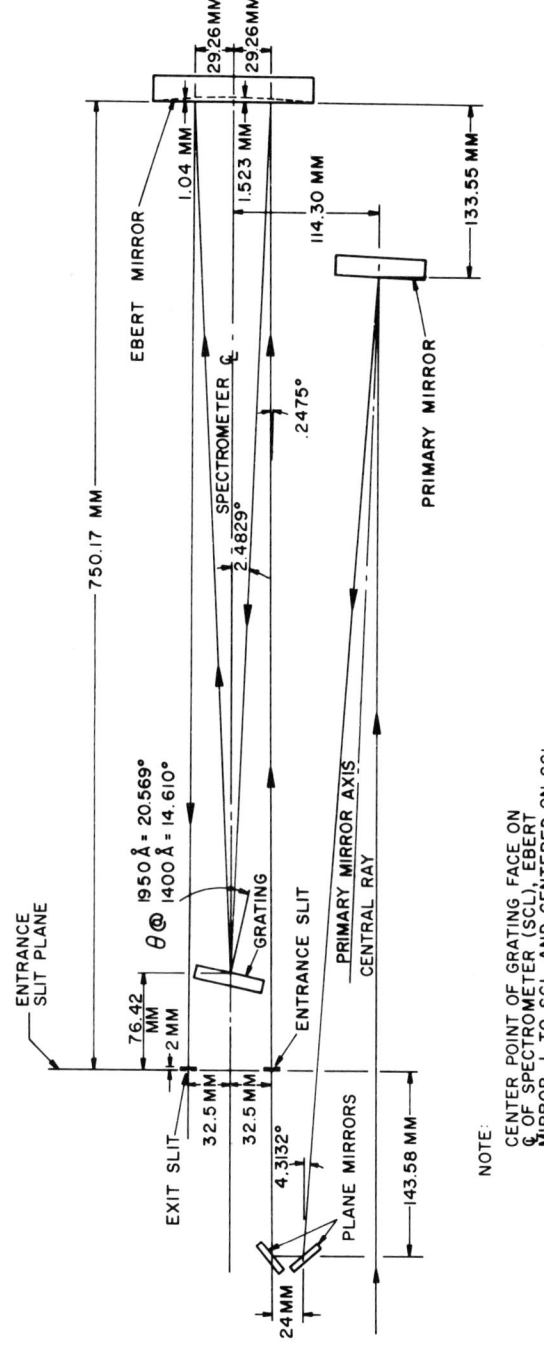

FIGURE 1  Schematic diagram of the rocket spectrometer and telescope optical arrangement.

sorbing wavelength band; this process ultimately results in an apparent difference between solar-model predictions and the observations.

The effect of the instrumental profile on the observations can be minimized with an accurate determination of the complete instrumental profile and the use of a deconvolution program [see, for example, De Jager and Neven, 1966].

The measured instrumental profile shown in Figure 2 was determined by using the $^{198}$Hg line at 253.651 nm. The profile extends over a 0.45-nm bandwidth and has a ratio of peak to minimum intensity over this range of $1 \times 10^4$. We also measured the instrumental profile at three other $^{198}$Hg lines near 265.3 nm and at several wavelengths throughout

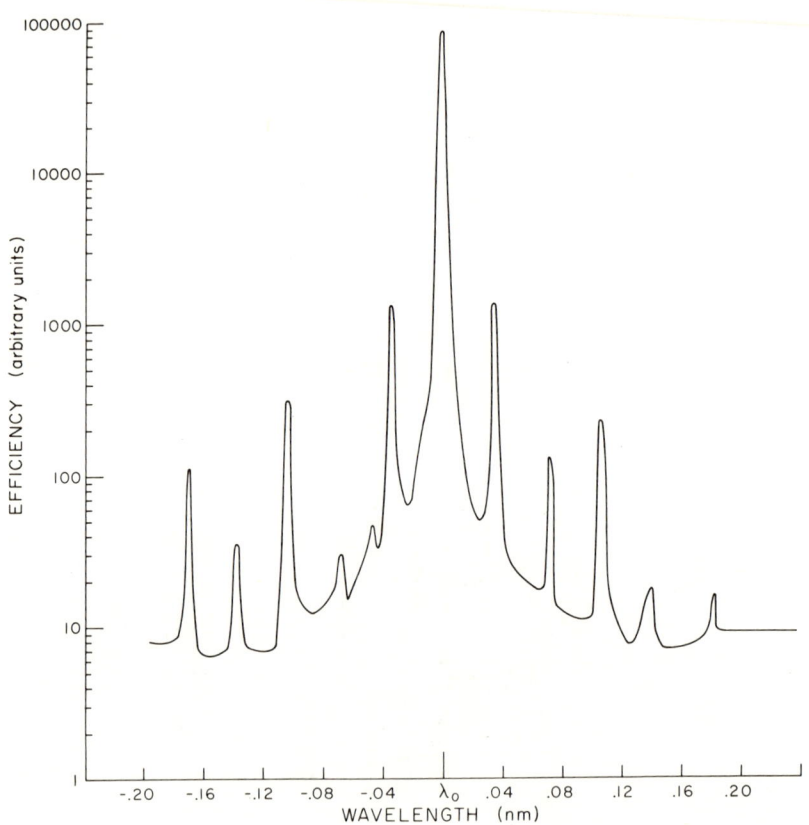

FIGURE 2  The instrumental line profile of the rocket spectrometer is indicated by the solid line, and $\lambda_0$ indicates the center of the profile. This curve is the average shape for several scans of the $^{198}$Hg line at 253.651 nm.

the flight range using the rotational lines of the CO emission spectrum excited in a Hunter-type discharge lamp. The profile was measured before and after an Aerobee-170 environmental-vibration test and after the actual flight and recovery of the instrument. There was no measurable change in the instrumental profile.

## Calibration Arrangement

A schematic diagram of the calibration arrangement is shown in Figure 3. The light source used for the calibration is a current-regulated Hunter discharge lamp [Hunter, 1962] containing a $CO_2$-He gas mixture. The predispersing instrument, a 1/3-m, folded Czerny-Turner spectrometer made by McPherson Instruments, used a 2.0-nm bandpass. The exit slit of this monochromator is the entrance aperture to the main calibration vacuum tank and is located at the focal length of a concave mirror 18 cm in diameter that collimates the light and fills the rocket telescope-objective. This mirror forms the image of the predisperser exit slit onto the entrance slit of the Ebert spectrometer. The telescope is masked to just underfill the grating.

The absolute intensity calibration of the rocket spectrometer is based on a special vacuum ultraviolet photodiode with a magnesium fluoride window and a cesium telluride cathode. This detector was calibrated at the National Bureau of Standards (NBS) before and after the rocket flight by direct comparison of photocurrents with an NBS-calibrated standard photodiode.

The location of the diode during calibration is shown in Figure 4. The light from the telescope objective is intercepted by a calibrated plane mirror coated with aluminum and overcoated with magnesium

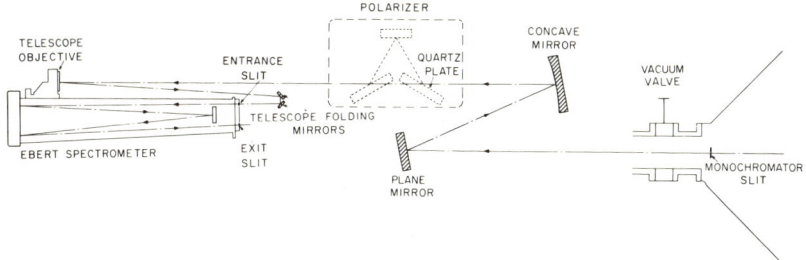

FIGURE 3 The calibration arrangement within the vacuum tank is shown in this schematic diagram. The polarizer, enclosed by the dotted lines, is present only during the polarization measurements.

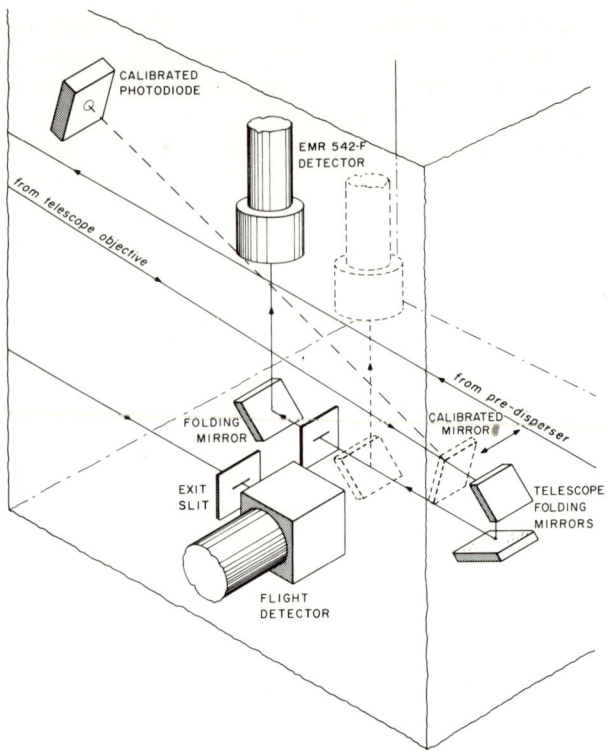

FIGURE 4  A closeup view of the calibration arrangement immediately in front of the spectrometer entrance slit is shown. The EMR 542-F photomultiplier and mirror, which are used to measure $T_s$, are indicated in the front position by the dashed line, and their position behind the slit is shown with a solid line.

fluoride that can be moved into and out of the light path in vacuum. This mirror folds the beam onto the diode, which is located at the focus of the telescope. The photodiode current is a measure of the light that is incident on the telescope folding mirrors within the bandpass of the predispersing optics.

To determine the calibration system function of the telescope folding mirrors, Ebert spectrometer, and EMR detector combination, the fractional amount of light lost on the entrance slit jaws of the spectrometer must also be measured. This is accomplished by means of an EMR 542-F photomultiplier and folding flat combination, also shown in Figure 4, that can be moved before or behind the entrance slit. The relative anode currents for the two positions yield the transmittance, $T_s$, at the slit.

Care must be taken at each position to ensure that the light beams cover the same area of the EMR 542-F photocathode and that this area has a uniform sensitivity. In practice, the EMR 542-F in the forward position was intercalibrated with the standard diode, so that it was possible to measure the slit transmittance in vacuum without moving the photomultiplier. The calibrated photodiode could not be used behind the entrance slit because of the low flux transmitted by the 6-$\mu$m slit.

## The System Function

Although we measured the efficiency of most of the elements of the rocket instrument individually, the total spectrometer and telescope are finally calibrated as a system so that the response of the complete instrument is known.

The system function as used here includes the reflectance of the two telescope folding mirrors, the transmittance of the Ebert spectrometer, and the quantum efficiency of the detector–amplifier–discriminator system but does not include the reflectance of the telescope objective or the solid angle and area factors. These quantities are determined separately.

The expression for the system function is given by:

$$S(\lambda) = i_d C_d(\lambda) \, T_s / \int_{\lambda_1}^{\lambda_2} \frac{K(\lambda', t) \, d\lambda'}{\Delta t \, \Delta \lambda}, \tag{1}$$

where $i_d$ is the photodiode current, $C_d(\lambda)$ is the sensitivity of the photodiode at wavelength $\lambda$ in photons per ampere, $T_s$ is the transmittance of the entrance slit, and the denominator of Eq. (1) represents the detector counts per second integrated over the bandwidth of the predisperser.

The system function evaluation is illustrated in Figure 5. A 2.0-nm segment of the CO emission spectrum from the Hunter lamp is selected by the monochromator and detected by the calibrated photodiode. The photocurrent, typically $5 \times 10^{-10}$ A, was measured with a Keithly model 640 electrometer, which was calibrated before and after the rocket instrument calibration by the Calcutron Corporation, with a current source traceable to the NBS. With the entrance slit transmittance measured, the CO spectrum was scanned and the single photon pulses from the flight detector were selected by the flight discriminator, counted during 40-msec gate times by our Lockheed Electronics Mac 16 on-line computer and stored on magnetic tape as a function of grating position. After the noise is subtracted out, the measured counts are integrated over the spectral bandpass of the predisperser by a CDC 6400 computer,

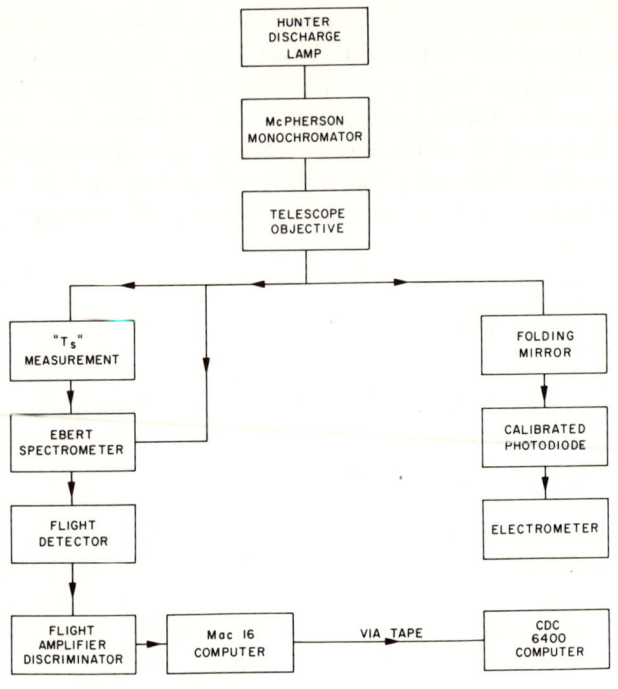

FIGURE 5 The calibration procedure is illustrated in this block diagram or flow chart.

and the integrated counts factor and system function are evaluated. To check beam nonuniformity, the system function was measured for the case in which only the central 25% of the telescope objective was illuminated. This value is within 5% of the value measured with full illumination.

**Calibration Equation**

The solar data are reduced by use of the basic calibration equation

$$\int_{\lambda_1}^{\lambda_2} I(\lambda)\, d\lambda = S(\bar{\lambda}) \int_{\lambda_1}^{\lambda_2} \frac{K(\lambda', t)\, d\lambda'}{\Delta t\, \Delta\lambda} \bigg/ R_t A_s \Omega_t, \qquad (2)$$

where $\int I(\lambda)\, d\lambda$ is the integrated intensity in photons $\sec^{-1}$ $cm^{-2}$ $sr^{-1}$ of any spectral feature, $R_t$ is the reflectance of the main telescope mirror,

$A_s$ is the area of the entrance slit, $\Omega_t$ is the solid angle subtended by the telescope mask at the entrance slit, $K(\lambda', t)$ is the detector counts in a quasi-digital output format, $\Delta t$ is the time interval, $\Delta \lambda$ is the spectral bandpass of the spectrometer exit slit, and $\lambda_2 - \lambda_1$ is the spectral extent of any feature of the solar spectrum. Notice that $\Delta \lambda$ is also contained in the expression for $S(\lambda)$ and, therefore, need not be known. The integration and conversion from the quasi-digital format is routinely accomplished by the CDC 6400 computer.

The entrance slit area was measured by three methods: electron micrographs with measurements from photographic prints, laser diffraction, and comparison of the fractional amounts of a visible light beam transmitted by the flight slits and by a 40-μm-wide reference slit whose width was measured by laser diffraction and a measuring microscope.

The third method was accepted as the most reliable and preferable to the laser diffraction measurement of the 6-μm slit, because the latter measurement was complicated by asymmetries in the diffraction pattern apparently due to the small offset of the slit jaws. The effect was negligible for the wider reference slit. Our electron micrograph area-determination was in agreement with laser diffraction within its larger error limits.

In addition to the factors given in Eq. (2), the data also contain minor corrections for scattered light and atmospheric absorption.

Equation (2) establishes the integrated intensity of a feature of the solar spectrum, such as a spectral line. The relative wavelength distribution of the integrated intensity—in other words, the "line shape"—is established either by the raw data $K(\lambda', t)$, or in some cases more accurately by the raw data deconvoluted from the instrumental profile.

## Errors

The sources of error are given in Table 1. The most probable error, assuming independent errors, is 11%. The largest sources of error are in the photodiode calibration and in the transmittance measurement of the entrance slit.

TABLE 1  Errors

| Quantity | $i_d$ | $C_d(\lambda)$ | $T_s$ | $\int_{\lambda_1}^{\lambda_2} \frac{K(\lambda', t)\, d\lambda'}{\Delta t\, \Delta \lambda}$ | $A_s$ | $\Omega_t$ | $R_t$ |
|---|---|---|---|---|---|---|---|
| Systematic error | ±1% | ±8% | ±5% | ±2% | ±4% | ±1% | ±2% |

## Polarization Studies

It is well known that grating spectrometers and large angle reflections have unequal efficiencies for light polarized parallel and perpendicular to the plane of incidence of the incoming light. Therefore, the apparent efficiency of a telescope–spectrometer system varies according to the degree of polarization of the incident radiation. These effects can lead to calibration errors when the light used during calibration differs in degree of polarization from the light source being investigated.

To study these effects and to determine the efficiencies of our instrument separately for the two components of polarization, we designed and constructed a reflecting polarizer that consists of an uncoated fused-quartz reflector plate, used at 60 deg incidence, and two aluminum-coated flat mirrors that provide an effectively straight-through optical path. A second quartz plate, also used at 60 deg incidence, is used as the polarization analyzer. At a wavelength of 150 nm the intensity ratio for crossed polarizers to aligned polarizers is 0.015. This fused-quartz reflecting polarizer is useful down to wavelengths of 130 nm, where this ratio was measured to be 0.12.

The polarizer was first used to qualify the calibrated photodiode by measuring its relative efficiency $E$ for plane-polarized light as a function of the orientation of the diode about the incident light direction. The photocurrent $i(a)$ of the diode in orientation $a$ due to plane-polarized light $I(a)$ from the predispersing optics, monochromator, and light source through the polarizer also in orientation $a$ is given by

$$i(a) = CE(a) I(a), \qquad (3)$$

where $C$ is a polarization-independent constant. If the polarizer is turned through 90° to position $b$, the photocurrent is given by

$$i(b) = CE(b) I(b). \qquad (4)$$

If the diode is turned through 90° to position $b$, the photocurrent for polarization $a$ is given by

$$i'(a) = CE(b) I(a) \qquad (5)$$

and

$$i'(b) = CE(a) I(b) \qquad (6)$$

for polarization $b$.

Dividing Eq. (3) by Eq. (4) and Eq. (5) by Eq. (6), and the results by each other yields an equation for the relative efficiency of the photodiode for the two components of polarized light, given by

$$\frac{E(a)}{E(b)} = \left[\frac{i(a) \cdot i'(b)}{i(b) \cdot i'(a)}\right]^{1/2}. \tag{7}$$

This assumes that the predisperser polarization does not change with time, but the light source intensity may change while the diode is being rotated. Most important, the measurement is independent of polarization of the light incident on the polarizer. It was determined that our photodiode has less than 1% change in efficiency with polarization at all orientations measured.

After the polarization sensitivity of the photodiode is established, the polarization of the light from the predisperser and feeding optics combination can be determined by substituting the result of Eq. (7) into the quotient of Eq. (3) divided by Eq. (4). The light emerging from our predispersing optics was found to have a polarization of less than 7% for all flight wavelengths.

The system function of the rocket instrument can also be measured for incident light that is plane-polarized perpendicular or parallel to the grating rulings, by simply illuminating the instrument with polarized light. The system function for nonpolarized light, which is the expected polarization of the solar radiation in our wavelength range, can be determined from this measurement. This may result in a slight correction to the system function measured without the polarizer since that value is for light with the 7% polarization of our predispersing optics. The postflight calibration, which is in progress at this writing, includes this polarization measurement.

The authors wish to thank all the engineering and technical staff of the Harvard College Observatory Solar Satellite Project. In particular, we acknowledge the contributions of J. J. Crawford, S. M. Diamond, L. Solomon, L. Lapson, and F. C. DeFreze. We also thank our Project computer programmer R. J. Freuder.

This program was supported by NASA Grant NGR 22-007-202.

## References

De Jager, C., and L. Neven, *Bull. Astron. Inst. Netherlands 18,* 306 (1966).
Griffin, R. F., *Mon. Notices R. Astron. Soc. 143,* 319 (1969).
Hunter, W. R., *Proc. 10th Colloquium Spectrosc. Intern.,* p. 247 (1962).
Parkinson, W. H., and E. M. Reeves, in *Symposium on Calibration Methods in Ultraviolet and X-Ray Regions of the Spectrum,* p. 219, ESRO, Paris (1968).
Parkinson, W. H., and E. M. Reeves, *Solar Phys. 10,* 342 (1969).

# Thin Films

G. HASS and W. R. HUNTER

# New Developments in Vacuum-Ultraviolet Reflecting Coatings for Space Astronomy

## I. Introduction

Since the end of World War II, vacuum-ultraviolet spectroscopy has become increasingly important as a tool in many fields of research. One of these fields is space astronomy in which rocketborne and satelliteborne spectrographs are used for studying emission and absorption features of the sun, stars, and interstellar medium. Most of these spectrographs use mirrors and gratings coated with reflecting films. This requirement has prompted a search for mirror coatings with high reflectance in the vacuum-ultraviolet spectral region. Important progress toward the development of reflecting coatings with improved efficiency and stability has been made in recent years.

There is a natural dichotomy in the wavelength characteristics of reflecting coatings that occurs approximately at 1000 Å and is caused by the optical properties of the coating materials. For wavelengths longer than 1000 Å, the intrinsic reflectance of aluminum is higher than that

---

G. Hass is in the USAECOM, Night Vision Laboratory, Fort Belvoir, Virginia 22060; W. R. Hunter is in the E. O. Hulburt Center for Space Research, U.S. Naval Research Laboratory, Washington, D.C. 20375.

of any other film material, but its high reflectance of about 90% can only be utilized if the formation of an oxide film on its surface can be prevented [Madden *et al.*, 1963]. A partial preservation of the high reflectance of aluminum has been achieved by overcoating freshly deposited Al with $MgF_2$ [Hass and Tousey, 1959; Canfield *et al.*, 1968; Hutcheson *et al.*, 1972b] or LiF [Angel *et al.*, 1961; Cox *et al.*, 1968; Hutcheson *et al.*, 1972b] films of precisely controlled thicknesses, which results in highly reflecting coatings from the visible region down approximately to the cutoff of the overcoating material: 1150 Å for $MgF_2$ and 1000 Å for LiF. Such films are very useful for extreme-ultraviolet stellar astronomy, where the absorption due to interstellar hydrogen increases rapidly from 1000 Å to the ionization limit at 912 Å and restricts observations to wavelengths longer than 1000 Å. This paper describes a new technique [Hutcheson *et al.*, 1972a] for precisely monitoring the thickness of the $MgF_2$ and LiF films on Al as the dielectric films are being deposited and discusses the evaporation conditions most suitable for preparing $MgF_2$ and LiF overcoated Al mirrors of optimum reflectance.

For the wavelength region below 1000 Å, where only the sun can be studied, the reflectance of all known mirror coatings is rather low. For a long time, platinum was extensively used because it was considered to be the best film material for this spectral region [Jacobus *et al.*, 1963]. New investigations have shown that films of iridium [Hass *et al.*, 1967], osmium [Cox *et al.*, 1972], rhenium and tungsten [Cox *et al.*, 1972] are more efficient mirror coatings for parts of this region. Data on the reflectance of Ir, Os, Re, and W films prepared under various conditions for the wavelength region from 300 Å to 2000 Å will be presented, and the effect of aging in air on the reflectances of these film materials will be discussed.

Finally, as space technology progresses and space observatories become more sophisticated, the very interesting possibility opens up of coating mirrors or gratings in an orbiting satellite. Some calculations have been made showing that Ir overcoated with 100–150 Å of unoxidized aluminum has a high reflectance over a very extended wavelength region [Hass and Hunter, 1967] and that for wavelengths between 500 Å and 800 Å, the Ir–Al combination is considerably more efficient than either Ir or Al. A brief résumé of these results will be presented, and the factors governing the usefulness of such coatings will be discussed.

## II. Reflecting Coatings for Wavelengths Longer Than 1000 Å— Al + MgF$_2$ and Al + LiF

In order to obtain a better appreciation of the problems associated with producing highly reflecting coatings for this region, a short review of the optical properties of Al will be presented.

Aluminum has a high intrinsic reflectance in the vacuum ultraviolet but it is a chemically active metal and combines readily with oxygen to form $Al_2O_3$. This oxide layer is highly absorbing and causes the reflectance of evaporated Al films to drop rapidly as it forms. Figure 1 shows the measured reflectance of evaporated Al coatings as freshly deposited and after exposure to air for 1 h, 1 day, and 1 month. The curve labeled UHV shows the results of a comparatively slow Al deposition, 10–20 Å/sec at $1 \times 10^{-9}$ to $3 \times 10^{-9}$ Torr, onto fire-polished glass [Feuerbacher and Steinmann, 1969]; and the curve labeled CONV. VAC. SYS. (conventional vacuum system) was obtained from reflectance measurements of Al films deposited at a high rate of about 500 Å/sec at about $3 \times 10^{-6}$ Torr onto optically polished glass [Madden et al., 1963]. The higher reflectance of the UHV curve is due to the smoothness of the fire-polished glass and not to the fact that the evaporation was performed in ultrahigh vacuum. Coatings deposited in ultrahigh vacuum and in conventional vacuum systems using high deposition rates have the same reflectance after exposure to air [Hutcheson

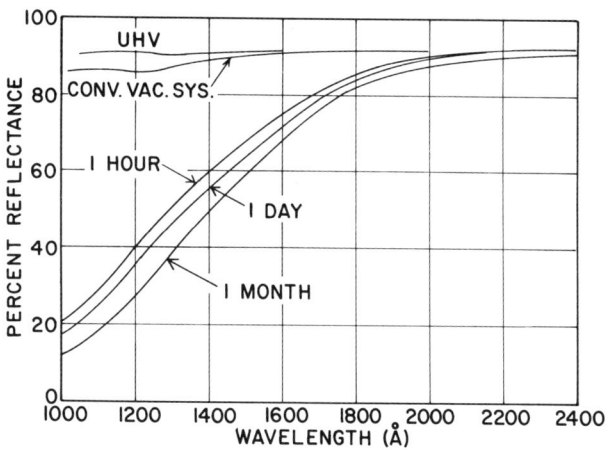

FIGURE 1 Reflectance of evaporated Al films before and after 1-h, 1-day, and 1-month exposure to air in the wavelength region from 1000 to 2400 Å.

*et al.,* 1971]. The loss in reflectance is not very great at wavelengths longer than 2000 Å but becomes even greater to shorter wavelengths.

A slight dip in the reflectance between 1200 Å and 1300 Å of the Al film freshly deposited in the conventional vacuum system on optically polished glass is caused by the surface plasmon oscillation [Feuerbacher and Steinmann, 1969]. This oscillation is coupled to the incident radiation through surface roughness. Therefore its effect is much less pronounced in the curve showing the reflectance of Al deposited on fire-polished glass because fire-polished glass is smoother than optically polished glass.

Feuerbacher and Steinmann [1969] have shown that this dip in reflectance shifts to longer wavelengths when the Al is overcoated with a dielectric film and that the dip is accentuated with increasing surface roughness. Under ordinary conditions the effect is small, but if the optical system has many reflections, there may be a significant loss of speed in the spectral region of the surface plasmon oscillation.

Figure 2 shows how rapid the loss in reflectance is on exposure to air for the wavelengths 1608 Å, 1216 Å, and 1026 Å. These measurements were made in a conventional vacuum system [Madden *et al.,* 1963] so that the reflectance was decreasing during the first 8 min even at pressures between $10^{-6}$ and $10^{-7}$ Torr. After 8 min in vacuum, the film was

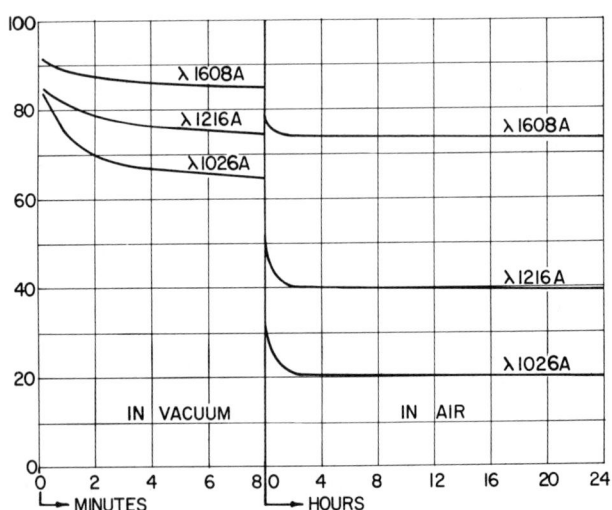

FIGURE 2  Effect of aging in vacuum and in air on the reflectance of evaporated Al at three different wavelengths in the vacuum ultraviolet.

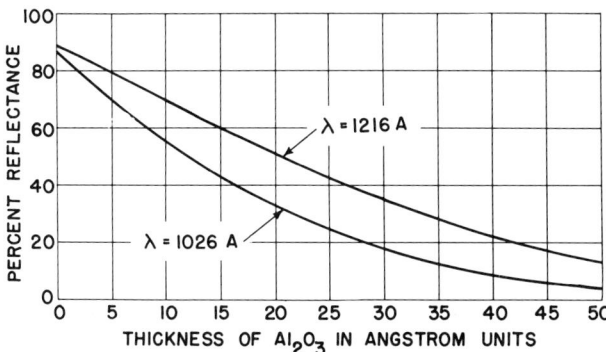

FIGURE 3 Calculated effect of oxide films of various thicknesses on the reflectance of Al at 1216 and 1026 Å.

exposed to air for a very short period before continuing measurements. During this short exposure the reflectance dropped from 75% at 1216 Å to about 50% and then continued to drop to 40% in the next 2 h. The loss in reflectance was even more pronounced at 1026 Å; from 65% to approximately 30% during the short exposure to air, and with a subsequent loss in 2 h to about 20%. At 1608 Å, the reflectance loss was much smaller.

Figure 3 shows the calculated reflectance of $Al + Al_2O_3$ as a function of oxide thickness at 1216 Å and 1026 Å. The optical constants used in the calculation were those reported by Hunter [1964] for Al and for $Al_2O_3$ values measured using evaporated films of $Al_2O_3$. According to the figure, an oxide thickness of 35 Å, approximately the terminal thickness of the natural oxide that forms on Al, reduced the reflectance to about 28% at 1216 Å, in agreement with Figure 1, and to about 12% at 1026 Å. This demonstrates that vacuum-ultraviolet reflectance measurements can be an extremely sensitive tool for studying the oxidation of Al and of other metals.

It has proven possible to preserve most of the high reflectance of Al coatings by overcoating them before formation of the oxide layer with a film material of high transparency in the vacuum ultraviolet [Hass and Tousey, 1959; Angel et al., 1961; Canfield et al., 1966; Cox et al., 1968; Hutcheson et al., 1972b]. The overcoating must be thick enough to prevent oxidation of the underlying Al surface by diffusion of oxygen through the coating. Overcoatings of $MgF_2$ and LiF have been used and result in coatings having high reflectance to wavelengths as short as 1200 Å and 1000 Å, respectively. The effect is more complicated than simply preserving the Al from oxidation because interference effects are involved; hence the thickness of the dielectric coating must be precisely controlled for optimum results, and the effect of evaporation conditions on the optical properties of the overcoating materials must be considered.

A special technique has been developed to monitor the thickness of the $MgF_2$ or LiF overcoating by measuring the reflectance of the Al + dielectric coating directly at 1216 Å as the coatings are applied [Hutcheson et al., 1972a]. Previously such monitoring was done using a quartz crystal microbalance [Feuerbacher et al., 1969] or by optical monitoring at longer wavelengths [Berning et al., 1960; Canfield et al., 1966]. Neither of these methods was really satisfactory because of the difficulty in relating the physical thickness indicated by the quartz crystal microbalance to the optical thickness of the film since the index of refraction of the film can change with evaporation conditions. Also, optical monitoring at longer wavelengths is rather insensitive because it is necessary to calculate and measure, for the particular monitoring wavelength, the loss in reflectance equivalent to a one-half wavelength thickness on Al at 1216 Å. The new technique became possible through the combined use of a specially designed hydrogen glow lamp (R. E. Ruskin, private communication, 1969) that emits mostly 1216 Å and an ion chamber [Byram et al., 1958] sensitive only to the wavelength region between 1140 Å and 1350 Å. The response of the detector is controlled on the short wavelength side by the cutoff of the $MgF_2$ window material and on the long wavelength side by the ionization limit of the NO gas with which it is filled.

Figure 4 is a line drawing of the hydrogen glow lamp, and Figure 5 shows its spectral intensity distribution as measured with an ion chamber

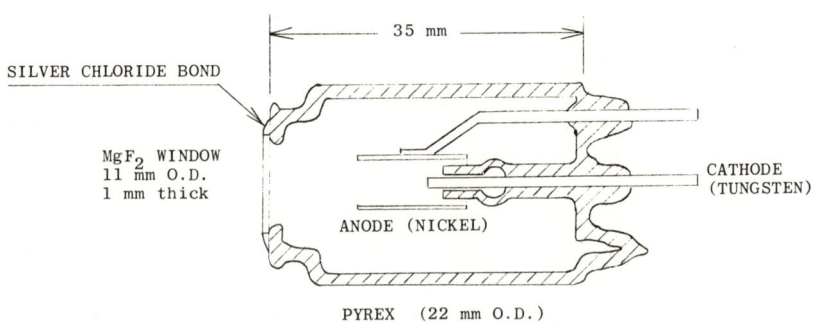

FIGURE 4 Cross-sectional drawing of the hydrogen dc glow discharge tube.

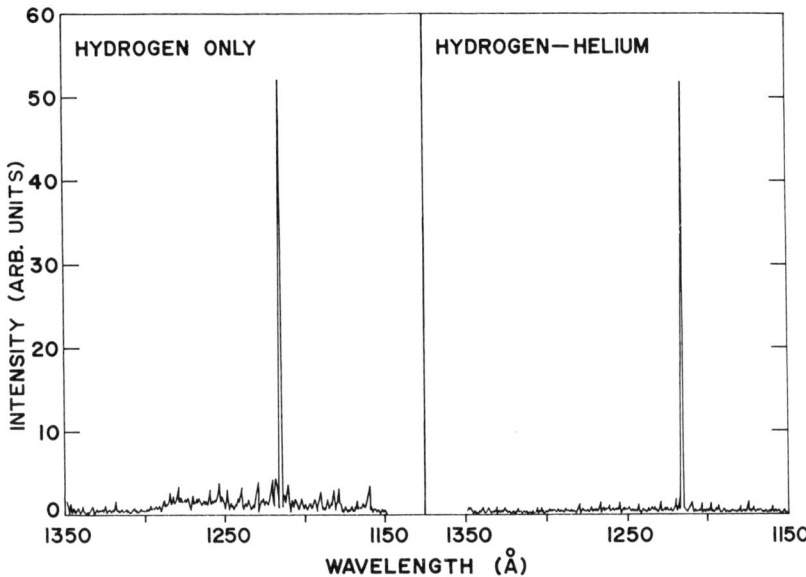

FIGURE 5  Spectral intensity distribution of glow discharge tubes filled with hydrogen only and hydrogen plus helium.

and a vacuum-ultraviolet monochromator that had a grating coated with Al + MgF$_2$. The discharge current in the lamp can be adjusted from about 0.2 to 5 mA, depending on the intensity desired and requires from 300 to 400 V during operation. A current-regulated power supply ensures excellent stability over periods of a half hour or more. Filling the lamp with a mixture of hydrogen and helium suppresses the many-lined hydrogen molecular spectrum below that obtained with only hydrogen in the lamp, as shown in Figure 5, and does not affect the intensity of the Ly-$\alpha$ line at 1216 Å. For this purpose, however, either filling proved satisfactory. Since the 1216 Å radiation from the source is many times more intense than the molecular spectrum within the sensitive region of the detector, the source–detector combination is essentially a monochromatic optical system for 1216 Å.

Figure 6 shows the monitoring signal during the fabrication of an Al + MgF$_2$ coating. Because the reflectance of the glass substrate was 13%, as determined by measurements using a reflectometer, the photometer-recorder was initially set at 13% of full scale. After deposition of the Al, the reflectance had risen to 88% of full scale. The monitoring signal was interrupted for about 15 sec by a shutter that shielded the fresh Al coating while the Al vapor source was cooling. During the delay after deposition of the Al and until the $MgF_2$ deposition started, the loss in reflectance of the Al coating was about 2%.

The $MgF_2$ deposition began at about 20 sec, and the curve shows the reflectance of the Al + $MgF_2$ coating as the thickness of the $MgF_2$ layer increased. Normally, to produce coatings with highest reflectance at Ly-$\alpha$, the deposition of $MgF_2$ would be stopped at the first half-wave thickness—the first maximum, as indicated in Figure 6. However, this deposition was deliberately continued to show the maxima and minima corresponding to the three, four, and five quarter-wave thicknesses.

The Ly-$\alpha$ monitor has also proven useful in monitoring the deposition of LiF layers over Al. Although the half-wave thickness of LiF at 1216 Å does not correspond to the thickness required for highest reflectance at 1026 Å, it is possible to obtain, routinely, coatings with optimum reflectance at 1026 Å by allowing the monitored reflectance to increase 3% beyond the quarter-wave thickness that is the first minimum.

Since use of the Ly-$\alpha$ monitor practically eliminates the variability of the thickness of the dielectric layer, it was possible to study the effect of substrate temperature and deposition rate of the dielectric coating on the reflectance of both $MgF_2$- and LiF-overcoated Al.

Figure 7 shows the effect of the $MgF_2$ deposition rate on the reflectance at 1216 Å of an Al + $MgF_2$ coating deposited at 40 °C. All the

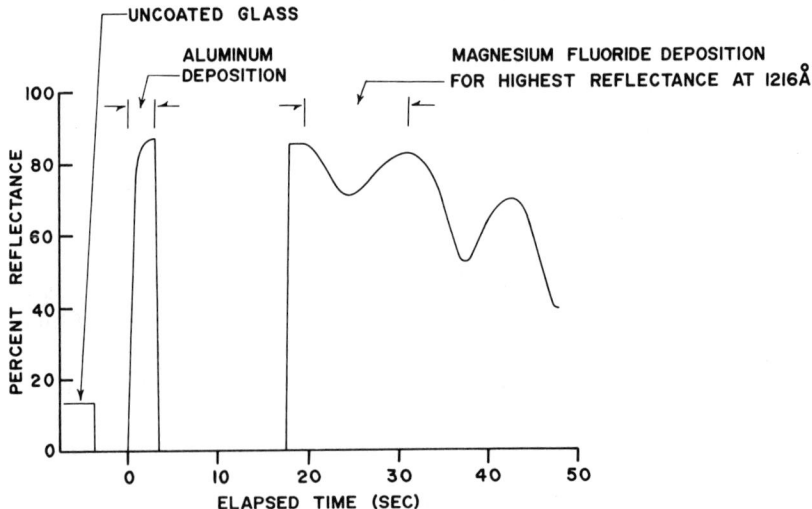

FIGURE 6  Trace of recorder output from Lyman-$\alpha$ film thickness monitor during Al and $MgF_2$ depositions.

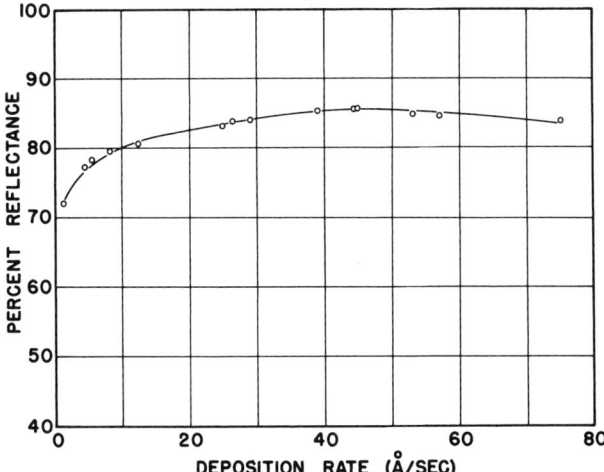

FIGURE 7 The effect of $MgF_2$ deposition rate on the reflectance at 1216 Å of Al films coated with 250 Å of $MgF_2$.

$MgF_2$ layers were one-half wavelength thick at 1216 Å. The reflectance rises with increasing deposition rate from 72% for an $MgF_2$ film deposited at a rate of 2 Å/sec to a maximum of 85.7% for an $MgF_2$ film deposited at a rate of about 45 Å/sec. For higher rates, the reflectance decreases slightly; at a rate of 75 Å/sec the reflectance was 84.1%. The initial increase in reflectance with increasing rate is attributed to the increasing purity and compactness of the $MgF_2$ films that can be obtained using rates up to 45 Å/sec. On the other hand, the higher temperatures needed to increase the evaporation rate beyond 45 Å/sec can also cause decomposition of the $MgF_2$ [Hacman, 1970], which is the probable explanation for the loss in reflectance at deposition rates higher than 45 Å/sec.

The increase in reflectance obtained by increasing the deposition rate of $MgF_2$ up to 45 Å/sec is not restricted to 1216 Å. Figure 8 shows the reflectance from 1000 Å to 2000 Å of two Al + $MgF_2$ coatings, deposited at 40 °C and at a rate of 8 and 45 Å/sec. At wavelengths longer than 1150 Å, the higher deposition rate resulted in significantly higher reflectances.

Using the optimum deposition rate of 45 Å/sec, no significant change in reflectance occurred as the substrate temperature was raised from 40 to 100 °C. When the substrate temperature was increased to 150 °C, however, the reflectance at 1216 Å was about 8% lower than that measured for a substrate held at 40 °C. This loss in reflectance was probably

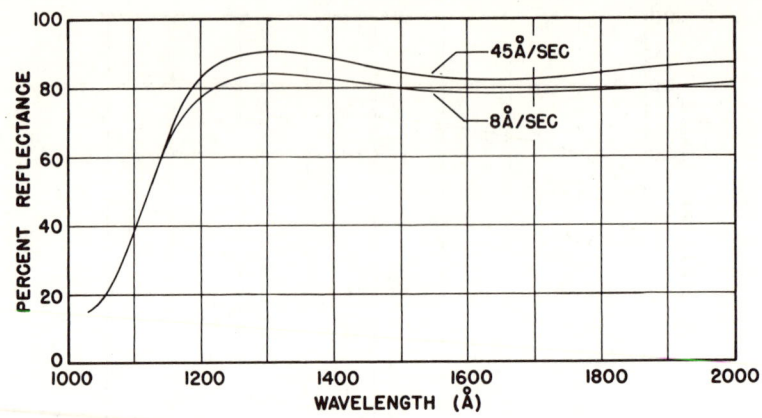

FIGURE 8 Reflectance in the wavelength region from 1000 to 2000 Å of Al coated with 250 Å of $MgF_2$ for $MgF_2$ deposition rates of 8 and 45 Å/sec.

caused by the increased crystal grain size and surface roughness known to occur in Al films deposited at substrate temperatures as high as 150 °C [Hass and Waylonis, 1961].

Coatings of Al + $MgF_2$ do not show any significant loss in reflectance, regardless of $MgF_2$ deposition conditions, for periods as long as 5 months. Previous work has shown that $MgF_2$-protected Al mirrors are very stable and are not generally affected by exposure to the atmosphere or even ultraviolet or charged-particle irradiation.

In contrast to the results obtained with $MgF_2$ overcoatings, the vacuum-ultraviolet reflectance of Al films coated with 140 Å of LiF did not show any significant dependence on LiF deposition rates over the range from 3.5 to 70 Å/sec; however, the reflectance of such coatings did depend strongly on the substrate temperature during deposition. At 1026 Å, for a deposition rate of approximately 15 Å/sec, increasing the substrate temperature from 40 to 100 °C caused the reflectance to increase from 74 to 81%. With a further increase in substrate temperature to 150 °C, the reflectance decreased to 77.5%. The improvement in reflectance with increasing substrate temperature up to 100 °C is probably due to increased crystallinity in the LiF film, which may result in a decrease in the absorption for wavelengths longer than 1000 Å and an increase in the index of refraction. The loss in reflectance for higher substrate temperatures is caused by the increase in roughness of the underlying Al film, just as with Al + $MgF_2$.

Figure 9 shows the reflectance of Al + LiF films, from 1000 to 2000 Å. The LiF deposition rate was 15 Å/sec, and the substrate tem-

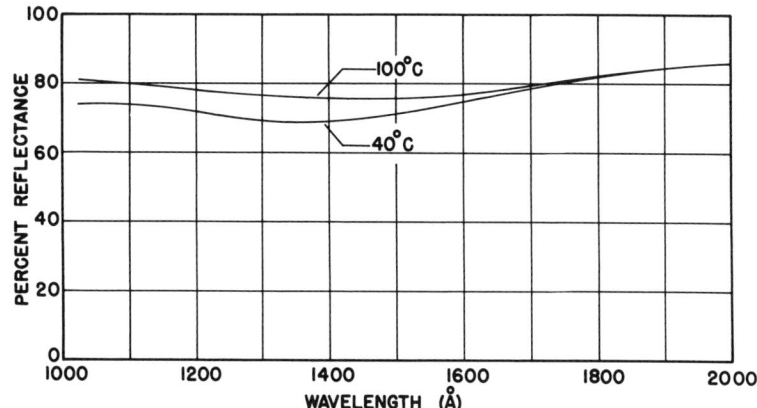

FIGURE 9 Reflectance in the wavelength region from 1000 to 2000 Å of Al coated with 140 Å of LiF deposited at substrate temperatures of 40 and 100 °C.

peratures were 40 and 100 °C. The reflectance obtained with a substrate temperature of 100 °C is higher than that obtained with a 40 °C substrate temperature at all wavelengths between 1026 and 1750 Å.

Although Al + LiF coatings have higher reflectances to shorter wavelengths than Al + $MgF_2$ coatings, the optical properties of LiF coatings are very sensitive to atmospheric water vapor, and the long-term stability of the vacuum-ultraviolet reflectances of Al + LiF coatings is strongly influenced by the environment in which the mirrors are stored. Infrared reflectance measurements made on Al coated with 5500 Å thick LiF films that had been exposed to atmospheric water vapor did not show the 3-$\mu$m water absorption band that is so evident for dielectric film materials such as $MgF_2$, $SiO_2$, and $CeO_2$ when they are deposited on Al to a thickness of a quarter-wave. This can be explained by the fact that the atmospheric water vapor reacts chemically with LiF to form a compound on the LiF film surface, which is absorbing throughout the entire vacuum ultraviolet [Patterson and Vaughan, 1963].

Experience has shown that the loss in reflectance of Al + LiF mirrors was retarded considerably by storing the mirrors in a desiccator even though the desiccator was opened frequently to remove mirrors for measurement during this aging period. The loss in reflectance of mirrors stored 2 months in ambient air was much greater than the loss in reflectance of mirrors stored 5 months in a desiccator, as shown in Table 1. The ambient air had 50 to 70% relative humidity (July–August).

Table 2 shows a comparison of the aging of Al + LiF films, which were deposited at three different substrate temperatures, after being

TABLE 1  Comparison of the Effects of Aging in Ambient Air and in a Desiccator on the Vacuum-Ultraviolet Reflectance of Al + LiF Coatings Made at 40 °C

|  | Ambient Air | | Desiccator | |
|---|---|---|---|---|
|  | Fresh | Aged 2 Months | Fresh | Aged 5 Months |
| $\lambda$ (Å) | Percent Reflectance | | | |
| 1026 | 74.0 | 41.6 | 74.4 | 63.4 |
| 1216 | 71.6 | 43.6 | 72.4 | 65.6 |
| 1608 | 74.5 | 46.9 | 73.0 | 62.3 |
| 2000 | 86.1 | 50.5 | 85.2 | 81.2 |

stored in a desiccator for 4 months. Although the actual change in reflectance with time was similar for all three films, the Al + LiF film that was deposited at a substrate temperature of 100 °C still had a higher reflectance than the films deposited at substrate temperatures of 40 and 150 °C, even after long-term aging.

Adriaens and Feuerbacher [1971] report that Al + dielectric coatings prepared in ultrahigh vacuum show an improvement in stability and a reflectance increase at wavelengths less than 1250 Å if they are annealed *in vacuo* ($10^{-7}$ Torr) at 300 °C for 60 h; however, there is also an accompanying decrease in reflectance for wavelengths longer than 1250 Å.

TABLE 2  Comparison of the Effects of Aging after 4 Months' Storage in a Desiccator on the Vacuum-Ultraviolet Reflectance of Al + LiF Coatings Made at Three Different Substrate Temperatures

|  |  | Substrate Temperature | | |
|---|---|---|---|---|
|  |  | 40 °C | 100 °C | 150 °C |
| $\lambda$ (Å) |  | Percent Reflectance | | |
| 1026 | Fresh | 74.4 | 81.0 | 77.5 |
|  | Aged | 63.4 | 76.7 | 72.1 |
| 1216 | Fresh | 72.4 | 77.6 | 75.0 |
|  | Aged | 65.6 | 66.0 | 59.9 |
| 1608 | Fresh | 73.0 | 77.0 | 76.2 |
|  | Aged | 62.3 | 68.1 | 63.2 |
| 2000 | Fresh | 85.2 | 84.4 | 81.5 |
|  | Aged | 81.2 | 82.3 | 77.7 |

Attempts to reproduce their results with coatings produced and annealed in conventional vacuum systems proved futile.

It would be incorrect to conclude that mirrors of Al overcoated with $MgF_2$ or LiF are useless for wavelengths less than 1000 Å. They are not so efficient as the Pt group metals, to be discussed in the next section, but they can be used to wavelengths as short as 500 Å [Hunter *et al.*, 1971]. Figure 10 shows the reflectance of an Al + $MgF_2$ coating, from 300 to 1600 Å, when the $MgF_2$ layer is 250 Å thick. It was measured at the three angles of incidence commonly encountered in vacuum-ultraviolet spectrometers: normal incidence, 35° corresponding approximately to the Seya-Namioka mounting, and grazing incidence. Practically, there is no difference between the reflectance measured at normal and 35° incidence; both average about 12–15% reflectance between 1000 and 500 Å. At wavelengths less than 500 Å, the reflectance decreases as the wavelength decreases, until at 300 Å the reflectance is about 1% or less.

Al + LiF coatings show the same general behavior as Al + $MgF_2$ coatings. When the LiF overcoating is 140 Å thick, the mirror has an average reflectance at normal incidence and 35° angle of incidence of about 10% between 1000 and 500 Å. Toward shorter wavelengths, the reflectance drops to about 1% or less at 300 Å.

At grazing incidence, both types of coating have reflectances of be-

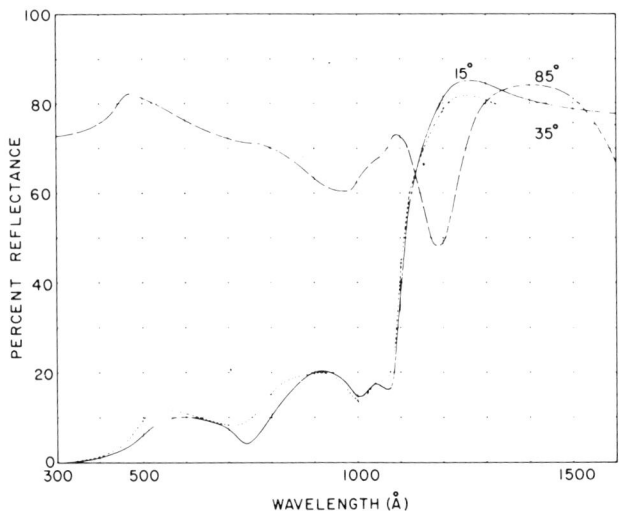

FIGURE 10 Measured reflectance of an Al + $MgF_2$ mirror from 300 to 1600 Å for three angles of incidence. The $MgF_2$ thickness is 250 Å.

tween 70 and 80%, which, although not as high a reflectance as that of aged Al, is still quite useful.

Al + MgF$_2$ mirrors with an effective half-wave thickness of MgF$_2$ at 1216 Å are also useful as detectors of contamination [Hass and Hunter, 1970; Gillette and Kenyon, 1971]. The effective thickness of the MgF$_2$ layer at 2000 Å, for such a mirror, is a quarter-wave. This means that at 2000 Å standing waves in the MgF$_2$ layer have a loop at the surface, and at 1216 Å they have a node at the surface. Hence any thin absorbing layer on the MgF$_2$ layer surface will cause very little loss in reflectance at 1216 Å but a large loss at 2000 Å. This property of Al + MgF$_2$ mirrors is being utilized in the design of a contamination monitor that measures the reflectance at 1216 and 2000 Å simultaneously to determine the onset and rate of contamination.

## III. Reflecting Coatings for Wavelengths Shorter Than 1000 Å—Pt, Ir, Os, Re, and W

A number of metals have been found to be useful as reflecting coatings for wavelengths less than 1000 Å. They are Pt, Ir, Os, Re, W, Rh, and Au, all second and third series transition metals. Some reflectance studies by Juenker *et al.* [1968], using solid samples of Ta and Mo, indicated that these metals may also be useful as reflecting films, but they have not yet been studied in that form. Au and Rh have been discussed elsewhere [Canfield *et al.*, 1964; Cox *et al.*, 1971] and will not be considered in this paper. Thus the metals to be discussed will be Pt, Ir, Os, Re, and W.

A general remark may be made about these five metals. They must be heated to quite high temperatures to obtain reasonable deposition rates (50–100 Å/sec). The vapor pressure of W remains quite low until it melts, which excludes the evaporation of W from hot filaments of itself. The other metals readily alloy with tungsten filaments or boats at high temperatures, therefore heating with a high-powered electron gun has been found to be the most convenient method for evaporating them.

### A. PLATINUM

Early measurements of the reflectance of Pt films [Gleason, 1929; Sabine, 1939] led to the conclusion that it was the best coating material for wavelengths less than 1000 Å. Surprisingly enough, despite its widespread use, systematic studies of the optical properties of Pt films have been reported for only two wavelengths less than 1000 Å: 736 and 584 Å [Jacobus et al., 1963].

Figure 11 shows the normal incidence reflectance of an opaquely reflecting film of Pt, deposited on a glass substrate at 300 °C, from 2000 to 150 Å. The reflectance decreases more or less uniformly from a value of about 27% at 2000 Å to about 14% at 700 Å. Toward shorter wavelengths there is a sharp increase in reflectance, reaching a maximum value of 24% at 570 Å. At even shorter wavelengths, the reflectance drops, goes through a second, small maximum of about 13% at 400 Å, and then drops to values of a few percent or less at wavelengths approaching 150 Å.

Jacobus et al. [1963] also studied the effect of evaporation conditions and aging on the reflectance of Pt films. Since their study has been the only one of its kind to date, their findings will be briefly summarized below.

The most important parameter in the preparation of Pt films is the substrate temperature. Pt films deposited on glass substrates at high substrate temperatures have higher reflectances than those deposited at room temperature (40 °C). Elevated substrate temperatures also improve adhesion of the film to the substrate and help to prevent the tendency of thick films to crack and peel off the substrate. Thin films deposited at room temperature do not show this tendency.

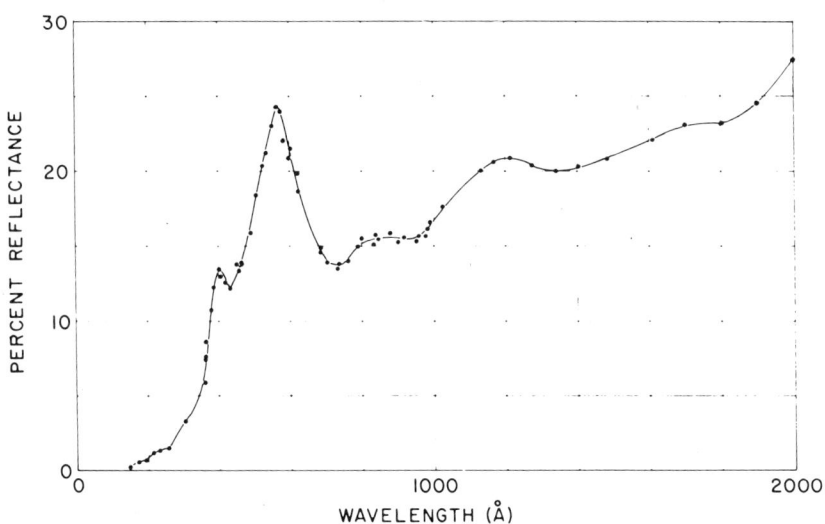

FIGURE 11  Reflectance of opaquely reflecting Pt deposited on glass at 300 °C in the wavelength region from 150 to 2000 Å.

Thin films of Pt on glass substrates have a higher reflectance than opaque films. This enhanced reflectance is caused by constructive interference between the wavefronts reflected from the metal–vacuum and the metal–glass interfaces. Later studies showed that this is a characteristic of all the second and third series transitions metals studied thus far.

The reflectance of Pt films is not sensitive to the pressure in the evaporator during deposition because of the inertness of Pt. Varying the pressure between $10^{-4}$ and $10^{-5}$ Torr had no effect on the reflectance.

There is no appreciable loss in reflectance with age of Pt films. Films stored in the laboratory did show a slight loss of 1 to 2% over the period of a year, but if they had been deposited on heated substrates the original reflectance could be restored by cleaning the surfaces with collodion. Collodion cleaning of Pt films deposited on unheated substrates may remove the film from the substrate because of the poor adhesion between film and substrate.

A study is now under way at the Naval Research Laboratory to measure more completely the optical properties of Pt in the vacuum ultraviolet.

B. IRIDIUM

The reflectance and optical constants of Ir have been reported by Hass et al. [1967] in the wavelength range from 500 to 2000 Å. Figure 12,

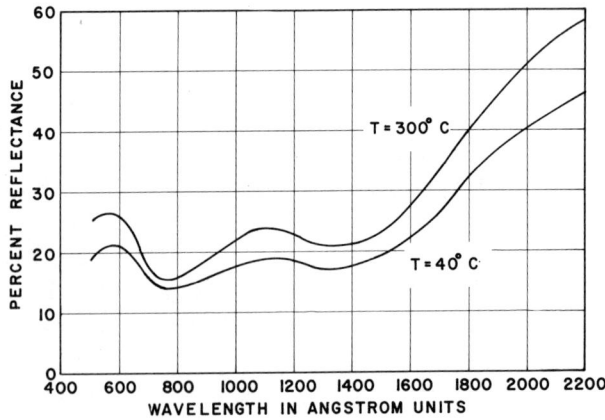

FIGURE 12  Reflectance of opaque Ir films deposited at two substrate temperatures as a function of wavelength from 500 to 2200 Å.

taken from their paper, shows the reflectance of opaquely reflecting Ir films deposited on a hot (300 °C) and on a cold (40 °C) substrate. The deposition rate was about 30 Å/sec, and the pressure was $10^{-5}$ Torr during deposition. The reflectance spectrum is similar to that of Pt, although the reflectance values are higher than those of Pt. The long wavelength drop in reflectance, from 2200 to 1600 Å, also occurs with Pt films but at longer wavelengths so that it is not shown in Figure 11. A broad, low maximum occurs at about 1100 Å and then, toward shorter wavelengths, there is a sharp increase in reflectance to a maximum value of 27% at 540 Å. Recent measurements made at wavelengths shorter than 500 Å have shown that a shoulder exists in the reflectance curve at 400 Å, where Pt has a small reflectance maximum, and that the reflectance drops to values of a few percent at wavelengths as short as 300 Å.

Over the wavelength region shown in the figure, the reflectance of the film deposited at 300 °C is considerably higher than the reflectance of the film depoisted at 40 °C. Furthermore, the increase in reflectance with increasing substrate temperature is not restricted to the vacuum-ultraviolet region but extends into the visible and infrared. In addition, films prepared on heated substrates show better adhesion.

Owing to the inert nature of Ir, the reflectance of Ir films shows only slight dependence on the deposition rate and pressure during evaporation. For depositions on substrates at 300 °C, an increase of deposition rate from 10 to 100 Å/sec and a decrease of the pressure from $5 \times 10^{-5}$ to below $10^{-5}$ Torr had practically no effect on the reflectances of the films. For room-temperature films, however, an increase of reflectance with increasing deposition rate and decreasing pressure was noticeable.

Figure 13 illustrates the enhancement of reflectance obtained by the use of thin, semitransparent Ir films. The semitransparent film was about 150 Å thick, and the other film was opaquely reflecting in the vacuum ultraviolet. Both were deposited at a rate of about 30 Å/sec on glass substrates at 300 °C. For all wavelengths less than 1700 Å, the reflectance of the thin film is higher than that of the opaque film; while at longer wavelengths, and in the visible region, opaque films are more efficient reflectors. For monitoring the film thickness during its deposition, it is important to know that a film thickness of 150 Å deposited at 300 °C corresponds to a transmittance at 5500 Å of about 8%.

Ir films show practically no change in reflectance in the vacuum ultraviolet during storage in air. For most films, reflectances measured a few hours after deposition were the same 6 months later. Any loss in reflectance during storage in air can be restored by rinsing with alcohol or cleaning with collodion.

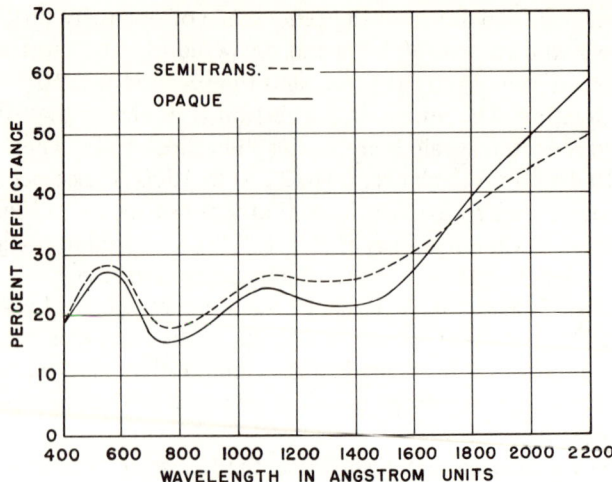

FIGURE 13 Reflectance of opaque and semitransparent ($t$ = 150 Å) Ir on glass as a function of wavelength from 400 to 2200 Å. Substrate temperature 300 °C.

C. OSMIUM

Recent studies have shown that evaporated films of Os [Cox et al., 1973] make excellent reflecting coatings for wavelengths less than 1000 Å. The Os was evaporated in an ion-titanium pumped system and deposited at a rate of 50 Å/sec at a pressure of about $10^{-6}$ Torr. Figure 14 shows the reflectance of two Os films from 300 to 2000 Å, one approximately 170 Å thick and the other opaquely reflecting in the vacuum ultraviolet. Both were deposited on superpolished fused quartz at a substrate temperature of 300 °C. As with Pt and Ir, there is a decrease in reflectance from 2000 to about 1400 Å, then a somewhat slower decrease to about 750 Å. For wavelengths below 750 Å, the reflectance increases rapidly to a value of 35% for the semitransparent film and 29% for the opaque film at 600 Å. Toward even shorter wavelengths, the reflectance spectrum differs somewhat from those of Pt and Ir in that the maximum and shoulder at 400 Å found in the first two metals is not present in Os, although there is an asymmetry to the reflectance maximum at 600 Å that suggests that, due to the different structure of Os, the feature at 400 Å in Ir and Pt has been shifted to about 500 Å, much closer to the large reflectance maximum.

Table 3 shows, in more detail, the dependence of the reflectance of Os films on the film thickness. The reflectance at different wavelengths

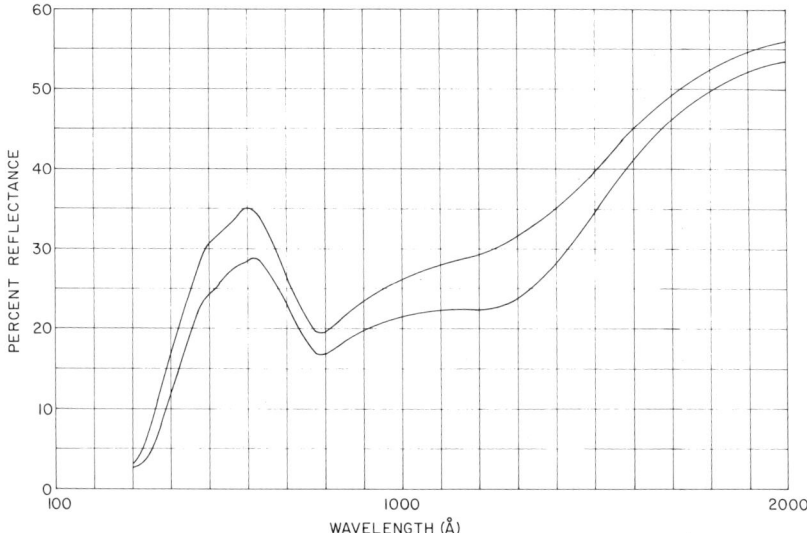

FIGURE 14  Reflectance of opaque (lower curve) and semitransparent (upper curve) ($t$ = 170 Å) Os deposited on super-polished fused quartz in the wavelength region from 300 to 2000 Å.

TABLE 3  Reflectance of Os Films Deposited on Glass at 300 °C as a Function of Thickness and Transmittance at $\lambda 5461$ Å for Various Wavelengths in the Vacuum Ultraviolet

|  | Percent Transmittance at $\lambda 5461$ Å | | | | | | |
| --- | --- | --- | --- | --- | --- | --- | --- |
|  | 0.2 | 0.5 | 1.0 | 2.0 | 5.0 | 10.0 | 15.0 |
|  | Thickness (Å) | | | | | | |
|  | 675 | 560 | 465 | 370 | 235 | 140 | 95 |
| $\lambda$ (Å) | Percent Reflectance | | | | | | |
| 584 | 26.0 | 26.7 | 27.7 | 29.1 | 31.4 | 32.2 | 30.4 |
| 1026 | 21.2 | 21.5 | 22.1 | 23.0 | 24.9 | 25.7 | 24.8 |
| 1216 | 21.5 | 22.0 | 22.6 | 23.9 | 26.4 | 27.8 | 27.5 |
| 1608 | 41.0 | 41.8 | 43.0 | 44.9 | 46.9 | 43.7 | 37.8 |
| 2000 | 53.3 | 54.5 | 55.6 | 57.3 | 59.2 | 54.0 | 47.0 |

in the vacuum ultraviolet is given, as well as the thickness and the transmittance at 5461 Å, which are useful in monitoring the film thickness during deposition. At all the wavelengths shown, there is a pronounced reflectance maximum that occurs at about 140 Å for wavelengths of 1216 Å and less and shifts to about 235 Å for wavelengths longer than 1216 Å and up to 2000 Å. The difference between the maximum value of 32.2% shown in the table for 584 Å and the value of 34% shown in Figure 14 may be caused by additional roughness of the glass substrate, used in obtaining the results for the table, over that of the superpolished fused quartz substrate used to obtain the values for the figure.

Unlike Pt and Ir, the reflectance of Os films is not very sensitive to substrate temperature. Increasing the substrate temperature from 40 to 300 °C during deposition causes an increase in reflectance of only 1 to 2%. Because of this property, Os is a very useful metal for coating replica gratings that cannot be heated much beyond 40 °C during the deposition.

Loss in reflectance of Os films during storage in air is also rather small. A loss of ½% in one month after deposition during storage in a desiccator, and of about 1% during storage in air, is characteristic of the films. This demonstrates that the oxide film that forms on Os must be very thin.

D. RHENIUM

Some earlier reflectance measurements by Juenker *et al.* [1968] on solid samples of Re and W indicated that coatings of these metals might be efficient reflectors. Accordingly a study of evaporated films of Re and W was undertaken [Cox *et al.*, 1972], and the results are summarized here.

Figure 15 shows the measured reflectances of two Re films, one 160 Å thick ($T_{5500 \text{ Å}} = 16\%$) and the other 480 Å thick ($T_{5500 \text{ Å}} = 2\%$). The 480 Å thick film can be considered opaque from 300 to 2000 Å. The figure also shows that there is a strong dependence of reflectance on film thickness but that the film thickness has very little effect on the reflectance versus wavelength characteristic.

At 2000 Å, the reflectance of Re is decreasing with wavelength and reaches a minimum at about 1200 Å. Toward shorter wavelengths there is a broad maximum between 1000 and 1100 Å, and at even shorter wavelengths the reflectance maximum has a value of 36% for a film thickness of 160 Å and is quite high for this region of the spectrum. For comparison, the highest reflectance of thin Os films occurs at 600 Å and is 35%. At wavelengths shorter than 630 Å, the reflectance decreases rapidly to values of only a few percent at 300 Å. In all respects, the re-

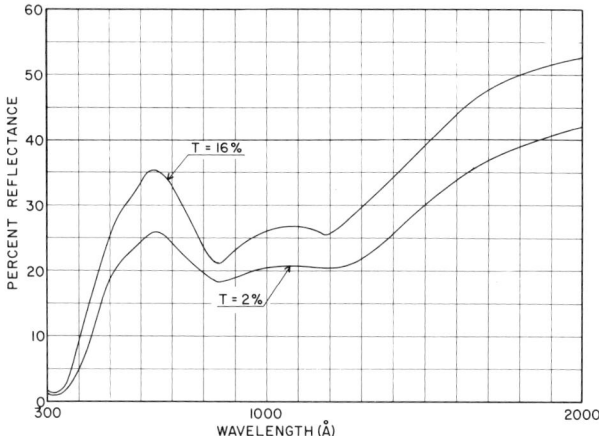

FIGURE 15  Reflectance of evaporated Re films approximately 160 Å thick ($T_{5500\,\text{Å}} = 16\%$) and 480 Å thick ($T_{5500\,\text{Å}} = 2\%$), deposited on glass substrates at 40 °C.

flectance spectrum of Re is similar to that of Os, including the asymmetry of the main reflectance maximum at 630 Å.

The dependence of reflectance on thickness is shown in some detail in Table 4 for different wavelengths in the vacuum ultraviolet. At 584 Å, the peak reflectance occurs for thicknesses between 120–150 Å and is approximately 33%. For an opaque film, the reflectance drops to 24%.

TABLE 4  Reflectance of Re Films Deposited on Glass at 40 °C as a Function of Thickness and Transmittance at $\lambda 5500$ Å for Various Wavelengths in the Vacuum Ultraviolet

|  | Percent Transmittance at $\lambda 5500$ A | | | | | | |
| --- | --- | --- | --- | --- | --- | --- | --- |
|  | 1 | 2 | 5 | 10 | 15 | 20 | 25 |
|  | Thickness (A) | | | | | | |
|  | 590 | 480 | 330 | 215 | 150 | 120 | 90 |
| $\lambda$ (Å) | Percent Reflectance | | | | | | |
| 584 | 24.0 | 25.0 | 28.0 | 31.0 | 33.0 | 33.0 | 29.0 |
| 736 | 22.5 | 23.0 | 25.0 | 27.0 | 30.0 | 30.0 | 27.0 |
| 1026 | 21.0 | 21.0 | 23.0 | 25.0 | 26.5 | 27.0 | 26.0 |
| 1216 | 21.0 | 22.0 | 23.0 | 25.5 | 27.0 | 28.0 | 27.0 |
| 1608 | 36.0 | 37.0 | 41.0 | 45.0 | 44.5 | 40.5 | 36.0 |
| 2000 | 43.5 | 44.0 | 48.0 | 53.0 | 54.0 | 49.5 | 43.5 |

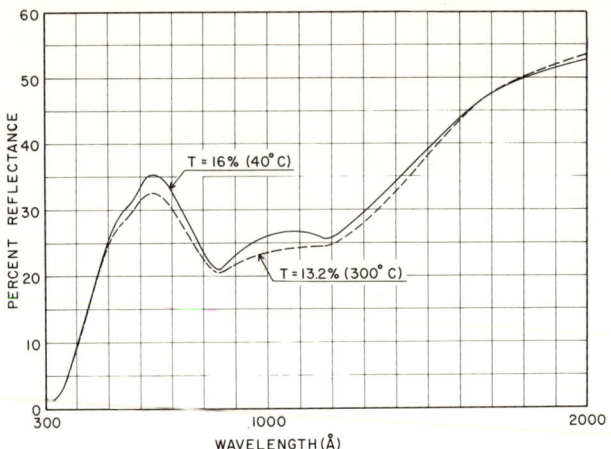

FIGURE 16 Reflectance of semitransparent Re films, approximately 140 Å thick, deposited on glass substrates at 40 and 300 °C.

This characteristic of Re films is common to all the wavelengths contained in the table, and the optimum thickness for Re on glass substrates lies between 120 Å and 150 Å.

The dependence of reflectance on substrate temperature is shown in Figure 16. Substrate temperatures of 40 and 300 °C were used, and the film thicknesses were about 140 Å, the optimum thickness according to Table 4. Over most of the wavelength region studied, the reflectance dropped as the substrate temperature rose. At wavelengths longer than 1800 Å, however, there was a slight increase in reflectance. This dependence is not very strong, amounting at most to a few percent. It appears unlikely that the loss in reflectance is caused by scattering due to increased roughening of the film as the substrate temperature is increased, because the loss in reflectance for wavelengths less than 500 Å, where scattering would be most effective, is negligible. As with Os, the fact that Re films deposited on room-temperature substrates have high reflectance makes Re a useful material for coating replica gratings. In fact, the recent successful photographs of the earth's geocorona [Carruthers and Page, 1972] obtained by a camera taken to the moon on Apollo 16 were obtained using Re-coated optics.

Re films show very little loss in reflectance on initial exposure to air. For example, an Re film deposited on a 40 °C substrate had an initial reflectance at 584 Å of 27.9%, which did not change until exposure to air when it dropped to 26.2% after 15-min exposure. After 4 days' exposure the measured reflectance was 24.6%. This change is probably due to oxidation, because the transmittance at 5500 Å increased from an initial value *in vacuo* of 14.5% to 15.9% after 4 days' exposure to air. Presumably the conversion of Re to its oxide decreased the optical depth of the film, which increased the transmittance.

On exposure to air containing traces of plasticizers or fumes of plastic cement, a hazy, bluish scattering coating accumulates very rapidly on Re films, and ellipsometric measurements show correspondingly large changes in the relative phase change. The nature of this layer is not known, but it may be caused by catalytic reaction on the Re surface.

Re films that have been stored in a desiccator do not acquire the scattering layer, although their reflectances decrease somewhat. For example, at 584 Å a film approximately 140 Å thick, which had an initial reflectance of 33%, decreased in reflectance in two months to 29%. At that time, a collodion cleaning caused the reflectance to increase to 30%. Similar behavior occurs for films of different thicknesses and at other wavelengths in the vacuum ultraviolet up to 2000 Å.

E. TUNGSTEN

The reflectance spectra of four W films are shown in Figure 17. The two solid curves represent the reflectance resulting from a substrate tempera-

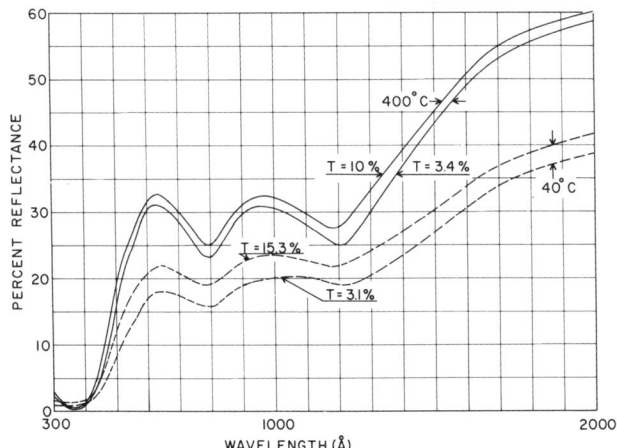

FIGURE 17 Reflectance of W films deposited on glass substrates at 40 and 400 °C. The approximate thicknesses of the W films are 120 Å ($T_{5500 Å}$ = 15.3%), 150 Å ($T_{5500 Å}$ = 10%), and 370 Å ($T_{5500 Å}$ = 3.4 and 3.1%).

ture of about 400 °C, and the two dashed curves are for a substrate temperature of 40 °C. Unlike Re films, low-temperature W films have considerably lower reflectance than those deposited on high-temperature substrates.

At 2000 Å, the reflectance is decreasing with decreasing wavelength and reaches a minimum at 1200 Å. To shorter wavelengths, the reflectance rises to a maximum at 970 Å, decreases, and again reaches a maximum at 630 Å. At shorter wavelengths, the reflectance becomes quite low. For wavelengths less than 1000 Å, the two reflectance maxima, at 970 and 630 Å, are almost equal and have values of about 33%. At 970 Å, this is the highest reflectance for an evaporated film yet reported.

Figure 17 also illustrates the effects of thickness on the reflectance of W films deposited at 40 and 400 °C. At both substrate temperatures, the reflectances of the thin films, 120 Å thick ($T_{5500 \text{ Å}} = 15.3\%$) and 150 Å thick ($T_{5500 \text{ Å}} = 10\%$), are higher than those of the thick films, which are about 370 Å thick. This dependence is less pronounced at the higher substrate temperatures.

Table 5 lists measured reflectances as a function of thickness at different wavelengths in the vacuum ultraviolet for a substrate temperature of 400 °C. At 584 Å, there is a reflectance maximum of 32% at a thickness of 190 Å that drops to 26.5% for an opaquely reflecting film. This type of behavior occurs at all the wavelengths listed, and an optimum thickness appears to be 190 Å for this spectral region.

TABLE 5  Reflectance of W Films Deposited on Glass at 400 °C as a Function of Thickness and Transmittance at $\lambda 5500$ Å for Various Wavelengths in the Vacuum Ultraviolet

| | Percent Transmittance at $\lambda 5500$ Å | | | | | |
|---|---|---|---|---|---|---|
| | 1 | 2 | 5 | 10 | 15 | 20 | 25 |
| | Thickness (Å) | | | | | |
| | 570 | 460 | 310 | 190 | 130 | 110 | 85 |
| $\lambda$ (Å) | Percent Reflectance | | | | | |
| 584 | 26.5 | 28.0 | 31.0 | 32.0 | 31.5 | 29.2 | 25.5 |
| 736 | 24.0 | 25.0 | 26.5 | 28.0 | 28.5 | 27.5 | 26.5 |
| 1026 | 27.5 | 29.0 | 31.5 | 32.5 | 33.0 | 32.0 | 31.0 |
| 1216 | 23.0 | 24.0 | 26.0 | 28.0 | 29.0 | 29.5 | 29.5 |
| 1608 | 46.0 | 48.0 | 51.5 | 51.5 | 50.5 | 45.0 | 38.0 |
| 2000 | 57.5 | 58.0 | 61.0 | 61.5 | 61.0 | 55.5 | 42.5 |

TABLE 6 Reflectance of W Films as a Function of Substrate Temperature for Various Wavelengths in the Vacuum Ultraviolet for Films with a Transmittance of about 12% at $\lambda = 5500$ Å

| | Substrate Temperature | | | | |
|---|---|---|---|---|---|
| | 40 °C | 150 °C | 300 °C | 400 °C | 500 °C |
| $\lambda$ (Å) | Percent Reflectance | | | | |
| 584 | 20.0 | 24.5 | 29.6 | 32.3 | 33.3 |
| 736 | 19.8 | 22.8 | 26.6 | 28.5 | 29.4 |
| 1026 | 24.0 | 27.8 | 31.7 | 33.7 | 34.1 |
| 1216 | 22.3 | 26.0 | 28.0 | 29.0 | 30.0 |
| 1608 | 35.0 | 41.0 | 48.0 | 51.6 | 53.0 |
| 2000 | 42.0 | 50.5 | 58.9 | 62.5 | 63.7 |

Substrate temperature has a much greater effect on the reflectance of W films than the thickness. The reflectances of the two thin films shown in Figure 17 clearly illustrate this effect. At 630 Å, these films have reflectances of 22 and 33% for substrate temperatures of 40 and 400 °C, respectively. Table 6 lists measured reflectance data that show this dependence in more detail for a W film approximately 170 Å thick. At 584 Å, a 40 °C film reflects 20%, but at 400 °C the reflectance has increased to 32.3%. At 2000 Å the corresponding reflectance increase is from 42 to 63.7%. Increasing the substrate temperature to values higher than 400 °C results in only a very slight increase in reflectance, so that 400 °C can be considered to be a limiting substrate temperature. Films deposited at low substrate temperatures (40 °C) are of no practical use because of their low reflectance.

Loss of reflectance of W films on initial exposure to air is approximately 1 to 1.5%. At 584 Å, the total decrease in reflectance is about 5% in ambient air and 3 to 4% in a desiccator during 2 months' storage. Reflectance losses decrease with increasing wavelength. Juenker *et al.* [1968] observed similar reflectance losses for their solid W samples—about 5% at 537 Å.

The initial reflectance loss and the loss occurring during storage in a desiccator are caused by the formation of an oxide film. Since the reflectance of aged W films reaches a steady value, the oxide film that forms must be stable and must reach a terminal thickness, in the same manner as the oxide that forms on Al films. It is much thinner, however, than the aluminum oxide film that reaches a terminal thickness of between 35 and 40 Å. For example, a loss in reflectance at 584 Å of 5%

corresponds to an oxide film thickness of about 10 Å. Most of this loss occurs during initial exposure to air, the loss incurred by subsequent exposure in a desiccator being rather small. If the tungsten oxide were to grow to the same thickness as the aluminum oxide, the reflectance at 584 Å would decrease to about 15%, a lower value than the reflectance of Pt or Ir at this wavelength, and would make W films useless as reflecting coatings for the vacuum ultraviolet.

## IV. Two-Layer Coatings—Ir + Al

Since it appears that no single film material exists that has a high reflectance over a very extended region in the vacuum ultraviolet, calculations were made to find a film combination that might have a high reflectance from about 500 to 2000 Å. Berning *et al.* [1960] made some calculations showing that the reflectance of Pt films could be increased at 584 and 736 Å by overcoating them with an Al film of the appropriate thickness. Hass and Hunter [1967] have calculated the effect of overcoating Ir with Al and have shown that efficient reflecting coatings are possible over the wavelength region from 500 to 2000 Å.

Figure 18 shows the calculated reflectance of Ir overcoated with Al from 500 to 2000 Å. The Al layers were 120 and 265 Å thick, the thicknesses necessary to optimize the reflectance at 584 and 736 Å, respectively. The reflectance of unoxidized Al and bare Ir are shown on the same figure as dashed curves for comparison. For wavelengths longer than 850 Å, unoxidized Al is unquestionably the best reflector. At shorter wavelengths, however, Ir coated with the proper thickness of Al

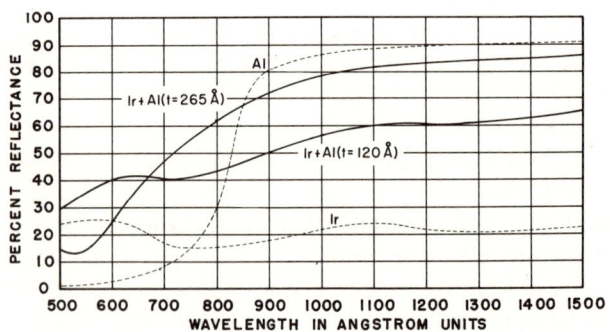

FIGURE 18 Calculated reflectance of Ir overcoated with different thicknesses of Al as a function of wavelength from 500 to 1500 Å. The dashed lines show the reflectances of Ir and Al for comparison.

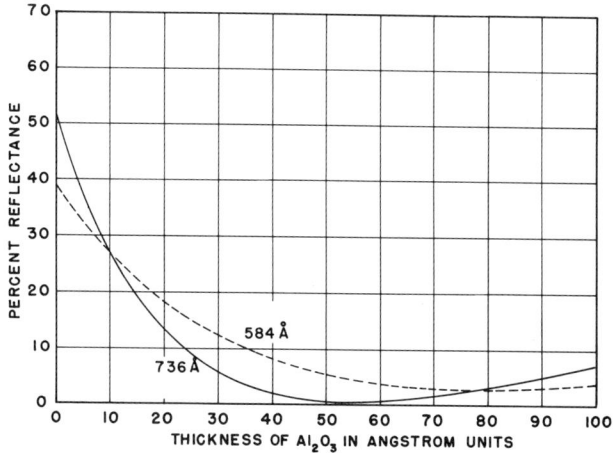

FIGURE 19 Calculated effect of a surface film of $Al_2O_3$ on the reflectance of Ir coated with Al ($t = 265$ Å) as a function of the $Al_2O_3$ thickness at 736 and 584 Å.

is considerably more efficient than plain Al or Ir and still shows a rather high reflectance at longer wavelengths. An Al overcoating of 120 Å gives a more uniform reflectance over the entire wavelength region, while a layer of Al 265 Å thick is superior for all wavelengths longer than 670 Å.

Both of these studies pointed out that if an oxide layer formed on the Al film, its high absorptance would cause a drastic reduction in reflectance. Figure 19 shows the calculated effect of thin surface films of $Al_2O_3$ on the reflectance of Ir + Al coatings at 736 and 584 Å. At 736 Å, the Al coating was 265 Å thick for maximum reflectance at this wavelength. As the thickness of the oxide film increases, the reflectance decreases until, at an oxide thickness of 35 Å, the reflectance is about 3%, which is much less than the reflectance of bare Ir. A smaller but still intolerable reflectance decrease is caused by the oxidation of Al at 584 Å on Ir + Al coatings designed for highest reflectance at this wavelength. This undesired behavior is also to be expected at all wavelengths from 500 to 1500 Å, because $Al_2O_3$ is highly absorbing over this wavelength region. Furthermore, film materials with low absorptance over this entire wavelength region that could be used as oxidation-preventing protective coatings are unknown.

Such film combinations must be *produced* and *stored* in an environment where the partial pressure of oxygen is low enough so that no appreciable oxide layer forms during the time of their use. Since orbiting satellites may exist in such an environment, the most likely application of highly reflecting Ir + Al coatings is in satelliteborne spectrographs or extreme-ultraviolet telescopes. An estimate of the altitude necessary to retard the Al oxidation to a negligible rate can be obtained in the following manner. If the oxygen partial pressure is $10^{-6}$ Torr and the sticking coefficient is unity, a monolayer of the oxide forms in about 1 sec, hence for the mirror to have a lifetime of 1 yr, where lifetime is arbitrarily defined as the time required for the formation of a monolayer, the oxygen pressure must be approximately $10^{-13}$ Torr. The altitude required for an oxygen pressure this low can be estimated by an extrapolation of published data [Roberts and Vanderslice, 1963] and is about 1500 km.

In the discussion of oxygen pressure and probability of oxidizing a fresh Al surface in the upper atmosphere, the space vehicle with the coating was assumed to be stationary. In reality, the spacecraft, especially in a low orbit, has a velocity considerably in excess of that of the average molecule, atom, or ion. Thus, the gas pressure at a given area of the satellite is highly directional, having a maximum when located normal to the forward movement and a minimum when pointing in the opposite direction. In addition, gas particles striking the surface have much higher velocity and, therefore, a greatly different sticking probability than the ones impinging on a stationary object. This makes it difficult to estimate the lifetime of Ir + Al mirrors at altitudes below 1500 km.

In principle, the vacuum-ultraviolet reflectance of various other second and third series transition metals can also be increased using an overcoating of Al. There is always the danger, however, that the two metals will diffuse into each other and spoil the coating. The interdiffusion of Au and Al layers, for example, is well known and, at room temperature, can be complete in from 6 months to a year with severe deterioration of the coating [Hunter *et al.*, 1972]. Hence, the compatibility of the two metals *vis-à-vis* interdiffusion must be determined before they are used to coat flight optics.

## References

Adriaens, M. R., and B. P. Feuerbacher, *Appl. Opt. 10*, 958 (1971).
Angel, D. W., W. R. Hunter, R. Tousey, and G. Hass, *J. Opt. Soc. Am. 51*, 913 (1961).
Berning, P. H., G. Hass, and R. P. Madden, *J. Opt. Soc. Am. 50*, 586 (1960).
Byram, E. T., T. A. Chubb, H. Friedman, J. E. Kupperian, and R. W. Kreplin, *Astrophys. J. 128*, 738 (1958).

Canfield, L. R., G. Hass, and W. R. Hunter, *J. Phys. (Paris)* 25, 124 (1964).
Canfield, L. R., G. Hass, and J. E. Waylonis, *Appl. Opt.* 5, 45 (1966).
Carruthers, G. R., and T. Page, *Science* 177, 788 (1972).
Cox, J. T., G. Hass, and J. E. Waylonis, *Appl. Opt.* 7, 1535 (1968).
Cox, J. T., G. Hass, and W. R. Hunter, *J. Opt. Soc. Am.* 61, 360 (1971).
Cox, J. T., G. Hass, J. B. Ramsey, and W. R. Hunter, *J. Opt. Soc. Am.* 62, 781 (1972).
Cox, J. T., G. Hass, J. B. Ramsey, and W. R. Hunter, *J. Opt. Soc. Am.* 63, 435 (1973).
Feuerbacher, B. P., B. Fitton, and W. Steinmann, ELDO/ESRO *Tech. Rev.* 1, 385 (1969).
Feuerbacher, B. P., and W. Steinmann, *Opt. Commun.* 1, 81 (1969).
Gillette, R. B., and B. A. Kenyon, *Appl. Opt.* 10, 545 (1971).
Gleason, P. R., *Proc. Natl. Acad. Sci. (U.S.)* 15, 551 (1929).
Hacman, D., *Opt. Acta* 17, 659 (1970).
Hass, G., and W. R. Hunter, *Appl. Opt.* 6, 2097 (1967).
Hass, G., and W. R. Hunter, *Appl. Opt.* 9, 2101 (1970).
Hass, G., G. F. Jacobus, and W. R. Hunter, *J. Opt. Soc. Am.* 57, 758 (1967).
Hass, G., and R. Tousey, *J. Opt. Soc. Am.* 49, 593 (1959).
Hass, G., and J. E. Waylonis, *J. Opt. Soc. Am.* 51, 719 (1961).
Hunter, W. R., *J. Phys. (Paris)* 25, 154 (1964).
Hunter, W. R., J. F. Osantowski, and G. Hass, *Appl. Opt.* 10, 540 (1971).
Hunter, W. R., T. L. Mikes, and G. Hass, *Appl. Opt.* 11, 1594 (1972).
Hutcheson, E. T., G. Hass, and J. K. Coulter, *Opt. Commun.* 3, 213 (1971).
Hutcheson, E. T., J. T. Cox, G. Hass, and W. R. Hunter, *Appl. Opt.* 11, 1590 (1972a).
Hutcheson, E. T., G. Hass, and J. T. Cox, *Appl. Opt.* 11, 2245 (1972b).
Jacobus, G. F., R. P. Madden, and L. R. Canfield, *J. Opt. Soc. Am.* 53, 1084 (1963).
Juenker, D. W., L. J. LeBlanc, and C. R. Martin, *J. Opt. Soc. Am.* 58, 164 (1968).
Madden, R. P., L. R. Canfield, and G. Hass, *J. Opt. Soc. Am.* 53, 620 (1963).
Patterson, D. A., and W. H. Vaughan, *J. Opt. Soc. Am.* 53, 851 (1963).
Roberts, R. W., and T. A. Vanderslice, *Ultrahigh Vacuum and Its Applications*, p. 180, Prentice-Hall, Englewood Cliffs, N.J. (1963).
Sabine, G. B., *Phys. Rev.* 55, 1064 (1939).

HIDEO IKEDA, HIDEKI AKASAKA, and
ZENJI WAKIMOTO

# MULTILAYER ANTIREFLECTION COATINGS FOR ULTRAVIOLET (200-400 nm) AND APPLICATION

## I. Introduction

In NASA planning, spectral photographic experiments have been proposed to photograph the earth's ozone layer in the range 255–285 nm and twilight airglow horizon in a broadband ultraviolet region such as 200–400 nm with a uv camera.

Practically available materials for making ultraviolet lenses are limited to $CaF_2$, $SiO_2$, LiF, and a few other materials. Since these materials have relatively low refractive indices and small differences in dispersion, a number of lens elements are required for faster lens speed. Multilayer antireflection coatings for the ultraviolet enable one to increase the number of lens elements without decreasing transmission.

For this purpose, we have developed design and manufacturing techniques for uv broadband multilayer antireflection coatings.

---

The authors are at Nippon Kogaku K.K., Tokyo, Japan; the first two named are in the Research Laboratory, and the third is in the Optical Designing Section.

## II. Design Concepts for Ultraviolet Antireflection Coatings

### A. SINGLE LAYER

At first, we consider the possibility of a broadband single antireflection coating of other than a quarter-wavelength, for the ultraviolet range between 200 and 400 nm. We assume $CaF_2$ or $SiO_2$ to be the substrate.

In this case, we can adopt as optical thickness of the layer $(\lambda_0/2)m$ instead of $(\lambda_0/4)m$ ($m = 1, 2, \ldots$), where $\lambda_0$ is the design wavelength in the ultraviolet range. The results obtained by the vector method clearly show that the most suitable layer substance should have a refractive index between 1.3 and 1.1.

Figure 1 indicates the optimum refractive indices obtained by the vector method for each wavelength. Figure 2 shows the reflectance of a single layer of the optimum refractive indices when the optical thickness is kept to $\lambda_0$. Of course, it is impossible to obtain a substrate with such an index and dispersion in the ultraviolet.

### B. SYMMETRIC LAYER COMBINATION

We already know the theory of "equivalent layer" introduced by Herpin [1947]. He has shown that a symmetric layer combination can be replaced by a single layer, called the equivalent layer, having a different

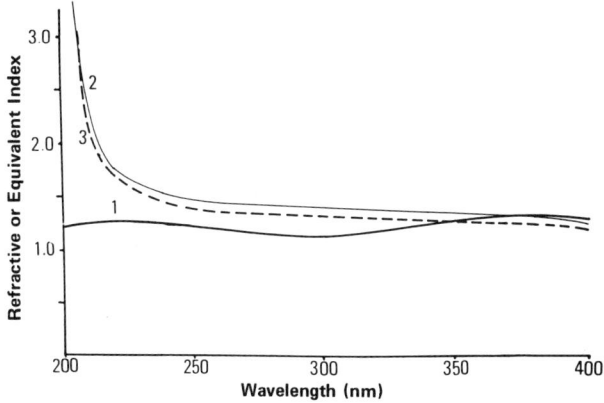

FIGURE 1 Refractive index as a function of wavelength for various designs. ━━━ Single layer (from vector method). ——— Equivalent layer (three-layer combinations). ------ Quasi-symmetric three-layer.

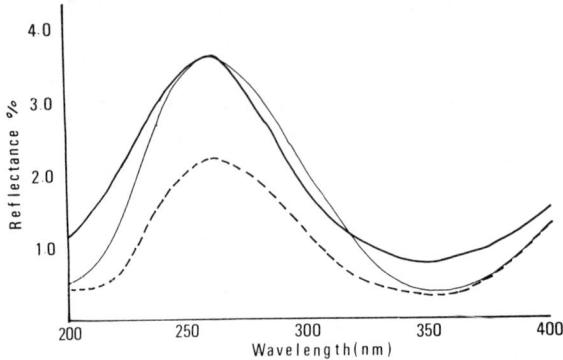

FIGURE 2  Reflectance as a function of wavelength for various designs using the optimum refractive index. ——— Single layer (from vector method). ——— Equivalent layer (three-layer combinations). - - - - - Quasi-symmetric three-layer.

refractive index for each wavelength. The equivalent layer has a nearly constant refractive index over some region of wavelength. The index changes remarkably outside this region.

It may seem likely that the single layer suggested in Sec. I.A can be approximately replaced by a symmetric three-layer combination if the combination can be suitably chosen. Unfortunately, however, this is not the case, as is shown by the curve 2 of reflectance in Figure 2 and by the curve 2 of equivalent index in Figure 1. It is apparent that the reflectance obtained is not satisfactory. This is due to the limited availability of substances that are transparent and physically and chemically stable.

C. QUASI-SYMMETRIC LAYER COMBINATION

We have found that if the equivalent refractive index could increase or decrease by just a small amount, say 5%, for each wavelength, the above symmetric three-layer coating would be greatly improved.

To do this, it is enough to convert the symmetric combination to a slightly asymmetric one, as will be discussed later.

Asymmetric three-layer combinations can be logically divided into the following two categories: The first is asymmetric with respect to refractive index but symmetric with respect to thickness. This is called a quasi-symmetric layer of the first kind. The second is specified as asymmetric with respect to thickness but symmetric with respect to the refractive index and is called a quasi-symmetric layer of the second kind.

*Quasi-symmetric Three Layers of the First Kind* A slightly asymmetric three-layer combination can be constructed from an original symmetric one that is characterized by equivalent index $N$ and equivalent optical thickness, and where $D$ is equivalent thickness. It is expressed in term of Herpin matrix $M_1$ :

$$M_1 = \begin{pmatrix} \cos g_1 & \dfrac{j}{n_1 + \Delta n} \sin g_1 \\ j(n_1 + \Delta n) \sin g_1 & \cos g_1 \end{pmatrix} \begin{pmatrix} \cos g_2 & \dfrac{j}{n_2} \sin g_2 \\ jn_2 \sin g_2 & \cos g_2 \end{pmatrix}$$

$$\begin{pmatrix} \cos g_1 & \dfrac{j}{n_1} \sin g_1 \\ j(n_1 + \Delta n) \sin g_1 & \cos g_1 \end{pmatrix},$$

where

$$g_i = 2\pi \dfrac{nd i}{\lambda_s} \cdot \dfrac{\lambda_s}{\lambda},$$

$nd/\lambda_s$ is optical thickness measured in design wavelengths, $n_i$ is the refractive index of the $i$th layer within the combination, $\Delta n$ is the difference in refractive indices of the outer layers, and $\Delta n \ll n$.

The matrix is approximately rewritten as

$$M_1 \simeq \begin{pmatrix} \dfrac{n_1}{n_1 + \Delta n} & 0 \\ 0 & \dfrac{n_1 + \Delta n}{n_1} \end{pmatrix} \begin{pmatrix} \cos g_1 & \dfrac{j}{n_1} \sin g_1 \\ jn_1 \sin g_1 & \cos g_1 \end{pmatrix} \begin{pmatrix} \cos g_2 & \dfrac{j}{n_2} \sin g_2 \\ jn_2 \sin g_2 & \cos g_2 \end{pmatrix}$$

$$\begin{pmatrix} \cos g_1 & \dfrac{j}{n_1} \sin g_1 \\ jn_1 \sin g_1 & \cos g_1 \end{pmatrix}$$

$$= \begin{pmatrix} \dfrac{n_1}{n_1 + on} & 0 \\ 0 & \dfrac{n_1 + on}{n_1} \end{pmatrix} \begin{pmatrix} \cos \Theta & \dfrac{j}{N} \sin \Theta \\ j \sin \Theta & \cos \Theta \end{pmatrix},$$

where $\Theta = 2\pi(ND/\lambda_s)(\lambda_s/\lambda)$, $ND/\lambda_s$ is the equivalent optical thickness for the original symmetric layer.

$$N = n_1 \left( \frac{n_1 n_2 \sin 2g_1 \cos g_2 + (n_2^2 \cos^2 g_1 - n_1^2 \sin^2 g_1) \sin g_2}{n_1 n_2 \sin 2g_1 \cos g_2 + (n_1^2 \cos^2 g_1 - n_2^2 \sin^2 g_1) \sin g_2} \right)^{1/2}.$$

This expression means that a slightly asymmetric three-layer combination can virtually be considered as a symmetric three-layer combination that has an equivalent index $N$ and equivalent thickness $D_1$, which are functions of $N_1^*$ and $D_1^*$, respectively, with a parameter of $\alpha = \Delta n/n_1$ as follows:

$$N_1^* \simeq N(1 + \alpha),$$

$$\cos \Theta^* \simeq \left[ 1 + \frac{\alpha^2}{2(1+\alpha)} \right] \cos \Theta,$$

where $\alpha = \Delta n/n_1$, $\Theta_1^* = 2\pi(N_1^* D^*/\lambda)$, $\Theta = 2\pi(ND/\lambda)$.

The asymmetric combination with different $\alpha$'s is equivalent to a single equivalent layer of different index and optical thickness. In other words, the idea of equivalent index or optical thickness of a quasi-symmetric layer can be obtained from that of symmetric layers. Introducing such a modification in the preliminary design, we get a reflectance curve shown by curve 3 in Figure 1.

It is to be noticed that the quasi-symmetric layer combination might be simplified by equalizing the refractive index of the innermost layer with that of the substrate material, for we can eventually exclude the innermost layer. The example will be given in Sec. III.B as Type 1 and Type 4.

*Quasi-symmetric Three Layers of the Second Kind* Similarly, a slightly asymmetric three-layer combination with respect to optical thickness can be constructed from an original symmetric one.

$$M_2 = M_E \begin{pmatrix} \cos \delta & \dfrac{j}{n_1} \sin \delta \\ jn_1 \sin \delta & \cos \delta \end{pmatrix},$$

where

$$M_E = \begin{pmatrix} \cos g_1 & \dfrac{j}{n_1}\sin g_1 \\ jn_1 \sin g_1 & \cos g_1 \end{pmatrix} \begin{pmatrix} \cos g_2 & \dfrac{j}{n_2}\sin g_2 \\ jn_2 \sin g_2 & \cos g_2 \end{pmatrix} \begin{pmatrix} \cos g_1 & \dfrac{j}{n_1}\sin g_1 \\ jn_1 \sin g_1 & \cos g_1 \end{pmatrix}.$$

This combination can be virtually considered as a symmetric three-layer combination of equivalent index $N_2{}^*$ and of thickness $D_2{}^*$, as follows:

$$N_2{}^* = N\left[1 - \frac{(a+b)k}{1+ak}\right]^{1/2},$$

$$\cos \Theta_2{}^* \simeq \cos(\Theta + \delta), \quad (\delta \ll \Theta),$$

where

$$a = (N - n_1)/n_1, \quad b = (N - n_1)/N,$$

$$k = \sin \delta \, \sin \Theta / \sin(\Theta + \delta).$$

In this case, the expression of the equivalent index and thickness contains three parameters $a$, $b$, and $\delta$ or $k$. With these parameters suitably chosen, this kind of quasi-symmetric layer can replace a layer with a dispersive index and dispersive optical thickness of a much wider range in magnitude and spectral variation than the first kind of quasi-symmetric layer.

## III. Computer Design Based on Quasi-symmetric Layers

A design of thin-film systems has to determine optical thickness and optical index (or material) for each layer. We call these the design parameters of a thin-film system. Many attempts have been made to use a computer to automatically search and find a solution for the design parameters. The concept of quasi-symmetric layers simplifies the automatic approach to thin-film systems design.

### A. COMPUTER PROGRAM FOR AUTOMATIC DESIGN

As is shown in Figure 3, our design system consists of three steps: first-order design, optimization, and refinement.

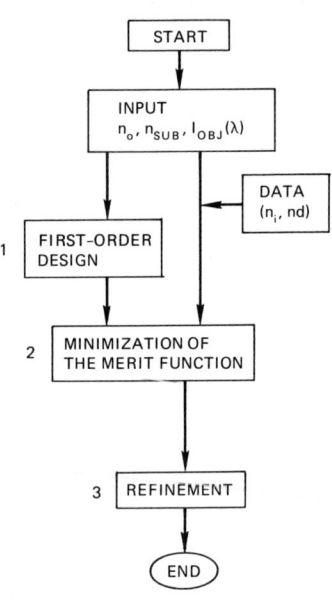

FIGURE 3 Block diagram of computer program for automatic design.

*1. First-Order Design Step* By utilizing an analytical synthesis method such as Kard's [1957], Delano's [1966], or the graphical technique on a computer display, we can determine a system of imaginary layers whose properties are specified to be in accordance with required characteristics. The simplest system may be a single layer, as shown in Sec. II.A. Then, we proceed to convert the required system of imaginary layers into an actual one. First, as shown in Figure 4, we choose several usable materials and make all possible quasi-symmetric combinations from three arbitrary materials with suitable thickness (but we neglect dispersion of the materials at this step). Then, the equivalent indices (EI) and equivalent optical thickness (ET) corresponding to those combinations represent so many points in the EI–ET plane or network. Out of these net elements, we then look for a set of the elements whose characteristics best approximate those of the required system of equivalent layers. This is easily done if we optimize the merit function for each set of the elements by the modified simplex method. After the best set of net elements has been found, it can be interpreted as a set of combinations of quasi-symmetric layers. These are the actual design parameters of the solution. The first-order design step may be programmed for the computer.

*2. Optimization Step* The design parameters of the first-order design

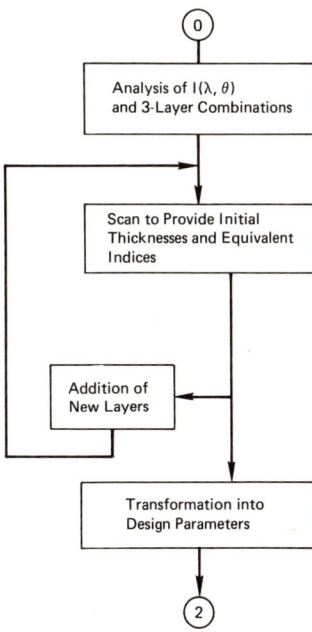

FIGURE 4 Block diagram of computer program for generating design parameters.

are the initial data for the optimization step. The optimization is based on the so-called parabolic approximation method [Meiron and Volinez, 1960], which uses a stored characteristic target function such as

$$\Phi = \sum_j \sum_i w(\theta_j, \bar{\lambda}_i) \Phi^{(t)}(\bar{\lambda}_i),$$

where

$$\Phi^{(t)}(\bar{\lambda}_i) = \frac{1}{(\lambda_{i_2} - \lambda_{i_1})} \int_{\lambda_{i_1}}^{\lambda_{i_2}} (I - I_{\text{opt}})^2 \, d\lambda,$$

that is to say, weighted mean of subtarget functions over the intended wavelength band.

At this step, we can take into consideration the dispersion of materials.

*3. Refinement Step* If we want to use only a few particularly useful materials such as $MgF_2$, $SiO_2$, or $NdF_3$ for some or all of the layers, we can utilize again the concept of the quasi-symmetric layer at this step.

B. OUTCOME OF DESIGN

Using the above-explained automatic synthesis program on the IBM 360/75I (or IBM 370/165), we systematically obtained the following $\lambda/4$ stacked antireflection coatings for the ultraviolet from 200 nm to 400 nm:

Type 1: S–$\lambda/4$–$\lambda/4$–air, S–$\lambda/2$–$\lambda/4$–air.
Type 2: S–$\lambda/2$–$\lambda/2$–$\lambda/4$–air, S–$\lambda/4$–$\lambda/2$–$\lambda/4$–air.
Type 3: S–$\lambda/2$–$\lambda/4$–$\lambda/2$–$\lambda/4$–air, S–$\lambda/4$–$\lambda/4$–$\lambda/2$–$\lambda/4$–air.
Type 4: S–$\lambda/4$–$\lambda/2$–$\lambda/2$–$\lambda/2$–$\lambda/4$–air.

The reflectance characteristic is plotted in the form of equireflectance contours in a plane defined by the angle of incidence and the wavelength. Figure 5 shows the average reflectance contour for Type 1 (two-layer coatings); Figure 6 its equireflectance contours for the $p$-component and $s$-component of Type 1 coatings; Figure 7 the equi-average-reflectance contour for Type 4 (five-layer coatings); and Figure 8 equireflectance contours for the $p$-component and $s$-component of Type 4 coatings. Figure 9 show the calculated reflectance curve of Type 1 and Type 4 coatings.

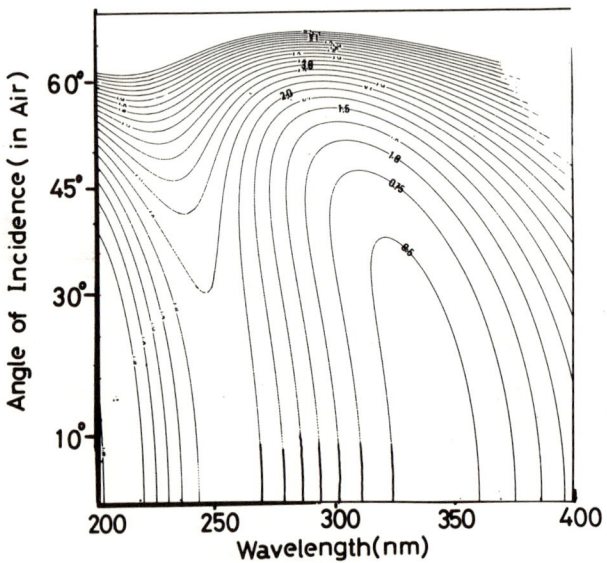

FIGURE 5  Equireflectance contours for two-layer uv antireflection coatings on $CaF_2$. $n_0 = 1.0$, $n_1 = 1.405$, $n_2 = 1.65$, $n_3 = 1.465$.

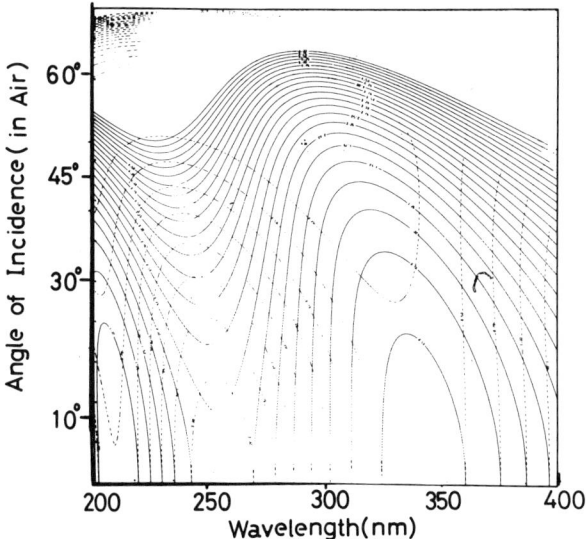

FIGURE 6  Equireflectance contours for two-layer uv antireflection coatings on $CaF_2$. $n_0 = 1.0, n_1 = 1.405, n_2 = 1.65, n_3 = 1.465$. ——— s-reflectance; - - - - - p-reflectance.

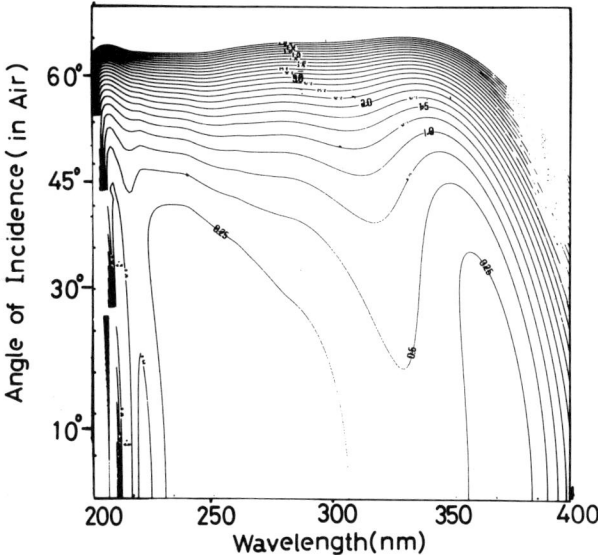

FIGURE 7  Equireflectance contours for five-layer uv antireflection coatings on $CaF_2$.

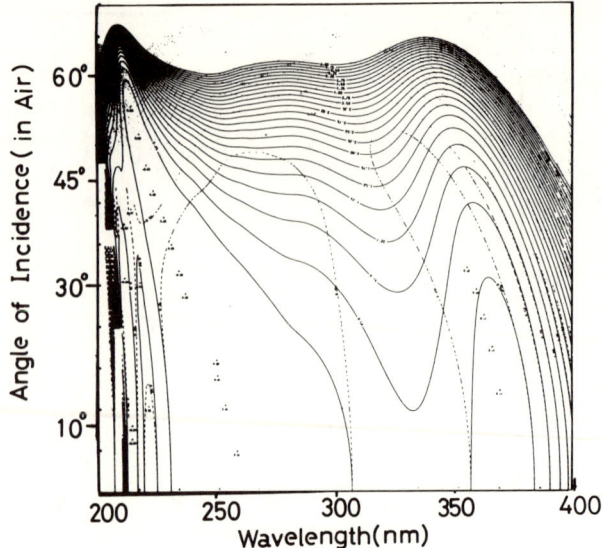

FIGURE 8  Equireflectance contours for five-layer uv antireflection coatings on $CaF_2$. —— $s$-reflectance; ---- $p$-reflectance.

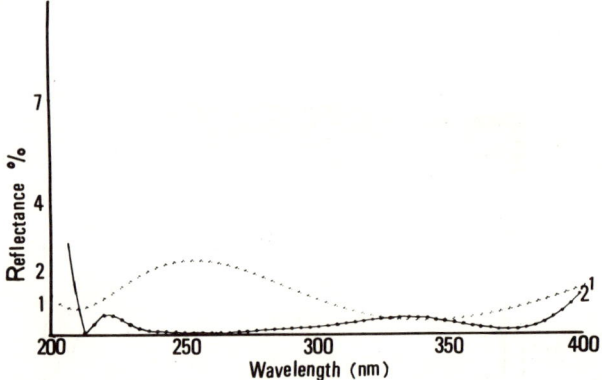

FIGURE 9  Calculated reflectance in air versus wavelength for Types 1 and 4 antireflection coatings on $CaF_2$. 1, Type 1; 2, Type 4.

## IV. Method for Controlling Optical Thickness

Taking advantage of the fact that our designs are all combinations of the $\lambda_0/4$ or $\lambda_0/2$ thick layers with designed wavelength $\lambda_0$ in the ultraviolet range, we carried out thickness monitoring by measuring the reflectance of evaporated layers for visible monochromatic light of $\lambda_M = 2\lambda_0$.

If we neglect the dispersion of materials, the optical thickness $\lambda_0/2$ corresponds to $\lambda_M/4$ and $\lambda_0/4$ to $\lambda_M/8$.

As is easily seen, the evolution of reflectance change in the course of evaporation manifests the steepest slope when the optical thickness is $\lambda_M/8$ (one eighth of the monitoring wavelength). This is true whenever the substrate is base or stacked with multiple $\lambda_M/4$ (quarter-wavelength) layers. Therefore, monitoring of the $\lambda_M/8$ layer is rather easy and accurate. We actually obtained a relative error of thickness monitoring within 6% for the Types 1–4. This accuracy was sufficient to maintain the characteristic reflectance of designed multilayer coatings.

## V. Application

### A. ULTRAVIOLET CAMERA USED FOR THE NASA PROJECT

In the NASA project, multispectral photographic experiments have been conducted by photographing the earth's ozone layer and twilight airglow with visible and ultraviolet light simultaneously.

Two cameras, one for the uv and the other for visible, were used. The uv camera is used to take a picture of ozone clouds of various thickness, which reflect ultraviolet sunlight, and the visible camera is used to take the usual color photograph of the area viewed by the ultraviolet camera. The uv light is filtered through a broadband filter corresponding to the absorption band of ozone (255–275 nm or ~320 nm). The photographs taken simultaneously with the visible camera permit us to evaluate the reflections from various terrain and cloud features beneath the ozone layers. These cameras are also intended for twilight and nighttime horizon airglow photography. The uv lens for the uv camera should have the same focal length, aperture, and field of view as the standard lens for the Nikon F camera.

### B. ANTIREFLECTION COATING AND ULTRAVIOLET LENS

Multilayer ultraviolet antireflection coatings are of value for increasing the number of lens elements in an optical system in order to obtain high optical performance.

This uv, 55-mm ($f/2$) lens is composed of 12 separate lens elements, made of quartz and fluorite. It is corrected for uv wavelength (200–400 nm) with an aperture of $f/2$, field of view 43°, and back focal length of about 40 mm for the 35-mm format. Figure 10 shows this uv lens section.

The T-number 3.5 without uv antireflection coatings for this lens

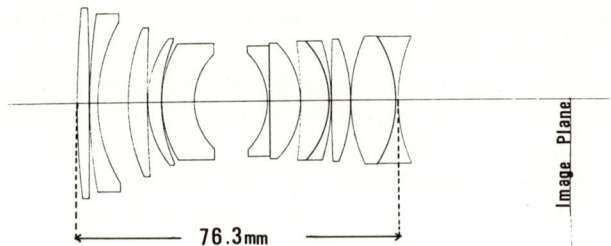

FIGURE 10  Ultraviolet lens section.

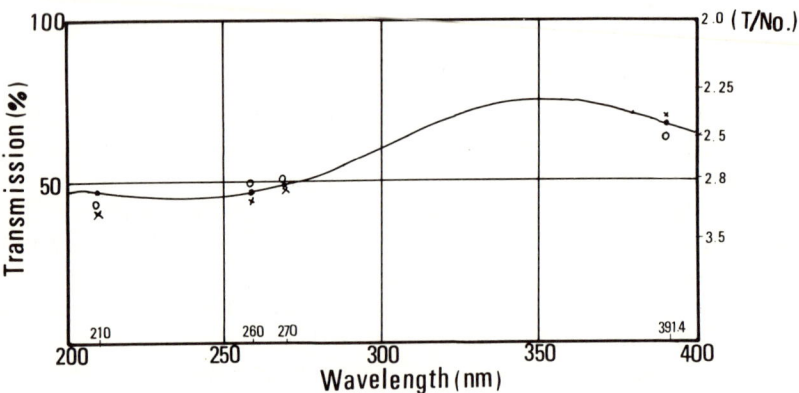

FIGURE 11  Comparison of theoretical (curve) and experimental (points) spectral transmission as a function of wavelength.

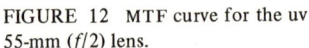

FIGURE 12  MTF curve for the uv 55-mm ($f/2$) lens.

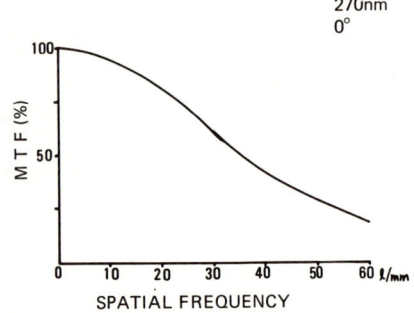

could be improved up to about 2.5 over the entire wavelength region with a $\lambda/2$ ($NdF_3$) + $\lambda/4$ ($MgF_2$) uv antireflection coating. Figure 11 shows the calculated spectral transmission curve and the dotted point measured by NASA. Figures 12–17 show the modulation transfer function. Figures 18–20 show the measured value of the resolving power test by NASA. This multilayer coating has an extremely low solubility in water, so it may be useful in a very humid environment.

The development of multilayer antireflection coatings for the ultraviolet has contributed greatly to ultraviolet optical system design.

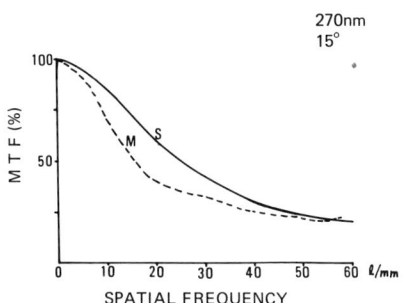

FIGURE 13  MTF curve for the uv 55-mm ($f/2$) lens.

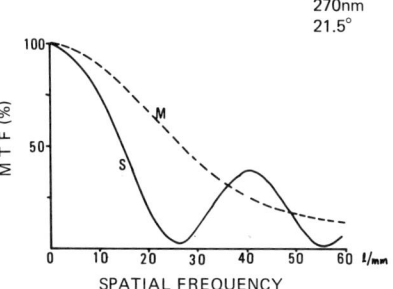

FIGURE 14  MTF curve for the uv 55-mm ($f/2$) lens.

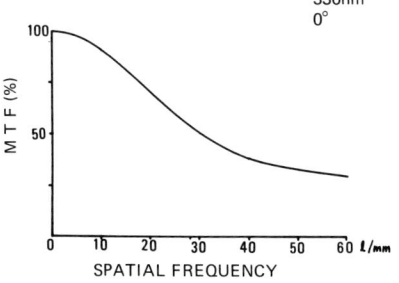

FIGURE 15  MTF curve for the uv 55-mm ($f/2$) lens.

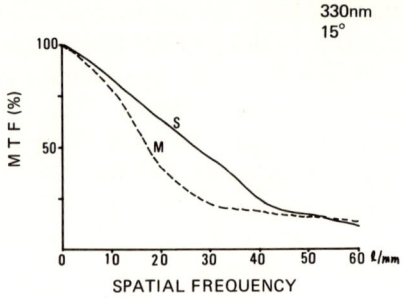

FIGURE 16  MTF curve for the uv 55-mm (f/2) lens.

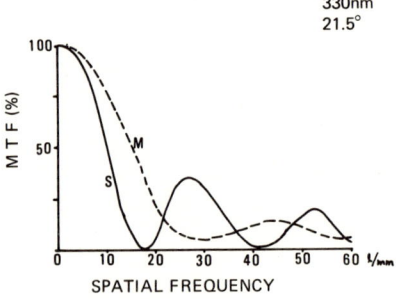

FIGURE 17  MTF curve for the uv 55-mm (f/2) lens.

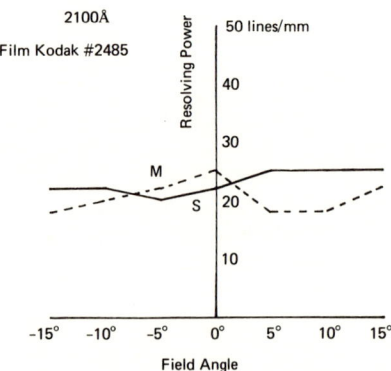

FIGURE 18  Resolving power of the uv 55-mm (f/2) lens.

FIGURE 19  Resolving power of the uv 55-mm (f/2) lens.

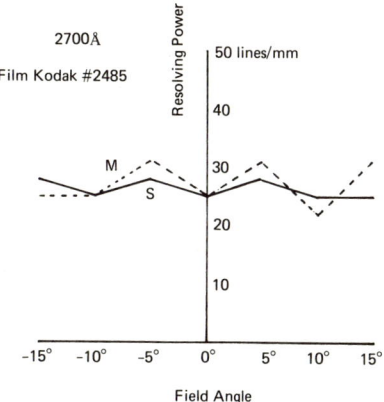

FIGURE 20  Resolving power of the uv 55-mm (f/2) lens.

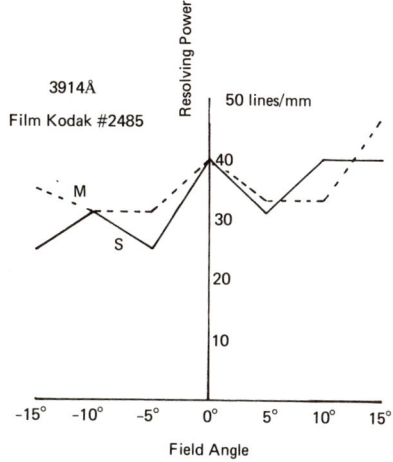

## References

Delano, E., PhD Thesis, University of Rochester (1966).
Herpin, M. A., *Compt. Rend. 225,* 182 (1947).
Kard, P. G., *Opt. Spectrosc. 2,* 236 (1957).
Meiron, J., and G. Volinez, *J. Opt. Soc. Am. 50,* 207 (1960).

J. A. DOBROWOLSKI

# AN AUTOMATIC THIN-FILM INTERFERENCE FILTER DESIGN PROGRAM BASED ON THE USE OF MINUS FILTERS

## Introduction

Optical filters are used in many instrument packages for the remote sensing of earth resources, pollution control and surveillance, and in a number of satellite-based scientific experiments [Katz, 1972]. Often simple bandpass or cutoff filters are all that is required. Present-day thin-film interference filters fill this need adequately.

For other applications, filters with rather complicated transmittance curves extending over broad spectral regions would be useful, but until recently there was no certainty that they could always be made. Filters of this type are generally constructed from colored glasses and other absorbing materials. However, at times satisfactory solutions cannot be found because of a lack of materials with suitable spectral absorption characteristics.

This paper describes a new automatic thin-film synthesis program for the design of filters with complicated spectral transmittance characteristics. Its chief advantage over previous thin-film design techniques is that it should yield a solution to any reasonable filter problem.

---

The author is in the Division of Physics, National Research Council of Canada, Ottawa, Ontario, Canada K1A OS1.

The simplest of the previous design methods are the refinement programs [Baumeister, 1958]. These improve the performance of a multilayer that is close to the desired performance by repeatedly making small changes in the thicknesses and refractive indexes of the layers. The procedure does not work well unless a reasonable starting design is available.

Several automatic thin-film synthesis programs have also been described in the past. In some of these, the performance of a starting design is improved by making random, drastic changes in one or more of its construction parameters and by refining the resulting system [Ermolaev et al., 1962]. In others, a suitable multilayer is gradually evolved by the repeated addition of appropriate layers to an existing multilayer [Dobrowolski, 1965]. Although the probability of finding a solution to a problem is greatly enhanced by these methods, there will be times when the programs will not yield a practical solution without an excessive expenditure of computer time.

## The Minus Filter Method

The present method is based on the use of minus filters (Figure 1). These are filters that transmit freely all the radiation incident upon them except in a narrow spectral region. They are characterized by the minimum transmittance, the wavelength at which it occurs, and by the half-width of the rejection region.

By placing in series a number of appropriately chosen minus filters, one can reproduce a desired spectral transmittance curve over a restricted range within any required tolerance (Figure 2). Should the desired curve be very complicated, or the tolerances very tight, the number of minus filters required will be large and the resulting system may not be very practical.

More than one minus filter can be deposited onto the same substrate surface without interference provided that only one of these filters has a transmittance significantly less than unity at any wavelength of interest.

To avoid errors in the overall transmittance it is necessary to prevent light that has undergone multiple reflections at the various surfaces of the component filters from falling onto the detector (Figure 3). Reflections originating at the air–substrate interfaces can be removed with antireflection coatings or by cementing. To remove reflections from the substrate–multilayer interfaces, the various components must be sufficiently inclined to one another to permit a spatial separation of the multiply reflected from the directly transmitted beams. This could be done by depositing the filters on wedged substrates.

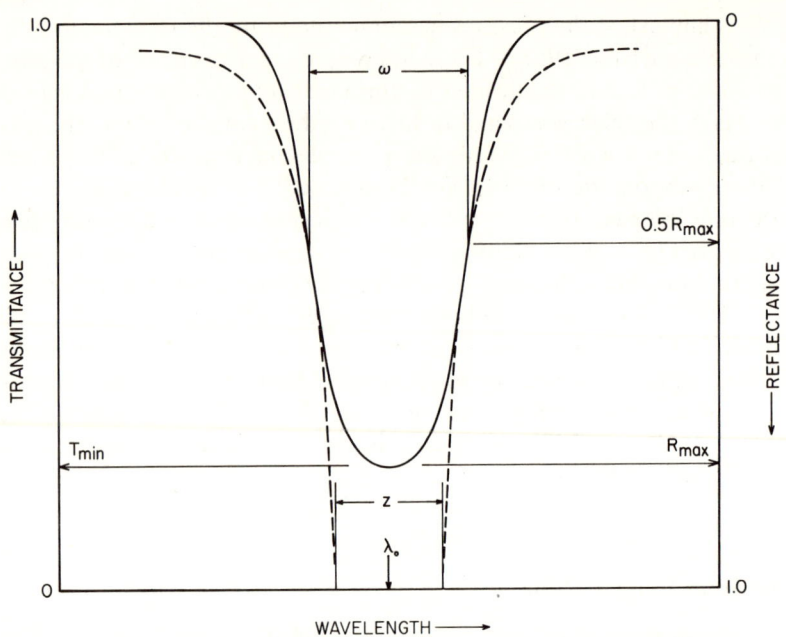

FIGURE 1  The chief characteristics of a nonabsorbing minus filter.

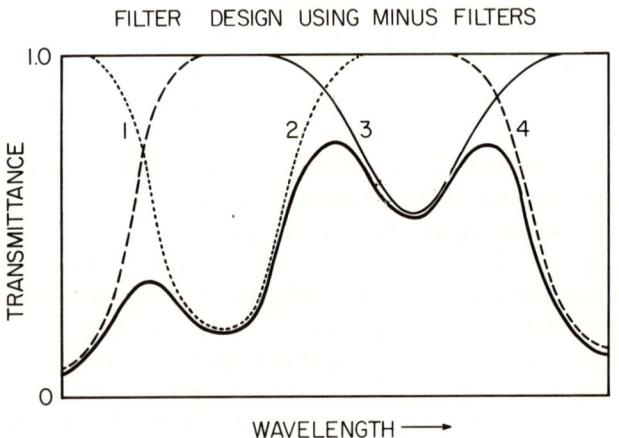

FIGURE 2  Principle of filter design with minus filter components.

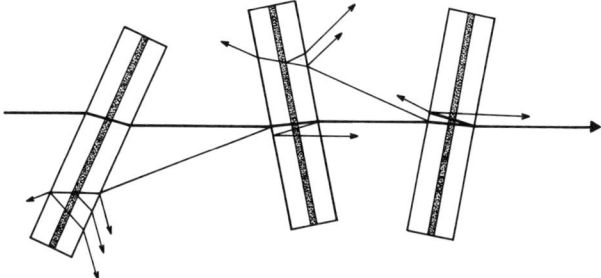

THE VARIOUS MINUS FILTERS

FIGURE 3 Typical interreflections occurring between the various minus filter components.

## Computer Program

The computer program that automatically finds the parameters of a set of minus filters necessary to achieve a certain spectral transmittance makes use of the type of minus filter described by Thelen [1971]. They consist essentially of a central two-material reflecting stack surrounded on both sides by special antireflection coatings that reduce the ripple in the transmission regions.

The only mandatory input to the program is the desired spectral transmittance (Figure 4). One can read in the performance or the construction parameters of several initial filters chosen by experience or obtained from previous calculations. Next the transmittance curve is calculated of the filter that, when placed in series with the existing filters, would yield the desired spectral transmittance. Unless otherwise instructed, by examining this "residual" transmittance curve the computer program will automatically find the most advantageous spectral region for placing of the next minus filter. A first estimate is now made of the rejection wavelength and the number and refractive indexes of the layers of an appropriate minus filter. This filter is optimized by modifying the rejection wavelength and the refractive indexes of the layers. A search is then carried out to see whether minus filters consisting of a smaller or larger number of layers would yield a better result. The best of the minus filters found is retained. A number of tests determine whether the cycle for the selection of another minus filter is to be repeated.

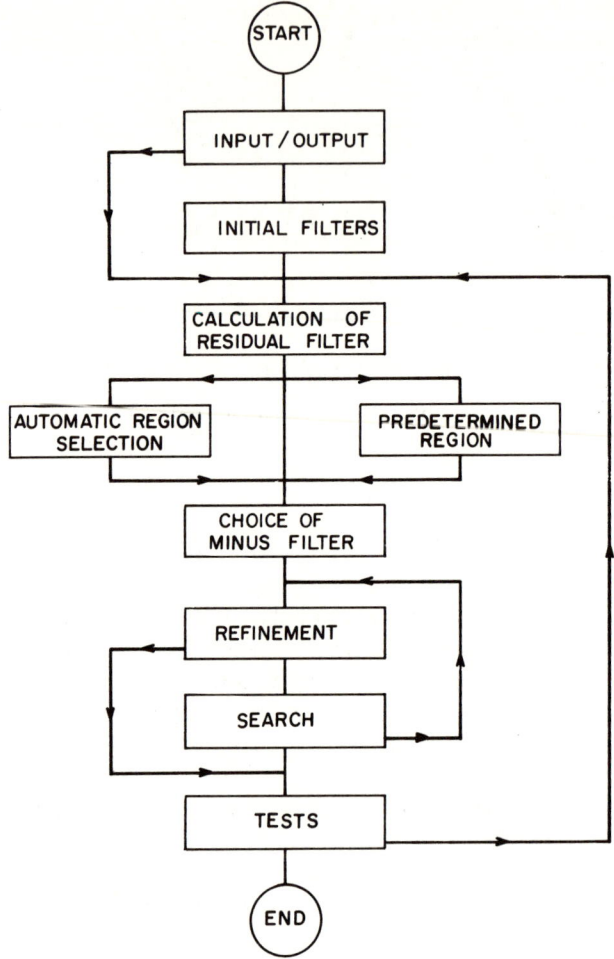

FIGURE 4  Flow of calculations in the computer program.

## Examples

It is possible to obtain very good solutions with minus filters only. But, in general, such solutions will consist of many components, and the total number of layers will be large. More practical solutions consisting of far fewer layers are obtained if the minus filter program is used to generate a comparatively rough solution that can serve as a starting design for refinement or for further calculations with one of the other existing automatic thin-film synthesis programs. Examples will now be given of

## J. A. Dobrowolski

both types of solution. Where filters were refined, the thicknesses of the layers were varied individually, but the refractive indexes were kept constant.

### A. A COMB FILTER

The desired spectral transmittance curve for this filter consisted of a number of regions of alternately high and low transmittance (Figure 5). Because the regions were equispaced on a wavelength scale, solutions based on multiples of quarter-wave layers centered at one wavelength would be difficult to achieve. The narrowness and high contrast of the spectral features indicate that the solution must consist of many layers. It is therefore unlikely that solutions to this problem would be found for a reasonable expenditure of computer time with any of the other known thin-film synthesis programs.

The solution obtained with the present program consisted of two 19- and one 21-layer minus filters centered at $\lambda = 1.343$, 1.647, and 1.947 $\mu$m. The rejection regions of these components do not overlap, and so they can be combined to form one 61-layer filter. The calculated trans-

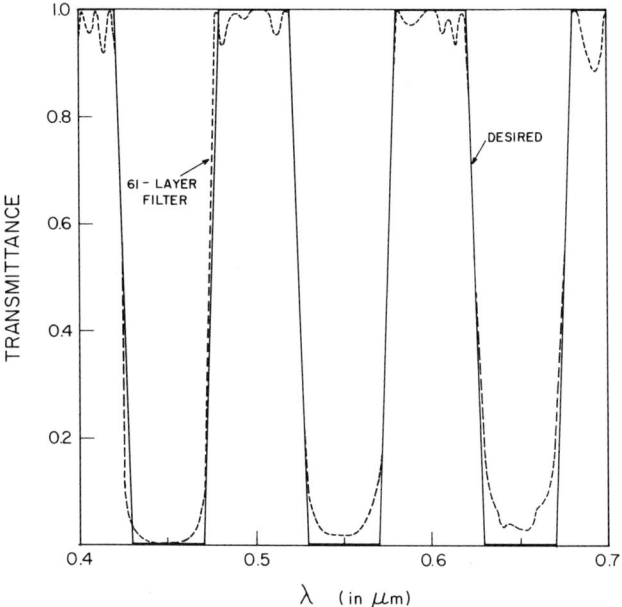

FIGURE 5 Calculated results for a comb filter.

mittance of this filter does not differ significantly from that of the three individual components placed in series. If necessary, the transmittance of the filter could be improved in the high transmittance region by further refinement.

### B. TRIANGULAR FILTER WITH ZERO TRANSMITTANCE AT 0.55 μm

The desired spectral transmittance curve for this filter is shown in Figure 6a. The difficult feature in this filter is, of course, the cusp at 0.55 μm. The performance of a system consisting of two minus filters of 13 and 9 layers each is inadequate (Figure 6c). By the addition of another minus filter of 13 layers, the desired curve can be met within ±7% (Figure 6b). But essentially the same performance can also be obtained by refining the two minus filters, and yet this solution consists of 12 fewer layers (Figure 6d)!

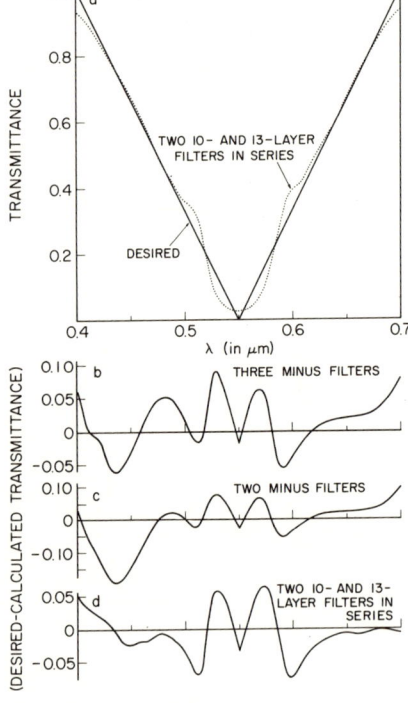

FIGURE 6 Triangular filter with zero transmittance at λ = 0.55 μm.

## C. TRIANGULAR FILTER WITH UNIT TRANSMITTANCE AT 0.55 μm

The desired spectral transmittance curve for this filter is shown in Figure 7a. The calculated transmittance of a four-component solution obtained with the aid of the computer program was within ±7% of this curve (Figure 7b). The filters can be combined in pairs and deposited on opposite sides of the same substrate. Refinement results in a system of two filters with a significantly improved performance of ±3% and a reduced number of layers (Figures 7a and 7c).

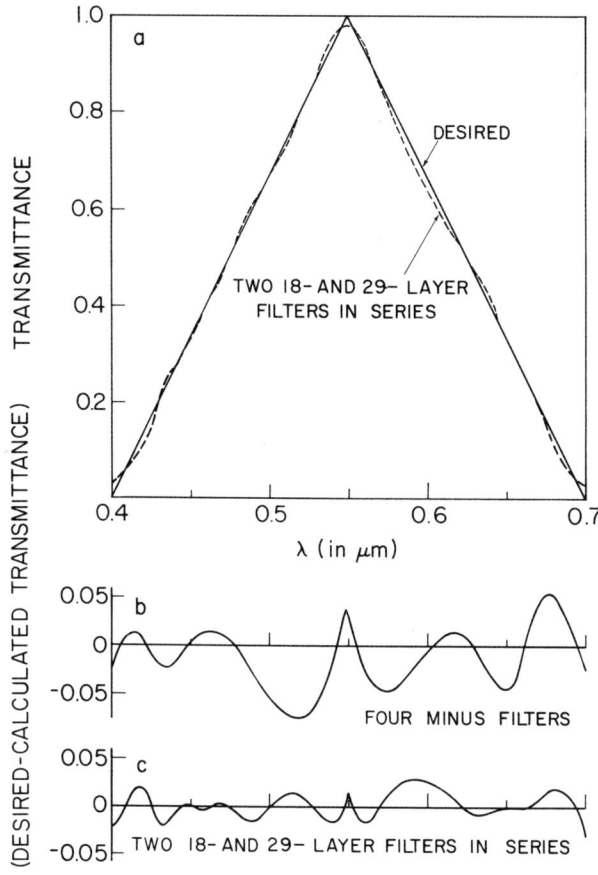

FIGURE 7  Triangular filter with unit transmittance at λ = 0.55 μm.

## D. FILTER WITH A TILTED SINE SHAPE

In the next example the desired transmittance curve was a sine curve drawn on the line joining the points ($T = 0$, $\lambda = 0.4$ μm) and ($T = 1.0$, $\lambda = 0.7$ μm), as shown in Figure 8a. This curve was chosen because it is rather difficult to obtain an extended region of constant transmittance of moderate value bordered on the side by a sharp transition to a high transmittance zone. The performance of a three-component solution to this problem is indicated by the curves of Figures 8a and 8b. The most serious departure from the desired curve occurs at the $\lambda = 0.7$ μm edge.

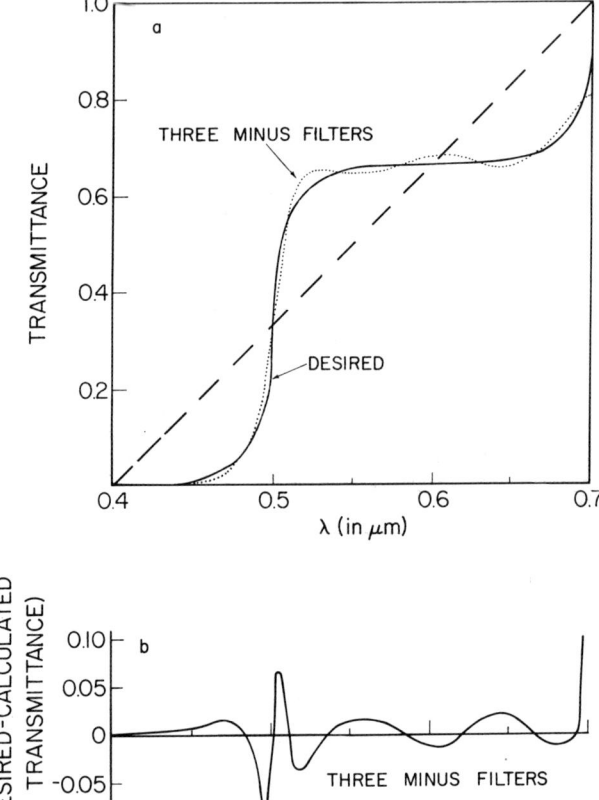

FIGURE 8 Filter with a tilted sine shape.

### E. A $\bar{y}_\lambda$ TRISTIMULUS FILTER

The last example is a $\bar{y}_\lambda$ tristimulus filter for use in a colorimeter (Figure 9a). The performance of a solution based on four minus filters with a total of 44 layers would be adequate for most applications (Figures 9a and 9b). But it is possible to obtain a more economical solution. Two minus filters whose performance was quite inadequate (Figure 9c) were combined and served as a starting design for calculations with a different thin-film synthesis program. The 24-layer filter that resulted had a performance that was comparable with that of the 44-layer solution (Figure 9d).

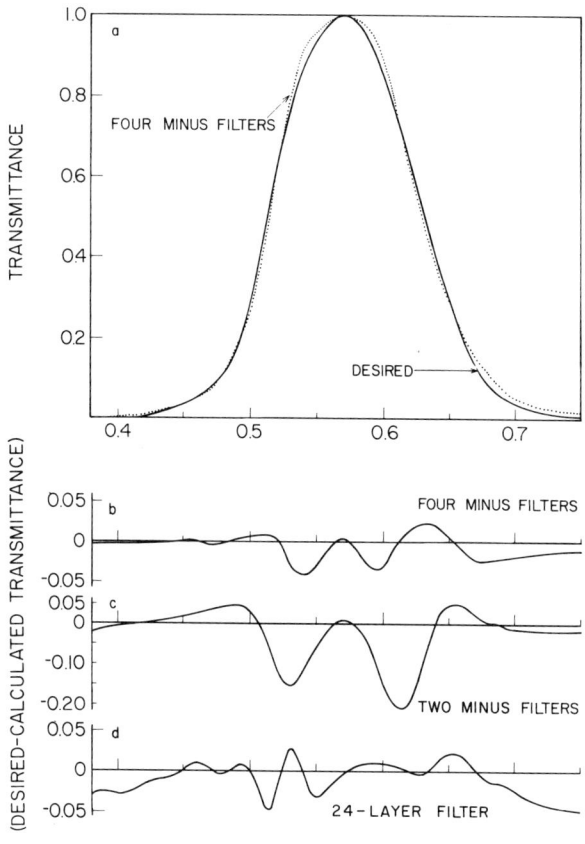

FIGURE 9 A tristimulus $\bar{y}_\lambda$ filter.

## Conclusion

To summarize, this paper describes a computer program for the automatic synthesis of optical filters with prescribed spectral characteristics that is based on the use of thin-film minus filters. The program is a useful complement to other existing methods of thin-film design. Theoretically it yields a solution to any problem providing that the spectral range over which the filter is defined is not excessive, even when the desired transmittance approaches unity in some parts of the spectrum. Naturally a complicated transmittance curve or tight tolerances will lead to a complex thin-film system. The method is very fast compared to other programs. For example, the time taken to obtain the four-component solution to the $\bar{y}_\lambda$ filter was about 1 min on an IBM 360/67 computer.

Disadvantages of the method are that the solutions often consist of more layers than those obtained by other methods and that sometimes the individual filters cannot be combined to form a single thin-film system.

The program has been demonstrated successfully on a number of rather difficult hypothetical problems. It seems now probable that with the aid of this and the other existing thin-film synthesis programs one should be able to find a satisfactory solution to any likely filtering problem in the visible and near-infrared spectral regions. It has been demonstrated before that filters of this type can be made in practice [Dobrowolski, 1970]. It is true that the solutions might at times be rather complicated and expensive to produce, but probably this would not be a deciding factor for any space applications.

I would like to acknowledge the contribution to this work of S. Cairns, who carried out much of the initial programming.

## References

Baumeister, P. W., *J. Opt. Soc. Am. 48,* 955 (1958).
Dobrowolski, J. A., *Appl. Opt. 4,* 937 (1965).
Dobrowolski, J. A., *Appl. Opt. 9,* 1396 (1970).
Ermolaev, A. M., I. M. Minkov, and A. G. Vlasov, *Opt. Spectrosc. 13,* 142 (1962).
Katz, Y. E., *Proc. Soc. Photo-Opt. Instrum. Eng. 27* (1972).
Thelen, A., *J. Opt. Soc. Am. 61,* 365 (1971).

EBERHARD SPILLER

# Multilayer Interference Coatings for the Vacuum Ultraviolet

## Introduction

It has been shown recently that low-loss coatings can be made using absorbing materials. A material with a large absorption index can have very small absorption losses if its environment is properly designed. This fact has been known since the last century. Wiener [1890] showed that light is not absorbed in the nodes of a standing wave in front of a mirror. In a perfect crystal, an anomalously high transmission has been observed for x rays incident at the Bragg angle [Borrmann, 1941]. This anomalous transmission is due to the fact that in this case a standing wave exists in the crystal with the atoms positioned in its nodes. The induced transmission filter [Berning and Turner, 1957] was the first application of these observations to the design of optical coatings; their filter can be described as a Fabry-Perot interferometer with a thin metal layer positioned in a node of the standing wave between the two Fabry-Perot mirrors.

We will describe in this paper multilayer coatings that use absorbing

---

The author is at the IBM Thomas J. Watson Research Center, Yorktown Heights, New York 10598.

TABLE 1  Applications of Multilayer Coatings of Absorbing Materials Discussed in this Paper

| Wavelength Region | Applications |
|---|---|
| 50–900 Å | Mirrors |
|  | Polarizers |
|  | Beam splitters |
| 1100–2500 Å | High-reflectivity mirrors |
|  | Beam splitters |
|  | Interference filters |
|  | Antireflection coatings |

materials. In the first section, we will discuss some general properties of these coatings and prove that, in theory, absorption-free coatings are possible. In the next two sections, we will discuss coatings for special applications in detail. Table 1 lists the applications that we consider important and that will be treated in the paper. We will also discuss some experimental results.

**Theoretical Considerations**

Every discontinuity of the complex refractive index $\hat{n}$ ($\hat{n} = n + ik$) causes a reflection of an incident wave at the boundary. This reflection is very small if the change in the optical constants is very small; however, if we add more boundaries in such a way that all add in phase to the reflected wave, then the reflected intensity increases proportional to the square of the number of boundaries $N^2$ until it reaches a final value due to either depletion of the incident beam or absorption. Without absorption, all the energy is in the reflected beam when the incident beam is depleted. A reflectivity $R = 1$ can be obtained in theory with arbitrarily small differences in the refractive index. If we use absorbing films, one generally assumes that additional absorption losses will limit the reflectivity obtainable to much lower values. This assumption, however, is wrong, and we now want to prove the following theorem:

*Reflectivities approaching $R = 1$ can be obtained with multilayer structures using absorbing films.*

Because we know already that a reflectivity approaching $R = 1$ can be obtained with multilayer structures of lossless films of arbitrarily small reflectivity, we only have to prove that the absorption losses of such a reflector using absorbing films can also be made arbitrarily small.

In front of and inside a reflector with $R = 1$ there exists a perfect standing wave with zero intensity in its nodes. Away from the node of the standing wave the intensity increases quadratically with distance. A very thin absorbing film (absorption index $k$, thickness $d$) positioned in the node of the standing wave has an absorption loss

$$A \propto \int_{-d/2}^{+d/2} k\, I(z)\, dz \propto \int_{-d/2}^{+d/2} k\, z^2\, dz \propto d^3. \tag{1}$$

The reflections from the front side and back side of this film cancel each other to a large extent with the remaining reflectivity

$$R \propto d^2. \tag{2}$$

Equations (1) and (2) show that the absorption losses of the film decrease faster than its reflectivity. Therefore, the absorption losses can be made arbitrarily small compared to the reflectivity if thinner and thinner films are used. In the limit, a very thin film is equivalent to an absorption-free film of low reflectivity.

A multilayer structure of arbitrarily small losses and high reflectivity can be constructed by using many of these films spaced in such a way that they all add in phase to the reflected wave. The space between the absorbing films contains the antinodes of the standing wave in this design and has to be completely absorption-free.

Curves that show how the reflectivity $R = 1$ is approached if the number of layers is increased and the thickness of each layer is decreased have been given previously [Spiller, 1972]. Figure 1 shows the enhancement of the transmission produced by the standing wave that is obtained from a periodic structure with the thickness of the absorbing layer optimized to produce maximum reflectivity. For structures with many layers, this enhancement becomes a very large effect; a transmission maximum occurs simultaneously with the reflection maximum. The curve in Figure 1 shows how fast the enhancements observed in x-ray diffraction (Borrmann effect) [Batterman and Cole, 1964] are approached if the number of periods is increased.

The ideas described give some guidance on how to reduce absorption losses in multilayer structures of absorbing materials. The method will work best for high-reflectivity mirrors where the standing wave is most pronounced. Other elements, for instance beam splitters or other partly reflecting mirrors, produce wave fields that have no planes with $I = 0$; for these coatings, the absorption cannot be reduced completely and an increase in the number of layers will produce only a smaller improvement. The same holds if a spacer layer that is completely absorption-free is not

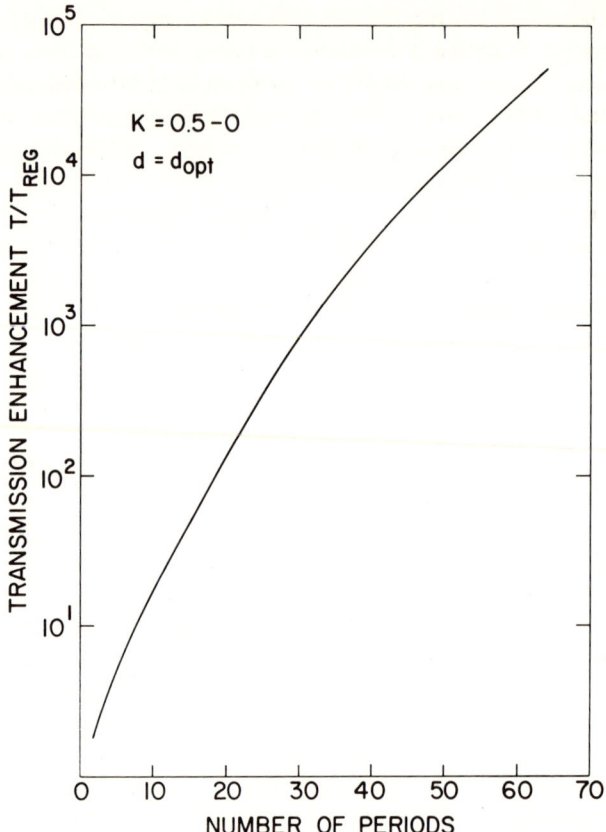

FIGURE 1  Transmission enhancement $T/T_{REG}$ for a periodic multilayer structure of two materials with $k = 0.5$ and $k = 0$ optimized to give maximum reflectivity as a function of the number of periods. $T_{REG} = (1 - R) \exp(-4\pi k D/\lambda)$ is the transmission that one would calculate if the standing wave in the coating were neglected, $D$ = total thickness of all absorbing layers. The refractive index for all layers is $n = 1$.

available. For each practical design, there exists an optimum performance that can be reached with a finite number of layers.

The equations for multilayer structures of absorbing materials are too long to be written out. However, the calculations are a trivial task for an electronic computer. We use the matrix method [Born and Wolf, 1965] with an IBM 360 time-sharing system. The available programs allow us to optimize every coating. As an example, we find a coating with a maximum (or minimum) for some quantity like reflectivity or absorption in certain layers by starting from a guessed design and proceeding along the gradient of this quantity with respect to some parameters (usually the thickness of each layer) until an extreme has been reached.

In general, the optimum design depends on the number of layers one wants to use and the optical constants of each layer. The reflectivity obtainable for a mirror increases with the number of layers and with the value of the Fresnel coefficient between adjacent layers; it decreases for increasing $k$ of the spacer layer. The maximum reflectivity obtainable with many layers ($N \rightarrow \infty$) depends only on the ratio $k_H/k_L$ for the case that all layers have the same refractive index and $k_H \ll 1$. ($k_H$, $k_L$ are the absorption indices of the absorber and the spacer layer.) This maximum reflectivity is plotted in Figure 2 to give a guide to what can be obtained without any change in the refractive index. While the maximum reflectivity obtainable for $N \rightarrow \infty$ depends only on $k_H/k_L$, the number of layers required to come close to this maximum reflectivity depends on the values of the absorption indices: for smaller values of $k$, more layers

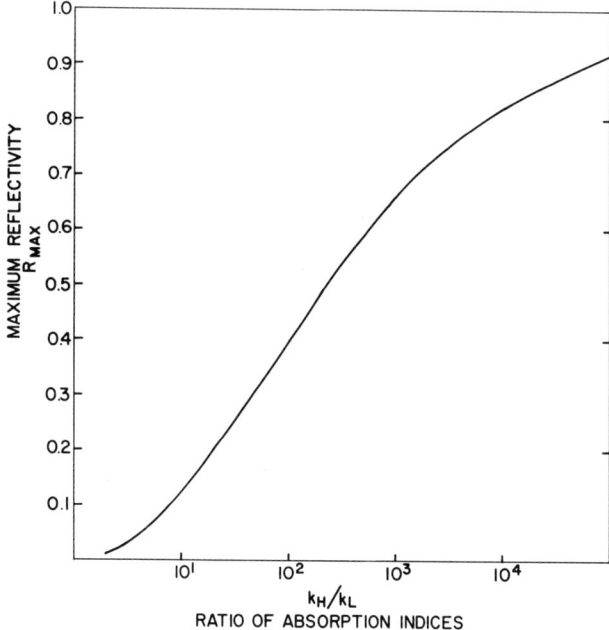

FIGURE 2 Maximum normal incidence reflectivity obtainable by alternating two materials of different absorption indices $k_H$, $k_L$ versus the ratio $k_H/k_L$ under the assumption that the number of periods is very large ($N \rightarrow \infty$), that $k_H \ll 1$, and that all refractive indices are one.

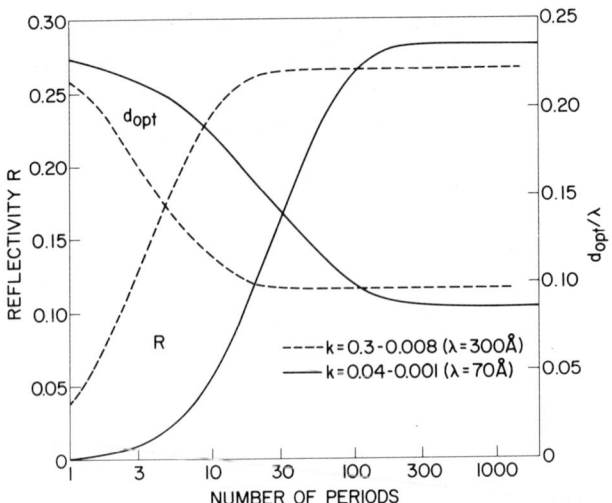

FIGURE 3 Normal incidence reflectivity for a periodic structure of period $\lambda/2$, where a material with high absorption index $k$ alternates with a material of low absorption index. The thickness $d_{opt}$ of the strong absorber is chosen to give the highest reflectivity and is also plotted. The $k$ values used are available at $\lambda = 300$ Å (dashed curves) and $\lambda = 70$ Å (solid curves).

are needed. This is demonstrated in Figure 3, where the maximum reflectivity obtainable is plotted versus the number of periods used. The values for $k$ chosen are typical values in the xuv region [Spiller, 1972]. For a wavelength $\lambda = 300$ Å only 10 periods are needed to obtain the largest possible reflectivity, while about 100 periods are necessary for $\lambda = 70$ Å.

## Interference Coatings for the xuv (50–900 Å)

### MIRRORS

In the xuv, no single material has a substantial reflectivity at normal incidence, the reflectivities available decrease with decreasing wavelength [Madden, 1963; Samson, 1967; Hass and Hunter, 1974]. The refractive indices of all materials are close to one in this region. The absorption indices show more variations. The highest absorption indices available at each wavelength are about a factor of 50 larger than the lowest [Spiller, 1972]. No absorption-free material is available. Figure 2 shows that the maximum possible reflectivity with multilayer coatings is around $R = 0.3$

for wavelengths shorter than 300 Å when the condition $k_H \ll 1$ is fulfilled; at longer wavelengths the highest values of $k$ are around $k \simeq 1$, and higher reflectivities are possible.

The best design for highest reflectivity is not the periodic structure but a structure where the thickness of the absorber layer increases from the top to the bottom (substrate) of the coating. The reason is that the standing wave is more pronounced at the top of the coating than in the deeper layers; for an optimum design, therefore, the first absorber layers will be thinner than the deeper ones; all the light that passes the last layer is transmitted and lost for reflection, therefore in a mirror for maximum reflectivity the last layer has the same thickness as that single film which gives the maximum reflectivity.

Possible materials for the xuv region are listed in Table 2. The best combination is the combination that combines the highest with the lowest value of the absorption index; however, some combinations will have to be excluded because the boundary between the two materials is not stable (for instance, Al–Au) [Hunter et al., 1972]. Many more possibilities for stable boundaries are expected if not only elements but also com-

TABLE 2 Selected Materials for xuv Coatings[a]

| Large $k$ | | | Small $k$ | | |
|---|---|---|---|---|---|
| Material | Wavelength (Å) | Reference | Material | Wavelength (Å) | Reference |
| Au | 200–900 | Canfield et al., 1964 | Mg | 250–900 | Hunter, 1964 |
| Pt | 100–900 | Dietrich and Kunz, 1972; Jacobus et al., 1963 | Al | 170–600 | Hunter, 1964 |
| Ir | 100?–900 | Hass et al., 1967 | Si | 130–350 | Hunter, 1964 |
| Re | 100?–900 | Cox et al., 1972 | | | |
| Os | 100?–900 | Hass and Hunter, 1973 | Be | 120–600 | Rustgi, 1965 |
| Rh | 100?–900 | Cox et al., 1971 | C | 50–300 | Samson and Cairns, 1965 |
| W | 100?–900 | Cox et al., 1972 | | | |
| $Al_2O_3$ | 100?–900 | Hass and Tousey, 1959; Madden, 1963 | | | |

[a] A question mark at a wavelength means that the optical constants at this wavelength have not been measured.

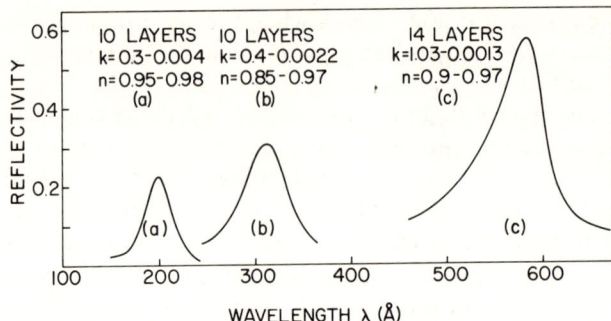

FIGURE 4 Reflectivity as a function of wavelength of mirror coatings for the xuv region. The thickness of the layers counted from the substrate are (in Å) (a) 61.4, 40.3, 66.6, 33.4, 71.9, 28.9, 75.2, 26, 77.5, 24.7; (b) 98.9, 51.9, 101.9, 45, 112.5, 38.3, 116.4, 34.9, 119.8, 34; (c) 198.4, 100.3, 237.9, 61, 250.3, 48.7, 257.7, 42.8, 262.8, 36.6, 268, 32.7, 269.4, 32.8. The optical constants used for the substrate are (a) $n = 0.95$, $k = 0.08$; (b) $n = 0.89$, $k = 0.092$; (c) $n = 0.85$, $k = 0.47$. The last layer of each coating (i.e., the first layer toward the incident light) is always a layer of the material with the high $k$.

pounds like metal oxides are used; however, not enough data on the optical constants of compounds are available to date.

Figure 4 gives the calculated reflectivity curves for three optimized designs in the xuv regions. The optical constants used correspond to the combination Si–Au for curve a, Mg–(Au, Pt, Ir) for curve b, and Mg–Pt, for curve c. No correction to the optical constants has been made to account for the fact that very thin films often have a lower density and therefore also different optical constants than thicker films. We see that the calculated reflectivities are about an order of magnitude larger than those obtainable with single films.

POLARIZERS

A mirror used at nonnormal incidence has different reflectivities for s- and p-polarization and can be used as a polarizer. The degree of polarization $R_s/R_p$ obtainable from one boundary depends on the optical constants and can be maximized by adjusting the angle of incidence [Hunter, 1964]. If a multilayer coating is used instead of a single boundary, the obtainable degree of polarization practically does not change; however, because the reflectivity can be increased, a multilayer structure gives a much larger intensity, and larger ratios $R_s/R_p$ become possible by using several reflections.

Figure 5 shows the calculated performance of such a multilayer reflector using optical constants available around $\lambda = 300$ Å. The structure optimizes the degree of polarization $R_s/R_p$ for an angle of incidence of 45°; it was found that the design is practically the same as that which maximizes the reflectivity for s-polarization, $R_s$.

BEAM SPLITTERS

The reflectivities possible in the xuv are too small and the absorption losses too high to make multiple-beam interferometers (Fabry-Perot interference filters) useful devices. If an element for wavelength selection is wanted, it is better to use a mirror coating of small bandwidth (i.e., with many layers) or to cascade several mirrors. For a two-beam (Michelson) interferometer some absorption losses can be tolerated. With a

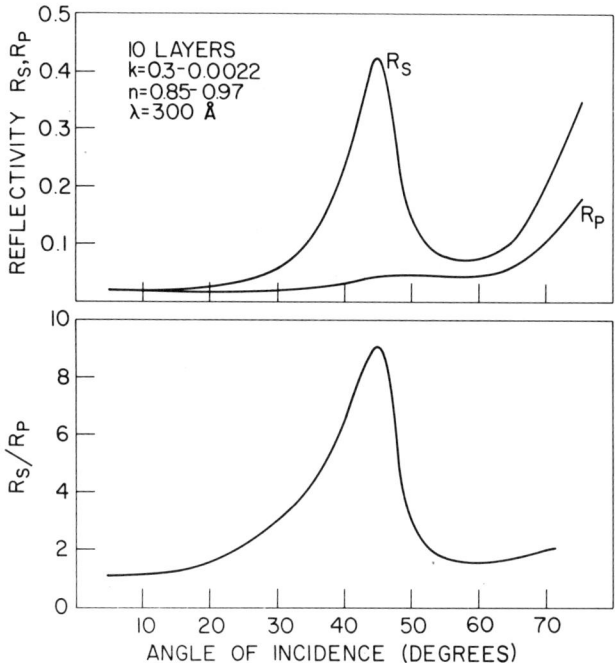

FIGURE 5  Reflectivities $R_s$ and $R_p$ for s- and p-polarization (top) and degree of polarization $R_s/R_p$ (bottom) versus the angle of incidence $\alpha$ for a coating useful as a polarizer at $\lambda = 300$ Å. Optical constants of the substrate: $n = 0.89$, $k = 0.092$. Thickness of each layer counted from the substrate (in Å): 157.7, 72.8, 168.1, 57.7, 178.9, 45.6, 186.9, 38.9, 191, 37.1.

multilayer structure the absorption losses can be reduced; however, the reduction is smaller than for a mirror of higher reflectivity because a beam splitter, which produces also a transmitted beam, always produces less pronounced standing waves than a mirror of higher reflectivity and no transmittance. A serious problem for any beam splitter is the absorption in the substrate; we ignore this problem completely by assuming that either an absorption-free substrate is available or unbacked films can be made.

Figure 6 shows the calculated performance of a beam splitter that might be used in a Michelson interferometer. This five-layer beam splitter for an angle of incidence of 45° has a transmission and reflection coefficient of 24%, while 52% of the incident light is absorbed. In contrast, a simple film of the high absorption material alone could have a transmission and reflection of only 18.5% with 63% absorption and for a three-layer coating (high $k$, low $k$, high $k$) 23% transmission and absorption are possible. As we had expected, the improvement over the single film is more modest than that obtained for the highest reflectivity mirror; practically no further reduction in absorption loss occurs when more than five layers are used.

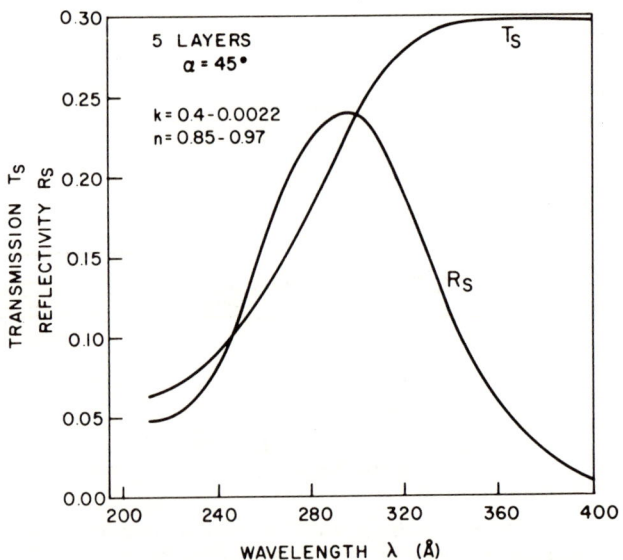

FIGURE 6 Transmission $T_s$ and reflection $R_s$ of a coating designed as a 50/50 beam splitter for 45° angle of incidence and $\lambda = 300$ Å. Layer thickness 27.65, 200.5, 27.65, 200.5, 27.65 Å.

## Interference Coatings for the Near-Vacuum Ultraviolet (1100–2500 Å)

A single film of aluminum has a much better optical performance in the wavelength region between 1100 and 2500 Å than any multilayer structure possible in the xuv. We will discuss in this section what further improvement can be obtained when the aluminum film is replaced by multilayer structures. Because nonabsorbing materials are available in this wavelength region, in theory ideal elements with no absorption can be produced. In practice, the required high quality of the spacer layer is the most important obstacle to obtain better and better elements. Spacer layers that are completely free of absorption and scattering are needed for the ideal element; we will discuss how much loss in the spacer layer can be tolerated.

### HIGH-REFLECTIVITY MIRRORS

High-reflectivity mirrors can be obtained by overcoating an opaque aluminum film with a multilayer structure of spacer layers and thin aluminum films. The process can be described as positioning very thin aluminum films into the nodes in front of the opaque film in such a way that they boost the reflectivity of this film. Figure 7 shows how the reflectivity increases if more and more layers are used. The curves have been calculated for spacer layers of $MgF_2$ with the values of optical constants for $\lambda = 1216$ Å; however, for other wavelengths in the region between 1100 and 2500 Å very similar curves are obtained. Figure 7 shows also the influence of losses in the spacer layer. If the spacer layer attenuates the light according to $k_{MgF_2} = 0.01$ (either due to absorption or to scattering) only a slight improvement is possible; for $k_{MgF_2} = 0.03$ no improvement occurs. It has been observed that $MgF_2$ films on Al can be described by $k_{MgF_2} = 0.03$ at $\lambda = 1216$ Å [Hutcheson et al., 1972]; therefore, improvements at 1216 Å can only be expected if better spacer films can be produced. At longer wavelengths where scattering and absorption losses are smaller, the task is easier.

### BEAM SPLITTERS AND INTERFERENCE FILTERS

Instead of using many layers to increase the reflectivity, it is also possible to maintain the reflectivity and increase the transmission of a mirror. Figure 8 shows as an example the transmission of an optimum periodic $(Al-MgF_2)^N$ structure with a reflectivity $R = 0.9$ versus the number of periods $N$. The absorption losses can be drastically reduced by using several layers. The most important application for these designs is their

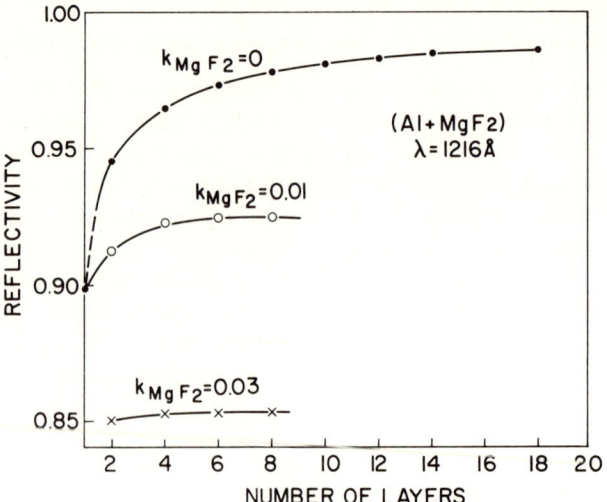

FIGURE 7 Maximum reflectivity obtainable from multilayer structures of Al and $MgF_2$ versus the number of layers used with $k_{MgF_2}$ as parameter. $n_{Al} = 0.056$, $k_{Al} = 1.04$, $n_{MgF_2} = 1.7$. All calculated points are for coatings terminated with $MgF_2$ except for the single Al film (one layer).

FIGURE 8 Maximum transmission of a periodic multilayer structure of Al and $MgF_2$ with a reflectivity $R = 0.9$ versus the number of periods in the structure. $n_{Al} = 0.12$, $k_{Al} = 2.12$, $n_{MgF_2} = 1.44$, $k_{MgF_2} = 0$.

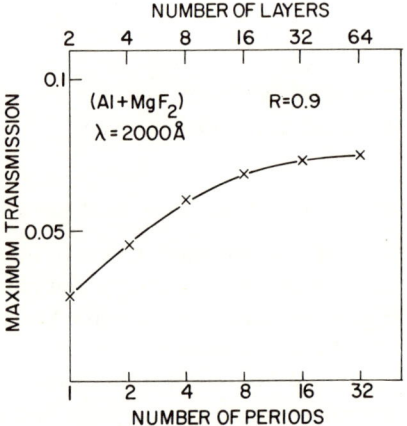

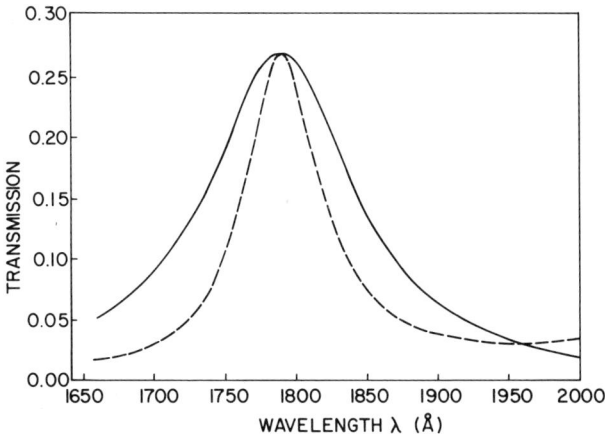

FIGURE 9 Calculated transmission curves for interference filters. *Solid curve*: Conventional interference filter, substrate Al–MgF$_2$– Al–MgF$_2$ with layer thicknessess of 282.5, 352.74, 282.5, 205.5 Å. *Dashed curve:* 12-layer structure of Al and MgF$_2$ with thickness of 109.8, 572.5, 109.8, 572.5, 109.8, 380, 78, 572.5, 78, 572.5, 78, 644 Å. Substrate is sapphire; the dispersion of all materials included and approximated by linear functions.

use in Fabry-Perot interference filters. Interference filters with single-film aluminum reflectors have been described before [Bates and Bradley, 1966]. In Figure 9 we compare the performance of these conventional interference filters (full curve) with the performance possible if the two Al–film reflectors are replaced by several layers (dashed curve). The multilayer filter has been designed to give the same peak transmission as a conventional filter, in this case the multilayer filter has a smaller bandwidth than the filter using single Al–film reflectors. It is also possible to design a filter with the same bandwidth but increased peak transmission. Theoretically a filter with an arbitrarily small bandwidth and a peak transmission close to one is possible; the practical limit is again determined by the losses in the spacer layer.

ANTIREFLECTION COATINGS

Antireflection coatings are useful to increase optically the efficiency of photodetectors. In the vacuum uv, however, not enough nonabsorbing materials are available to use the standard designs. If absorbing materials are used, part of the incident light is absorbed in the coating and not in the material of the detector. Because our design principle allows us to reduce

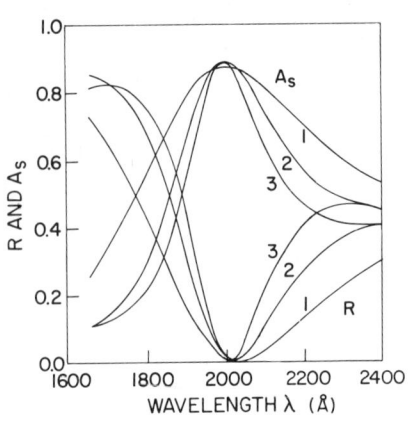

FIGURE 10 Calculated performance of antireflection coatings on silicon made of Al and $MgF_2$ and using one, two, or three Al films. The reflectivity $R$ and the fraction of the incident light absorbed in the silicon substrate $A_s$ are plotted as a function of wavelength. $n_{SI} = 1, k_{SI} = 2.5, n_{MgF_2} = 1.44, k_{MgF_2} = 0, n_{Al} = 2.12$.

the absorption in an absorbing film, we want to investigate now how far it is possible to reduce the reflection with absorbing materials in such a way that most of the incident light is absorbed in the substrate and a minimum amount in the coating. We choose as a practical example silicon with $n = 1$ and $k = 2.5$ for $\lambda = 2000$ Å. More than 60% of the incident light is reflected, only 40% is absorbed in silicon.* The simplest antireflection coating is a thin film of aluminum with thickness and distance from the silicon adjusted such that the reflections from the substrate and the aluminum cancel each other. Curves 1 in Figure 10 show the reflectivity of this design and the fraction of the light $A_s$ absorbed in the silicon substrate as a function of wavelength. The reflectivity has been reduced to zero, 87% of the incident light is absorbed in the substrate for $\lambda = 2000$ Å. $MgF_2$ has been used as the spacer layer and as a protective overcoating of this design. Curves 2 and 3 in Figure 10 show the performance obtainable if two and three Al films with the corresponding $MgF_2$ spacer layers are used. There is only a very small increase in substrate absorption possible if one goes from one Al film to two films; the difference between two and more layers is unnoticeable in the scale of Figure 10. We conclude that for our special problem a system with a single Al film gives the best performance. The differences between the system with a single Al film and those with more films become larger if substrates with higher initial reflectivities are used. For this case, a system with higher reflectivity is needed to cancel the reflectivity; this results in better pronounced nodes of the standing wave with larger payoff for thinner films of the absorbing material. If for some reason Al cannot be used in the overcoating and a material with not so well suited optical constants has to be selected for the coating, the step from a single absorb-

*I thank Peter Cone for bringing this problem to my attention.

*Eberhard Spiller* 595

ing film to many thin films will also have a larger effect on the absorption of the substrate.

EXPERIMENTS (1800-3000 Å)

Transmission measurements on $MgF_2$ films deposited on sapphire substrates showed that these films have $k < 0.001$ for $\lambda > 1800$ Å. The thickness of the films used for these measurements was around 5000 Å, the deposition time varied between 2 and 6 min, and the pressure in the chamber was $1 \times 10^{-5}$ Torr or better. The substrate was at room temperature; higher substrate temperatures (200 °C) or heating of the film in air to 200 °C for 2 h after the deposition produced more lossy films. From these measurements we conclude that the designs described in the previous section can be realized experimentally for $\lambda > 1800$ Å.

Figure 11 shows that this conclusion is correct. In this figure we compare the measured reflectivity of a single thick Al film and of an Al film overcoated with $MgF_2$ to a four-layer system ($Al-MgF_2-Al-MgF_2$) designed to give maximum reflectivity at $\lambda = 2000$ Å. We see that indeed

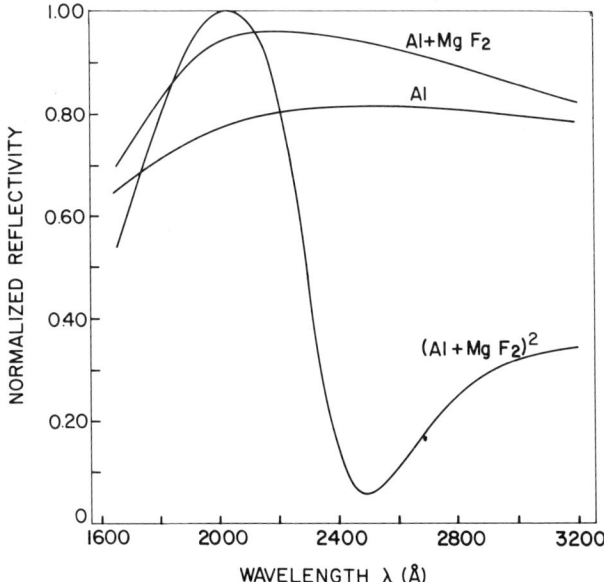

FIGURE 11 Measured reflectivity of a single opaque fresh Al film, an Al film overcoated with $MgF_2$, and a four-layer Al-$MgF_2$ coating designed to give maximum reflectivity for $\lambda = 2200$ Å. Angle of incidence 10%. The thicknesses desired were 1000, 635, 140, and 665 Å; we estimate the calibration of the quartz thickness monitor to have an error of 10%

higher reflectivities can be obtained with multilayer systems. The reflectivity values plotted are normalized to the maximum reflectivity of the four-layer system. The reflectivity values have been measured in a Beckman DK2a spectrophotometer with an aluminum mirror in the reference beam. The reflectivity of a sapphire surface was used as a standard to correct all measured reflectivity values in such a way that the measured reflectivity for sapphire agreed to its calculated values obtained from extrapolated values of the published refractive index for sapphire [Malitson et al., 1958]. This procedure resulted in a peak reflectivity of $R_{max}$ = 1.02; the scale in the plot was then changed to produce $R_{max}$ = 1.00. In the scale of Figure 11, therefore, all measured reflectivity values for sapphire would appear 2% lower than the calculated ones.

The films have been prepared by evaporation of the materials with an electron gun. The pressure measured with an ionization gauge in the chamber close to the electron gun was $2 \times 10^{-7}$ Torr between evaporations and $2 \times 10^{-6}$ Torr or less during evaporations. The deposition time for the Al films was about 0.5 min, and for the $MgF_2$ films about 1 min. The thickness of the films was monitored with an Edwards quartz thickness monitor.

The calculated maximum reflectivity obtainable with an optimized four-layer system is $R = 0.947$, while $R = 0.928$ is possible with an aluminum film overcoated with $MgF_2$. The measured ratio of the peak reflectivities for the two coatings is in reasonable agreement with these values.

## Conclusion

We have shown that multilayer structures of absorbing materials can be useful in the vacuum uv. Ideal elements with no absorption loss at all can be realized if one material free of absorption and scattering is available that can be used as the spacer material between the layers of an absorber. Absorption-free materials are available for wavelengths above 1100 Å. A problem for further research is how to deposit thin films of these materials in such a way that neither absorption nor scattering occurs. The problem is most severe at the shorter wavelengths; our experiments show that $MgF_2$ films are of sufficient optical quality to allow multilayer coatings with high performance for wavelengths longer than 1800 Å.

In the xuv region no absorption-free material is available—the absorption in the spacer layer makes ideal lossless elements impossible. In spite of that, mirror reflectivities that are an order of magnitude higher than those of single films seem possible. In order to be able to select the best material combination, data on the optical constants of more materials are necessary. Only little is known at present regarding the optical constants in the xuv region for the very thin films required for multilayer coatings.

I thank A. Neureuther for helpful discussions.

## References

Bates, B., and D. J. Bradley, *Appl. Opt. 5*, 971 (1966).
Batterman, B. W., and H. Cole, *Rev. Mod. Phys. 36*, 681 (1964).
Berning, P. H., and A. F. Turner, *J. Opt. Soc. Am. 47*, 230 (1957).
Born, M., and E. Wolf, *Principles of Optics*, 3rd ed., p. 51, Pergamon, New York (1965).
Borrmann, G., *Phys. Z. 42*, 157 (1941).
Canfield, L. R., G. Hass, and W. R. Hunter *J. Phys. 25*, 124 1964).
Cox, J. T., G. Hass, and W. R. Hunter, *J. Opt. Soc. Am. 61* 360 (1971).
Cox, J. T., G. Hass, J. B. Ramsey, and W. R. Hunter, *J. Opt. Soc. Am. 62*, 781 (1972).
Hass, G., and W. R. Hunter, this volume, p. 525 (1974).
Hass, G., and R. Tousey, *J. Opt. Soc. Am. 49*, 593 (1959).
Hass, G., G. F. Jacobus, and W. R. Hunter, *J. Opt. Soc. Am. 57*, 758 (1967).
Hunter, W. R., *J. Phys. 25*, 154 (1964).
Hunter W. R., T. L. Mikes, and G. Hass, *Appl. Opt. 11*, 1594 (1972).
Hutcheson, E. T., J. T. Cox, G. Hass, and W. R. Hunter, *Appl. Opt. 11*, 1590 (1972).
Jacobus, G. F., R. P. Madden, and L. R. Canfield, *J. Opt. Soc. Am. 53*, 1084 (1963).
Madden, R. P., in *Physics of Thin Films*, Vol. 1, p. 123, G. Hass, ed., Academic Press, New York (1963).
Malitson, I. H., F. V. Murphy, Jr., and W. S. Rodney, *J. Opt. Soc. Am. 48*, 72 (1958).
Rustgi, O. P., *J. Opt. Soc. Am. 55*, 630 (1965).
Samson, J. A. R., and R. B. Cairns, *Appl. Opt. 4*, 915 (1965).
Spiller, E., *Appl. Phys. Lett 20*, 365 (1972).
Wiener, O., *Ann. Phys. 40*, 203 (1890).

# IMAGE PROCESSING AND HOLOGRAPHY

S. DEBRUS, M. FRANÇON, and
C. P. GROVER

# A NEW METHOD OF OPTICAL PROCESSING APPLIED TO DETECTION OF DIFFERENCES BETWEEN TWO IMAGES

## I. Introduction

The phenomenon of interference in diffused light was observed for the first time by Newton [1931]. A mirror, with reflecting back surface, was illuminated by a point source of light placed at its center of curvature. A beautiful system of colored rings, centered on the source, was then observed. Continuing the experiment of Newton, De Chaulnes [1755] showed that the visibility of the rings could be increased by using a mirror with partially diffusing front surface. Further studies on this subject were made by Young [1802] and Herschel [1830]. Fabry and Perot [1897] observed a similar system of rings in transmitted light. Their system consisted of a semireflecting plate with plane-parallel faces, one of which was diffusing.

Burch [1953] for the first time applied the interference phenomenon in diffused light for the construction of an interferometer.

After the discovery of the laser and holography, the problem of interferometry in diffused light drew attention of many workers. This resulted in a great number of publications on various aspects of the prob-

---

The authors are in the Institute of Optics, University of Paris, France.

lem. One of the important aspects of the problem is the study of the interference phenomenon at infinity. In an experiment by Burch and Tokarski [1968] a number of speckle patterns, displaced laterally relative to one another and recorded simultaneously on the same photographic plate, give at infinity a system of fringes.

We use a new method of optical processing that is based on the speckle interference phenomenon at infinity. After describing the general principle of the method, we shall give an application to the detection of differences between two images.

## II. General Principle

The main feature, common to all the applications of our method, is the modulation of the signal to be processed by a random diffuser. The diffuser contains high spatial frequencies that give in the focal plane of a lens a wide spread-out field. The spectrum of the signal, which is normally given by a delta function, is, then, carried away from the center due to this modulation. Let $A$ be the signal to be processed. Generally it is a transparency with $A$ as its irradiance distribution. $D$ represents the factor of transmittance of the diffuser. Figure 1 shows the recording geometry: the transparency, A, is imaged by a lens system, O, on the diffuser, D. A. high-resolution photographic plate, H, placed immediately behind D, records the product $AD$. The distance between D and H is so small that for all practical purposes they can be considered to lie in the same plane. Let this plane be represented by the coordinates $(\eta, \zeta)$. The photographic plate is exposed twice to this irradiance distribution, and between the exposures it is translated through $\zeta_0$ in a direction parallel to $\zeta$ axis. The total irradiance recorded is

$$A(\eta, \zeta) D(\eta, \zeta) + A(\eta, \zeta - \zeta_0) D(\eta, \zeta - \zeta_0). \qquad (1)$$

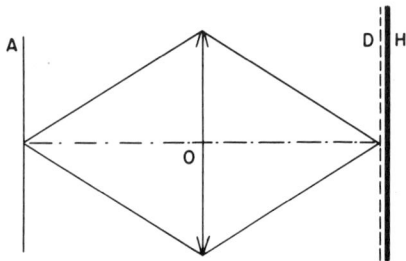

FIGURE 1 Optical arrangement for signal recording.

S. Debrus et al. 603

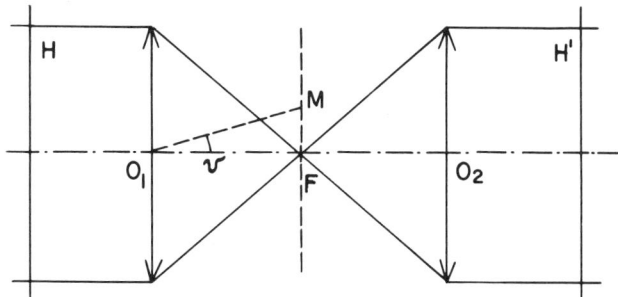

FIGURE 2  Optical arrangement for signal retrieval.

This can be rewritten as

$$A(\eta, \zeta) D(\eta, \zeta) \oplus [\delta(\eta, \zeta) + \delta(\eta, \zeta - \zeta_0)] \qquad (2)$$

because a translation is equivalent to a convolution by a delta function. The photographic plate is processed under the usual conditions of linearity. It is illuminated by a parallel beam of light as shown in Figure 2. The transmitted amplitude is given by

$$t(\eta, \zeta) = t_0 - \beta_0 \{ A(\eta, \zeta) D(\eta, \zeta) \oplus [\delta(\eta, \zeta) + \delta(\eta, \zeta - \zeta_0)] \}, \qquad (3)$$

where $t_0$ and $\beta_0$ are constants.

The amplitude distribution in the focal plane of the objective $O_1$ is given by the Fourier transform of $t(\eta, \zeta)$, that is,

$$U(u, v) = t_0 \delta(u, v) - \beta_0 [\tilde{A}(u, v) \oplus \tilde{D}(u, v)] (1 + e^{jkv\zeta_0}) \qquad (4)$$

($k = 2\pi/\lambda$, $\lambda$ = wavelength) the symbol $\sim$ represents the Fourier transform and $(u, v)$ are the coordinates in the focal plane. The first term on the right-hand side of Eq. (4) represents the direct image of the source that is located at the focus of $O_1$. It can be neglected due to its small size. In the second term, apart from a constant, the spectrum of the product $AD$ is modulated by the factor $1 + e^{jkv\zeta_0}$, which represents a system of Young's fringes. These fringes are perpendicular to the direction of translation. In intensity, the irradiance of these fringes is proportional to $\cos^2(kv\zeta_0/2)$.

Let us consider a grid consisting of an array of slits having the same spatial frequency as that of the fringes. Such a grid, when placed in the focal plane and so positioned that the slits coincide with the fringe maxima, let pass the signal to give its image in plane H' (Figure 2). If the grid is translated in its plane through half a period so that the slits now fall on the fringe minima, no information from the signal will be transmitted. We see that the system of fringes that modulates the spectrum of the signal provides a sort of filter for the extraction of the required information. Depending on the experiment, the modulating fringes can be shaped accordingly.

Let us consider the case when the photographic plate H is exposed $(N + 1)$ times to the irradiance $AD$. Between two successive exposures, H is translated through $\zeta_0$. Under these conditions, the total irradiance recorded is

$$(AD) \oplus \sum_{n=0}^{N} \delta(\eta, \zeta - n\zeta_0). \qquad (5)$$

For $N \to \infty$, the spectrum of H is given by

$$t_0 \, \delta(u, v) - \beta_0'(\widetilde{A} \oplus \widetilde{D}) \, [\text{comb} \, (v/v_0)], \qquad (6)$$

where $v_0 = \lambda/\zeta_0$. In Eq. (6) the convolution $\widetilde{A} \oplus \widetilde{D}$ is modulated by a Dirac comb (Figure 3); the period of the comb is determined by the amount of translation given to the plate.

As $N$ is always limited, the intensity of the modulating fringe system is given by

$$\left[\frac{\sin (N + 1) kv\zeta_0/2}{\sin kv\zeta_0/2}\right]^2. \qquad (7)$$

Between two primary maxima, there are $(N - 1)$ secondary maxima (Figure 4). These secondary maxima can be suppressed if the exposures are in the ratio of the binomial coefficients and if the plate is given equal displacements between them [Burch and Tokarski, 1968].

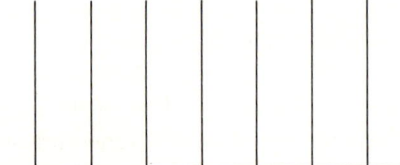

FIGURE 3 Dirac comb-type modulation.

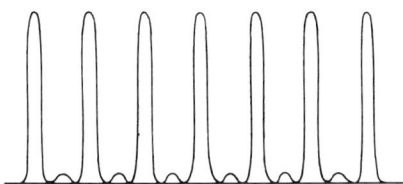

FIGURE 4 Multiple-exposure fringe-type modulation for finite value of $N$.

## III. Detection of the Difference between Two Images [Debrus *et al.*, 1971]

Let $A$ and $B$ be the two signals to be compared. In actual practice the signals are derived from two black-and-white transparencies. There are certain regions in the two photographs that are identical. The problem consists in the detection of the difference $b = B - A$.

The optical arrangement of Figure 1 is used to record the signals. In the first exposure, $A$ modulated by the diffuser D is recorded. The plate is translated through $\zeta_0$, and then the second exposure is made replacing $A$ by $B$. It is absolutely necessary that the identical regions of the two transparencies coincide in the plane of the diffuser. The total irradiance recorded in the two exposures is

$$(AD) \oplus \delta(\eta, \zeta) + (BD) \oplus \delta(\eta, \zeta - \zeta_0). \tag{8}$$

This can be rewritten as

$$(AD) \oplus [\delta(\eta, \zeta) + \delta(\eta, \zeta - \zeta_0)] + (bD) \oplus \delta(\eta, \zeta - \zeta_0) \tag{9}$$

and in a symmetrical form as

$$(AD) \oplus \{\delta[\eta, \zeta + (\zeta_0/2)] + \delta[\eta, \zeta - (\zeta_0/2)]\} + (bD) \oplus \delta[\eta, \zeta - (\zeta_0/2)]. \tag{10}$$

The spectrum of such a recording after processing can be written as

$$2(\tilde{A} \oplus \tilde{D}) \cos \varphi + (\tilde{b} \oplus \tilde{D}) e^{-j\varphi}, \tag{11}$$

where $\varphi = \pi \nu \zeta_0/\lambda$; the direct image of the source has been neglected. If a grid, with the same period as that of the fringe system, is placed in the focal plane of $O_1$, with slits coinciding with the fringe minima, the first term of Eq. (11) is stopped and the second term is transmitted. One more Fourier transformation restores the difference $b$ of the two images modulated by the diffuser D. Let us note that the spatial frequencies of D are high enough (size of the structure of our diffuser: ~3μm) not to alter the quality of the reconstructed image b. The grid can be replaced by a single slit to act as a filter.

The amplitude modulation of the type described above is represented by curve 1 in Figure 5. As the slope of the curve at $M_0$ is nonzero, it is necessary to use a fine slit. This increases the noise level. This can be avoided by modifying the shape of the fringes.

If, instead of two, three exposures are made, we have the results shown in Table 1. The total irradiance is given by

$$(AD) \oplus \left[ \delta(\zeta) + \frac{1}{2} \delta(\zeta + \zeta_0) + \frac{1}{2} \delta(\zeta - \zeta_0) \right] + (bD) \oplus \delta(\zeta) \qquad (12)$$

and the spectrum by

$$2(\tilde{A} \oplus \tilde{D}) \cos^2 \varphi + \tilde{b} \oplus \tilde{D}. \qquad (13)$$

The modulating factor in amplitude is given by $\cos^2 \varphi$, which is represented by curve 2 in Figure 5. The resulting flat minimum at $M_0$ permits one to use a comparatively wider slit, and, consequently, the signal-to-noise ratio is improved. The gain is still higher if the number of exposures is increased. For a series of exposures with equal amounts of time, the secondary maxima obtained increase the noise level. These can be suppressed, as described earlier. By assigning even values to $N$, the modulating factor becomes an even power of $\cos \varphi$. Equation (13) becomes

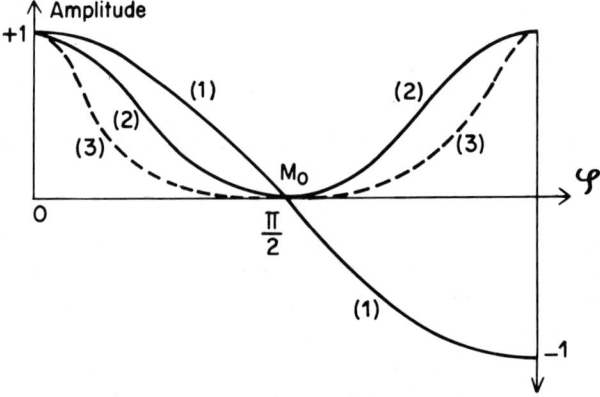

FIGURE 5  Amplitude variation for different types of exposition.

TABLE 1  Results for Various Exposures

| Exposure | Transparency | Translation | Time of Exposure (sec) |
|---|---|---|---|
| 1st | $B = A + b$ | 0 | 2 |
| 2nd | $A$ | $+\zeta_0$ | 1 |
| 3rd | $A$ | $-\zeta_0$ | 1 |

$$2^N(\widetilde{A} \oplus \widetilde{D}) \cos^N \varphi + \overset{N/2}{\underset{N}{C}} (\widetilde{b} \oplus \widetilde{D}), \qquad (14)$$

where $\overset{N/2}{\underset{N}{C}}$ are the binomial coefficients.

As an example, for $N = 4$ and $N = 6$, respectively, we have for Eq. (14)

$$16(\widetilde{A} \oplus \widetilde{D}) \cos^4 \varphi + 6(\widetilde{b} \oplus \widetilde{D}) \qquad (15)$$

and

$$64(\widetilde{A} \oplus \widetilde{D}) \cos^6 \varphi + 20(\widetilde{b} \oplus \widetilde{D}). \qquad (16)$$

The corresponding translations and exposures times are given in Table 2. Curve 3 in Figure 5 shows the case for $N = 6$.

This type of study is important from the point of view of comparing the photographs taken by an artificial satellite at two different times. The comparison consists of the detection of perturbations in the atmosphere or the changes occurring on the earth surface due to natural phenomena.

Results obtained with this technique are shown in Figure 6 at 11. Figure 8 is the difference between the photographs shown in Figures 6

TABLE 2  Translations and Exposure Times

| Total Number of Exposures | Transparency | Translation | Time of Exposure (sec) |
|---|---|---|---|
| 5 | $B = A + b$ | 0 | 6 |
| 5 | $A$ | $\pm \zeta_0$ | 4 |
| 5 | $A$ | $\pm 2\zeta_0$ | 1 |
| 7 | $B = A + b$ | 0 | 20 |
| 7 | $A$ | $\pm \zeta_0$ | 15 |
| 7 | $A$ | $\pm 2\zeta_0$ | 6 |
| 7 | $A$ | $\pm 3\zeta_0$ | 1 |

FIGURE 6  Aerial photograph of sea coast.

and 7, which corresponds only to the rectangles along the sea coast. The theory shows that the reconstructed difference is always modulated by the diffuser. Due to the fine structure of the grains ($\sim 3$ $\mu$m), such modulation is not visible in Figure 8. Similarly, Figure 11 represents the difference between Figures 9 and 10. Between the photographs to be compared, the profile of the sea has changed, as shown in Figure 11. Apart from this, the fields by the side of the sea are also shown in Figure 11 because of their unequal irradiance on Figures 9 and 10.

FIGURE 7  Photograph to be compared with the one shown in Figure 6.

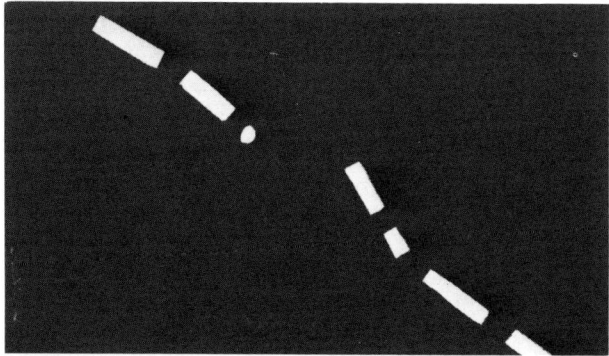

FIGURE 8  Difference between Figures 6 and 7. The difference is reconstructed with reversal contrast.

FIGURE 9  Aerial photograph of sea coast taken at certain time.

FIGURE 10  Photograph to be compared with that of Figure 9.

## IV. Decoding of a Message

The preceding method can also be applied to the decoding of a message. If $A$ is a message and $B$ is a random distribution that conceals it, the photograph to be processed has the irradiance $A + B$. In this case, another photograph with irradiance $B$ acts as a "key" for the decoding. Two exposures are made on the same photographic plate, as before, so that the irradiances $(A + B)$ and $B$ are recorded with a displacement $\zeta_0$ between them. In the process of filtering, $B$ can be eliminated leaving behind the required signal $A$. It can be seen that without the irradiance distribution $B$, which constitutes the "key," it is impossible to extract the signal $A$.

*S. Debrus et al.*

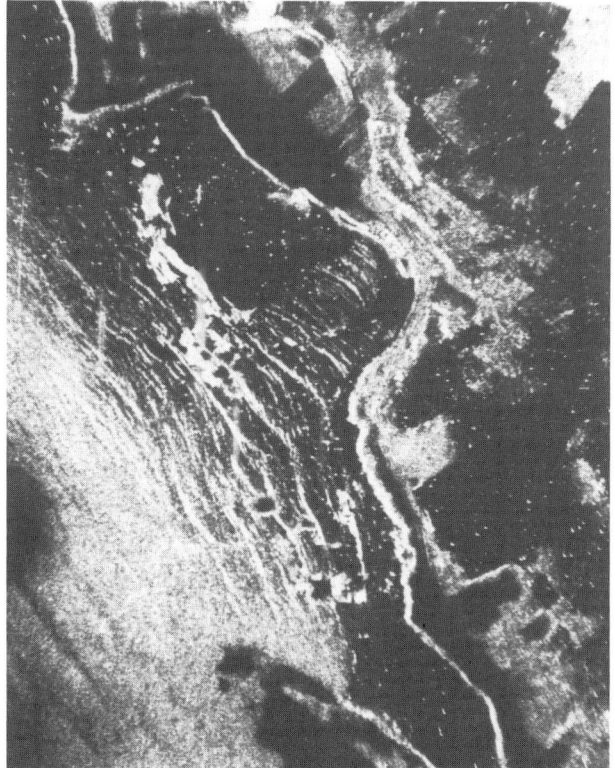

FIGURE 11  The difference between Figures 9 and 10.

## V. Other Applications–Conclusion

The same method has been applied to the multiplexing of images, the contrast reversal of the reconstructed images in carrier frequency photography, and the reconstruction of color images from black-and-white recordings.

The originality of this extremely simple method lies in the fact that the signal is modulated by a random diffuser to give a spread-out spectrum in the focal plane of a lens. The inconvenience caused by such modulation does not affect the quality of the images because we use diffusers of very high spatial frequencies. An easy recording operation and the use of simple filters give the method an additional advantage over the others.

We conclude by mentioning an additional way of modulating the spectrum of the recording. Hitherto, we have considered the modulation by two-beam Young's fringes and multiple-beam fringes. It is also possible to modulate the spectrum by the Fraunhofer diffraction pattern of a slit. This is brought about by displacing continuously the photographic plate H during the exposures. The width of the central maximum is determined by the displacement given to H.

## References

Burch, J. M., *Nature 171*, 889 (1953).
Burch, J. M., and J. M. J. Tokarski, *Opt. Acta 15*, 101 (1968).
Debrus, S., M. Françon, and C. P. Grover, *Opt. Commun. 4*, 172 (1971).
De Chaulnes, Mem. l'Anc. Acad. Sci. Paris, p. 136 (1755).
Fabry, C., and R. Perot, *Ann. Chim. Phys. 12*, 7th ser. (1897).
Herschel, T., *Encyclopedia Metropolitana*, Vol. 2, pt. 2, p. 473, Baldwin and Crodock (1830).
Newton, I., *Optics*, 4th ed., p. 289, Bell and Sons, London (1931).
Young, T., *Philos. Trans.*, p. 41 (1802).

Y. BELVAUX and S. LOWENTHAL

# Subtraction (or Addition) of Illuminance

## I. Introduction

We describe a method using a birefringent prism to subtract image illuminances.

Amplitude subtraction can be obtained using, for example, holographic methods [Gabor *et al.*, 1965; Bromley *et al.*, 1969; Lee *et al.*, 1970; Collins, 1968]. Nevertheless, it is often of interest to perform intensity subtraction in order to determine differences that exist between two photographic records or scenes. Such a method, which avoids phase effects and correlative experimental difficulties, yields numerous applications. For example, automatic surveillance, analysis of meteorological photographs, and transmission of the only interesting information in a TV system (modifications of the picture).

To realize such an illuminance subtraction, Debrus *et al.* [1971] use a ground glass that randomly modulates the images, and they give a translation to the photographic plate between the two exposures. Another method uses two complementary gratings to modulate the objects that are to be subtracted. For that purpose, Pennington, who described

---

The authors are at the Institut d'Optique, Faculté des Sciences, 91 Orsay, France.

another method, uses a Ronchi ruling in contact with the object (or its image). Between the two exposures, he gives a half-period translation to the ruling. The processed photographic plate obtained is an image-hologram [Lohmann and Paris, 1966], which, after reconstruction through a spatial filtering device, shows the difference between the two scenes.

A difficulty of this method is that the ruling must be translated exactly half a period to obtain a 180° phase shift of the equivalent reference wave. Another difficulty is the noise that appears during the reconstruction due to the coherent stray light.

The method we shall describe avoids these difficulties. Two complementary gratings are used as in Pennington's method, but the gratings are not material and no mechanical translation is needed to perform the 180° phase shift. The images are actually modulated by a set of interference fringes that are obtained by means of a birefringent prism. Translation of the ruling is replaced by giving a 90° rotation to a polarizer to obtain a complementary interference pattern.

We shall describe two devices: The first one uses a Wollaston prism and can be used to process transmitting objects. The other one uses a Savart plate and is more adequate to treat incoherent objects, for example, three-dimensional scenes like landscapes, since extended spatially incoherent sources can be used. It will be shown that inconveniences related to the coherent reconstruction can be greatly reduced when using partially coherent light.

## II. Recording

A. PRINCIPLE

For illuminance subtraction, a double exposure is performed. Let $I_1(x,y)$ and $I_2(x,y)$ be the illuminances of two images that are to be subtracted so that $I_1(x,y) - I_2(x,y)$ is obtained.

During the first exposure, the image $I_1$ is modulated by a sinusoidal grating of frequency $\nu$ and the illuminance on the plate is $I_1(x,y) \times (1 + \cos 2\pi\nu x)$. Before the second exposure, the grating is shifted by half a period. During the second exposure, the illuminance is $I_2(x,y) \times (1 - \cos 2\pi\nu x)$. The total recorded illuminance is then $I_T = I_1 + I_2 + (I_1 - I_2) \cos 2\pi\nu x$, where the difference $I_1 - I_2$ appears, modulated by a carrier. Thus, it can be separated using adequate techniques of bandpass filtering.

B. FIRST DEVICE

The first device we describe can be used to perform subtraction of transmitting objects, for example, scenes recorded on slides.

*1. Description* This device uses a Wollaston prism located in an optically conjugate plane of the object.

The Wollaston prism W (Figure 1), placed between crossed or parallel polarizers P and A, yields two images $S_1$ and $S_2$ of the source S. The two images have perpendicular polarizations: interference occurs after analyzer A in plane H. The contrast is 1 for a 45° orientation of P with respect to the Wollaston axes.

Since the object O is imaged on the prism W, the final image at plane H is not sheared. The single image obtained is 100% intensity modulated by a sinusoidal fringe pattern.

*2. Operation* Suppose that to the intensity transmittance $T_1(x,y)$ of the object corresponds the complex amplitude $T_1(x,y)^{1/2} \exp i\phi_1(x,y)$, where $\phi_1(x,y)$ includes the phase due to the object and to the optical system.

The Wollaston prism yields at plane H the image amplitude

$$A_1(x,y) = T_1(x,y)^{1/2} \exp i\phi_1(x,y) \cos 2\pi\nu x , \qquad (1)$$

where we suppose, for example, parallel polarizers and the two images of the source symmetrically set with respect to the optical axis.

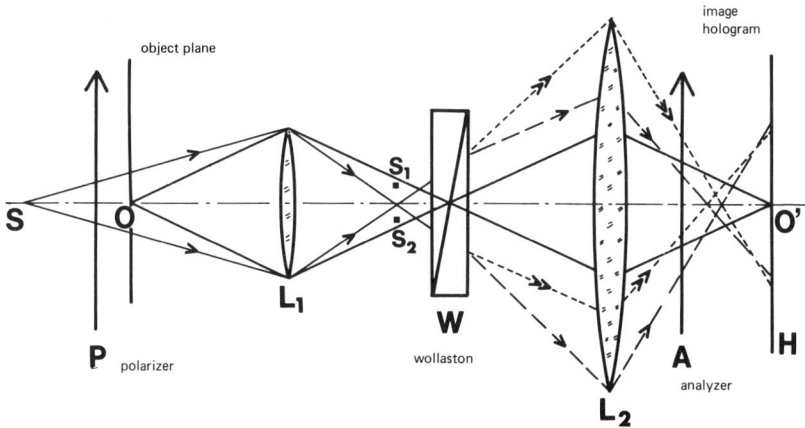

FIGURE 1 Optical system for the subtraction of two recorded scenes.

In the illuminance recorded by the plate,

$$I_1 = |A_1|^2 = T_1(x,y) \cos^2 2\pi\nu x = (T_1/2)(1 + \cos 4\pi\nu x), \qquad (2)$$

the phase factor disappears.

For the second exposure $T_2$ is substituted for $T_1$ and P is rotated through a 90° angle. The illumination is then

$$I_2 = |A_2|^2 = T_2(x,y) \sin^2 2\pi\nu x = (T_2/2)(1 - \cos 4\pi\nu x). \qquad (3)$$

If the two recordings are made on the same suitably pre-exposed plate [Biedermann, 1968] (satisfying the linearity conditions), the complex amplitude transmittance of the processed plate is of the form

$$t = a + b(T_1 + T_2) + b(T_1 - T_2) \cos 4\pi\nu x. \qquad (4)$$

Equation (4) represents the image hologram of differences $T_1 - T_2$ recorded with a $2\nu$ spatial carrier frequency.

It can be noted that the phase shift introduced between the two exposures is exactly $\pi$, bright fringes being substituted for dark fringes and conversely.

From an experimental point of view, the polarizers are not perfect flats and introduce some phase terms. These terms have no influence on the phase shift, provided that polarizer P (and not analyzer A) is rotated. Since P is located before beam splitter W, the fringe patterns are always perfectly complementary.

*3. Partially Coherent Recording* The image hologram obtained by the method we have just described suffers from speckle, spurious interference fringes due to stray light, etc. Since in an image hologram there is a point-to-point correspondance between the plate and the reconstructed image, the image is drastically degraded by these defects, and it is then very interesting to use incoherent or partially coherent light. A first possibility is to use a line source set parallel to the fringes. We used such a source synthesized by a vibrating mirror, and we obtained improved results. Moreover, it can be shown that, in the third-order approximation, the previous device gives linear fringes located on the prism when using a spatially incoherent source.

## C. SECOND DEVICE—INCOHERENT RECORDING

Another method is to use an experimental configuration giving Young's fringes with spatially incoherent light [Lowenthal et al., 1969; Aspect, 1971].

Let S be a broad monochromatic source and let $S_1$ and $S_2$ be two images of S given by a beam splitter BS in such a way that $S_1$ and $S_2$ are separated by a shear due to a pure translation (Figure 2).

Then, let us consider two homologous points $M_1$, $M_2$ corresponding to $S_1$ and $S_2$ that emit two parallel rays. These rays are focused at point P in the focal plane H of lens L. The state of interference at P depends on the path difference $M_2 K$. Since the vector $M_1 M_2$ is constant in space (pure translation), the path difference is independent of the pair $M_1 M_2$ under consideration. Thus the interference pattern is a set of Young's fringes 100% modulated, even when an extended quasi-monochromatic source is used. If the object is imaged in the focal plane H, the image is intensity modulated with the sinusoidal fringe pattern.

To combine both the extended source advantages and the possibility of obtaining exact complementary fringe patterns, a Savart plate can be used.

Such a plate placed between polarizer P and analyzer A (Figure 3) in front of a camera lens L produces the desired fringes in the focal plane of the lens.

Since the fringes of the Savart plate are located at infinity, an auxiliary lens $L_0$ is needed when the object O is at a finite distance from lens L.

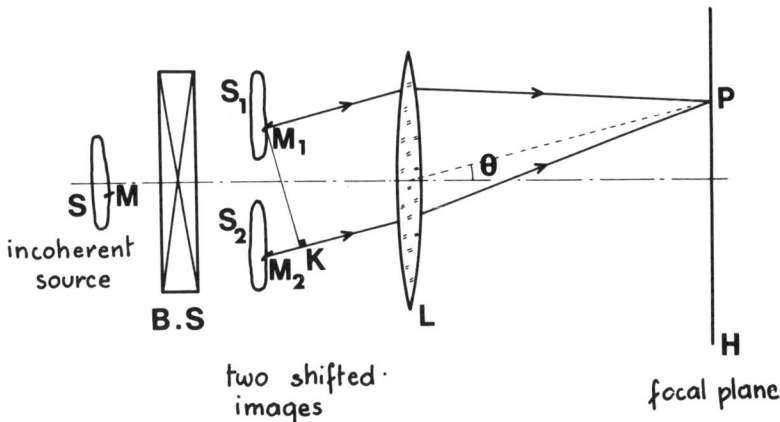

FIGURE 2  Optical system for the generation of Young's fringe with spatially incoherent light.

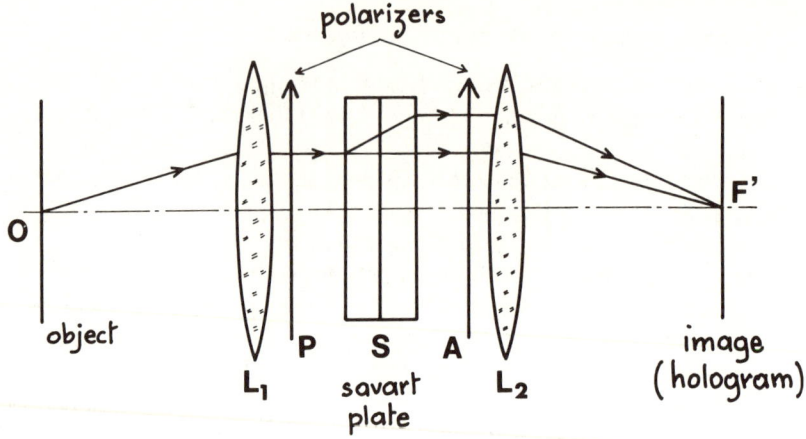

FIGURE 3  Interferometer using a Savart plate.

The modified camera can be used to carry out subtraction of outdoor scenes, for example, provided that a monochromatic filter is used.

### III. Reconstruction

For any type of recording device, the processed plate acts as an image hologram whose amplitude transmittance is given by Eq. (4). This hologram can be reconstructed using a converging beam (Figure 4). A stop E located in the focal plane of lens $L_1$ just selects the +1 (or −1) diffracted order.

The final image $H'$ given by $L_2$ has an amplitude that is proportional to the difference between the illuminances of the recorded images. Of course, $L_2$ can be the eye and $H'$ the retina of the observer.

To avoid the noise effects due to coherent reconstruction, different reconstruction methods with partially coherent light can be used. In any case, one can employ a line source parallel to the fringes.

Moreover, when the carrier frequency is high enough when compared with the spatial frequency content of the images, the three spectra plotted in plane E (Figure 4) are widely separated. Thus, an extended incoherent source can be used for reconstruction, in the limit of no spectrum overlapping. In the same manner, a finite spectral bandwidth source can be used.

*Y. Belvaux and S. Lowenthal* 619

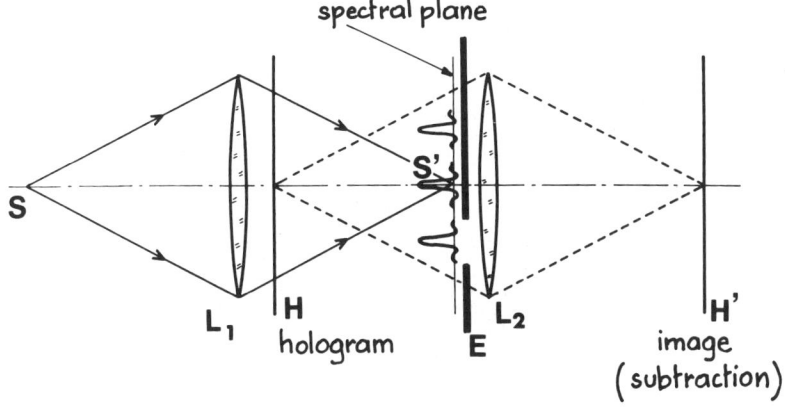

FIGURE 4  Schematic of system for image reconstruction.

## IV. Experimental Results

Figure 5 shows the first results obtained with a fairly simple object made of a set of geometrical drawings, some with a high contrast, the others with a lower contrast. Figure 5(a) shows the original object; Figure 5(b) shows the result obtained by subtracting the two upper lines of the object.

As a test of quality, we compared the fluxes diffracted in the first order with and without subtraction. The object was a free pupil, and we obtained a ratio of 20 to 1 that corresponds to an attenuation factor of 13 dB.

Figure 6 corresponds to a much more complicated object. It is a book

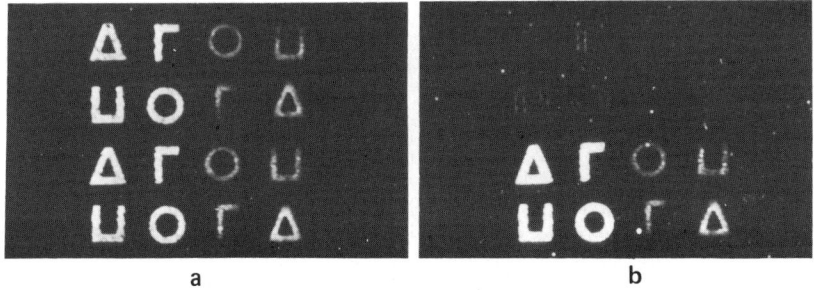

FIGURE 5  Experimental results: (a) the original object; (b) image after subtraction of upper two lines.

FIGURE 6 Illustration of subtraction of (a) from (b) to yield the difference (c).

shelf from which a book has been removed between the two exposures. Figure 6(c) shows the difference between Figures 6(a) and 6(b). These three pictures correspond to the images obtained when letting through the first-order spectrum only. The relatively poor quality of these images is due to bandpass limitation, which occurs when filtering the spectra because the spacing of the interference fringes was not small enough with the birefringent prism we had at our disposal.

## V. Conclusion

It has been shown that illuminance subtraction can be obtained using birefringent prisms. However, the technique must still be improved, and for now a Savart prism giving a fairly high-frequency set of fringes is being made.

It must be noticed that, whatever birefringent prism is used, exact subtraction occurs even if fringes are not straight lines. A set of fringes being substituted for the complementary one, the fringe distortion involves only spectral distortion, but the image illuminance is not modified.

## References

Aspect, A., Thesis, Orsay (1971).
Biedermann, K., *Optik 28*, 160 (1968).
Bromley, K., M. A. Monahan, J. F. Bryant, and B. J. Thompson, *Appl. Phys. Lett. 14*, 67 (1969).
Collins, L. G., *Appl. Opt. 7*, 203 (1968).
Debrus, S., M. Françon, and C. P. Grover, *Opt. Commun. 4*, 172 (1971).
Gabor, D., G. W. Stroke, R. Restrike, A. Funkhouser, and D. Brumm, *Phys. Lett. 18*, 116 (1965).
Lee, S. H., S. K. Yao, and A. G. Milnes, *J. Opt. Soc. Am. 60*, 1037 (1970).
Lohmann, A. W., and D. P. Paris, *J. Opt. Soc. Am. 56*, 537A (1966).
Lowenthal, S., J. Serres, and C. Froehly, *C. R. Acad. Sci. (Paris) 268B*, 1481 (1969).

J. T. WINTHROP and R. F. VAN LIGTEN

# HOLOGRAPHIC MICROSCOPY IN EXOBIOLOGY: TRANSMISSION OF A HOLOMICROGRAM OVER A LIMITED TELEMETRY CHANNEL

## Introduction

The remote detection of extraterrestrial life involves special problems arising from the need to transmit data over great distances. In the case of remote microscopy, the electronic transmission system limits the amount of image data that can be sent to the earth-bound observer. This paper argues that, under such conditions, holographic microscopy is the most practical method of acquiring images of exobiological structures. The feasibility of the method is demonstrated by the results of a computer simulation.

Our study relates specifically to exobiology on the planet Mars. The microscope is to provide images of a cylindrical object volume 500 μm in diameter and 500 μm deep (see Figure 1). The transverse resolution is required to be 0.5 μm at the best-resolved part of the volume and no worse than 1.0 μm elsewhere. The microscope images are telemetered to earth at a maximum rate of $1.9 \times 10^7$ bits/day.

It is convenient to distinguish two methods of microscopy, conven-

---

The authors are in the Research Division, American Optical Corporation, Framingham Centre, Massachusetts 01701.

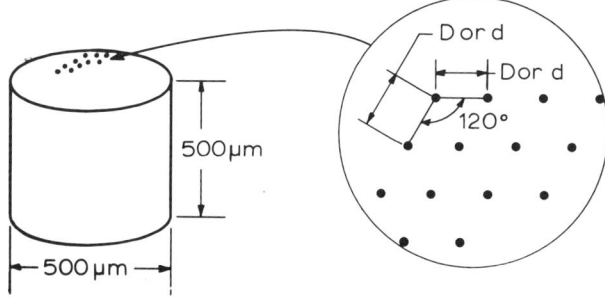

FIGURE 1 Image volume (normalized to unit magnification) and hexagonal sampling lattice.

tional and holographic. The relative merits of the two methods for exobiology can be summarized as follows:

1. Conventional microscopy, using incoherent or partially coherent illumination, has the potential to provide images of high quality. Holography has the disadvantage that the reconstructed images suffer from speckle noise caused by coherent illumination of the object volume.

2. Conventional microscopy requires automatic refocusing for each distinguishable object plane in depth. Holography, on the other hand, records the entire object volume with one focal setting.

3. During the time needed for the conventional microscope to focus through the object volume, the object may have moved. This may result in a distorted or incomplete three-dimensional image. Holographic microscopy with pulsed illumination has the advantage that it stops object motion.

4. Conventional microscopy requires two images for each focal setting, one showing absorption objects and the other showing phase objects. With holography, images of either type of object are available from one hologram of the object volume.

5. The number of on–off pulses (bits) needed to describe the volume image formed by conventional microscopy is three orders of magnitude greater than the number needed to describe a holographic image of the same volume.

We conclude that, apart from the question of image quality, holographic microscopy is preferable to conventional microscopy for the problem of exobiology. Because of the importance of the fifth factor, data economy, we will discuss it in detail.

## Information Handling

The discussion assumes water-immersion objectives (numerical aperture = $n \sin \theta = 1.33 \sin \theta$) and illumination at vacuum wavelength $\lambda = 0.488\ \mu m$.

The number of bits needed to describe an image of the object volume formed by *conventional microscopy* can be found as follows. An image of the object volume is constructed by focusing on a succession of object planes in depth. The spatial Fourier spectrum of the intensity distribution in one plane of the image (normalized to unit magnification) occupies a circular disk of diameter $D^*$, where [O'Neill, 1963]

$$D^* = 4n \sin \theta / \lambda. \qquad (1)$$

According to sampling theory [Petersen and Middleton, 1962], the image in this plane can be described most efficiently in terms of samples of the intensity taken on a hexagonal sampling lattice, where the spacing $D$ between nearest-neighbor sampling points is given by (Figure 1)

$$D = 1.15/D^*. \qquad (2)$$

The diameter $D^*$, and hence $D$, can be related to the resolution limit $\Delta r$ through the Rayleigh criterion for incoherent illumination [Born and Wolf, 1964],

$$\Delta r = 0.61 \lambda / n \sin \theta. \qquad (3)$$

From Eqs. (1), (2), and (3),

$$D = 0.47 \Delta r. \qquad (4)$$

If $\Delta r = 0.5\ \mu m$, then $D = 0.24\ \mu m$, and one plane of the object volume 500 $\mu m$ in diameter contains $3.9 \times 10^6$ independent sampled data. To get the number of refocusing steps, we note that in the axial direction, the spatial Fourier spectrum of the image (normalized to unit magnification) is limited to an interval $\Delta Z^*$, where

$$\Delta Z^* = 2n(1 - \cos \theta)/\lambda, \qquad (5)$$

and $\cos \theta$ is determined by the Rayleigh criterion. Then, according to sampling theory, the object can be specified in depth by refocusing at intervals $\Delta Z$, where

$$\Delta Z = 1/\Delta Z^* \qquad (6)$$

[Frieden, 1966]. The known values of $n$, $\lambda$, and $\Delta r$ give $\Delta Z = 1.75$ µm, and the number of refocusing steps in the object cylinder of depth 500 µm is $2.9 \times 10^2$. Thus, the total number of intensity values needed to specify the volume image is $1.1 \times 10^9$. If the intensity distribution is quantized into $16 = 2^4$ gray levels, corresponding to 4 bits of information per data point, then there will be $4.5 \times 10^9$ bits of information in the image. Finally, since two images per specimen are required in conventional microscopy, the total amount of information to be transmitted per specimen becomes *$9 \times 10^9$ bits.*

In *holographic microscopy,* it is only necessary to describe the amplitude and phase at one plane of the image field. The amplitude and phase at all other planes can be obtained by propagation. We describe the field in the plane containing the image of the end of the object cylinder nearest the objective. The spatial Fourier spectrum of this image (normalized to unit magnification) is confined to a circular disk of diameter $d^*$, where [O'Neill, 1963]

$$d^* = 2n \sin \theta / \lambda. \qquad (7)$$

The field can therefore be specified most efficiently in terms of sample values taken on a hexagonal sampling lattice, where the spacing $d$ between nearest-neighbor sampling points is given by

$$d = 1.15/d^*. \qquad (8)$$

The criterion for resolution $\Delta r$ with coherent imagery is [Born and Wolf, 1963]

$$\Delta r = 0.83 \lambda / n \sin \theta, \qquad (9)$$

and, therefore, from Eqs. (7), (8), and (9),

$$d = 0.69 \Delta r. \qquad (10)$$

Since $\Delta r = 0.5$ µm, we have $d = 0.35$ µm, and the image area 500 µm in diameter contains $1.9 \times 10^6$ sampling points. This corresponds to $3.8 \times 10^6$ real data, because both amplitude and phase have to be specified. If the amplitude and phase are quantized into $16 = 2^4$ levels each, then the total amount of information to be transmitted per specimen in holographic microscopy is *$1.5 \times 10^7$ bits.*

Thus, the amount of information that has to be transmitted in holographic microscopy is about 600 times less than the amount that has to be transmitted in conventional microscopy. On this basis, holographic microscopy has a distinct advantage over conventional microscopy.

## Hologram Bandwidth Reduction

The hologram contains a record of the amplitude and phase of the image formed by the microscope. Before the hologram transmittance can be telemetered to earth, it has to be sampled, and the magnitude of each sample has to be quantized into a finite number of levels. Because of the limited capacity of the telemetry channel, it is essential that the number of sampling points and quantization levels be kept as small as possible.

To minimize the number of sampling points, the hologram is recorded by the lensless Fourier transform method, according to which the reference field diverges from a point in the plane of the exit pupil of the microscope objective. As is well known, the hologram produced by this method has minimum spatial bandwidth. Since the number of sampling data needed to describe the hologram transmittance depends on the area occupied by its Fourier spectrum, the lensless Fourier transform hologram requires the least number of samples for its complete description.

As shown in Figure 2(a), the Fourier spectrum of the hologram (amplitude) transmittance consists of three parts: two image-bearing sidebands of diameter $d^* = 2n \sin \theta / \lambda$ and a central area of diameter $2d^*$ containing the Fourier spectrum of the image intensity. The latter is sometimes called the intermodulation term of the hologram. It does not contribute to the formation of the holographic image. If the intermodulation term were not present, the number of sampling data needed to describe the hologram transmittance would be the same as the number of real data needed to describe the holographic image ($3.8 \times 10^6$ real data, as calculated above). The presence of the intermodulation term more than dou-

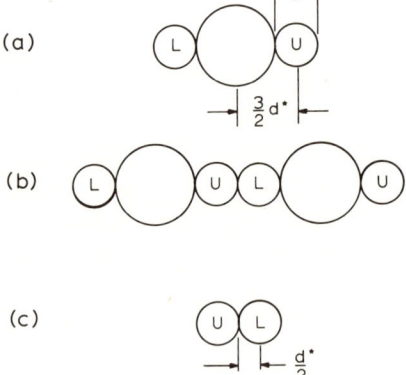

FIGURE 2 Stages in the reduction of hologram bandwidth. (a) Fourier spectrum of original hologram. U and L denote upper and lower sidebands. (b) Replication of hologram spectrum that results when the hologram transmittance is multiplied by a sinusoid. (c) Spectrum of bandwidth-reduced hologram, produced by low-pass filtering of spectrum (b).

FIGURE 3 Coherent optical processor used to generate bandwidth-reduced hologram. A pair of mutually coherent plane waves illuminates the original hologram placed at the input plane. A low-pass filter transmits a field whose spectrum at the output plane occupies the region indicated in Figure 2(c). The addition of a plane wave normally incident on the output plane produces a real, non-negative field amplitude suitable for measurement.

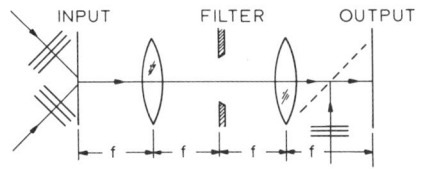

bles the number of data needed to specify the hologram. Thus, the intermodulation term needs to be eliminated before the telemetry stage. We refer to the elimination of this term as bandwidth reduction.

Two possible methods of bandwidth reduction will be discussed. In the first method, suggested to us by J. W. Goodman (private communication, 1972), the transmittance of the original hologram is converted to an electrical signal by means of a scanning pinhole and photodetector. The lines of scan run parallel to the direction of spatial-frequency offset. The output of the photodetector is multiplied by a sinusoid, resulting in a signal whose spectrum consists of the sum of upshifted and downshifted replicas of the original hologram spectrum. The frequency of the sinusoid is chosen so that the spectrum of the product signal occupies the domain indicated in Figure 2(b). Note that the lower sideband L of the upshifted spectrum touches, but does not overlap, the upper sideband U of the downshifted spectrum. A low-pass filter then discards all but the two adjacent sidebands, resulting in a bandwidth-reduced signal whose spectrum is indicated in Figure 2(c). It is this signal that when sampled and quantized is telemetered to earth.

An optical method that achieves the same reduction of bandwidth is shown in Figure 3. The original hologram is placed at the input plane of a coherent optical processor and is illuminated by a pair of mutually coherent plane waves. The directions of the plane waves are chosen so that the light distribution in the filter plane consists of two contiguous hologram spectra, as shown in Figure 2(b). A mask in the filter plane blocks all but the two adjacent sidebands. The wave field transmitted to the output plane has the Fourier spectrum shown in Figure 2(c). The complex amplitude of the field is real valued, but to provide a non-negative amplitude for measurement, the field must be added coherently to a wave of constant amplitude and zero phase, as indicated in Figure 3. The resultant amplitude at the output plane is then scanned by a pinhole and detector, and the electrical signal is stored for transmission to earth.

FIGURE 4 Derivation of optimum periodic sampling lattice. (a) Closest packing of repeated hologram spectra generates reciprocal lattice with unit cell of area $A^*$. (b) Fourier transform of reciprocal lattice gives optimum periodic sampling lattice with unit cell of area $1/A^*$.

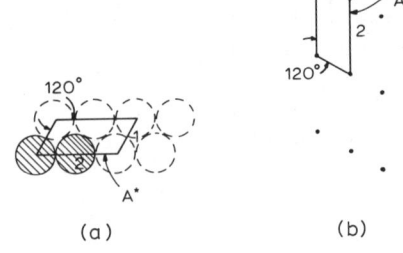

(a)   (b)

## Sampling

An optimally efficient scheme for periodic sampling of the bandwidth-reduced hologram can be derived from the condition of closest packing of repeated hologram spectra. As shown in Figure 4(a), the unit cell of the array of repeated spectra is a parallelogram with sides in the ratio 2:1 and with an included angle of 120°. The array of corners of repeated unit cells comprises the reciprocal lattice. The optimally efficient sampling lattice is the Fourier transform of the reciprocal lattice [Figure 4(b)]. The two Fourier-conjugate lattices are related by a 90° rotation; the area of the unit cell of the sampling lattice is the reciprocal of the area of the unit cell of the reciprocal lattice. The periodic sampling lattice defined in this way is optimally efficient in the sense that it provides a complete description of the hologram with the minimum number of sampling points.

## Experiment

A test object was prepared consisting of a collection of 0.5-μm latex spheres. A hologram of the object was made in 4880 Å light, using a lensless-Fourier-transform microscope equipped with a water-immersion objective of numerical aperture 0.75. The hologram was then processed and reconstructed on a computer system at the Visibility Laboratories of the University of California at San Diego. The computer simulation enables us to judge the effects of bandwidth reduction, sampling, and quantization under controlled, noise-free conditions.

Figure 5(a) shows a direct image, using 4880 Å coherent illumination, of a collection of the 0.5-μm spheres used for the experiment. In order to limit the number of data points presented to the computer, it was necessary to mask off all but a small portion of the object field. Figure 5(b) shows a direct image of the object field masked by a square aperture 33 μm × 33 μm. The mask lies somewhat outside the plane of focus. Thus, a set of regular fringes, caused by interference between the coherent background and light diffracted from the edges of the aperture, covers the field of view.

A hologram was made of the object field of Figure 5(b). The transmittance values of the hologram were then sampled in accord with the sampling theorem, and the sample values were stored in the computer. The stored hologram, consisting of a 256 × 256 array of sample values, is shown in Figure 6.

Next, a computation was made of the Fourier transform of the hologram transmittance, the result of which is shown in Figure 7. The computer was then instructed to discard all but the signal sidebands and to shift them to the positions indicated in Figure 2(c). An inverse Fourier transform then provided the bandwidth-reduced hologram shown in Figure 8. The coarse fringes show that a reduction of bandwidth has taken place.

Reconstruction of the hologram of Figure 8 was performed by computer. The reconstructions of various planes of focus are shown in Figures 9(b)–9(g). [Figure 9(a) is a direct image of the original object.] Figure 9(c) represents the plane of best focus for the pair of spheres circled in Figure 9(a). In spite of the overlay of noise due to the square mask, the spheres can be seen as resolved.

We have so far demonstrated the feasibility of the sampling and bandwidth-reduction processes. However, the reconstructions of Figure 9 were obtained from an essentially unquantized hologram. To test the

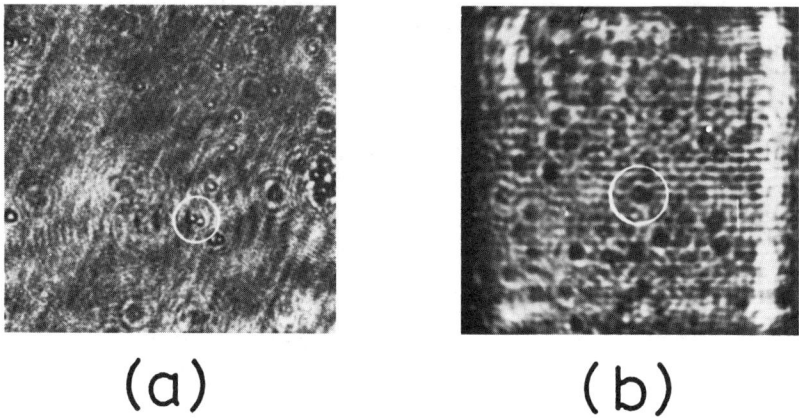

FIGURE 5 The holographic object. (a) Collection of 0.5-μm latex spheres, viewed directly through microscope using 4880 Å coherent illumination. (b) The object field of (a) masked by a 33-μm-square aperture.

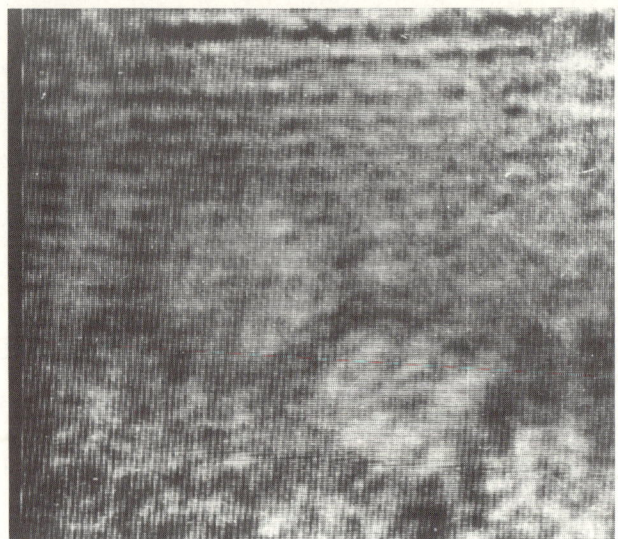

FIGURE 6  256 × 256 array of sample values of an optical hologram of the object of Figure 5(b).

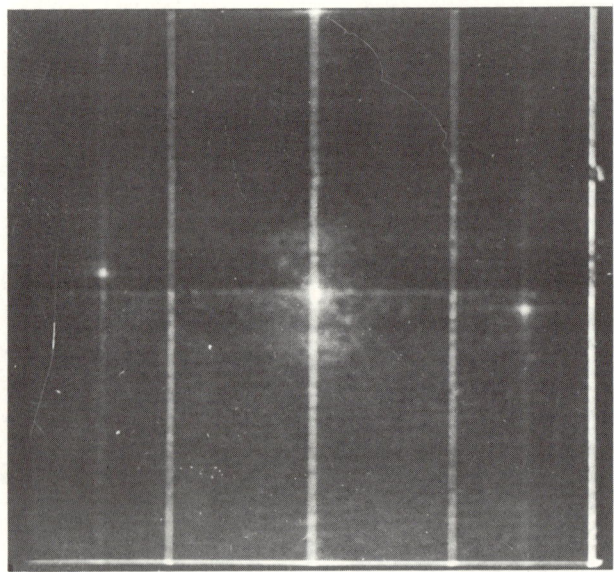

FIGURE 7  Computer-generated Fourier transform of the hologram of Figure 6.

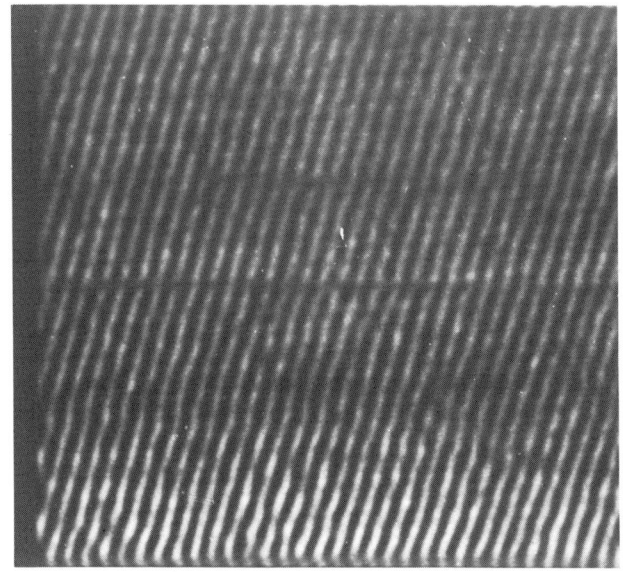

FIGURE 8  Bandwidth-reduced hologram.

effect of quantization, reconstructions of the same object were made from holograms quantized in various numbers of equally spaced levels of transmittance. The results are shown in Figure 10. To the eye, it makes little difference whether the hologram is quantized in 128 or 16 levels. At 8 levels, however, the reconstruction deteriorates markedly.

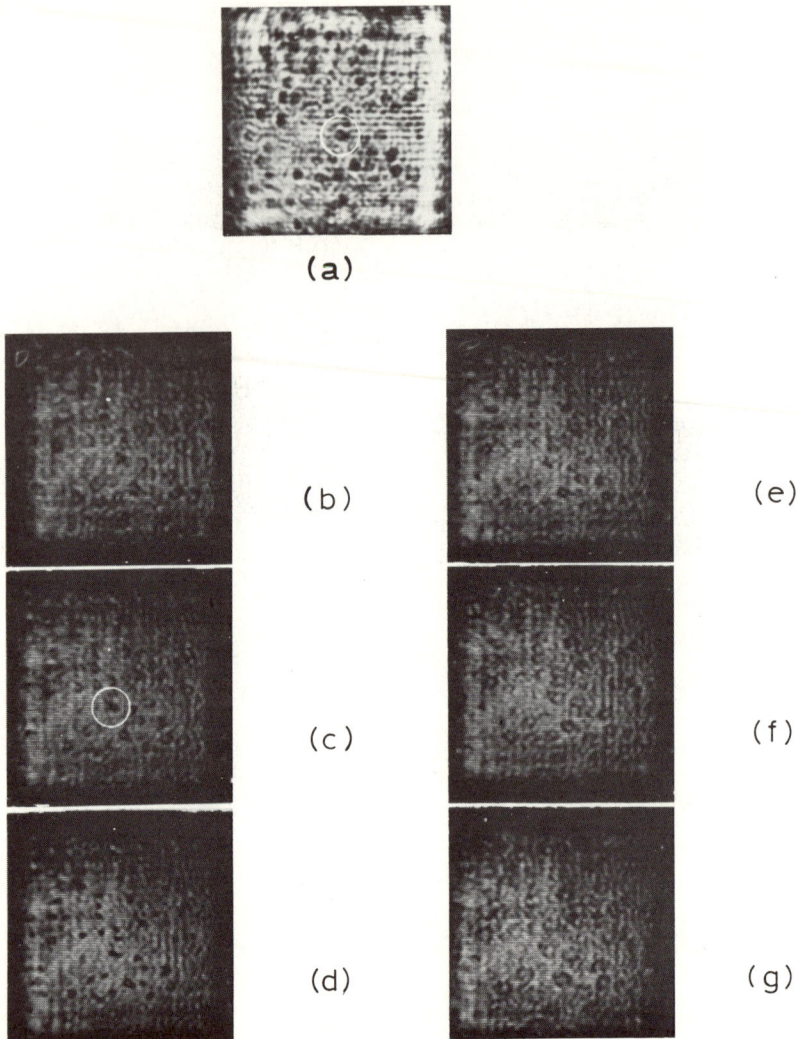

FIGURE 9 Computer reconstruction of hologram of Figure 8. (a) Direct image of original object. (b)–(g) Various planes of focus. Figure (c) represents the plane of best focus for the pair of spheres circled in (a).

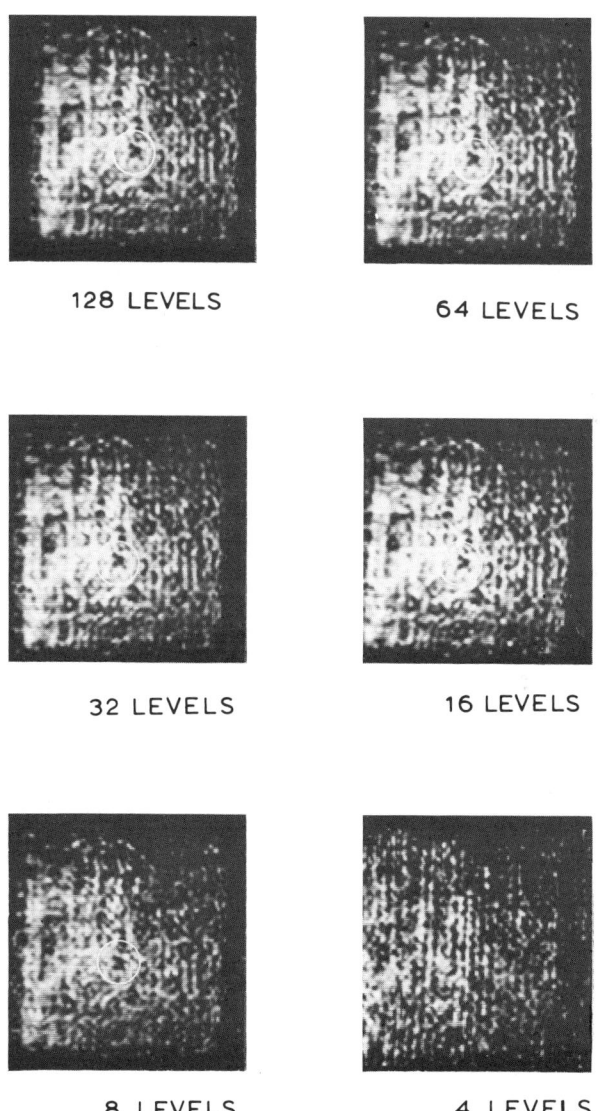

FIGURE 10 Reconstructions from holograms that have been quantized to various equally spaced levels of transmittance.

## Conclusion

We have shown that it is possible to prepare an ordinary hologram for electronic transmission by these steps:

1. Scan the hologram, converting transmittance variations into an electrical signal.
2. Process the electrical signal to obtain a new hologram of reduced bandwidth.
3. Sample the bandwidth-reduced hologram.
4. Quantize the sample values into 16 equally spaced values.

In terms of the problem of exobiology described in the Introduction, this procedure enables us to transmit a hologram of the cylindrical object volume using $1.5 \times 10^7$ bits, or about one hologram per day.

After the hologram has been transmitted and received, it can be reconstructed by computer, as was demonstrated here, or by conventional optical means.

It may be possible to reduce even further the amount of data that have to be sent by making use of special sampling and coding methods [Huang et al., 1971]. For example, with a scheme of nonuniform quantization, three instead of four bits of quantization per data point may suffice. Experiments have been planned to investigate these possibilities.

We are indebted to J. A. Levitt, J. W. Goodman, and A. W. Lohmann for their contributions to the conceptual aspects of this work; to K. C. Lawton for preparation of the test object and the optical hologram; and to J. L. Harris, Sr., for assistance in the computer simulation.

This work was done under NASA Contract NAS2-6472, "Holographic Microscopy in Exobiology."

## References

Born, M., and E. Wolf, *Principles of Optics*, 2nd ed., p. 524, Pergamon, New York (1964).
Frieden, B. R., *J. Opt. Soc. Am. 56*, 1495 (1966).
Huang, T. S., W. F. Schreiber, and O. J. Tretiak, *Proc. IEEE 59*, 1586 (1971).
O'Neill, E. L., *Introduction to Statistical Optics*, p. 83, Addison-Wesley, Reading, Mass. (1963).
Petersen, D. P., and D. Middleton, *Inform. Control 5*, 279 (1962).

N. BALASUBRAMANIAN

# CONTOURING FROM HOLOGRAPHIC STEREOMODELS

Introduction

During the last few years, there has been considerable interest in the application of modern coherent optical techniques to achieve storage and subsequent processing of data obtained through photogrammetric and remote-sensing techniques. The theoretical considerations by Real [1969], as well as the recent studies at Bendix [Krulikoski *et al.*, 1968, 1970] were directed toward the demonstration of the capability of coherent optical processing techniques to obtain instant elevation profiles and contours through optical correlation. The second area of optical research in photogrammetry is the application of holography to meet the problems of storage, retrieval, and display. Recently, Kurtz *et al.* [1971] produced photogrammetric holograms of stereomodels composed of pairs of overlapping photographs. It is possible to extend the optical-processing techniques to obtain contour information directly from the holographic stereomodel. This would facilitate not only the storage of raw photogrammetric data in a highly processed form but also would

---

The author is in the Institute of Optics, University of Rochester, Rochester, New York 14627.

permit automated retrieval of contour information from the holographic stereomodel as and when required.

## Basic Concept

The concept of holographic stereomodel involves replacing the overlapping stereophotographs or their projections with their overlapping holographic virtual images. The holographic stereomodel is produced by recording, on a single hologram, the overlapping images of two photographs of a common area having proper perspective for subsequent stereoviewing. The optical system used for recording the holographic stereomodel is shown in Figure 1. The transparencies are projected onto the rear projection screen using coherent illumination derived from a laser source. First, the double projection system is relatively oriented to remove differences in scale and angular tilt at the overlapped image plane. While recording, only one projection system is illuminated at a time. Each of the projected transparencies is recorded on the holographic plate, and the two reference beams used for the two exposures allow late image separation upon reconstruction. The optical system used for viewing the holographic stereomodel is shown in Figure 2. Here the polarization of one reconstruction beam is rotated by 90° using half-wave retarder; orthogonally oriented polarizing filters are then used for viewing the reconstructed holographic stereomodel. A simple approach to mapping the information contained in the holographic stereomodel is to use a self-illuminated dot attached to an $XYZ$ coordinate measuring device and placed in the virtual image space of the hologram. A detailed description of such an instrument is given by Kurtz *et al.* [1971]. The purpose of this paper is to show the advantages of the holographic stereomodel from the viewpoint of further data processing.

For the purposes of this analysis, a modified version of the arrangement used for constructing the holographic stereomodel is shown in Figure 3. The amplitude and phase distribution on the rear projection screen is $[u_{10} + u_1(x,y)] \exp[i\phi_1(x,y)]$ when transparency $t_1(x,y)$ is projected on the screen. $u_0$ represents the undiffracted transmission of the transparency and $u_1(x,y)$ is due to the structure information on the transparency. The corresponding amplitude and phase distribution for the transparency $t_2(x,y)$ is $[u_{20} + u_2(x,y)] \exp[i\phi_2(x,y)]$. It is assumed here that the transparencies do not introduce phase distortions and that $\phi_1(x,y)$ and $\phi_2(x,y)$ are essentially a function of the double projection systems. The reference beam 1 reconstructs $[u_{01} + u_1(x,y)]$

$\exp [i \phi_1 (x,y)]$, and the reference beam 2 reconstructs $[u_{02} + u_2 (x,y)]$ $\exp [i \phi_2 (x,y)]$. At the superimposed image plane, the two reconstructed reference beams interfere, and the resultant intensity distribution is given by

$$I(x,y) = (E_1 + E_2)(E_1^* + E_2^*), \quad (1)$$

where

$$E_1 = [u_{01} + u_1 (x,y)] \exp [i \phi_1 (x,y)]$$

and

$$E_2 = [u_{02} + u_1 (x,y)] \exp [i \phi_2 (x,y)].$$

$$I(x,y) = [u_{01} + u_1 (x,y)]^2 + [u_{02} + u_2 (x,y)]^2$$
$$+ 2[u_{01} + u_1 (x,y)] [u_{02} + u_2 (x,y)]$$
$$\cos [\phi_1 (x,y) - \phi_2 (x,y)]. \quad (2)$$

This can be expanded to give the expression

$$I(x,y) = u_{01}^2 + u_{02}^2 + 2u_{01} u_{02} \cos (\phi_1 - \phi_2)$$
$$+ 2u_{01} u_1 + 2u_{02} u_2 + 2u_{01} u_2 + u_{02} u_1 \cos (\phi_1 - \phi_2)$$
$$+ u_1^2 + u_2^2 + 2u_1 u_2 \cos (\phi_1 - \phi_2). \quad (3)$$

The terms containing $\cos (\phi_1 - \phi_2)$ represent a fringe modulation whose shape and spatial frequency are dependent on the relative distributions of functions $\phi_1$ and $\phi_2$.

When a dc block is placed at the focal plane of the imaging lens $L_1$ in Figure 3, Eq. (3) becomes

$$I(x,y) = u_1^2 + u_2^2 + 2u_1 u_2 \cos (\phi_1 - \phi_2). \quad (4)$$

Regions having zero $x$-parallax between the two projections on the rear projection screen exhibit maximum correlation. Hence in the regions of zero $x$-parallax, where conjugate images of the stereotransparencies are superimposed, $u_1 (x,y)$ is approximately equal to $u_2 (x,y)$. Intensity distribution at the regions where conjugate images are coincident can be expressed as

$$I_c(x,y) = 2u_1^2 [1 + \cos (\phi_1 - \phi_2)]. \quad (5)$$

Equation (5) shows that coincident regions in the superimposed image plane would exhibit maximum fringe modulation.

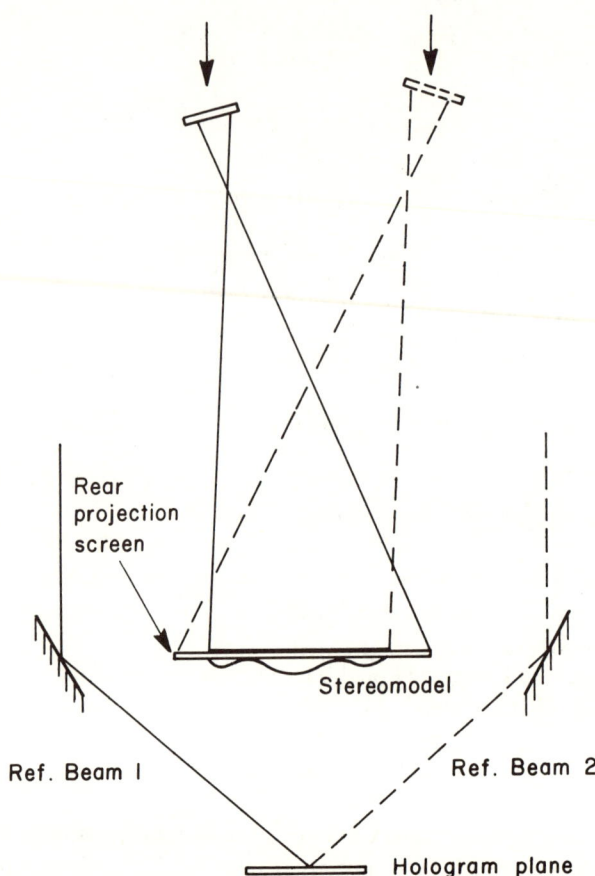

FIGURE 1  Recording a projected stereomodel.

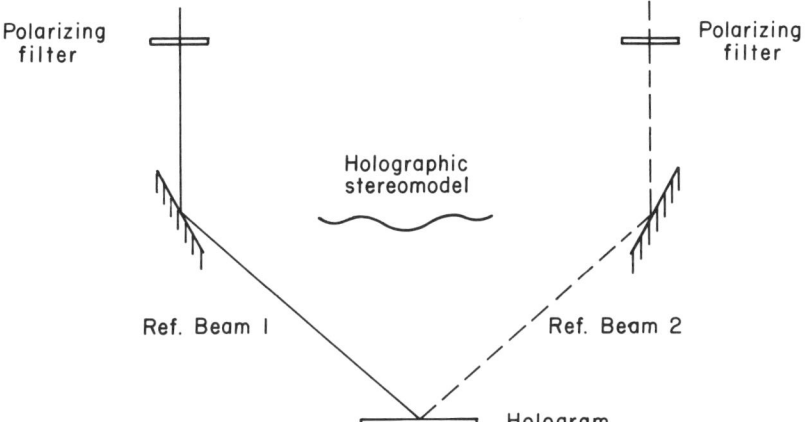

FIGURE 2  Viewing system for holographic stereomodel.

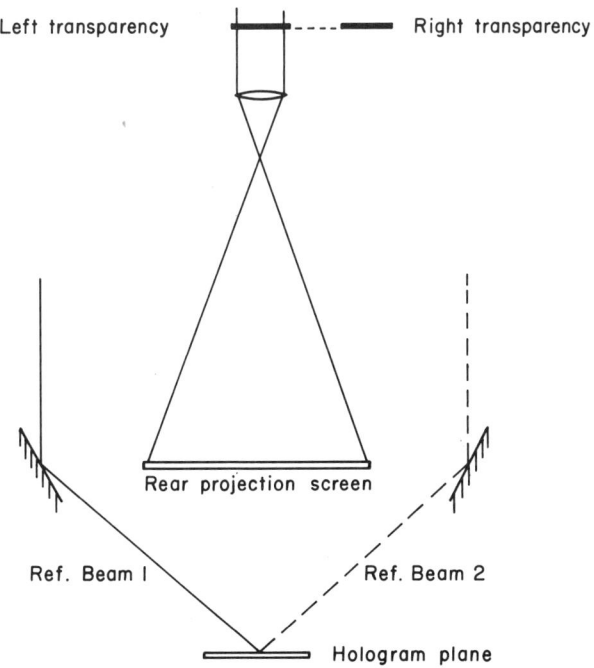

FIGURE 3  Recording holographic stereomodel (vertical case).

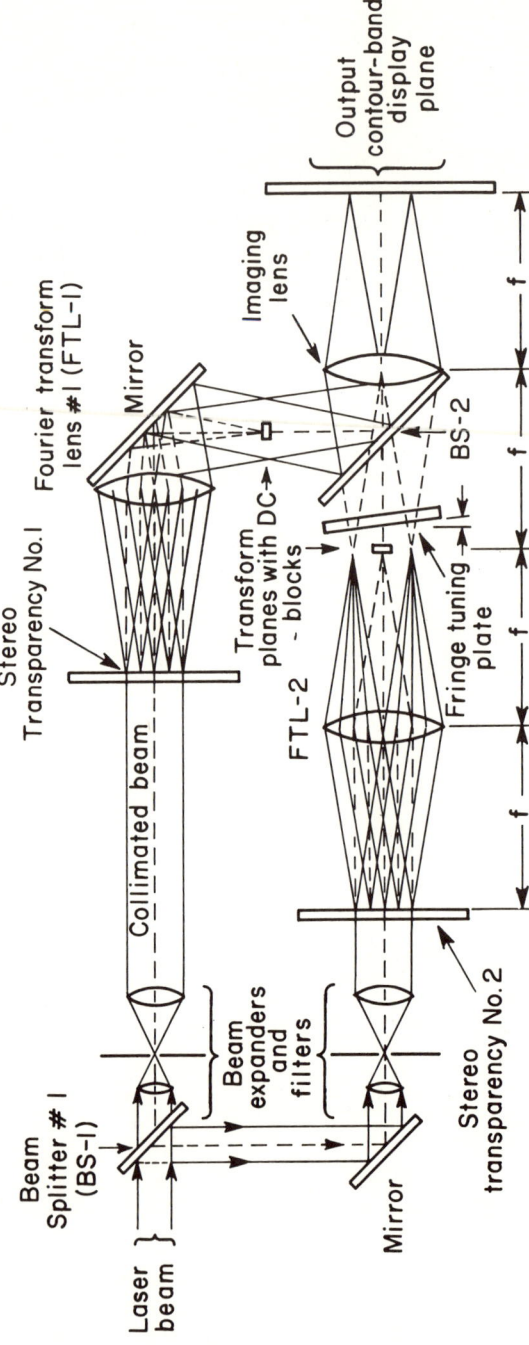

FIGURE 4 Contouring interferometer (adopted from Krulikoski et al. [1968]).

## Comparison with Contouring Interferometer

The holographic stereomodel configuration can be considered as a modified version of the contouring interferometer proposed earlier by Krulikoski et al. [1970]. The optical arrangement used by Krulikoski et al. is shown in Figure 4. It essentially consists of a Mach-Zehnder interferometer and two $x$-$y$ photocarriages that hold the stereotransparencies in each channel of the interferometer. The average transmittences of the transparencies are removed through spatial filtering, and the filtered images are then superimposed at the output plane of the interferometer. The regions of zero $x$-parallax at the superimposed image plane are detected through the existence of interferometric fringe modulation. Different height contours are generated by translating one of the transparencies in the $x$-direction (to provide different $x$-parallaxes).

It is clear from above considerations that the holographic stereomodel arrangement can be considered equivalent to a Mach-Zehnder interferometer arrangement in which fringes are localized at the rear projection screen. At the reconstructed virtual image space of the holographic stereomodel we have essentially the equivalent of two relatively oriented rectified stereophotographs. The intensity distribution at the plane of the rear projection screen (at the virtual image) is similar to the distribution at the output plane of the contouring interferometer. Once the holographic stereomodel has been made, there is no provision to translate the projections with respect to each other (to obtain different $x$-parallax contours). Hence, one holographic stereomodel provides information regarding only one height contour for different contour intervals; multiple recording of different holographic stereomodels representing different $x$-parallax differences at the plane of the rear projection screen must be made. Because of the ease with which the holographic stereomodels can be recorded once the relative orientation has been obtained, this poses no serious problem.

## Advantages

Visual detection of fringe visibility limits considerably the sensitivity that can be obtained. However, if the holographic stereomodel is reconstructed in such a manner that the real image is superimposed on a scanning detector plane, the output of the detector can be used to obtain the signal information. Furthermore, if the phase of one of the reference beams is modulated, the amplitude distribution corresponding to one of the projections also gets phase modulated. The time-varying intensity at the superposed real-image plane would give rise to an ac beat signal at the detector, the amplitude of which is proportional to the contrast of the fringe modulation at that point. Because of this, the actual fringe spacing at the superposed image plane would have no effect on the spatial resolution that could be obtained.

In the holographic stereomodel, the detection of coincidence of conjugate images is made on the stereomodel space, and, hence, the contours obtained represent directly orthoscopic contours. Unlike the case of interferometric contouring system, there are no stringent requirements on the quality of the optics used, and the system itself is basically simple. The direct contouring approach can be extended to a general holographic stereomodel configuration shown in Figure 1; however, the limitation on the detectability of the fringe spacing on the rear projection limits the range of base-to-height ratios to which this contouring system can be applied.

## Conclusion

The holographic stereomodel not only provides a means of storing and displaying photogrammetric data in a processed form but also permits direct processing of the stored information to obtain contour information. Research is presently continuing on the instrumentation of the system described, as well as on the extension of this technique of coincidence detection of conjugate images directly on the stereomodel.

Most of the work reported here has been supported by the Research Institute, U.S. Army Topographic Laboratories, Fort Belvoir, Virginia.

## References

Krulikoski, S. J., D. C. Kowalski, and F. R. Whitehead, *Bendix Tech. J. 1*, No. 2, 50 (1968).
Krulikoski, S. J., *et al.*, Coherent Optical Mapping Techniques, RADC-TR-70-62 (May 1970).
Kurtz, M. K., *et al.*, Study of Potential Application of Holographic Techniques to Mapping, Final Tech. Rep., DAAK-02-69-C-0563, U.S. Army Topographic Labs., Fort Belvoir, Va. (1971).
Real, R. R., *Appl. Opt. 8,* 411 (1969).

J. C. WYANT

# HOLOGRAPHIC TESTING OF ASPHERIC OPTICAL ELEMENTS

## Introduction

The high performance requirements of modern optical systems have made the inclusion of aspheric surfaces in the design increasingly advantageous. This is especially true for systems such as the Large Space Telescope, solar telescope, and associated spectrographic instruments where for the first time it will be possible to utilize the resolution capabilities of large aspheric often nonsymmetric, optical components. Thus, the ability to test aspheric surfaces in both fabrication and final stage is extremely important. In an effort to satisfy these new testing requirements, an investigation has been made of new interferometric testing techniques made possible by the use of holography.

A common arrangement for testing spherical surfaces is a Twyman-Green interferometer, which compares the surface under test with a flat or spherical reference surface. Often when the surface under test is aspheric, the difference between the reference surface and test surface is so large that the resulting interferogram is too complicated to analyze. The most common method of solving this problem when the surface de-

The author is at the Itek Corporation, Lexington, Massachusetts 02173.

parts only a fringe or so from the desired shape is to make a second optical system (null lens or null mirror), which converts the wavefront produced by the element under test into either a spherical or a plane wavefront. This wavefront can then be interferometrically compared with a spherical or plane reference wavefront. Often in early fabrication stage testing the surface is not known accurately enough to perform a null test. Even if a null test is attempted, the resulting interferogram will still contain too many fringes to analyze. Since high accuracy is neither needed nor desired in early fabrication stage testing a longer wavelength light source can be used in the interferometer to reduce the number of fringes. Using a longer-wavelength light source in the interferometer also creates problems because film cannot be used to record the interferogram directly and the inability to see the radiation causes considerable experimental difficulty. As shown in recent papers [Hildebrand and Haines, 1967; Zelenka and Varner, 1968, 1969; Wyant, 1971], two-wavelength holography (TWH) provides a means of using visible light to obtain an interferogram identical to the one that would be obtained if a longer nonvisible wavelength were used. This paper will review these concepts, and details will be given of a special interferometer built for the TWH testing of aspheric elements. Experimental results will be given for the testing of both polished and ground-glass optical surfaces.

In final stage testing, high accuracy is needed, so TWH cannot be used to reduce the number of fringes. Since the surface departs only a fringe or so from the desired shape, null optics can be used to convert the aspheric wavefront under test into a spherical wavefront. However, the null optics required is often very difficult and expensive to produce. The difficulty and expense becomes even more severe when nonsymmetric wavefronts are tested. Other papers have indicated that in many cases computer-generated holograms (CGH) provide a method of either eliminating or reducing the complexity of the required null optics [Pastor, 1969; Pastor et al., 1968; MacGovern and Wyant, 1971]. In this paper, the concepts of CGH testing will be reviewed, the sources of error will be discussed, and an interferometer built for CGH testing will be described. Some experimental results will also be given.

### Two-Wavelength Holography (TWH)

There are two basic TWH techniques useful for early fabrication-stage testing of optical elements. One technique consists of first photographing the fringe pattern obtained by testing an optical element using a wavelength $\lambda_1$ in an interferometer such as the modified Twyman-Green shown in Figure 1. This photographic recording of the fringe pattern (hologram) is then developed and replaced in the interferometer in the exact position it occupied during exposure, and it is illuminated with the fringe pattern obtained by testing the optical element using a different wavelength, $\lambda_2$. It can be shown [Wyant, 1971] that the moiré pattern obtained is identical to the interferogram that would have been obtained if the optical element were tested using a wavelength $\lambda_{eq}$, where

$$\lambda_{eq} = \frac{\lambda_1 \lambda_2}{|\lambda_1 - \lambda_2|}. \tag{1}$$

See Table 1 for various values of $\lambda_{eq}$ that can be obtained using various pairs of wavelengths from an argon and He-Ne laser.

This moiré pattern will not have high contrast if the two fringe patterns giving the moiré pattern do not have high contrast. If desired, the contrast of the final interferogram can be increased by spatial filtering. If this filtering is to be effective, the angle between the two interfering beams in the interferometer should be such that only the object beam, and not the reference beam, passes through the spatial filter (aperture) shown in Figure 1. The spatially filtered moiré pattern is a result of the interference between the wavefront produced by illuminating (with wavelength $\lambda_2$) the hologram recorded using wavelength $\lambda_1$ and the wavefront obtained from the optical element using wavelength $\lambda_2$.

The fringe pattern (hologram) must be recorded in the image plane of the exit pupil of the optical element under test, since the interfero-

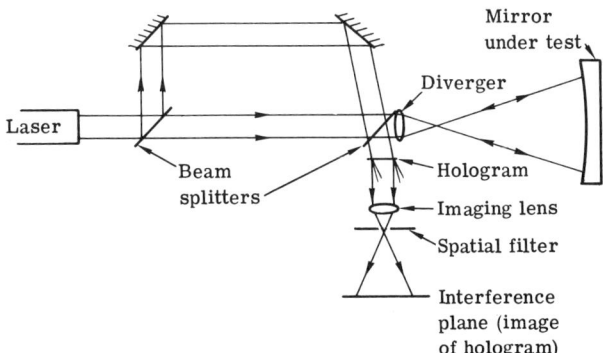

FIGURE 1 Modified Twyman-Green interferometer used for holographic testing.

**TABLE 1** Possible Equivalent Wavelengths, $\lambda_{eq}$, Obtainable Using an Argon and an He–Ne Laser

| | $\lambda_1, \mu m$ | | | | | |
|---|---|---|---|---|---|---|
| $\lambda_2, \mu m$ | 0.4765 | 0.4880 | 0.4965 | 0.5017 | 0.5145 | 0.6328 |
| 0.4765 | – | 20.22 | 11.83 | 9.49 | 6.45 | 1.93 |
| 0.4800 | 20.22 | – | 28.5 | 17.87 | 9.47 | 2.13 |
| 0.4965 | 11.83 | 28.5 | – | 47.9 | 14.19 | 2.30 |
| 0.5017 | 9.49 | 17.87 | 47.9 | – | 20.16 | 2.42 |
| 0.5145 | 6.45 | 9.47 | 14.19 | 20.16 | – | 2.75 |
| 0.6328 | 1.93 | 2.13 | 2.30 | 2.42 | 2.75 | – |

gram obtained using TWH correctly gives the difference between the two interfering beams only in the plane of the hologram. The final photograph of the interferogram should be recorded in the image plane of the hologram, i.e., in the image plane of the exit pupil of the optical element under test.

Figure 2(a) shows an interferogram of an optical element tested using a wavelength of 0.4880 μm. The other interferograms shown in the figure were obtained using TWH to test the same optical element. The interferograms shown in Figures 2(b), 2(c), 2(d), and 2(e) were ob-

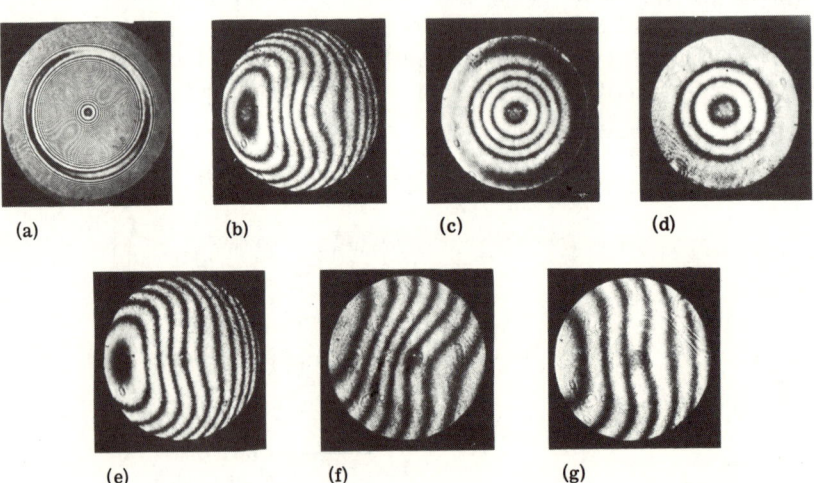

FIGURE 2 Interferograms of an optical element (a) $\lambda$ = 0.4880 μm, (b) $\lambda_{eq}$ = 6.45 μm, (c) $\lambda_{eq}$ = 6.45 μm, (d) $\lambda_{eq}$ = 9.47 μm, (e) $\lambda_{eq}$ = 9.47 μm, (f) $\lambda_{eq}$ = 20.22 μm, (g) $\lambda_{eq}$ = 28.5 μm.

tained by first recording an interferogram (hologram) using a wavelength of 0.5145 μm and then illuminating the recording with a fringe pattern obtained using a wavelength of 0.4765 μm for Figures 2(b) and 2(c) and 0.4880 μm for Figures 2(d) and 2(e). The interferograms were spatially filtered. The amount of tilt shown in the interferograms was adjusted in real time by changing the angle at which the reference wavefront was incident upon the hologram during the reconstruction.

The interferograms shown in Figures 2(f) and 2(g) were obtained by first recording an interferogram using a wavelength of 0.4880 μm and then illuminating this recording with a fringe pattern obtained using a wavelength of 0.4765 μm and 0.4965 μm, respectively.

In the method of TWH described above, the final interferogram gives the difference between a fringe pattern recorded at one instant of time and a fringe pattern existing at some later instant of time. If the two fringe patterns are different for reasons other than wavelength change, e.g., air turbulence, incorrect results are obtained. For example, if air turbulence causes one fringe change between the fringe pattern obtained using $\lambda_1 = 0.4880$ μm and the fringe pattern obtained using $\lambda_2 = 0.5145$ μm, the moiré interferogram will contain one fringe error, which, as Table 1 indicates, corresponds to an error of 9.47 μm.

The effect of air turbulence can be reduced by recording simultaneously the two interferograms resulting from the two wavelengths. If the recording process is sufficiently nonlinear and the interferograms have sufficiently high contrast, this interferogram shows the same moiré pattern described above. Generally, this moiré pattern is too low in contrast to be useful. However, when this interferogram (hologram) is illuminated with a plane wave, spatially filtered, and reimaged in the same manner as shown in Figure 1, one obtains a high-contrast interferogram, identical to that obtained using the first method of TWH described above. Since both fringe patterns are recorded simultaneously, and air dispersion is small, the sensitivity of the interferometer to air turbulence is essentially the same as if a long-wavelength light source were used in the interferometer.

One problem in using double-exposure TWH, as just described, is that the amount of tilt in the final interferogram cannot be adjusted after the hologram is recorded. When desired, this problem can be solved by using the procedure described later in the paper.

Since our initial TWH results obtained using a rather crude laboratory setup proved very useful, it was decided to build a special Twyman-Green interferometer for TWH testing of aspheric wavefronts in the shop. To reduce the turbulence problems mentioned above, it was decided that the simultaneous double-exposure TWH technique had to be used in the shop interferometer. Furthermore, it was decided to design the interferometer to use principally the 0.4880-$\mu$m and 0.5145-$\mu$m wavelengths. These two wavelengths were selected because they are the two strongest lines from an argon laser and because our initial work indicated that an equivalent wavelength of 9.47 $\mu$m gave in most cases the desired sensitivity. A 2-W Ar laser was used, with an etalon included to obtain coherence with large optical path differences.

Figure 3 shows two views of the completed interferometer. The optical components include one 4-in.-diameter beam splitter (Ar coated), two 3-in.-diameter flat mirrors (hologram reference beam mirrors), one 1-in.-diameter flat folding mirror, one 6-in.-diameter flat folding mirror, one 6-in.-diameter parabolic mirror (mounted off-axis), and two 2-in.-diameter flat mirrors (holographic playback beam mirrors).

As can be seen from the photographs of the interferometer, the entire optical component housing was mounted on translation stages to allow precision positioning of the optical component housing along three orthogonal axes.

Direct radiation from the laser first enters the line filtering section. Since a filter could not be obtained to pass only the 4880 Å and 5145 Å lines and block all the other Ar lines, the laser beam had to be split into two beams, and a filter passing only the 4880 Å line was placed in one beam and a filter passing only the 5145 Å line was placed in the other beam. The two beams were then recombined. After the filter section, the light is reflected to the laser beam expander/collimator, which produces a 50-mm-diameter collimated wavefront. This size was selected for two reasons: First, since the whole purpose of the interferometer is to test aspheric wavefronts, the optics had to be made large enough to accept large blur circles. Second, this enabled the use of a diverger having a relatively long focal length so the image of the piece under test would be at a position where the photographic plate could be conveniently placed.

The collimated beam is split by a beam splitter. Part of the light travels from the beam splitter to mirror 1, mirror 2, back through the beam splitter, and illuminates the photographic plate as the reference wavefront for the image plane hologram. The other portion of the beam goes through the beam splitter to the off-axis parabolic mirror. The parabolic mirror is used as the diverger so there are no chromatic aberration problems. After focus, the light strikes mirrors 3 and 4 and goes out to the optics under test. Light returning from the test optics surface follows a path back to mirrors 4 and 3, the off-axis parabola, the beam splitter, and, finally, to the photographic plate. The optical distance from the photographic plate to the off-axis parabola is adjusted to lo-

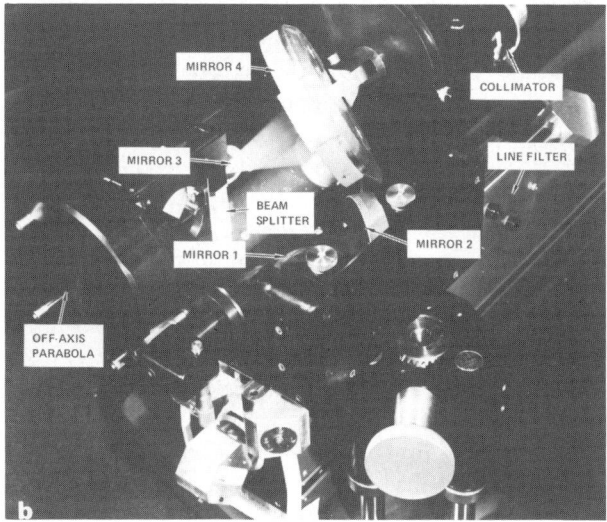

FIGURE 3  Two-wavelength holographic interferometer (a) side view, (b) front view.

cate a real image of the optics under test in the plane of the photographic emulsion. The two-color real image and two-color reference wave produce two distinct sets of fringes at the photographic plate. The two sets of fringes are stored photographically and constitute the image-plane hologram.

The holograms are recorded on Agfa 10E56 photographic plates. Using this interferometer, the exposure times are approximately 1/100 sec, even for the testing of the ground surfaces described later.

After the holograms are processed, they could be illuminated with a plane wavefront and spatially filtered as described earlier to obtain the final interferogram. However, since it is often desirable to be able to adjust in real time the amount of tilt in the final interferogram, the setup shown in Figure 4 and described below was used in the hologram reconstruction process. The interferometer is easily arranged to perform the holographic reconstruction process by inserting mirrors 5 and 6 [Figure 4(b)] in front of mirror 1 and the off-axis parabola, respectively. A collimated reconstructing beam of only one color is split into two portions by the beam splitter. The reconstructing beam leaving mirror 6 is reflected again by the beam splitter and illuminates the hologram at approximately the same angle as the reconstructing beam from mirror 5, which passes through the beam splitter onto the hologram.

Two reconstructing waves illuminating two sets of holographic fringes produce four reconstructed images at the image plane hologram. The reconstructing beam from mirror 6 produces one image from the green-light fringes and one image from the blue-light fringes. These images are designated M6G and M6B in Figure 4(a). Similarly, the reconstructing beam from mirror 5 forms two images, designated M5G and M5B.

Mirror 5 has real-time tip, tilt adjustment and is used to obtain a slight tilt between the reconstructed images M6B and M5G or M6G and M5B. The other reconstructions are spatially filtered at the focal plane of lens L to select out either M6B and M5G or M6G and M5B. The interferogram obtained by interfering a proper set of reconstructions shows one fringe per 4.735-$\mu$m surface departure from a sphere, i.e., the same results as would be obtained if 9.47-$\mu$m wavelength were used in a regular Twyman-Green interferometer.

Figure 5 shows the results obtained using the interferometer to test a 32-in.-diameter off-axis nearly hyperbolic polished mirror. Also shown in the figure is an interferogram of the same surface obtained using a wavelength of 0.5145 $\mu$m. As can be seen, the interferogram obtained at a wavelength of 0.5145 $\mu$m is too complicated to analyze, while the interferogram obtained using the TWH at an equivalent wavelength of 9.47 $\mu$m can easily be analyzed.

Just as 10.6 $\mu$m from a $CO_2$ laser can be used to obtain interferograms of ground-glass surfaces [Munnerlyn and Latta, 1968], so can TWH. Figure 6 shows a TWH interferogram of approximately one half of an $f/12$, 7.5-cm-diameter ground-glass mirror. As can be seen, the fringes have amazingly good contrast. However, there are two problems in using TWH for testing ground-glass surfaces. First, since the hologram

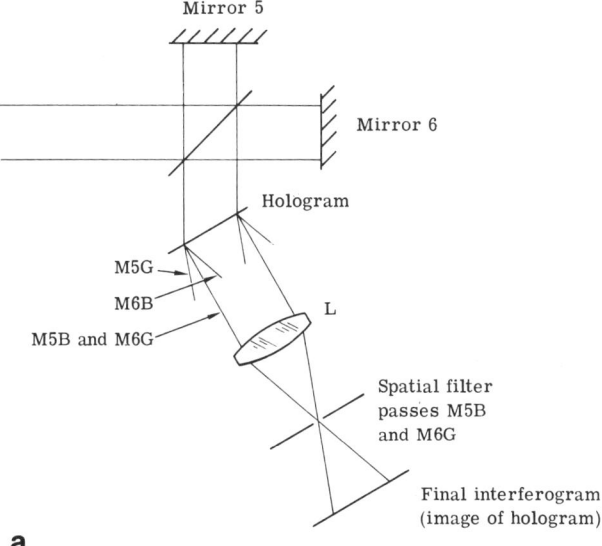

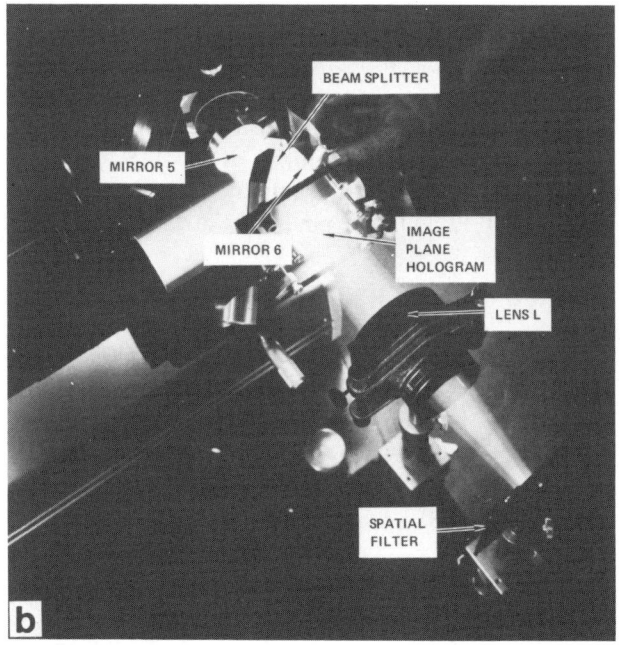

FIGURE 4  TWH reconstruction setup: (a) schematic diagram, (b) reconstruction of image-plane hologram.

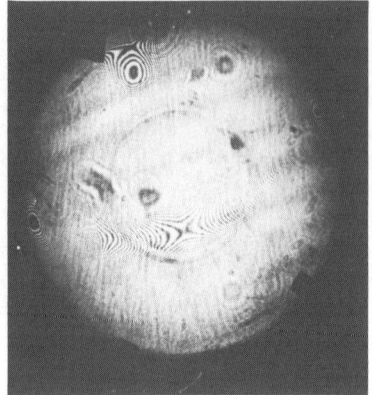 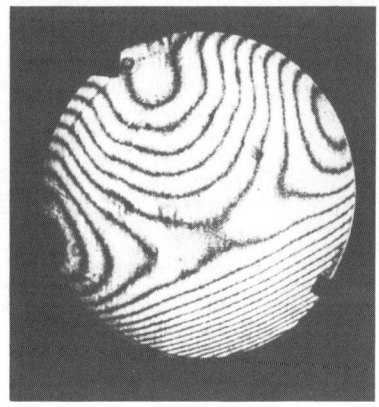

FIGURE 5  Interferograms of off-axis hyperbolic mirror: *left*, $\lambda = 0.5145$ μm; *right*, $\lambda = 9.47$ μm.

is made using visible light, the ground-glass surface scatters the light so much that very little light gets back through the imaging lens onto the hologram. Thus, long exposures are required. The second problem is the difficulty involved in setting up an interferometer when the piece under test does not give a specular reflection.

To get around this problem, various methods of waxing the ground-glass surface to obtain a specular reflection have been suggested [Moreau and Hopkins, 1969]. We have tried many of these techniques, and the method that appears to work best is a method first suggested and tried by Paul Remijan of Itek. If the ground-glass surface to be tested is first

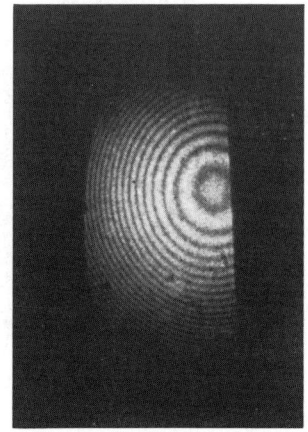

FIGURE 6  Interferograms of a portion of a ground-glass mirror, $\lambda_{eq} = 9.47$ μm.

FIGURE 7 Interferogram of spray-varnished ground-glass hyperbolic mirror ($\lambda_{eq}$ = 9.47 μm).

sprayed lightly with a varnish (Hyplar spray varnish by Grumbacher was used) normally intended for use as a coating for oil paintings, a specular reflection suitable for TWH testing is obtained. Figure 7 shows the results for testing the off-axis hyperbola mentioned earlier after it has been ground with M-5 grit. Repeated tests have shown that the uniformity of the spray coating is good to within a 3-μm peak-to-peak error, which should be adequate for fabrication stage testing. Since polishing a large surface takes several hours and the spraying takes but a few minutes, the spraying technique can be very time saving. Thus, it is felt that the combination of TWH and this spraying technique should prove very useful for early fabrication-stage testing.

## Computer-Generated Holograms (CGH)

The CGH used for final-stage testing of aspheric wavefronts as described in this paper are basically a binary representation of the actual interferogram that would be obtained if the ideal aspheric wavefront being tested were interfered with a tilted-plane wavefront. This will become clear as the procedure for making and using a hologram is looked at.

A CGH can be used with a wide variety of experimental setups used to test aspheric wavefronts. A convenient setup for testing aspheric mirrors is the same setup that was shown in Figure 1 for the TWH testing. The hologram is placed in the image plane of the mirror under test, i.e., in the same position film would be placed if a recording were to be made of the interference of the aspheric wavefront produced by the mirror under test with the tilted plane reference wavefront.

To make the desired CGH, the system must first be ray-traced. A computer program was written to obtain the position of the fringes in the hologram plane that result from the interference of the tilted-plane wave and the aspheric wavefront that would be obtained if the mirror under test were perfect. The program locates the fringe position by iteration to any desired accuracy and outputs the sequential positions along a fringe onto a tape, which is used to drive the plotter. For all the results shown in this paper, a 28-in.-diameter hologram was plotted using a Calcomp Model 736 plotter. The hologram is plotted one fringe at a time, and parabolic interpolation is used to produce smooth fringes. To achieve wide fringes, each one is traced a number of times, usually five, with a small lateral shift introduced. The resultant plot consists of wide dark fringes against a white background. A typical computer plot is shown in Figure 8. The total computer time on a CDC 3300 used to produce one hologram is approximately 20 min, and plotter time is about 10 h. The 28-in.-diameter computer plot is photoreduced to the correct size with an $f/3$, 39-mm focal-length Nikon lens.

When the CGH is placed in the interferometer as shown in Figure 1, the CGH and the interference fringes produced by the interference of the reference wavefront and the wavefront produced by the mirror under test produce a moiré pattern, which gives the difference between the CGH and the interference fringes. Spatial filtering can be used to improve the contrast of the moiré pattern if in the making of the CGH the tilt of the plane reference wavefront was selected to be at least as large as the maximum slope of the aspheric wavefront along the intersection of the plane of incidence of the plane wave and the aspheric wavefront. Then, spatial filtering is accomplished by reimaging the hologram with an appropriately placed small aperture in the focal plane of the reimaging lens. This aperture is placed so that it passes only the wavefront from the mirror under test and the corresponding wavefront produced by illuminating the hologram with a plane wavefront. Thus, in the interference plane shown, an interferogram is produced that gives the difference between the wavefront produced by the mirror under test and the corresponding wavefront produced by the hologram.

Although there are obviously many places in the interferometer where a CGH could be placed, when it is placed as shown, the thickness variations in the hologram plate have no effect on the results, and, thus, what could be a very serious source of error is eliminated.

The above ray-tracing procedure used to make the holograms can be

FIGURE 8  Typical computer plot for CGH.

used for any general optical system. The only requirement is that all the optics in the interferometer be known so the system can be ray-traced. An important consequence of ray-tracing the entire interferometer is that even though the diverger may be corrected only for spherical wavefronts and may introduce additional aberrations in the aspheric wavefront being passed through it, the hologram automatically corrects for these aberrations when a null test (or for all practical purposes, a near-null test) is performed.

Besides the usual errors produced by interferometer misalignment, there are five main sources of error in a CGH test: emulsion movement, plotter distortion, photoreduction lens distortion, incorrect hologram size, and misalignment of hologram in interferometer. As the errors are looked at below, it will be seen that all five errors are proportional to the maximum slope of the departure of the aspheric wavefront from a spherical wavefront.

To determine the error produced by emulsion movement, nine 25-mm-diameter holograms of two collimated wavefronts were made on Kodak 649-F plates. The spatial frequencies of the holograms were 40 lines/mm, 330 lines/mm, and 1000 lines/mm. The holograms were developed in Kodak HRP for 5 min, after which they were put in an acetic stop bath for 15 sec and Kodak fixer for 3 min. They were washed in running water for 5 min and rinsed for 30 sec in Yankee Instant Film Dryer and Conditioner and air dried.

After processing, the hologram was replaced into the original setup and one of the original collimated wavefronts was interfered with the corresponding wavefront produced by the hologram. Interferograms were recorded, and the average rms and peak wavefront errors measured for the three spatial frequencies investigated are shown in Table 2.

It is doubtful that the wavefront errors shown in Table 2 were predominately a result of emulsion movement, since the magnitude of the error does not appear to be largely dependent on the spatial frequency of the hologram fringes. Other possible sources of error are noise produced by dust in the collimated wavefronts, small error in repositioning of the hologram, turbulence, and what is believed to be the largest source of error, noise in the data-reduction process. The important conclusion is that the rms error produced by emulsion movement is certainly less than $1/40\lambda$.

The next source of error to be investigated is distortion in the hologram plotter. To show how the CGH wavefront accuracy depends on the number of plotter resolution points and the maximum slope of the aspheric wavefront being tested, let us assume that the plotter has $P \times P$ resolution points. Thus, there are $P/2$ resolution points across the radius of the hologram. Since the maximum error in plotting any point is one half a resolution unit, any portion of each line making up the

TABLE 2 Average rms and Peak Error in Wavefront Produced by Hologram

| Spatial Frequency of Hologram Fringes (lines/mm) | Average rms Error | Average Peak Error |
|---|---|---|
| 40 | $0.025\lambda$ | $0.073\lambda$ |
| 330 | $0.021\lambda$ | $0.061\lambda$ |
| 1000 | $0.023\lambda$ | $0.065\lambda$ |

hologram could be displaced from where it should be a distance equal to $1/P$ the radius of the hologram. Let the maximum difference between the slope of the aspheric wavefront and the tilted plate wave be $S$ waves per hologram radius. Thus, the phase of the plane wave at the hologram lines can differ from that of the required wavefront at the same lines by as much as $S/P$ waves. Therefore, in the hologram plane the error in the reconstructed wavefront can be as large as $S/P$ waves. Since the final interferogram is recorded in the image plane of the hologram, the quantization due to the finite number of resolution points causes a peak error in the final interferogram of $S/P$ waves. Figure 9 is a log-log graph of peak wavefront error versus number of plotter resolution points for various amounts of maximum difference between the slope of the aspheric wavefront and the slope of the tilted plane wave.

As mentioned above, to maximize the contrast of the final interferogram, the hologram should be made such that it is possible to spatially filter the hologram to select out the first-order reconstruction. This requires that the slope (tilt) of the plane reference wavefront be at least as large as the maximum slope of the aspheric wavefront along the intersection of the plane of incidence of the plane wave and the aspheric wavefront. Since increasing the slope of the plane reference wavefront decreases the accuracy of the aspheric wavefront produced, advantage should be taken of the fact that a smaller reference wavefront tilt can be used in the testing of nonsymmetric wavefronts if the plane of incidence of the reference wavefront is along the direction of minimum slope of the aspheric wavefront.

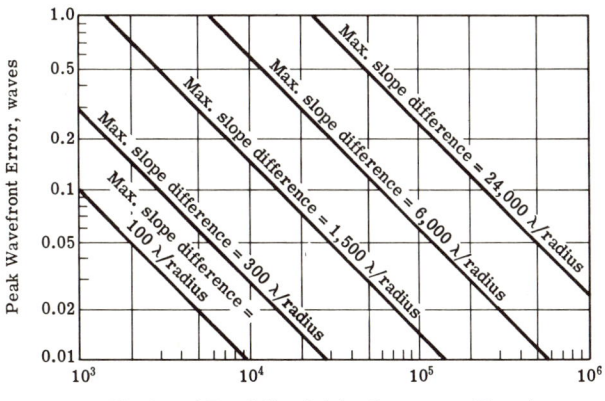

FIGURE 9 Peak wavefront error versus number of plotter resolution points.

The third source of error to be investigated is the error due to incorrect hologram size. Let $\phi(r, \theta)$ be the aberrated wavefront being tested in the plane of the hologram. If the hologram is the correct size, then this is the wavefront that the hologram produces. If the hologram has the incorrect size by magnification factor, $M$, then the test gives the difference $\phi(r/M, \theta) - \phi(r, \theta)$. Now, by a Taylor's expansion

$$\phi(r/M, \theta) = \phi[r + (1/M - 1)(r, \theta)] = \phi(r, \theta) + \{[\partial\phi(r, \theta)]/\partial r\}[(1/M) - 1]r + \cdots . \quad (2)$$

Terms higher than first order in the expansion can be neglected if $M$ is sufficiently close to 1 and a small region is looked at.

Thus, the error in the test results caused by the hologram having the incorrect size is given by

$$\phi(r/M, \theta) - \phi(r, \theta) = \{[\partial\phi(r, \theta)]/\partial r\}[(1/M) - 1]r + \cdots . \quad (3)$$

A distortion error is analyzed the same as a magnification error except that the magnification, $M$, is a function of position $r$. Let $\alpha(r)$ be the distortion as a function of radius $r$. That is, a point that is supposed to be at the radius $r$ is at a radius $r[1 + \alpha(r)]$. Since often in a test both the distortion and wavefront slope are a maximum at the maximum value of $r$, the error due to distortion can be reduced by adjusting the magnification to balance out the distortion error at the edge of the plot. That is, the photoreduction should be demagnified by a factor $1 + \alpha(r_{max})$ from what it would be if no distortion were present. Thus, the magnification error due to distortion at any radius $r$ is

$$M(\alpha) = \frac{1 + \alpha(r)}{1 + \alpha(r_{max})}. \quad (4)$$

Error due to misalignment of the hologram in the interferometer could be due to either an off-center error or, in the case of testing nonsymmetric wavefronts, a rotation error. The off-center error will be looked at first.

Let $\phi(x, y)$ be the wavefront being tested. Let the center of the aberrated wavefront and the center of the hologram be displaced a distance $\Delta x$ in the $x$ direction. Then the result of the interference test gives $\phi(x + \Delta x, y) - \phi(x, y)$. Just as before, a Taylor series expansion leads to

$$\phi(x + \Delta x, y) - \phi(x, y) \approx \frac{\partial \phi(x,y)}{\partial x} \Delta x. \quad (5)$$

Equation (5) gives the error resulting from off centering the hologram a distance $\Delta x$. $\partial \phi / \partial x$ is just the slope in the $x$ direction.

When nonrotationally symmetric wavefronts are being tested, an error will result when the hologram has an incorrect rotational position. Again writing $\phi$ in polar coordinates, i.e., $\phi(r, \theta)$, the test gives the difference $\phi(r, \theta + \Delta\theta) - \phi(r, \theta)$, where $\Delta\theta$ is the angular error. The Taylor series expansion gives

$$\phi(r, \theta + \Delta\theta) - \phi(r, \theta) \approx \frac{\partial \phi(r, \theta)}{\partial \theta} \Delta\theta. \tag{6}$$

Thus, Eq. (6) is the error that results from an angular error. $[\partial \phi(r, \theta)]/\partial \theta$ is the angular slope.

Several aspheric wavefronts have been tested using CGH. To verify the analysis given above, CGH were made to test the known aspheric nonsymmetric wavefronts produced by the experimental setup shown in Figure 10. The amount of aberration in the wavefront being tested was selected by tilting the plane-parallel plate placed between the diverger and spherical mirror.

For all our tests, the spatial filter shown in Figure 10 was positioned so that the final interferogram showed the interference between the aberrated wavefront produced by the optical system and the aberrated wavefront produced by the hologram. The same results would have been obtained if the spatial filter were positioned to pass the plane reference wave and the plane wavefront produced by the hologram when the aberrated wavefront is used as the reconstructing beam.

Figures 11 and 12 show typical interferograms resulting from the

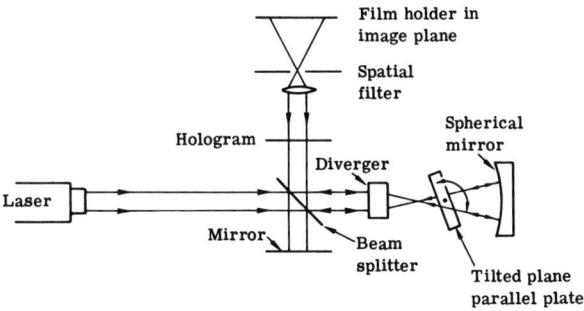

FIGURE 10 Experimental setup to obtain nonsymmetric aspheric wavefront.

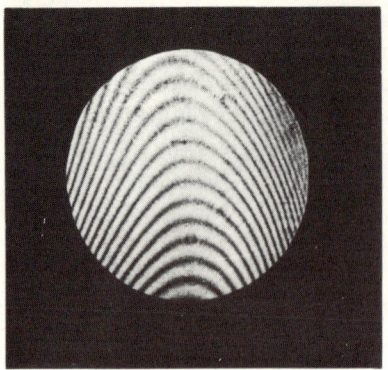 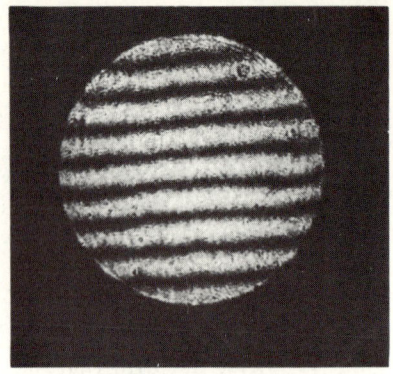

Results of CGH test | Interference of aberrated wavefront and plane wave

FIGURE 11 CGH test of aspheric wavefront having a maximum slope of 35 waves per radius and 19 waves departure.

CGH testing of the nonsymmetric aspheric wavefronts obtained using the setup shown in Figure 10. The ideal result would, of course, be equispaced straight fringes. Also shown in the figures are the interferograms obtained by interfering the aspheric wavefronts with a tilted plane wave. Table 3 summarizes the experimental results. Using the analysis given above, it was found that all the results shown in Table 3 are within the estimated errors in the experiment.

Several parabolic mirrors have been tested using CGH. Figure 13(a) shows the result for testing an $f/3$, 40-cm-diameter parabolic mirror. Figure 13(b) shows the fringes obtained by testing this mirror in autocollimation using a 61-cm-diameter flat mirror. Allowing for the fact that when the mirror was tested in autocollimation it was tested in double pass, it is seen that the two resulting interferograms closely agree. The agreement is to within the errors calculated using the analysis given above.

Figure 14 shows the results for testing a lens having 50 waves of third- and fifth-order spherical aberration. An analysis of this interferogram shows that the wavefront produced by the lens and the wavefront produced by the hologram differed by less than 1/15 wave rms.

In all the above results, the limiting factor in the accuracy was the distortion in plotter. The errors due to incorrect hologram size and incorrect positioning of the hologram are determined by the smallest dis-

(a) **Results of CGH test**

FIGURE 12 CGH test of aspheric wavefront having a maximum slope of 126 waves per radius and 64 waves departure.

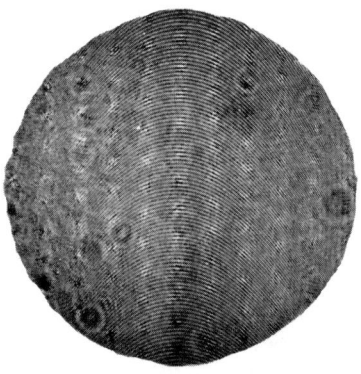

(b) **Interference of aberrated wavefront and plane wave**

TABLE 3  Summary of Experimental Results

| Tilted Plate Angle | Maximum Slope of Aspheric Wavefront in Units of $\lambda$/Radius | Maximum Departure of Aspheric Wavefront from a Spherical Wavefront | rms Error of CGH Test | Peak Error of CGH Test |
|---|---|---|---|---|
| 20° | 24  | 12.54$\lambda$ | 0.04$\lambda$ | 0.13$\lambda$ |
| 25° | 35  | 19.16$\lambda$ | 0.05$\lambda$ | 0.15$\lambda$ |
| 30° | 56  | 30$\lambda$    | 0.05$\lambda$ | 0.15$\lambda$ |
| 35° | 77  | 40$\lambda$    | 0.07$\lambda$ | 0.20$\lambda$ |
| 40° | 100 | 51.6$\lambda$  | 0.06$\lambda$ | 0.17$\lambda$ |
| 45° | 126 | 64$\lambda$    | 0.07$\lambda$ | 0.22$\lambda$ |

## f/3 PARABOLIC MIRROR

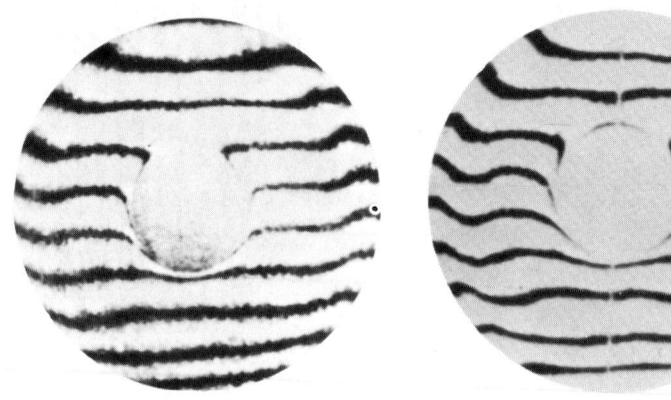

a. Holographic test (single pass)

b. Test in autocollimation (double pass)

FIGURE 13   Result of test of $f/3$ parabolic mirror.

tance that can be measured in the hologram plane. If the hologram size were increased, the smallest distance that could be measured in the hologram plane would remain essentially constant. Thus, the errors due to incorrect hologram size and incorrect hologram position decrease as the size of the hologram increases. Also, these errors due to incorrect hologram size and position are random errors and can be reduced by repeating the experiment many times and averaging. However, error due

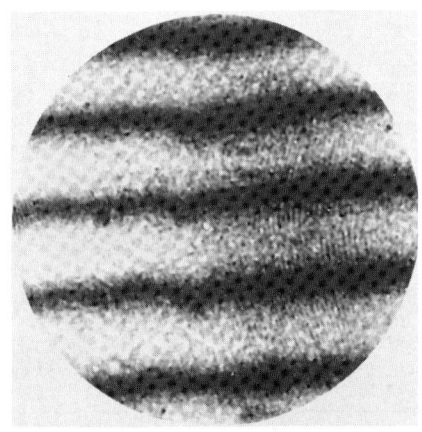

FIGURE 14   Result of holographic test of lens having 50 waves of third- and fifth-order spherical aberration.

to plotter inaccuracies can be reduced only by using a more accurate plotter. We are presently investigating the use of a laser beam recorder, in which, although it has only approximately the same number of distortion-free resolution points as the presently used plotter, most of the distortion is repeatable, so the distortion can be removed in the software and it is hoped that almost an order of magnitude more resolution points will be obtained. If it becomes necessary, the errors due to distortion in the photoreduction lens can also be removed in the software.

Since the requirements for an interferometer built for TWH testing and CGH testing are similar, it makes sense to use the same interferometer for both tests. This thought was kept in mind when the TWH interferometer described earlier was designed. Only two changes need be made in the TWH interferometer to convert it to a CGH interferometer. First, the off-axis parabolic mirror diverger must be replaced with a diverging lens. This is necessary because the parabolic mirror is not of sufficient quality for final-stage CGH testing. To satisfy the requirements for positional accuracy of the CGH, a second change must be made. The hologram must be mounted on a stage that provides accurate and smooth $xy$ translations and rotation.

Although CGH can be designed for any optical system, and one can think of large interferometrically controlled plotting devices that could at least in principle produce the CGH required to test an arbitrarily complicated optical system, one soon reaches a point where the time and expense required to make a CGH are unreasonable. Also, given enough time and money, null optics could probably be designed and built to test almost any arbitrarily complicated optical system. Again a point is soon reached where the time and expense required are not practical. The question arises, can the complicated CGH or the complicated null optics required to test complicated aspherics be replaced with the combination of relatively simple null optics and relatively simple CGH? In many cases the answer is yes, and that is probably the approach that should often be taken. At the present time, we are in the process of using simple null optics and a CGH to test a nonsymmetric mirror having hundreds of waves of departure and hundreds of waves per radius of slope from the best fitting spherical surface. The simple null optics (spherical mirror) reduces the departure to about 60 waves and the slope to about 100 waves per radius. A CGH is used to convert this wavefront to a spherical wave. Our analysis indicates that the accuracy of the test should be about 0.1 wave peak to peak.

## Conclusions

The results given in this paper show that two-wavelength holography provides a good method of using visible light to obtain an interferogram identical to the one that would be obtained if a longer wavelength were used. A wide range of equivalent wavelengths can be obtained using commercially available lasers. Ground-glass surfaces can easily be tested if the surface is first sprayed lightly with varnish. It is felt that the combination of TWH and this spraying technique should prove useful for early fabrication-stage testing.

It is seen that computer-generated holograms can in many instances be used to either eliminate or reduce the complexity of null optics for the final-stage testing of aspheric optical elements. Probably their greatest use will come in replacing complicated null optics with simple null optics and a CGH. The error analysis given can be used to calculate before the test the approximate errors in the experiment to determine if adequate accuracy will be obtained. The interferometer described can be used to do all the testing from early fabrication-stage TWH testing up through final-stage CGH testing.

The author would like to thank P. W. Remijan and V. P. Bennett for performing much of the experimental work shown in the paper, and he would like to especially thank J. Smith, who did the mechanical design and construction of the TWH and CGH interferometer shown. Special thanks should also go to P. K. O'Neill, whose contributions to this paper are too numerous to mention.

## References

Hildebrand, B. P., and K. A. Haines, *J. Opt. Soc. Am. 57*, 155 (1967).
MacGovern, A. J., and J. C. Wyant, *Appl. Opt. 10*, 619 (1971).
Moreau, B. G., and R. E. Hopkins, *Appl. Opt. 8*, 2150 (1969).
Munnerlyn, C. R., and M. Latta, *Appl. Opt. 7*, 1858 (1968).
Pastor, J., *Appl. Opt. 8*, 525 (1969).
Pastor, J., J. S. Harris, and G. E. Evans, *J. Opt. Soc. Am. 58*, 1556A (1968).
Wyant, J. C., *Appl. Opt. 10*, 2113 (1971).
Zelenka, J. S., and J. R. Varner, *Appl. Opt. 7*, 2107 (1968).
Zelenka, J. S., and J. R. Varner, *Appl. Opt. 8*, 1431 (1969).

# OPTICAL TECHNOLOGY

R. PETIT, D. MAYSTRE, and M. NEVIÈRE

# Practical Applications of the Electromagnetic Theory of Gratings

## I. Introduction

In our universities, the theory of gratings is generally taught in the optics course after interference and diffraction phenomena. Teaching this way, students might believe that gratings properties do not depend on the polarization of the light. It is true for geometrical properties, and nobody contests that the famous "grating formula" is able to give the diffraction directions. But it is wrong for what we call "energy properties," and it is now well known that the distribution of diffracted energy among the different spectral orders depends strongly on polarization if the wavelength $\lambda$ and the grating period $d$ are of the same order of magnitude. Therefore, an electromagnetic theory of gratings has been developed during the last ten years in order to determine the efficiencies of a modern grating when both its profile and working conditions are known. Two of our recent papers [Petit and Maystre, 1972; Nevière et al., 1973] have been devoted to this theory; this short paper is for experimentalists in order for them to have an idea of the help they can expect from applied mathematics when they are confronted with plane gratings problems.

---

The authors are at the Université de Provence, Centre de St-Jérôme, Optique Electromagnétique, 13013 Marseille, France.

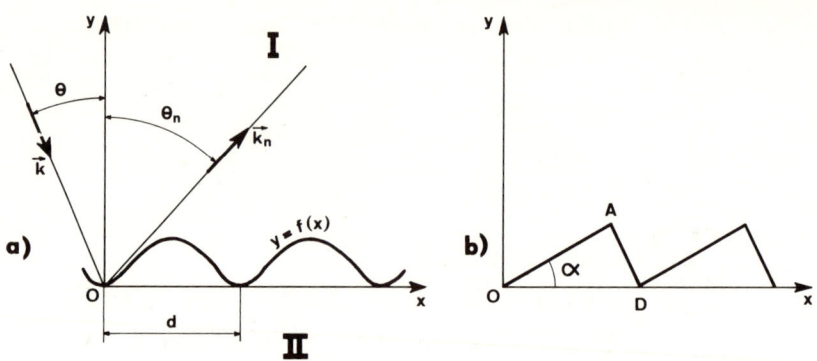

FIGURE 1 (a) Notations for classical working conditions. (b) Echelette grating profile. All through the paper $\hat{A}$ is supposed to be $90°$.

## II. Notations

Because we are not dealing with the resolving power, we always suppose the grating to be infinitely wide and describe it by the cylindrical surface $y = f(x)$ [Figure 1(a)]. The period $d$ of $f(x)$ is the grating period. If the graph of $f(x)$ is obtained by translations from a triangle [Figure 1(b)] the grating is called an "echelette grating." In the other cases, it is called a grating of arbitrary profile or, sometimes, for obvious reasons, a holographic grating.

Region I $[y > f(x)]$ is a vacuum or air. Depending on whether region II $[y < f(x)]$ is filled with a conducting or infinitely conducting metal, we speak of a conducting or infinitely conducting grating.

The plane and monochromatic incident wave is described by the wave vector **k**, the modulus of which is $2\pi/\lambda$. If **k** is parallel to the $xOy$ plane (orthogonal to the grooves) we say that we are under classical working conditions, and **k** is then located only by the incidence angle $\theta$. In the other cases (**k** not orthogonal to the grooves) we say that we are working with "conical diffraction" because it is known that **k** and the different diffracted wave vectors $\mathbf{k}_n$ are then located on a cone the axis of which is $Oz$. As usual, the efficiency $e_n$ in the $n$th order is the fraction of incident energy diffracted in this order.

## III. Efficiency of an Infinitely Conducting Echelette Grating under Classical Working Conditions

The problem of the theoretical determination of efficiencies when **k** is orthogonal to the grooves is now perfectly resolved [Petit and Maystre,

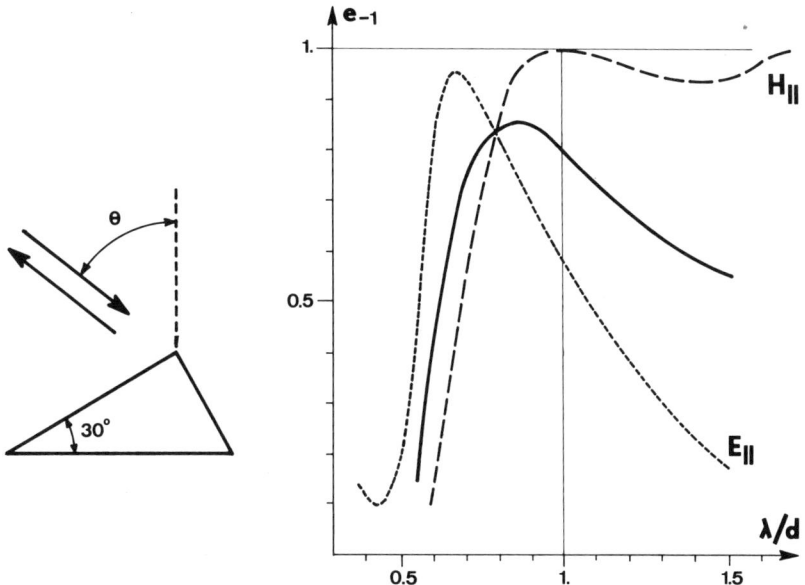

FIGURE 2 Efficiency of an echelette grating: Littrow's mounting, −1 order, $\lambda/d = 2 \sin\theta$. ---- $H_\|$ case, ------ $E_\|$ case, ——— natural light (unpolarized).

1972] for any values of the $AOD$ triangle angles, of the incidence angle $\theta$, and whatever the polarization. Accuracy is about 1% whenever the ratio $\lambda/d$ is greater than 0.2. Numerous curves have been drawn for the constant deviation mountings that are often used. Particular attention has been devoted to Littrow's mounting (zero deviation). Figure 2, which corresponds to a 30° blaze angle, comes from an important paper already published [Maystre and Petit, 1971a] in which we study systematically two fundamental cases of polarization called $E_\|$ and $H_\|$ according to whether the electric or magnetic field is parallel to the grooves. Figure 3 recalls an older work done to examine the influence of a bad ruling on efficiency.

## IV. Efficiency of Infinitely Conducting Gratings for Classical Working Conditions

No theoretical complication appears since some of the different methods described by Petit and Maystre in the *Revue de Physique Appliquée* are well adapted for arbitrary profiles. The main difficulty is to know the exact profile shape of the commercial holographic gratings. For this

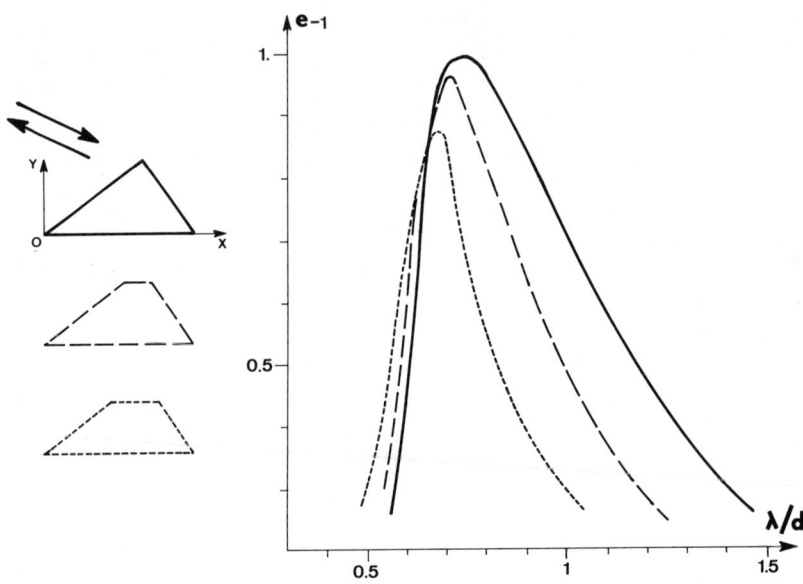

FIGURE 3 Influence of a bad ruling (Littrow's mounting, −1 order, $E_\parallel$). The profile and the corresponding efficiency curve have been drawn using the same type of line.

purpose several methods have been employed: stereoscopic electron micrographs of the grooves (Figure 4), grazing incidence electron micrograph [Bousquet et al., 1969], and the "sandwich method" [Chauvineau et al., 1967] with the help of either an ordinary or a scanning electron microscope. The results obtained by these methods are not in perfect agreement. Nevertheless, they allow us to determine a profile shape the accuracy of which is good enough for a first quantitative study. In fact, Figure 5 shows us that the mathematical representation used to describe the profile (obtained here by grazing incidence method) has only a very slight influence on efficiency curves. For instance, using the profile of Figure 5 the grooves' depth influence has been investigated. Results have already been given in an earlier paper [Maystre and Petit, 1970], from which Figures 6 and 7 have been chosen. Note that, for −1 order and Littrow's mounting at least, both "blaze efficiency" and "blaze wavelength" are increasing with the depth $h$.

Another study, in which the ratio $h/d$ has been fixed to the highest value (0.2) that seemed to be possible one year ago, has been carried out to determine, for a given period, the groove width that we must choose to get the best efficiency for Littrow's mounting and −1 order [Maystre and Petit, 1971b]. It is the trapezoidal profile (Figure 5a) that has been

R. Petit et al.  671

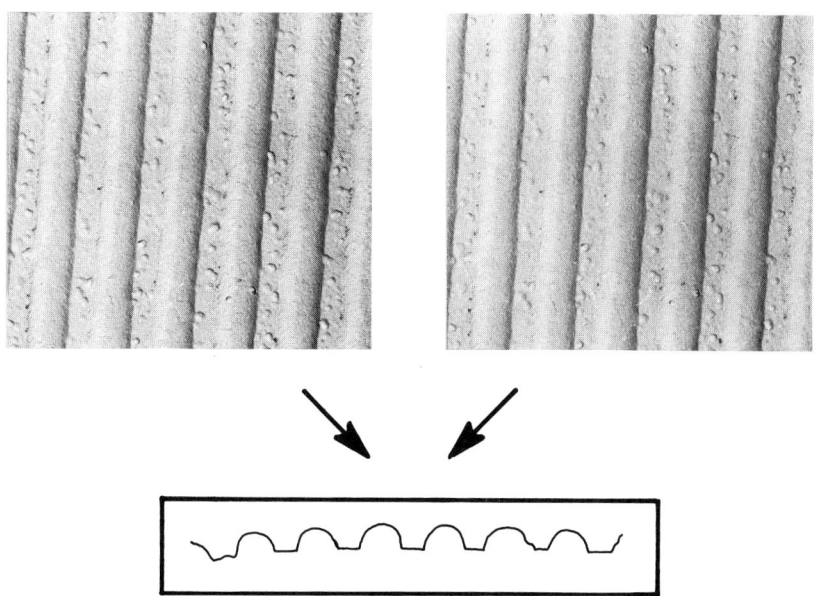

FIGURE 4  Stereoscopic electron micrographs (Karl Zeiss, Oberkochen) and associated restitution (Institut géographique national, Paris).

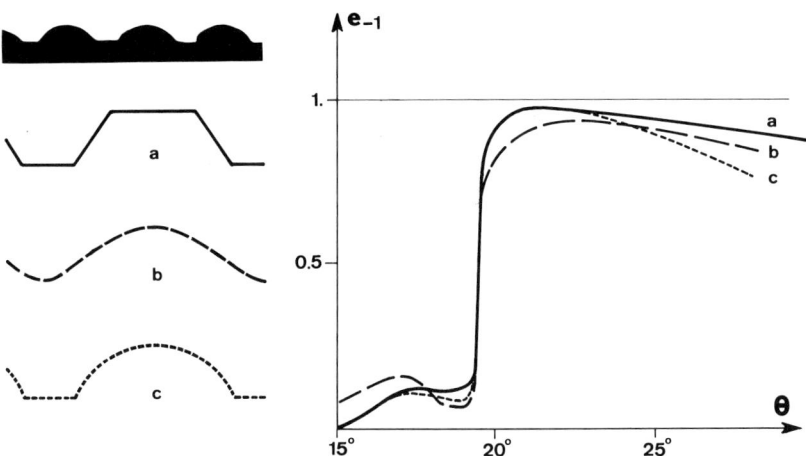

FIGURE 5  Different mathematical representations can be used to interpret the photograph. The curve b is the graph of $x = d[u - 0.04 \sin(2\pi u)]$, $y = 0.115d[1 - \cos(2\pi u)]$. For each representation, the corresponding efficiency curve is given for Littrow's mounting.

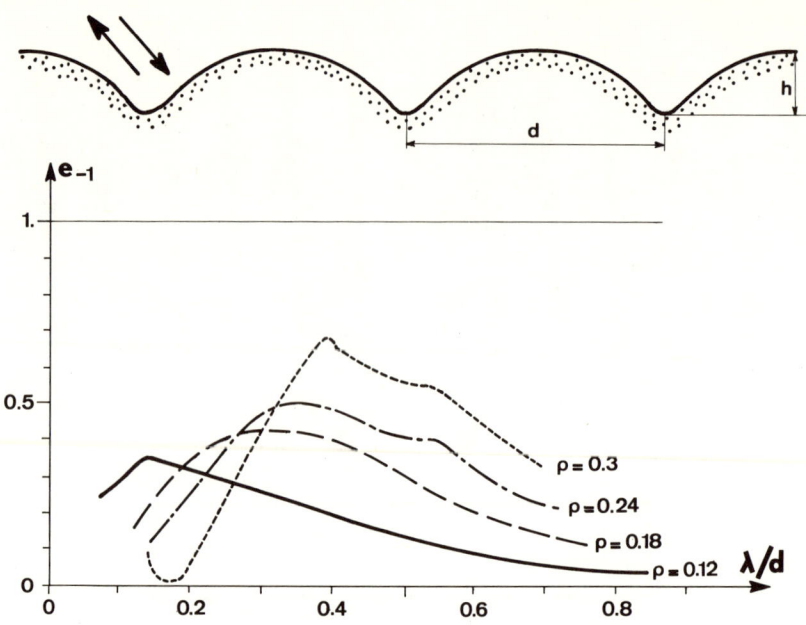

FIGURE 6  Influence of groove depth (Littrow's mounting, $E_\|$, $\rho = h/d$).

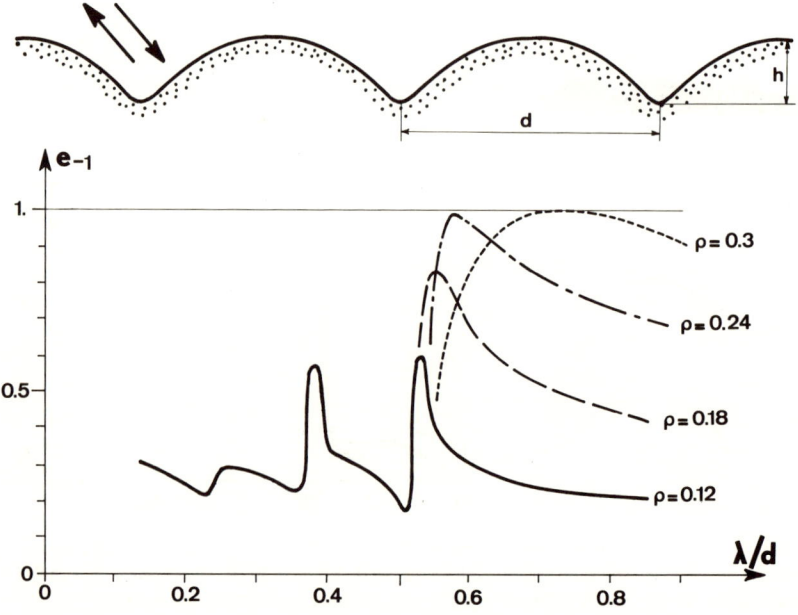

FIGURE 7  Influence of groove depth (Littrow's mounting, $H_\|$, $\rho = h/d$).

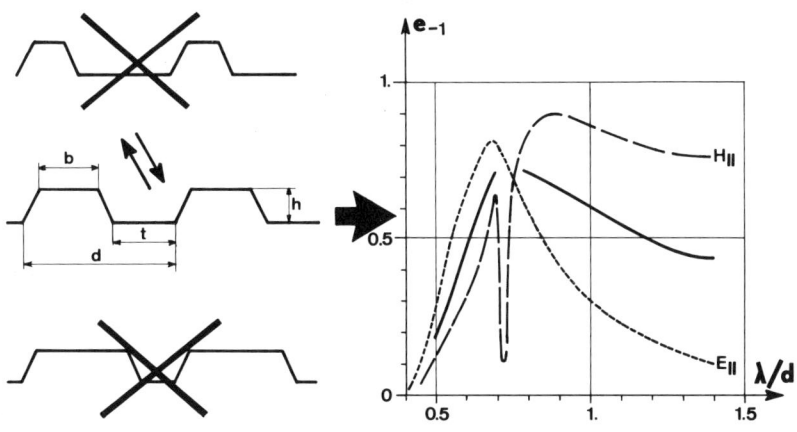

FIGURE 8 Influence of groove width. There is a cross on bad profiles. For the good one $b = t$, $\rho = h/d = 0.2$.

used for this purpose, and Figure 8 exhibits the best solution; in spite of an important but narrow Wood anomaly in the $H_\parallel$ case, an efficiency of 0.75 is expected with unpolarized light provided $\lambda/d$ is near 0.75.

People who want to use holographic gratings with ultraviolet radiation often ask the following question: Is it possible to get good efficiency when using, for example, a 3000 lines/mm grating and a 0.1-$\mu$m wavelength? Figure 9 allows us to reply in the affirmative. The groove depth must be small, and in any case $h/d$ must be less than 0.2; the efficiency is about 0.4 when $\lambda/d$ is 0.3 and $h/d$ is 0.08. Unfortunately, this conclusion is not valid for ultraviolet radiation, because in this spectral region aluminum (and, of course, oxidized aluminum) is not perfectly conducting.

## V. Efficiency of Conducting Gratings in Classical Working Conditions

The theory of conducting gratings is now in the process of being elaborated. We have recently proposed an integral method [Maystre and Vincent, 1972; Maystre, 1972], but numerical data are available only for large values of the conductivity $\sigma$. Figures 10 and 11 allow us to compare for visible radiation the ideal efficiency curve ($\sigma = \infty$) with the more realistic curve relative to a complex index $\nu + i_\chi$. There is nothing surprising in the $E_\parallel$ case; we get the second curve from the first one by multiplying the values of the ordinate by a coefficient that roughly corresponds to the reflectance of an aluminum mirror (Figure 10). On the contrary, in the other fundamental case of polarization (Figure 11), the

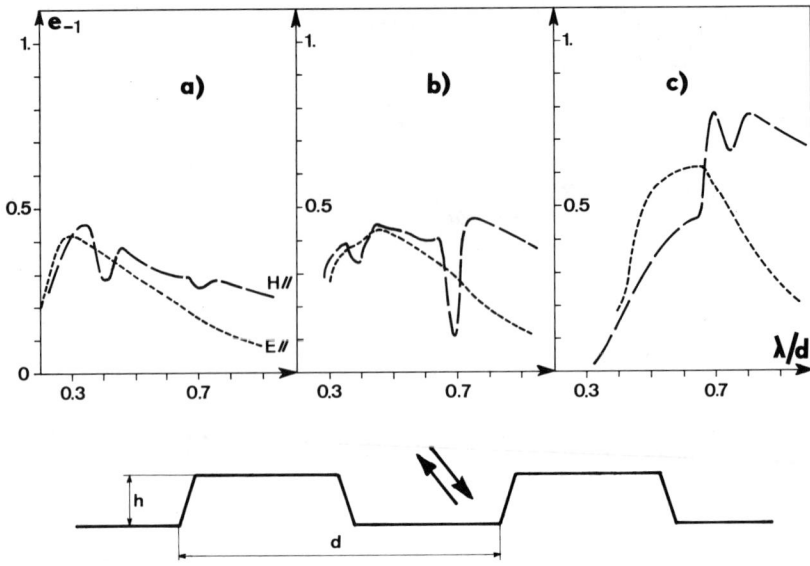

FIGURE 9 Obtention of a suitable efficiency for small values of $\lambda/d$ with Littrow's mounting. (a) $h/d = 0.082$. (b) $h/d = 0.104$. (c) $h/d = 0.157$.

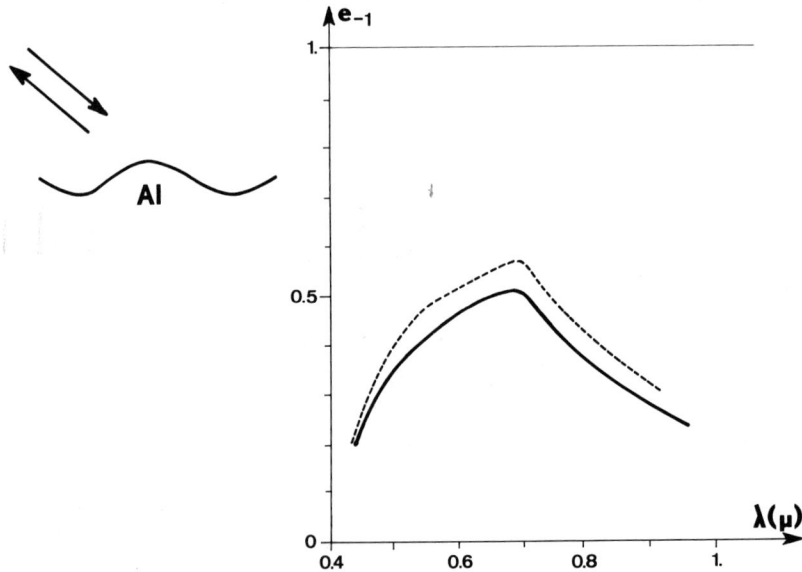

FIGURE 10 Efficiency of a 1000 lines/mm aluminum grating illuminated with visible radiation. $E_\parallel$ case. ——— $n = \nu + i_\chi$. ------ $\sigma = \infty$.

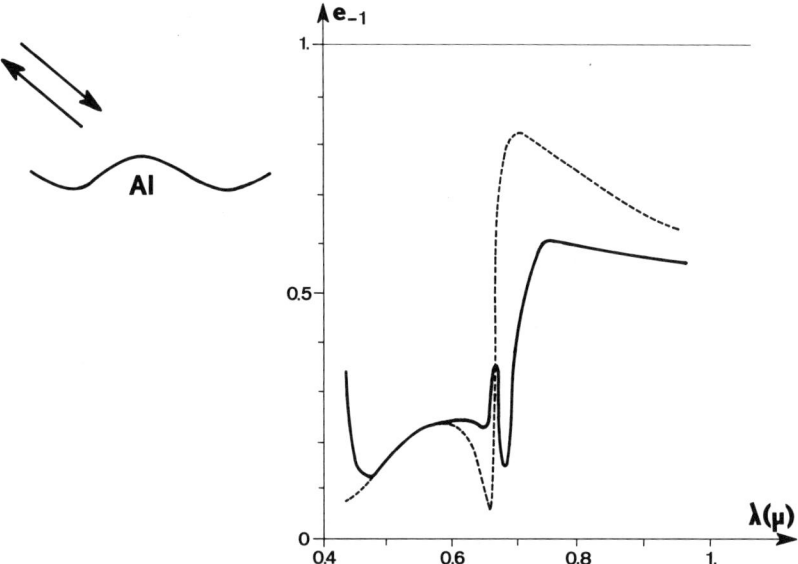

FIGURE 11 Efficiency of a 1000 lines/mm aluminum grating illuminated with visible radiation. $H_\parallel$ case. ——— $n = \nu + i_\chi$. ------ $\sigma = \infty$.

conductivity influence seems to be important (we say "seems" and not "is" because now only a few numerical data are available; a more systematic work is needed to draw a definite conclusion). Thus we hope we will be able to solve practical gratings problems for ultraviolet radiation very soon; but first we have to understand the effect of a dielectric coating (a natural aluminum oxide coating or an artificial protecting one).

## VI. Efficiency of a Coated Infinitely Conducting Grating

The differential method that we have published [Nevière et al., 1972] is not quite applicable to the problem we spoke of at the end of the preceding paragraph because the method uses an infinite conductivity. But it can predict the efficiencies of a dielectric-coated grating illuminated by visible radiation because, in this case, aluminum can be considered as infinitely conducting. Figure 12, extracted from Nevière et al. [1972], shows that it seems to be possible, even with unpolarized light, to double the "blaze width" using an adequate deposit of magnesium fluoride.

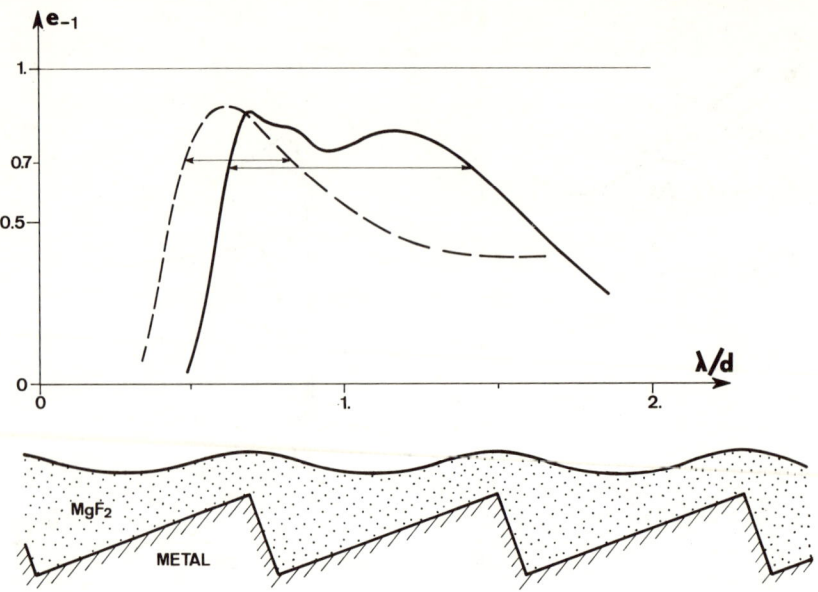

FIGURE 12 Influence of a dielectric coating on Littrow's mounting. ———— Coated grating. ------ Uncoated grating. The metal conductivity is infinite; the index of $MgF_2$ is 1.38. For the coated grating the blaze width corresponds to an octave interval.

## VII. Efficiency of a Holographic Grating Working in "Conical Diffraction"

If, in classical working conditions, the efficiencies are *known for any incidence and for the both fundamental cases of polarization*, it is very easy, as we mentioned [Petit and Maystre, 1972], to obtain the efficiency $e_n$, which, for a given polarization, corresponds to a wave vector that is not orthogonal to the grooves. It is not necessary to write new computer programs; a very simple formula is sufficient [Petit and Maystre, 1972], and this is why both the $E_\parallel$ and $H_\parallel$ cases are called *fundamental cases*. This formula shows that what we called classical working conditions perhaps are not always the best ones in order to have at the same time a great efficiency and a great blaze width. Because reading theoretical papers dealing with "conical diffraction" is a tedious chore from which we wish to save the reader, we would like to point out only one important but not intuitive theorem:

$\mathcal{E}$ being the set of k vectors the projection of which on the $x0y$ plane is a fixed vector, then for a given $n$:

(a) The projection on the same $xOy$ plane of the $n$th order diffracted wave vector $\mathbf{k}_n$ is also a fixed vector;

(b) The efficiency $e_n$ (for the $n$th order) is constant provided the incident-wave polarization is fixed.

If we remember [Maystre and Petit, 1971a] that, for classical working conditions, $e_{-1}$ is very close to unity for both fundamental cases of polarization, provided that in Littrow's mounting $\mathbf{k}$ is perpendicular to the large facet of an echelette grating, the preceding theorem allows us to state the following proposition:

If, as it is assumed on Figure 13, $\mathbf{k}$ is parallel to the small facet plane and, moreover, if $\lambda/d = 2 \sin \alpha \cos \phi$, there is a specular reflection of the incident wave on the large facet plane. This plane works as a plane mirror the high reflectance of which is very close to unity and does not depend on the wavelength. Therefore, it seems to be possible to imagine grating *spectrometers or monochromators the transmission of which is very good and does not depend on the wavelength.* Figure 14 describes

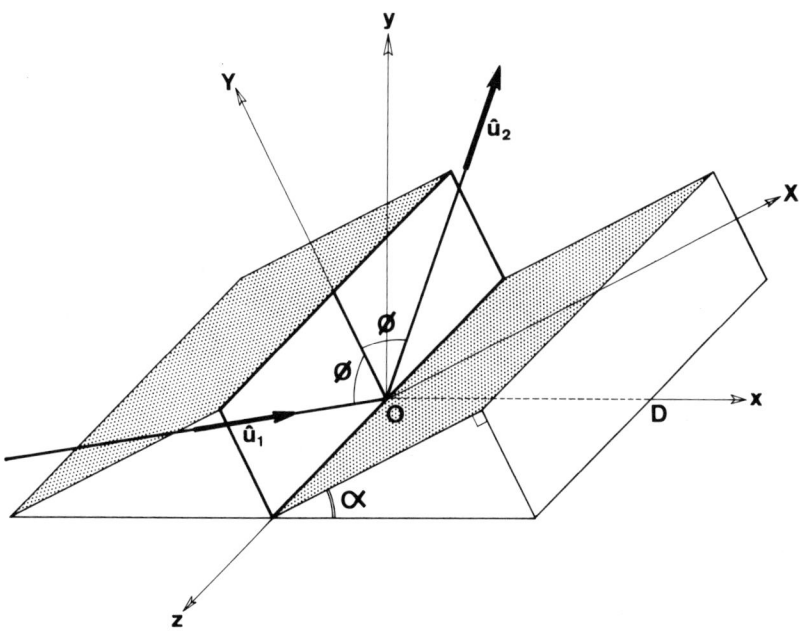

FIGURE 13 Grating configuration with $\mathbf{k}$ parallel to the small facet plane.

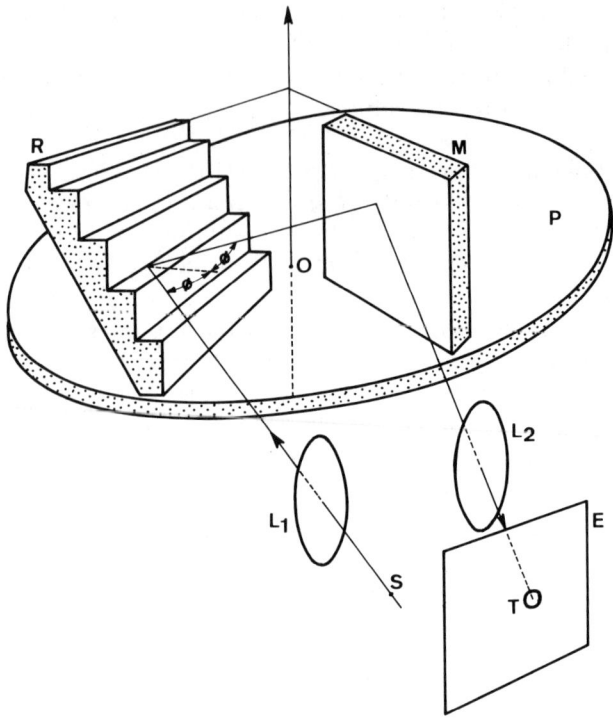

FIGURE 14 Basic scheme of a spectrometer the light transmission of which is not dependent on wavelength.

such a grating arrangement. The lenses $L_1$, $L_2$ and the opaque screen E are fixed, S is a polychromatic light source situated at the focus of lens $L_1$. The grating R and the mirror M are fixed to the stand P. When P rotates about the vertical axis, $\phi$ varies and the hole T is illuminated by different wavelengths. Slight modifications are necessary if we wish to use a slit source rather than a point source [Maystre and Petit, 1972a].

For a long time we thought the arrangement we have just described to be a new one; but we have been told recently that it was described more than ten years ago (U.S. Patent 3,069,967, December 9, 1959). To our knowledge it is not yet often used, and we think it is a great pity because both its geometrical and energy properties indeed look excellent. Moreover, it also works for a holographic grating if **k** is parallel to a convenient plane that we are able to specify.

## VIII. Grating Bandpass Filter

In optics, gratings are mainly dispersing agents, but it is known that they can be used also as wavelength filters. We suggested [Maystre and Petit, 1972b] a grating arrangement that, with polarized light, works as a bandpass filter the transmission of which is very close to unity and the bandwidth of which is tunable. It is an application of numerical data reported on Figure 15 and of some theoretical reciprocity relations [Maystre and Petit, 1972b]. Its basic principle is explained in Figure 16 after the curve of Figure 15 has been idealized. We mentioned [Maystre and Petit, 1972b] how the device described in Figure 17 avoids angular dispersion, which exists with the simplest device of Figure 16.

## IX. Conclusion

We hope that this short paper exhibits the practical utility of applied mathematics in electromagnetic optics. Theoretical considerations have already contributed to improve gratings utilization and also gratings production through a friendly collaboration with the gratings department

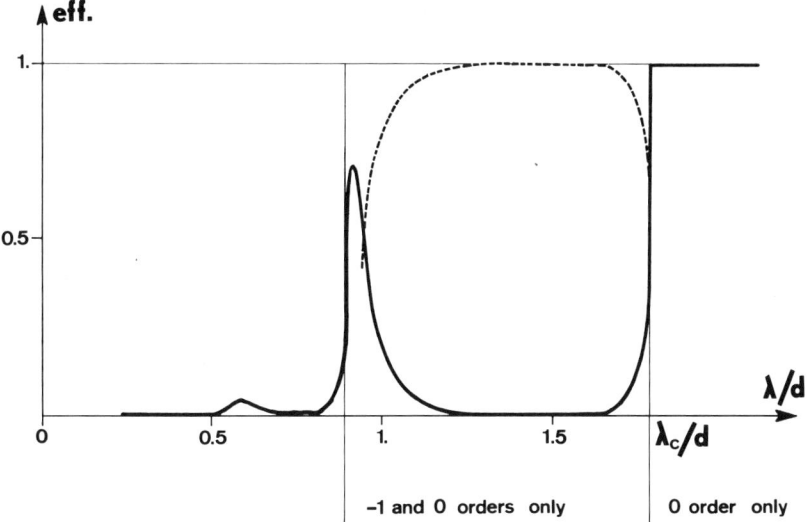

FIGURE 15 Efficiencies for a fixed incidence ($\theta = 52°$) in the $H_\parallel$ case: ——— zero order; ------ $-1$ order. $\lambda_c/d = 1 + \sin \theta$, $= 38°$. Notice the important Wood's anomaly when $-2$ order appears.

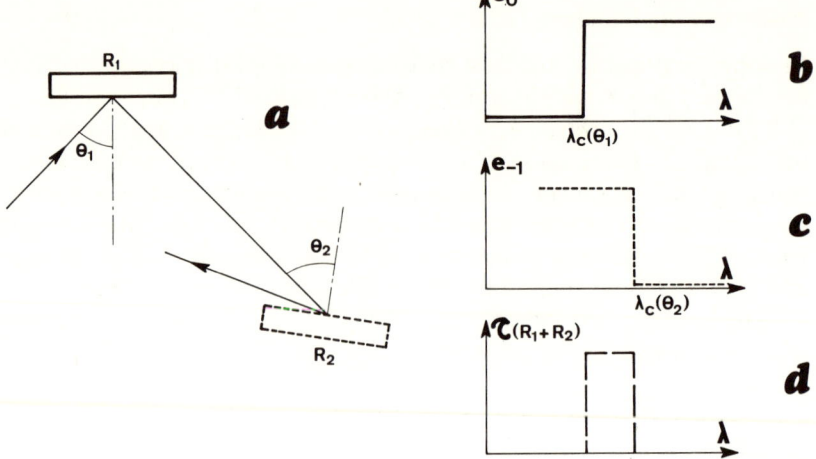

FIGURE 16 (a) basic scheme, (b) efficiency of $R_1$ working in zero order, (c) efficiency of $R_2$ working in $-1$ order, (d) transmission of $R_1 + R_2$. The direction of transmitted wave is dependent on $\lambda$.

of the French firm Jobin-Yvon. Either for ruled or holographic gratings arrangement, important help can be expected from electromagnetic theory by everyone needing great luminosity. Unfortunately, we must point out an important restriction: gratings must be plane gratings because, to the authors' knowledge, the electromagnetic theory of concave gratings has not yet been elaborated.

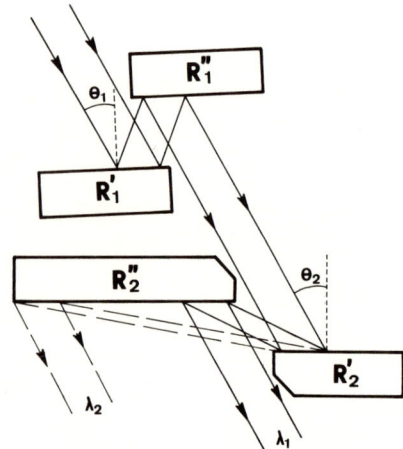

FIGURE 17 With this arrangement we have proposed that the transmitted wave direction is not dependent on $\lambda$.

## References

Bousquet, P., L. Capella, A. Fornier, and J. Gonella, *Appl. Opt. 8,* 1229 (1969).
Chauvineau, J. P., L. Constanciel, A. Marraud, and R. Petit, *Rev. Opt. 46,* 417 (1967).
Maystre, D., *Opt. Commun. 6,* 50 (1972).
Maystre, D., and R. Petit, *Opt. Commun. 2,* 309 (1970).
Maystre, D., and R. Petit, *Nouv. Rev. Opt. Appl. 2,* 115 (1971a).
Maystre, D., and R. Petit, *Opt. Commun. 4,* 25 (1971b).
Maystre, D., and R. Petit, *Opt. Commun. 5,* 35 (1972a).
Maystre, D., and R. Petit, *Opt. Commun. 4,* 380 (1972b).
Maystre, D., and P. Vincent, *Opt. Commun. 5,* 327 (1972).
Nevière, M., M. Cadilhac, and R. Petit, *Opt. Commun. 6,* 34 (1972).
Nevière, M., M. Cadilhac, and R. Petit, *IEEE Trans. Antennas Propag. AP-21,* 37 (1973).
Petit, R., and D. Maystre, *Rev. Phys. Appl. 7,* 427 (1972).

G. C. RIGHINI, V. RUSSO, S. SOTTINI, and
G. TORALDO DI FRANCIA

# Thin-Film Geodesic Lenses

## I. Introduction

Integrated optics has many attractive features for applications in laser-beam guidance and optical signal processing in compact form. Active and passive components of two-dimensional optical circuitry are obtained by means of thin films, capable of guiding light.

Some authors have already suggested different techniques to build two-dimensional lenses, which are one of the basic components in two-dimensional processors. Shubert and Harris [1968, 1971] suggested shaped structures of different refractive indexes either inserted in the film or deposited on top of the main film. Ulrich and Martin [1971] tested thin-film lenses in which the velocity of the guided light was varied locally by properly shaping the thickness profile of the film. In all these cases the curved boundary of the lens must be sufficiently sharp.

Optical systems for guided waves can also be obtained by extending to thin-film optics the principle of configuration lenses already studied for application to microwave antennas.

---

The authors are at the Istituto di Ricerca sulle Onde Elettromagnetiche del C.N.R., Firenze, Italy.

Here we present some geodesic (or configuration) lenses constructed and tested in the optical region for application to integrated optics.

## II. Configuration Lenses

Let us briefly recall the working principle of a "configuration" lens.

It is well known that Maxwell's fisheye, shown in Figure 1(a), represents a perfect optical system. It is a variable-index lens with spherical symmetry. The Luneberg lens, shown in Figure 1(b), is another perfect

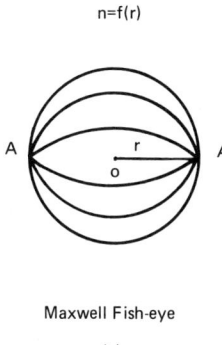

Maxwell Fish-eye

(a)

FIGURE 1 (a) Maxwell fisheye $[n = 2/(1 + r^2)]$ : a point source A is perfectly imaged at A'. (b) Luneberg lens $[n = (2 - r^2)^{1/2}]$ : a point source A located at infinity is perfectly imaged at A.

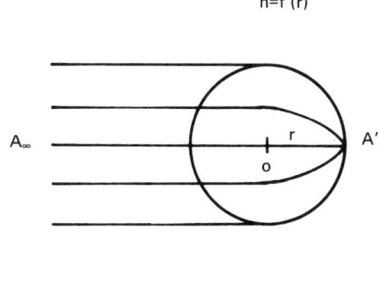

Luneberg Lens

(b)

FIGURE 2 Rinehart lens: the point source A is perfectly imaged at infinite distance on the plane rim.

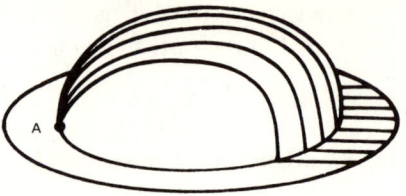

Rinehart Lens

optical system with different properties, due to a different distribution of refractive index. Consider now a plane where the refractive index distribution corresponds to one of the preceding optical systems. By recalling Fermat's principle, it is possible to find a two-dimensional non-Euclidean space having the same optical properties as the preceding variable-index planar lens. The rays follow the geodesics of this space. Focusing depends on the relative paths along the different rays.

It is readily found that the two-dimensional analog of the fisheye is a spherical surface. Here all great circles through a point intersect each other at the diametrically opposite point.

The two-dimensional analog of the Luneberg lens is the so-called Rinehart lens [Rinehart, 1948], shown in Figure 2.

A generalization made independently by Kunz [1954] and Toraldo [1955a] showed that a family of perfect configuration lenses exists, of which the Rinehart lens is only a particular case.

## III. Thin-Film Lenses

Most configuration lenses were built in the microwave region for application to high-speed scanning.

A two-dimensional Riemann space (surface) can be easily achieved in the case of microwaves. Two parallel metal plates suitably bent constrain microwaves to travel along a given surface.

Now geodesic lenses can be made also in optics and can find application in thin-film optical circuitry.

A dielectric thin film deposited on a curved substrate of different refractive index can constitute a two-dimensional Riemann space. If the thickness of the film is sufficiently small, the propagation can be considered to occur along the mean surface of the film. The rays are the geodesics of the mean surface.

At first, the simplest geodesic lens was constructed and tested [Righini et al., 1972]. It is a quarter of a spherical surface, which can be used as a two-dimensional focusing element.

The lens is made up of a spherical glass covered by a thin epoxy film (Araldite MY 757 CIBA), doped with Rhodamine B in order to make the path of the guided light evident. The film was deposited from a liquid solution by slow evaporation of the solvent.

Figure 3 shows a plane laser beam (2-cm diameter) perfectly focused at a point located a quarter of the great circle apart from the input edge. Figure 4 shows two plane-parallel laser beams focused at the same point.

The coupling at the input edge is easily obtained by roughly tapering the dielectric film. More efficient coupling would be obtained by means of a curved holographic grating. The only disadvantage of the spherical lens is that it cannot be easily inserted in planar circuits. Then we started investigating lenses having planar input and output.

The lens introduced by one of us [Toraldo, 1957; Scheggi and Toraldo, 1960] has been considered; it has the same properties of the Rinehart

FIGURE 3 A geodesic lens, constituted by a quarter of a spherical surface, focusing a plane laser beam.

FIGURE 4  Two plane and parallel laser beams focused at the same point by the geodesic lens.

lens without presenting any discontinuity for the tangent plane. This lens is perfect on almost the entire aperture.

Figure 5 shows the meridional curve of the lens with the parameters $a$ and $b$. By choosing different values for $a$ and $b$, it is possible to obtain lenses with different apertures and different curvature radii of the toroidal junction.

Figure 6 shows the shaped glass substrate of the lens we have built. The linear aperture is 6 cm.

Figure 7 shows a plane beam focused at the opposite point.

Figure 8 shows two parallel beams focused on the predicted focal line. Here the conventional technique of the prism-film coupling is used.

A lens of this type can be easily inserted in planar film circuits. It is evident that the lens can be constructed as a protrusion with respect to a planar film or a depression in the planar film (Figure 9). In this connection we may mention that Van Duzer [1970], dealing with acoustic waves, described with a different approach the focusing properties of a depression in the substrate.

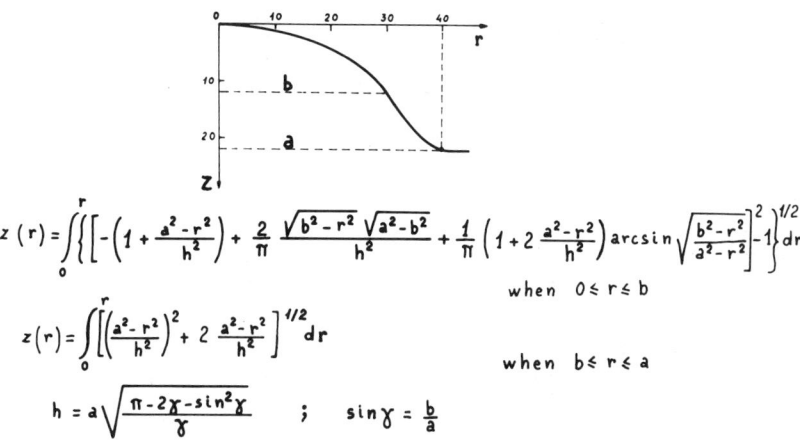

$$z(r) = \int_0^r \left\{ \left[ -\left(1 + \frac{a^2-r^2}{h^2}\right) + \frac{2}{\pi} \frac{\sqrt{b^2-r^2}\sqrt{a^2-b^2}}{h^2} + \frac{1}{\pi}\left(1 + 2\frac{a^2-r^2}{h^2}\right)\arcsin\sqrt{\frac{b^2-r^2}{a^2-r^2}} \right]^2 - 1 \right\}^{1/2} dr$$

when $0 \leq r \leq b$

$$z(r) = \int_0^r \left[ \left(\frac{a^2-r^2}{h^2}\right)^2 + 2\frac{a^2-r^2}{h^2} \right]^{1/2} dr$$

when $b \leq r \leq a$

$$h = a\sqrt{\frac{\pi - 2\gamma - \sin^2\gamma}{\gamma}} \quad ; \quad \sin\gamma = \frac{b}{a}$$

FIGURE 5  Meridional curve of the Toraldo lens.

Optical systems of very simple construction can also be designed, having as a basic element a *conflection lens.*

In principle, a conflection lens is constituted by two coaxial cones joined together along a common section. The particular case of a converging lens is shown in Figure 10, where the two cones have been developed on the same plane. When meeting the section, the rays will undergo a change of direction. The quantitative law of conflection, which can be obtained most readily by applying the Fermat principle, is very simple. It states that the angle of incidence and the angle of conflection are equal.

FIGURE 6  The shaped and polished glass substrate of the Toraldo lens. Radius $a$: 4 cm, radius $b$: 3 cm, linear aperture: 6 cm.

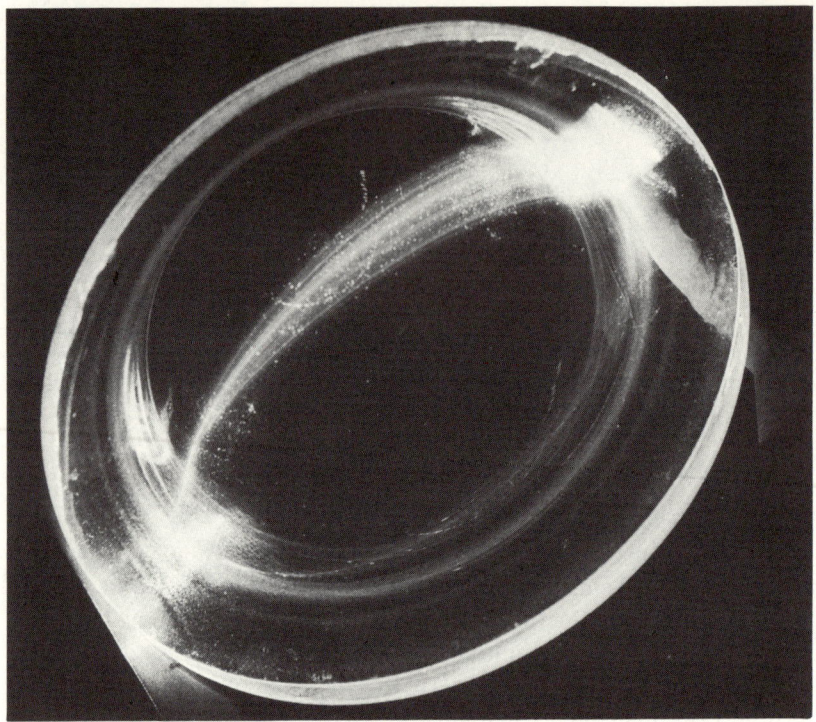

FIGURE 7  Toraldo lens focusing a plane beam.

It was found by Toraldo [1955b] that this very simple system behaves like a thin lens of classical optics with refractive index $n = 0$. The third-order spherical aberration of a conflection lens was also evaluated. Then it was possible to combine one converging lens and one diverging lens in such a way as to obtain an optical system corrected for third-order spherical aberration. It was called a *conflection doublet*.

Figure 11 shows a particular conflection doublet developed on a plane. A conflection doublet with the specifications shown in Figure 12 has been constructed and tested. Figure 13 shows a plane beam impinging on the outer disk; the beam is focused on the inner disk. The system is corrected for third-order spherical aberration.

The glass substrate of the lenses has been shaped and polished in the laboratory. Surface irregularities due to an imperfect polishing are mainly responsible for the losses that are evident in some lenses.

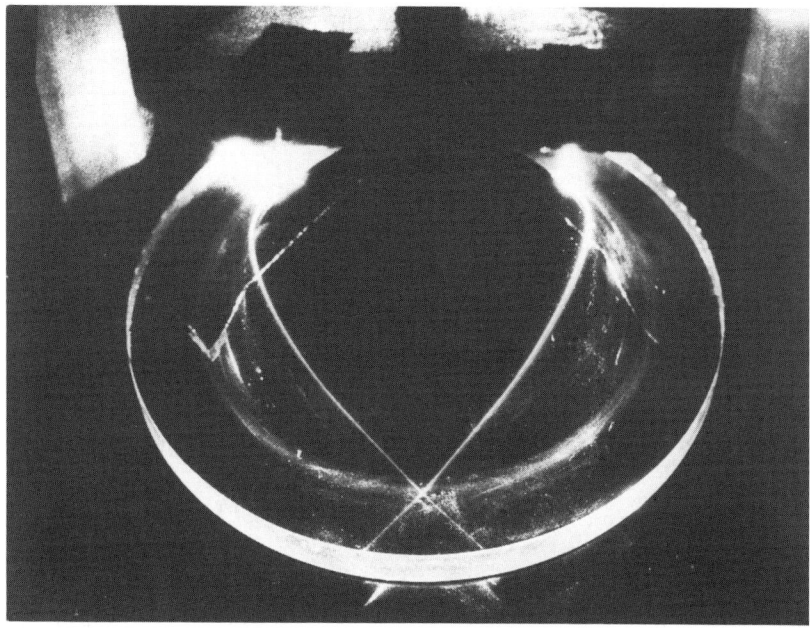

FIGURE 8  Two parallel beams focused on the predicted focal line. The prism–film coupling is used.

FIGURE 9  The same geodesic lens constructed either as a protrusion with respect to a planar film (a), or as a depression in the planar film (b).

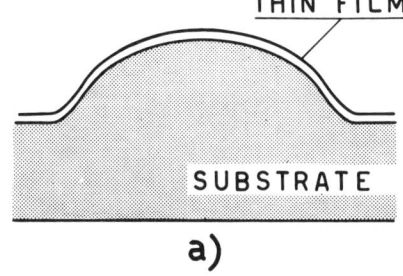

a)

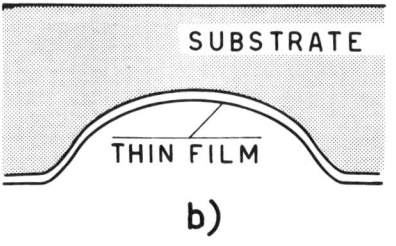

b)

690  OPTICAL TECHNOLOGY

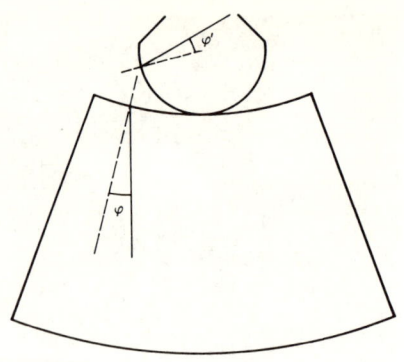

FIGURE 10 Two coaxial cones having a common section, developed on the same plane. The system behaves like a thin lens of classical optics with refractive index $n = 0$.

Conflection Law
(Converging Lens)

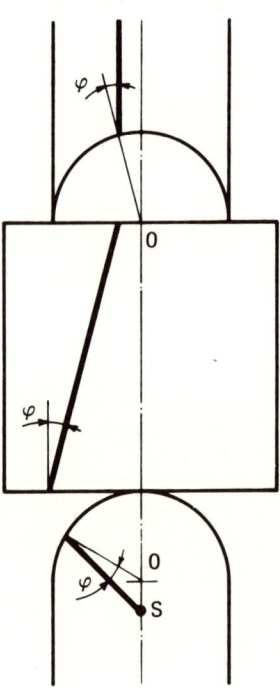

FIGURE 11 Development of a conflection doublet on a plane.

Conflection Doublet

FIGURE 12 Cross section and perspective view of a conflection doublet where a small amount of negative third-order spherical aberration has been left. The linear aperture is $\frac{4}{3}R$ with a maximum aberration of $\sim 1°$. $AB$ is the input plane (collimated beam). $F$ is the focus.

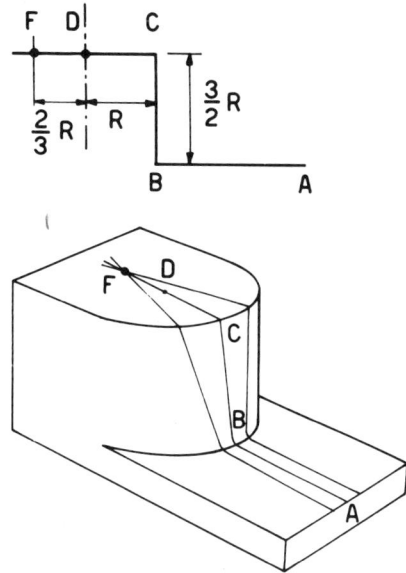

FIGURE 13 A conflection doublet ($R$ = 2 cm) focusing a collimated beam.

## IV. Conclusions

Geodesic lenses have been constructed and tested that represent an alternative to the more conventional techniques of building thin-film lenses. In addition, they have the advantage that the substrate can be prepared in advance and with the accuracy of glass optics. Then the dielectric deposition can be made with the same method as for the more conventional components of the optical circuit.

The authors would like to thank Dr. Raimondi of CIBA-GEIGY for providing the epoxy material and D. Pucci of the Laboratorio di Elettronica Quantistica for construction of the lenses substrate.

### References

Kunz, K. S., *J. Appl. Phys. 25,* 642 (1954).
Righini, G. C., V. Russo, S. Sottini, and G. Toraldo di Francia, *Appl. Opt. 11,* 1442 (1972).
Rinehart, R. F., *J. Appl. Phys. 19,* 860 (1948).
Scheggi, A. M., and G. Toraldo di Francia, *Alta Freq. 29,* 438 (1960).
Shubert, R., and J. H. Harris, *IEEE Trans. Microwave Theory Tech. MTT-16,* 1048 (1968).
Shubert, R., and J. H. Harris, *J. Opt. Soc. Am. 61,* 154 (1971).
Toraldo di Francia, G., *Opt. Acta 1,* 157 (1955a).
Toraldo di Francia, G., *J. Opt. Soc. Am. 45,* 621 (1955b).
Toraldo di Francia, G., *Atti Fondaz. Ronchi 12,* 151 (1957).
Ulrich, R., and R. J. Martin, *Appl. Opt. 10, 2077* (1971).
Van Duzer, T., *Proc. IEEE 58,* 1230 (1970).

R. A. SCHOWENGERDT and P. N. SLATER

# Determination of the In-Flight Optical Transfer Function of Orbital Earth Resources Sensors

## I. Introduction

The past few years have seen an increasing interest in the worldwide assessment of natural resources and the detection of environmental pollution. A common characteristic of many instruments used for such purposes is that they monitor radiation reflected from, or emitted by, large areas of the earth's surface, in different parts of the electromagnetic spectrum. Frequently, the output from these instruments, which are referred to as multispectral remote sensors, is converted into a photographic image for analysis purposes. For example, the density function of the image may be digitized with a microdensitometer. The resulting values are related to ground reflectances (not a straightforward task), which are used as an aid in the production of thematic maps from the imagery [Park, 1972].

One subject of practical interest to those analyzing imagery from orbiting spacecraft is the quality of the imagery, which is expressed in terms of spatial resolution and spectroradiometric accuracy, quantities

---

The authors are at the Optical Sciences Center, University of Arizona, Tucson, Arizona 85721.

that are related and equally important in remote-sensor imagery. The blurring of the object, which occurs in any image, decreases the accuracy of spectroradiometric calculations on microimage areas, particularly when the image modulation is decreased to the point where it becomes indistinguishable from noise.

We are concerned here with techniques for measuring the quality of operational imagery and in particular with a method that is uniquely suited to the characteristics of orbiting multispectral sensors. In the next section, several of these characteristics will be discussed from the viewpoint of their importance to the image-evaluation problem.

## FACTORS INFLUENCING THE SELECTION OF AN IMAGE-EVALUATION METHOD

In selecting an in-flight image-evaluation method we first have to take into account the unique characteristics of both orbital multispectral sensors and the imagery they produce. Multispectral sensors form several images of the ground scene simultaneously through broadband spectral filters or dispersive elements. Now, in general, the spatial distribution of scene radiance will be different from band to band. Thus, the edge between two fields may be a good step function in a red band, but, because of sparse vegetation near the edge, it may be a poor step function in a green band. Consequently, a given object, particularly a naturally occurring one, may not be suitable for evaluating the image in all bands of the sensor. In addition, wavelength-dependent scattering of light in the atmosphere will reduce the modulation of the image by different amounts in each band. The signal-to-noise ratio will therefore vary from band to band even if the image recording components in each band are identical. Moreover, the optical system(s) used in the sensor will generally have different imaging characteristics in each band because of the dependence of aberrations on wavelength.

The low ground resolution typical of these sensors bears directly on the choice of an image evaluation method. Table 1 compares the resolution of low-contrast, three-bar ground targets for past, current, and future systems [Colvocoresses, 1972; Slater, 1972]. As we will discuss further in the next section, these values generally rule out the possibility of utilizing man-made test targets.

Finally, in all earth remote-sensing programs involving spaceborne sensors, for example, those onboard NASA's Earth Resources Technology Satellite (ERTS), simultaneous underflight photography is scheduled regularly. The imagery from these underflights is used as an aid for calibration of spacecraft data in terms of ground measurements. The air-

TABLE 1  Resolution of Earth-Orbiting Remote Sensors

| Sensor | Approximate Ground Resolution | |
|---|---|---|
| | m/line pair | line pair/km |
| Apollo 9 S 065 experiment | | |
| (4 Hasselblad cameras) | 100 | 10 |
| ERTS-1 (Earth Resources Technology Satellite) | | |
| RBV (Return Beam Vidicon) | 180–280 | 3.5–5.5 |
| MSS (Multispectral Scanner) | 300 | 3.5 |
| Skylab S-190 experiment | | |
| (6-lens Itek camera) | 20–100 | 10–50 |

craft sensors usually use the same spectral bands as those in the spacecraft and in some cases duplicate systems are under construction [Forkey and Womble, 1972]. Simultaneous underflights are flown from low altitudes of a few hundred meters to very high altitudes of 15 to 20 km. The imagery from these underflights is necessary for the image-evaluation technique discussed in this paper.

REVIEW OF CURRENT IN-FLIGHT IMAGE-EVALUATION TECHNIQUES

Sensor imaging capabilities can be predicted at the design stage and measured in the laboratory for complete systems. However, sensor performance cannot be predicted accurately and reliably for an extended operational period in the space environment. Imaging systems carried by aircraft are often evaluated in flight by the use of the three-bar resolution type of ground target. In this discussion we are concerned, however, with a more complete analysis that extends to the measurement of the optical transfer function (OTF), which is symbolized by $\tau(\bar{f})$, where $\bar{f}$ is a (possibly) two-dimensional spatial frequency variable.

Measurement of $\tau(\bar{f})$ for in-flight sensors has been achieved with the use of special objects, such as man-made edges [Roetling et al., 1969] or lines [Hendeberg and Welander, 1963] and their naturally occurring counterparts in the form of coast lines, field boundaries, lunar crater edges [Mazurowski and Kinzly, 1969], etc. The use of naturally occurring targets has several limitations. Ideal edges and lines do not occur in nature, and reasonable facsimiles are often of unknown quality. As mentioned earlier, a given target may not be suitable for the evaluation of all the bands in a multispectral sensor. Furthermore, the low ground resolution typical of many of these sensors sets a severe requirement on the minimum size of both natural and man-made targets. Consider a sensor with a 100-m/cycle ground resolution, and let that distance correspond roughly to the half-width of the central lobe in the sensor spread function. Then, if we want to measure the first or second side lobes of the spread function, the length of the target in any given direction must be at least 200 to 300 m and at least that long in the perpendicular direction. Naturally occurring objects that are large and straight over that length would be difficult to find, and deployment and maintenance of such large man-made targets would be difficult if not impossible. Even if such an object was used, its position and orientation in the field of view would be unique, and, consequently, its use would be limited.

The technique we will describe can be applied to any imagery for which there is simultaneous underflight coverage, and it does not have any direct dependence on the nature of the object. Consequently, it is of more practical value than an analysis using isolated targets.

## II. Theory

The fundamental imaging equation for linear, stationary optical systems is

$$I(\bar{f}) = \tau(\bar{f})O(\bar{f}),$$

where $I(\bar{f})$ and $O(\bar{f})$ are the image and object spatial spectra, respectively. In general, all quantities in this equation are complex.

To measure $\tau(\bar{f})$ it is necessary to know $I(\bar{f})$ and $O(\bar{f})$. As discussed above, $O(\bar{f})$ is not known for naturally occurring objects. Man-made targets are often used because $O(\bar{f})$ is then known and $I(\bar{f})$ can be measured from the imagery. Now the simultaneous underflight imagery obtained in multispectral sensor experiments gives us a good measure of $O(\bar{f})$ for any part of a scene. The scale factor between the underflight imagery and the spacecraft imagery indicates that we need measure only very low spatial frequencies in the underflight image and then scale these up to the correspondingly higher frequencies in the spacecraft image to evaluate $\tau(\bar{f})$. For example, if the cutoff frequency (assuming noiseless imagery) of the spacecraft sensor OTF is 50 cycles/mm and the aircraft underflight sensor is of the same focal length and flown at an altitude 1/10 of that of the spacecraft, frequencies up to only 5 cycles/mm need to be measured in the underflight image. To determine $O(\bar{f})$, the OTF for the aircraft sensor should be divided into the underflight image spectrum, but the highest frequency of interest, which in the above case is 5 cycles/mm, may be so low that this correction is unnecessary.

In practice, the two sets of images from the spacecraft and the simultaneous underflight can be scanned and digitized with a microdensitometer in either one or two dimensions. The same ground area is scanned in each set of images, and the scanning aperture size and sampling rate are scaled by approximately the scale between the images. Because of the scale factor, the aperture size is large for the underflight image. In the previous example, the aperture size would be about 100 to 200 $\mu$m. Photographic grain noise is thus a minor problem in the measurement of $O(\bar{f})$. For one-dimensional scans, a slit aperture can be used to reduce the grain noise even further.

Now, the spacecraft image scan should not be longer than the size of an isoplanatic, or stationary, region to ensure that $\tau(\bar{f})$ is essentially constant over the scan length. Because the same ground area is scanned in each of the two images, the length of the underflight image scan is longer than the spacecraft image scan. Thus, the underflight image scan may extend over a significant part of the field (say 5° to 10°), and care should be taken that this scan also does not extend outside an isoplanatic region. However, the restriction to low frequencies in this image means that the underflight sensor OTF, in this frequency range, will likely be constant over the scan length.

In addition to the sensitometric conversion from film density to effective image irradiance for all data, the underflight image data should be corrected for $\cos^4$ falloff in irradiance off-axis.

Distortion in the underflight image owing to topographic elevation differences on the ground should be considered. The positional distortion $\Delta r$ for an image point at a distance $r$ from the center of the image is given by

$$\Delta r = r(\Delta H/H),$$

where $\Delta H$ is the difference in ground elevation of the on-axis object point and the point imaged at $r$, and $H$ is the aircraft altitude. For $H = 20$ km, $\Delta H = 100$ m, and $r = 10$ mm, we have $\Delta r = 0.05$ mm, which is less than the required microdensitometer aperture size mentioned earlier and would be considered negligible. For each scan, however, it would be prudent to check topographic maps of the area, estimate the distortion from elevation differences, and, if necessary, apply a correctional transformation to the data.

Because it is unlikely that the aperture size and sample interval could be scaled exactly on the microdensitometer, correction for aperture and microdensitometer OTF and exact scaling of the data must be done on the digitized data in a computer. A technique for scaling that has been successful is to start at the same ground point in both the spacecraft and underflight image, take the same number of points in each set of data but with the sample interval on the underflight image chosen as close as possible to the scale factor times the sample interval on the spacecraft image, and stretch or shrink the underflight image in consecutive steps by a linear interpolation scheme, which keeps the number of points constant. The integrals

$$\text{mean squared difference} = \int [o(x) - i(x)]^2 \, dx \Big/ \int [o(x)]^2 \, dx,$$

$$\text{correlation factor} = \int o(x) \, i(x) \, dx \Big/ \int [o(x)]^2 \, dx$$

are evaluated for each step of the stretching or shrinking process. A minimum will appear in the mean-squared difference between object and image at some scale factor, and a maximum will appear in the correlation factor, usually at the same scale factor. We thus have two independent criteria for determining the scale factor. In addition, by using this procedure, the same number of real points is obtained in each set of data, which allows us to use a fast Fourier transform (FFT) routine that performs two real transforms simultaneously, an efficient use of the FFT algorithm.

After correction for microdensitometer OTF, sensitometry, and scaling, the data are Fourier transformed, and the ratio of corresponding spectral values gives the OTF of the spacecraft sensor. Now, in any procedure that involves sampled data and calculation of spectra, the spectra are replicated in the frequency domain at intervals of $1/\Delta x$, the sample interval. If $\Delta x$ is too large, overlap of the spectra may occur, which results in aliasing [Blackman and Tukey, 1959], i.e., high frequencies appearing as lower frequencies. We would expect aliasing to be most severe in the underflight image data, where large values of $\Delta x$ are used. However, the microdensitometer aperture is also large and consequently serves to reduce the modulation of higher frequencies and thus also the aliasing. Using underflight data from the Apollo 9 S 065 experiment, we have determined the aliasing errors in Table 2 for *one particular* image spectrum. The same set of data was used but was sampled at different intervals. The error was measured only for frequencies below the first zero, $f_c$, of the scanning aperture OTF. In this example, the phase errors occurred only in the region of 2.5 to 5 cycles/mm.

Finally, we note that the low ground resolution and the large final product format sizes (S 065, 70 mm; ERTS, 24 cm) typical of orbital multispectral images means that the requirements placed on microdensitometry by the above technique are not severe. For example, in evalu-

TABLE 2  Aliasing Error

|  |  |  | Maximum Aliasing Error | |
|---|---|---|---|---|
| $f_c$ | $\Delta x$ (mm) | $1/\Delta x$ (cycles/mm) | Modulus | Phase |
| 5 cycles/mm | 0.012 | 83 | Assumed zero | Assumed zero |
|  | 0.024 | 42 | 5% | 10% |
|  | 0.048 | 21 | 5% | 50% |
|  | 0.096 | 11 | 5% | 50% |

ating the S 065 system, aperture sizes of 0.02 mm × 0.1 mm and 0.2 mm × 1.0 mm and sample intervals of 0.006 mm and 0.06 mm were used on the spacecraft and underflight imagery, respectively.

### III. Examples of Data from Apollo 9 S 065 Evaluation

Figure 1 shows microdensitometer scans of the image of the same ground area in each of three bands: BB (green filter, Pan-X film), CC (near-ir filter, black-and-white ir film), and DD (red filter, Pan-X film). The curves illustrate some of the statements made earlier. For example, the modulation in the BB band is the lowest of the three, which is due to atmospheric scattering and to low modulation of the object in the green band (the image was of southern Arizona). Also note that grain noise in the ir band is more prominent than in the other two bands because of the high granularity of the ir film.

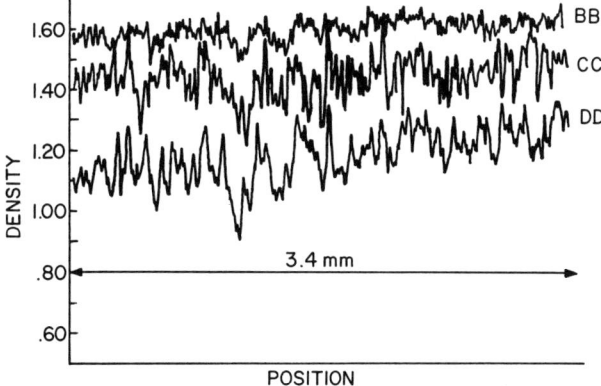

FIGURE 1  Image scans, spacecraft imagery.

Figure 2 is a plot of the mean-squared difference and correlation factor between the underflight (object, $o$) and spacecraft (image, $i$) scans (DD band) as a function of scale factor. It can be seen that a scale factor of about 10.7 gives the best match between object and image. The curves indicate that the two criteria for matching are sensitive to the scale factor, and it is expected that an accuracy of ±2.5% can be obtained in determination of the scale factor.

Figure 3 shows the image function (DD band) and the object function as originally sampled and at the correct scale factor. The same number of points is represented in each curve.

Figure 4 illustrates the effects of aliasing. The modulus and phase of the spectrum of a set of underflight data, sampled at two different intervals, are shown. At the greater sample interval, the modulus has a positive error increasing at higher frequencies, and the phase shows varying error, also increasing at higher frequencies.

Figure 5 is the OTF for the DD band and represents the average of OTF's obtained from several portions of one scan. The real and imaginary spectral components of the OTF determined from each set of data were weighted by the strength of the image spectrum modulus and then averaged to obtain the final OTF. The dashed bounds on the lower sections of the MTF represent relative uncertainty based on the strength of

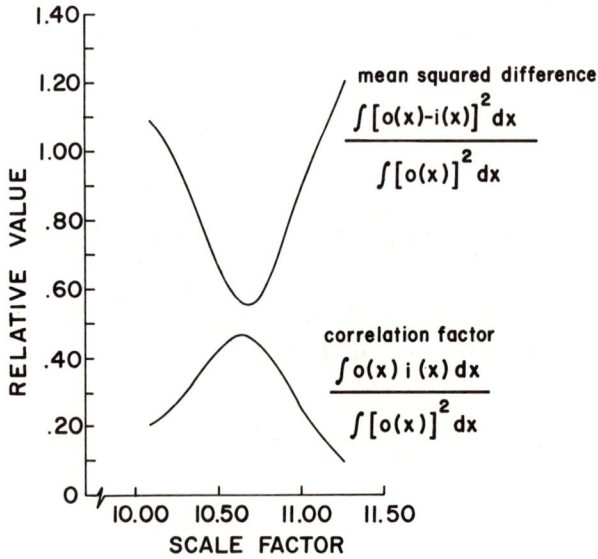

FIGURE 2  Criteria for scale-factor determination.

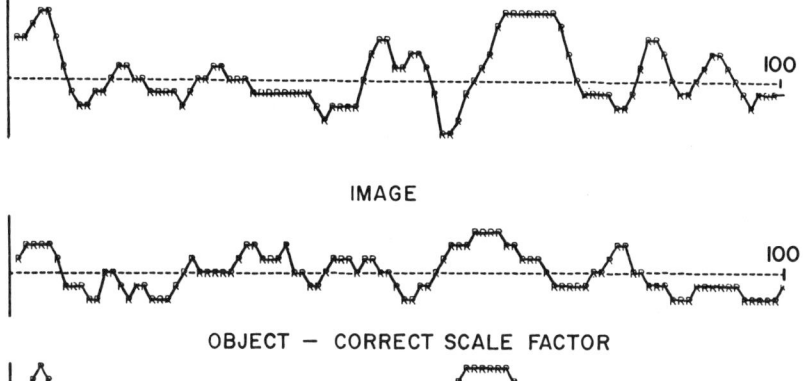

FIGURE 3  Scaling of object to image.

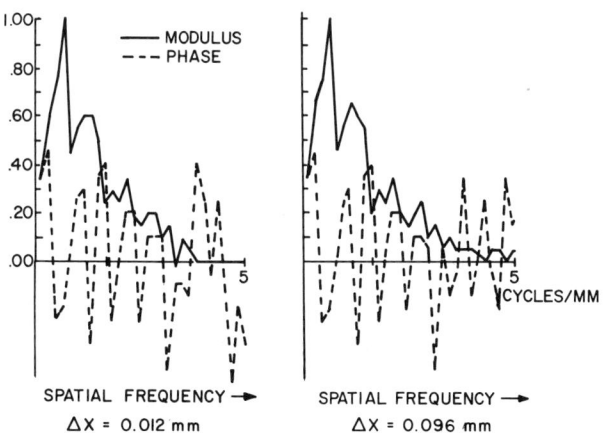

FIGURE 4  Aliasing error.

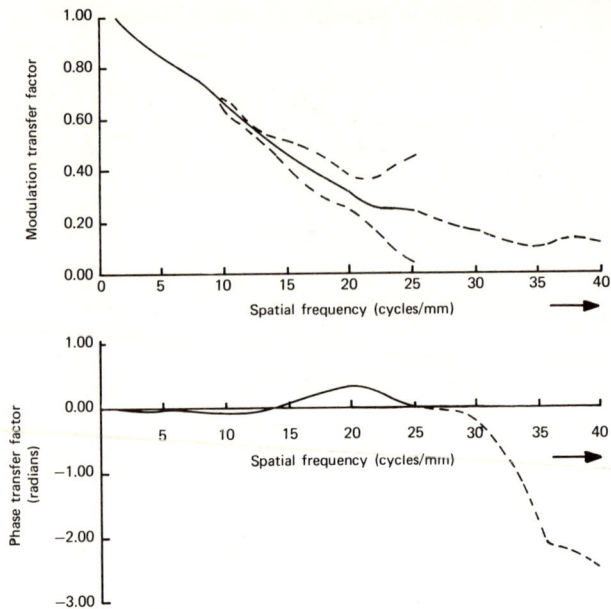

FIGURE 5  S 065 sensor (DD band) OTF.

the image modulation at each frequency. Additional smoothing of the OTF was achieved by eliminating negative lobes in the corresponding spread function and by convolving the OTF with a Gaussian function. The curves are dashed above 25 cycles/mm because they have not been corrected for aliasing in this region. The effect of aliasing is particularly evident in the phase transfer factor.

## IV. Discussion of the Technique

One of the difficulties in using natural terrain for image evaluation as discussed in this paper is the low modulation of the ground as seen from above the atmosphere. The recorded images are of even lower modulation, and the signal-to-noise ratio, i.e. (image modulation)$^2$/grain noise variance, which is a function of spatial frequency, can easily be as low as 5:1 and decrease rapidly with increasing spatial frequency. With edge analysis, multiple scans are usually averaged to increase the signal-to-noise ratio, but this is not possible with the general technique described here. However, it is possible to decrease the uncertainty in the OTF by averaging OTF's obtained from several scans within an isoplanatic region.

Locating exactly the same ground area and determining the scale between the two images are problems with this approach, but they can be handled satisfactorily by mean-square difference and correlation matching.

In spite of these difficulties, our approach possesses several unique assets. The orbiting sensor OTF can be determined from any imagery (and in any portion of the field of view) that is covered by simultaneous underflights. There is no need for special targets or reliance on natural objects of unknown quality as test objects. Indeed, the use of natural terrain for image evaluation provides additional information about the usefulness of the imagery. Those analyzing remote-sensing data can use the statistical results of visual or machine-aided photointerpretation to establish relationships among the quantity and quality of data extractable from the imagery, the spatial frequency content of the imagery, and the sensor OTF. These relationships would not only be useful for determining the value of given imagery but also for specifying requirements on future sensors [Slater and Schowengerdt, 1973].

The technique has been applied to evaluation of the Apollo 9 S 065 photography [Schowengerdt and Slater, 1972] and is currently being used at the Optical Sciences Center for quality evaluation of the ERTS-1 RBV and MSS sensors.

We wish to acknowledge NASA's continuing support of this effort under contract NAS 9-9333 for the Apollo 9 studies and contract NAS 5-21849 for the ERTS-1 investigation.

## References

Blackman, R. B., and J. W. Tukey, *The Measurement of Power Spectra,* p. 31, Dover, New York (1959).
Colvocoresses, A. P., *Photogramm. Eng. 38,* 33 (1972).
Forkey, R. E., and D. A. Womble, "Unique Lens Design for Multiband Cameras," presented at ICO-IX Congress on Space Optics, Santa Monica, Calif. (1972).
Hendeberg, L. O., and E. Welander, *Appl. Opt. 2,* 379 (1963).
Mazurowski, M. J., and R. E. Kinzly, "The Precision of Edge Analysis Applied to the Evaluation of Motion-Degraded Images," in *Evaluation of Motion-Degraded Images,* NASA SP-193 (1969).
Park, A. B., "Earth Resources Program," presented at ICO-IX Congress on Space Optics, Santa Monica, Calif. (1972).
Roetling, P. G., R. C. Haas, and R. E. Kinzly, "Some Practical Aspects of Measurement and Restoration of Motion-Degraded Images," in *Evaluation of Motion-Degraded Images,* NASA SP-193 (1969).
Schowengerdt, R. A., and P. N. Slater, Final Post-flight Calibration Report on Apollo 9 Multiband Photography Experiment S 065, NASA Contr. NAS9-9333 (May 1972).
Slater, P. N., *Photogramm. Eng. 38,* 543 (1972).
Slater, P. N., and R. A. Schowengerdt, *Photogramm. Eng. 39,* 197 (1973).

MASAHARU KAWAI

# FABRICATION OF AN ASPHERIC SURFACE FOR OBSERVING AIRFLOW IN A ROCKET NOZZLE

## I. Introduction

There are two cases of introducing an aspheric surface to an optical system. One is to improve an image performance of an optical system that has been designed by spherical surface. Many attempts at this case have been made, such as applying an aspheric surface to an astronomical objective. In the other case, the aspheric surface plays a spherical role that can never be replaced by any combination of spherical surfaces. One example of this case is a cylindrical lens. This study is concerned with the latter case.

The optical observation techniques through plane-parallel windows, such as a shadowgraph or the Schlieren method, are widely used for aerodynamic study in a wind tunnel. However, for a phenomenon enclosed in the bore of a tube, the conventional method cannot be used, since the shape of the inner surface greatly affects the flow field.

For the purpose of observing the phenomena, a tubelike lens that has a conical inner surface and an aspheric outer surface is fabricated. The latter is one of the ruled surfaces, i.e., surfaces generated by the motion

---

The author is with Nippon Kogaku, K.K., Tokyo, Japan.

of the straight line. It enables a light beam to be collimated not only in the bore but outside of the lens. It is used for observing phenomena taking place in the bore of the tube by the shadowgraph method and may be called a wind tunnel pipe.

In this paper, the fabrication method of such aspheric lenses is described.

## II. Determination of the Outer Aspheric Surface

The above-discussed lens has a property that incoming parallel light rays traverse the bore parallel and emerge parallel (Figure 1). The relation for determining a shape of the outer surface is described in the following.

In Figure 2, the position vector **S** of a point p on the inner surface is given by

$$S(x, y, z); x = f(t) \cos \theta,$$

$$y = f(t) \sin \theta, \tag{1}$$

$$z = t,$$

where $f(t)$ represents the inner radius of the pipe at $z = t$, and $t$ are curvilinear coordinates of the surface.

The analysis is confined to the half-path of a light ray for simplicity.

In Figure 2, **r** is a unit vector that is parallel to the incident light ray at p. **r**′ is a unit vector showing the direction of the refracted ray at p. Let $S_n$ be the normal vector at the point p on the inner surface; **r**, $S_n$, and **r**′ must be in the same plane. The relationship is obtained as follows:

$$[\mathbf{r} \times \mathbf{S}_n] \times [\mathbf{S}_n \times \mathbf{r}'] = 0. \tag{2}$$

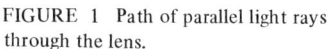

FIGURE 1  Path of parallel light rays through the lens.

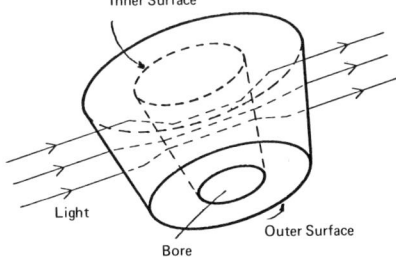

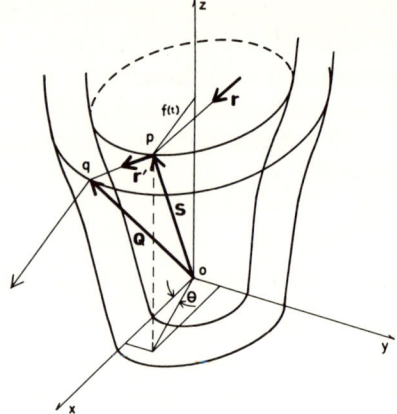

FIGURE 2 Illustration of the coordinate system.

Snell's law determines the relationship among **r**, $S_n$, and **r'**.

$$n = \sin \beta / \sin \gamma, \qquad (3)$$

where $\beta$ and $\gamma$ are angles between **r** and $S_n$, and $S_n$ and **r'**, respectively, and $n$ is a refractive index of the lens.

The refracted light ray parallel to **r'** passes a point q on the outer surface and emerges parallel to **r**. Therefore, the normal vector $Q_n$ at the point q on the outer surface is parallel to $S_n$.

$$S_n \cdot Q_n = 0. \qquad (4)$$

Using differential geometry and vector analysis for Eqs. (2), (3), and (4), the shape of outer surface can be determined.

Finally, the position vector **Q** of q is expressed by

$$Q = S + \delta r', \qquad (5)$$

where $\delta$ is a distance between p and q.

Assuming the inner surface to be conical, we know from Eq. (5) that the outer surface consists of straight lines, each parallel to the generating line of the inner cone. The fabrication method of the outer surface based on this property is developed.

## III. Fabrication Method

In Figure 3, the inner surface of the lens is indicated by a conical surface whose vertex lies at V. Here, $z$ is the cone axis and the $x$ and $y$ axes are perpendicular to the $z$ axis. The outer surface, which has an optical property as shown in Figure 1, is a ruled surface generated by a line $QQ'$, which is parallel to and deviated by a certain distance from a generating line of the inner cone $SV$. The relative position of $QQ'$ and $SV$ is determined by a vector $SQ$ on a plane $(x, y)$. This vector is a function of $\theta$. Its $x$ and $y$ components are given by

$$SQ_s = X(\theta),$$
$$SQ_y = Y(\theta).$$
(6)

Using this geometrical property of the outer surface, a grinding and polishing machine was designed and constructed as shown in Figure 4. A work turns around the $z$ axis and a tool T moves along the line $QQ'$. The relative position of the work to the tool is controlled by two cams C1 and C2 satisfying Eq. (6). The movement of the tool along $QQ'$ and the rotation of the work are independent of each other. By setting the tangent line of the cutting edge of tool parallel to line $QQ'$, there are no limitations to the tool radius.

The machine is shown in Figure 5. The machine can grind and polish not only the outer surface but the inner surface. It has four parts: a work table, a tool table, a polishing apparatus, and a base.

The work table consists of a work spindle and $x$ and $y$ guides. A work

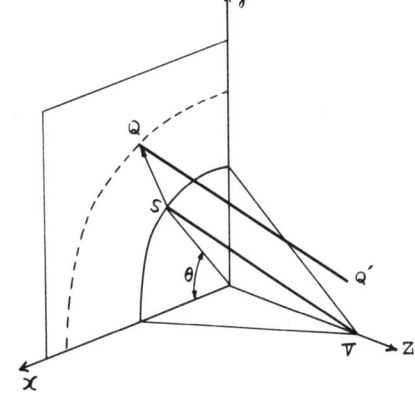

FIGURE 3  Definition of parameters used in fabrication.

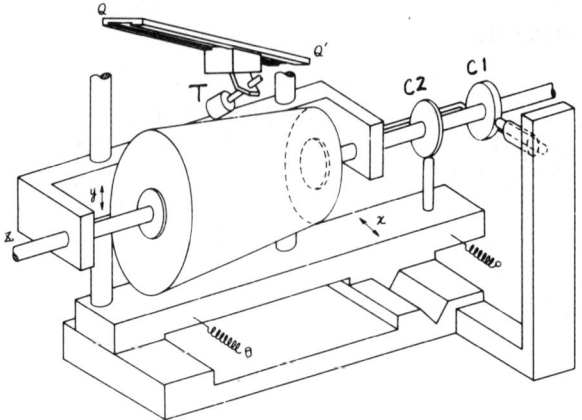

FIGURE 4  The grinding and polishing machine.

FIGURE 5  Photograph of the grinding and polishing machine.

of glass and two cams are held on the work spindle, which rotates at 0.7 rpm. During grinding and polishing the outer surface, the work table is moved following two cams along the $x$ and $y$ guides, and its motions determine the relative position of the work to the tool. For the grinding and polishing of the inner surface, these cams are set free and the work table is fixed to the base of the machine.

A front part of the machine, shown in Figure 5, is the tool table, which is composed of a tool spindle and a motor. For grinding the outer surface, metal bonded diamond cup wheels of 30-mm diameter are used at 9000 rpm; and for the inner surface, wheels 28 mm in diameter are used. The tool table traverses along the generating line with speeds of 0.4–1.8 mm/min. Until the work has a required diameter, the grinding is repeated. After each grinding, the work is removed from the work spindle and measured by a profile projector. Grinding takes two steps— rough and fine.

After the fine grinding, the tool spindle is removed and the polishing apparatus is set on the tool table. The polishing apparatus consists of flexible and slender polishers and a reciprocation mechanism. Before the polishing, in order to remove the grinding marks, the work is smoothed by the flexible laps, which are made of aluminum foil. The flexible slender polishers, which have the same length as the work and are 20 mm in width, are used for polishing. They are made of polyurethane and are reciprocated along the generating line. Reciprocation speed is 129 stroke/min.

Fabrication process of the lens is as follows: (1) preparative shaping, (2) outer surface grinding, (3) outer surface polishing, (4) inner surface grinding, (5) inner surface polishing, (6) side edge finishing. In process 1, a glass blank is shaped to fit a chuck of the work spindle. Processes 2–5 are put in practice with the machine. Process 6 is a final shaping process to fit the experimental equipment.

## IV. Inspection

For performance evaluation, a finished lens was inspected by the shadowgraph method, as shown in Figure 6. Three types of chart were used. One is orthographic projection of the cone, and the other two are of crossing lines of 2-mm spacing. One of the latter charts was placed at the position 1 of Figure 6, and the other and the orthographic chart were placed at the position 2. Shadowgraphs were taken with parallel light and are used for calibration of the image coordinate.

710                                                                OPTICAL TECHNOLOGY

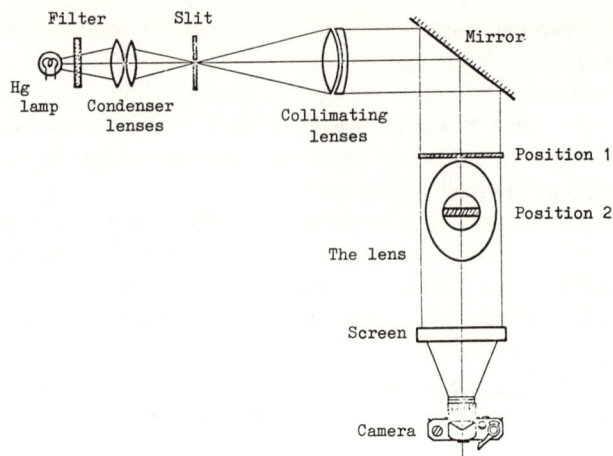

FIGURE 6  Shadowgraph system for performance evaluation of finished lenses.

The profile of the outer surface is measured mechanically by a universal measuring microscope.

## V. Result

One of the finished lenses is shown in Figure 7. It has a half cone angle of 15°, minimum radius of inner cone of 17.47 mm, and is 52.0 mm in

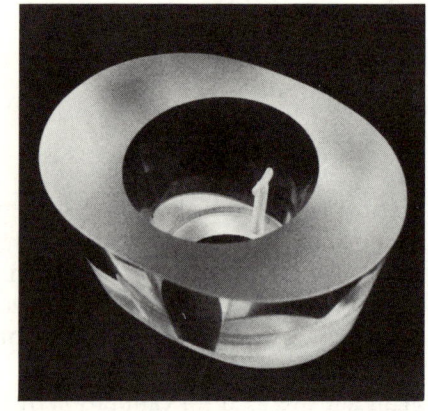

FIGURE 7  A finished lens.

length. Figure 8 shows deviation from the calculated profile of the outer surface. Figure 9 shows a shadowgraph of the crossing line chart placed in the bore. The finished lenses proved satisfactory.

Experiments were made using this lens at the National Aerospace Laboratory (Tokyo) for the purpose of analyzing the flow field at a rocket nozzle, especially in the study of SITVC (Secondary Injection Thrust Vector Control), which is one of the guiding methods of solid propellant rockets. Shadowgraphs of the airflow phenomena that are caused by collision of the main and the secondary injection flows in a rocket nozzle model were taken by the lenses, which are called models S, SL, R, and R2. The secondary flow was injected through a small orifice at the wall of the lens (Figure 7). Figure 10 is one of the shadowgraphs, showing a shock wave caused by the collision. The shape of the shock wave is important for the analysis.

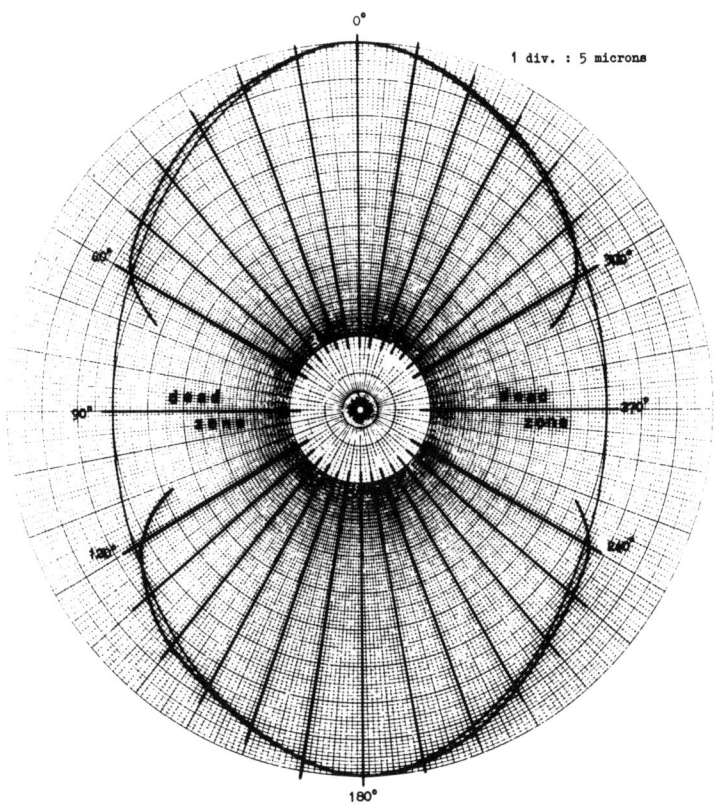

FIGURE 8   Deviation of the actual profile of the outer surface from that calculated.

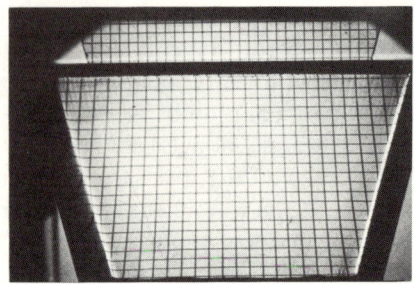

FIGURE 9 Shadowgraph of the crossing lines chart placed in the bore.

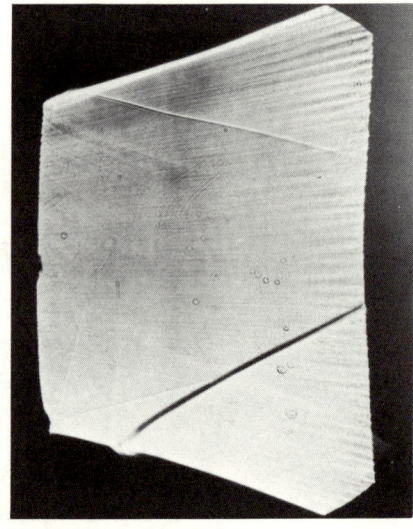

FIGURE 10 Shadowgraph showing the shock wave produced by the collision of the main and secondary air flows.

## VI. Conclusion

An aspheric tubelike lens, which has a property that collimated incident light rays traverse its bore parallel and emerge parallel, was fabricated with sufficient accuracy by a newly developed machine. By shadowgraph technique, the lens proved useful for the observation of flow-field phenomena in the conical bore. The lens is expected to make a considerable contribution to developments of a guiding method of the solid propellant rocket.

This machine can fabricate other types of ruled surface.

The author would like to thank Dr. Yamanaka, National Aerospace Laboratory, for many stimulating suggestions and helpful discussions.

## Bibliography

Yamanaka, T., *Proc. of the 8th International Symposium on Space Technology and Science,* Tokyo, 1969, p. 370.

Yamanaka, T., and H. Azuma, *J. Spacecraft Rockets 4,* 1272 (1972).

# OPTICAL METHODS

H. E. BENNETT, J. L. STANFORD, and
J. M. BENNETT

# SCATTERING FROM MIRROR SURFACES USED IN SPACE APPLICATIONS

## Introduction

Outside the earth's atmosphere, optical signals can be transmitted over very long distances without serious attenuation. However, if they are to be detected, these signals must be separated from background radiation, which also can travel over long distances. A similar problem arises in extraterrestrial astronomy. In order to utilize the superior resolving power that can be achieved if seeing limitations imposed by the atmosphere are removed, the effect of background radiation from extraneous sources must be minimized. Since the intensity of light reaching the earth from astronomical bodies ranges over a factor exceeding $10^{19}$, levels of scattered light within telescopes in a satellite observatory may well be the limiting factor in their performance. A particularly difficult situation arises when the stellar object is angularly near a strong natural emitter such as the sun or the earth's limb. To detect signals under these situations, the scattering levels in the optical system must be extremely low.

---

The authors are in the Michelson Laboratory, Naval Weapons Center, China Lake, California 93555.

Light scattering by optical components may be divided into two categories, surface scattering and scattering in the bulk of the material. Bulk scattering phenomena such as Rayleigh, Brillouin, and Raman scattering have been extensively studied in transparent materials, and some work has been done on metals. Surface scattering, although orders of magnitude larger than bulk scattering for most mirror components, has received less theoretical attention. This scattering from mirrors arises from two sources: (1) macroirregularities such as scratches, the scattering from which is governed by geometrical optics, and (2) surface microirregularities, which produce scattering governed by diffraction theory, an aspect of physical optics. We have found that, contrary to popular conception, high-quality optical surfaces are usually sufficiently scratch-free so that (2) dominates.

Unlike scattering from macroirregularities, scattering from microirregularities is strongly wavelength-dependent. In the visible region, typical hemispherical scattering levels from metal-coated glass mirrors are a few parts in $10^3$. Scattering levels an order of magnitude lower than these values can be obtained by superpolishing some types of mirrors. On the other hand, scattering levels over an order of magnitude higher have been observed on mirrors that could not be distinguished visually from those with low scattering levels.

The above values are for hemispherical scattering. Near the specular direction, scattering levels will be one or more orders of magnitude lower than the hemispherical values, depending on the character of the particular surface and the instrumental acceptance angle used. In this paper, we will discuss (1) the theoretical relationships between scattered light from mirror surfaces and their roughness characteristics, (2) some experimental techniques for determining relevant surface parameters and scattering levels, and (3) the results obtained and how they correlate with theoretical predictions.


## Classical Theory of Scattering by Diffraction

The scattering resulting from surface microirregularities depends on how these irregular surfaces modify the surface currents that are induced by the incident light wave within the "skin depth" of the metallic coating. Various calculational procedures have been used to determine the properties of the resulting scattered light. Scalar theories [Davies, 1954; Porteus, 1963] based on the Kirchhoff diffraction integral predict the total amount of light scattered from irregular surfaces and the angular dependence for scattering near the specular direction. The polarization and angular dependence of light scattered at arbitrary angles has more recently been obtained from vector theories [Hunderi and Beaglehole, 1970; Kröger and Kretschmann, 1970; Elson, to be published], which consider the radiation from induced surface currents modulated by surface irregularities. At least one calculation of scattered light based on quantum theory has been carried out [Elson and Ritchie, 1971].

In all the above theories, the surface irregularity structure is described statistically. A simple statistical model of the surface is one in which the surface irregularities have a Gaussian height distribution about the mean surface level. This model was apparently first used by Chinmayanandam [1919], who published a paper describing the ratio of specular to diffuse reflectance from such a surface. The exponential relationship he obtained has been rediscovered numerous times by authors working in different wavelength regions. One of the best known derivations was published by Davies [1954] in connection with the reflection of radar waves. Davies showed that expressions for the amount of light scattered near the specular direction and its angular dependence can be obtained for a perfectly reflecting surface under the restrictions that (1) both the height distribution and autocovariance functions are Gaussian, and (2) the resulting rms roughness $\delta$ and autocovariance length $a$ are small compared to the wavelength $\lambda$ of light incident on the surface. Davies' results were analyzed, and the spectral dependence of scattered light at normal incidence that they predicted was experimentally verified by Bennett and Porteus [1961]. Porteus [1963] subsequently developed a general theory for scattering at normal incidence from a surface of arbitrary joint density function. In the special case of a Gaussian height-distribution function, he showed that if the slope of the surface irregularities is not excessive, the coherent reflectance $R_c$ (which is equivalent to the specular reflectance when $\delta/\lambda \ll 1$) is given by

$$R_c = R_o \exp[-(4\pi\delta/\lambda)^2] \tag{1}$$

without restriction on the value of $\delta/\lambda$. Here $R_o$ is the total reflectance of the surface, specular plus diffuse. The fraction of the incident light that is incoherently reflected (i.e., the scattered light) is then $R_i$, where

$$R_i = R_o \left\{ 1 - \exp\left[(4\pi\delta/\lambda)^2\right] \right\} \tag{2}$$

$$\cong (4\pi\delta/\lambda)^2 R_o, \quad \text{when } \delta/\lambda \ll 1. \tag{3}$$

It should be emphasized that Eqs. (1–3) hold strictly only for surface irregularities that have a Gaussian height distribution about the mean surface level. There is no reason to expect that actual surface irregularities should even be symmetric about the mean surface level. However, the true height-distribution functions can often be approximated reasonably well by a Gaussian function, particularly when $\delta/\lambda$ is small. The statistical parameters describing the surface are then easily understood, and a comparison of the parameters for different surfaces is a reasonably valid way of comparing the surfaces themselves. It is common practice in statistics to assume when an rms value is quoted that a Gaussian distribution function is implied. If this practice is followed when relating optical scattering to surface roughness, "effective roughness" values describing the surface may be computed. The effective roughness of a surface at a given wavelength is then a measure of the scattered light to be expected from that surface at that wavelength. The effective roughness values for a surface will change when the wavelength changes if the surface does not in fact have a Gaussian height distribution. This variation in effective roughness with wavelength in no way implies a breakdown in the theory; rather, it serves as a measure of how close to Gaussian the actual height distribution of the surface really is.

The above scattering theories based on the Kirchhoff diffraction integral are necessarily scalar theories, hence they do not treat effects that give rise to variation in the polarization properties of the scattered light. In addition, in considering the angular distribution of scattered light, simplifying approximations have been made that limit the validity of these theories to small angles about the specular direction.

In order to treat the polarization and angular dependence of diffuse scattering, recent authors [Hunderi and Beaglehole, 1970; Kröger and Kretschmann, 1970; Elson, to be published] have employed perturbational methods to calculate the radiation from induced surface polarization currents whose properties depend on surface structure. These approaches are limited by the condition $\delta/\lambda \ll 1$; as has been indicated, this condition is satisfied for many cases of interest. For unpolarized light incident from vacuum normally onto a metal surface of rms roughness $\delta$, the results of the above theories are identical; the bidirectional reflectance–distribution function is given by

$$\frac{dR_i}{d\Omega} = \frac{1}{2} \left(\frac{2\pi}{\lambda}\right)^4 \left(\frac{\delta}{\pi}\right)^2 \left\{ \left|\frac{(\epsilon - \sin^2 \theta)^{1/2}}{(\epsilon - \sin^2 \theta)^{1/2} + \epsilon \cos \theta}\right|^2 + \left|\frac{1}{(\epsilon - \sin^2 \theta)^{1/2} + \cos \theta}\right|^2 \right\}$$

$$\times \left|1 - \sqrt{\epsilon}\right|^2 (\cos^2 \theta) \left[g\left(\frac{2\pi}{\lambda} \sin \theta\right)\right] \quad (4)$$

In the above expression, $\epsilon$ is the wavelength-dependent dielectric constant $\epsilon = \epsilon_1 + i\epsilon_2$. The function $g[(2\pi/\lambda)\sin\theta] = g(\kappa)$ is the Fourier transform of the autocorrelation function $G(\rho)$, defined by

$$g(\kappa) = 2\pi \int_0^\infty G(\rho)e^{i\kappa\rho}d\rho, \qquad (5)$$

$$G(\rho) \equiv \lim_{L\to\infty} \frac{1}{L\delta^2} \int_0^L z(x)z(x+\rho)dx, \qquad (6)$$

for an isotropic surface structure. The quantity $z(x)$ is the height of the surface at a point $x$ on an axis in the mean surface plane, and $\rho$ is a finite displacement along $x$.

The first term in the curly brackets in Eq. (4) corresponds to light polarized in a direction parallel to the scattering plane ($p$-polarized), and the second to $s$-polarized scattered light. For most metals in the visible and near infrared, $\epsilon_1 < 0$, and $|\epsilon_1| \gg \epsilon_2$. Therefore the intensity of $p$-polarized scattered light is somewhat greater than $s$-polarized scattered light at all angles except $\theta = 0$, where the two components are equal. The angular dependence of the two components may be strongly affected by the quantity $g[(2\pi/\lambda)\sin\theta]$. For the case that $G(\rho)$ is a Gaussian,

$$G(\rho) = \exp(-\rho^2/a^2), \qquad (7)$$

then from Eq. (5),

$$g[(2\pi/\lambda)\sin\theta] = \pi a^2 \exp(-\pi^2 a^2 \sin^2\theta/\lambda^2). \qquad (8)$$

As $a/\lambda$ increases, the scattered light near the specular direction ($\theta = 0$) increases. In order to predict the angular distribution of scattered light from a given surface, the autocorrelation function must therefore be known.

## Anomalous Scattering

As we have just seen, surface polarization currents induced by a light wave incident on an irregular surface radiate light diffusely (i.e., scatter the incident light). Surface microirregularities and the resulting modification of induced currents are also indirectly responsible for at least one other type of diffuse radiation. This kind of scattering is considered anomalous in that it is not predicted by conventional scattering theories.

If the fields associated with the modulated surface currents have the proper spatial phase variation, they excite traveling surface polarization waves [Elson and Ritchie, 1971]. In the case of a metal, these waves consist of a collective motion of the relatively free conduction electrons. They are commonly called surface plasma waves, the quanta of which are surface plasmons. The effects of surface plasmon excitation are (1) a strong resonance-type absorption of the incident light near the surface plasma wavelength $\lambda_{sp}$ [$\epsilon_1(\lambda_{sp}) = -1$] and (2) diffuse radiation by these "anomalous" currents at wavelengths near $\lambda_{sp}$. As we shall see shortly, the intensity of this anomalous, resonance-type scattering may be an order of magnitude greater than that of the light directly scattered by the surface irregularities.

A second type of resonance scattering is associated with interference effects produced by thin dielectric films on irregular metal surfaces. The strength of the electric field at the surface of the metal may be increased significantly by the addition of a dielectric layer of suitable thickness. Since it is this field that determines the amplitude of the induced surface currents, one may therefore expect to see increased scattering at wavelengths where this field enhancement is greatest. This effect is not anomalous in the same sense as the indirect effects discussed above, since it is predicted in a straightforward way by classical scattering theories [Kröger and Kretschmann, 1970].

## Experimental

Several instruments are being used in our laboratory to measure scattering and/or surface irregularities that produce scattering. These include instruments to measure both scattered light and optical absorption, a FECO interferometric system to measure surface roughness, an electron microscope, and an optical microscope with a Nomarski attachment to photograph surface irregularities under 100 Å rms in height. We will describe the instrumentation in this section.

### SCATTERED-LIGHT APPARATUS

Figure 1 shows a schematic diagram of an instrument that can be used to study the incoherent re-emission of optically excited surface plasmons, as well as light scattered by simple diffraction. Light from xenon arc Xe or tungsten lamp W is focused on the entrance slit $E_1$ of a monochromator. Light of the desired wavelength passes through the exit slit $E_2$ and is focused on pinhole aperture $A_1$. Alternatively, low-pressure

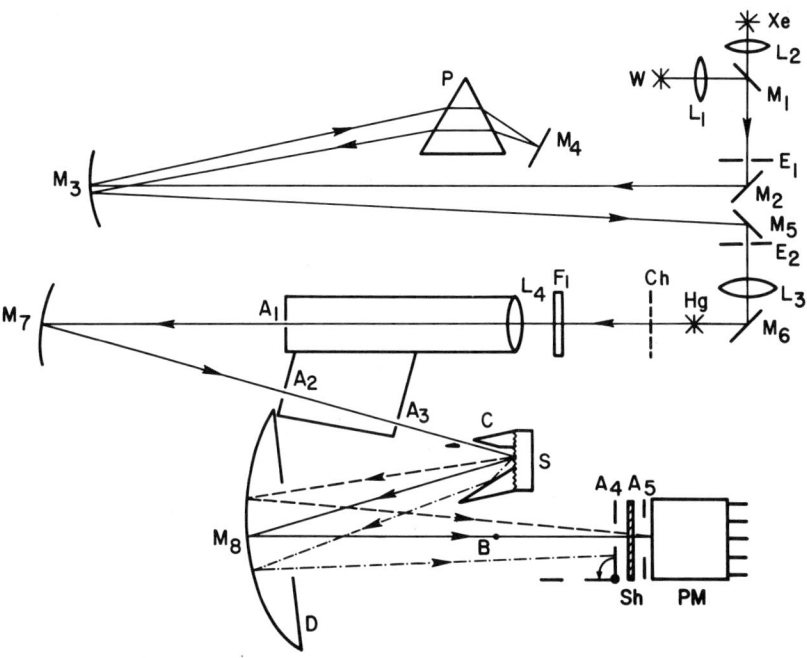

FIGURE 1 Schematic diagram of roughness analyzer used to measure total scattered light and its angular dependence on plane 3.86-cm-diameter samples.

mercury arc Hg may be inserted in the beam, and, by using filter $F_1$, monochromatic light may be obtained. This source is particularly useful at 2536 Å in the ultraviolet, where sufficient intensity is difficult to obtain with our monochromator.

In any case, the narrow pencil of light from $A_1$ is collimated by mirror $M_7$ and passes through aperture $A_2$ and baffle $A_3$ to strike the optically flat sample S. Light specularly reflected from the sample passes along the solid line to mirror $M_8$, which images the sample surface on $A_5$. The specular beam may be blocked from striking the photomultiplier PM by blocking mask B, which is at the image of aperture $A_2$. The mask, whose dimensions are several times that of the specular beam at that point, cuts out light within a semiangle of 37 min of arc to the specular direction. Light scattered within 20° of the specular direction is focused by mirror $M_8$ on $A_5$ directly. By closing down diaphragm D, the acceptance angle about the specular direction can be reduced. Light scattered into angles larger than 20° strikes cone C, which redirects it to mirror $M_8$ but introduces a virtual object, so that the image at $A_5$ of this light is in the form of a ring about the specular beam. It can be blocked out by aperture $A_4$. Nearly all of the light scattered at angles larger than 20° is picked up by the cone. Scattered light from the surface of $M_8$ has a minimal effect since it is not focused on the detector, whereas that from the sample is. Nevertheless, $M_8$ has been given a supersmooth polish to make doubly sure that no serious systematic error arises from this source.

By manipulating B, D, and $A_4$, one can measure (1) the total light reflected from the sample, (2) the scattered light only, and (3) that scattered into a range of angles from 20° to 1½° from the specular direction. Since the amount of light scattered in the ultraviolet, visible, or near infrared by a good optical surface may be orders of magnitude less than that specularly reflected, some care must be taken to ensure that the electronics can handle a wide dynamic range linearly. In our case, the photomultiplier, which has been checked for linearity using a three-polarizer method, feeds the signal through a lock-in amplifier to a voltage-to-frequency converter and counter. The resulting signal levels can then be read to five significant figures. Very small relative signal levels can thus be measured with precision.

OPTICAL EVALUATION FACILITY

Only optically flat samples 3.86 cm in diameter can be accommodated in the Scattered Light Apparatus described above. With a new instrument, the Optical Evaluation Facility, which is now nearing completion, we will be able to measure the total amount of scattered light and to some degree its angular dependence on mirrors up to 40 cm in diameter with more or less arbitrary radius of curvature, provided it is larger than 115 mm. A schematic diagram of the instrument is shown in Figure 2. A light beam from source L, which may be either a krypton, helium–neon, helium–cadmium, or carbon dioxide laser, is spatially filtered by SF, chopped at 104 Hz by $C_1$, and then strikes 13-Hz mirror chopper $C_2$. It is either deflected directly to pyroelectric detector $D_2$ or passes through to strike the sample mirror $M_s$, which is mounted on an optical dividing head in a class 100 clean area. By translating, tilting, and rotating $M_s$, the entire mirror surface can be examined. The light is reflected from $M_s$ to plane mirror $M_1$ and thence to detector $D_2$. Alternately, it is scattered, strikes Coblentz sphere $M_c$, and is imaged on domed pyroelectric detector $D_1$. In either case, the signals generated go to one of the lock-in amplifiers LI. The low-frequency envelope seen by $D_2$ is indicated by the plot of intensity versus time in Figure 2. Detector $D_1$ sees a modulated signal whose minimum goes to zero. The incident intensity $I_0$ that strikes $D_2$ is controlled by attenuator A and monitored

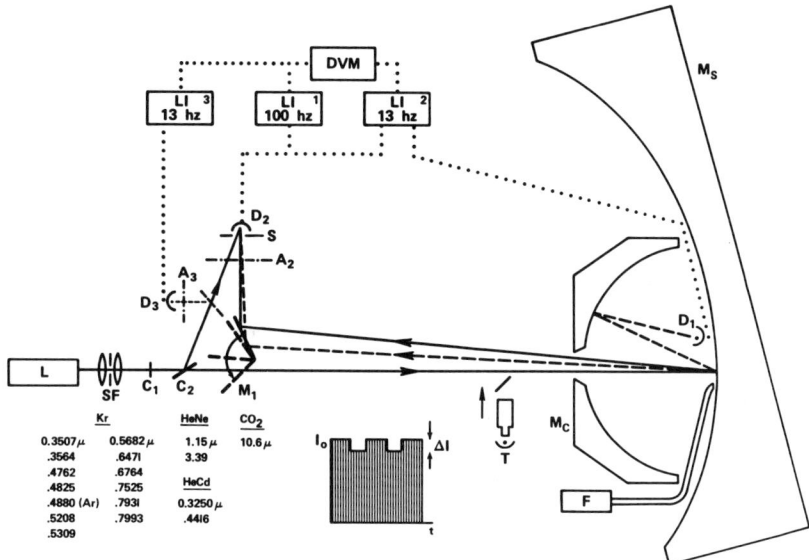

FIGURE 2 Schematic diagram of Optical Evaluation Facility used to measure total scattered light and small-angle scattering from plane and concave mirrors.

by one LI operating at the 104-Hz frequency. The second LI, operating at 13 Hz, can monitor either $D_1$ or $D_2$, so that by taking the ratio of signals from the lock-ins using digital voltmeter DVM, both scattering and absorption can be determined. Since the angle of $M_1$ can be changed, light scattered within 2° of the specular direction, indicated by the dashed line, can also be determined. Mirror $M_s$ is positioned using fiber optic microprobe F and autocollimator T. The reflectance of $M_1$ can be calibrated by rotating it into the incident beam as shown.

An advantage of this instrument is that the lock-in signal is proportional to mirror absorption or scattering, not mirror reflection. Very small scattering levels and reflectances very close to unity should thus be measurable. We expect to measure scattering levels of $10^{-7}$ and reflectances with an uncertainty of ±0.001 or better for high reflectance samples using this instrument.

FECO INTERFEROMETRIC SURFACE SCANNER

An alternate approach to the surface evaluation problem is to use interferometric analysis. Fringes of equal chromatic order (FECO fringes) [Bennett and Bennett, 1967] give direct information about the roughness of optical surfaces and can be used to distinguish between different finishes even if the rms roughnesses are the same. Figure 3 shows a schematic diagram of the FECO system, which is currently being modified to include the slow scan TV camera and data-analysis system. White light from zirconium arc Z is collimated by lens $L_1$ and directed by beam splitter B to the interferometer I. The interferometer consists of a very smooth reference flat coated with a nearly opaque (95% reflectance) silver film and the plane sample coated with an opaque silver film. The sample is almost in contact with the reference surface, with only about five wavelengths of light separating the two surfaces. Lens $L_2$, an 89-mm $f/3.1$ macrolens for the Vickers Projection Microscope, focuses a magnified image of I on the entrance slit S of a Bausch & Lomb constant-deviation spectrograph. The resulting spectrum containing the FECO fringes is formed at the focus of lens $L_4$ and can be viewed with an eyepiece (not shown) or photographed.

Figure 4 is a group of such photographs showing FECO fringes of a superpolished fused-quartz surface, a good-quality polished glass surface, and a very smooth polished metal surface. The latter two surfaces have the same rms roughness but different finishes, as is clearly shown by the different character of the FECO fringes. The fringe to the right in all three photos occurs in the red portion of the spectrum and has

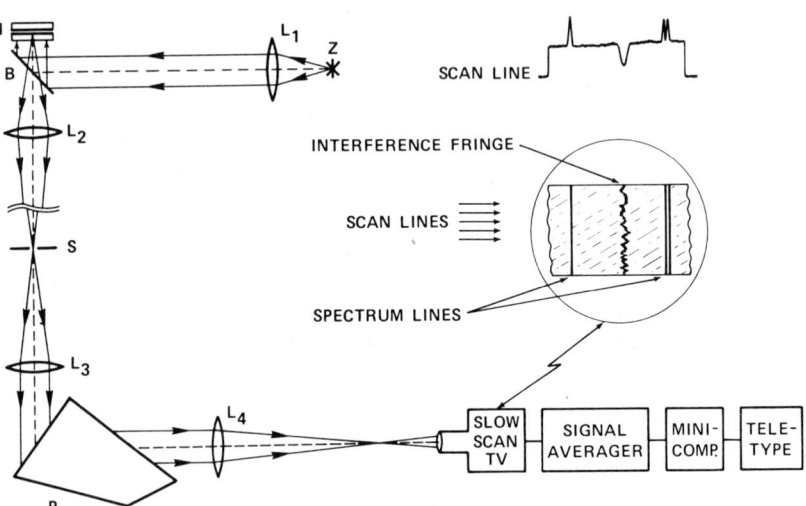

FIGURE 3 Schematic diagram of FECO Interferometric Surface Scanner used to measure surface roughness of plane samples. The image detected by the slow scan TV is shown at right center, and a single scan line is shown above it.

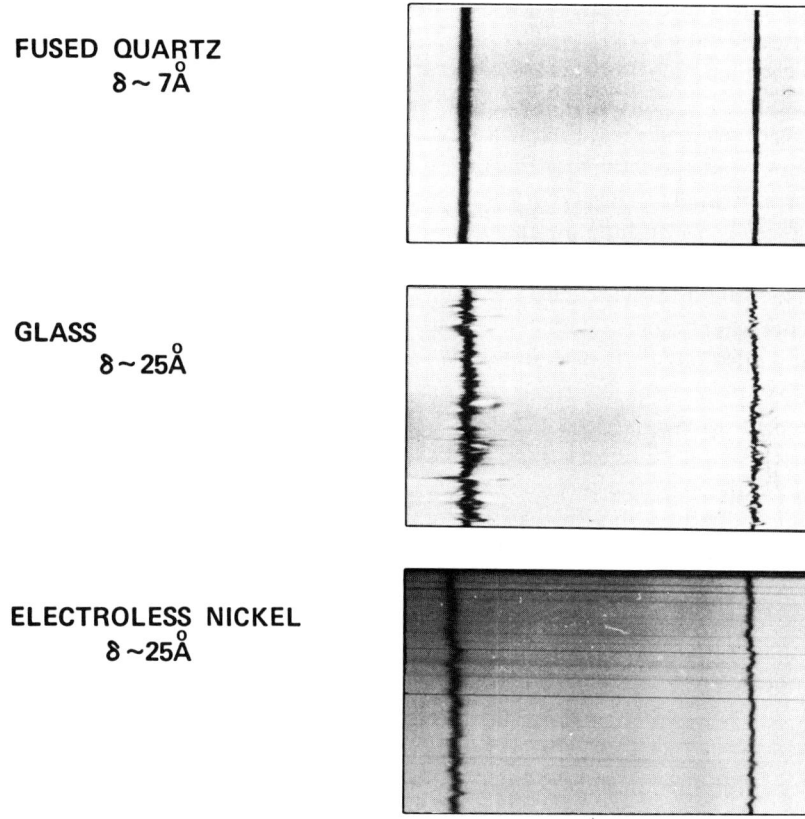

FIGURE 4  FECO fringes obtained from (a) a superpolished fused-quartz optical flat, (b) a good-quality polished-glass surface, and (c) a very smooth polished metal surface. The reference surface in all cases was a superpolished fused-quartz optical flat.

order of interference 7 (i.e., 7 half-wavelengths of red light separate the reference and sample surfaces), while the fringe to the left is order of interference 8 and occurs in the blue-green spectral region where the reflectance of silver is lower and the dispersion of the spectrograph is larger. Even though the appearance of the two FECO fringes in each spectrum is somewhat different, both give the same roughness profile of a 1 mm $\times$ 3 $\mu$m rectangular area of the interferometer.

Prior to the installation of the TV system, the peak-to-peak roughness of the pair of surfaces (reference surface plus sample) was obtained visually by setting a crosshair on the extremities of a FECO fringe and converting the measured wavelength difference into a roughness value.

With the new interferometric surface scanner, an image of the FECO fringes and reference spectral lines (Figure 3 at right) is detected by a slow scan TV camera employing an ITT image-dissector tube. The camera is operating in a line scan mode in which a single line is repetitively scanned until the command is received to proceed to the next line. There are 500 scan lines from the top to the bottom of the picture. A single line is repetitively scanned with the output going to the signal averager until the signal-to-noise level is adequate and the contour (intensity versus position) appears as in the upper part of Figure 3. Then the information on the scan line is digitized and fed into a minicomputer, which calculates and stores the wavelength of the segment of the interference fringe contained in the scan line. The camera now shifts down to the next scan line and repeats the process. The significant output information is the wavelength of the interference fringe at 500 equally spaced points along its length. From these data, a statistical description of the surface including the autocovariance function, true rms height, and height-distribution function can be developed. When operational, this instrument will provide a unique, direct, and nearly indisputable measure of the surface irregularities on optically polished surfaces and will be able to detect irregularities under 10 Å peak-to-valley provided that the lateral scale of the roughness is greater than the 3-$\mu$m lateral resolution limit of the optical system. Thus, the instrument will be complementary to the electron microscope and will furnish a check on the power spectrum results obtained from Nomarski micrographs.

### NOMARSKI MICROGRAPHS

An alternate technique for obtaining the autocovariance function of the surface is to scan superimposed Nomarski micrographs of an optical surface using a densitometer such as our Moll with a reliable screw motion. Such micrographs also are qualitatively useful in that a picture of the surface microirregularities can be obtained. It is difficult, however, to obtain quantitative values of heights of surface irregularities such as are furnished by the FECO technique. It is at first surprising that irregularities a few tens of angstroms in height can be examined using an optical microscope. However, the Nomarski technique [Nomarski and Weill, 1954; Allen *et al.*, 1969] depends on interference between beams initially polarized at right angles to each other, which are reflected from the optical surface and then recombined. Thus, its sensitivity to height differences arises from the same source as that of the FECO technique.

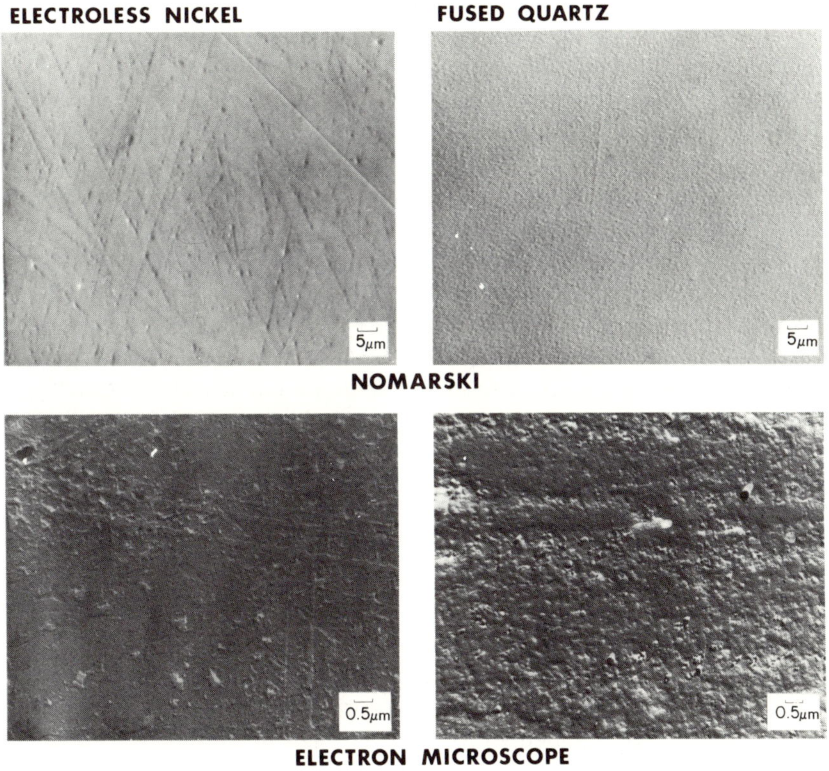

| ELECTROLESS NICKEL | FUSED QUARTZ |

NOMARSKI

ELECTRON MICROSCOPE

FIGURE 5  Nomarski and electron micrographs of an optically polished electroless nickel surface and a superpolished fused-quartz surface. Note the difference in magnification for the two types of micrograph.

Figure 5 shows micrographs of two very smooth optically polished surfaces, one fused quartz and the other superpolished electroless nickel, taken using the Nomarski technique. Although the effective rms roughnesses of the two surfaces were comparable, differences in character between them are readily apparent. These micrographs were taken using a Zeiss universal research microscope with a Nomarski 40X plan objective and a total magnification of 530X.

ELECTRON MICROGRAPHS

Although very small height differences can be observed using the FECO or Nomarski techniques, the lateral resolution that can be obtained is of the order of micrometers. Studies of optical scattering from microcrystallite irregularities in evaporated films indicate that more closely spaced irregularities are also important. To see such closely spaced irregularities, an electron microscope is required. The lower part of Figure 5 shows electron micrographs of surfaces comparable with those on which the Nomarski micrographs were taken. To enhance depth sensitivity, the specimen was shadowed at an angle of 15° using platinum–carbon. Since the side of the 0.25% Formvar replica in contact with the sample was shadowed, these micrographs differ from those obtained using standard techniques in that the depressions in the original surface appear as asperities in the replica, and vice versa. Maximum surface fidelity is, however, achieved in this way. The magnification was 26,500X, so that the total width of the electron micrograph corresponds to a distance of about 10 $\mu$m on the sample, about three times the minimum lateral separation observable using FECO and Nomarski methods. By combining the information from these various techniques, we then have a powerful tool for determining the character of the surface in detail.

## Results

Table 1 gives a comparison of the roughness values obtained from scattered light and FECO measurements on fused-quartz samples coated with different thicknesses of calcium fluoride to obtain various roughnesses and then overcoated with metal. The agreement between these two independent techniques is quite good. For thick calcium fluoride films, the values obtained from scattered-light measurements are slightly larger than those obtained interferometrically, possibly because the calcium fluoride crystallites, many of which are too close together to be resolved by the optical system of the interferometer, play an increasingly important role in scattering from the surface as film thickness increases.

TABLE 1 Comparison of Effective rms Roughness Values of Calcium Fluoride Films Measured Interferometrically and by Scattered Light

| Thickness $CaF_2$ (Å) | Effective rms Roughness | |
|---|---|---|
| | $\delta_{int}$(Å) | $\delta_{scat}$(Å) |
| 0 | 11 | 10 |
| 875 | 12 | 13 |
| 1750 | 16 | 19 |
| 2625 | 19 | 26 |

The wavelength dependence of the scattered light predicted by Eq. (2) is frequently well obeyed. Figure 6 shows the scattering level observed for an aluminum-coated superpolished fused-quartz surface. Over the wavelength range studied, the agreement between the theoretically predicted scattering from a 12.8 Å rms surface, shown by the solid line, and that experimentally observed is excellent. Similar agreement is found in the infrared, as shown in Figure 7.

The levels of scattered light predicted theoretically at various wavelengths from surfaces having various effective rms roughness values are given in Figure 8. The dashed lines give typical roughness values for conventionally polished high-quality metal mirrors, glass mirrors, and fused-quartz or metal mirrors that have received a superpolish. The diagonal lines show the scattered light levels to be expected. For example, at a wavelength of 1 μm the light scattered from a superpolished mirror would be about $10^{-4}$ of that reflected by the mirror. The light scattered by such a mirror into a small angle near the specular direction would be significantly less than $10^{-4}$. Note, however, that the bidirectional reflec-

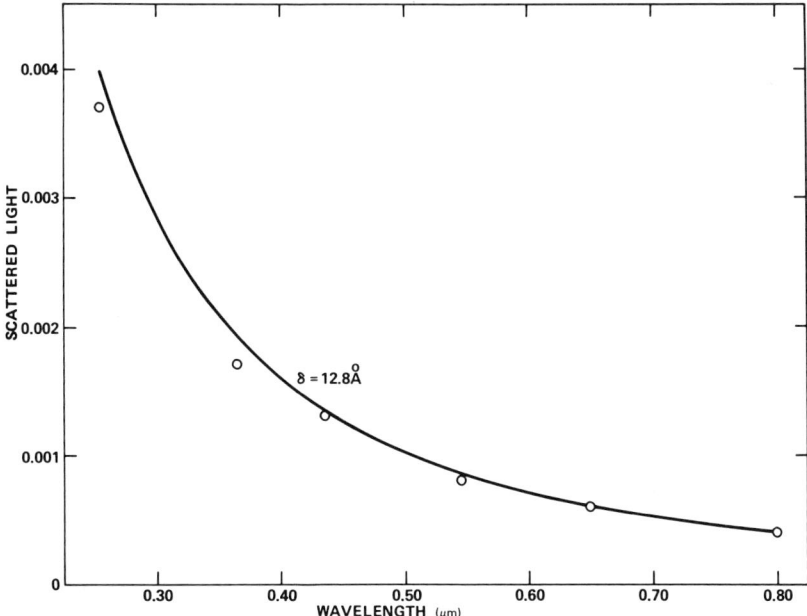

FIGURE 6 Wavelength dependence of the total scattered light from an aluminum-coated superpolished fused-quartz sample. Circles are measured values, and the solid curve is calculated from Eq. (2) assuming a value of 12.8 Å for δ.

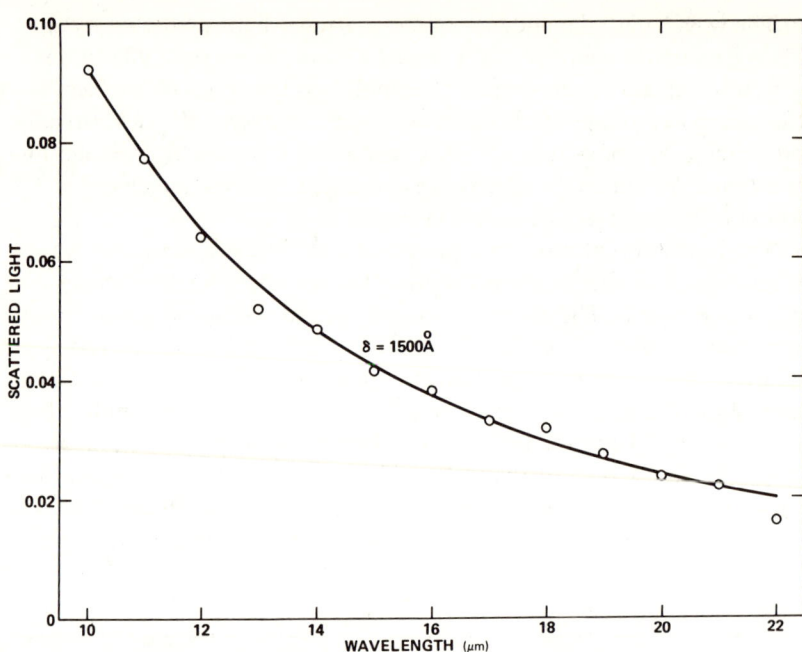

FIGURE 7 Wavelength dependence of the total scattered light from an aluminum-coated ground-glass surface. Circles are measured values, and the solid curve is calculated from Eq. (2) assuming a value of 1500 Å for $\delta$.

tance–distribution function [Nicodemus, 1970; Ginsberg et al., to be published], i.e., the fraction of the light diffusely reflected (scattered) or specularly reflected into a small solid angle and expressed in terms of the reflected light *per steradian,* may be much larger than $10^{-4}$ $sr^{-1}$ for this mirror near the specular direction.

Scattering levels from the mirrors shown in Figure 8 differ at a given wavelength by well over an order of magnitude. However, the superpolished mirror could probably not be distinguished from the metal mirror having a normal polish simply by visual inspection. A slight haze can sometimes be observed on still rougher mirrors by comparing them very carefully with a smooth sample. Also, the "scratch and dig" test commonly used by the optics industry would not reveal the difference between a superpolished and a conventionally polished mirror. In fact, the latter could easily be judged higher in quality than the former on the basis of this test. A better testing procedure is needed if differences in mirror quality are to be recognized. One suggestion is the "veiling glare" test in which near-forward scattering is measured directly [Baird, 1949; McLeod, 1955; MIL-STD-150A, 1959].

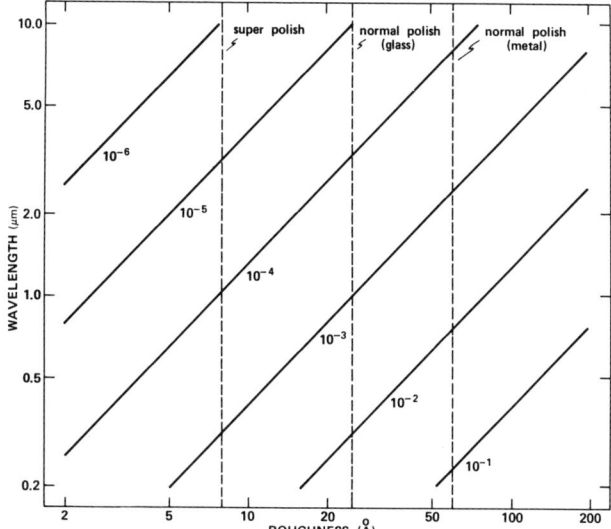

FIGURE 8 Scattered light levels predicted theoretically (diagonal lines) for surfaces having rms roughnesses from 2 to 200 Å. The dashed lines indicate typical roughnesses of various kinds of polished surface. Wavelengths from the ultraviolet to the infrared are plotted logarithmically on the ordinate.

Not all surfaces scatter according to Eq. (2). As was mentioned previously, if the surface irregularities do not have a Gaussian height distribution, an "effective roughness" can be used to describe their behavior. The value of this effective roughness will then be wavelength-dependent. At nonnormal incidence or when the slopes of the surface irregularities become too large, more elaborate theories are required. This is also true if the polarization dependence or the angular dependence at large angles is required. To illustrate, we show in Figure 9 a comparison of experimental results obtained by Hunderi and Beaglehole [1970] with the predictions of the perturbation-type theories [Eq. (4)]. The sample was prepared by first coating a smooth substrate with polycrystalline calcium fluoride by vacuum evaporation to generate an irregular surface, then overcoating this with an opaque deposit of silver. The scattering of normally incident light from this silver sample was determined as a function of scattering angle and polarization at a wavelength of 5500 Å. The experimental data are indicated by data points through which smooth, solid curves have been drawn. Shown are data for scattered light that is polarized parallel to the scattering plane (P) and light polarized in a direction perpendicular to the scattering plane (S). The dashed lines are the theoretical predictions from Eq. (4) assuming

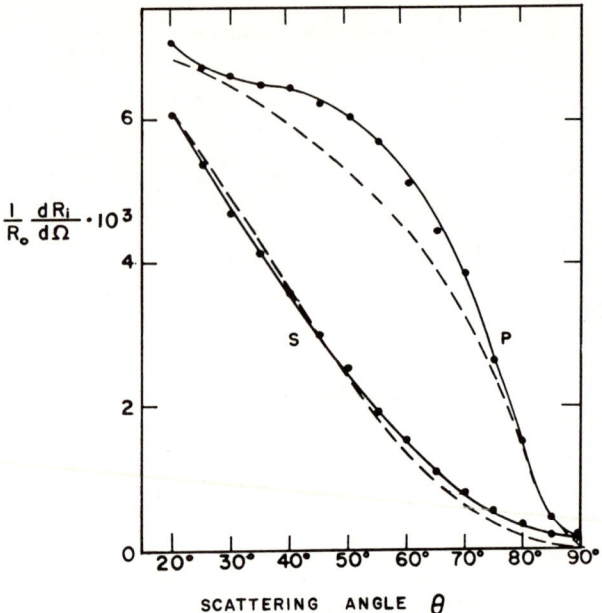

FIGURE 9 Bidirectional reflectance distribution from "rough" silver surface at a wavelength of 5500 Å for scattered light polarized parallel (P) and perpendicular (S) to the scattering plane. Solid curves are drawn through experimental data points; dashed curves are theoretical predictions [Hunderi and Beaglehole, 1970].

a Gaussian autocorrelation function with an autocorrelation length of 1100 Å. The scalar theory does not distinguish light of the two polarizations, which in this case differ in intensity by more than a factor of 3 over a fairly large range of scattering angles.

In some cases additional phenomena are involved. Figure 10 shows such an example—scattered-light measurements made on a silver-coated mirror having an effective roughness of 21.5 Å, as determined from scattering measurements made in the ultraviolet. This "rough" mirror was also made by depositing calcium fluoride onto a superpolished fused-quartz flat prior to the deposition of an opaque layer of silver. For this mirror, the ratio of scattered light to incident light is plotted instead of the ratio of scattered to reflected light as shown in the previous graphs. The level of scattered light predicted by Eq. (2) and the reflectance of the material is indicated by the long-dashed curve. The *difference* between what was actually observed for the silver-coated surface and the predicted level is shown by the solid curve. This curve has

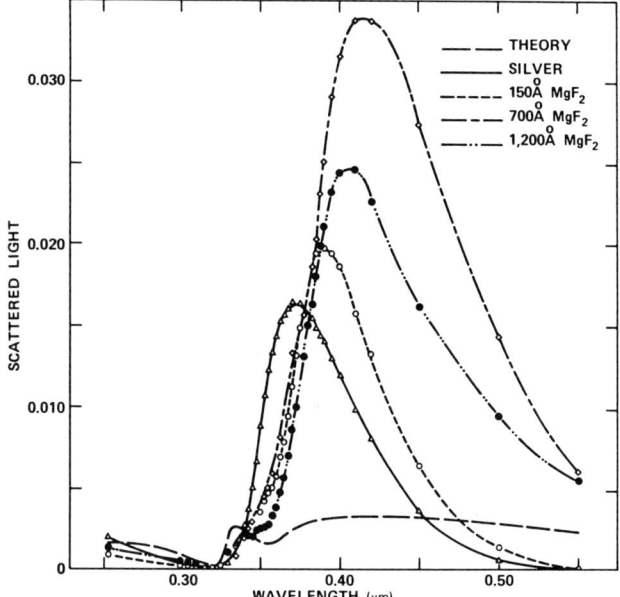

FIGURE 10 The long-dash curve is the wavelength dependence of the ratio of the total scattered light to the incident light for a "rough" (21.5 Å rms) silver surface calculated from Eq. (2). Other curves are the *difference* between measured values and the calculated (long-dash) curve: solid curve is for a bare silver surface, and other dashed curves are for silver surfaces coated with the indicated thicknesses of magnesium fluoride.

its maximum near 3700 Å and shows that at this wavelength the surface scatters over seven times the amount of light predicted by theory. When layers of magnesium fluoride of various thicknesses are added to the silver mirror, the scattered light level increases further and the peak of the scattering shifts to longer wavelengths. The maximum amount of scattering occurs with a 700 Å thick overcoating layer of magnesium fluoride; for this case, the mirror scattered over 11 times as much light as is predicted by Eq. (2). As the overcoating layer becomes thicker, the scattered light level drops and the peak shifts back to shorter wavelengths.

The scattering behavior of the silver-coated mirror shown in Figure 10 can be explained in terms of optical excitation of surface plasmons, which subsequently decay causing incoherent re-emission to occur. The shift in the peak of the scattering curve to longer wavelengths and its growth when magnesium fluoride is deposited on the sample is caused by the effect of a dielectric other than air on the dispersion curve of the surface plasmons and by an interference effect in the magnesium fluoride film itself. This example illustrates that scattered light can arise in more complicated ways than would be predicted from conventional diffraction theory.

The results presented thus far have been for mirrors having specially prepared rough surfaces. However, similar results have been observed on commercially available samples. The dielectric-overcoated gold mirror whose scattering characteristics are shown in Figure 11 was obtained from a commercial source. In the ultraviolet region, the scattered light level indicates an effective roughness of about 10.6 Å, i.e., a very smooth surface. However, in the visible region, the scattering level increases to over four times the theoretical value. It then drops to nearly the value predicted from the ultraviolet data. Sufficient information is not available in this case to determine the origin of the observed effect. This example illustrates, however, that unexpected scattering properties do sometimes occur in mirrors used in operational systems and indicates that there is considerable work still to be done on the deceptively simple appearing problem of scattered light from metal-coated mirror surfaces.

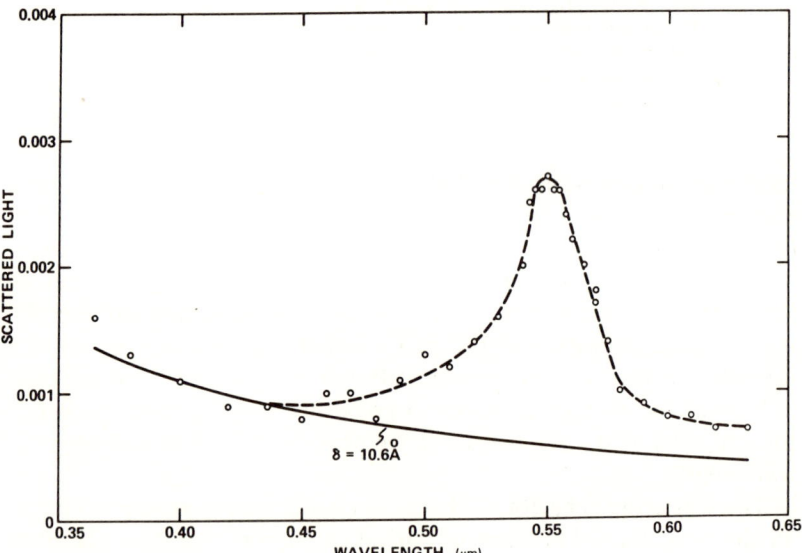

FIGURE 11 Measured wavelength dependence of the total scattered light from a dielectric-overcoated gold mirror (circles) and calculated curve for a 10.6 Å rms surface.

## Conclusions

The scattered light from high-quality mirror surfaces arises primarily from diffraction effects not from geometrical optics. Therefore, to reduce scattering levels further, it is the surface microirregularities that must be controlled not scratches and other macroirregularities. Several theoretical approaches to the scattering problem have been proposed. For light normally incident on a surface whose statistical description is comparable with that of many actual optical surfaces, the theoretical predictions are in approximate agreement with each other and with the experimental results. For nonnormal incidence, the situation is less clear but is being actively investigated. A reliable theory would make it possible to characterize a surface either from scattered-light measurements in a limited spectral range or from measurements of surface irregularities using interference and electron microscope techniques. The scattering properties of that surface in arbitrary wavelength regions should then be accurately predictable. However, besides additional theoretical work, special instruments will be required to make such measurements. These instruments are being developed; their existence will make it possible to develop better and more reproducible polishing techniques. Such techniques are badly needed, particularly for metal mirrors. Mirrors of this type are vitally necessary for many space applications, but at present polishing techniques for making these mirrors are seriously inadequate.

Finally, scattering may arise from mechanisms other than conventional diffraction. In these cases, scattering levels may be several times those that would be predicted from the above theories. This anomalous scattering is strongly material-dependent, so that the mirror coating material, as well as the surface microirregularities, is important when minimum scattered light is required.

We are grateful for the assistance of John Dancy, who took the electron micrographs, Marian Hills, who helped us to use the Nomarski technique, and James Rogers of Hughes Aircraft Company, who sent us the commercially obtained gold-coated sample for testing.

## References

Allen, R. D., G. B. David, and G. Nomarski, *Z. Wiss. Mikrosk. 69,* 193 (1969).
Baird, K. M., *Can. J. Res.* 27A, 130 (1949).
Bennett, H. E., and J. M. Bennett, in *Physics of Thin Films,* Vol. 4, pp. 31-37, G. Hass and R. E. Thun, eds., Academic Press, New York (1967).
Bennett, H. E., and J. O. Porteus, *J. Opt. Soc. Am. 51,* 123 (1961).

Chinmayanandam, T. K., *Phys. Rev. 13*, 96 (1919).
Davies, H., *Proc. Inst. Elec. Eng. 101*, 209 (1954).
Elson, J. M., and R. H. Ritchie, *Phys. Status Solidi*, to be published.
Elson, J. M., and R. H. Ritchie, *Phys. Rev. B 4*, 4129 (1971).
Ginsberg, I. W., T. Limperis, F. E. Nicodemus, and J. C. Richmond, NBS Tech. Note, to be published.
Hunderi, O., and D. Beaglehole, *Phys. Rev. B 2*, 321 (1970).
Kröger, E., and E. Kretschmann, *Z. Phys. 237*, 1 (1970).
McLeod, J. H., *J. Opt. Soc. Am. 45*, 402A (1955).
MIL-STD-150A (May 1959).
Nicodemus, F. E., *Appl. Opt. 9*, 1474 (1970).
Nomarski, G., and A. R. Weill, *Bull. Soc. Franc. Miner. Crist. 77*, 840 (1954).
Porteus, J. O., *J. Opt. Soc. Am. 53*, 1394 (1963).

A. MARÉCHAL and G. FORTUNATO

# Recent developments in selective modulation spectrometry

## I. Introduction

The performance of spectrometers mainly depends on two factors: resolution limit and luminosity. In many types of spectrometer, these two parameters are not independent of each other. Jacquinot [1954] has shown that the relation, luminosity × resolution = constant, does not allow sufficient luminosity and resolution to be obtained simultaneously. However, several systems make it possible to avoid this limitation. This is particularly the case for spectrometers using a confocal Fabry-Perot [Connes, 1958], as well as the Girard's [1967] grid spectrometer.

Other systems without totally escaping from this limitation, make it possible to obtain a better compromise between these two parameters. This is the case, for example, of SISAM [Connes, 1959, 1960], of Fabry-Perot [Chabbal, 1953/54], and in Fourier transform spectroscopy [Connes, 1961]. The last one, however, in spite of its advantage of being multiplex, necessitates the use of a computer.

Our goal is to show the possibilities of a new type of interference spectrometer with selective modulation, which associates an interfer-

---

The authors are at the Institut d'Optique and ENSET, Paris, France.

ometer with very high luminosity to a periodic grid [Fortunato, 1972; Fortunato and Maréchal, 1972]. The luminosity of this system depends only on its geometry. The resolution increases with the number of the grid slits, so that these two parameters, luminosity and resolution, are not directly dependent and hence can be adjusted at will.

## II. Principle of the System

This principle has been set by Prat [1965, 1971], who, while developing the concept of spatial resonance, has indicated several new arrangements, the simplest of which is the following (Figure 1):

An extended source S is used to form two images $S_1$ and $S_2$ by means of splitting system D. Under these conditions, the images $M_1$ and $M_2$ of a point M on the source are coherent, and the optical path difference between the two vibrations corresponding to two parallel rays coming out of the interferometer is

$$\Delta = T \cos \phi,$$

where $T$ is the distance $M_1 M_2$ and $\phi$ is the angle between the exit beam direction and the vector $\mathbf{M_1 M_2}$. If $T$ is a constant, then when M is displaced over the source, $\Delta$ depends only on $\phi$ and the fringe pattern is observable at infinity. Its contrast is independent of the source extent. These fringes can be localized in the image focal plane of a lens L, and their luminosity depends only upon the geometry of the splitting system D and the quality of the lens L. If now, we introduce in the fringe plane a periodic grid that we vibrate, the flux going through this grid will be

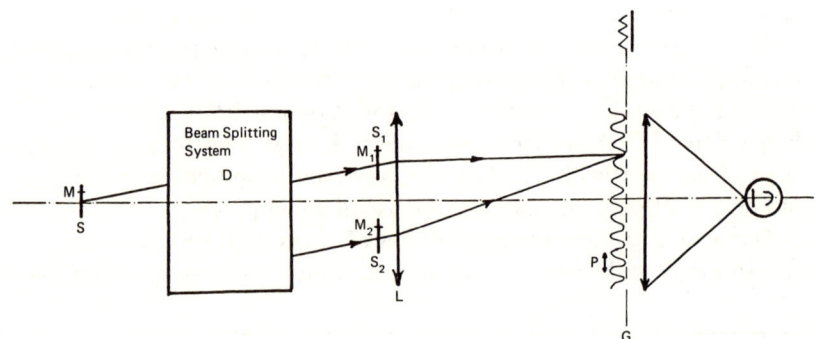

FIGURE 1  Analysis of rectilinear fringes.

modulated only for the wavelength λ such that: $T \times P = F \times \lambda$ ($F =$ focal length of the lens, $P =$ grid spacing). A study of the resolution limit of this system shows that this limit depends only on the number of grooves, $G$, on the grid and hence does not obey a relationship of the type $L \times R = C$.

We have arrived at the same kind of arrangement as Prat by studying the variations in the spatial degree of coherence in interferometers and by asking the following question: How can we obtain a temporal modulation of only one wavelength among others, while keeping the etendue as large as possible? The simplest solution seems to be the following. An interferometer, made of plane mirrors, splits an image point M into two object points $M_1$ and $M_2$ of which the relative distance $T$ does not depend on the position of M. The degree of partial coherence $\gamma(M_1, M_2, t)$ must depend only upon the relative positions of $M_1$ and $M_2$. We can then write it in the form $\gamma(M_1 - M_2, t)$. This is obtained if the source is localized at infinity in the object space, that is to say, in fact, in the object focal plane of a lens (Figure 2).

The Van Cittert-Zernike theorem shows that the degree of partial coherence is related to the Fourier transform of the intensity distribution of the source. In order to obtain a nonnegligible value of $\gamma$ when the source is large, it is necessary to modulate the light intensity of the source with a given periodicity. We can do that by putting a periodic grid in the focal plane of lens L (Figure 2). The degree of partial coherence between two points, $M_1$ and $M_2$, at distance $T$ from one another has then a maximum if the grid spacing $P$ satisfies the relation:

$$\lambda F = TP.$$

If we displace the grid by a quantity $u$, in its plane, the degree of

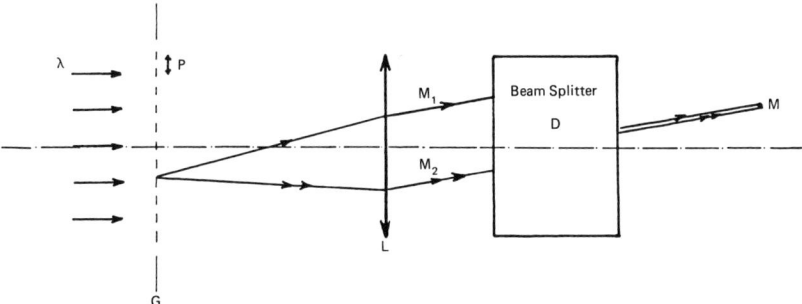

FIGURE 2 Selective modulation interference spectrometer.

partial coherence is multipled by $\exp[j(2\pi u/P)]$, and, hence, the flux impinging on every point M of the detector is modulated with respect to the grid's position and this modulation is obtained only for the wavelength that satisfies the above relation $\lambda F = TP$.

In conclusion, we can see that this process allows us to selectively modulate one and only one element $\Delta\lambda$ in the spectral range of the source.

## III. Practical Choice of the Components

### A. THE INTERFEROMETER

The choice of the interferometer is very important, and practical considerations made us choose an interferometer of the Sagnac type (Figure 3). Its main advantages are the following:

The apparatus is easily made; it consists simply of a beam splitter and a compensating plate, along with two mirrors. The tolerances upon the flatness and parallelism are of the order of $\lambda/4$ in order to get a resolution of $10^4$.

The spectrum is explored by simply translating one of the two mirrors.

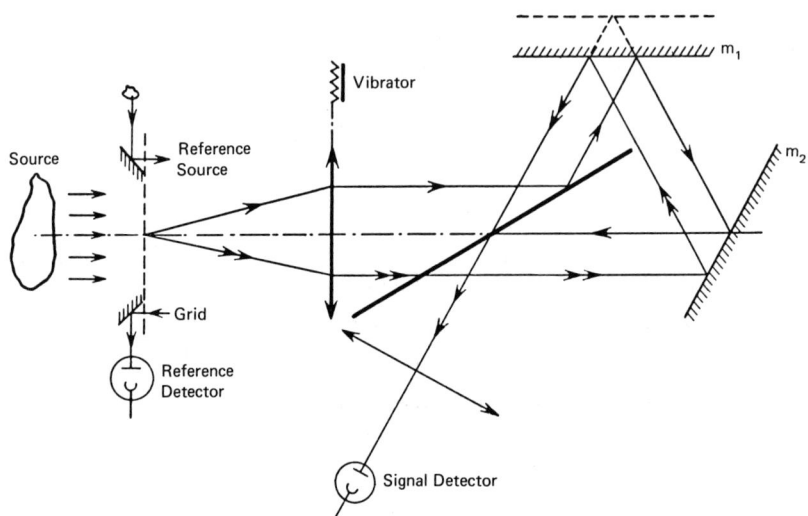

FIGURE 3  Sagnac-type interferometer.

The interferometer is quite insensitive to perturbations, vibrations, etc. (since the two beams follow the same path).

Finally, it is easy to set it up.

The main disadvantage of this type of interferometer is that its geometry limits the etendue of the beams. Indeed, it is clear from Figure 3 that if the system is unfolded the beam splitter acts like several apertures that may limit the etendue of the system when the splitting becomes important. However, there is no limitation, in the direction perpendicular to the plane of the figure.

B. METHOD OF PRODUCING AND DETECTING THE SIGNAL

Among several possibilities, we have chosen a sawtooth modulation of the grid's position, the amplitude being equal to a large number of the grid's spacings. The reference signal is obtained by forming the image of a small part of the moving grid upon a fixed part of the same grid; we are then able to use synchronous detection.

## IV. Systems Performance

A. RESOLVING POWER

For a given wavelength, the degree of partial coherence is a function of the distance $T = M_1 M_2$. If $N$ is the number of the grid's spacings and if the source intensity is uniform, $\gamma$ has a maximum for $T = \lambda F/P$ and its representation versus $T$ has the shape indicated by Figure 4. The full width at half-maximum is $\lambda F/NP$. Two wavelengths $\lambda$ and $\lambda + \Delta\lambda$, with comparable intensities, will be separated if the two values of $\gamma$ for the same value of $T$ are different. For this to be so, it is necessary for $\Delta\lambda$ to be of the order of $\lambda/N$. The resolving power $R = \lambda/\Delta\lambda$ will then be of the order of $N$. Apodization can be obtained by suitably modulating the intensity distribution on the grid or by putting a suitable aperture function in the plane of this grid, for example, in the form of a Gaussian. This apodization can also be determined by putting the aperture function in the conjugate plane of the grid.

B. LUMINOSITY

This quantity depends on the etendue of the beams, that is to say, on the area of the grid and the solid angle with which the beams propagate inside the interferometer. As we have noticed above, the only theoreti-

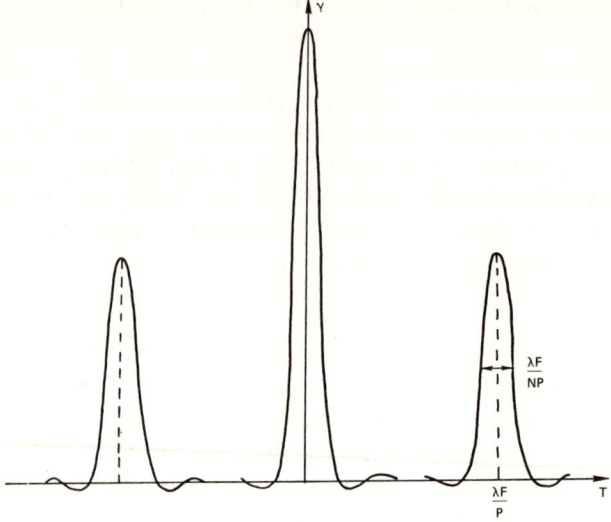

FIGURE 4  Spatial variation of the coherence.

cal limitation is imposed by the geometry of the interferometer (limitation of the aperture angle in the plane of Figure 3). If we define luminosity $L$ as being the ratio between the flux incident on the detector and the source radiance, we get the following relation:

$$L = \tau S \alpha \beta,$$

where $\tau$ is the transmission factor of the system. $S$ is the area of lens $L$. $\alpha$ and $\beta$ are the angular dimensions of the grid.

Let us compare this luminosity to that of a classical grating spectrometer. We know that in this case, if we define $\tau$ as being the spectrometer transmittance, $S'$ the grating area, $\beta'$ the angular width of the slit, and $R'$ the effective resolving power, the luminosity can be written as

$$L' = \tau' \, (S'\beta'/R').$$

The gain in luminosity of our spectrometer (SIMS) is then

$$g = L/L' = (\tau S \alpha \beta / \tau' S' \beta') \, R'.$$

This shows that we can simultaneously increase the luminosity and the resolution of our interferometer. Indeed, the resolution $R$ is equal to $N$, the number of the grid's grooves, which is proportional to $\alpha$. On the other hand, the gain in luminosity may be important as is shown in the following example.

The experiments have been done until now with

A circular grid, 40 mm in diameter, 70 grooves per mm;
A lens, 36 cm$^2$ in area, 500-mm focal length.

To make a valid comparison, we consider a grating spectrometer with the following characteristics:

Grating area: 50 cm$^2$;
Angular width of slit: $5 \times 10^{-2}$ rad.

With the same resolution for both these instruments, that is to say, $R' = 2800$, the gain in luminosity of ours is of the order of 250.

We are now setting up an interferometer, which will consist of an 80-mm-diameter grid, a lens area of 100 cm$^2$, and 500-mm focal length. This should provide a gain of 9000 over the same grating spectrometer.

C. SIGNAL-TO-NOISE RATIO

This ratio is fundamental in the exact evaluation of the qualities of any spectrometer. The noise arises from different phenomena and depends on several parameters. In the ultraviolet region, photon noise is predominant; in the infrared, it is detector noise. When photon noise is predominant, our system is still interesting if the number of spectral elements or the intensity ratio between two lines is less than the luminosity gain that has been previously defined. When the noise source is the detector, the gain in signal-to-noise ratio is equal to the gain in luminosity.

## V. Experimental Study

The experimental results obtained with a preliminary system (which is very modest as far as the quality of the optical and mechanical components are concerned) have permitted us to verify some of the theoretical predictions that were previously announced. The resolving power is equal to the theoretical one of the grid. The gain in luminosity, mea-

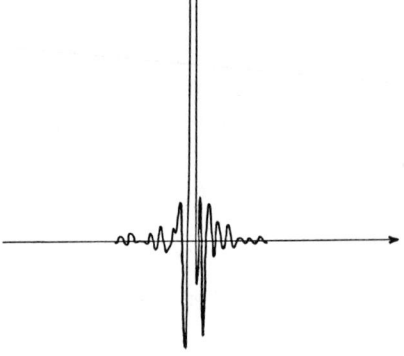

FIGURE 5 Selective modulation spectrometer instrumental function. $\lambda = 5460$ Å; $\Delta\lambda = 5$ Å.

sured by comparing our spectrometer to a commercial prism spectrometer having the same resolving power, is of the order of $3 \times 10^4$. This measurement has been done, on the one hand, by using a monochromatic line (Figure 5) ($\lambda = 5460$ Å from a low-pressure mercury vapor lamp) and, on the other hand, on the line spectrum of a neon lamp (Figure 6). We also performed an experiment in absorption spectroscopy by using holmium oxide [Figure 7(a)]. The source we used in this last experiment (tungsten filament lamp) has the spectrum shown in Figure 7(b). In every experiment, the detectors are photomultipliers.

## VI. Conclusion

This first theoretical and experimental study indicates that in the visible range if the spectrum is not too dense, this spectrometer will allow

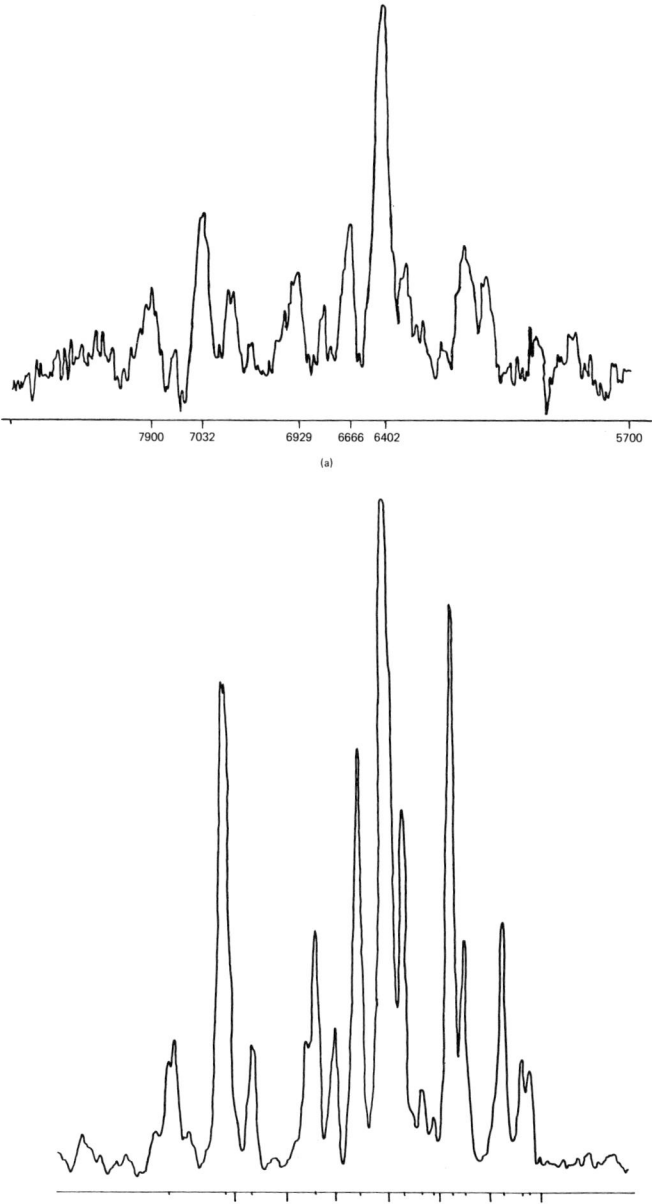

FIGURE 6 (a) Neon spectrum obtained with a slit spectrometer; (b) neon spectrum obtained with the interferometric spectrometer.

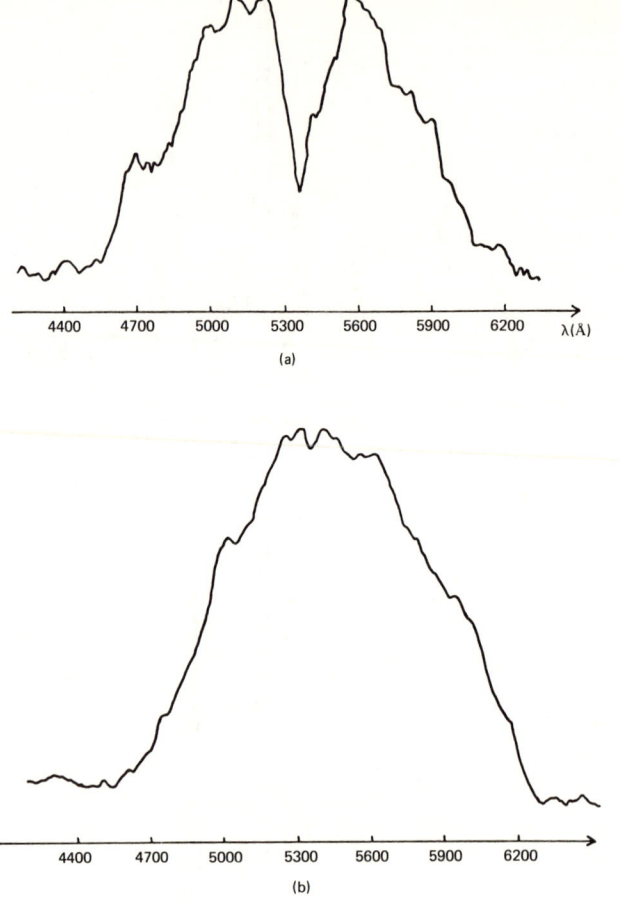

FIGURE 7 (a) Absorption spectrum, holmium oxide; (b) emission spectrum, tungsten.

either important gains in speed or the possibility of studying very weak sources. Until now, we have been limited only by the response time of the recorder, and we have been able to record low-density spectra, as that of neon, at the speed of 4000 Å/sec, without notably diminishing the signal-to-noise ratio. Experiments now in progress will allow us to test more completely this apparatus in the visible, especially in absorption and fluorescence spectroscopy. Furthermore, these experiments in the visible, the most unfavorable region for selective modulation interference spectroscopy, allow us to hope for interesting results in the infrared region.

The editors wish to thank J. L. Meysonette for preparing the English translation of this paper.

## References

Chabbal, R., *J. Rech. CNRS 5,* 138 (1953/54).
Connes, J., *Rev. Opt. 40,* 45, 116, 171, 231 (1961).
Connes, P., *J. Phys. Rad. 19,* 262 (1958).
Connes, P., *Rev. Opt. 38,* 157, 416 (1959).
Connes, P., *Rev. Opt. 39,* 402 (1960).
Fortunato, G., *Compt. Rend. Ser. B 274,* 688 (1972).
Fortunato, G., and A. Maréchal, *Compt. Rend. Ser. B 274,* 931 (1972).
Girard, A., Thesis, Paris (July 1967).
Jacquinot, P., *J. Opt. Soc. Am. 44,* 761 (1954).
Prat, R., *Japan. J. Appl. Phys. Suppl. 1,* 448 (1965).
Prat, R., *Opt. Acta 18,* 213 (1971).

O. S. HEAVENS and S. K. SHARMA

# USE OF LATERAL WAVES IN THE STUDY OF SURFACE FILMS

## I. Introduction

In the study of surface films by total internal reflection spectroscopy, use is made of the penetration into the less dense medium of a light wave at angles greater than the critical angle. The penetration is generally to a depth of the order of a wavelength, and reflection is accompanied by a shift—the Goos-Hänchen shift—which is of the same order. Thus the volume of the medium sampled is small, and a large number of reflections is required in order to obtain a measurable amount of absorption. This necessitates a parallel-sided specimen of reasonable size— generally of the order of 100 mm in length. When this can be realized, then very high sensitivities are obtainable in favorable circumstances— such that a monolayer of adsorbed gas may be detected. This paper describes a possible alternative method for cases for which the above form of specimen preparation may not be possible. It makes use of the lateral wave that is propagated along the interface between the two media when light is incident *at* the critical angle. Such waves have been extensively studied [Maeker, 1949; Ott, 1949; Acloque and Guillemet,

---

The authors are in the Department of Physics, University of York, England.

1960; Osterberg and Smith, 1964; Tamir and Felsen, 1965] in a problem in which the properties of a thin layer in the neighborhood of a surface are to be investigated. The exact theory for the case of propagation of a lateral wave in a medium in which the optical properties vary in the direction normal to the plane of the interface has not been solved. For the case of a uniform film at the surface, however, expressions may be obtained that are capable of yielding numerical solutions.

The method is illustrated by application to the study of color centers formed by bombardment of the surface of a crystal by low-energy electrons.

## II. Analysis of Lateral Wave

We may first note that the displacement of a beam of finite width on reflection at a surface is a normal consequence of diffraction theory and is not restricted to the case of total reflection. The problem has been examined by Brekhovskikh [1960]. For the case in which a narrow pencil beam is reflected at an angle $\theta_0$ at a surface with reflection coefficient $R(\theta)$, the displacement may be written as $-x^1(\alpha)$, where $\alpha = k \sin \theta_0$ and $x(p)$ is the phase of the reflection coefficient, expressed in terms of the variable $p = k \sin \theta$. The value of $\Delta$ depends on the state of polarization of the incident light, but for angles close to the critical angle the values for the $p$- and $s$-components, $\Delta_p$ and $\Delta_s$ are approximately equal and are given by

$$\Delta_p = \Delta_s = (\lambda/\pi) \tan \theta_0 / (\sin^2 \theta_0 - n^2)^{\frac{1}{2}}, \tag{1}$$

where $n$ is the refractive index of the medium of incidence or the ratio of indices of the first to second medium. The expansion on which the above result is based is not strictly valid for $\theta_0 = \arcsin n$, but it is clear that as the critical angle is approached the displacement can become very large. Viewed in this way, the displacement on a beam incident in the neighborhood of the critical angle may be regarded as being associated with a lateral wave that traverses the region of the surface between an incident ingoing bundle and an emerging bundle at a distance $\Delta$.

In order to calculate the flux in the portion of the wave emerging from the surface, it is necessary to consider a point source (Figure 1) emitting spherical waves and to make a Fourier expansion using plane waves. In order to encompass the singularity associated with the point source, the expansion must necessarily involve plane waves with complex wavenumber. In the integration required to determine the reflected

FIGURE 1 Lateral wave flux at $P$.
$$F_{\text{lat}} = \frac{\lambda}{\pi^2 n^2 (n^2 - 1) r l^3}.$$

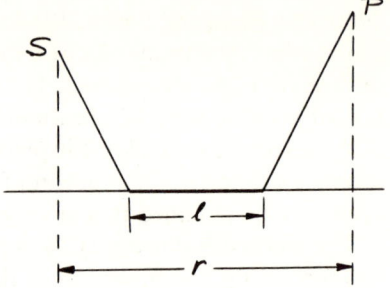

amplitude, due attention must be paid to the branch points and singularities in the complex plane. For the case for which the distances involved are large compared with the wavelength, the Hankel functions involved in the expression for the reflected field may be replaced by asymptotic exponential forms. The result of the analysis leads to the following dependence of the lateral wave flux on the experimental parameters, for the case in which the media either side of the boundary are isotropic and nonabsorbing:

$$F_{\text{lat}} = \frac{\lambda^2}{\pi^2 n^2 (n^2 - 1) r l^3}, \tag{2}$$

where the symbols are as shown in Figure 1.

Equation (2) is obtained for uniform, isotropic media as a result of integrating the expression for the field due to a wave reflected at a surface. This integral takes the form

$$\psi \sim e^{i\pi/4} \left(\frac{k}{2\pi r}\right)^{1/2} \int_{-(\pi/2)+i\infty}^{(\pi/2)-i\infty} \exp ikR_1 \cos(\theta - \theta_0) R(\theta) \left(1 + \frac{1}{8ikr \sin\theta}\right) \sin^{1/2}\theta \, d\theta \tag{3}$$

(see Figure 2), where the integral is evaluated over the appropriate path in the complex plane.

In Eq. (3), $R(\theta)$ is the reflection coefficient that in the case of uniform media is simply the Fresnel reflection coefficient. In the case of a lateral wave traversing a surface carrying a uniform film, an analytic expression for $R(\theta)$ can easily be obtained [Heavens, 1955].

For the more complicated case of a surface film in which the optical properties vary with depth, an analytical expression for $R$ is not nor-

*Method of steepest descents:*

$$\psi_{lat} = \frac{-2i \exp[ikR_1 \cos(\theta_c - \theta_0)]}{nk\sqrt{r\cos\theta_c}\,[R_1 \sin(\theta_c - \theta_0)]^{3/2}}$$

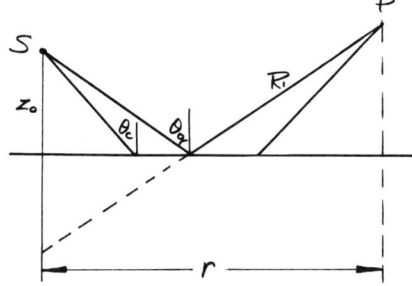

FIGURE 2  Definition of the parameters $z_0$, $R_1$, $\theta_C$, and $\theta_0$.

mally obtainable. In this case, numerical methods would need to be used.

## III. Application of Lateral Waves to Study of Absorbing Layers

The experiments to be described arose in connection with an examination of the behavior of surfaces of alkali halides that had been subjected to slow electron bombardment. It was found that the nucleating properties of the surfaces of such crystals were profoundly modified by the action of such electron bombardment.

It was suspected that this might be due to the formation of damage centers (e.g., $F$-centers), and the presence of such centers had been inferred by taking an absorption spectrum of the damage layer. When this spectrum is obtained by the use of light at normal incidence, detection is possible only if the density of centers is extremely high. This is because the depth of penetration of the slow electrons is very small, amounting perhaps to a few tens of nanometers. In order to be able to determine the density of color centers under modest bombardment, a technique alternative to normal incidence transmission is needed, and for this purpose use may be made of lateral waves.

The convenient experimental arrangement is shown in Figure 3, in which convergent polychromatic light is incident on one prism face in

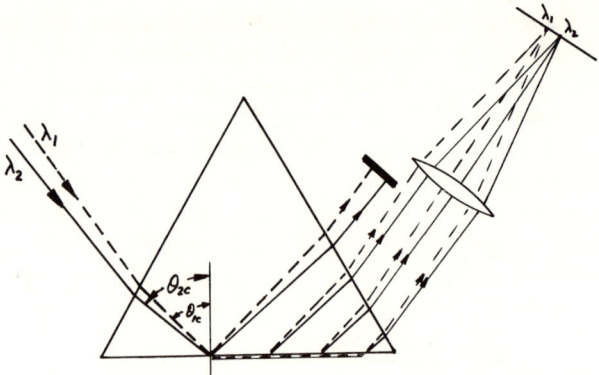

FIGURE 3  Experimental arrangement. Convergent polychromatic light is incident on one prism face so that lateral waves transverse the lower face.

such a way that lateral waves traverse the layer face. On account of the dispersion of the prism, the critical angle for different wavelengths varies so that the lateral wave spectrum is dispersed and may be focused in the way shown by means of a lens. The experimental arrangement for examining bombarded crystals is shown in Figure 4. In this case, monochromatic radiation whose wavelength corresponds to the peak of the ab-

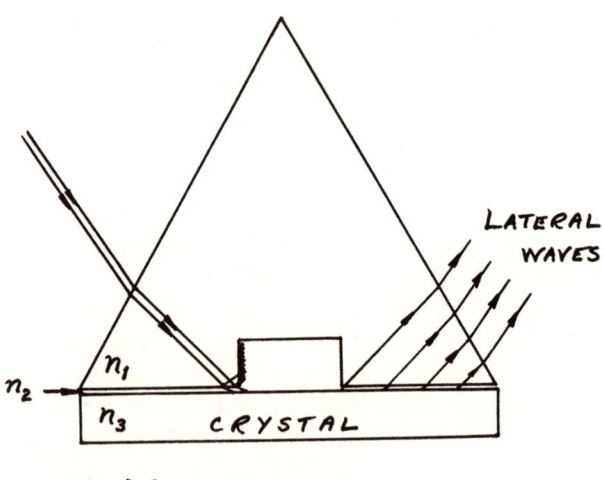

FIGURE 4  Experimental arrangement for examining bombarded crystals.

sorption spectrum of the *F*-centers is used. By arranging that only strips of crystal (Figure 5) be exposed to electron bombardment and by traversing the crystal in a direction perpendicular to that of the lateral waves, a comparison of the lateral wave intensities for bombarded and unexposed surfaces is possible. A typical result of these experiments is shown in Figure 6, in which the normal-incidence transmission curve (dotted) is compared with the lateral-wave curve. It is immediately seen that the depth of modulation is enormously larger for the lateral waves.

The quantitative exploitation of this technique requires that the crystal surface be polished to a high degree of smoothness. This is so because the penetration of the lateral-wave field into the second medium extends only to a distance of the order of the wavelength. Thus the presence of scratches, even of a depth a fraction of a wavelength would interfere seriously with the propagation of the lateral wave. Experience has shown that conventional polished crystal flats are unsuitable for lateral-wave studies on account of the effects of surface scratches.

The present position is that lateral waves are demonstrably sensitive enough to probe very thin absorbing surface layers. Exact theory is not yet available for real systems, although an approximate theory, invoking the known exponential decay of field in the second medium, can yield results with a reasonable degree of accuracy.

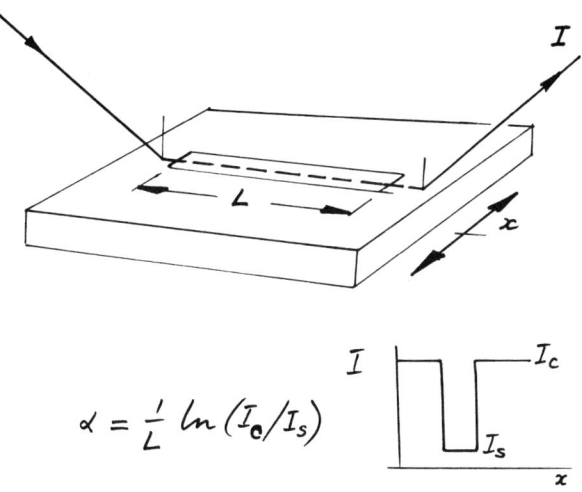

FIGURE 5 Strip of crystal exposed to electron bombardment. $\alpha = \frac{1}{L} \ln(I_c/I_s)$.

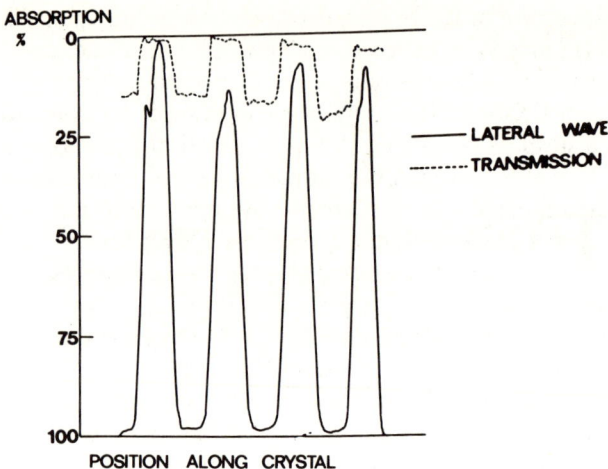

FIGURE 6  Experimental result. The normal-incidence curve ( · · · ) is compared with the lateral curve (———).

## References

Acloque P., and C. Guillemet, *Compt. Rend. 250*, 4328 (1960).
Brekhovskikh, L. M., *Waves in Layered Media,* Academic Press, New York (1960).
Heavens, O. S., *Optical Properties of Thin Solid Films,* Butterworths, London (1955).
Maeker, H., *Ann. Phys. 4*, 28 (1949).
Osterberg, H., and L. W. Smith, *J. Opt. Soc. Am. 54*, 1073 (1964).
Ott, H., *Ann. Phys. 4*, 432 (1949).
Tamir, T., and L. B. Felsen, *IEEE Trans. Antennas Propag. AP-13*, 410 (1965).

R. BOULAY, J. W. Y. LIT, and R. TREMBLAY

# Focusing Systems Adaptable to Integrated Optics

## Introduction

Different focusing systems have been studied. The basic elements are free irises, metallic cylinders, and annular lenses. All the systems have a few things in common, among which is the dependence of their focusing capabilities, at least in part, on diffraction effects. In this paper, we shall summarize the principal results that we have obtained. All, except one, of the experiments were done with microwaves and in the three-dimensional case. A brief mention on the possibility of constructing similar focusing systems in thin films will also be given.

## Focusing Systems

Consider an ellipsoid of revolution whose cross section (an ellipse) is shown by the broken line in Figure 1(a). S and S' are the foci of the ellipsoid. The ellipsoid is specified by the distance $L$ between S and S'

---

The authors are in the Laboratoire de recherches en optique et laser, Departement de Physique, Université Laval, Québec 10, P.Q., Canada.

and by the Fresnel number $N$, which is the path difference (expressed in units of $\lambda/2$, where $\lambda$ is the wavelength for which the system is designed) between the distance from S to S' via a point on the surface of the ellipsoid and the direct distance $L$. If free irises ($A_1, A_2, \ldots$) are placed between S and S' in such a way that the rims of all the irises touch the surface of the reference ellipsoid, then any wave that is emitted by S, diffracted by the edge of one of the irises, and then arrives at S' (such as $SCS'$) in Figure 1(a) will have traveled the same physical path length.

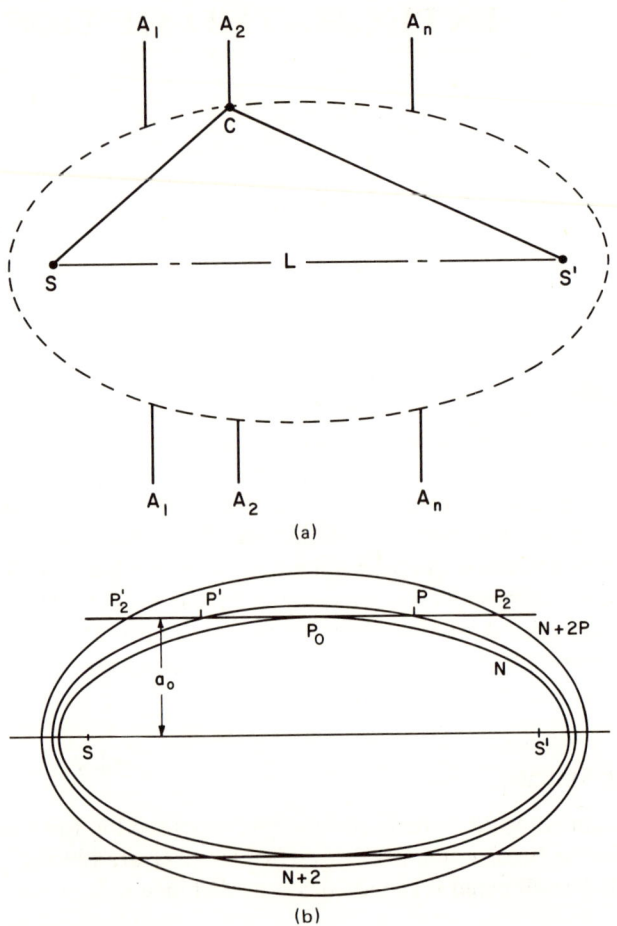

FIGURE 1 Geometry of iris systems. (a) System with one reference ellipsoids—radii of irises different. (b) System with a set of reference ellipsoids—radii of irises all equal.

Thus, if S is a point source, all the singly diffracted waves will be in phase at S'. Such an iris system will possess focusing properties [De et al., 1968].

Instead of making use of a single ellipsoid, consider a family of ellipsoids, having common foci S and S' [Figure 1(b)]. The difference between the distance from S to S' via a point on the ellipsoid and the direct distance from S to S' is $(N + 2p)(\lambda/2)$, where $N$ is any positive number, and $p$ is equal to 0, 1, 2, etc., with $p$ for the smallest chosen ellipsoid being zero. As shown in the cross-sectional diagram [Figure 1(b)], a straight line parallel to $SS'$ (at a distance $a_0$) is drawn tangent to the smallest chosen ellipse, at the point $P_0$, cutting the other ellipses at points $P_{-1}$, etc. The system studied is then formed by placing free irises with radii all equal to $a_0$ at positions $P_0$, $P_{-1}$, etc. Following the same reasoning as before, it can easily be seen that with a point source at S, such a system will possess focusing capabilities [Lit et al., 1970]. Pushing the concept of equal-radii iris systems a little further, it is expected that a system of suitably positioned cylinders of appropriate lengths will also possess focusing properties [Lit et al., 1971]. Finally, instead of irises, annular lenses can be used; systems similar to those shown in Figure 1 will also be expected to possess focusing properties [Boulay et al., 1971].

## Theory

Since all the systems have diffraction effects, we shall now first summarize the diffraction theory that is used to find the diffracted fields of the systems [Lit et al., 1969].

Consider a point source S situated on the perpendicular axis of an iris (Figure 2). The boundary-diffraction-wave theory says that the field at an observation point P is given by the sum of a geometric-optical wave, which comes directly from S, and a boundary-diffraction wave, which is given by

$$U_B(P) = \frac{1}{4\pi} \int_\Gamma U(Q) \frac{\exp(ik\rho)}{\rho} \frac{\vec{r} \times \vec{\rho}}{1 - \vec{r} \cdot \vec{\rho}} \cdot \vec{l} \, dl, \tag{1}$$

where Q is a typical point on the diffracting edge $\Gamma$; $r$ and $\rho$ are, respectively, the distance from the source to Q and the distance from Q to P, with corresponding unit vectors $\vec{r}$ and $\vec{\rho}$.

For the situation shown in Figure 2, Eq. (1) can be evaluated, for the case with P on the axis (i.e., $x = 0$),

$$U_B(P) = -a^2 Z \exp[ik(r + \rho_0)]/2r\rho_0(r\rho_0 + a^2 - zs). \tag{2}$$

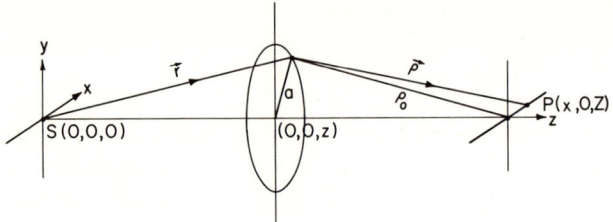

FIGURE 2 Geometry of diffraction by an iris.

When P is away from the axis, assuming that $x$ and $a \ll z$ and $Z$,

$$U_B(P) = C[C_0 J_0\,(kax/s) - i\, C_1 J_1\,(kax/s) - C_2 E], \tag{3}$$

where

$$E = \beta^{-1} \sum_{m=0}^{\infty} \epsilon_m (-i)^m\, \alpha^m\, J_m\,(kax/s), \tag{4}$$

$$C = \frac{-a\, \exp\{ik[r+s+(a^2+x^2)/2s]\}}{2rs\{rs[1+(a^2+x^2)/2s^2]+a^2-zs\}},$$

$$C_0 = \{C_1 + zx[1-(a^2+x^2)/2s^2] - Za^2 x/s^2\}/B,$$

$$C_1 = zax^2/s^2 B,$$

$$C_2 = C_0 - aZ[1-(a^2+x^2)/2s^2],$$

$$B = ax(1+r/s)/\{rs[1+(a^2+x^2)/2s^2]+a^2-zs\},$$

$$\alpha = (1-\beta)/B,$$

$$\beta = +(1-B^2)^{1/2},$$

and the Neumann factor $\epsilon_m = 1$ or 2 according to whether $m$ is equal or not equal to zero. $J_m(v)$ is a Bessel function of the first kind and of integral order $m$. It can be shown that the absolute value of $B$, and thus the value of $\alpha$, is less than unity. Consequently, the series in Eq. (4) converges reasonably fast.

### Results and Discussions

Figure 3 shows the transverse irradiance distribution in the focal plane of an iris system with a single reference ellipsoid whose Fresnel number $N$

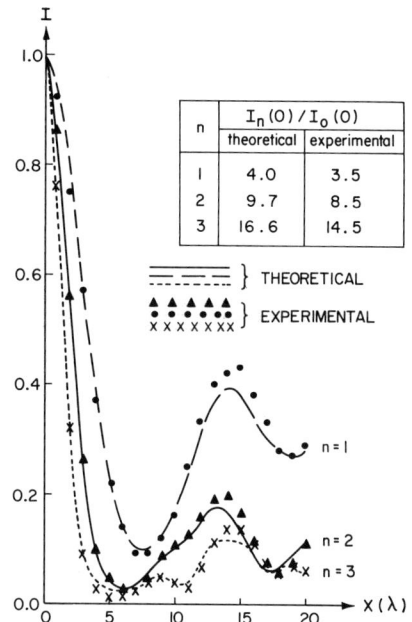

FIGURE 3  Transverse irradiance distributions in principal focal plane of one-ellipsoid system. Fresnel number of system $N = 5$. $n$ = number of irises. Table shows irradiances at principal focus as a function of $n$. $I_0$ = irradiance produced by free-space propagation.

is equal to 5. The accompanying table shows that the irradiance at the principal focus increases as the number $n$ of irises increases. This shows that the system possesses focusing capabilities that increase with increase of irises. It is also evident that as $n$ increases, both the width of the central lobe and the irradiance of the secondary maximum decrease, indicating an increase in focusing capabilities. This effect is again strongly borne out by the table in Figure 4. It can be noted that for the same number $n$ (say, 3), the irradiance values given in the tables in Figures 3 and 4 are different. This is because the axial spacings of the irises in the two cases are different. For a system with a given $N$, the axial spacings can be optimized so that, for a given $n$, the maximum irradiance is produced at the principal focus [Lit et al., 1969).

The Airy pattern of an equivalent perfect lens is also shown in Figure 4. The shapes of the central lobes of the two curves are very similar. Although the secondary maxima of the iris system are generally higher than those of the perfect lens, the reverse is true for the first side maximum. Consequently, it can be concluded that the iris systems can focus reasonably well.

Figure 5 shows the axial irradiance distribution for a system with $N = 6$ and $n = 11$. (A system with $N = 5$ produces essentially the same features.) From Figures 4 and 5, it can be conjectured that the principal

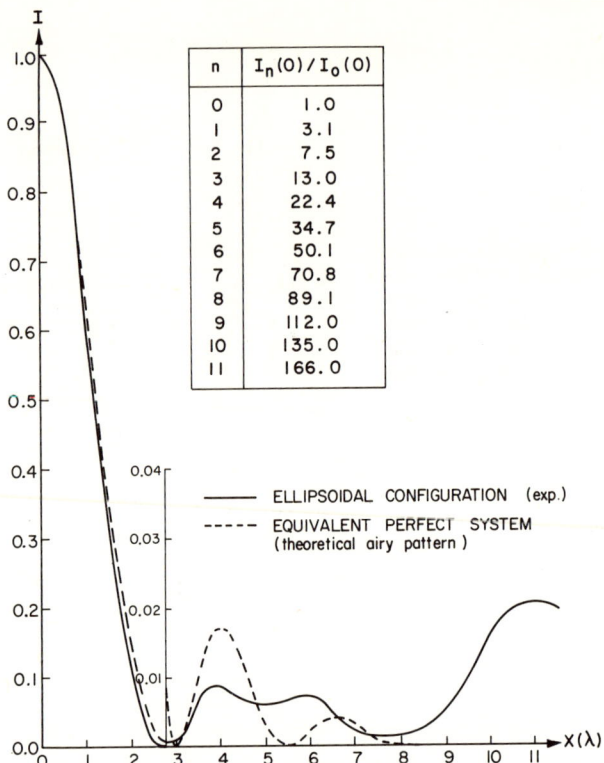

FIGURE 4 Transverse irradiance distributions in principal focal plane. Full curve: one-ellipsoid system with $N = 5$, $n = 11$. Dotted curve: equivalent perfect lens. Table shows irradiances at principal focus of iris system as a function of $n$.

focus of the iris system is marked by a small tubular region of relatively high irradiance, a phenomenon produced by a conventional lens.

If an iris system is constructed in the way shown in Figure 1(b), the radii of all the irises are equal. Figure 6 shows the transverse irradiance distribution in the focal plane of such a system with Fresnel numbers of the reference ellipses equal to 7 and 5. The axial irradiance distribution of the same system is shown in Figure 7. The general characteristics of these two curves are the same as the corresponding ones of the single-ellipsoid systems previously shown, except that, in Figure 7, the main lobe is narrower but the secondary maxima are higher than those in Figure 5.

In Figure 1, it can be easily seen that, with a point source at S, the singly diffracted waves in the one-ellipsoid system [Figure 1(a)] will

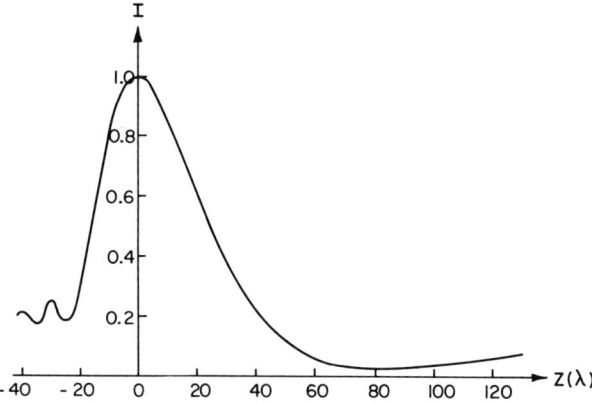

FIGURE 5 Axial irradiance distribution of one-ellipsoid system with $N = 6$, $n = 11$. Point $z = 0$ is the principal focus. Negative values of $z$ means toward system.

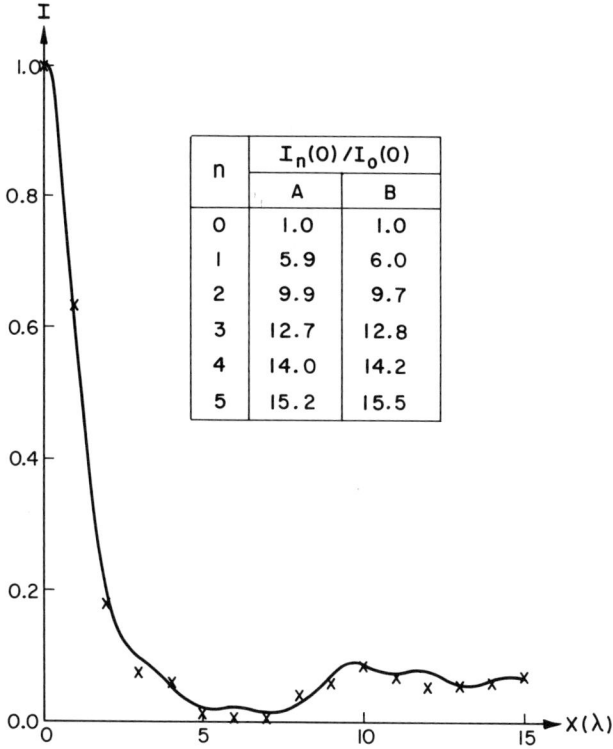

FIGURE 6 Transverse irradiance distribution of two-ellipsoid iris system. Fresnel numbers of reference ellipsoids are 5 and 7. $n = 3$. Table shows irradiances at principal focus as a function of $n$.

| n | $I_n(0)/I_0(0)$ | |
|---|---|---|
| | A | B |
| 0 | 1.0 | 1.0 |
| 1 | 5.9 | 6.0 |
| 2 | 9.9 | 9.7 |
| 3 | 12.7 | 12.8 |
| 4 | 14.0 | 14.2 |
| 5 | 15.2 | 15.5 |

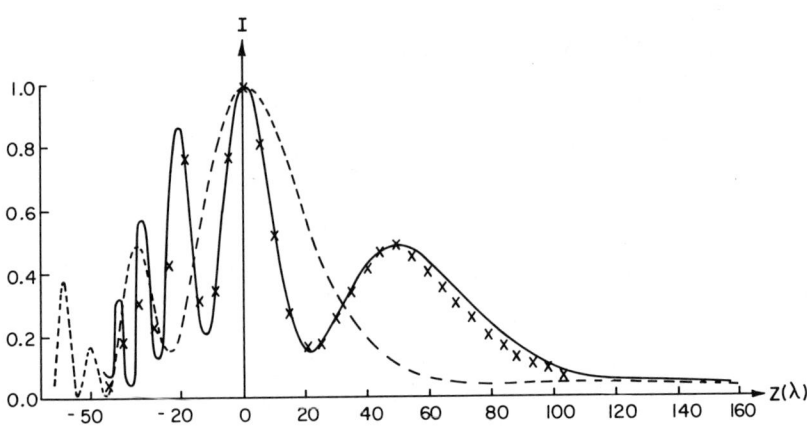

FIGURE 7 Axial irradiance distributions. Full curve: three-ellipsoid system with $N = 5$ and 7, $n = 3$. Dotted curve: one-ellipsoid system with $N = 5$, $n = 3$. Curves are theoretical results. Crosses are experimental data.

always arrive at S′ in phase, independent of the wavelength used. But this is not true with the case shown in Figure 1(b). Here, whether the different singly diffracted waves will arrive at S′ in phase depends on the value of $p$, which in turn depends on the wavelength $\lambda$ of the wave used. Consequently, we should expect that a single-ellipsoid system will have a frequency dependence much less than a multiellipsoid system. This is shown in Figure 8, which shows the theoretical results of the irradiance at the principal focus as a function of frequency. Both systems have five irises and have the same $L = 318.75\lambda$. The Fresnel number of the single-ellipsoid system is 5. The multiellipsoid system has three reference ellipsoids with $N = 5$ and $p = 0, 1$, and 2.

So far we have considered only irises that are far apart. We shall now present some results obtained with two irises separated by a small distance $\delta$ and placed midway between a point source and a detector. Figure 9 shows the variation of the irradiance at the detector as a function of $\delta$. The variation is periodic, with a period of $1\lambda$. The irises were made in metallic plates, and curve (a) in Figure 9 gives the results when the two surfaces of the plates facing each other were not covered by absorbers. When the two surfaces were separated by an absorber, the results are shown by curve (b). The curves show that the periodic variations of the irradiance were due to the interaction of the edges of the two irises.

To simulate a number of closely spaced irises, we have made a metallic cylinder with grooves inside it. In fact, this is a kind of cylindrical grating that makes use of the edge diffraction effects to achieve focusing phe-

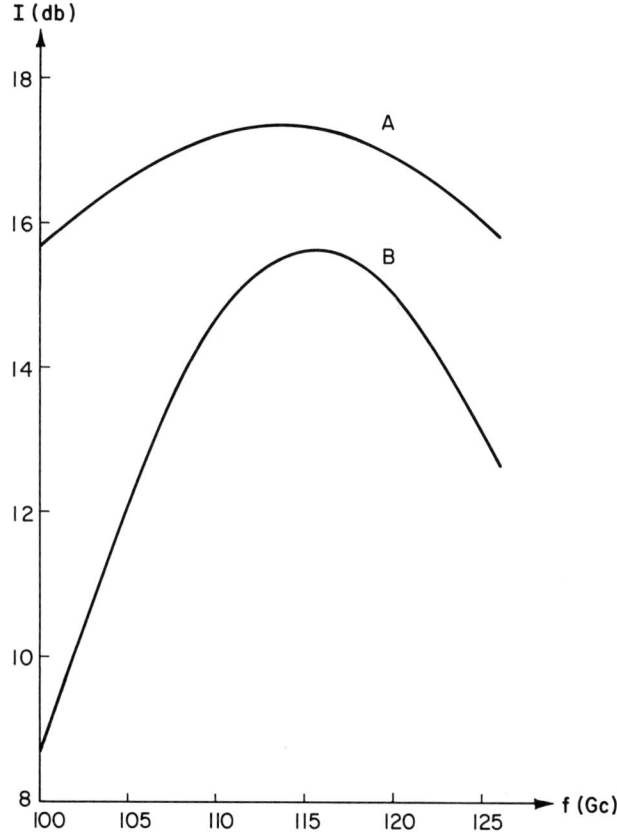

FIGURE 8 Irradiance at principal focus as a function of frequency. Curve A: one-ellipsoid system with $N = 5$; $n = 5$. Curve B: three-ellipsoid system with $N = 5, 7$, and $9$; $n = 5$. Irradiances are normalized with respect to that produced by free-space propagation.

nomena. The width of each groove was $1\lambda$ and the radius was $20\lambda$. The cylinder was placed midway between a point source and a detector $318.75\lambda$ apart. Figure 10 shows the axial irradiance distributions of two cylinders, identical in all respects, except that the grooves of one had a depth of $0.5\lambda$ and that of the other, $1\lambda$. The two curves are nearly identical. The transverse irradiance distributions in the plane $z = 0$ and $z = -70\lambda$ (position of maximum in Figure 10) are shown in Figure 11. In plane $z = -70\lambda$, the half-width of the central lobe was approximately $2\lambda$ and the relative height of the secondary maximum was less than 0.1. The amount of energy concentrated in the central lobe was high. By as-

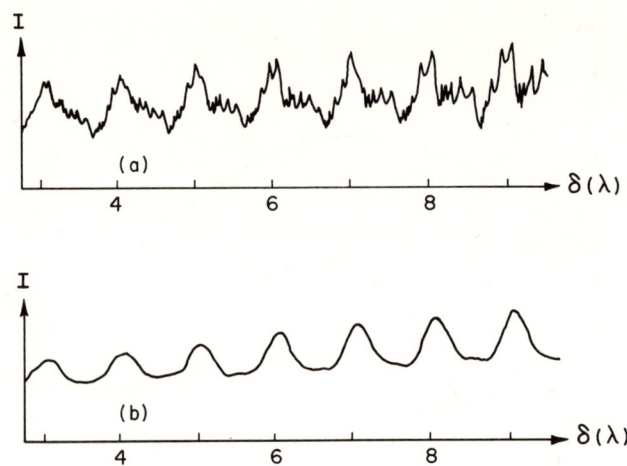

FIGURE 9 Irradiance at a chosen axial point of two closely spaced irises as a function of iris spacing $\delta$. (a) No absorber between irises. (b) Irises separated by absorber.

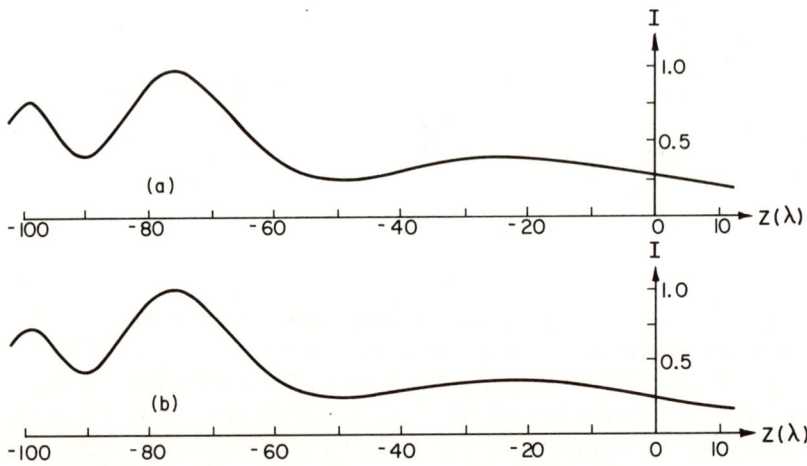

FIGURE 10 Axial irradiance distribution of a cylinder with 13 grooves. The zero point on the $z$ axis is the point that was at the same distance from the cylinder as was the source. A gain of approximately 13 dB with respect to free-space propagation was recorded in each case at $z = 0$. (a) Each groove had a depth of $0.5\lambda$. (b) Each groove had a depth of $1\lambda$.

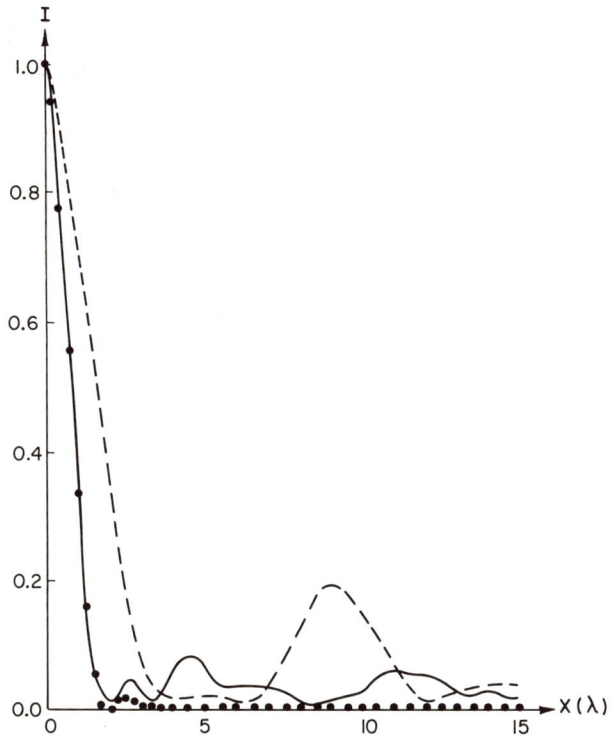

FIGURE 11 Transverse irradiance distributions. Broken curve: in plane $z = 0$ of cylinder with grooves of depth $0.5\lambda$. Full curve: in plane $z = -70\lambda$ of the same cylinder. Points: in focal plane of equivalent perfect lens.

suming the cylinder to be a lens, we calculated its equivalent focal length. For comparison, the transverse irradiance distribution in the focal plane of a perfect lens with the equivalent focal length is also shown in Figure 10. We note that the central lobes of the two curves coincide very well, showing that the cylinder with grooves possesses good focusing effects.

Extending the idea of irises with equal radii [Figure 1(b)] a little further, it is expected that systems built with smooth cylinders may possess certain focusing effects. Moreover, if iris systems are to be constructed for visible light, due to the finite thicknesses of the screens, the irises will appear to be short cylinders. So the properties of systems of cylinders have also been studied. We have placed a source and a detector at a distance of 318.75λ apart. Cylinders of radii 20λ but of different lengths were then placed, one by one, midway between the source and the detector. From these results, the cylinder (60λ long) that gave the maximum irradiance (12 dB) was chosen. On the two sides of the chosen cylinder, two more short cylinders of various lengths were placed, one on either side of the chosen cylinder. The irradiance at the detector as a function of the length of the two side cylinders is shown in Figure 12 (curve A). The two cylinders (10λ long) that gave a total maximum irradiance (17 dB) were again chosen. The process was then repeated for the next two additional cylinders (curve B in Figure 12). From these results, it is clear that focusing systems can be built with cylinders.

Instead of irises, annular lenses were used. As expected, the irradiance at the focus depends very much on the width of the annular lens (here width means the difference between the radius of the whole annular lens and the radius of the iris at the center). However, for a system of given annular lenses, the irradiance gain also depends on the longitudinal

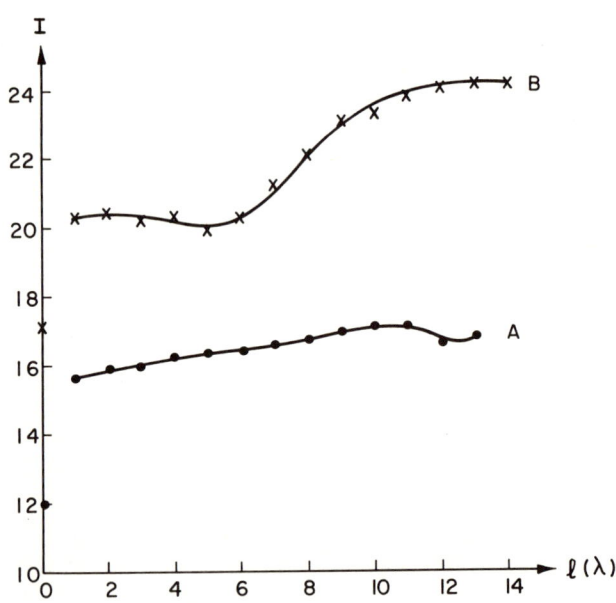

FIGURE 12 -Irradiances produced at a chosen axial point of cylinder systems. $l$ = length of cylinder. Curve A: three cylinders. Curve B: five cylinders.

positions of the annular lenses; this means that the diffraction effects are still significant.

Figure 13 shows the transverse irradiance distributions of two annular-lens systems, one with three lenses and the other with one lens. The central lobe of the former is narrower than that of the latter. Also, the former has secondary maxima that are very much lower than those of the latter. These facts, together with the data given in the accompanying table in Figure 13, show that the focusing properties of the system, like those of the other systems, increase with the number of the basic elements used. Finally, Figure 14 shows the axial irradiance distribution of an annular-lens system with three lenses. The general features are the same as the iris cases, and so the comments given before can be repeated here.

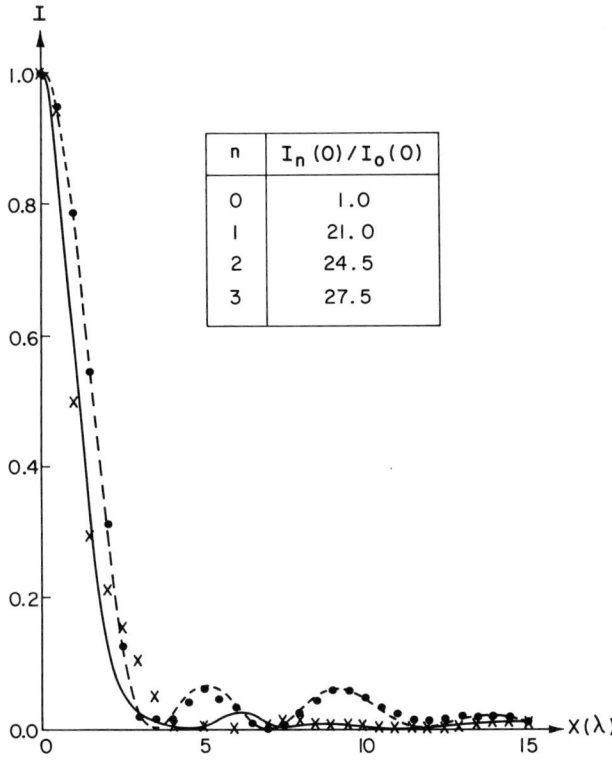

FIGURE 13 Transverse irradiance in focal planes of annular-lens systems. Full curve: three lenses. Broken curve: one lens. Points and crosses are experimental data. Table shows irradiance at principal focus as a function of $n$, the number of lenses.

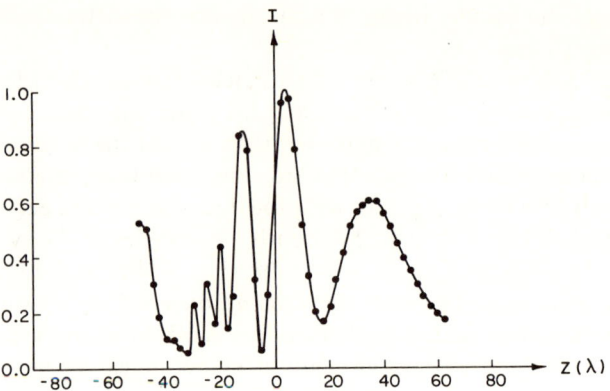

FIGURE 14   Axial irradiance distribution of a three-annular-lens system. Curve was drawn to pass through the experimental points.

**Visible-Light Systems**

Since all the results given above are those of microwave systems, the question is asked whether the systems will work in the visible region. Here the alignment problem should be much more critical. In the case of irises, because they must have finite thicknesses, they will appear, in the visible region, to be short cylinders. Nevertheless, short-cylinder systems have been shown to possess focusing properties, so the problem of finite thickness should not be too serious. Indeed, an iris system has already been constructed for visible white light. The diffraction pattern at the principal focus was found to be a central white spot, surrounded by much weaker colored rings [Lit *et al.*, 1969]. This shows that focusing systems similar to those built in the microwave region can also be built in the visible region.

In trying to build similar focusing systems, using thin films, as in integrated optics, there are a few ways to make the basic elements cited above. Among the ways are (1) to monitor the film thickness to produce an annular lens and (2) to use thin lines to imitate diffracting screens. Both of these can be achieved. There is, therefore, a good possibility of adapting the systems discussed above to integrated optics systems.

The focusing systems have also been studied because of the possibility of using them to build beam waveguides and new forms of open resonators [Tremblay and De, 1966; Lit and Van Rooy, 1972; Lit *et al.*, 1972]. Similar applications of this kind in the field of thin films are not inconceivable.

## References

Boulay, R., J. W. Y. Lit, and R. Tremblay, *Opt. Commun. 4,* 163 (1971).
De, M., J. W. Y. Lit, and R. Tremblay, *Appl. Opt. 7,* 483 (1968).
Lit, J. W. Y., and R. Tremblay, *J. Opt. Soc. Am. 59,* 559 (1969).
Lit, J. W. Y., and D. Van Rooy, "Annular-Lens Optical Waveguide," Communication presented at the VII International Quantum Electronics Conf., Montreal (May 8-11, 1972).
Lit, J. W. Y., R. Boulay, and R. Tremblay, *Opt. Commun. 1,* 280 (1970).
Lit, J. W. Y., R. Boulay, and R. Tremblay, "Longitudinally Zoned Optical Elements," *Proc. Symp. on Submillimeter Waves,* p. 487, Microwave Res. Inst. Symp. Series, Vol. XX, J. Fox, ed., Polytechnic Press, Brooklyn, N.Y. (1971).
Lit, J. W. Y., P. Lavigne, and R. Tremblay, *J. Opt. Soc. Am. 62,* 718 (1972).
Tremblay, R., and M. De, *Appl. Phys. Lett. 9,* 136 (1966).

G. L. WEISSLER and SANTOSH K. SRIVASTAVA

# A WALL-STABILIZED DOUBLED ARC AS A STANDARD SOURCE IN THE VACUUM ULTRAVIOLET

## Introduction

The measurement of absolute intensity in the vacuum uv (vuv) region of spectrum is of importance to astrophysics, among other areas, in order to provide more insight into our understanding of hot and cool plasmas, including planetary atmospheres. Particularly, such absolute intensity measurements can be used to produce secondary standards or to calculate the response of vuv spectrometers. To the several methods [Schreider, 1965] that have been employed for this purpose, there must be added the use of the wall-stabilized arc [Maecker, 1956], which gives values of absolute intensities with a high degree of accuracy.

This arc when operated in argon, hydrogen, nitrogen, krypton, etc., may emit optically thick spectral lines that reach a saturation value, and their peak intensity is given by the Planck function. This property was used by Boldt [1970] to develop a radiation standard in the 1100–3100 Å region. In this arc, high temperatures above 10,000 K can be produced by flowing high currents of more than 100 Å through different

---

The authors are in the Department of Physics, University of Southern California, Los Angeles, California 90007.

gases at approximately atmospheric pressure. Under these conditions, a state of local thermodynamic equilibrium (LTE) can be shown to exist in the hot plasma of many gases, and the emitted intensity, $I_\lambda$, is given by the modification of Kirchhoff's law of emissivity being proportional to absorptivity, namely,

$$I_\lambda = B_\lambda(T)(1 - e^{-\tau_\lambda}), \tag{1}$$

and

$$B_\lambda(T) = [(2hc^2)/\lambda^5] \ [\exp(hc/kT\lambda) - 1]^{-1}, \tag{2}$$

where $B_\lambda(T)$ is the Planck function of the source; $\tau_\lambda$ the absorption coefficient of the plasma at wavelength $\lambda$; and $h$, $c$, $k$ represent the well-known constants. If $\tau_\lambda$ is large, then $I_\lambda = B_\lambda(T)$ and the radiation at $\lambda$ becomes optically thick. The intensity of an optically thick spectral line can thus be accurately calculated, if the temperature of the source is known.

In the visible and near ultraviolet region of spectrum, there are two methods to determine if the center of a spectral line is optically thick or not. In the first [Jürgens, 1952], the profile of a line is scanned end-on as well as side-on. The length of the emitting plasma is quite different for these two observations. If at a certain distance from the line center the ratio of intensities obtained in this fashion is very much less than the geometric ratio, $l_{(end\text{-}on)}/l_{(side\text{-}on)}$, then the intensity at the center of the line is optically thick. In the second method, a mirror is placed behind the plasma. If the intensity at the center of a line does not change, then the line is optically thick.

The above methods cannot be used in the vuv region of the spectrum for the following reasons. First, most of the lines in the vuv are resonance lines, and side-on and end-on observations may give a false value of their intensities because of absorption due to colder regions of the arc. Second, reflectivities of mirrors may be too small in the vuv.

In the vuv, there have been two approaches to evaluate optically thick conditions of spectral lines. First, one may plot the intensity of a spectral line with respect to the flow rate of a gas in the plasma. That rate at which the intensity at the center of a line reaches a saturation value gives the optically thick condition. Second, one can use the LTE properties of the plasma and calculate the number density of atoms emitting a particular line. Then, the optically thick condition can be estimated [Stuck and Wende, 1972].

In the present work, we describe a third approach, which not only permits the accurate evaluation of the optically thick condition of emission lines but which also utilizes emission or resonance continua (free-bound transitions) to obtain $B_\lambda(T)$, the blackbody radiation limit.

### Analysis and Method of Measurements

For this purpose, a double arc has been constructed, which consists of two arcs in tandem. Its use can be explained in the following way. The intensity output $I_\lambda$ of the first arc (closest to the spectrometer) is given by Eq. (1). If behind this arc No. 1 a second arc No. 2 emits radiation $I_\lambda^{in}$ which is incident on arc No. 1, then the resulting output intensity $I_\lambda'$ from both arcs is given by

$$I_\lambda' = I_\lambda + I_\lambda^{in} e^{-\tau_\lambda}, \tag{3}$$

and eliminating $\tau_\lambda$ from Eqs. (1) and (3), we obtain

$$B_\lambda(T) = (I_\lambda \, I_\lambda^{in})/(I_\lambda^{in} + I_\lambda - I_\lambda'). \tag{4}$$

If radiation $I_\lambda$ from the arc No. 1 is optically thick, then the introduction of radiation $I_\lambda^{in}$ into this arc will not change its value and $I_\lambda$ will be equal to $I_\lambda'$. On the other hand, when radiation $I_\lambda$ is not optically thick, passage of radiation $I_\lambda^{in}$ through the arc will change its value to $I_\lambda'$. The measurement of these three quantities then permits the calculation of $B_\lambda(T)$ from Eq. (4).

In the double arc, the first one serves as a source of $I_\lambda$, and the second provides $I_\lambda^{in}$. In both arcs, identical spectra are produced and the spectrum of one is superimposed over that of the other. Thus we find that such an arrangement is not only useful in the vuv for checking optically thick conditions, but it can also be used in the visible and near-uv regions.

In actual practice, for these measurements the intensity of a line or a continuum is measured in terms of its height, recorded photoelectrically by a paper recorder, which can be related to the actual intensity at the source by the following relation:

$$I_\lambda = K_\lambda \, h_\lambda, \tag{5}$$

where $K_\lambda$ is an instrumentation constant that depends on the geometry of the apparatus and the response of the optical recording system. Substitution of Eq. (5) into Eq. (4) gives

$$B_\lambda(T) = K_\lambda \, [(h_\lambda \, h_\lambda^{in})/(h_\lambda^{in} + h_\lambda - h_\lambda')] = K_\lambda \, h_\lambda^t, \tag{6}$$

and

$$h_\lambda^t = (h_\lambda\, h_\lambda^{in})/(h_\lambda^{in} + h_\lambda - h_\lambda'), \qquad (7)$$

where $h_\lambda^t$ is the height of the recorded spectrum at wavelength $\lambda$ for an optically thick radiation. If $K_\lambda$, the instrumentation constant, is known, then the absolute value of intensity $I_\lambda$ of a source, recorded as $h_\lambda$, can be calculated from Eq. (5). The instrumentation constant, $K_\lambda$, is obtained from Eq. (6) by measuring the height, $h_\lambda^t$, of the optically thick spectral lines and calculating $B_\lambda(T)$ from Eq. (2). The optically thick height, $h_\lambda^t$, can also be calculated from Eq. (7) if the heights $h_\lambda$, $h_\lambda^{in}$, and $h_\lambda'$ are known.

We have made use of the double arc to calibrate the spectrum in the 600 to 1216 Å region. Optically thick hydrogen lines, Ly-$\alpha$, Ly-$\beta$, Ly-$\gamma$, and Ly-$\delta$, nitrogen NI lines at 1200, 1134, and 1100 Å, and a CI line at 1194 Å were used between 945 and 1216 Å. In addition, the carbon emission continuum, which begins at about 1100 Å, and the hydrogen continuum at 900 Å were used to calculate $h_\lambda^t$ from Eq. (7) in those regions where optically thick lines were absent.

In the 600 to 800 Å region, the argon resonance continuum was obtained in emission by operating the arc in a He–Ar mixture, and it was employed to obtain relative intensities, which were then converted to absolute values by extrapolating via Eq. (2) the $h_\lambda^t$ value at 950 Å to shorter wavelengths.

For the absolute value of $B_\lambda(T)$, an accurate knowledge of the plasma temperature is required. This was determined in all cases reported here from the line intensity ratio of Ar I 4300 Å to Ar II 4806 Å, measured in the visible region. The temperature, so obtained, was accurate to within ±2%, giving rise to an error of about 20% in $B_\lambda(T)$.

## The Double Arc

Figure 1 shows a double arc consisting of 19 copper plates with a central bore of 4.8 mm. Each of them has one inlet for gases and one inlet and one outlet for cooling water (not shown). Plates number 1 and 19 serve as anodes, with two tungsten inserts and a central hole 1.65 mm in diameter. The plates themselves are 7.65 mm thick and 91 mm in diameter. Plates number 9 and 11, the cathodes, are also fitted with tungsten inserts with 2.4-mm-diameter holes. Plate number 10 is 12.7 mm thick and has a rod as a light shutter, which can be moved in or out of the light path. All plates are electrically insulated by Teflon sheet spacers and are pressed together by four stainless steel screws (not shown). One end of the two arcs is optically coupled to a vacuum spectrograph, and the other end via a window of glass to a monochromator operating in the visible region.

### Experimental Arrangement

Electrical connections are shown in Figure 2. Power to the two arcs is supplied by two separate and electrically independent welding power supplies, which each can deliver a current of 325 A at 320 V dc. Inductances $L_1$ and $L_2$ are the windings of two electromagnets that are cooled by water, and resistances $R_1$ and $R_2$ are water-cooled stainless steel tubes of about 0.9 $\Omega$ each. These two arcs are thus electrically isolated, and any change in the current of one does not affect the other. Each arc is started by a high-voltage condenser discharge of about 3 kV between anode and cathode at a pressure of 5 Torr in argon.

Our spectroscopic arrangement, also shown in Figure 2, is the same as reported earlier in detail by Hofmann and Weissler [1971].

### Calibration between 900 and 1216 Å

The double arc was operated at a current of 100 A, and argon was flown into it at the rate of 40 $cm^3$/sec at plate number 10 and at 20 $cm^3$/sec at plates number 1 and 18. After passing through the arc channel, the gases were allowed to exit into the surrounding atmosphere via a 50-cm-long plastic tube to prevent backdiffusion. These exit ports were located at plates number 3 and 17 of arc No. 2 and arc No. 1, respectively. Helium was flown into the arc at plate number 19 at the rate of 11 $cm^3$/sec and into plate number 20 in such amounts that the pressure in the fine hole joining the arc with the differential pumping system was measurably greater than 1 atm.

Commercial grades of argon and helium were used. When these unpurified gases were flown into the arc, impurity lines of hydrogen, carbon, and oxygen appeared in the vuv spectra. Therefore, argon was passed through a cold trap containing zeolite, cooled by a dry ice and methyl alcohol mixture, and helium was passed through a liquid-nitrogen-cooled trap. Under these conditions, impurity lines of oxygen and carbon disappeared, but hydrogen lines were still present, though at much lower intensity.

The following procedure was adopted for recording optically thick

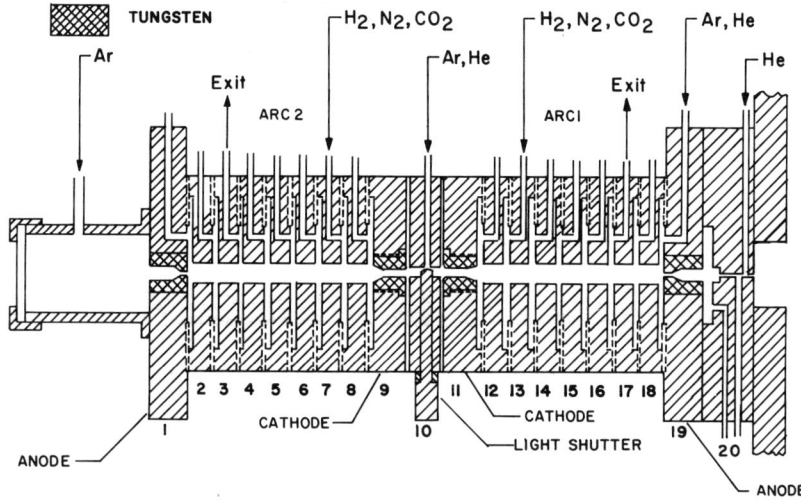

FIGURE 1  The wall-stabilized double arc.

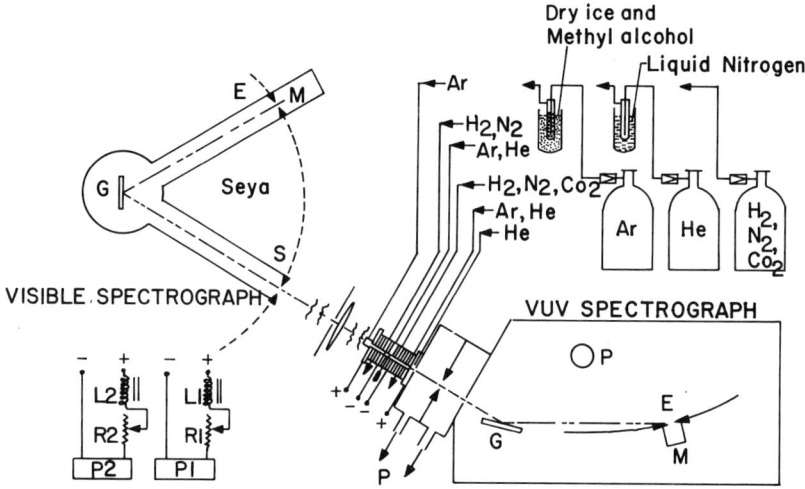

FIGURE 2  The spectroscopic vuv and visible arrangement. The primary slits, gratings, exit slits, and photomultiplier detectors are marked by S, G, E, and M, respectively. Pumps are indicated by P.

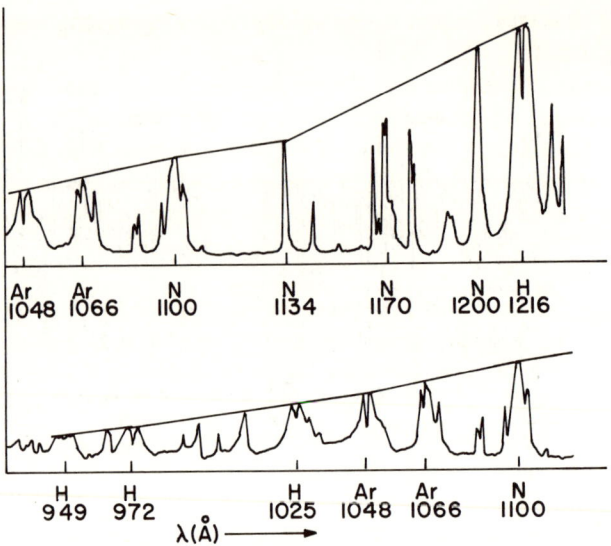

FIGURE 3  A recorder trace showing the optically thick hydrogen Lyman lines, N I lines, and Ar I lines. Solid curve gives the value of $h_\lambda^t \propto B_\lambda(T)$ obtained by joining the peak intensities of these optically thick lines.

lines. In arc No. 2, hydrogen and nitrogen were flown in at plate number 7 at the rates of 1.5 cm$^3$/sec and 2 cm$^3$/sec, respectively. Then the intensities of hydrogen lines, Ly-$\alpha$, Ly-$\beta$, Ly-$\gamma$, and Ly-$\delta$, and nitrogen lines at 1200, 1134, and 1100 Å were recorded. This provided the value of $h_\lambda^{in}$ in Eq. (7), for these lines. The mixture of these gases was then also flown into arc No. 1 through plate number 13, and the output radiation provided the value of $h_\lambda'$. Following this, the light shutter at plate number 10 was closed, so that light from arc No. 2 could not pass through arc No. 1, and the spectrum was again recorded and provided the value of $h_\lambda$. The flow rate of hydrogen and nitrogen was then increased in arc No. 1 to such a value that closing or opening the light shutter did not make any difference in the value of $h_\lambda$, i.e., $h_\lambda = h_\lambda'$. When this condition was fulfilled, then the intensity at the maximum of the line was optically thick and its height was given by $h_\lambda^t$. A typical recorder trace of the optically thick hydrogen and nitrogen lines is shown in Figure 3. The peak intensities of these lines are joined by a solid line, which gives the measured value of $h_\lambda^t$ in this region. It is seen that while hydrogen lines are self-reversed in the center, nitrogen lines are not. This is due to the fact that argon and helium both contained hydrogen as an impurity that could not be removed by the cold traps and thus was present in the colder region between the arc and the vuv spectrometer.

In order to check that this self-reversal did not give a false saturation level, the height of the optically thick nitrogen line at 1200 Å was extrapolated to Ly-α at 1216 Å by using Planck's function with the temperature obtained from visible diagnostics. This extrapolated value agreed very closely, within ±1%, with the height of the observed Lyman line.

One more test was applied to check that the saturated level given by the Lyman lines was true. For this purpose, they were recorded for two plasmas at two different temperatures, $T_1$ and $T_2$. Using one set of data at a temperature $T_1$, the instrumentation constant, $K_\lambda$, given by Eq. (6), was determined between 950 and 1216 Å. Since $K_\lambda$ does not depend on temperature, its value should remain the same when determined from the optically thick line heights at a new temperature $T_2$. It was found that the $K_\lambda$ values obtained from these two temperatures were in close agreement. Thus, we have further support that the self-reversal of the Lyman lines did not falsify our results of $h_\lambda^t$.

In order to evaluate $h_\lambda^t$ between lines and beyond Ly-δ toward shorter wavelengths, we used the CI resonance continuum (1120 to 980 Å) and the hydrogen continuum at 910 Å, which were *not* optically thick. The double arc was operated in argon, and carbon dioxide and hydrogen were flown in at plates number 13 and 7. Thus, values of $h_\lambda^{in}$, $h_\lambda$, and $h_\lambda'$ for these continua were obtained as described above, and Eq. (7) was then used to calculate $h_\lambda^t$. Figure 4 shows a typical recorder

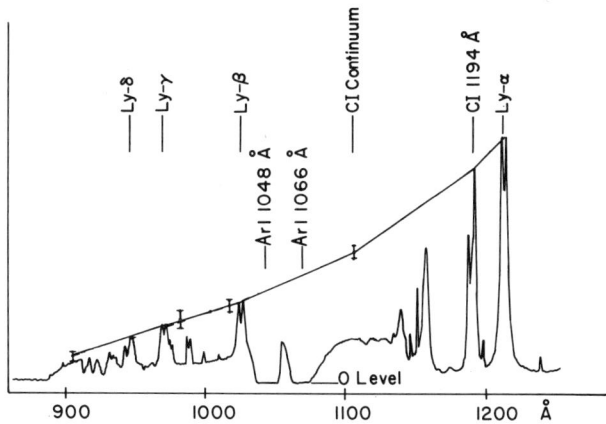

FIGURE 4 A recorder trace showing the optically thick Lyman lines and the carbon and hydrogen continua in the 900–1216 Å region. Solid line gives the experimentally obtained value of $h_\lambda^t \propto B_\lambda(T)$.

trace of these continua, together with optically thick Lyman lines. (As is seen there, the vuv radiation is completely absorbed by the two resonance lines of Ar I at 1048 and 1066 Å due to the presence of cold argon in the light path.) The points with error bars are the values of $h_\lambda^t$ obtained from the underlying carbon and hydrogen continua and the solid curve shows the values of $h_\lambda^t$ for this region, obtained by joining the optically thick Lyman lines, the optically thick CI line at 1194 Å, and the above-mentioned points for the continua.

This method can lead to large errors in the calculated values of $h_\lambda^t$, if the absorption coefficient for the carbon atoms in the arc No. 1 is not large. We found that the best results were obtained when the absorptivity, $a_\lambda = 1 - \exp(-\tau_\lambda)$, was between 0.3 and 0.4. In our measurements, errors in the $h_\lambda^t$ values varied from ±5% in the 1100 Å region to ±16% in the 900 Å region.

### Calibration between 600 and 800 Å

A pure argon arc has a short-wavelength transmission limit at its ionization potential corresponding to about 800 Å. Therefore, the arc was operated in helium as a carrier gas, which transmits down to 504 Å, and small amounts of argon were flown into it. In actual operation, the arc was first started in argon, which was slowly replaced by helium, since a pure helium arc was found to be electrically unstable. A small admixture of argon in helium, when flown in from the cathode end, completely removed this instability. Therefore, a gas-flow scheme was chosen in which a mixture of argon and helium entered into the arc at plate number 10, and pure helium was flown through plates number 19 and 20. These gases were then allowed to exit at plate number 17. Only arc No. 1 was used, and all the inlets of arc No. 2 were closed.

A typical recorder trace of the argon resonance continuum and some Ar II lines between 650 and 950 Å is shown in Figure 5. Since it is known that LTE does not exist in a helium arc [Uhlenbusch, 1969], the question arises as to whether this Ar continuum can be assumed to be in LTE and can be described by Kirchhoff's law given by Eq. (1). It seems that in a mixed plasma of this nature argon atoms can be in LTE, if the electron density and temperature are each above a certain minimum value. Therefore, both electron density and temperature, which would describe argon in LTE in a helium plasma, were determined experimentally. For this purpose, the temperature of plasma was obtained by two methods:

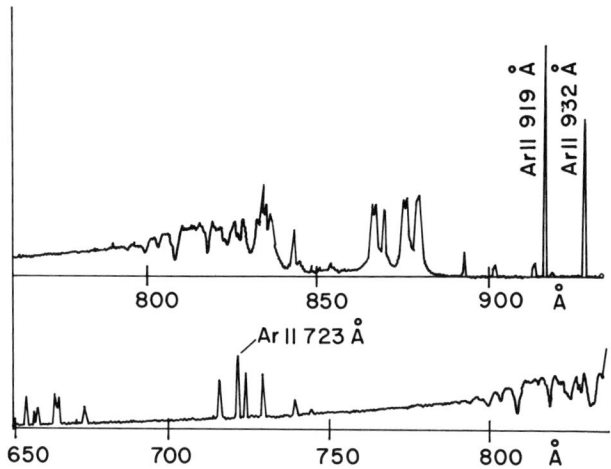

FIGURE 5  A typical spectrum of Ar II lines and the Ar I continuum between 650 and 932 Å emitted along the axis of an He–Ar arc.

SPECTROSCOPIC TEMPERATURE DETERMINATION IN THE VISIBLE REGION

Here, the temperature was obtained, as before, from the line intensity ratio of Ar I 4300 Å to Ar II 4806 Å, while the corresponding electron density was calculated from the Stark-broadened H-$\beta$ line. The details have been given by Stuck and Wende [1972], who showed that temperatures thus determined are accurate to ±2%.

TEMPERATURE DETERMINATION FROM VUV MEASUREMENTS

In this case, it is assumed that the Ar I and Ar II atoms are in LTE, and that the radiation emitted by them in the vuv is optically thin. Under such conditions, the intensity, $I_\lambda$, of the argon recombination continuum is given by

$$I_\lambda/B_\lambda(T) = N_g\, \sigma_\lambda L, \tag{8}$$

where $N_g$ is the number density of neutral argon atoms in the ground state, $\sigma_\lambda$ the photoionization cross section at wavelength $\lambda$, and $L$ the length of emitting plasma. Similarly, the intensity of a resonance line of an argon ion is given by

$$I_\lambda^+/B_\lambda^+ = (\pi e^2/m_e C^2)\, \lambda_+^2\, f_{mn}^+\, N_g^+\, L, \tag{9}$$

where $I_\lambda^+$, $\lambda_+$, and $f_{mn}^+$ are the intensity, wavelength, and oscillator strength, respectively, of the Ar II line. $N_g^+$ is the ground-state density of the argon ions, and $L$ is again the length of the emitting plasma. The ratio of Eqs. (9)/(8) is

$$\frac{I_\lambda^+}{I_\lambda} \cdot \frac{B_\lambda}{B_\lambda^+} = \left(\frac{\pi e^2}{m_e C^2}\right) \lambda_+^2 \cdot \frac{f_{mn}^+}{\sigma_\lambda} \cdot \frac{N_g^+}{N_g} . \tag{10a}$$

In addition, we have from the Saha equation

$$\frac{N_g^+}{N_g} = \frac{2}{N_e} \cdot \frac{g_0^+}{g_0} \cdot \exp\left(-\frac{(E_\infty - \Delta E_\infty)}{kT}\right) \cdot \frac{(2\pi m_e k)^{3/2}}{h^3} \cdot T^{3/2} , \tag{10b}$$

where $N_e$ is the electron density, $g_0$ and $g_0^+$ are the statistical weights of the ground states of the Ar I and Ar II atoms, respectively. $E_\infty$ is the ionization energy of an isolated neutral argon atom, $\Delta E_\infty$ the lowering of the ionization energy due to neighboring charges, and $T$ the temperature of the plasma. Quantities with a + sign correspond to argon ions, and all other symbols represent customary quantities.

In the present experiment, the intensity of the Ar II line at 919 Å was compared with the intensity of the Ar I continuum at 750 Å. The oscillator strength, $f_{mn}^+$, of this line is given by Wiese et al. [1969] and was obtained from the lifetime measurements of Lawrence [1968], and the photoionization cross sections, $\sigma_\lambda$, for the argon continuum are given by Samson [1964]. The intensity of 919 Å Ar II line was measured from the area enclosed by the line, and the intensity of continuum at 750 Å was given by its height. The temperature of the plasma was thus measured using Eqs. (10a) and (10b).

It is important to point out that in the derivation of Eqs. (10a) and (10b) it is assumed that the 919 Å Ar II line and the argon continuum at 750 Å are both optically thin and that the response function $K_\lambda$, Eq. (5), of the vuv recording system is constant in this region. This response depends on the quantum efficiency of the sodium salicylate radiation converter placed in front of the photomultiplier and shown by Samson [1967] to be substantially constant. In addition, the response depends on the reflectivity of the grating. Since our grating was a Siegbahn type, lightly ruled on glass, and was used to grazing incidence of about 80°, its reflectivity and therefore its response in this arrangement did not change much from 950 to 750 Å.

In order to check whether the Ar II line at 919 Å was indeed thin, it was compared with another Ar II line at 932 Å, the oscillator strength of which is also known accurately [Wiese et al., 1969]. If the argon atoms were in LTE and these lines were optically thin, then the ratio of their intensities should be given by the ratio of their oscillator strengths, multiplied by the ratio of their statistical weights. This was found to be true within ±10%.

The following criterion was used to prove that the intensity of the Ar I continuum was optically thin. Near 750 Å (Figure 5) the spectrum consisted of an Ar II line at 723 Å on the shorter wavelength side and an Ar II line at 919 Å on the longer wavelength side. If their peaks were joined by a straight line, then we obtained at 750 Å the minimum possible height for optically thick radiation. A comparison of the height of the Ar I continuum at 750 Å with this extrapolated optically thick height showed unambiguously that it was optically thin. Thus, the ratio $I^+_{919}/I_{750}$ could be used in Eq. (10a) for the determination of the temperature of the plasma. This ratio is sensitive to small changes in temperature, and therefore $T$ should be accurate to ±2%. Figure 6 shows this ratio for two electron densities of the plasma.

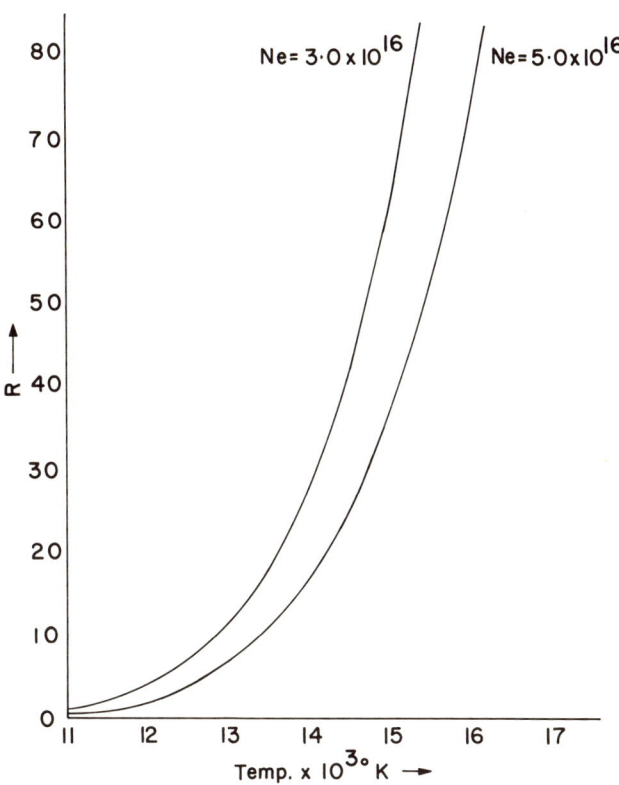

FIGURE 6 Variation of $R = I^+_{919}/I_{750}$ with temperature $T$ in kelvins.

# INSTRUMENTATION

PIERRE WEBER and PIERRE DURRENBERGER

# Method and Equipment for Localizing Satellites by Laser Range-and-Direction Finding

## I. Introduction

For 15 years, the existence of artificial satellites has been the source of scientific developments, some of them quite spectacular. Not only do the measurements made on board the satellites provide data of a fundamentally new nature, but also their very presence around the globe offers a precious advantage: the observation of their movements from the ground provides information that may be useful for many applications.

The precise localization of satellites has been used in geometrical geodesy (setting up of large bases with a 1-m precision), in dynamic geodesy (study of terrestrial gravitation, etc.), in practical applications (calibration, synchronization, navigation).

Improvements on satellite localization from the ground concerned both the observation means and the satellites themselves. As regards optical means, the first ones, like radar, had mostly a role of surveillance (Moonwatch American network, AT-1 Russian detectors). Very early visual observation was supplemented by photography: when a satellite

---

The authors are in the Office National d'Etudes et de Recherches Aérospatiales (ONERA), 92320 Chatillon (France).

is illuminated, at night, by the sun, its crossing through the camera field creates a line on the photograph. If the satellite is bright enough, the camera may be rotated along with the diurnal movement so that the stars appear as fixed points; a shutter provokes, at known instants, breaks in the satellite trace, and the plotting of these breaks relative to known stars provides the satellite angular positions.

If the satellite is not bright enough, its movement may be followed for some time by the camera in order to accumulate the light on one point of the photographic plate; it is along these principles that the Baker-Nunn, Antares, and AFU-75 cameras were designed. The angular precisions are of the order of a few seconds of arc [Lambeck, 1968].

The invention of the laser revolutionized the range-measuring techniques. The laser range finder, which works as an optical radar, imposed itself as one of the most precise means for satellite localizing by trilateration. A number of stations were developed, in particular by the Smithsonian Astrophysical Observatory [Lehr et al., 1970], the NASA Goddard Space Flight Center [Premo, 1970], the Aeronomy Service of CNRS (the French Center of Scientific Research) [Bivas, 1968], and ONERA (the French Institute for Aerospace Research) [Moreau and Véret, 1969].

Some satellites have been specially equipped to play the role of passive targets. They carry retroreflectors (cube corners), which raise considerably the range of laser range finders and improve their precision, presently to the order of 1 m for most of them.

Among reflector-equipped satellites, some follow a particular trajectory: with a view to geodetic applications, PEOLE, launched in 1971 by CNES (the French Space Agency) from its Guiana Space Center, flies over equatorial regions with an inclination of only 15°. Others (Anna 1 B launched in 1962, GEOS-A in 1965, GEOS-B in 1968) carry powerful lamps delivering very brief and intense flashes at known moments; contrast and resolution power of the images are improved; angular precision reaches 1 sec of arc. If the same flashes are photographed by several stations, the synchronization of their observations is obviously quite easy.

The laser range finders for which the satellites are acquired visually present the drawback of a time-limited operation: for a given station, only three passages of a satellite are visible by night, as a rule, for three consecutive weeks; then the satellite is no longer visible for about three more weeks. In order to ensure the observation permanency, some organizations made use of programmed laser range finders, capable of acquiring the satellites by day also (Goddard Space Flight Center, Air Force Cambridge Research Laboratories).

Whatever the acquisition means, the photographs of satellite traces are taken against a star background. The flash satellites may be photographed even if not illuminated by the sun. But the satellites equipped this way are not many (only GEOS-B remains, to our knowledge), the flashlamps have a limited lifetime and function only at predetermined moments. Moreover, neither the laser range finders nor the photographic cameras make up autonomous devices capable, from only one point on earth, of locating a satellite; results obtained from several observation stations have to be put together so that triangulation and/or trilateration operations may be performed.

ONERA developed and experimented with a process for locating satellites from a single station. The range-and-direction finder provides polar coordinates through radial ranges and angular positions, at known moments: distances are given by a laser range finder and angles by a photograph of the satellite while illuminated by a high-energy pulsed laser.

In parallel, the Air Force Cambridge Research Laboratories performed a study on a process whose principle is almost the same as that of ONERA but with different methods and means [Iliff, 1970].

After presenting the principle of laser range-and-direction finding, the paper will describe the ONERA station, present the experiments performed, and analyze the results obtained.

## II. Principle of the Apparatus

The range-and-direction apparatus is made of a photographic camera, a tracking turret, and measuring means. The turret carries two lasers (Figure 1): the first is a $Q$-switched ruby laser for range measuring; the second is also a ruby laser, but working on its natural mode it emits more energy than the first, and its purpose is to illuminate the satellite on its background of stars.

Emissions of the two lasers are synchronized by a precise clock according to which all measurements are timed.

The turret also carries the sighting telescope for visual pointing and tracking and the receiving telescope for picking up the range-finder echo.

A chronometer measures the duration of the light round trip between station and satellite.

The camera is fixed on an equatorial mounting, which compensates for the apparent movement of the stars. The camera axis is pointed beforehand toward the zone expected to be crossed by the satellite at the proper time. The operator, sitting on the turret, acquires the satellite

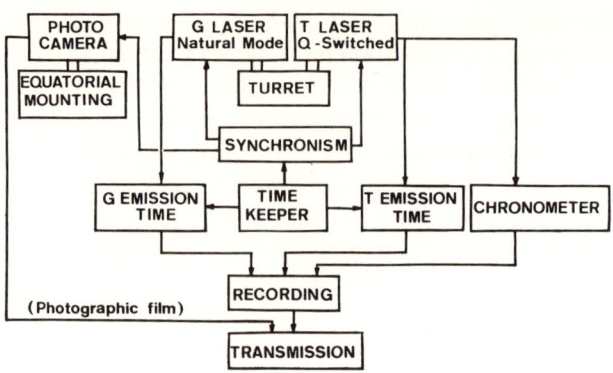

FIGURE 1  Principle of the laser range-and-direction finding system.

through a rendezvous procedure; during the whole tracking period he keeps the range finder working; when the satellite is in the receiving telescope field, the illuminating laser is also made to work, and the echoes are photographed.

The characteristics of the equipment are as follows:

1. The turret (Figure 2) is of the "elevation-azimuth" type; it is servo-controlled in speed or in position by local or remote control.

2. The sighting telescope (field 3°, magnification 23 ×, aperture 125 mm) is articulated in elevation so that the eyepiece remains fixed relative to the operator.

FIGURE 2  The tracking turret.

3. The range-finger telescope is of the Cassegrain type (field 1.5 mrad); the aperture is large enough (600 mm) to make it easy to pick up the echoes by night.

4. The range-finder laser is switched by a rotating prism. A 1-J energy is emitted in a single pulse of 28-nsec duration, at the rate of 1/sec. The natural divergence of the laser beam is reduced to 1 mrad by an afocal Galilean telescope (magnification 10X).

5. The illuminating laser (Figure 3) emits 30-J energy in a series of brief pulses lasting about 650 $\mu$sec. The ruby (length 203 mm, diameter 16 mm) is placed along the axis of a cylindrical tubular diffuser containing magnesion powder. Every 2 min it emits a train containing up to ten successive emissions, 4 sec apart. The natural divergence is reduced to 0.7 mrad by an afocal Galilean telescope (magnification 8X).

6. Timekeeping is ensured by a cesium clock, compared with very-low-frequency emissions.

7. The range-finder chronometer has a resolution of 10 nsec (1.5 m): calibration and some tests are performed with a 1-nsec chronometer.

The photographic camera (Figure 4) is a telescope of the Schmidt type, as this is the astronomical instrument that best fills the following requirements:

1. The field must be large enough so that many stars, used as references, and as many echoes as possible are recorded.

2. The aperture must be large enough for the received light flux to be at a sufficient level.

3. The aberrations must be properly corrected within the whole field so that defining power and sensitivity remain the same in the whole field.

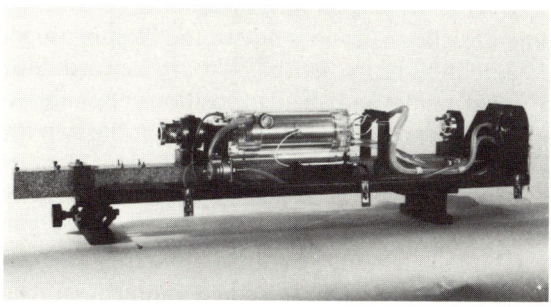

FIGURE 3  The 30-J illuminating laser.

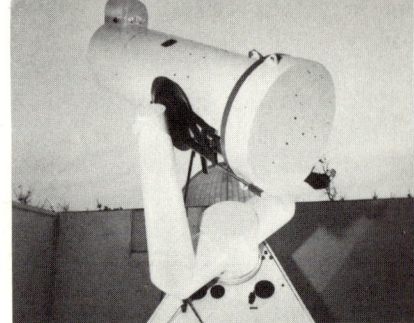

FIGURE 4  The Schmidt telescope.

The telescope used has a 300-mm aperture, a 600-mm focal length, a 10° field, and a 0.5 sec of arc resolving power. The shutter is remotely controlled. The film (useful diameter 110 mm) is maintained by suction on a porous spherical mounting. The exposure time is a function of the field crossing time of the satellite (25 to 55 sec).

The equatorial mounting enjoys two degrees of freedom, in hour angle and declination.

## III. Particular Points

The principle retained by ONERA presents several peculiarities as regards the synchronization of the two lasers—the echo recognition and the placing of the Schmidt telescope relative to the illuminating laser.

### A. LASER SYNCHRONIZATION

The two lasers may operate simultaneously or successively. Simultaneous emissions present a drawback, as the range-finding echo detection is hindered by the presence of echoes due to the illuminating laser, and very elaborate detection processes then have to be used. Moreover, it is not mandatory that ranges and angular positions be simultaneous to complete the satellite trajectory. For this reason, the two lasers emit successively. In the particular case of GEOS-B, their synchronization (Figure 5) is ensured in relation to the spacecraft-borne lamp flashes; in this way it is possible to distinguish, on the same photograph, the positions of the satellite made visible by the illuminating laser; sometimes, its positions while illuminated by the range-finding laser, which happens in clear weather; and the GEOS-B lamp flashes, if any.

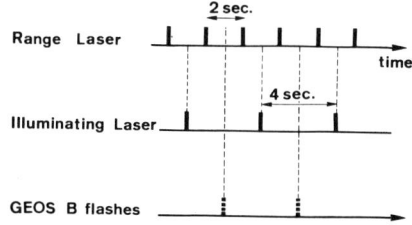

FIGURE 5 Synchronization of the laser emissions.

### B. ECHO RECOGNITION

For the range finder, echo recognition consists in distinguishing them in time from the noise, in order to detect low-energy echoes and to improve the maximum range. The latter is obtained by a relation between emitted and received energies:

$$R = \left( \frac{16}{\pi^2} \cdot \frac{W_t}{W_r} \cdot \frac{\rho_r A_r \alpha_s A_s T^2}{\omega_t^2 \cdot \omega_s^2} \right)^{1/4}, \tag{1}$$

where

$R$ is the maximum range,
$W_t$ is the energy emitted by the range-finding laser,
$W_r$ is the minimum detectable energy, at receiver input,
$\rho_r$ is the transmission factor of the receiving telescope,
$A_r$ is the receiver area,
$\alpha_s$ is the reflecting factor of the satellite retroreflectors,
$A_s$ is the retroreflector area,
$T$ is the atmospheric transmission factor,
$\omega_t$ is the angular aperture of the beam emerging from the range-finder laser telescope,
$\omega_s$ is the angular aperture of the beam returned by the satellite.

The theoretical maximum range of the ONERA range-and-direction finder is in the order of 7000 km. Practically it is about half this value.

The means used to reduce the noise and to detect weak echoes are the following:

1. 0.5- and 1-nm interference filters, when necessary—in dark night they are not needed;
2. A very-low-noise, ITT/FW 130, uncooled photomultiplier—high voltage is permanently applied, echoes are detected at single electron level;
3. An electronic system of temporal gate, open for a short time

around the expected echo return—this system may be controlled during operation by the range value previously obtained;

4. Various electronic filtering systems may also be used [Besson and Weber, 1966; Boileau and Weber, 1970].

For the photographic process, echo recognition consists in distinguishing them from the stars and from the defects on the films. No filter is used for photographing retroreflector-equipped satellites.

The choice of emulsion is essential. The lumination $L_f$ received by the film placed in the focal plane is given by [Véret, 1971]:

$$L_f = \left(\frac{4}{\pi}\right)^3 \cdot 0.9 \cdot \frac{W_g}{\omega_g^2} \cdot \frac{\alpha_s A_s}{\omega_s^2} \cdot \frac{A_p \rho_p}{a^2} \cdot \frac{T^2}{R^4}, \qquad (2)$$

where

$W_g$ is the energy emitted by the illuminating laser,
$\omega_g$ is the angular aperture of the beam emerging from the illuminating laser telescope,
$A_p$ is the area of the camera aperture,
$\rho_p$ is the camera transmission factor,
$a$ is the diameter of the satellite image in the focal plane.

The emulsion used is Kodak 2475, deposited on a Mylar support (0.1

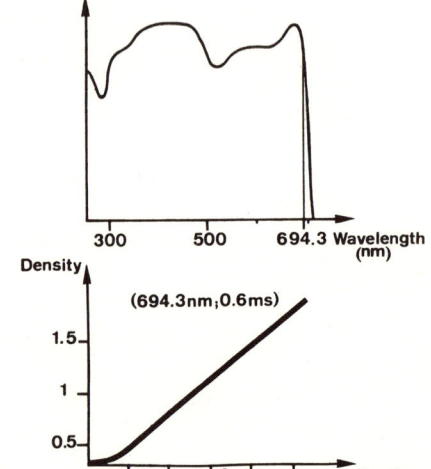

FIGURE 6 Sensitivity of the Kodak 2475 emulsion.

mm thick); this emulsion was chosen for its sensitivity in the red. At 694.3 nm it is in the order of $3 \times 10^{-4}$ erg/cm² for the minimum detectable darkening and for a nonpunctual image (Figure 6). To compute the theoretical maximum range, it is necessary to take the whole fog into account; if a minimum detectable lumination of $2 \times 10^{-2}$ erg/cm² is chosen (giving a density of 0.3 above the fog), the theoretical maximum range is of the order of 4500 km.

In order to distinguish the illuminated satellite from the stars, a double exposure is performed. Moreover, at the end of the second exposure, which lasts as long as the first one, a small trail is obtained by stopping the equatorial mounting, more or less according to the displayed declination. The trail is useful to orient the film during data reduction and to recognize the star images corresponding to the first exposure.

The images of the satellites and the graininess of the film are not doubled; however, the atmospheric backscattering phenomenon and the choice of the camera location help to separate the ones from the others, as shown below.

### C. LOCATION OF THE CAMERA

The location of the camera relative to the illuminating laser depends on three main factors: the aperture of the beam reflected by the satellite, the velocity aberration, and the aperture of the light beam emitted by the laser and backscattered by the atmosphere.

The retroreflectors carried by the satellite are made of silica trihedra (cube corners) having the property of sending the light back in the direction of the source. The retroreflected light is concentrated within a cone, which defines on the ground an ellipse whose half minor axis has a minimum value of about 100 m (for the lowest aperture and satellite altitude).

But as the satellite is moving, the beam reflected by the corners is deviated by an angle $\phi$ in the direction of the satellite velocity vector:

$$\phi = \frac{2v}{c} \left[ 1 - \left( \frac{r}{r+1} \right)^2 \cos^2 s \cos^2 \omega \right]^{1/2}, \qquad (3)$$

where

$v$ is the satellite velocity relative to earth,
$c$ is the speed of light,
$r$ is the terrestrial radius,
$h$ is the satellite altitude,
$s$ is the elevation under which the satellite is seen from the station,

$\omega$ is the angle between the orbit plane and the vertical plane containing the line of sight.

The reflected light ellipse on the ground is shifted in the same direction; the maximum value of the deflection angle is of the order of $5 \times 10^{-5}$ rad, corresponding to a 50-m displacement for a 1000-km range.

Relative to a fixed illuminating laser, the position of the ellipse center on the ground varies in direction and distance according to the satellite altitude and the trajectory orientation. The best solution, quite unrealistic, would be to locate the camera according to the configuration of each satellite pass. In practice, a preliminary study of the ephemerides of all the satellites to be observed during a testing campaign permits one to determine a mean orientation of the laser–camera line. The maximum distance between the two is chosen according to the most unfavorable case (maximum aberration, 90° elevation, and velocity vector opposed to the laser–camera vector), in such a way that the camera is then on the edge of the retroreflection circle. The latter having at least a 100-m radius for a 50-m aberration, the distance must not exceed 50 m (Figure 7).

The light emitted by the laser is scattered by aerosols and molecules; experience shows (Figure 8) that the films receive an image from the light scattered within a beam emitted up to a distance of about 30 km. If the camera is too near the laser, the satellite image appears within the trace left by the backscattered light and may be immersed in it. The minimum distance $d_m$ corresponds to the half major axis of the ellipse formed by the intersection of the emitted light cone with the horizontal plane, which is supposed to be the upper limit of the backscattering phenomenon at altitude $H$ (Figure 9). The minimum distance is 41 m for $H = 30$ km and $s = 30°$, a value under which satellites are rarely photographed.

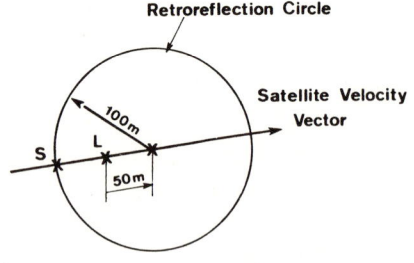

FIGURE 7 Velocity aberration in the retroreflection circle.

S: Schmidt telescope
L: Laser

FIGURE 8  Backscattering traces on a photograph.

As a conclusion, the camera should be at a distance from the illuminating laser of between 40 and 50 m.

In these conditions, the satellite images are, on the film, in line with the backscattering lines (Figure 10). As they are also on the trace of the satellite, when it is illuminated by the sun, the echoes are easily distinguished from the film defects.

FIGURE 9  Minimum distance between laser and camera.

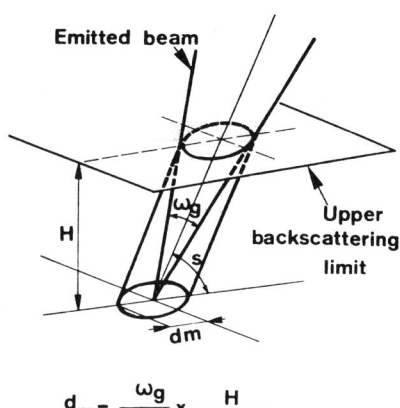

$$d_m = \frac{\omega_g}{2} \times \frac{H}{\sin^2 s}$$

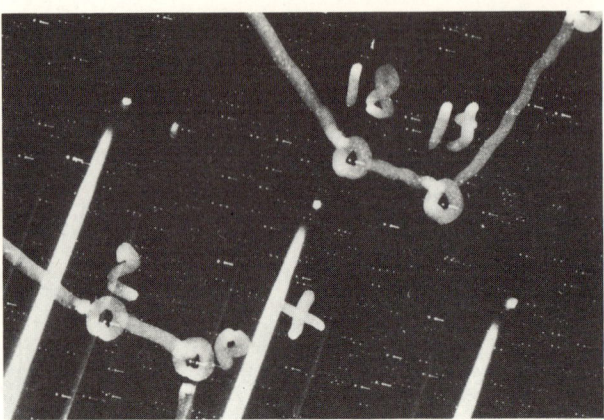

FIGURE 10 Partial enlargement of a photograph.

## IV. Experiments and Mode of Operation

Feasibility experiments on range-and-direction finding were performed in 1967 and 1968 [Müller *et al.*, 1969, 1970]. Later, ONERA took part in operation ISAGEX (International SAtellite Geodesy EXperiment), with its range-and-direction finder installed at the Saint-Michel-de-Provence Observatory (S.E. France). The ISAGEX experiment was directed by CNES, with the collaboration of many organizations. A network of observation stations (laser range finders, cameras, radars) covering the whole earth was used for the precise localization of satellites equipped with retroreflectors. All the results were gathered, to be treated by CNES and the Smithsonian Astrophysical Observatory.

During four campaigns, lasting three weeks each, for a total of 49 nights of observation in good conditions, the ONERA range-and-direction finder followed 137 satellite passes and provided nearly 10,000 range measurements and 260 direction measurements.

The functioning of the station was organized as follows. By a preliminary operation, the station was localized by geodetic survey, and the range finder was calibrated on fixed targets. The energy of the target-obtained echoes was made about equal to that received from satellites by introduction of optical densities in the receiver, so that the detection probability was only 50%.

The station clock was synchronized by the flying-clock method, and the time was kept during the 6 months of the experiment within ±1 $\mu$sec, as compared with Coordinated Universal Time.

Every day, the passes foreseen for the next night were prepared from the ephemerides: rendezvous for acquisition, pointing of the camera. If the satellite was not acquired at the first rendezvous, several others were attempted; the minimum interval between two attempts was 30 sec. The opening of the Schmidt telescope was remotely controlled at the time the satellite covered the field. The exposure duration was automatically adjusted. Measurements were punched on tape.

At the end of the test, while the film was developed, the tape was completed with data concerning the identification and meteorological conditions during tracking. A rapid survey of the film and the measurements was performed before transmission to the organizations responsible for reduction and orbit calculations.

The film data reduction was performed as follows:

1. The stars whose magnitude is between 7.8 and 8.3 (SAO catalog) and placed near the center of the photograph are traced, by a computer, on a graph at the same scale as that of the film.
2. The film is superposed to the star tracing.
3. A polygon surrounding the satellite images is made, with 24 stars as an average (Figure 11).
4. The rectangular coordinates of the traced stars and those of the satellite images are measured by means of a comparator.
5. The star positions measured on the film are compared with those given by the SAO catalog; adjustment is made if necessary.
6. The satellite positions are eventually converted into equatorial coordinates, taking account of their timing.

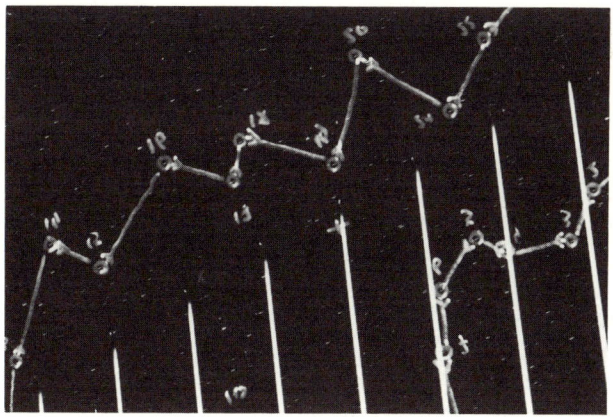

FIGURE 11   Star polygon around satellite images.

## V. Analysis of Results

The analysis of the results obtained during operation ISAGEX leads to several remarks.

Range measurements were obtained up to 3200 km. The average number of measurements for each pass was 70; this point is interesting, as it is desirable to have enough points at each pass to reduce the random error to a value lower than that of the systematic error, while computing the orbits. The overall false alarm ratio was 3%. Satellite passes were classed into three categories, according the measurement quality: 93% are satisfactory, 6% doubtful, and 1% eliminated because of false alarms (Table 1).

Among passes followed by the range finder, 77% gave rise also to directional measures. Satellite images were obtained up to 2600 km. The number of images per pass does not seem to depend on the satellite distance (this number reaches its maximum for 2200 km as well as for 1000 km). Other factors act on the number of images and the film quality: the tracking precision, the Schmidt telescope adjustment and the film positioning in the focal plane, the meteorological conditions, the thickness of the atmospheric layer crossed by the light beam, and the cleanliness and absolute darkness during the film development.

The images of the laser-illuminated satellites are as round as and sometimes brighter than those of the GEOS-B flashes (in fact, GEOS-B had only two lamps working out of four during the ISAGEX operation).

Accuracy range is 1.2 m (standard deviation). A better accuracy may be obtained with ultrafast detector and electronic circuits, with narrower pulses, and through an analysis of the echo shapes [Lehr et al., 1970; Gaignebet, 1971; Weber, 1972]. In these conditions, an accuracy of about 0.40 m is reached; this may still be improved by treating a large number of measures. Such an accuracy is necessary for certain studies, such as that of ground tides.

The accuracy of angular results, obtained from the photographs, depends mainly on

1. The positioning error of the film relative to the stars taken as reference in the SAO catalog: the mean error, with 24 stars, is $3 \times 10^{-6}$ rad (0.6 sec of arc); this error takes into account the diffusion within the film gelatin, the error made in measuring the star position, the star flicker, the film surface defects, and its positioning error in the telescope focal plane.

2. The error made in reading the rectangular positions of the satellite image, 2.5-$\mu$ average; this is equivalent to $3 \times 10^{-6}$ rad in direction.

TABLE 1  Distribution of the Satellite Passes Followed by the Range Finder

| Satellite | GEOS-A | D1C | GEOS-B | BEC | D1D | BEB |
|---|---|---|---|---|---|---|
| Correct | 54 | 36 | 13 | 10 | 8 | 7 |
| Uncertain | 5 | 1 | 0 | 0 | 1 | 1 |
| Eliminated | 1 | 0 | 0 | 0 | 0 | 0 |

3. The error due to random image displacement whose estimated value is lower than $3 \times 10^{-6}$ rad.

Image timing is ensured with a 0.1-msec resolution; the satellite velocity being of the order of 8000 m/sec, the corresponding angular error is at most $8 \times 10^{-7}$ rad, or 0.2 sec of arc.

In order to obtain an experimental estimate of the direction finder angular accuracy a comparison was made with flashes from GEOS-B recorded with a Baker-Nunn camera placed in the same part of the earth.

Orbit calculations are made according to a potential model (Standard Earth 1969), taking account of the sun and moon perturbations, and making use of laser range finders distributed all over the earth. Residues of the GEOS-B angular measures by the laser direction finder are of the same order as those taken by the flash recording camera (Table 2). The value of these residues is about 3 sec of arc; this value is less good than the direction finder accuracy, as new errors are introduced in the orbit calculation. Even though only part of the results was treated, it can be asserted that the angular accuracy of the laser direction finder reached 1 sec of arc.

To improve the accuracy, the number of measurements should be increased, either by taking several successive photographs or by using an

TABLE 2  Standard Deviations[a] of Residues Relative to Orbit Calculations

|  | $\delta$ | $\alpha \cos \delta$ |
|---|---|---|
| Illuminating Laser | $14.5 \times 10^{-6}$ rad (2.9 sec of arc) | $18 \times 10^{-6}$ rad (3.6 sec of arc) |
| GEOS-B Flashes (Baker-Nunn) | $20.5 \times 10^{-6}$ rad (4.1 sec of arc) | $17.1 \times 10^{-6}$ rad (3.4 sec of arc) |

[a] $\alpha$: right ascension; $\delta$: declination.

illuminating laser with a higher rate. The minimum time interval between two photographs taken with the same telescope is 2 min; experiments were attempted with two Schmidt telescopes pointed on contiguous parts of the trajectory.

## VI. Conclusions and Prospects

During the ISAGEX experiments, the ONERA range-and-direction finder provided range and angular measurements on satellites equipped with retroreflectors. The tests were performed by night, on sunlit satellites in order to ensure visual tracking. A study is under way at ONERA with a view to control the tracking turret movements according to the satellite ephemerides [Staron, 1972]: in this system the angular deviation between the pointing direction and the calculated direction is displayed on a screen; the operator ensures tracking by keeping this deviation to a minimum. In this way, it will be possible to localize precisely nonvisible satellites.

The range-and-direction finder may be used by itself or in association with other means. As an autonomous localization instrument it should permit one to check calibration methods of other systems, such as space-borne altimeters (radar echoes on the sea surface). Integrated into a station network it may contribute to the measure of orbit planes in inertial systems. Knowing almost simultaneously the angles and distances from a single station makes it easier to calculate orbit arcs.

As a last remark, let us mention the advantage of using such an instrument instead of having to resort to flash-carrying satellites such as GEOS. There seems to be a tendency to abandon the latter, in particular because of their limited lifetime and their operating cost. With the illuminating laser procedure, the active equipment remains on the ground. Moreover, it is possible to associate with the illuminating laser a high-precision range finder for advanced studies, such as ground tides or earth pole shift. It appears that the laser range-and-direction finder should be of some use not only in geodesy but in geophysics research fields.

## References

Besson, J., and P. F. Weber, Système de télémétrie laser à double impulsion, French Patent No. 1,509,017.

Bivas, R., La télémétrie spatiale par laser—Applications géodésiques, PhD thesis, Paris (1968).

Boileau, J., and P. F. Weber, Système de télémétrie à chronométrie multiple, French Patent Appl. No. 70/32,910 (1970).

Gaignebet, J., "Réalisation d'une nouvelle génération de télémètres laser," in *Bull. du Groupe Rech. de Géodésie Spatiale, Paris,* No. 1 (1971).

Iliff, R. L., "AFCRL Laser Satellite Geodesy and Future Plans," presented at GEOS-II Rev. Conf. GSFC (May 1970).

Lambeck, K. "Comments on the Accuracy of Baker-Nunn Observations," presented at Conf. on Photographic Astrometric Techniques, Tampa, Fla. (Mar. 1968).

Lehr, C. G., M. R. Pearlman, J. L. Scott, and J. Wohn, "Laser Satellite Ranging," in *Laser Applications in the Geosciences,* McDonnell Douglas (1970).

Moreau, R., and C. Veret, La télémétrie laser. Participation de l'ONERA aux opérations de géodésie au moyen des satellites Diadème," ONERA Tech. Note 156 (1969).

Müller, P., and C. Veret, "Photographie des échos laser sur satellites," in *Space Research X,* North-Holland, Amsterdam, Holland (1970).

Müller, P., R. Moreau, and C. Veret, "Expériences de photographie d'échos laser sur satellites," in *Space Research IX,* North-Holland, Amsterdam, Holland (1969).

Premo, D. A., "Goddard Mobile Laser—System Description," in *Proceedings of the GEOS-II Program Review Meeting,* Vol. III, Computer Sciences Corp., ed. (1970).

Staron, M., "Dispositif de poursuite semi-automatique pour satellites," La Recherche Aérospatiale No. 1972-2 (1972).

Weber, P. F., "Adaptation d'un lidar à la télémétrie des avions et des missiles," La Recherche Aérospatiale No. 1972-1 (1972).

A. MONFILS, J. P. MACAU, and S. GARDIER

# Spectrometric Device for the Analysis of Light Emission from a Venus Orbiter

## I. Astronomical Considerations

The spectroscopic observation of a planetary atmosphere by an orbiting probe is a problem likely to retain for the future a high scientific interest with regard to the exploration of the solar system.

In the case of Venus, several problems may be mentioned: (1) identification and determination of the concentration of minor components of the atmosphere, (2) scale height measurements, (3) determination of the temperature of the upper layers, (4) identification of light-emission mechanisms, and, (5) mapping of cloud cover at various uv wavelengths.

Points (1) and (3) may require a resolution approaching the angstrom, especially at lower wavelengths. The others may be satisfied by resolutions of several tens of angstroms.

According to the relative position of the sun, the planet, and the spectrograph, the observation modes are various and permit the detection and the study of atmospheric constituents in emission or absorption.

---

The authors are at the Institut d'Astrophysique, Universite de Liège, Cointe-Sclessin, Belgium.

Five major configurations are to be envisaged for the definition of the instrumentation. They are given on Figure 1, where the periaster is supposed as being in the 24-h meridian. The corresponding distances of observation are assembled in Table 1 as a function of the inclination of the orbit and the periaster altitude. The probe is supposed to be spinning around an axis that is perpendicular to the ecliptic plane, and the optical axes of the telescope(s) are parallel to the spin axis.

For profile scanning, it is accepted that a scale height should be roughly separable. The scale heights in Venus are easy to compute (Table 2).

For position (2) with an inclination of 60° and periaster altitude of 1000 km, a resolution of 20 km necessitates a beam limited to 1/200 rad angular diameter.

Another specification arises from the very short time available for profile scans. As the speed of the probe is of the order of 4 to 5 km/sec, a 200-km profile requires only 40 to 50 sec.

Scanning of the twilight zone is somewhat more favorable, but the period does not exceed 5 min.

As far as the wavelength range is concerned, it is well known that the

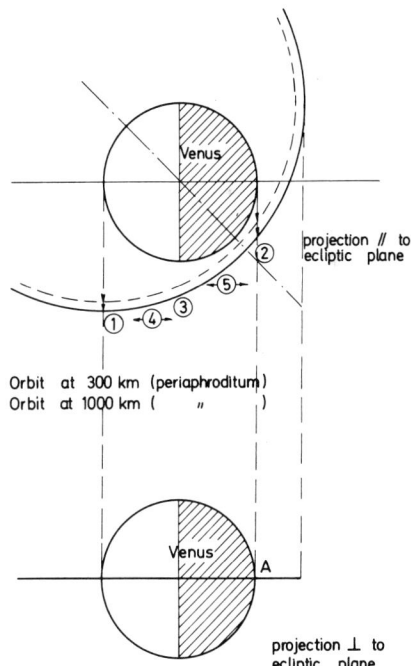

FIGURE 1  Major observation configurations.

TABLE 1  Distance of Observation as a Function of Inclination of Orbit and Periaster Altitude

| Inclination | Periaster (km) | (1) (km) | (2) (km) | (3) (km) | (4) (km) | (5) (km) |
|---|---|---|---|---|---|---|
| 90° | 300 | 8000 | 2600 | 1400 | 4500 | 700 |
|     | 1000 | 9000 | 4000 | 2200 | 5200 | 1700 |
| 60° | 300 | 7100 | 3000 | 1900 |   |   |
|     | 1000 | 7700 | 4300 | 2700 |   |   |
| 45° | 300 | 6500 | 3100 | 2900 |   |   |
|     | 1000 | 7000 | 4100 | 4100 |   |   |
| 30° | 300 | 5800 | 3400 | — |   |   |
|     | 1000 | 6200 | 4300 | — |   |   |
| 0°  | 300 | 3800 | 3800 | — |   |   |
|     | 1000 | 4300 | 4300 | — |   |   |

TABLE 2  Scale Heights in Venus

| $T$ (K) | Masses (km) | | | |
|---|---|---|---|---|
|   | 1 | 4 | 28 | 44 |
| 200 | 190 | 47.5 | 6.8 | 4.3 |
| 400 | 380 | 95 | 13.6 | 8.6 |
| 600 | 572 | 143 | 20.4 | 13.2 |

1000 to 3000 Å region is very interesting: one can briefly mention among molecular features [Barth *et al.*, 1971] the $N_2$ Lyman-Birge-Hopfield system (1200–2300 Å), the CO fourth positive system (1400–2400 Å), the $N_2$ Vegard-Kaplan bands (1400–3000 Å), the CO Cameron bands (1800–2800 Å), the $CO^+$ first negative system (2000–2700 Å), and the $CO_2^+$ *B-X* system at 2900 Å; and among atomic lines, the O transitions at 1304 and 2972 Å and the C line at 1657 Å. Longer wavelengths are far from being devoid of interest: one may cite the $CO^+$ comet tail bands (up to 7000 Å) and the $N_2$ first positive system that extends as high as 1.2 μm. The 6300 Å O doublet must also be mentioned.

## II. Experimental Considerations

A spectrographic system designed for the observation of planetary atmospheres must be strictly adapted to its mission. The weight, telemetry, and power will be critical, especially if the spectrograph is to be mounted on a Pioneer-type probe.

It has already been shown how the astronomical aspects of the mission primarily impose a series of specifications:

1/200 rad angular diameter view angle,
Repetitivity of short scanning periods,
Wide wavelength range,
Wide resolving power range (±2 to ±20 Å).

Two other conditions appear to be very important: as the luminosity-to-weight ratio must be maximum, the design must have a high throughput; furthermore, as the probe is supposed to be spinning, observation along the axis is the only solution for continuous scans. This, in turn, makes the planet image rotate around the optical axis, which prohibits the use of long slits: square, or preferably, circular holes are best suited.

A high throughput coupled with wide wavelength range points to a coupling of an interferometer with a spectrograph.

Now, it is well known that, for spectrographs, the luminosity resolution product $LR$ is proportional to the focal distance $F$. The weight $W$ is proportional to the third power of $F$. This means that

$$LR/W \sim 1/F^2.$$

It is, of course, possible to increase the luminosity without modifying the focal distance: when the source is extended, which is the case here, the spectrograph may simply be multiplied; the total luminosity is then directly proportional to the number of spectrographs, i.e., to the total weight

$$LR/W \sim C^t.$$

The optimum solution, consequently, is to be found by decreasing the focal distance down to a value limited by other considerations and by multiplying the number of spectrographs in order to recover some luminosity (a factor of 4 is proposed).

Finally, the resolution is optionally increased by the insertion of a Fabry-Perot interferometer. This permits a further gain of roughly 5 on the $LR$ product. Other advantages will be detailed later.

### III. Description of the Spectrograph–Interferometer Coupling

As the basic mounting we have adopted the Czerny-Turner monochromator, which has been widely adopted for the recording of spectra between 1200 Å and the infrared. The main advantages are the ease of wavelength scanning and high angular acceptance. Below 1200 Å, unfortunately, the impossibility at the present stage of the technique of obtaining high reflectivities prevents its use because of the number (3) of reflections inherent to the mounting.

The shape of the entrance aperture (a circular hole) permits, in this case, the use of off-axis parabolic mirrors, which, in turn, guarantee parallel beams with very low aberrations.

The parallel beams are very well suited for the installation of a Fabry-Perot interferometer. The latter is, in fact, a thin plate that may be rotated and is close to the tilting filter concept.

Figure 2 shows the dimensions and how the elements are disposed.

The dispersion angle is chosen to be 53°, which corresponds to a mounting free from stray light (multiple light diffraction). The focal distance is very short (9 cm). This value has been chosen on the basis of two criteria: resolution and weight. These will be developed later in the text.

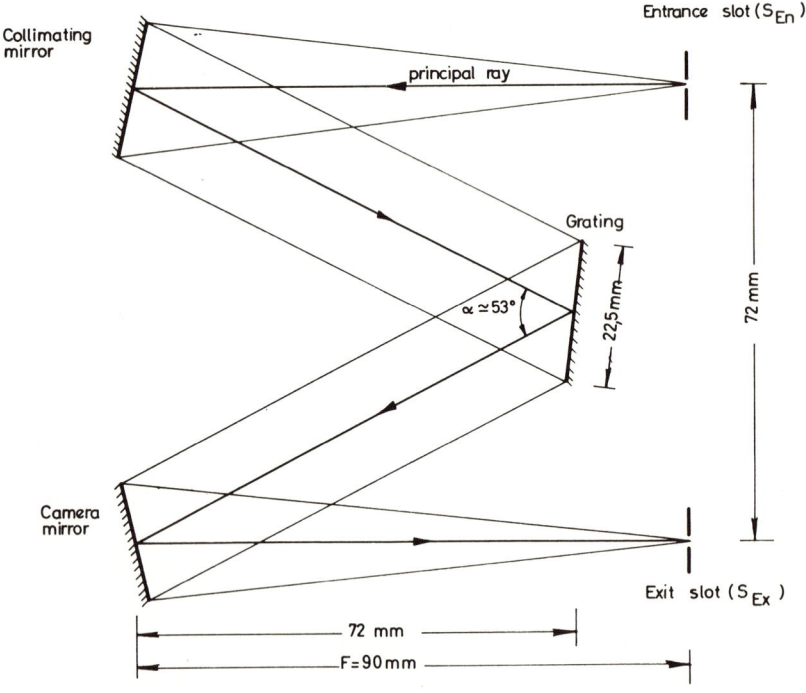

FIGURE 2 Czerny-Turner mounting.

## IV. Gratings and Bandpasses

The choice of the number $N$ of lines per millimeter of a grating is basic. In principle, this number conditions the throughput, which is proportional to $N$. Unfortunately, the wavelength sets a drastic upper limit through the well-known formula

$$d(\sin i + \sin r) = k\lambda.$$

The sum of the sinus is to be lower or equal to 2 in theory, and in practice to some lower value that we accept here as being 1.2 for luminosity reasons:

$$d = \lambda_{max}/1.2 \text{ (for the first order)},$$

so that

a 1500 lines/mm grating is useful up to 8000 Å,
a 2500 lines/mm grating is useful up to 5000 Å,
a 3600 lines/mm grating is useful up to 3300 Å,
a 6000 lines/mm grating is useful up to 2000 Å.

We immediately see then a further advantage to the multiplication of the monochromators: for each of them the grating may be chosen to be optimized for a limited wavelength range:

(1) 1000 to 2000 Å for the 6000 lines/mm grating,
(2) 1650 to 3300 Å for the 3600 lines/mm grating,
(3) 2500 to 5000 Å for the 2400 lines/mm grating,
(4) 4000 to 8000 Å for the 1500 lines/mm grating.

The wavelength ranges so determined cover one octave, which facilitates the separation of the higher orders by filters [LiF, $SiO_2$, and suitable glasses, respectively, for monochromators (1), (2), (3), and (4)].

The bandpass is, in a Czerny-Turner monochromator, the convolution of the equivalent entrance and exit slit widths.

$$L_{En} = dS_{En} \cos i/F,$$

$$L_{Ex} = dS_{Ex} \cos r/F,$$

where $S_{En}$ and $S_{Ex}$ are the entrance and exit slot widths and $F$ the focal distance.

TABLE 3  Resolution and $K$ for Selected Wavelengths for Four Monochromators

| | Wavelengths (A) | Resolution (A) | $K$ |
|---|---|---|---|
| Monochromator 1 | 2000 | 17.8 | 112 |
| | 1500 | 20.6 | 73 |
| | 1000 | 22.1 | 45 |
| Monochromator 2 | 3300 | 29.3 | 112 |
| | 2475 | 34.0 | 73 |
| | 1650 | 36.3 | 45 |
| Monochromator 3 | 5000 | 44.5 | 112 |
| | 3750 | 51.5 | 73 |
| | 2500 | 55.0 | 45 |
| Monochromator 4 | 8000 | 71.0 | 112 |
| | 6000 | 82.0 | 73 |
| | 4000 | 88.5 | 45 |

It is well known that the convolution gives a triangular profile if the slits are rectangular and if $L_{En} = L_{Ex}$. Circular holes do not give a very different solution. If $L_{En} \neq L_{Ex}$, the profile is a trapezium.

The bandpass concept differs according to the application: for the separation of two spectral lines, it is the width at half-height that matters. For the isolation of a spectral range to be scanned by the Fabry-Perot, it is the full width at the base.

The focal distance being fixed at 9 cm, the bandpasses may be computed as a function of $\lambda$ if $S_{En}$ and $S_{Ex}$ are chosen. First, there is no reason to make $S_{En} \neq S_{Ex}$. The values will consequently be equal. Strictly speaking, there is no precise figure to be computed. We have taken 0.7 mm as the dimensions of the square holes,

$$S_{En} = S_{Ex} = 0.7 \text{ mm}$$

for various reasons:

Such a value happens to correspond to $1/129$ rad with $F = 9$ cm, i.e., somewhat less than 30 min of arc. It will be seen that this value is very suitable for the Fabry-Perot scanning.

The luminosity, which depends on the square of the hole dimensions, is just sufficient. Finally, the necessary geometrical separation ($1/200$ rad) is attainable with a telescope of acceptable dimensions. The value $S_{En} = S_{Ex} = 0.7$ mm corresponds to the resolutions shown in Table 3.

## Fabry-Perot Interferometers

The aim of the introduction of these devices is to increase by a factor of the order of 5 the resolution while keeping the throughput constant.

This multiplies the $LR$ product by the same factor, but implies a finesse of at least 10.*

It is well known that the instrumental function (3) of an interferometer may be expressed as

$$W(\sigma) = \int W'(\sigma')F(\sigma-\sigma')d\sigma' = R(\sigma)*T(\sigma)*D(\sigma),$$

where $R$, $T$, and $D$ are the elementary instrumental functions linked, respectively, to the reflection finesse, the thickness finesse, and the diaphragm finesse. We may expect that, due to the limitations imposed to $\mathcal{F}_T$ by the low-wavelength limit and to $\mathcal{F}_D$ by the dimension of the hole, the reflection finesse will be easily higher than the two first cited, and that according to Jacquinot [1960], the total instrumental function $W$ will be conditioned by

$$T(\sigma) \sim D(\sigma) \gg R(\sigma).$$

In this case

$$W = (W_T^2 + W_D^2)^{1/2}.$$

The attainable thickness finesse depends on technical possibilities quite difficult to determine here. We may, however, accept as a working hypothesis that

$$\mathcal{F}_T = 50 \text{ at } 5000 \text{ Å}$$

and is proportional to the wavelength.

It is easy to show that the diaphragm finesse may be written

$$\mathcal{F}_D = F/K \tan i.$$

In this formula, $K$ is the order of interference. It must be chosen so as to leave free the wavelength ranges $\Delta\lambda$ not resolved by the spectrograph, i.e., those illustrated in Table 3. The $K$ values are easily obtained by the well-known formula

$$K = \lambda/\Delta\lambda$$

*The factor of 2 existing between the finesse and the resolution gain is linked with the difference of the resolution concept mentioned earlier.

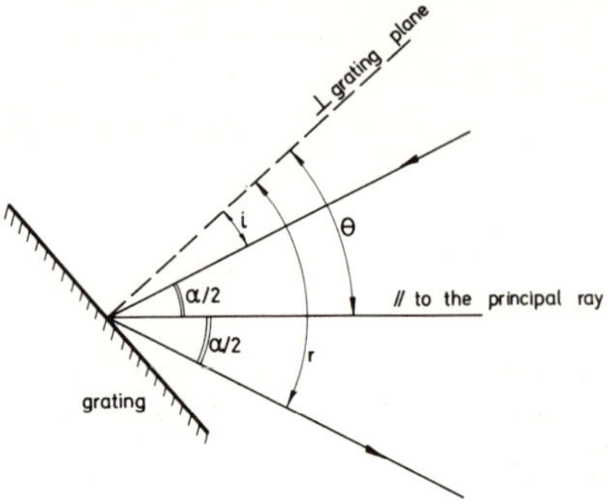

FIGURE 3  Definition of the monochromator angular parameters.

and may be found in the last column of Table 3; $F$ is the focal distance (9 cm); $i$ is the angle of incidence. (See Figure 3.) Its minimum value is 0, as the scanning is supposed to start with the plate perpendicular to the beam. Its maximum value may be obtained by the formula

$$\Delta(\cos i) = \cos i/K \sim 1/K.$$

Here, a mechanical limitation occurs: we must choose a single angular scanning to avoid undue mechanical complications. We must then adopt the lower $K$ value: 45 or, in order to ensure a security margin, 40. $\Delta(\cos i)$ is then equal to 0.025, which corresponds to an angular scan of $12°50'$, and a maximum value of $\tan i$ of 0.228. Thus

$$\mathcal{F}_D^{\min} = 14.$$

Of course, when the Fabry-Perot is perpendicular to the beam, $\tan i = 0$ and $\mathcal{F}_D = \infty$.

We can now write in Table 4 the evolution of $\mathcal{F}_T$ and $\mathcal{F}_D$. Supposing that a plate is only used for half the spectral ranges of each monochromator, the minimum value of $K$ is 27, which corresponds, respectively, to $15°38'$ and 0.28 for the scan and $\tan i$.

These figures confirm the validity of the option not to consider the reflecting finesse. The reflectivity is in fact more and more easy to control as $\lambda$ increases, making it possible to keep $\mathcal{F}_R \gg (\mathcal{F}_T, \mathcal{F}_D)$ for the whole spectral range.

The next quantity to compute is the resolving power (Table 5):

$$R = \lambda/\Delta\lambda = KF.$$

TABLE 4  Evolution of $\mathcal{F}_T$ and $\mathcal{F}_D$

|  | Monochromator 1 | | | Monochromator 2 | | | Monochromator 3 | | | Monochromator 4 | | |
|---|---|---|---|---|---|---|---|---|---|---|---|---|
| λ(Å) | 1000 | 1500 | 2000 | 1650 | 2475 | 3300 | 2500 | 3750 | 5000 | 4000 | 6000 | 8000 |
| $\mathcal{F}_T$ | 10 | 15 | 20 | 16.5 | 25 | 33 | 25 | 37.5 | 50 | 40 | 60 | 80 |
| $\mathcal{F}_D{}^a$ | ∞ to 14 | ∞ to 21 | ∞ to 19 | ∞ to 14 | ∞ to 21 | ∞ to 19 | ∞ to 14 | ∞ to 21 | ∞ to 19 | ∞ to 14 | ∞ to 21 | ∞ to 19 |
|  |  | ∞ to 14 |  |  | ∞ to 14 |  |  | ∞ to 14 |  |  | ∞ to 14 |  |
| $\mathcal{F}^a$ | 10 to 8 | 15 to 12 | 20 to 14 | 16.5 to 11 | 25 to 16 | 33 to 17 | 25 to 12 | 37 to 18 | 50 to 18 | 40 to 13 | 60 to 20 | 80 to 18 |
|  |  | 15 to 10 |  |  | 25 to 12 |  |  | 37.5 to 13 |  |  | 60 to 14 |  |

$a$ The limits corresponding, respectively, to tan $i$ = 0 and 0.32.

## Telescopes

The specifications of each telescope are simple: spatial resolution $1/200$ rad; exit diaphragm dimensions = 0.7 mm; angular aperture, $f/5$.
The focal distance is then

$$200 \times 0.07 \text{ cm} = 14 \text{ cm}.$$

The sides of the mirror are 3.5 cm wide.
The mirror is, here again, an off-axis paraboloid.

## Luminosity

The number of photons recorded per second with a symmetric Czerny-Turner is

$$N_P = \mathcal{L} A_F \Omega T \epsilon,$$

where $\mathcal{L}$ is the luminance expressed in photons cm$^{-2}$ sr$^{-1}$ of the observed object, $A_F$ is the surface of the entrance slot, $T$ the transmission coefficient of the optics, and $\epsilon$ the quantum efficiency. The product $A_F \Omega$ is the throughput of the mounting and

$$\Omega = (A_R/F^2) \cos i,$$

where $A_R$ is the grating area, $F$ the focal distance of the mirrors, and $i$ the angle of incidence on the grating.
Admitting values such as

$$\mathcal{L} = 1 \text{ kR},$$
$$T = 3\%,$$
$$A_F = 4 \times 10^{-3} \text{ cm}^2,$$
$$\Omega = 6 \times 10^{-2} \text{ sr},$$

the photon number recorded per second is

$$N_P = 600.$$

Consequently, 1.66 sec are necessary to record a 1-kR signal with a statistical error of 3%.
As the total spectrograph scanning takes 256 steps and that 4 Fabry-

Perot scannings (with 64 steps each) amounting to 256 other ones are foreseen, a complete mixed scan where 4 Czerny-Turner bandwidths are analyzed by the Fabry-Perot will take 854 sec, i.e., a little more than 14 min. Of course, this recording time may be split (and must be) into a large number of elementary ones. The example taken corresponds to a limit where a very faint source will have to be completely analyzed from 2000 to 8000 Å, a total of 16 regions being examined at high resolution. In practice, dayglow emissions are more intense than 1 kR, and a 3% precision is not always required. Furthermore, the scanning will be reduced to parts of the spectrum. If the situation is difficult, because of the very limited amount of available light (nightglow), the fact that the probe is in orbit around the planet must be used and the information added progressively in order to increase the statistical signal-to-noise ratio. The true limit will be set by the dark current of the detector and the lifetime of the satellite.

## General Description

Figures 4 and 5 show how the monochromators are designed and disposed on top of each other.

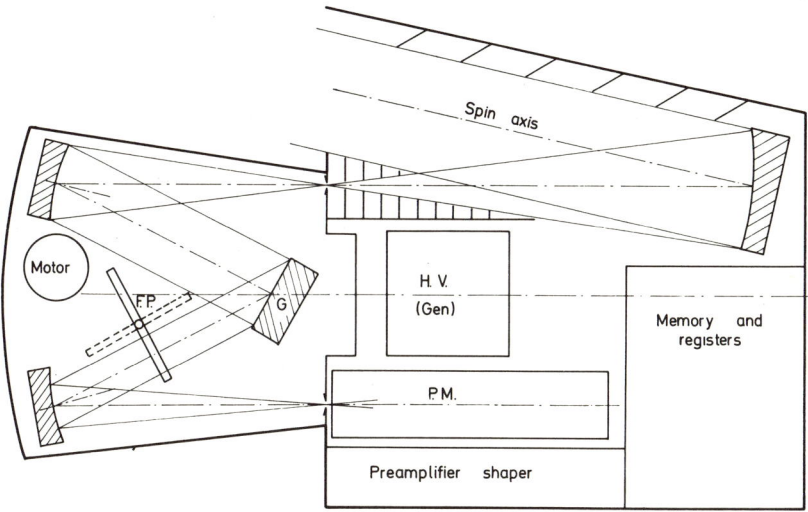

FIGURE 4  General design of a spectroscopic assembly. G is the grating, HV is the high-voltage power supply, and PM is the photomultiplier. The power supply and the memory-register are common to the set of four assemblies.

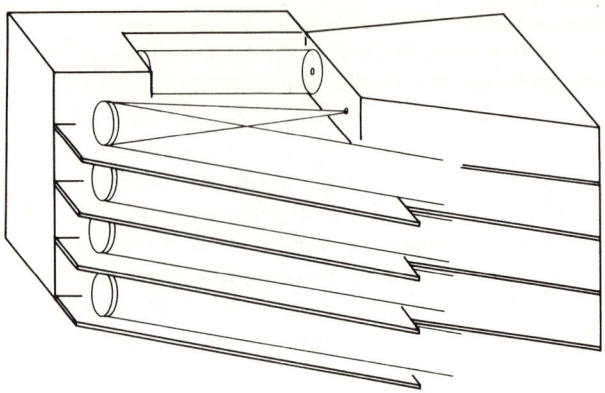

FIGURE 5  Sketch of the total package.

The two sets of Fabry-Perots may be interchanged in the exit parallel beam or removed completely. The telescopes and the monochromators leave a roughly cubic space free, which is used for the detectors and electronics.

**Detectors**

We intend to use photomultiplier tubes. Their technique has been widely developed during the last ten years. Very light and efficient tubes with very low dark current are at present available. The shape of the exit slots allows very small cathodes to be used, with very low dark current counts.

For CsI and RbTe cathodes, these counts may reach the order of one count per second. The detectability limit drops then down to a few Rayleighs, which permits the detection of nightglows.

Long-wavelength channels will, of course, be less sensitive as a result of higher dark currents, although the cooling of the cathode may be considered in space. The wavelength splittings between the four monochromators correspond to a very favorable series of cathodes:

CsI up to 1900 Å,
Rb-Te up to 3000 Å,
Bialkali up to 6000 Å,
Trialkali up to 8000 Å.

Here again, each monochromator is optimized.

A small problem arises between 1900 and 2000 Å and between 3000 and 3300 Å. It is covered by the range superpositions.

## Data Handling

The four photomultipliers (Figure 6) are each to be followed by a threshold amplifier and pulse shapers (A–PS). Four counters (C1 to C4) are in action during a $\Delta t$ period corresponding to one Czerny-Turner (C.T.) or Fabry-Perot (F.P.) channel (viz., one step duration of the C.T. stepping motor or F.P. stepping motor). At the end of the counting period, the results of the four counters are added to the content of the corresponding memory shift register (MSR).

An oscillator, OS, followed by a frequency divider (FD) delivers a signal at the channel frequency for the command of the stepping motors, a signal for the sequencer (SA) and to the control unit of the (digital) telemetry.

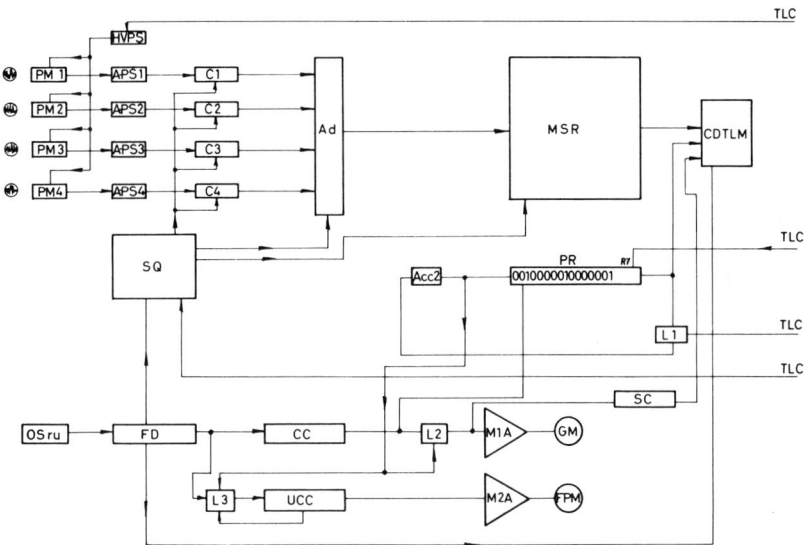

FIGURE 6 Diagram of the data-handling system. ACC1 and Acc2 are, respectively, the count and program accumulators; CDTLM is the control of digital telemetry; CC the channel counter; UCC the under channel counter; GM and FPM the gratings and Fabry-Perot motors; TLC are the telecommands; L1 to L3 the logic controls; AM the motor amplifiers; and SC is the spectrum counter.

In order to allow the simultaneous use of the low and high resolution, the different programs are stored in an R.O.M. (Read Only Memory). The program is selected by a telecommand supplying the power to the corresponding part of the R.O.M.

In low-resolution mode, only the grating motor is in action, and an accumulation in the corresponding memories is obtained.

In high-resolution mode, the R.O.M. loads with a series of 0 and 1 controls the working program. It contains, in principle, 4"1" for 252 "0." Each 1 corresponds to a channel to be subscanned by the F.P. Of course, this program is preadapted to the observation necessities. The necessary capacity of the memory is, for the C.T.,

$$4 \times 256 \times 12 = 12{,}288 \text{ bits;}$$

for 4 counting chains,

12 bits per word,
256 channels per chain.

As the F.P.'s are considered in parallel, this number is to be doubled.

If we record 10 channels per second, the dynamic range is 40,960 pulses per second. For planetary atmosphere profiles, it is advantageous to record at the highest speed compatible with the equipment: the elementary spectra will be added channel per channel; the total number of bits will only be 24,576 if the F.P. and the C.T. are used in parallel at full capacity and up to the dynamic range given here above, which corresponds to a precision of 1.56%.

**Weight**

The whole package is composed of

Four sets comprising
    1 telescope of 140-mm $F$
    1 monochromator C.T.
    1 F.P.
    1 detector chain (1 PM + 1 preamplifier + 1 pulse shaper)
1 HV power supply
1 grating stepping motor
2 F.P.
1 container for electronics

It appears possible to build four sets within a total weight of a little over 4 kg.

This weight breakdown corresponds roughly to the equation

$$W_D = W_T + W_E \text{ and } W_T \sim W_E,$$

where $W_D$ is the weight of the dispersing device (C.T. + F.P.). $W_T$ is the weight of the telescope and $W_E$ is the weight of the electronics.

Here lies the reason for the choice of the focal distance: as the weight of the C.T. is proportional to the cube of the focal distance, a higher or lower focal distance leads to an unbalanced weight of the total package.

## Power Consumption

Three stepping motors command, respectively, the gratings, the F.P. scanning, and the switching of the two F.P. blades corresponding to the half-ranges.

It is to be recalled that the three gratings are rigidly locked to each other. Likewise for the F.P.'s, which are scanned together and switched together. Each motor, including the command electronics, is consuming around 3 W. As they are never used simultaneously, the total power necessary remains at 3 W for the motors.

The HV power supply is only requiring 0.1 W or so.

The counting and logics electronics are difficult to evaluate with some precision as the circuits are not fully designed. It may be estimated, however, that 100 to 200 integrated circuits will be necessary, which amounts to 2 to 4 W.

The experiment will consequently require 5 to 7 W.

It is to be stressed that the power does not depend critically on the number of spectrographs, which is another argument in favor of their multiplication.

## Conclusions

We propose to develop a spectrometric device composed of four 9-cm focal distance Czerny-Turner monochromators, each of them optimized as far as gratings, detectors, and reflectivities are concerned, and each of them preceded by a 13-cm focal distance telescope. Only square or circular holes are used as entrance apertures. Fabry-Perot interferometers allow a gain of a factor of 5 to be realized on the resolution for 16 spectral regions.

The resolutions are written in Table 5 and correspond roughly to a constant resolving power of 500 from 1000 Å to 8000 Å. The weight and power are thought to be kept as low as 4.1 kg and 5 to 7 W. The telemetry requirements depend on the observation program. The 200 kbits to be stored (corresponding to 16 independent complete spectra) will need typical telemetry rates of 2 bits/sec.

The basic specifications such as the slot width are, of course, to be considered as a parameter commanding the resolving power, the luminosity, and the recording time.

It must be stressed, however, that the ultimate resolution depends on the slot width according to an approximately square law: the luminosity $L$ is, obviously, proportional to $1/S^2$. Consequently, $RL \sim C^t$.

This means that it is, in principle, possible to increase the resolving power, but at the direct expense of the luminosity. A second consequence is the linear increase of the number of resolved channels (at constant luminosity). The observation time for a given number of spectra will consequently increase with the square or the cube of the resolving power, for bands or lines, respectively. An increase of the resolving power will then be almost certainly conditioned by a restriction on the wavelength range to be scanned.

## References

Barth, C. A., C. W. Hord, J. B. Pearch, K. K. Kelly, G. P. Anderson, and A. I. Stewart, *J. Geophys. Res.* **76**, 2213 (1971).
Jacquinot, P., *Rev. Prog. Phys.* **23**, 267 (1960).
Marette, G., Article sur *Bull. Acad. R. Belg. Cl. Sci.* **56**, 310 (1970).
Roig, J., *Optique Physique I*, Masson, Paris (1967).
Shannon, D. C., *Bell Syst. Tech. J.* **3** (1948).

R. HOEKSTRA, K. A. VAN DER HUCHT,
TH. KAMPERMAN, and H. J. LAMERS

# THE UTRECHT ORBITING STELLAR SPECTROPHOTOMETER S 59

## I. Introduction

The Utrecht Orbiting Stellar Spectrophotometer S 59 is one among the seven scientific astronomical instruments aboard the European Space Research Organization (ESRO) satellite TD-1A. The successful launch of the spacecraft was accomplished by the 88th Delta vehicle of NASA on March 11, 1972, from the Western Test Range in California. The satellite is the largest and most complicated ever built in Europe; ESRO was responsible for the realization of the spacecraft, while the scientific instruments were developed by a number of universities and space-research centers. A detailed description of the spacecraft and its scientific package is given by Tilgner [1971]. Notwithstanding this, a brief introduction to the spacecraft orbit and attitude (Section II) will precede the description of the S 59 instrument (Section III), because S 59 is adapted extensively to the dynamics of the spacecraft. The calibration of S 59 before and during the integration of the satellite will be summarized in the Section IV. Finally, a survey of the first astronomical results deduced from the preliminary S 59 data will be given in Section V.

---

The authors are in The Astronomical Institute at Utrecht, The Netherlands.

## II. Orbit and Attitude Control of the Spacecraft

The orbit of the spacecraft is a nearly polar orbit, and due to its inclination of 97.5° the orbit is retrograde and the orbital plane precesses 1° per day, i.e., the precession rate of the orbital plane is equal to the rotation rate of the earth around the sun. In this way, it is possible to keep the orbital plane approximately perpendicular to the direction earth–sun. The altitude of the orbit is 550 km above the earth's surface. Its revolution time is 96 min, implying that the precession of the orbital plane per orbit amounts to 4 min of arc.

The spacecraft achieves a three-axis stabilization with a 1 min of arc accuracy by using reaction wheels and gas nozzles: the spacecraft $Z$-axis points to earth and the $X$-axis points to the sun. Any instrument pointing along the $-Z$-axis thus scans the celestial sphere along ecliptic meridians, and a complete coverage of the sphere is obtained in half a year.

## III. The S 59 Instrument

The S 59 instrument uses the scanning advantage of the spacecraft and is in this way able to measure more than 200 bright early-type stars in half a year in a highly automatic mode of operation. The instrumental concept of S 59 is shown in Figure 1 and in the description of the instrument given below, the numbers in parentheses refer to that figure.

When the star of interest is located on the S 59 optical axis, the parallel light beam originating from the star (1) is reflected by the ellipsoidal primary telescope mirror (2), subsequently by the spherical secondary mirror (3), and is then focused in the center of the entrance diaphragm of the spectrometer box (4). This entrance diaphragm acts at the same time as a field-of-view aperture for the telescope, with limiting dimensions of 10 × 50 min of arc, the 50 min of arc long side being perpendicular to the scanning movement of the spacecraft. Inside the spectrometer box the off-axis paraboloidal mirror (5) provides a collimated beam to the off-axis paraboloidal diffraction grating (7), which produces the spectrum (8) alongside one of the edges of the primary telescope mirror.

This unconventional grating mounting was applied here in order to minimize the dimensions and moment of inertia of the instrument. As shown in the figure, the heaviest components of the telescope–spectrometer combination (i.e., the primary mirror and the package housing the five photomultiplier tubes) could be located close to each other near the instrument gimbal axes (16, 17) minimizing in this

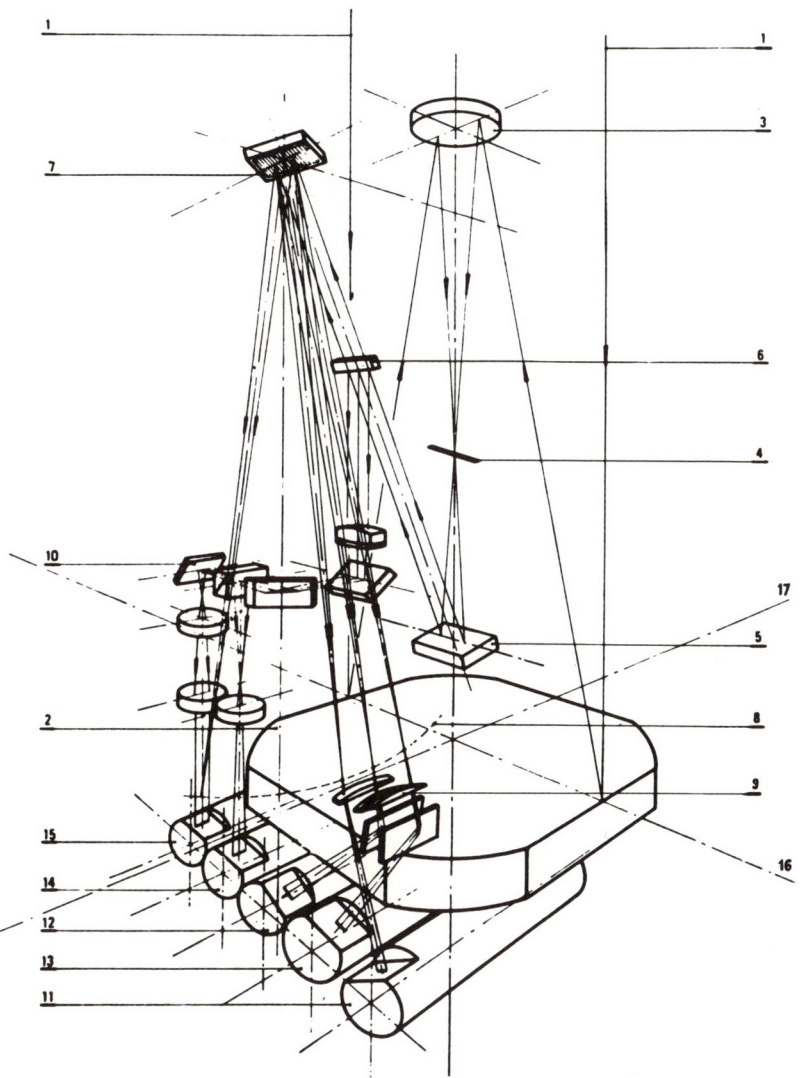

FIGURE 1 Optical system of S 59: 1, Incident beam; 2, primary mirror; 3, secondary mirror; 4, entrance diaphragm of spectrometer box; 5, collimator mirror; 6, small mirror reflecting 10% of the light for the pointing system; 7, grating; 8, spectrum; 9, condenser system; 10, optical elements of the pointing system; 11, 12, and 13, photomultipliers for measuring the spectrum; 14 and 15, photomultipliers for the pointing system; 16, main axis; 17, cross-axis.

way the moment of inertia of the moving part of the instrument. The grating mounting is one of the outcomes of a theoretical study of Werner [1970] at the Institute for Applied Physics in Delft, The Netherlands, about "imaging properties of diffraction gratings."

The spectrum is scanned by means of a curved mask with three 25-$\mu$m-wide slits. The spectrum scanner makes a stepwise movement with steps of 15 $\mu$m, corresponding to 0.45 Å in the spectrum. One step takes 1.2 sec. During a spectral scan, 200 of these steps are made, so the total time needed for a scan is 4 min, and the spectral coverage consists of three spectral bands of a width of ~90 Å. The ultraviolet detectors used (11, 12, 13) are EMR 641 F photomultiplier tubes, used in a pulse-counting mode. The photocathode material of the tubes is $Cs_2Te$; the window material is $MgF_2$ for the 2110 Å tube and 9741 glass for the 2540 and 2820 Å tubes.

In order to keep the telescope pointed to the star during the 4 min of observation time, the whole telescope–spectrometer combination can be rotated around two axes: the main axis (16) and the cross axis (17). The rotation around the main axis compensates for the roll motion of the spacecraft, and therefore the rotation around the main axis amounts to 15° during the 4-min tracking period. The cross-axis rotation has a much smaller range—only 50 min of arc corresponding to the length of the field-of-view aperture of the telescope—but the cross-axis pointing must be highly accurate (better than 1 sec of arc) because it acts in the direction of the dispersion of the spectrum.

The internal pointing is servo driven, with an optical sensing obtained in the star-pointing optics. Nearly 10% of the parallel beam going to the grating is fed into the pointing optics by means of a small mirror (6). By means of a doublet and a beam splitter, two images of the star to be measured are obtained. A vibrating knife edge, located in the image point of the "fine pointing" channel, provides the signal for the cross-axis movement, while a fixed knife edge in the other channel provides the adequate information for the main-axis servo loop. The spectral region from 4000 up to 5000 Å is used in the star-pointing system. The photomultipliers of the pointing system are also EMR tubes, but now with trialkali or multialkali cathode material and 7056-glass windows.

The operation of the instrument in orbit is as follows:

1. The telescope scans the sky due to the scanning movement of the spacecraft. In this "standby" mode, the small measuring slits in the ultraviolet channels are replaced by much wider so-called acquisition slits in order to have a higher sensitivity in these channels. When a star with a sufficient ultraviolet flux enters the field of view of the tele-

scope, this is detected by the ultraviolet photomultiplier tubes and causes the instrument to change its standby mode into the "tracking" mode.

2. In this tracking mode, the internal star-pointing system locks the telescope during 4 min on the star. Meanwhile, the spectrum scanner makes its 200 steps. Having completed the spectrum scan, the instrument changes its mode into "reverse."

3. In the reverse mode the telescope and the scanner move back to their standby positions, and the instrument is then ready for catching the next star.

This method of operation has the important advantage of being fully automatic. Only incidentally are telecommands from the ground required, for instance, for optimizing the star-presence levels (i.e., threshold levels for the ultraviolet channels in the standby mode) for certain parts of the sky. Another example of the need for telecommands is the monthly crossing of the scan plane by the moon, in which case the high tensions of the photomultiplier tubes are switched off temporary.

Dimensions and optical parameters concerning the S 59 telescope-spectrometer combination are given in Table 1.

TABLE 1  Information Concerning the S 59 Optics

| | |
|---|---|
| *Telescope* | |
| Outer dimensions of primary mirror | $22.0 \times 22.0$ cm$^2$ |
| Net collecting surface | 290 cm$^2$ |
| Focal length telescope | 106.5 cm |
| *Spectrometer* | |
| Wavelength ranges | 2064–2158 Å |
| | 2497–2591 Å |
| | 2777–2868 Å |
| Grating | Off-axis paraboloidal grating, ruled by Bausch & Lomb |
| Grating ruling | 1200 lines/mm |
| Ruled area | $44 \times 33$ mm$^2$ |
| Blaze angle | 8° 10' |
| Blaze wavelength | 2286 Å |
| Wavelength of stigmatism | 2387 Å |
| Spectral dispersion | 0.033 mm/Å |
| Angular dispersion | 2.4 sec of arc/Å |
| Spectral resolution | 1.8 Å |
| *Materials* | |
| Grating | BK 7 optical glass |
| All other optical components | Fused silica |
| Reflection coatings | Al + MgF$_2$ |

## IV. Calibration

The S 59 instrument has been tested and calibrated using a parallel test beam, generated by the equipment shown in Figure 2. Because the wavelength bands of the S 59 ultraviolet channels lie above 2000 Å, the whole calibration configuration and the S 59 instrument could remain simply in air during the measurements. The light emitted by a 450-W xenon light source (1) is filtered by a Czerny-Turner monochromator (2) and is then focused onto a 10-$\mu$m-diameter pinhole (3) in the focal point of a paraboloidal mirror (4) of 2-m focal length. In this way, a parallel monochromatic beam of ultraviolet radiation (5) is obtained. The parallelism of the beam is $\sim$1 sec of arc FWHM, and the spectral bandpass of the monochromator a few tenths of an angstrom. The wavelength of the test beam can be chosen by means of the monochromator setting, and the absolute flux of the beam can be measured by means of an absolute-calibrated detector (e.g., photomultiplier tube). Because the S 59 star-pointing system uses starlight in the visible spectral region, some visible radiation is added to the ultraviolet beam from a second Xe lamp (6).

During the calibration measurements, the S 59 instrument was located in the test beam, and information could be gathered about the wavelength scale under a wide variety of conditions, the spectral resolution, and the absolute sensitivity of the ultraviolet spectrophotometer channels. An example of results obtained is presented in Figure 3, which shows the instrumental response for a test beam composed of visible light and an ultraviolet component of 2861 Å. Line profiles ob-

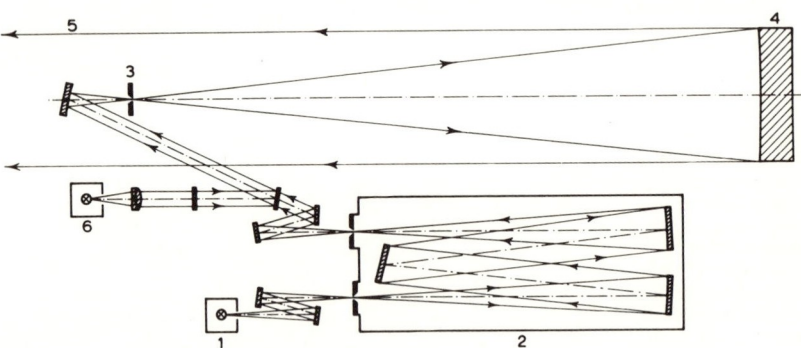

FIGURE 2 Instrumentation used for the S 59 calibration. 1, 450-W xenon lamp; 2, Czerny-Turner monochromator; 3, pinhole, diameter 10 $\mu$m; 4, paraboloidal mirror, diameter 40 cm, focal length 2 m; 5, test beam produced; 6, 150-W xenon lamp providing visible light for the pointing system.

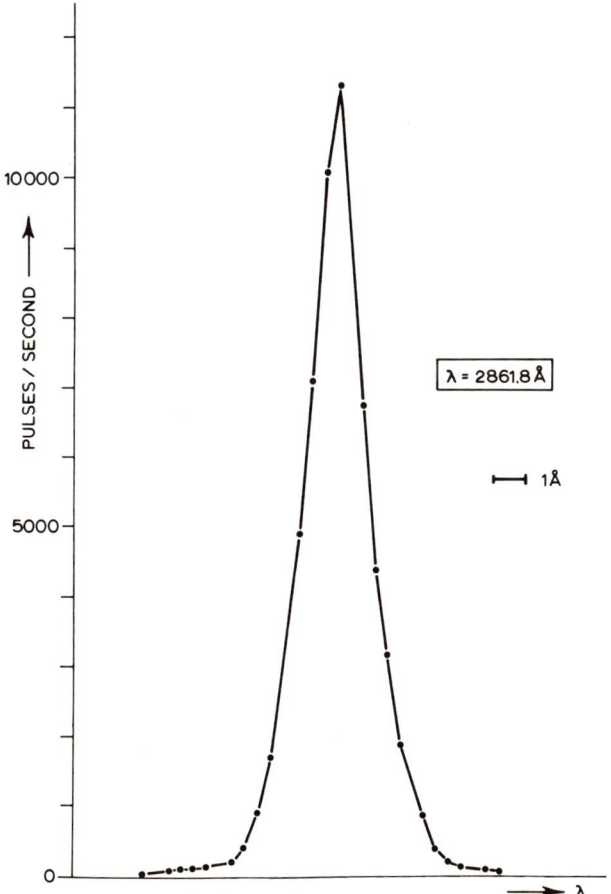

FIGURE 3 Observed line profile for λ = 2861.8 Å, obtained during laboratory tests.

tained in this way show a line width of ~1.8 Å FWHM in the three spectrophotometer channels.

## V. Results

At present, S 59 has fulfilled its half-year mission and has measured spectra of more than 200 stars. Because the scan plane precesses 4 min of arc per orbit, and the field of view perpendicular to the scan plane amounts to 50 min of arc, 12 scans are obtained for each bright ecliptic star. Stars at higher ecliptic latitudes are observed more times: ζ Dra, near the ecliptic pole has been observed in more than 140 consecutive orbits. Unfortunately, the tape recorders of the spacecraft failed after two months of operation in orbit, and since then the telemetric data received from the spacecraft were restricted to real-time data during ground-station passes only. Thanks to a considerable increase of the number of used ground stations from 5 to 31, a coverage of 50 percent could nevertheless be attained.

An example of a stellar spectrum obtained is given in Figure 4, which shows the three wavelength bands of β CMa, averaged over only three orbits. These orbits were available on "quick-look tapes," intended for a technological survey. The final data tapes, however, will provide 15 orbits of this star, so the final presentation of this spectrum will have an improved signal-to-noise ratio. The S 59 β CMa spectrum will be investigated in detail by C. de Jager at the Astronomical Institute of Utrecht.

The spectrum of β CMa shows the general nature exhibited by many of the S 59 spectra—a continuum on which absorption lines are superposed. The majority of the absorption lines are formed in the stellar atmospheres, and, accordingly, the study of these lines may give information about stellar atmospheres and the physical processes involved.

Striking lines are the Mg II resonance lines at 2795 and 2803 Å. The behavior of these lines along a sequence of stellar spectral types has already been studied from preliminary S 59 data [Lamers *et al.*, 1973]. This study shows a discrepancy between observed equivalent widths and theory for middle- and early-B stars: observed lines, corrected for interstellar contribution, are too strong by a factor of 1.4 (B5–B6) to 4 (B0–B1). It is hoped that these observations will contribute to a better understanding of the line-formation processes in stellar atmospheres.

Absorption lines also can be of interstellar origin, in which case the lines contain information about interstellar matter. A number of interstellar lines have already been identified in the preliminary S 59 data of ζ Pup [de Boer *et al.*, 1972]. Another striking spectral feature observed is a very much broadened emission line in the spectrum of the Wolf-Rayet star $\gamma^2$ Velorum, visualizing the enormous mass transport that takes place from the Wolf-Rayet star into space with velocities of the order of 1000 km/sec [van der Hucht and Lamers, 1973].

The examples given above only constitute a small fraction of the astrophysical content of the data. A detailed analysis of all spectra will

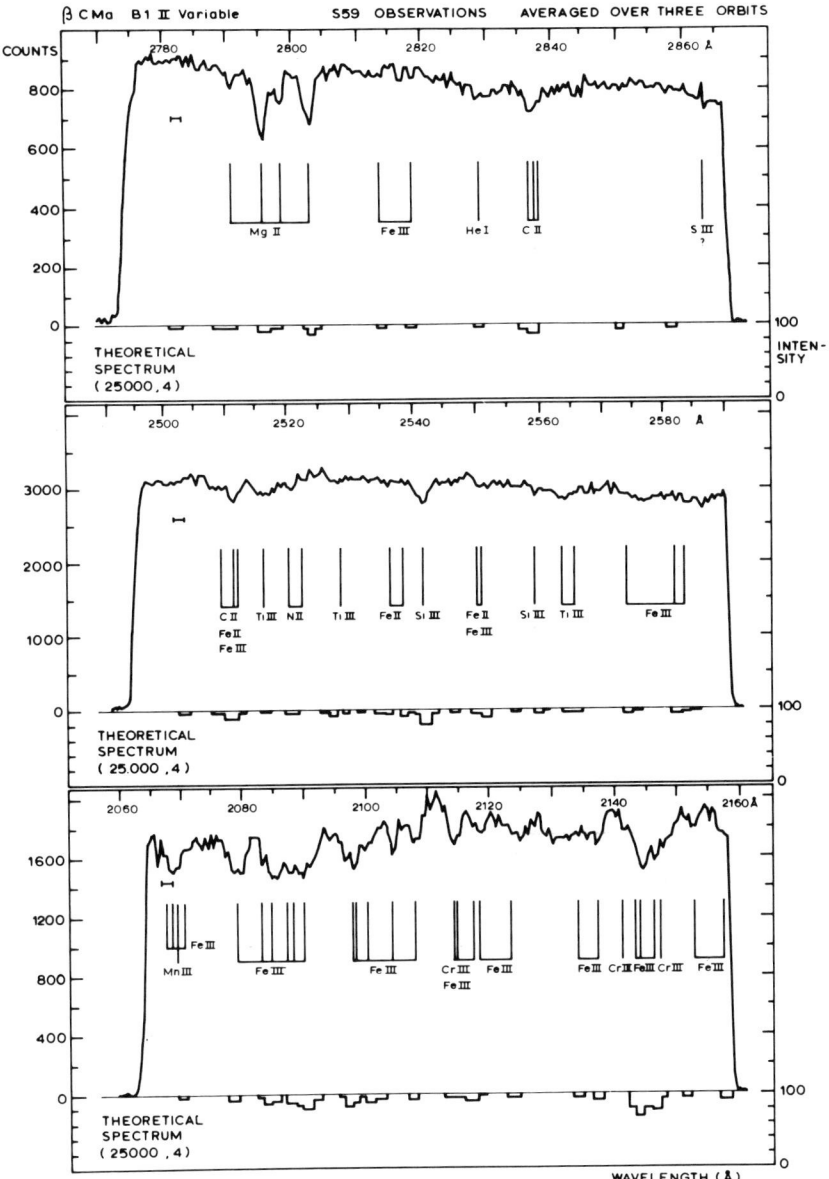

FIGURE 4 Spectrum of β Canis Majoris in the three wavelength bands, compared with theoretical predictions.

be carried out by the Utrecht Astronomical Institute and by more than 30 research groups from all over the world.

The authors wish to acknowledge A. Hammerschlag, W. Werner, and their collaborators at the Institute for Applied Physics TNO-TH (TPD), Delft, The Netherlands, for their outstanding contribution to the realization of the S 59 instrument.

## References

de Boer, K. S., R. Hoekstra, K. A. van der Hucht, T. M. Kamperman, H. J. Lamers, and S. R. Pottasch, *Astron. Astrophys. 21,* 447 (1972).
Lamers, H. J., K. A. van der Hucht, M. A. J. Snijders, and N. Sakhibulin, *Astron. Astrophys. 25,* 105 (1973).
Tilgner, B., *ELDO/ESRO Sci. Tech. Rev. 3,* 567 (1971).
van der Hucht, K. A., and H. J. Lamers, *Astrophys. J. 181,* 537 (1973).
Werner, W., Thesis, Delft, The Netherlands (1970).

# APPENDIXES

# Appendix A
# ICO BUREAU
# 1969-1972

*President*

H. H. HOPKINS, The University of Reading, England

*Vice Presidents*

B. HAVELKA, Université Palacky, Czechoslovakia
KOREO KINOSITA, Gakushuin University, Japan
R. M. SCOTT, The Perkin-Elmer Corporation, United States of America
W. H. STEEL, National Standards Laboratory, Australia

*Secretary-Treasurer*

J. CH. VIÉNOT, Université de Besancon, France

# Appendix B
# U.S. NATIONAL COMMITTEE FOR ICO

*Chairman*
F. DOW SMITH, Itek Corporation

*Vice Chairman*
S. Q. DUNTLEY, University of California, San Diego

IWAO P. ADACHI, Minolta Camera Company
HAROLD E. BENNETT, U.S. Naval Weapons Center
BRUCE H. BILLINGS, American Embassy, Taiwan
MICHAEL HERCHER, University of Rochester
JOHN N. HOWARD, Air Force Cambridge
    Research Laboratories
LLOYD G. MUNDIE, The RAND Corporation
DONALD S. NICHOLSON, The Aerospace Corporation
ARTHUR L. SCHAWLOW, Stanford University
ROBERT R. SHANNON, University of Arizona
BRIAN J. THOMPSON, University of Rochester

ICO IX PROGRAM COMMITTEE

*Chairman*
R. R. SHANNON, University of Arizona
K. M. BAIRD, National Research Council of Canada

P. L. BENDER, Joint Institute for Laboratory Astrophysics
H. L. BENNETT, U.S. Naval Weapons Center
B. H. BILLINGS, American Embassy, Taiwan
R. H. CHASE, Headquarters, NASA
J. G. CONWAY, University of California, Berkeley
S. P. DAVIS, University of California, Berkeley
S. Q. DUNTLEY, University of California, San Diego
M. HERCHER, University of Rochester
D. R. HERRIOTT, Bell Telephone Laboratories
D. S. NICHOLSON, The Aerospace Corporation
F. PAUL, Goddard Space Flight Center, NASA
R. D. RAWCLIFFE, The Aerospace Corporation
N. ROMAN, Headquarters, NASA
B. J. THOMPSON, University of Rochester

## LOCAL ARRANGEMENTS COMMITTEE

*Chairman*

D. S. NICHOLSON, The Aerospace Corporation

*Finance*

I. C. SANDBACK, Hughes Aircraft Corporation

*Operations*

L. G. MUNDIE, The RAND Corporation

*Publications*

E. G. BROCK, North American Rockwell Corporation

*Social*

VICTOR BEELIK, Hughes Aircraft Corporation
R. D. RAWCLIFFE, The Aerospace Corporation

# Appendix C
# SPONSORS

Aerojet General Corporation
The Aerospace Corporation
American Optical Company
Amersil, Inc.
Baird-Atomic, Inc.
Ball Brothers Research Corporation
Barnes Engineering Company
Bausch & Lomb Inc.
Boller & Chivens Division of
    The Perkin-Elmer Corporation
Corning Glass Works
Dynasil Corporation
EG&G, Inc.
Eastman Kodak Company
Edmund Scientific Company
Fairchild Space & Defense Systems
Farrand Optical Company, Inc.
Ford Motor Company
Gamma Scientific, Inc.
General Scientific Corporation
Herron Optical Company
Hughes Aircraft Company
International Business Machines
    Corporation
Itek Corporation
Ernst Leitz (Canada) Ltd.
Minolta Corporation
McDonnell Douglas Corporation
Muffoletto Optical Company, Inc.
Optical Coating Laboratory, Inc.
Optical Instrument Corporation
Optical Publishing Company
Optical Research Associates
Optical Sciences Group, Inc.
Optical Society of America
Optic-Electronic Corporation
Optovac, Inc.
Owens-Illinois Technical Center
Owens-Illinois Development Center
The Perkin-Elmer Corporation
RCA Laboratories
Rogers & Clarke Manufacturing
    Company
Santa Barbara Research Center

*Sponsors*

Schott Optical Glass, Inc.
Servo Corporation
Society of Photo-Optical Instrumentation Engineers
Tinsley Laboratories, Inc.
Tropel, Inc.
TRW, Inc.
Xerox Corporation
Carl Zeiss Oberkochen
Carl Zeiss, Inc.

SEP 27 1974

QB
86
I
59